The Elements of Botany

Adrien de Jussieu
Translated by
James Hewetson Wilson

Cambridge University Press

CAMBRIDGE UNIVERSITY PRESS

Cambridge, New York, Melbourne, Madrid, Cape Town,
Singapore, São Paolo, Delhi, Tokyo, Mexico City

Published in the United States of America by Cambridge University Press, New York

www.cambridge.org
Information on this title: www.cambridge.org/9781108037310

This edition first published 1849
This digitally printed version 2011

ISBN 978-1-108-03731-0 Paperback

Life Sciences

Until the nineteenth century, the various subjects now known as the life sciences were regarded either as arcane studies which had little impact on ordinary daily life, or as a genteel hobby for the leisured classes. The increasing academic rigour and systematisation brought to the study of botany, zoology and other disciplines, and their adoption in university curricula, are reflected in the books reissued in this series.

The Elements of Botany

The author of pioneering works on the plant families *Euphorbiaceae* and *Malpighiaceae*, French botanist Adrien de Jussieu (1797–1853), from a famous family of scientists, became professor of botany at the Paris Museum National d'Histoire Naturelle in 1826. The author of several specialised monographs, he is best known for his *Botanique: Cours élémentaire d'histoire naturelle* (1842), an exhaustive introduction to botany originally intended for use in French schools. This much acclaimed book went through twelve editions between 1842 and 1884, and was translated into many languages. This English translation, completed while he was still an Oxford student by the young British botanist James Hewetson Wilson (1826–50), was first published in 1849. It contains additional descriptions of the different systems and structures that clarify de Jussieu's terminology, as well as an appendix on geology.

Cambridge University Press has long been a pioneer in the reissuing of out-of-print titles from its own backlist, producing digital reprints of books that are still sought after by scholars and students but could not be reprinted economically using traditional technology. The Cambridge Library Collection extends this activity to a wider range of books which are still of importance to researchers and professionals, either for the source material they contain, or as landmarks in the history of their academic discipline.

Drawing from the world-renowned collections in the Cambridge University Library, and guided by the advice of experts in each subject area, Cambridge University Press is using state-of-the-art scanning machines in its own Printing House to capture the content of each book selected for inclusion. The files are processed to give a consistently clear, crisp image, and the books finished to the high quality standard for which the Press is recognised around the world. The latest print-on-demand technology ensures that the books will remain available indefinitely, and that orders for single or multiple copies can quickly be supplied.

The Cambridge Library Collection will bring back to life books of enduring scholarly value (including out-of-copyright works originally issued by other publishers) across a wide range of disciplines in the humanities and social sciences and in science and technology.

THE

ELEMENTS OF BOTANY.

BY

M. ADRIEN DE JUSSIEU,

MEMBRE DE L'INSTITUT, PROFESSEUR AU MUSÉUM D'HISTOIRE NATURELLE, ETC.

TRANSLATED BY

JAMES HEWETSON WILSON,

F.L.S., F.R.B.S.,

MEMBER OF THE BOTANICAL SOCIETY OF LONDON, &c., &c.

LONDON:

JOHN VAN VOORST, PATERNOSTER ROW.

MDCCCXLIX.

OXFORD:
PRINTED BY I. SHRIMPTON.

PREFACE.

The celebrity, which the Elements of Botany by M. De Jussieu has attained in his own country and even in England, is a sufficient excuse for the Translator's presenting it to the notice of the Public in the following shape. It forms part of The Elementary Course of Natural History adopted by the *Conseil Royal de l'instruction publique* for the use of the colleges in France.

To render the work more complete, an Introduction, to which De Jussieu refers, has been taken from the Treatise on Zoology in the same Course. In the Appendix the plates are copied from the Geology of M. Beudant, but descriptions of the different Systems and Formations have been added by the Translator to elucidate the terms used by De Jussieu.

Wadham College, Oxford,
Dec. 8th, 1848.

TABLE OF CONTENTS.

(The figures refer to the number of the paragraph).

THE INTRODUCTION.

THE ELEMENTS OF BOTANY.

THE ORGANS OF VEGETATION.

ORGANS OF REPRODUCTION.

FAMILIES.

SYNOPTICAL TABLES OF THE FAMILIES,

ARRANGED ACCORDING TO THEIR PRINCIPAL CHARACTERISTICS.

THE PRINCIPLES OF CLASSIFICATION.

THE DETAILS OF THE CHARACTERISTICS OF A FEW FAMILIES.

BOTANICAL GEOGRAPHY.

APPENDIX.

INTRODUCTION.

THE OBJECT AND UTILITY OF NATURAL HISTORY.

§ 1. We designate by the name of Natural History the science which treats of the structure of the bodies scattered over the surface of the globe, or united to form its mass, of the phenomena of which these bodies are the seat, of the characteristics adapted to enable us to distinguish them from one another, and of the part they play in the Creation. Its province, as we see, is immense, and its importance does not yield to its extent. A person not familiar with science might fancy it to be only a collection of facts and anecdotes, better suited to excite curiosity than to exercise the mind, or else a dry study of technical names and of arbitrary classifications; but such an opinion can only arise from ignorance, and whoever possesses the slightest knowledge of Natural History cannot refuse to recognise its immense utility. This great and harmonious spectacle of nature, by shewing how much the real beauty of the creation excels the ideal perfection of the inventions of man, elevates the soul, and unceasingly raises in the mind high and salutary thoughts; the knowledge of ourselves and of the objects that surround us is not gained only to satisfy that craving for information which is always increased in proportion as the mind expands; it is a necessary basis to several other studies, it is eminently adapted to give to the judgment that correctness without which the most brilliant qualities lose their value, and in the course of life destroy their possessor rather than conduct to a useful end. The practical importance of the natural sciences is too apparent for us to venture to demonstrate it. In order to convince ourselves of it, we have only to cast our eyes around us; to think of the riches buried in the bosom of the earth, and of the services which geology and mineralogy render each day to our manufactures; to see the varied and beautiful plants which supply our wants with such prodigal magnificence, and consider that it is Natural History which serves as a guide to agriculture; to enumerate those animals which furnish us with wool, silk and honey, which lend us the strength we want or which instead of being useful to us like the preceding, destroy our harvests; to recall to ourselves lastly the numerous evils with which the human frame is at times afflicted, and to bear in mind this truth, that medicine always acts in the dark, whenever she does not lean for support on the scientific study of the nature of man. The practical importance of these studies, we repeat, has no need of proof, and is felt whatever may be our pursuits; but their utility is not limited here, and the influence which they exercise over our faculties, also merits the most serious attention. Indeed the Natural Sciences, from the peculiar manner of prosecuting them, accustom the mind to look from the effect to the cause, and at the same time continually to subject the results deduced from preceding observations to the proof of fresh facts: they lead us to the most speculative ideas, but never allow the imagination to be spoiled, for they always place the material proof by the side of hypothesis. Lastly, better than any other study, that of Natural History exercises the mind to *method*, a part of logic without which all investigation is laborious, and every exposition obscure.

Natural History ought therefore to constitute one of the elements of every system of a liberal education; but this does not mean that it is necessary to make every young man a naturalist. A science so vast to be studied deeply would require time, which other classical studies would not allow him to employ in that manner, and comprehends a crowd of details useful only to those who wish to make it a subject of especial research. The knowledge of an enlightened man need not extend to that of the characteristics by means of which such a genus of plants or animals is distinguished from such another, nor of the exact course of each artery or nerve in the human body: to load his memory with these things would be to impose upon himself a labour which would leave traces neither durable nor useful; but it is important for him to have correct notions of the great questions, the solutions of which are the pursuits of the natural sciences; of the formation of the globe and the physical revolutions which have succeeded one another on its surface; of the nature of plants and animals; of the manner in which the functions of these beings are performed, and of the principal modifications which are observable in their structure, according to the kind of life to which they are destined. This is knowledge which once acquired, is hardly ever forgotten, and will not only serve as the basis of the special studies of him who wishes to become a naturalist, but is also sufficient for men whose occupations do not connect them intimately with the sciences. These are consequently the general notions with which we wish to impress the mind of those who are receiving a liberal education, and this is the course which we propose to adopt in this book.

DIVISION OF NATURAL BODIES INTO THREE KINGDOMS.

§ 2. Natural History, as we have already mentioned, treats of all the bodies scattered over the face of the globe, or assembled in the interior of the earth; and these bodies, as every one knows, are of two kinds, *inanimate bodies* or *minerals*, and *animate* or *organic bodies*. The latter are in their turn divided into two groups, which no one can mistake: *vegetables* and *animals*. So that in scientific language, as well as in common parlance, we distinguish three great divisions, or KINGDOMS, known under the names of MINERAL, VEGETABLE, and ANIMAL.

On commencing the study of Natural History, we are necessarily led to ask in the first place on what these apparent divisions are founded, and to enquire what are the fundamental differences which distinguish an inanimate body from an animate being, a plant from an animal.

DIFFERENCE BETWEEN INANIMATE BODIES AND ANIMATE BEINGS.

§ 3. These points of dissimilarity are numerous, and arise in whatever point of view we compare minerals with organic beings; the origin, the kind of existence, the duration, the method of destruction, the general form, the intimate structure, and even the very elementary composition: all are dissimilar, as very few words will suffice to demonstrate.

§ 4. Thus the *origin* is not the same in inanimate bodies and living beings. Now, when a mineral body is formed, it springs immediately from the union of two or several substances, which in their nature differ essentially from its nature, and which combine with one another on account of the chemical affinities with which they are endowed. A living being on the contrary is never produced by these spontaneous combinations of matter; it can only be formed under the influence of a living body similar to itself, and the vital force essential to its existence is transmitted in an uninterrupted succession of individuals springing from and resembling one another. Common salt, for instance, will be formed, whenever two particular substances, soda and hydrochloric acid, which in no way resemble this product, happen to unite; and these substances to be thus combined will not need the presence of a salt similar to that which they are going to form. A plant or an animal on the contrary is never thus created from the beginning, and, in order to exist, must necessarily participate in some way in the life of a *parent*, that is to say, of a living body previously developed, and from which it proceeds. These beings, in order that they may exist, seem to require a foreign impulse, and this impulse they can only receive from a body similar to that which they themselves will be.

§ 5. The *manner of existence* of living beings, compared with that of inorganic bodies, is equally characteristic. Inanimate bodies, such as stones and minerals, are in a state of permanent internal repose; the molecules of which they are composed are not renewed; if they increase it is only because fresh bodies similar to them happen to be deposited on their surface; and if they lose a part of their own substance, it is quite accidentally, and by the action of some outward force completely independent of the cause of their existence. Every living body on the contrary is the seat of an internal and unceasing movement of molecular composition and decomposition, as a consequence of which a part of its component substance is insensibly renewed. It is unceasingly occupied in incorporating into its own substances foreign molecules, which it derives from without, and also perpetually restores to the exterior a portion of its constituent substance. This species of circulation constitutes the phenomena of *nutrition*, and its continuity is a condition of the life of every organic being. It is also on this movement that the changes of volume, which living bodies undergo, depend; when their mass diminishes, it is because the quantity of substances expelled exceeds that of the fresh molecules which they assimilate; and when they increase, it is by *intussusception*, and not by *juxtaposition*, as in minerals; for the fresh matter added to their mass is not deposited on the external surface, but penetrates into their substance, so as to be interposed between the already existing molecules, at the same time that they replace those which the nutritive process had thrown to the outside.

§ 6. Lastly, after having existed in this way during a certain time, the *limit* of which is determined in each species, living beings infallibly perish, whilst inanimate bodies, once formed, exist so long as they are not destroyed by some foreign force; their duration has no necessary limit, and they contain in themselves no principles of destruction. In organic beings, we repeat, *death* is every where a necessary consequence of life; and as they cannot spring up spontaneously, they would soon disappear from the surface of the earth, if, besides the faculty of supporting themselves, they had not also the power of propagating their species; but this power is equally accorded to every living being, and thus constitutes one of the characteristics which essentially distinguish organic beings from inorganic bodies.

§ 7. The differences which are observable between inorganic bodies and living beings, when considered with regard to their form and size, also deserve to be mentioned. Every living being is to a certain extent predestined to acquire some general determinate form, which it has not when it commences to exist, but which is gradually developed; and this form has nothing of the geometrical simplicity which we find in minerals, when their molecules are united in crystals. Each living being is also subjected to certain limits in size which it cannot exceed; and an internal force tends to determine its increase, until it approaches these limits, which vary according to the species. Among inanimate bodies, it is quite different; their mass has no necessary limits; marble, for instance, may exist under the form of a microscopical fragment equally well as under that of an entire mountain; a plant, an insect, a bird could not live without they attained their determinate size, and could never exceed certain limits, which nature has assigned to their growth. An inanimate body also may be always divided mechanically, without the parts thus separated changing their nature, and losing their essential properties; the different parts of the same mass are not necessarily connected with one another, and it is only in the mind that we can admit of an indivisible mineral *individual.* Among plants and animals, on the contrary, different parts united by nature constitute a whole necessary to the existence of each of them, one single whole, an *individual being*, distinct from every thing which surrounds it, and not capable of being mutilated beyond a certain degree without ceasing to exist.

§ 8. Other characteristics peculiar to living bodies are furnished by their internal structure. They are always formed by the union of solid and liquid parts; the latter are spread in a greater or less proportion throughout every part of the mass, and the solid parts to contain these liquids assume the form of thin laminæ or filaments, arranged so as to surround interstices or cavities. A similar arrangement is met with in every living body, and the name of *organisation* has been given to this general structure; but in the mineral kingdom we never find an analogous texture. This mode of conformation is a condition of the existence of every living being, and we shall easily understand that this is necessary, if we reflect for an instant on what we have said concerning the nutritive process, which constitutes the most constant and characteristic phenomenon of life. Indeed, to give these bodies any form whatever, solid parts are evidently necessary; and in order to cause the foreign substances, destined to be incorporated with them, to penetrate into their innermost tissue, and to reject the particles which cease to belong to it, fluids are also necessary, for fluids alone present in their molecules sufficient mobility to impart such a motion. These fluids can penetrate to whatever part there may be life to support, into the thickness of the solids as well as to their surface, and therefore these solid parts must consequently have a spongy and areolary texture. It is therefore impossible to conceive the existence of a movement similar to the nutritive work without a structure such as that of which we have just spoken, and as we have already said we are taught by observation that this organisation is found in all living beings, in vegetables as well as in animals, so we give to these beings the name of *organised bodies* (*corps organisés*), in opposition to minerals, which are termed *inorganic bodies* (*corps inorganiques*).

§ 9. Lastly, we shall find important differences in the elementary or chemical composition of the matter in the mineral kingdom compared with that in the great division of animate beings.

An inanimate body, such as a stone or mineral, may be formed by molecules alone of the same simple or elementary substance, iron or sulphur for instance, or else result from the combination of two or more of the chemical elements, the list of which is now raised to more than fifty: nature has imposed on herself in this respect no restriction, and in every compound mineral body it has associated the constituent elements only in very simple proportions.

In living bodies, it is by no means so; they are always very complex in their chemical composition, and in order to gain a correct knowledge of the nature of these component parts of their bodies, we must divide these substances into three classes. Indeed, among these substances, some are also met with in the mineral kingdom, and present in plants and animals nothing peculiar; water and several salts may be mentioned among these, and will enter into the class of *inorganic bodies.* Other substances, which we may term *organic substances,* sugar and urea for instance, bear a great resemblance to the former in their manner of formation, but are only formed in nature by the action of life. Others, again, such as albumen, fibrine and cellulose, for which it is convenient to reserve the name of *organised substances,* (*matières organisées*), resemble the latter in their origin, but are distinguished from them as well as from inanimate bodies by chemical characteristics of great importance; they always result from the combination of three or four determinate elements; namely carbon, hydrogen and oxygen, either alone or combined with a fourth principle, azote; they are remarkable for their slight stability and for the manner in which they are destroyed by putrefaction, when they are exposed for a certain time to the influence of the warm and humid atmosphere; lastly, they differ from inanimate bodies in their manner of molecular constitution, for as chemistry informs us every atom of an organised substance results from the union of a very large number of atoms of the different elements assembled together in order to form it, whilst in the mineral kingdom each atom of a compound body contains only a very small number of elementary atoms[a].

[a] Thus an atom of carbonic acid is formed by one atom of carbon united to two atoms of oxygen, whilst an atom of the species of fat known as *stearine*, seems to contain 140 atoms of carbon, 134 atoms of hydrogen, and 5 atoms of oxygen.

Now it is these organised substances, which form the essential basis of all the living parts of animals and plants, which constitute in some kind the woof; and organic or mineral substances only play a part more or less secondary in the economy of these beings. Every living body is, consequently, chemically characterized by the presence of these peculiar compounds of carbon, of hydrogen and of oxygen, or else of azote combined with the three elements we have just named; for in the mineral kingdom we are not acquainted with any similar compound.

§ 10. Thus living bodies differ from inorganic bodies in their chemical composition, in their internal structure, in their general conformation, in the manner of their origin, of their existence, and of their destruction. But in order to characterize them, it is not necessary to enumerate all these differences; it is sufficient to say that they are *beings which support and reproduce themselves;* for these are the most remarkable and general phenomena by which life is manifested. The essential characteristic of plants and animals, considered collectively, is therefore *life*, which these beings mutually enjoy, and life itself, reduced to its simplest definition, is the faculty of supporting itself; but, as we shall soon see, it is only very rarely that we find it in such a simple phase, and it is generally the origin of a multitude of other phenomena.

§ 11. Science possesses no certain facts concerning the *principle of life*, but, as we personify in physics the source of heat under the name of caloric, although we know nothing of its nature, so in the same way in physiology, in order to facilitate the expression of facts, we admit the existence of a special force as the cause of the phenomena peculiar to living beings, at the same time inexplicable by the common laws of chemistry or of physics; this cause has been termed the *vital force*, but we are ignorant of the laws which regulate it. We know nothing more than that it is developed only in organised bodies, and that in order to be there found, these bodies must be placed in certain determinate conditions of existence. Thus one of the circumstances indispensable to the manifestation of the vital phenomena, is the presence of a certain quantity of water in the bodies of the organised beings. There are both plants and animals in which life is completely suspended by the effect of dessication, and restored again as soon as the moisture necessary to its existence is once more imbibed by the apparently dead being; but in the majority of cases this deprivation of water immediately causes death. Another condition of the existence of living beings is the influence of a certain temperature. Lastly, to all the influence of the air is necessary.

ORGANS.

§ 12. Besides, the vital force is manifested only by the intermediate agency of the *organs* or instruments, more or less numerous, which conjointly form the body of the living being. Each of the phenomena which are developed in a plant or in an animal, is the result of the action of a determinate part of its body, and there always exists a necessary connexion between the conformation of this part and the nature of the actions with the performance of which it is entrusted. Thus man can execute his movements only by the agency of certain *organs* or instruments termed muscles, and can only gain the knowledge of what is passing around him by the assistance of the *organs* of the senses, and the conformation of each of the organs varies according to its functions.

METHOD OF STUDYING LIVING OR ORGANIC BODIES.

§ 13. The study of the way in which the organs of an animal or of a plant are formed constitutes that branch of Natural History known under the name of *anatomy*. The study of the *functions* of these beings bears the name of *physiology*.

ANATOMY therefore is the science *which treats of the structure of organised bodies*, and PHYSIOLOGY is the *science of life*. But between these two sciences there exists the closest connexion, for physiology depends upon anatomy, and anatomy in its turn would lose the whole of its interest, were it separated from physiology. In short, to comprehend the mechanism, by which a vital phenomenon is produced, we must first be well acquainted with the arrangement and structure of the organs that cause it, and on the other hand, the knowledge of the structure of these organs would be of little importance if we did not, at the same time, attempt to discover their uses.

Anatomy and physiology constitute the base of the Natural History of organised beings; these two sciences are insufficient, and we must consequently study them in other connexions in order to gain a knowledge of plants and animals. Thus, in order to distinguish this almost innumerable mass of bodies from one another, we must observe the peculiarities they present, and employ them as *characteristics* to recognise each of these with exactness. We must also, to assist the memory, *class* them, so as to facilitate these distinctions, and group them according to the different degrees of resemblance or dissimilitude remarkable in them; for by the aid of such classifications, we state in a few words the most important points in the history of living beings. It is equally interesting to consider the way in which plants and animals are scattered over the surface of the globe, as well as the laws which preside over this distribution: so also the varied uses of these bodies. Lastly, Natural History is not only occupied with the beings which to-day sport around us, but also researches into the darkness of antiquity, and, seeking the traces of those which time has destroyed, and examining the fossil remains left in the bosom of the earth by its ancient inhabitants, arrives at the knowledge of what existed before man himself had drawn his first breath. These various studies are naturally divided into two branches, according as their object is animals or plants. The name of ZOOLOGY is therefore given to the history of the animal kingdom, and that of BOTANY to the science which treats of the vegetable kingdom. In this treatise Botany will alone engage our attention.

DIFFERENCE BETWEEN PLANTS AND ANIMALS.

§ 14. On comparing organised beings with inorganic bodies, we have pointed out the principal characteristics which distinguish the animal from the mineral kingdom; but these characteristics also belong to the vegetable kingdom, for they are inherent in every living thing, and on commencing the history of plants we must also point out the differences which separate them from animals.

The boundary between the animal and vegetable kingdoms is not always as easy to recognise as one would at first suppose, for there exist bodies of great simplicity of structure which seem to establish a kind of connexion between these two groups, and sometimes embarrass the naturalist when he attempts to classify them; but in the majority of cases, nothing is easier than to distinguish a plant from an animal, and the uncertain cases of which we have just spoken are perhaps referable to our imperfect knowledge rather than to the nature of things; cannot we therefore stop here, and state in a general manner that plants are distinguished from animals by very important characteristics, derived at the same time from the nature of the phenomena by which life is manifested in these beings, from their structure, and from the chemical composition of the principal constituent substances of their bodies.

§ 15. The object of the acts of vegetables are merely the support of the individual or the reproduction of the species. In animals life appears under a more complicated form; to the faculty of supporting and reproducing itself is added that of making, under the influence of an internal will, motions tending to a determinate end, and also that of receiving and perceiving external impressions. Hence is derived the name of *animate beings* given to animals in opposition to vegetables which are called *inanimate beings.*

Thus VEGETABLES are *bodies which can support and reproduce themselves, but which can neither feel nor change their position from any will of their own;* and ANIMALS are *bodies which support and reproduce themselves, which feel and move voluntarily.*

There exists also a considerable difference in the manner in which plants and animals exercise the same functions; thus the two grand divisions of living bodies do not perform the functions of nutrition in the same manner; this diversity of appearance can only be known by studying these functions, and it would be premature to enter upon it now.

§ 16. There is no less variety in the conformation of the organs which constitute a vegetable or an animal, consequent on the different functions. Animals, being endowed with a greater number of faculties than vegetables, necessarily possess more varied and complicated organs. But it is not only in this that they differ anatomically, the structure of the constituent tissues of their organs is not the same. The parts which compose these tissues, and which are, as it were, the organic materials of a vegetable, are essentially formed in the shape of *cells* or *utricles,* furnished with walls and hollow in the centre; it is not so in animals, the tissues are for the most part composed of filaments or laminæ, which intercross one another so as to circumscribe imperfect lacunæ and to constitute masses or membranes more or less spongy, but not divided into a number of independent cells as in vegetables; animal tissues in the process of formation frequently, it is true, appear to be composed of cells, but generally, this structure, which is permanent in the plant, is merely transitory in the animal, and remains only in the least active parts of the economy, as in the epidermic membrane.

§ 17. Lastly, to the characteristics derived from the formation and structure of plants and animals, we must add those furnished by the chemical nature of these beings. In short, the organic substances which form the base of the living tissues are composed of carbon, hydrogen, and oxygen, only in plants, whilst azote or nitrogen enters along with these three elements into the composition of that of animals. There exists, however, azotised matter among vegetables, as well as substances which contain no azote among animals; but the organic matter essential to the formation of living parts presents in the two kingdoms the chemical combination we have just mentioned.

THE

ELEMENTS OF BOTANY.

---

§ 1. BOTANY is that science which treats of the vegetable kingdom.

We cannot at present give a more correct notion of the distinction between the three kingdoms at the commencement of this book, than that which is found in the introduction. A stricter and more correct definition will arise from the facts which will be stated in the course of this book.

To the word plant we commonly attach the idea of a tree or herb, and we may at first be satisfied with this general notion. This plant commonly has roots, a stem and branches, leaves, flowers, and afterward fruits and seeds. This is what every one knows, and those who have investigated the subject more deeply, know nothing more than that these parts, the flowers for instance, are themselves composed of several smaller parts.

If we divide these last in their turn, and then endeavour by a series of analyses to divide into still more minute particles those which remain, we shall at last find bodies which cannot again be divided. We must consider these as the elements of the body which we are examining, and give them the name of *elementary organs* (*organes élémentaires*). The parts resulting from their union, which themselves form a clearly limited whole uniting in the performance of some act or function of life, receive the name of *compound organs* (*organes composées*).

ELEMENTARY ORGANS.

§ 2. Now although we have given to these small parts the name of elementary organs, we must by no means consider them as such definitely and absolutely, for the human mind can conceive no body that has no parts, or is so small as to be incapable of being divided. Reason, however, tells us that there is a certain limit beyond which our senses, aided by the most powerful auxiliaries of science, can shew us nothing clear, nothing certain, and then we are apt to wander through the obscure and boundless path of hypothesis. The improvements in the manner of investigation and the instruments of observation, especially in the microscope, have greatly extended our knowledge of facts, and thereby enabled us to test with greater accuracy the value of theories[a].

When we employ these means to examine some portion of a vegetable, we find that it is composed of a multitude of cavities of various shapes and sizes. Some are surrounded by a wall, as it were, by which each is closed up from those which surround it, and may consequently be compared to a bladder or vesicle: the rest are only the interval between the former empty spaces which these bladders, piled one above another, leave between themselves, whenever their walls or sides do not immediately touch.

The shape of the bladders with walls may be reduced to three principal modifications. Sometimes they are almost equally distended on every side, or at least they do not extend one way more frequently than another. The bladders which assume this shape are called *cells* (*cellules*)[b], or *utricles* (*utricules*) (*fig.* 1).

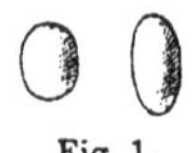
Fig. 1.

Sometimes they are lengthened in one direction, so that this diameter equals the transverse diameter a certain number of times. They are then commonly sharpened at their ends; if they are short, they have the form of a spindle, whence M. Dutrochet called them *closteres* (κλωστὴρ) (*clostres*) (*fig.* 2). When they are a little longer than the latter, their form is that of a tube pointed at each extremity. Since these cavities constitute the greater part of the wood, and as in this case they are commonly known under the name of ligneous or woody fibres, botanists have agreed to give them the generic name of *fibres* (*fig.* 3).

[a] Without the aid of the microscope, the parts of which we shall first treat cannot be well seen, and it is to be regretted that the pupil cannot verify our statements by his own observation. He must, therefore, obtain the assistance of some one, familiarized with the use of the instruments, and the preparation of the tissues, to shew him under the microscope the principal modifications. To facilitate this, our examples have been taken from those plants most easily procured.

[b] The French word for any botanical term will always be given, except when it is spelt alike in both languages.

Lastly, these bladders may assume the shape of long tubes whose ends are so far apart that both cannot be seen in the field of the microscope at the same time. We then term them *vessels* (*vaisseaux*) (*fig.* 4).

Between these three forms, cells or utricles, fibres, and vessels, the boundaries are by no means clearly defined. Fibres, for instance, may be so short as to receive the name of utricles, or so long as to merit that of vessels: but this confusion has very little inconvenience, since in reality we always treat with the same class of organs, diversely modified. Let us now examine each one of these forms successively, and the secondary modifications of which it is susceptible.

2 3 4

UTRICLES OR CELLS.

§ 3. When the cells are not pressed against one another, when they are equally developed in every part, without meeting with any obstacle on any side (*fig.* 8) to stop their growth, their surface is round, and their form is sometimes that of a sphere (*fig.* 5), sometimes of an ellipsoid (*fig.* 6). When, on the contrary, during their developement they meet and press against one another, the faces thus brought in contact are necessarily flattened, and the bladder assumes the form of a solid with many angles or polyhedron (*fig.* 7).

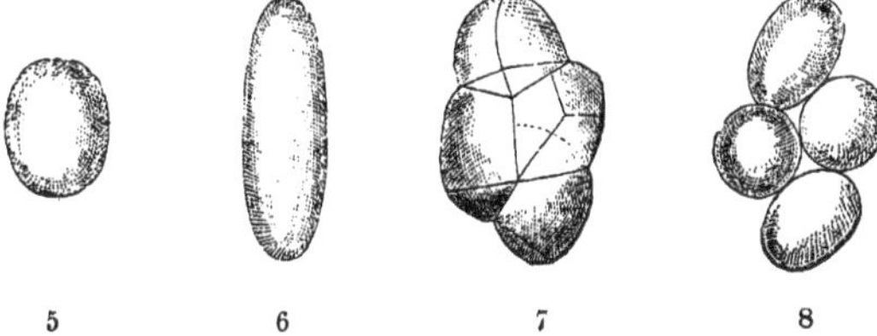

5 6 7 8

In this state their appearance, somewhat similar to that of a honeycomb, gave them (*figs.* 14 & 15) the name of cells, which is now indiscriminately employed with that of utricles. The tissue which results from their union is designated by the adjective *utricular* (*utriculaire*), or cellular (*cellulaire*), or else by the substantive PARENCHYMA (*Parenchyme*). Some writers have proposed to term the close tissues formed of angular or polyhedric cells alone (*fig.* 7), Parenchyma, and to distinguish the loose tissues composed of spherical or ellipsoidal cells (*fig.* 8), from them by the use of the word MERENCHYMA (*merenchyme*) for the latter.

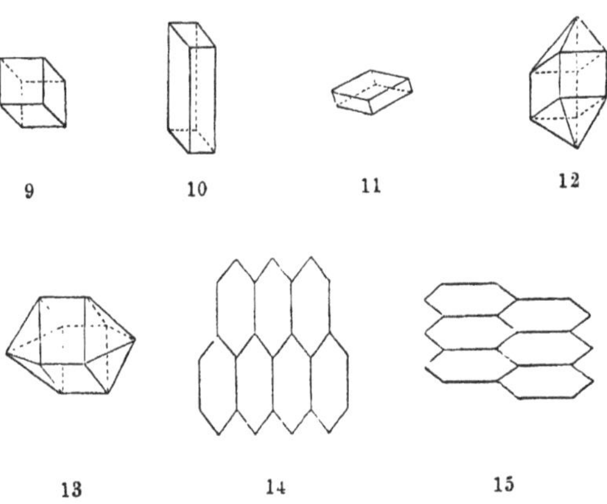

The commonest forms of polyhedric cells are the following: 1st, the cube or die (*fig.* 9): 2ndly, the prismatic column with four sides, in which the height exceeds the other dimensions (*fig.* 10): 3rdly, the tabular form, that is, a prism in which on the contrary the height does not equal the other dimensions (*fig.* 11): 4thly, the dodecahedron (*figs.* 12 & 13). We can pretty nearly determine the shape of the cells by comparing the horizontal and vertical sections of the tissue, without taking the trouble to isolate them. It is, for instance, apparent that the cubical cells, cut either horizontally or vertically, would give equal squares in either section; and that the dodecahedron (*figs.* 12 & 13) in one section shews a square, and in the contrary a hexagon, (*figs.* 14 & 15), &c. We must, however, guard the reader from supposing that these cells have the strict regularity of the geometrical figures to which we have compared them. They are generally far from it. The angles are rounded, the sides of the same square are not quite equal, the lines are not quite straight. By having represented a regularity that exists not in nature, the ancient plates of vegetable anatomy have partly owed their dissimilitude to the object for which they are intended.

The cells may be found on one part of their surface and flat on the other. This combination may be perfectly regular; as in the form of a cylinder (*fig.* 16 *a*), or of a barrel (*b*).

They may, lastly, be branched, that is, they are elongated at several points of their surface and in different directions. In this case the cells commonly touch one another by the points or ends of these elongations, and are frequently also very irregular (*fig.* 17); and yet even here some little regularity may be observed: thus, this form of cell often gives a very good idea of the trunks of fluted columns, &c.

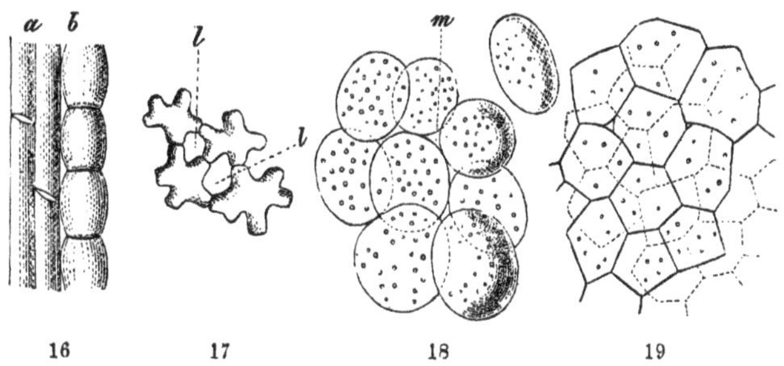

16 17 18 19

§ 4. In compact tissues, where the cells are, as it were, mortised into one another, and, touching only by flat surfaces, fit exactly, we may easily conceive that there will be no empty space left between them (*figs.* 7 & 19). In loose tissues, where the round surfaces can only touch in a small number of points, there must on the contrary remain between them greater or less intervals (*figs.* 8, 18): these intervals are called ***intercellular meati***, or ***passages*** (*méats intercellulaires* (*fig.* 18 *m*). These meati exist however in the generality of tissues, because on account of that slight degree of irregularity of which we have spoken as a general fact, the arrangement of the parts is not rigorously exact: but still, at the same time it must be remembered that these intervals are fewer in number, and smaller in size, in proportion to the compactness of the tissue.

These meati necessarily occupy a larger space between the

17. Branched cells from the Bean (*Fève de Marais*) (VICIA FABA).—*ll* Lacunæ.
18. Loose cellular tissue or merenchyma from a young leaf of House-Leek (*Joubarbe*) (SEMPERVIVUM TECTORUM).—*m* Intercellular meati.
19. Cellular tissue from the pith of the Elder (*Sureau*) (SAMBUCUS NIGRA). The cells are dotted in this as well as in the preceding figure.

branched cells, which touch, it will be remembered, at the extremity of elongations radiating from a common centre, and are then called *lacunæ* (*lacunes*) (*fig.* 17 *ll*). This name is generally given to every interval rather larger than ordinary, contained between several cells, and having no other wall than those of the surrounding cells. We frequently find great regularity in lacunæ, whether considered as to themselves or as to their position with regard to one another (*fig.* 20).

20

§ 5. Cells may be placed together without any appreciable order in their arrangement: and this happens most frequently when they are irregular in their shape, and unequal in their dimensions. But when they are regular and equal, we sometimes observe a certain degree of system in their arrangement, and they are disposed in rectilineal lines, sometimes horizontally, sometimes vertically. The cells of the two adjoining series may either be opposite, i. e. situate at the same height (*fig.* 21), or else alternate, i. e. at different heights, so that the middle of the cells of one series nearly corresponds with the extremities of those of the adjoining series (*fig.* 22). The latter series is almost necessarily found, when the cells are larger in the middle than at the ends; as in the dodecaheron (*figs.* 14 & 15).

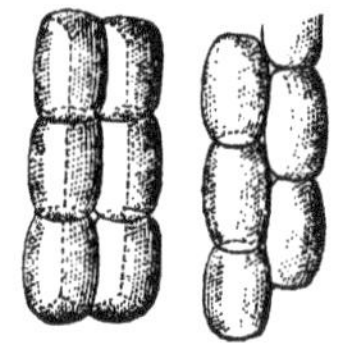
21 22

§ 6. The walls of the cells do not always present the same appearance. They are sometimes formed of one simple homogeneous membrane (*figs.* 5, 6): sometimes this membrane is marked by a certain number of little points (*fig.* 23), or of short transverse or

23

24

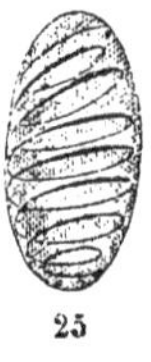
25

26

27

20. Lacunæ in the tissue of the Water Crowfoot (*Renoncule aquatique*) (RANUNCULUS *aquatilis*).

23 and 24. Dotted and striped cells from the tissues of the Elder (SAMBUCUS NIGRA).

25, 26, and 27. Spiral, annular, and reticulated cells from the tissue of the Misseltoe (*Gui*) (VISCUM ALBUM).

oblique lines (*fig.* 24): sometimes it appears to be doubled at certain intervals by little threads or bandlets; these threads commonly describe a spiral coil from one extremity of the cellule to the other (*fig.* 25); these bandlets, also, either follow one spiral line, or else separate into small, nearly horizontal rings (*fig.* 26); or lastly, form on the surface a kind of net with larger or smaller meshes (*fig.* 27). Now, observation has assured us that these divers appearances do not constantly characterize cells of different kinds, but that the same cell may present several of them successively, according to the time at which we examine it. It is necessary then to follow their developement closely, in order to explain clearly their different appearances and the causes which produce them.

This examination teaches us that the cell, directly we begin to perceive it as a distinct organ, is a small bladder formed by a simple membrane, perfectly continuous and homogeneous, the substance of which, at first soft and humid, gradually dries and hardens. We sometimes find that it remains in this state with the exception of a change in its form and volume only. But sometimes, over the whole of the interior surface of the bladder, a second is formed at a certain subsequent period. This new membrane does not appear to be identical with the first in its manner of developement; for, instead of extending in one continuous sheet perfectly corresponding to the former, it breaks in several places, perhaps, because the interior membrane, younger and softer than the exterior, is not developed with sufficient speed to follow its increase. At these places, the first is evidently not doubled by the second, and hence results that inequality of thickness at different parts of its surface. We may suppose, therefore, that the internal membrane, when thus distended, is frayed, as it were, in a great number of points, and thus are determined the dots which we perceive on several cells; but most frequently a marvellous degree of regularity seems to preside over these solutions of the continuity of the interior envelope, which is wound from the bottom upwards like a spiral thread or ribbon. If the edges of the revolutions of this spire do not touch, but are at an appreciable distance from one another, we have two parallel spiral zones, one where the external membrane is doubled, and the other where it is left bare. If, on the contrary, the edges touch exactly, the interval is indicated merely by an extremely fine line, which sometimes even ceases to be perceptible. But they are frequently separated a little at regular distances, leaving the exterior membrane naked in spaces which to our eye, do not exceed a point or a short line in their extent. Hence, perhaps, result the regularity and direction which are observable in these points and lines with which the cell is every where dotted. The annular and the reticulated or net-like bands appear to be susceptible of an analogous explanation, which we will resume in treating of the vessels, where the phenomena become less obscure, on account of the larger scale on which we can observe them.

The thickness of the walls of the cellule may be successively augmented by the formation of a third bladder, which is deposited on the interior of the second, of a fourth which is deposited on the interior of the third, and so on continually. If the third is not moulded exactly on the second, but happens to cover the parts where it has left the primitive membrane naked, we may conceive that it may be perceived in these spaces and hidden in all the other parts where it has already been doubled. Some authors have thus explained the compound appearance, as it were, of certain cellules; of those, for instance, which seem to be covered with a net, the meshes of which are dotted: the net would belong to the second, the dots to the third membrane.

But commonly the second membrane serves as a mould for those which are successively developed in the interior; they follow its shape and break where it breaks. This will at once be seen by looking at the vertical (*fig.* 29), or horizontal (*fig.* 28), section of a cell composed of a certain number of superposed membranes. We here see very plainly several concentric circles round a central cavity, which is small in proportion as the number of layers deposited increases; from this cavity small canals spread horizontally, which are closed by the external membrane, and correspond to the solutions of continuity of the secondary membranes. It is clear, that if they were not exactly moulded on one another, their revolutions would not correspond so as to form these continuous canals.

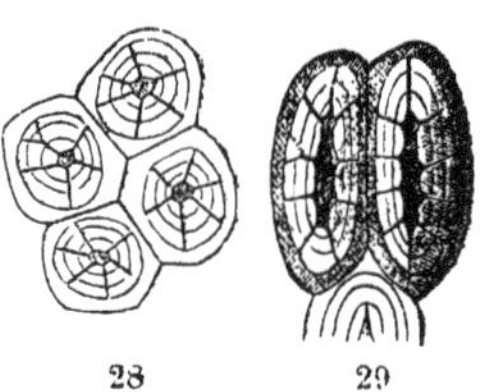

Fibres.

§ 7.—The history of the formation of cells, which has just been given, will save us from entering so much into detail on that of

28. Horizontal section of cells from the flesh of the Pear.
29. Vertical section of the same.

fibres, since they only differ in their form, and, their developement being the same, the appearance of their surface will present analogous modifications.

It has already been mentioned that the length of fibres is variable; that some are so short as to have received from certain authors the name of lengthened cells: some so long, that they resemble vessels, and have, therefore, been often classed with them under the name of fibrous vessels. The tissue which is formed by the union of fibres is called (PROSENCHYMA) (*prosenchyme*). Those which happen to be placed nearly at the same height, touch one another at their sides; but at their pointed ends necessarily leave between them spaces, in which are inserted the analogous extremities of the fibres, situated above and below (*fig.* 30). In the parenchyma, on the contrary, the upper and lower cells are placed upon one another on plain surfaces, (*figs.* 19, 21), which terminate them, (*cellulæ parenchymatis sibi extremitatibus impositæ sunt, prosenchymatis appositæ* [c]).

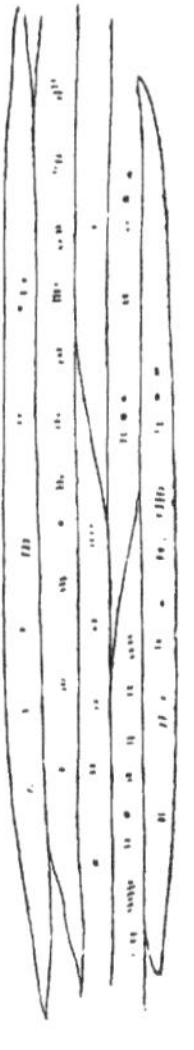

30

Their wall is generally thick and hard: it is at first, a short continuous membrane, which may acquire, without the addition of any other, a certain degree of thickness. But generally several layers grow successively from the exterior to the interior, so that the fibre, whose hollow axis is gradually contracted, and is at last reduced almost to nothing, may appear to be full or entirely solid.

Hence the section of a prosenchyma presents a generally compact mass, in which the proportion of the full and solid parts exceeds the hollow ones; the internal cavity of the fibres is at most a thin oblong canal; whilst their external surfaces touch so exactly as to render the intercellular meati almost imperceptible. Another consequence of this juxtaposition is the flattening of the sides in contact, so that the fibre is prismatic externally, though internally it re-

30. Fibres from the common Clematis (*Clématite commune*) (CLEMATIS VITALBA).

[c] The cells of the Parenchyma are placed over one another at their extremities; of the Prosenchyma, by the side of one another.

mains cylindrical. The horizontal section of a prosenchyma, sufficiently developed, shews this very clearly (*fig.* 31).

It has already been asserted, that the developement of the fibres is the same as that of cells. The increase of the primitive utricule, or of the external membrane, determines their length and breadth: the later formation of the internal layers determines their thickness, and the definite appearance of their surface, which may, therefore, present the same modifications as that of cells. Yet, it frequently happens that the internal layer covers the external without any interruption of continuity, so that the fibre remains as smooth as it was at first. The second layer may appear also, either as a spiral-thread, or as a number of little bands, forming the meshes of a net; but this condition is not often met with. Dotted fibres (*fig.* 32), on the contrary, are very common. These dots shew, as in cells, where the external membrane is not doubled by the internal, and where the little canals resulting from this interruption terminate. This is particularly remarkable in the wood of the fir and of the other analogous trees, commonly called evergreens, forming the family of the Coniferæ, which we shall consider in the course of the work. Here the dots are large enough to be mistaken for real holes; they are arranged in two rectilinear series, occupying the two opposite sides of the fibre, and each of them is often surrounded by a circle or areola (*fig.* 33). If we examine with a good microscope a very thin slice of the fibre cut down the middle of the dots, we shall see at these places the wall of the fibre bend towards the interior, thus determining the little hollow, the circumference of which is circular or elliptical (*fig.* 34) It is this hollow which forms the circle or areola, and at its centre a short lateral canal, analogous to those of cells or dotted fibres, ends and forms the central point, and, since the dots of two adjacent fibres

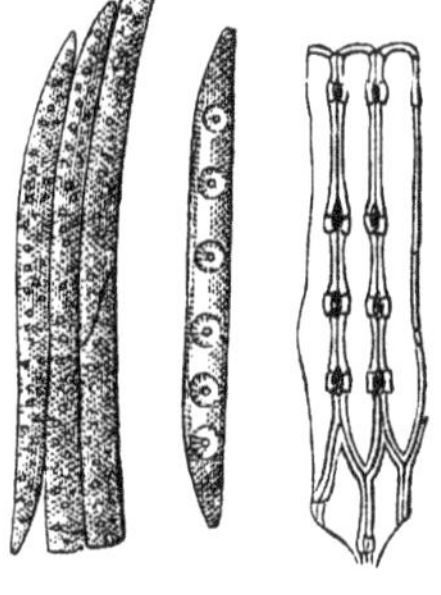

31. Horizontal section of the same.
32. Dotted fibres from the winged edge of a seed of a Bignonia (*Bignone*).
33. Fibres from the wood of the common Pine (*Pin*) (PINUS SYLVESTRIS).
34. Longitudinal section of the same.

ordinarily correspond, there is a little empty space of a lenticular form between them, as if two watch-glasses had been applied to one another at their circumferences.

Vessels.

§ 8. (*Vaisseaux*).—We have not as yet attempted to distinguish vessels or tubes from fibres, otherwise than by their greater length. This is sometimes considerable, and almost equal to that of the entire vegetable. One may easily shew this by passing through a long slip of the plant a fine hair, which, if introduced at one end of a canal, will come out at the other, and thus prove the continuity of the passage. When the plant is very thick and straight, as for instance, the branch of a vine, by applying the eye to one end, we may perceive the light at the other.

If one of these long vessels be laid bare and closely examined, when magnified by a lens of sufficient power, we shall instantly observe two characteristics: in the first place, its surface is never smooth, as that of cells or fibres frequently is, but always presents those inequalities, which we have seen in the latter at a certain age, under the appearance of dots, lines, rings, &c.; secondly, the cylinder formed by the vessel is not perfectly regular throughout its whole extent, but presents at intervals knot-like contractions or strangulations. These knots are sometimes at regular distances, and very near one another. But again, they may be found wide apart, or separated by unequal intervals. On observing attentively that portion of a vessel between two successive knots, we shall be struck with its resemblance to a cell or a fibre. This resemblance becomes still more evident, when the vessel is acted upon by dilute nitric acid at a boiling temperature. Whilst this portion is thus isolated, we can only distinguish it from a cell or a fibre, by its being pierced at the two extremities by which it is continued with the rest of the vessel.

We are led from these observations to conclude, that a vessel is formed by a series of cells or fibres, placed end to end, communicating uninterruptedly with one another by means of openings in the ends. If it be composed of cells in a series, then the strangulations will be near to one another, and the line which traces them will be horizontal or slightly oblique, like that formed by the surfaces on which cells are commonly placed on one another (*fig.* 22). If it be formed by a series of fibres, the contractions will be rather more distant from one another, and the line which points out their places will be very oblique, since they are joined at the side of the taper end of the cone (*fig.* 35).

On admitting these remarks, we shall see that vessels are organs, which cannot so justly be termed elementary as cells or fibres, since they are produced by the union of several of these latter. We ought not therefore to be astonished at finding upon their surface the same appearances of dots, lines, bands forming a continuous spiral fibre, or detached in the form of rings, or joined in a net, &c., which we have remarked on cells and fibres. There is, however, this difference, that in vessels we can constantly find these marks, whilst in cells or fibres we have seen that they do not exist in the primitive state, but are the result of a successive series of fresh internal layers, increasing both in number and size with age. We may, therefore, suppose that vessels, as we have described them, have already a certain age, that they are not thus formed of one continuous piece, but that they have previously passed through other forms.

Indeed, if we take a vegetable, or a piece of a vegetable at its first appearance, we cannot find the least trace of a vessel, but only utricles enclosed in a smooth and homogeneous membrane. It is only at a later period of their existence, that we see certain of these cells lengthening into fibres; and the plant is still older, when we find the walls losing their homogeneity and vessels shewing themselves. These vessels will have passed through the same periods of formation as the cells and fibres. A membranous bladder, at first single and continuous, is thickened by the addition of other bladders fitting one to another, and embroidered, if we may use the expression, with open-work: at the same time, it becomes closely connected with bladders of the same kind, the one placed above, the other beneath it; but the part of the walls thus united, instead of growing thick with the rest of the vessel, becomes thinner and partly disappears. The diaphragms which ought to remain at these points of junction, if not completely effaced, are represented, either by a small wrinkle following the circumference, or by open net-work. Thus a continuous duct is formed, closed externally by a membrane continuous in itself, single in a great number of points differently arranged, doubled, tripled, &c., in the rest of the internal surface.

Various kinds of vessels have been distinguished according to the general form of their tube and the various modifications of their surface. We will now name them in succession, and as briefly as possible. In this series of cells or fibres, we have only to treat of the combination of forms, of the repetition of appearances already well known. Yet, to the details already given, we may add something else; for the modifications of vessels, on account of their larger size, exhibit themselves more plainly, and hence have been more frequently and better studied than those of cells or fibres.

We have just stated that the surface of vessels is always uneven, and marked by points or lines, naturally arranged in the same manner as those of cells, i. e., generally in a spiral direction; thus we find them mentioned in the majority of modern works, under the generic name of spiral vessels or tubes, (*vasa spiralia, tubuli spirales,*) to distinguish them from vessels with smooth walls, or from those vessels called fibrous, of which we have spoken under the head of fibres, or from proper or lacticiferous vessels, of which we shall treat presently.

Spiral vessels themselves we separate into two kinds, the true vessels, or tracheæ; the false, which comprise the annular, the reticulated, striped, dotted, &c.

§ 9. TRACHEÆ[d], (*Trachées*),—are formed by a membranous cylinder, within which is wound a spiral fibre. This cylinder, without any change of form or surface, is of considerable length, and terminates in the apex of a cone at its two extremities, to which are frequently applied those of other tracheæ, which thus continue the first at the top and at the bottom. The tracheæ therefore are really composed of very long fibres (*fig.* 35).

The spiral fibre of a trachea extends from one end to the other without interruption. It may be compared to the brass wire which forms the spring of braces; and this gives rather a correct idea of its arrangement. Its colour generally resembles that of mother of pearl. With regard to its form authors have differed: some thought that the thread itself was a hollow tube; others that it was grooved into a kind of channel on the inner side; others have assigned to it other different forms too numerous to be enumerated here. From the most exact observations, made with the assistance of the most perfect instruments we now possess, it seems that the

35

36

[d] Henshaw discovered them in the Hazel in 1661, a year after Hook perfected the microscope. De Cand.—TRANS.

thread is always solid, but varies in form according to the places and the parts from which it is taken. It is sometimes flattened like a ribbon, and its section presents the form of a circle, an ellipse, or a quadrilateral figure. If we pull the broken trachea gently, the coils of the spiral separate one from the other, just like that of the elastic spring of a brace (*fig.* 36). When we break a young branch carefully, (of Elder, for instance,) the lower fragment is sometimes suspended to the upper by threads so fine that the eye can scarcely perceive them, these are the unrolled tracheæ; and this property has gained for these vessels the appellation of unrollable *tracheæ*, which is opposed to that of the other spiral vessels which do not possess it. This does not however appear at all ages indifferently; in the extremely young trachea, the tissue of which is still rather soft, the thread will not yet have acquired that elasticity it acquires at a later period, and will break with the tube without unrolling. It may also lose it again, as it grows old, doubtless by connecting itself intimately with the neighbouring parts.

The distance between the revolutions of the spiral varies, and each coil immediately touches the two nearest, both above and below (*fig.* 36). Then the external membrane cannot be perceived, for the interval between them is reduced to nothing: but united, doubtless, to the thread, it is torn and follows it in the process of unwinding. At other times the coils leave a perceptible interval between them, and even much larger than the thickness of the thread; and it is only in these cases that the external membrane can be seen at all plainly (*figs.* 37, 40).

As to the directions which the spire follows, one may be remarked more frequently than any other; that from right to left, supposing the vessel to be placed opposite to the observer in its natural position, that is, the extremity growing farthest from the ground turned upwards. We frequently also imagine the observer to be placed in the axis of a cylinder, around which the spiral line rises; and it is clear, that in this case, it seems to be in an inverse direction, from left to right (*fig.* 38).

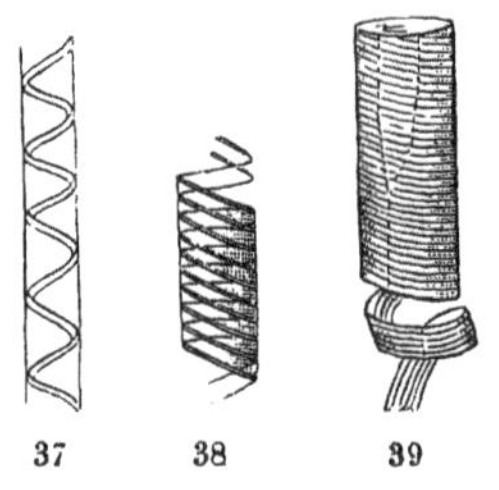

37. Trachea with separated convolutions from the stem of the Pumpkin (*Potiron*).
39. Trachea with several parallel fibres from the stem of the Banana tree (*Bananier*).

Under the microscope the face of the vessel turned towards the observer will be found with respect to him in the first position, and the face on the other side will be in the second, and the two lines of direction cross each other. If the vessel is so small that its two faces are almost placed together in the field of the microscope, it will then appear to be traversed by two threads which follow different directions, and describe a net-work of small lozenges (*fig.* 39). Some botanists have been deceived by this appearance, for which they could not account. But whatever may be the direction of the spire, it never changes from one extremity of the tracheal fibre to the other.

Most commonly the spiral fibre is single; but it is not unusual for it to be double (*fig.* 38). Sometimes there is a much larger number of them (*fig.* 39); and in the Banana (*Bananier*) more than twenty have been counted. These threads, close and parallel, are similar to a spiral ribbon. It is evident in this case that the obliquity of coils of the spiral increases in proportion to the number of the juxtaposed threads, since between two coils of each thread, there must always be the whole breadth of the ribbon. When on the contrary the thread is single, and the two coils touch, as in the spring of a brace, they are only separated by the thickness of the thread itself, and are situated in an almost horizontal line, so gentle is their ascent.

40

The single thread does not always remain so throughout the whole of its passage; but sometimes it forks and doubles itself, so as to present two finer fibres running parallel with one another, instead of only one (*fig.* 40). It may fork again, all the branches following the direction of the spire. This is a transition to reticulated vessels.

§ 10. ANNULAR AND RETICULATED VESSELS, (*Vaisseaux annulaires et réticulés.*)—The name tracheæ, if we go back to its meaning, would seem to be better adapted to the vessels we are about to describe, than to the preceding. Indeed, composed of a membranous tube, which is supported internally by the rings or circles placed one above the other (*fig.* 41), they may with greater justice be compared to the tracheal artery or windpipe of animals. They

40. Trachea, with a spiral fibre single below and double above, taken from the *Beet*, (BETA VULGARIS), (*Betterave*).

are generally thicker and much less uniform from one extremity to the other than true tracheæ. The rings of the same tube are not, in fact, always perfectly similar (*fig.* 42); although commonly horizontal, they may also be irregularly inclined to one side or the other; nor are they separated from one another by equal and regular intervals; and, lastly, they may be reduced to annular fragments, or represent some other kind of curve than a circle. Thus it is not unusual to see between the rings fragments of a spiral, which sometimes connects them (*fig.* 43), sometimes remains independent of them.

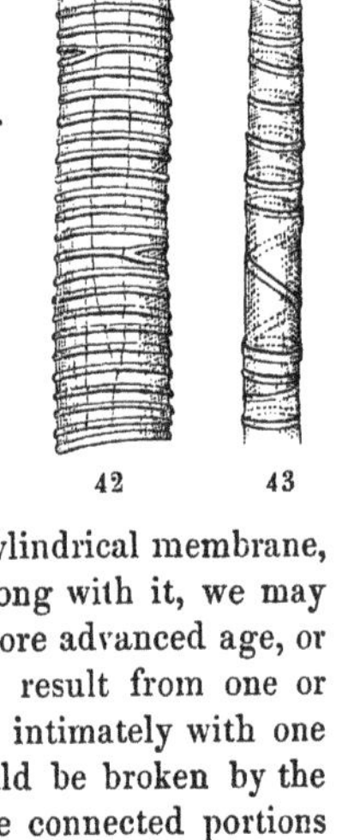

These portions of the spiral fibre have been naturally compared to those of a trachea; and when we have observed the only distinctive characteristic, that they are not unrollable, and, that adhering more or less to the surrounding cylindrical membrane, the spiral fibre in tearing draws the shreds along with it, we may conclude that these vessels are tracheæ of a more advanced age, or rather, stage of formation. The rings would result from one or more convolutions of the spiral fibre adhering intimately with one another: the continuity of the spiral fibre would be broken by the lengthening of the vessels, which separates the connected portions from one another; according to others, because the intermediate portion between two successive rings would frequently be absorbed. This hypothesis is supported by the fact that the rings are continuous with the spiral fibres, by the frequent existence in the middle of the annular ribbons, of lines, of furrows, or even of complete splits parallel at their edges, which seem to point out the line of union between two or more parallel fibres.

Other botanists think, that the internal thickening of the membrane of annular vessels does not commence by forming a continuous spiral, but that they are formed, as we see them, some portions in circles, some in spirals, some in a mixed form. These ob-

41 & 43. Annular vessels from the common Balsam (*Balsamine.*)

servers could not perceive at any time the continuity of the spiral fibre, and afterwards its dislocation and disappearance. Besides, according to the hypothesis which they oppose, the intermediate twisted portion between two rings ought always to point in the same direction; its breadth ought always to be some aliquot part of that of the ring, generally half, since this results from the union of several (generally of two) convolutions; the lines or slits, frequently observable on the ring, ought to mark the direction of the spiral fibre. But this is not the case. The intermediate portion is sometimes twisted in opposite directions, and that, even in the same interval; it is almost always as wide as the ring: the lines found on it, are parallel to the edges, and not obliquely directed from the lower to the upper (*fig.* 44).

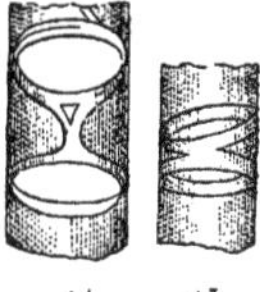

Sometimes we find two rings following one another, or even, when blended at one point of their margins, inclined towards each other in two inverse directions (*fig.* 45). This fact, along with the preceding, may be shewn to have arisen from the rings at first adhering so slightly to the membranous tube, that they are sometimes partly or completely detached, so as to be at the point of falling into the cavity of the tube; and then, abandoning their former position, they assume a fresh one, which to a certain extent approaches the vertical.

From all that has been mentioned, we can easily see with what facility the transition from annular to reticulated vessels takes place. These rings, at various degrees of inclination, connected with each other, either directly, or by means of little bands variously twisted, some of them embroidered with open-work by slits, even now frequently represent a kind of loose net-work. Let these elements multiply and draw nearer to one another, and we then have a closer and more complicated net, so that it is not uncommon to see a vessel annular in one part and reticulated in another (*fig.* 46).

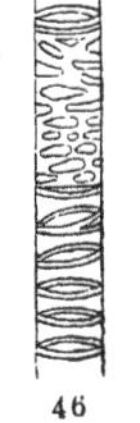

The termination of these vessels is a slender cone: and the length of the interval between these two ends proves that they, as well as tracheæ, are generally composed of fibres.

44 & 45. Fragments of vessels from the COMMELINA TUBEROSA.

§ 11. STRIPED VESSELS (*Vaisseaux rayés*),—instead of spiral fibres, circles, or irregular areolæ, exhibit transverse parallel lines which only occupy a part of the circumference of the tube, and are generally placed regularly above one another (*fig.* 47). The shape of the vessel frequently assumes that of a prism, the lateral faces of which are thus furrowed with lines which stop near the angles (*fig.* 48). This arrangement of these lines and their intervals has been compared to that of the rungs of a ladder, and, consequently, vessels presenting this appearance have received the name of *Scalariform*. The lines or bars are not, however, constantly developed in this manner, but sometimes take the form of small button-holes situated in the same places, but much more numerous, so that it seems as if several were formed at the expense of one single line broken at intervals.

47

48

It is now agreed to suppose that these vessels are, like the other organs previously examined, composed of a membranous tube, doubled in the interior by open net-work; the lines are the spaces which answer to the meshes, and in which the vessel is only enclosed by the external membrane. Some authors suppose that they have observed that the position and the shape of the lines agree to a certain extent with the parts by which the vessel is surrounded and compressed; that the lines remain where the surfaces are in contact with one another; that the membrane is thickened where it is in contact with an angle; that thus in those points where the vessel is contiguous to other similar vessels, there will be a degree of regularity which generally disappeared where it is contiguous to shorter and more numerous cells.

Striped vessels have been confounded by several authors with annular and reticulated vessels; when there is in the latter some regularity in the arrangement, when the meshes are narrow and lengthened in a transverse direction, it is difficult to distinguish between them. They have been considered also as a metamorphosis of the trachea, the convolutions of whose spiral fibre, dis-

47. A fragment of a striped vessel from the Vine (*Vigne*).

48. A fragment of a prismatic striped vessel from the Osmund-royal or Flowering Fern (*Osmund, Fougère-fleurie*) (OSMUNDA REGALIS).

tinct and somewhat separated in the spaces which answer to the stripes, would be confounded together in the intermediate spaces. This opinion seems to obtain some degree of probability in a certain number of cases, where the obliquity of the stripes follows a spiral line, and the vessel, when torn in two, is unrolled like a helix. But even then the fibre is not a single spiral contained between superposed stripes, but a zone comprising several parallel series of stripes (*fig.* 49). Besides, the stripes are for the most part perfectly horizontal, and seem to be so in the youngest vessels; the first change which we perceive on the simple homogeneous membrane, being the appearance of greyish parallel lines (*fig.* 50), and the earliest and most minute observation shews that in no way does it at all resemble a spiral fibre.

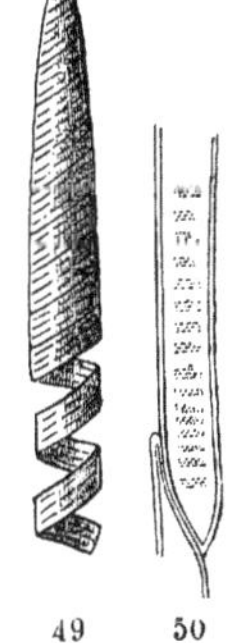

Striped vessels are formed by a series of fibres with cone-like ends, or else of elongated cells, terminated and adjusted one to the other by an oblique wall.

§ 12. Dotted Vessels (*Vasa punctata*),—which attain the greatest size in the vegetable, and whose internal canal may even be frequently seen with the naked eye, seem to be punctured with little holes generally arranged in parallel horizontal lines, sometimes, though more rarely, in oblique ones (*fig.* 51).

Botanists have assumed with regard to these series of opaque dots the same hypothesis as they did relative to the stripes, and the same objections may be adduced against it.

These vessels present the form of a cylinder, on whose surface are described circles without points, slightly oblique, or more frequently horizontal, placed sometimes at different, but

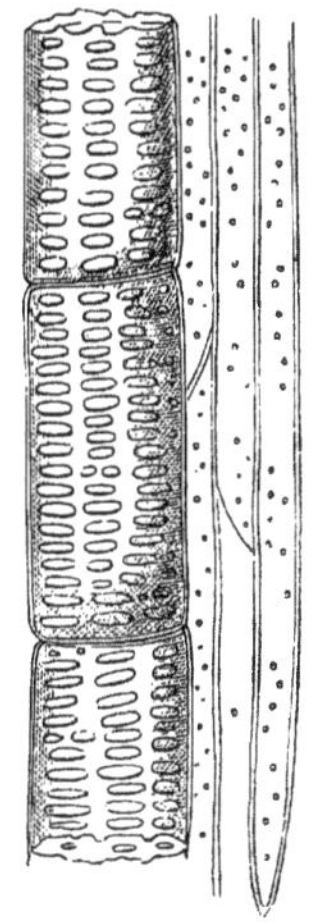

49. The end of a striped vessel from a Lycopodium; the lower part is unrolled like a ribbon.

50. A scalariform vessel at the commencement of its developement, from the root of a Date in germination.

51. A fragment of a dotted vessel from the Vine. Adjoining are some dotted fibres.

most commonly at equal distances. These circles have the same diameter as the rest of the tube, though sometimes it is rather less, and in that case a kind of strangulation or contraction is observable at intervals (*fig.* 52). The vertical section of the vessel shews us that in the interior, a circular fold or wrinkle corresponds with these strangulations; and sometimes also the vessel, when heated in aqua-fortis (nitric acid), breaks into several fragments following these lines, each of these fragments clearly representing a utricle, in the form of a barrel or tub with open ends. It is therefore apparent, that dotted vessels are formed by a series of utricles united together: and the dots are the places where the external membrane is left bare, not being covered by the internal ones.

If these strangulations which result from the union of a series of cells, swollen more in the middle than at the ends, are very apparent, the appearance of the vessel will remind us of a necklace of beads pressed against one another (*fig.* 53). These vessels, termed also necklace-like or worm-like vessels (*vaisseaux en chapelet*, or *vermiformes*), because they may be also compared to the body of a worm composed of a number of rings (*fig.* 54), are therefore only a modification of a more general form; and this modification appears not only in dotted vessels, but also in others. Generally, at the origin of fresh organs, as a twig springing from a branch, a leaf from a twig, where the vessels, in order to pass from one into the other, depart from their rectilinear direction, we find their elements in this place assisting this deviation by becoming shorter, more irregular, and uniting by smaller surfaces. The straight line breaks into a series of short lines so as to form a bend. It is thus that the tracheal fibres or annular or reticulated vessels, commonly very long when in the stem of the plant, contract at the knots, and even take the form of cells (*fig.* 54).

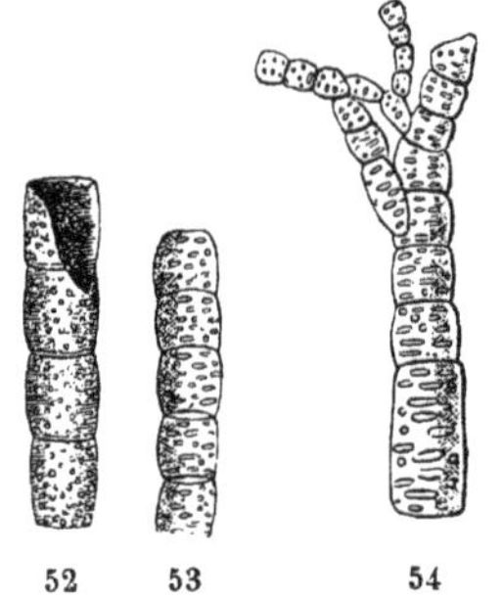

52. A fragment of a dotted vessel from the Common Clematis.
53. A fragment of a dotted vessel from the Miseltoe; it is passing into the appearance of a worm-like vessel (*Vaisseau vermiforme*).
54. A dotted vessel from the Balsam, assuming the same form in its upper part.

§ 13. Metamorphosis of Spiral Vessels.—We have now found annular vessels presenting long fragments of a spiral thread; we have found them passing in another part into striped vessels by reticulated ones; we have found the continuous line forming the stripes broken so as to be changed into a line of dots, and thus establishing the passage of the striped to the dotted vessel; we have sometimes seen them unrolled into little spiral bands by traction. We cannot, therefore, wonder much that the metamorphosis of vessels from one to another, is admitted by a great number of botanists, who see in all the different modifications previously explained only the different degrees of the developement of an organ, and think that a vessel may pass successively through all these forms, and present several of them at once in the different parts of its body. This hypothesis might even be carried farther, since it is impossible to distinguish precisely between vessels and fibres, fibres and cells: that generally, vessels are composed of the two others, and that all three originally presented the same appearance, a cell covered by a single homogeneous membrane. Let us also add, that the substance which forms the wall of a cell, of a fibre and of a vessel, is identically the same, chemically composed of the same elements and in the same proportion. We should therefore be tempted to conclude, that in a vegetable there exists only one organ, which, Proteus-like, in a series of transformations puts on the several modifications which have been taken for different organs.

But let us not take appearances for facts. Are not the different modifications of vessels the different successive degrees of developement in the same organ, and, according as this developement is arrested at such or such a point, shall we have such or such a modification? Observation, which proves the truth of theories, determines the contrary. On watching the developement of a vessel, we do not find any one, which in its different phases, would have represented all the other kinds of vessels, and the same thing may be said of cells. Remark moreover: 1st. That in each part of a plant, such and such modifications of cells, of fibres, of vessels, are found; we have, for instance, in certain places unrollable tracheæ; though in others we never meet with them. In the whole of such a plant, or in such a part of it, we shall never meet with striped vessels: in such another they will abound; 2nd. That, in spite of the similarity of the chemical composition of the walls, that of their contents is quite different; and, like the shape, constant in appearance, and agreeing with the place which the cavity occupies in the vegetable. Thus, therefore, if all the elementary organs of vegetables commence their growth as utricles, amongst which we cannot discover any appreciable differences, it is no less true that each utricle appears destined from the beginning to assume in its ulterior developement such a form and not such another, to contain or to elaborate such a substance. It is not, therefore, always the same organ. In passing from its original to its destined form, it may undergo several changes, and it is the limit or end of these transformations, which must be determined, before we can describe accurately the elementary organs of vegetables. In this research, care must be taken not to be misled by the illusions to which the extreme simplicity of the vegetable tissue gives rise; a simplicity, which, by stimulating differences and increasing resemblances, gives an appearance of a factitious unity, which we find so much the oftener and more necessarily as we study the smaller parts with imperfect instruments.

§ 14. LACTICIFEROUS VESSELS.—We have postponed to the end of this discussion, the description of an order of vessels which are so different from all the rest, that they have never been confounded with them: these are those vessels which have received the name of *milk* or *Lacticiferous vessels* (*vaisseaux lacticifères*), because they contain the juice of the plant or LATEX [e].

They are membranous tubes, communicating freely with one another by transverse branches, so that they resemble a large piece of net-work (*figs.* 56, 57). They resemble, therefore, more than all the other vessels previously described, the arteries of animals, one characteristic of which is to ramify throughout the body. This word ramification would, however, be improperly applied to milk vessels, which we do not find diminishing by successive divisions, as a trunk is divided into branches, these branches into twigs, and so on. Here the branches are almost equal to the canals with which they communicate, and from which they spring at either right or acute angles.

The membranous wall is transparent and homogeneous; for, if it is thicker at certain places than at others, these irregularities do not seem to be of any other kind than the rest of the vessel. The tubes

[e] The tissue composed of these vessels is called CINENCHYMA.—TRANS.

are at first very thin, slender, and regular cylinders (*fig.* 55 *a*); afterwards, as they grow, they either preserve their original cylindrical form or swell in certain places (*figs.* 55 *b*, 56); swellings, which during life may, perhaps, be temporary, and result, as in the veins of animals, from a momentary stagnation of the contained fluid, but which may sometimes be rendered permanent. The internal cavity may also be irregular and present strangulations or contractions at various distances, even when the exterior is smooth and regular.

M. C. H. Schultz, who has furnished the most complete memoirs of these vessels that we possess, admits that with age at the points corresponding to these contractions, at that branching below the swelling, a strangulation commences, followed by a complete division; then the lacticiferous vessels, instead of a continuous cavity, present a series of cavities separated from one another by so many articulations: he then terms them articulated (*vasa lacticis articulata*, *fig.* 57); in a state of expansion (*expansa*, *fig.* 56, 55 *b*), in that which preceded and seemed to constitute the highest degree of their vital activity; in a state of contraction (*contracta*, *fig.* 55 *a*) in the first stage of developement.

55 & 56. Lacticiferous vessels from the Celandine (*Éclaire*) (CHELIDONIUM MAJUS).
57. Lacticiferous vessels from the EUPHORBIA DULCIS.

On comparing them with the other vessels, we find something of opposite character; since, instead of cells, primarily distinct, and afterwards joining and communicating with one another to form a continuous tube, we find tubes, at first continuous, and then interrupted and broken, as it were, into utricles. Lacticiferous vessels are, therefore, eminently distinguished by this characteristic, if its truth be not disputed, and, at all events, by the numerous communications between the tubes forming a net, as well as by the imperforation of their wall, in whose substance we do not observe the small thinner intervals, leaving the primitive membrane naked in the form of bands, stripes, or dots, mentioned in other vessels, fibres and cells. This wall is thin when it is young; but as it grows old it is thickened, as we have already said: some authors have even supposed that they recognised a succession of layers.

§ 15. Means by which Elementary Organs are united.—After having shewn the principal modifications which the elementary organs of vegetables present, we cannot pass over in silence a problem which has lately occupied the attention of botanists; this is the research into the means or the force by which these vessels, which we have hitherto examined separately, are held together. As all may be reduced to the cell from which they began, as cells frequently form the greater part, sometimes even the whole of the vegetable, the problem is reduced to the determination of the mode of connection between cells.

According to some authors, their union is immediate; the walls of the cells, at first semifluid, preserve for some time a certain degree of softness, which is still sufficient when the walls of several neighbouring cells meet and touch one another in their developement, to glue them together: and even, as they dry, they remain agglutinated with various degrees of force, according to the nature and form of the tissue which they constitute.

The theory of mediate union, (i. e. by means of a medium,) which has never had many partisans, was revived a few years ago by a great authority, M. Hugo Mohl. He thinks, that between the cells there exudes a kind of paste, differing from them in its nature, which binds them, and which he calls intercellular matter. In certain vegetables of a very simple structure, such as aquatic plants, and chiefly those sea weeds known commonly as Varech, and which we shall hereafter study under the name of Fucus, the utricles (*fig.* 58 *aa*), of which the whole plant is composed, are very distant, leaving between them an interval very often greater than their diameter, and the whole of this interval is filled by this intercellular matter (*b*), which, consequently, forms the greater part of the mass. In more complicated vegetables, on the contrary, in the trees and herbs which cover the earth, the cells (*figs.* 7, 28) touch, and between their faces in contact, the matter which unites them is reduced to a layer so thin, that it escapes our eye; but, in the intercellular passages it sometimes becomes more visible, and may even fill them completely. It is distinguished by its different colour and structure; and, because its composition is not identical with that of the cellular membrane, the cells may be separated by means of chemicals, which dissolve the intercellular substance without destroying the membrane.

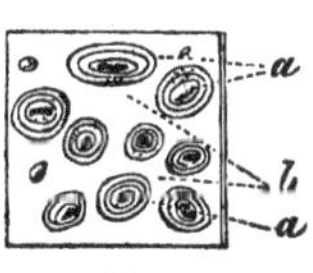

58

§ 16. M. Mirbel's theory does not touch upon either of the preceding. According to him, the vegetable tissue is at first a kind of mucilage, like a solution of gum arabic, which gradually thickens, and though at first continuous and solid, is at last broken into a great number of small holes, which will be the cavities of the cells. The neighbouring cells would, therefore, at first be separated by a common wall, which might remain as such, but which, still more frequently, would at last be doubled, sometimes in the whole of its extent, sometimes only in part, and always beginning towards the angles. According to this theory, the developement of the cells would then be perfectly inverse to that specified in the other theories; they would tend to separate and not to unite, and their union would only be the normal and original state, and would be progressively effaced with age. When this tissue (*fig.* 59 *b b*) remains in this state, and thus forms a piece of continuous net-work, the meshes of which are each doubled by a distinct cell (*aa*), M. Mirbel calls it interposed cellular tissue

59

58. A portion of the tissue of a Sea-weed (HIMANTHALIA LOREA).—*a a* Cells.—*b* Intercellular matter.

59. Central part of a young root of the Date tree.—*a a a* Cells.—*b b b* Interposed cellular tissue of M. Mirbel.

(*fig.* 59). It is clear that it here plays the part of the intercellular substance of M. Mohl, although its origin is quite different.

§ 17. The means of communication between the Elementary Organs.—If the means by which elementary organs are united give rise to some uncertainties, their mode of communication with one another is very evident. We have seen, indeed, that they are surrounded by a thin and simple membrane, which does not thicken uniformly over the whole of its internal surface, but always remains bare in a great number of places. Now, the permeability of such a membrane is put beyond a doubt by numerous decisive experiments. The gases or liquids, therefore, contained in the cavities of vessels and cells, always find, in order to pass from one to the other, a number of lateral canals, intercepted only by a membranous diaphragm. Several botanists have even denied the existence of these diaphragms : what we have called dots and stripes, they term pores and slits. It is very probable, that the membrane frequently disappears from the space left bare by the second membrane ; we have seen this taking place at the extremities of fibres and cells in contact, which by this arrangement form vessels. Sometimes the membranous tube of a vessel at last completely disappears, and leaves the rings supported by the neighbouring parts, which in the intervals serve as the wall of the cylinder. The existence of real holes has been proved beyond a doubt in the cells of certain vegetables.

In two contiguous cells, the lateral ducts of the one correspond to those of the other, so that two ducts seem to be only one (*fig.* 60 *a*), forming the means of communication between the two cavities, its continuity from one to another being uninterrupted. The passage, then, of a fluid from one vessel to another is always either entirely free, or easy, or at least possible.

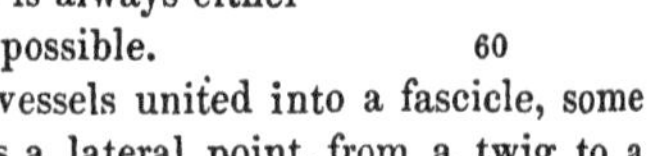

§ 18. When, from a group of vessels united into a fascicle, some detach themselves to go towards a lateral point, from a twig to a

60. Lengthened cells from the root of the Date tree.—*aa* Canals of communication.

leaf, for instance, and thus abandon their original rectilinear direction to form a curve, the continuity of the tubes appears interrupted at the bend, the vessels which follow this fresh direction join at their extremity (*fig.* 61 *aaa*) the ends of the vessels which form the first fascicle. Nearly the same thing, then, takes place here as at each point of the union of cells or fibres whose rectilinear series forms a vessel, viz., the partial or total perforation of the two united adjacent surfaces; only there is a little deviation of the ends which are thus joined.

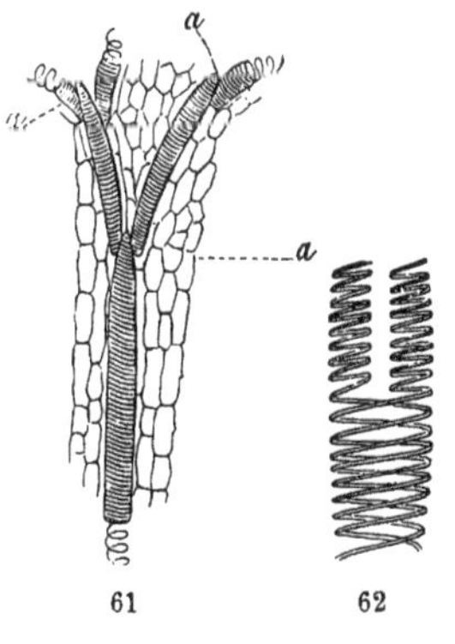

Although this is the almost constant origin of the vascular branches, yet we have examples of spiral tubes ramifying without articulation. Thus, we sometimes see a trachea with a double spiral fibre, fork out into two tracheæ with single spires, (*fig.* 62,) by the angular separation of the two spires whose coils were before completely parallel. The characteristic of ramification with continuity of tube, by which lacticiferous vessels have been distinguished from all others, is not, therefore, so absolute as to be proclaimed general, although they are in other points too distinct to render it difficult to define them clearly.

§ 19. The Contents of the Organs.—We have seen the vegetable tissue perforated with cavities of various forms, which occupy the interior of the cells, fibres or vessels, or else the intercellular passages. It now remains for us to determine whether these cavities are empty or full, and, if they are full, what they contain.

They frequently appear to be absolutely empty; but then, on opening them under water, a number of little bubbles commonly arise, thus demonstrating the presence of a gas. All the intermediate degrees of condition, from this aëriform state to the solid, are presented to us by the substances contained in the cavities of the vegetable tissue; they may be either a gas, a limpid or thick fluid, a jelly, a paste, or granules, either scattered or united into

62. A Trachea from a Pumpkin (*Potiron*) (Cucurbita pepo).

a mass of stone, or crystal. Now, it is clear, that as they approach to the solid state, it becomes easier to prosecute our researches into their nature: when in the gaseous state, we require, in order to determine them, the aid of chemistry; when liquid, they are apt to evaporate as soon as the tissue has been exposed; and the cell, which, during life, was extended by a liquid, under the microscope appears shrivelled or empty, or else to contain only a few traces of a solid remainder, if the evaporation has caused some bodies previously held in solution, as gum or sugar in water, or a resinous gum in a volatile oil, to be deposited. Chemical changes may also take place during the examination of the contents, which, separated from the vital parts, are necessarily brought into contact with fresh agents, as air, water, &c. This enquiry, the results of which are of so much importance in giving us clear ideas of the life of a vegetable, requires greater precautions, less direct and more varied processes, and, consequently, is much less advanced than those which have hitherto occupied our attention.

§ 20. If we commence with the observation of the solids, since that is the easiest, we shall see that they may occupy different positions in the cavity which encloses them. Thus, they sometimes line the whole of the internal surface, and thus form, as it were, another layer, only distinguishable from the others when reactives are employed, which make their different chemical composition apparent: this is the case in strongly azotised substances, which are found so placed in almost all very young cells, and are tinged red by the protonitrate of mercury, which does not colour the rest of the walls. That, which is called *lignine* (*ligneux*) and is found in wood of ligneous fibres, is not an internal layer on the walls only, but it also impregnates them, like some substance which, absorbed by a sponge in a liquid state, is afterwards solidified and forms along with it one whole. In these cases, the contents lying close to the wall or even introduced into its substance, are liable to be confounded to a certain extent with the containing wall.

It is the same with regard to the silex or flint which rather frequently incrusts the vegetable tissue. This may be seen by burning carefully on a piece of glass a small piece of one of these tissues. The fire destroys the vegetable tissue, but not the mineral substance with which it was penetrated, and which, deposited on the glass, represents the silicious skeleton of the organic tissue which has disappeared. This is shewn in a remarkable degree in the stems of the Gramineæ, commonly known by the name of straws, and is the origin of those lumps of vitrified matter that are sometimes found in the *débris* of a burnt corn-stack.

§ 21. The substance contained in the interior of the cells often appears under the form of granules, sometimes dispersed, sometimes collected into a ball. They frequently line the walls, to which they even seem to adhere; and then they are like spots or dots, which, on the outside of the cell, may be confounded with those resulting from the perforations we have previously mentioned, as existing in the substance of the membranes of which this wall is composed.

In most very young cells, and even in all, according to some botanists, there may be observed a granular mass, in the shape of a ball or a double convex lens, attached to the interior surface, or even buried in the membrane (*fig.* 63). It generally becomes less and less apparent as it grows older, at last disappearing entirely. It remains, nevertheless, in certain parts and in certain vegetables. It has been termed *nucleus*, or germ of the cell, for which name M. Schleiden has proposed to substitute that of *Cytoblast* (*Cytoblaste*) (χυτὸς, *cavity, cell;* βλάστος, *bud, germ*), on account of the functions which he attributes to it, those of producing, in its ulterior developement, the cells, of which it is, as it were, the embryo.

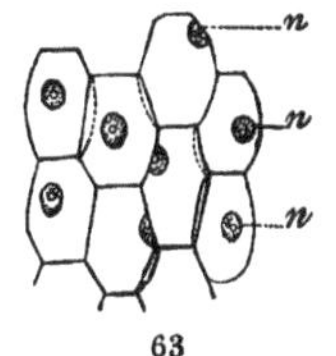

The granules are sometimes numerous and close enough to unite (especially after the evaporation of the liquids) into a compact mass, which fills almost the whole cellular cavity; at other times they are separate and distinct. Their nature is various, as may be supposed from the variety of their form, though principally from the different results produced by certain substances, especially by the solution of iodine, commonly employed in these observations. When a drop is poured upon the cells, whose contents we are examining, the granules are coloured, at one time, brown or yellow, more or less pale; at another, blue or violet, more or less deep. We may generally conclude, that, in the first case, the granules are composed of substances more or less azotised, of albumen or of vegetable casein; in the second, of a substance entirely free from azote, which is widely spread in vegetation, the fecula. If we examine the latter granules without colouring them, we shall notice certain characteristics also in their form: it is generally that of an irregular spheroïd,

63. Young cells from the Beet, with their nuclei *n n n*.

or of a polyhedron, on which are described several concentric circles around a point commonly placed on its circumference. This point or HILUM (*hile*) corresponds to that of the internal wall of the utricle, and, around it, is placed the first scale of the granule of fecula; then a second, which pushes the first to the inside of the cell; then a third, which forces the second in like manner, and so on: so that the granule may be considered as composed of a succession of scales or disks, piled one upon the other, sometimes horizontally, sometimes obliquely; so much the younger and softer as they approach the hilum, so much the older and harder as they draw nearer to the opposite extremity. It sometimes happens, that two or three granules, when in close juxtaposition, are united into one, in which we then remark two or three distinct hila; but the ulterior growth always takes place on one of them, on which the scales are produced after the union. If we wish to enter more into detail, we find the form of the fecula vary in different plants, but sufficiently constant in the same species to know whence it was taken. Thus, to take the best known instances, let us examine a very thin slice of a potato (*fig.* 64), of a grain of wheat (*fig.* 65), and of a grain of maize (*fig.* 66), and we shall find that the granules of the fecula in the three plants differ from one another sufficiently to enable us to recognise them with facility. If, in the little drop of water, in which the slices are generally put to prevent their drying, we mix a little of the solution of iodine, we shall immediately see the granules coloured blue, and shewing clearly the cells which contain them (*fig.* 64.)

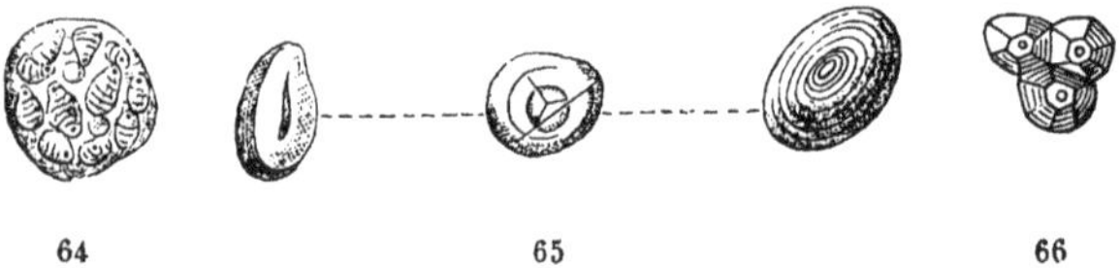

64 65 66

The circulation of the latex in the vessels to which it has given its name (*figs.* 55, 56, 57) is discernible from the presence of numberless granules. They are generally very small, like dust, but unequal in size, and in the midst of them are seen some of a much greater magnitude, and of forms sometimes very extraordinary, which iodine demonstrates to be granules of fecula.

64. A Cell of a Potato (*Pomme de terre*) filled with granules of fecula.
65. Granules of the fecula of Wheat (*Blé*).
66. Granules of the fecula of Maize (*Maïs*).

§ 22. It is very common to meet with crystals in the interior of the cells. This is by no means astonishing, for there are formed in the organs of the plant by the very act of vegetation several peculiar acids (as oxalic, malic, citric, &c.); and at the same time carbonic acid is attracted from the earth and air; on the other hand, the soil contains in solution inorganic alkalies, such as chalk, potassa, silica, &c., which are absorbed and circulate with the sap. These different solutions will frequently come in contact in the cavities of the plant; and, if the bodies have chemical affinity for one another and the property of forming an insoluble salt, crystals of various natures and shapes will be found. Hence, it would appear at first sight, that this is purely a chemical operation, which takes place in the interior of the cells, in the same way that it would, if the substances were to meet in any other vessel, in which the solutions were mixed and in a state of quiescence; and, when we find that the crystals grow more numerous as the parenchyma grows older and its vital powers are weakened, we shall be confirmed in the idea, that their formation depends upon inorganic forces, and not on those of life. There are several considerations, however, which will lead us to think the contrary, and especially the recent observations of M. Payen. According to him, the crystals are not formed nor do they float as detached substances in the utricle, but there exists a well-organized peculiar apparatus which produces and contains them. From a point in the wall of these cells starts a cord composed itself of smaller cells suspending a mass, which turns out to be a small cellular tissue just beginning to form (*fig.* 67). Afterwards, in the interior of the little cells of the mass thus suspended, there is deposited the crystallized mineral substance; and these seem to determine the size and form; so that one salt, oxalate of lime, for instance, may crystallize in vegetables in totally different forms, owing to the difference of the apparatus in which the operation is taking place. We have seen the cellular tissue before the deposition

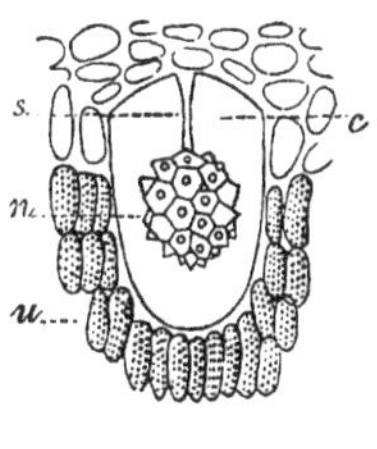

67

67. A mass of cellular tissue observed in the leaf of an India-rubber tree (*Figuier*) (FICUS ELASTICA), in which crystals are developed, the united mass of which represents a bristled nucleus *n*. It is suspended by a kind of tube *s* in the interior of an enlarged cell *c*, situated under the epidermis and surrounded by smaller utricles *u*, filled with granules which are coloured green by the chlorophyll.

of the salt; we may afterwards see it in a more developed and fixed form by dissolving the salt by means of a reactive, which does not hurt the tissue. Without the employment of this agent, the organic tissue which envelopes the crystals will escape from our notice, on account of the fineness of the membrane, which fits the crystals very closely.

Sometimes a cell contains only one or a very few crystals, and then their size enables us to determine their shape very clearly. But, more commonly, there is a great quantity, and, consequently, their smallness renders it very difficult to determine the form. Then they are generally arranged in two different ways, either radiating from a common centre, or parallel to one another. In the first case, they are commonly thicker and shorter, and their whole represents a kind of spherical or egg-shaped core or nucleus covered with dots (*figs.* 68, 69); in the second instance, they seem like a bundle of fine needles (*figs.* 70, 71). These needles, taken at first for a ve-

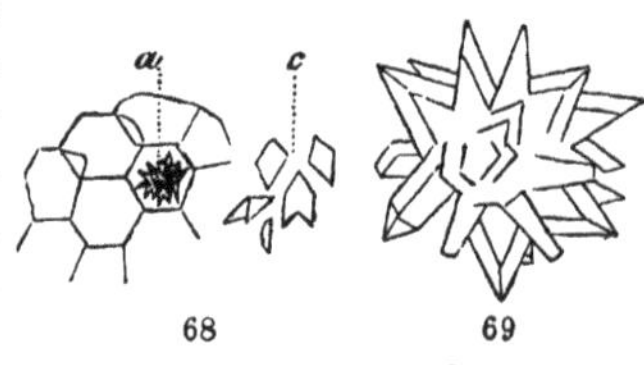

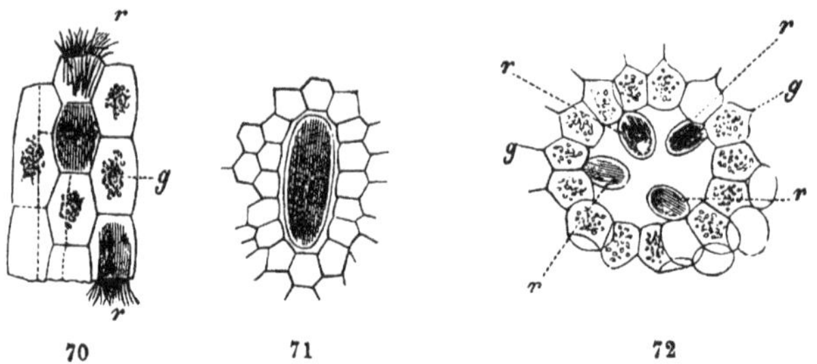

68. Cellular tissue of the Beet; in one of the cells is an accumulation *a* of bristled crystals.—*c* Separate crystals.

69. Accumulated crystals found in a cell of the petiole of the Rhubarb (*Rhubarbe*) (RHEUM UNDULATUM).

70. Cellular tissue of Lords and Ladies (*Pied-de-veau*) (ARUM VULGARE). Several cells are filled with granules of chlorophylle *g*, others by bundles of raphides *r r*.

71. Bundles of raphides in an enlarged cell which is surrounded by other smaller cells. These also come from the ARUM.

72. Portion of the tissue of the COLOCASIA ODORA. Cells, filled with granules of chlorophylle *g g*, leave between them a lacuna, into which project four other cells containing bundles of raphides *r r r r*.

getable organ, a kind of hair, have received the name of *Raphides.* We may remark that these crystals necessarily exclude the organic granules of which we have just spoken.

The crystals, and especially the raphides, have been sometimes described as placed outside the cell in the meati or lacunæ. This mistake may arise from several causes: the most common and natural is the dispersion of these bodies by the very act of dissection, which requires much care and skill so as not to tear the walls of the cells; the instrument also often penetrates the cells, and, separating the raphides, throws them promiscuously into the neighbouring cavities. But, even when the tissue has been operated upon with the greatest care, the bundles of raphides have been supposed to occupy some portion, if not the whole, of a lacuna. The cells in this case may have been extraordinarily developed, so that their cavity appears to be a lacuna (*fig.* 71); or, placed round the wall of a true lacuna, they extend into its interior (*fig.* 72), which then appears to contain the crystals. We may safely affirm, that up to this period they have been discovered only in the interior of the cells.

It is true that we sometimes see in the lacunæ and meati a kind of mineral substance formed of silex, either alone, or in combination with an acid: but it is in irregular masses, and not crystallized.

§ 23. The cells which form the greater part of the grains of wheat and other cereal plants, so full of the fecula whence flour is extracted, contain also a soft and very elastic mass, which fills their cavity and glues the granules together. This substance has received the name of *gluten.*

§ 24. Other substances present a consistence less and less solid without being quite liquid; they seem like cloudy flakes or jelly: of this kind is that which performs such important functions in the life and appearance of a vegetable, and which is called green matter, or *chlorophyll* (*chlorophylle*). It floats in the cellular fluid, and has a tendency to deposit itself on the solid parts which are in contact with it, on the internal walls of the cell and on the granules which it may happen to enclose. This is the reason, why some botanists have often described it as a green liquid, others as a mass of granules of the same colour. The liquids and the granules are in themselves colourless, but borrow their tints from the layer which encloses them. The so called granules of chlorophylle may, therefore, like those of which we have treated, occupy different situations: they may be fixed on the wall of the cell; or, floating in the liquid which fills it, present different forms and dimensions, and, doubtless, different natures. We find several of them coloured blue by the solution of iodine, which shews that they are nothing else than granules of fecula. Others are tinged with brown, but it is uncertain whether this arises from the green envelope being so thick as to prevent the action of the iodine on the nucleus, or from the substance being different. Sometimes several nuclei are pasted together with a sort of green jelly, and thus form a compound granule. Any green part of a vegetable, kept sometime in alcohol, loses its colour; we thus easily determine the solubility of the chlorophyll in this liquid, and thence also deduce its analogy to the resins.

The substance, which sometimes colours the cells yellow, appears to resemble it in its nature and properties, whilst that which colours them blue, red, or violet, presents the consistence of an aqueous fluid, in which it is completely dissolved. More commonly the cellular liquid is colourless, or nearly so, as well as that, which under the name of sap, flows through the spiral vessels. This already contains in solution some of the substances which are deposited in the cells, or the elements which combine to form them. We also find in the cells or in the lacunæ acting as reservoirs liquids of other natures, such as solutions of gums in water, fixed and volatile oils, which hold in solution resinous substances. Gases exist, especially in lacunæ, sometimes close to the surface of the plants, sometimes deeply buried in its interior.

§ 25. We have seen from what has preceded, that substances of different natures and consistences fill the cavities of the different organs of which the vegetable is composed, not indifferently, but placed regularly in such or such of these organs. We must not, however, expect to find the same organ constantly filled with the same substance; the progress of its life causes almost continual changes, and it is only when this is suspended or extinct that these changes cease. The cell, which becomes green by the action of light, was formerly colourless, and, as it grows older, will assume various tints, some of them quite different; before being filled with solid granules it was swelled with liquid only; it does not contain crystals till later; the vessels, which at some seasons convey the sap, are at others distended with air alone. An examination made at one period, would only tend to give a false impression of the functions of all these organs which we ought to study in all the phases of their life; and it is, doubtless, because the observations have been thus isolated, that the theories on this subject are so much opposed to one another.

COMPOUND ORGANS.

§ 26. The elementary organs, whose principal modifications we have just explained, form in combination *compound organs*. These, more or less complicated, more or less numerous, combine in their turn to form the vegetable. To observe this vegetable from its origin, from its simplest state, to follow it afterwards during its developement, to take notice of all the changes which it undergoes, and to analyze all the parts as they grow, are the surest means of becoming acquainted as completely and as clearly as possible with all those organs, which, when united, form its method of existence, whose action constitutes its life.

§ 27. The first state under which a vegetable presents itself is that of a utricle or cell (*figs.* 74, 75 E[1]), filled with a granular matter (*fig.* 75 E[2]). There are some plants which are hardly developed beyond this degree of extreme simplicity during the whole course of their existence, and all present it at the commencement, even those which attain the highest degree of vegetable organization. The first phase of the life of an organized being was that during which it still made a part of the being like itself, in which it was formed and which gave it birth. It then bears the name of *embryo* (*embryon*), and this period of its life is called the embryonary.

The vegetable embryo is, therefore, at first a single utricle, with granules in its cavity (*fig.* 73). Some changes in its coverings and in the contents of the cells are the only ones which result from the developement of certain embryoes; sometimes, also, other cells arrange themselves around the first, but without its being possible to distinguish several different parts, several different regions in this small homogeneous mass.

73.

§ 28. Frequently, on the contrary, the plant, which, in the state of an embryo, has not only acquired a much greater size by the accumulation of a rather large number of cells, but has also assumed very well determined forms, and two extremities very different from one another, may be distinguished very early. The one follows the direction of the axis of this ovoïd or egg-shaped body; the other deviates a little from it, forming, as it were, a kind of nipple projecting

73. An acotyledonous embryo, that of the Common Hepatica (*Hépatique commune*) (MARCHANTIA POLYMORPHA). Embryoes of this kind are also termed *spores* (*spores*).

laterally (*fig.* 74, E^4 *c*), or else two symmetrical nipples (*fig.* 75, E^4 *c c*), between which the axis would pass. These nipples will form

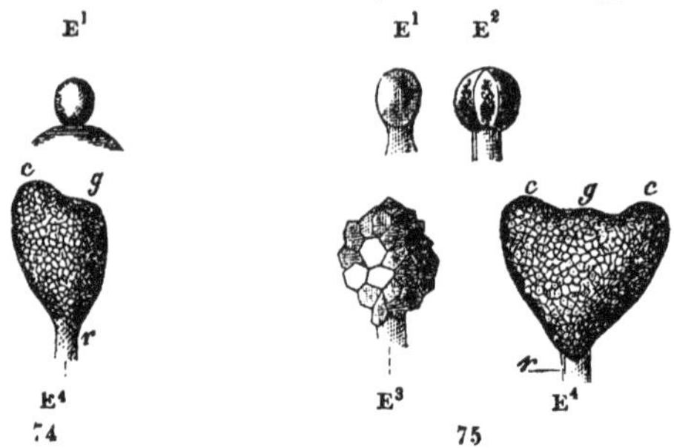

what we call the *cotyledons* (*cotylédon*)[f], and we have in this first period of the developement of the life of the plants, three modifications of the embryo: that which is homogeneous without distinction of parts, without cotyledon (*fig.* 73), that which has one (*fig.* 74, E^4), and that which has two (*fig.* 75, E^4). The first is called *acotyledonous* (*acotylédoné*), the second *monocotyledonous* (*monocotylédoné*), the third *dicotyledonous* (*dicotylédoné*).

§ 29. In general, cotyledonous embryoes do not stop at this first developement of the organs which compose them; but, enclosed in the seed still attached to the mother plant, they continue to grow in every part, principally in their single cotyledon (*fig.* 76 *c*) or double (*fig.* 77 *c c*), which form a large, and sometimes even the greatest, portion of the perfect embryo. The end opposite to the cotyledon has received the name of (*figs.* 76 *r*, 77 *r*) *radicle* (*radicule*) or little root, because roots will afterwards spring from it. Above the radicle, and in continuation with the axis, we remark, between the cotyledons, if there be two (*fig.* 75 *g*), concealed at the base of the cotyledon, if there be one (*fig.* 76 *g*), a much smaller body than those we have hitherto described. At first sight it has the appearance of a simple nipple; but a closer examination by means of the

74. A monocotyledonous embryo of the POTAMOGETON PERFOLIATUM, at different stages of its developement.—E^1 at its first appearance whilst it is still in the state of a utricle.—E^4 when its different parts, the radicle *r*, the gemmule *g*, the cotyledon *c*, begin to become distinct.

75. A dicotyledonous embryo of one of the Onagraceæ (ŒNOTHERA CRASSIPES), at different periods of its developement.—E^1 at its first appearance, whilst it is still in the state of a utricle.—E^2 E^3 when it is formed of three united utricles, or afterwards of a larger number.—E^4 when its different parts, the radicle *r*, the gemmule *g*, the cotyledons *c c*, begin to become distinct.

[f] From κοτυληδὼν, a cup-shaped hollow.—TRANS.

microscope, shews that it is itself composed of several small lobes (*fig.* 78 *g*), situated laterally, with respect to the axis, like the cotyle-

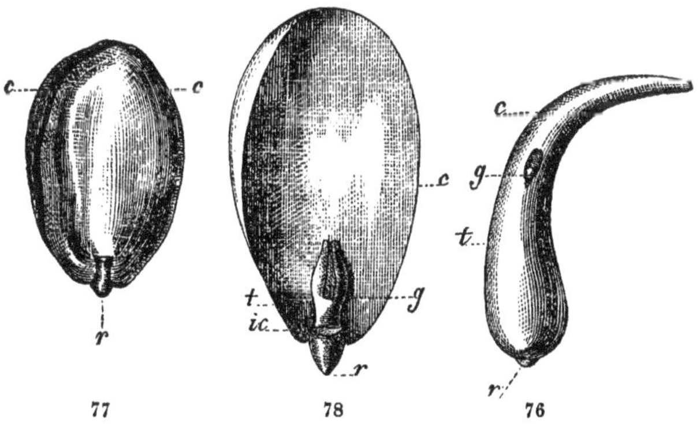

don, whose first developement is thereby, as it were, brought before our eyes. These small lobes will in a little time develope themselves into leaves; and, as a whole, they are termed *plumule*, *gemmule* (*gemmule*), or little bud, because we apply the word bud to an assemblage of undeveloped leaves placed on a short axis about to develope itself into a branch. The cotyledons themselves are only one or two of the first leaves of the young plant, but in general differing from those which succeed in their form and functions. The embryo, therefore, presents to us a continuation of lateral organs or leaves on an axis, the leafless extremity of which will form the root, and all the rest the stem. The latter, in the embryo, has received the diminutive appellation of *tigelle*.

§ 30. FUNDAMENTAL ORGANS.—These three parts, which are already very distinct in the cotyledonous embryo, may be called *Fundamental Organs*. All those, which the subsequent developement of

76. A monocotyledonous embryo of the POTAMOGETON PERFOLIATUM, nearly ripe.—*r* Radicle.—*t* Tigelle.—*c* Cotyledon.—*g* Gemmule.

77. A ripe dicotyledonous embryo of the common Almond (*Amandier commun*).—*r* Radicle.—*c c* Cotyledons.

78. The same, when the parts concealed by the cotyledons have been exposed by taking away one of them.—*r* Radicle.—*t* Tigelle.—*c* One of the cotyledons which has been left.—*ic* Wound at the place whence the other cotyledon has been taken.—*g* Gemmule composed of several small leaves.

the vegetable presents, in spite of their striking differences, in spite of the various names which have thence been given to them, are, however, now generally considered only as modifications of these primary organs.

§ 31. We recognise them by their relative forms and positions: for their elementary composition is identical: it is a mass of utricles, more or less compact. When the embryo is completely formed, the cells of its cotyledons abound more or less in fecula, especially if they are very thick; when they commonly fill the whole of the grain or seed, and are exclusively laden with the fecula which nourishes the young plant in the first stages of its life, which follow the embryonary period, and when it commences to exist detached from the parent plant. It is this accumulation of fecula in the embryo which causes such a quantity of seeds to be employed for the food of men and animals.

When the cotyledons are already largely developed in the embryo, and especially when they are in the form of leaves, we may observe in certain directions fascicles of lengthened cells, the first sketch, as it were, of the vessels.

§ 32. After the perfection of the vegetable embryo, or, to speak more familiarly, after the vegetable egg is laid, if certain conditions favourable to its ulterior and independent developement exist around it, (among which are first to be mentioned moisture and heat; conditions which are usually furnished by the earth, when it is buried to a certain depth,) it is hatched, as it were, and thus begins the second period of the life of the plant commonly called the *Germination.*

§ 33. At first, it draws its nourishment from itself, from the principles accumulated either in its integuments or its cotyledons: these principles undergo divers chemical changes on account of the fresh circumstances in which they are placed, and thus become fit to serve as food for the nascent plant. Thence results the enlargement of all the other parts, of the radicle, and of the gemmule. The developement of the radicle commences the first, and proceeds with the greatest activity; and we may then be sure, that in the portion of the axis beneath the cotyledon, comprehended under the name of radicle, the extremity only belongs to the root, and descends, whilst all the rest ascends, and belongs to the stem. The root is easily recognised by the small filaments with which its surface is covered (*fig.* 120, *r r′ r′*), and by means of which it begins to extract nutriment from the earth. The gemmule also is soon lengthened, and comes to light by pushing back the cotyledon or cotyledons which concealed it. It expands the leaves of which it is composed (*figs.* 79, 80

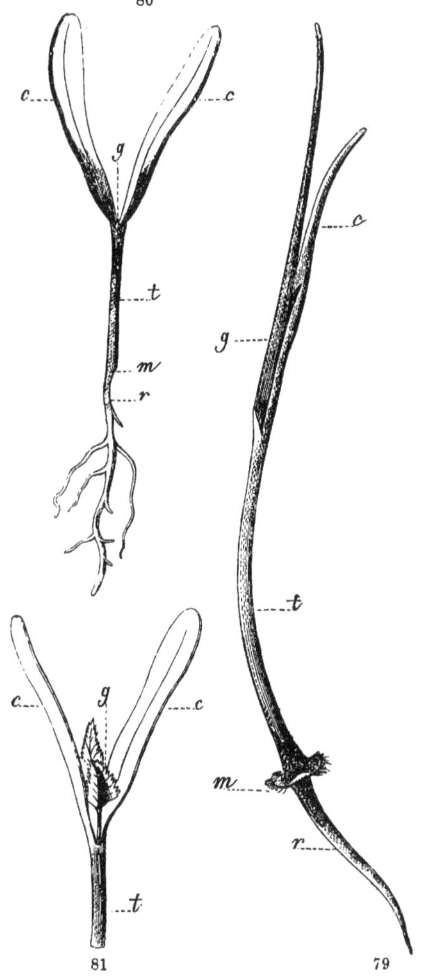

79. The Germination of the monocotyledonous embryo of the ZANICHELLIA PALUSTRIS, which is almost similar to the POTAMOGETON.—*m* Collum, the intermediate point between the stem *t* and the root *r*. We see that the latter results from the

81 *g*). These leaves are developed successively from bottom to top on the axis, i. e., they display themselves so much the sooner as they are lower on the axis. In proportion as these leaves increase in size and number, the cotyledons fade, and mostly finish by falling off. The new leaves generally differ from them in their shape (*fig.* 81), and tend more and more to assume that of the leaves of the fully developed plant. The first, however, especially those which immediately follow the cotyledons, frequently differ from the rest.

§ 34. When the little vegetable is freed from its now useless coverings, when it no longer draws its nutriment from the shrivelled or detached cotyledon, but by means of its roots absorbs it from the surrounding earth, then its germination may be said to be complete. It has created no fresh parts, but has rendered them more evident, by developing those which already existed in the embryo, viz., the fundamental organs, the stem, the root and the leaves. These organs continue to grow, and, in proportion as the axis is elongated, fresh leaves are produced in a lateral direction.

§ 35. The vegetation of some plants is merely this evolution continued for a certain length of time, and they only consist of one axis laden with leaves diversely modified. But, very often, and especially in dicotyledonous plants, at certain parts of the stem we find small excrescences, which we term buds, and which, in their turn, reproduce by their developement what we have just remarked in that of the gemmule, the production of leaves on a lengthening axis. The bud, and afterwards the branch, which is only a further developement of it, do not differ from the gemmule and the stem covered with leaves, except in the fact of their being planted in the stem instead of in the ground. This manner of growth, which may be repeated several times, and whence we obtain the ramification of the vegetable, is merely the reproduction of what we have observed in the evolution of the first axis, existing in the embryo, and to describe the one is to describe the others. In all the branches we shall only find leaves on axes, different in age and rank, it is true,

developement of the terminal projection (*fig.* 76 *r*), quite at the base of the embryo, beneath a dilation, here manifested by a kind of swelling in the form of a collar *m*.—*c* Cotyledon.—*g* Gemmule, the first leaf of which projecting out of the sheath of the cotyledon conceals the others.

80. Germination of the dicotyledonous embryo of a species of Sycamore (*Érable*) (Acer Negundo).—*m* Collum.—*r* Root.—*t* Stem—*c c* Cotyledons.—*g* Gemmule.

81. Upper part of the same, more developed.—*c c* Cotyledons.—*g* Gemmule, the first leaves of which are now expanded.

but the same in nature. Let us, then, examine the changes which the first axis progressively undergoes in its structure, and in that of its leaves.

§ 36. This research will naturally form three divisions; the stem, the root, and the leaves; but these three have one part common to all of them: this is a thin envelope extending over the whole surface of the vegetable called the EPIDERMIS (*épiderme*), upon which we will now enter.

EPIDERMIS.

§ 37. (*Epiderme*).—For a long time, the epidermis was thought to be a portion of the cellular tissue which it covers, to be only the external part, which, to the depth of one or more layers, is hardened and slightly modified by its exposure to the air. This is true in a certain number of vegetables of a very simple structure; but in others, the cells which form the epidermis are generally so different from those of the subjacent tissue in their shape, their dimensions, their manner of union, their contents, that it is now agreed to consider the epidermis as quite a distinct system.

§ 38. Let us examine, in the first place, those parts which are exposed to the air on the stems and leaves. This is that tissue, which is detached with more or less facility from the surface of these parts, when young, most frequently under the form of a colourless and transparent membrane; in some cases, this may be done without any previous preparation; in others, it can be done only after a prolonged maceration by which the subjacent cellular tissue is destroyed. If the maceration be continued for a long time, the epidermis itself is attacked, and it is then seen to be composed of two parts: the one, the external and most durable, is a thin continuous pellicle which extends over the whole surface (*fig.* 82 *p p*); the other, the internal, is the epidermis properly so called, composed of cells in close juxtaposition (*fig.* 82 *e e*).

§ 39. These cells, of almost equal dimensions and of the tabular form, are generally arranged in one layer of uniform thickness (*fig.* 83 *ee*). They are mostly larger than those of the subjacent tissue, although there are exceptions to this rule; for example, in the India-rubber tree (*Figuier élastique*), in the ORNITHOGALUM SYLVATICUM, where they are, on the contrary, much smaller. If we lay the transparent lamina of the epidermis flat on the object-glass

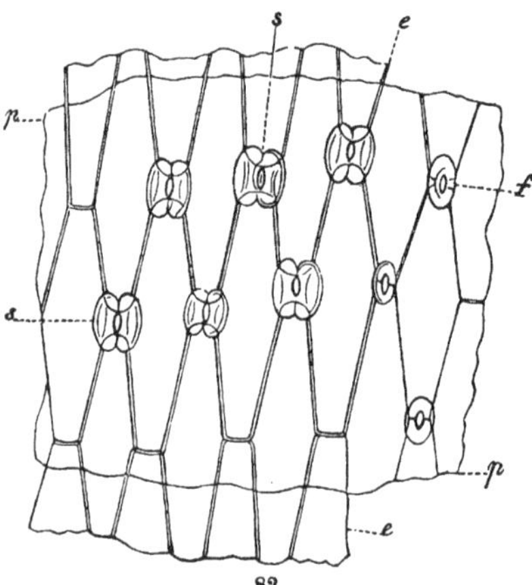

82

of the microscope, we may very readily observe the cells and the outline of their upper side, sometimes regular (*fig.* 82), sometimes irregular (*fig.* 84), often surrounded by double lines (*fig.* 85). In the first case, the quadrilateral and hexagonal figures are those which we most commonly find.

In the cells of the epidermis, besides the lateral walls, there is an

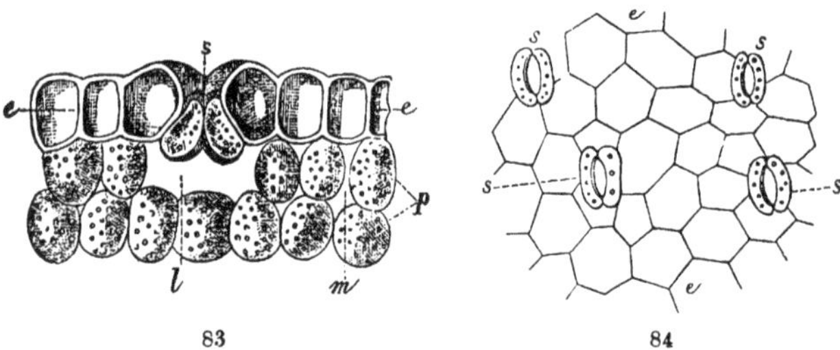

83 84

82. An horizontal piece of the epidermis of a leaf of the Iris (IRIS GERMANICA).—We here find an epidermal pellicle *pp* pierced with slits, somewhat resembling button-holes *f*, lying on a portion of the epidermis properly so called *e e*, composed of long hexagonal cells.—*s s* Stomata.

83. A vertical section of the epidermis of the same leaf, shewing the close union between the epidermal cells *ee*, and the lax junction with the green subjacent parenchyma *p*, broken by lacunæ *l* and meati *m*.—*s* Stomate.

84. A horizontal piece of the epidermis of the upper side of a leaf of the Water Crowfoot, growing out of the water.—*ee* Epidermal cells.—*ssss* Stomata.

upper as well as a lower one. The lateral walls adhere strongly to the analogous walls of the neighbouring cells, and from this close union results the absence of intercellular meati or passages, as well as the solidity of the whole membrane. The cells, although thinner at the sides than at the top and bottom, still present rather a remarkable degree of thickness. Their cavities are thus separated by rather large zones circumscribed by two lines, which appear to be a little deeper when examined under a microscope; and this is the reason, why several botanists formerly admitted the false notion of a plexus of small vessels in the intervals of the cells of the epidermis, especially when their shape is sinuous (*fig.* 85). These pretended vessels are only the walls of the cells.

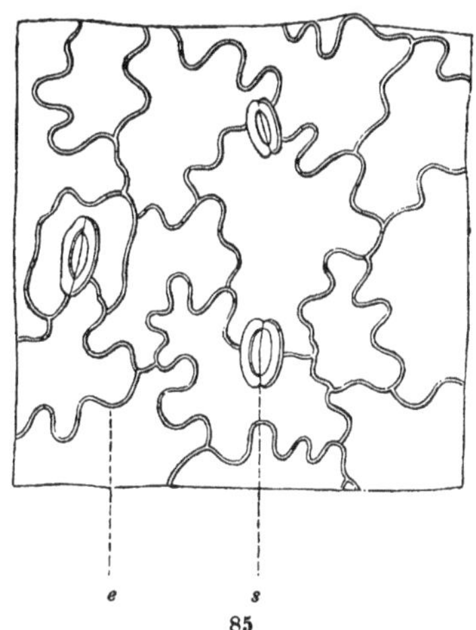

85

The lower wall, in the ordinary cases in which the epidermis is formed of only one layer, rests on the cells of the subjacent tissue, though it adheres to it very feebly. Thence, and from the adherence of the epidermal cells to one another, arises the ease with which they are detached in laminæ from the subjacent tissue.

The external wall, which is in communication with the air, is often much thicker than the others, so much so, that in some cases its

85. A piece of the epidermis of the lower side of a leaf of the Madder (*Garance*) (Rubia tinctorum).—*c* Epidermal cells—*s* Stomata.

substance takes up even the half of the cell (*fig.* 91 *e*). This all is generally smooth, and, consequently, the epidermis is so. But, sometimes, each cell is rounded at the top, and then the surface of the epidermis, examined by a lens, appears to be bristled, as it were, with a series of mamillæ (*fig.* 86 *e'*). When these grow a little longer, a hair or some other analogous organ commences (*fig.* 87 *p*). But we must postpone the detailed examination of them, which would interrupt the general observations with which we are at present engaged.

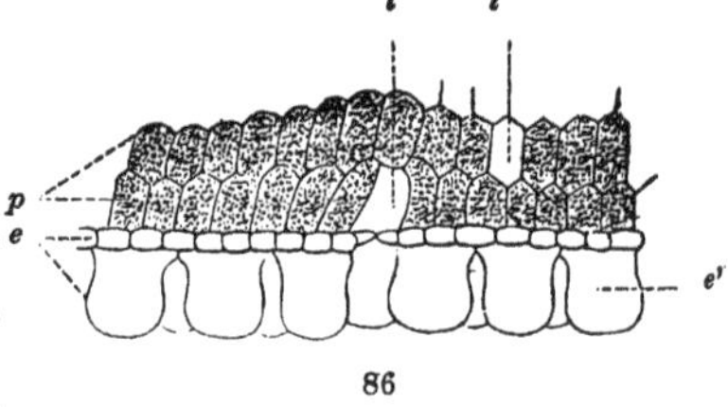

§ 40. STOMATES, STOMATA, STOMATIA (*Stomates*).—The external surface of the epidermis, at all its parts exposed to the atmosphere, is marked here and there by little spots, which a further examination by means of stronger lenses, shews to be so many breaks or interruptions in the continuity of the tissue surrounded by a peculiar swelling. When we examine a small fragment of a leaf of the common Iris (*fig.* 82), we shall find it composed of cells forming on the epidermal surface hexagons lengthened the same way as the leaf, and arranged in very narrow rectilinear series the other way. Between the small sides of the hexagons which follow one another in one and the same series, at short intervals are interposed little bodies (*fig.* 82 *ss*) of an oval form, which in their centre are pierced by an oblong hole surrounded by their pro-

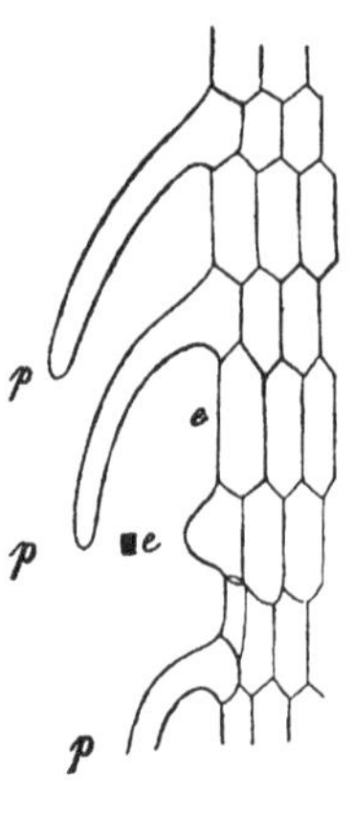

86. The horizontal section of the epidermis of the leaf of the ROCHEA FALCATA.—Its epidermis *e* is composed of two layers of cells: those of the exterior *e'*, very large and much swollen; those of the interior, perforated at *s* with a stomate, very small and even less than the green subjacent parenchyma *p*.—*ll* Lacunæ, one of which corresponds to a stomate.

87. A fragment of the epidermis of a young root of Madder. Several cells *p* are being lengthened so as to form hairs. The rest *e* have remained flat.

jecting swelling. This swelling is not one single piece; it is composed itself of two bodies slightly arched, the concave part of which is turned towards the hole, the convex to the outside, and united at their ends to one another. They have been compared to two lips, and the hole which they surround to the mouth. Thence the name of *Stomates* (*στόμα, a mouth*), by which name we agree to designate these organs, whose form in most plants resembles more or less that we have just described in the Iris. We will not stop here to mention all those little intermediate modifications of form between a circle and a very long and narrow oval we may find among stomata. Moreover, the same stomate will frequently vary according to the condition in which we may find it, according to its state of humidity or dryness. Of this we may be assured by comparing under the microscope two halves of the same fragment, one of which is moist and the other dry. In the former, the lips are swelled, and leave between them a large interval, thus augmenting their arch; they are shrivelled, narrow and contiguous in the second. We may suppose, then, that during life the afflux of the liquids tends to produce the former of these two appearances, i. e., to keep the stomates open, and the communication free between the exterior and the parts enveloped by the epidermis.

§ 41. The stomates are not found indifferently over all parts of the vegetable exposed to the air: it is only on the leaves that they are most abundant, and commonly on their lower surface; their number varies much according to the plants, and of course there are more as they are smaller. Let us cite some examples to shew this great diversity; they are the approximate numbers which certain botanists have found in the leaves of a few vegetables.

The stomata counted on a square inch of the epidermis of the leaf, give us the following numbers:—

	UPPER FACE.	LOWER FACE.
1. Miseltoe (*Gui*) (VISCUM ALBUM)	200	.. 200
2. Iris (*Iris*) (IRIS GERMANICA)	11,572	.. 11,572
3. Clove pink (*Œillet des jardins*) (DIANTHUS CARYOPHYLLUS)	38,500	.. 38,500
4. Water-plantain (*Plantain d'eau*) (ALISMA PLANTAGO)	12,000	.. 6,000
5. COBŒA SCANDENS	0	.. 20,000
6. Lilac (*Lilas*) (SYRINGA VULGARIS)	0	..160,000

These few examples, chosen from a large number, are sufficient to shew the enormous difference we may find in the absolute number of the stomata on various leaves, in their relative number on various parts of the same leaf. The last proportion is the exact expression of the degree of difference which may be found between the two sides of a leaf, as we shall see a little lower.

§ 42. The arrangement of the stomates is as variable as their number. Sometimes they seem to be scattered quite irregularly (*fig.* 84 *ssss*), sometimes they are placed in rectilineal lines: this generally takes place, when the cells of the epidermis are themselves arranged in the same way (*fig.* 82). Sometimes the series are separated from one another by equal spaces, at others they approach one another by twos, by threes, &c.; then comes rather a large zone quite bare of them; then again, another zone quite covered with them. In these several cases, and, in fact, generally, the stomatia are slightly separated from one another; but at other times, though very rarely, they approach and press against the sides of one another, and, if we except these small groups, the surface which bears them does not exhibit any stomata (*fig.* 88). The family of the Protæaceæ, of the Begonaceæ and of the Saxifrageæ, present several examples of this peculiar arrangement.

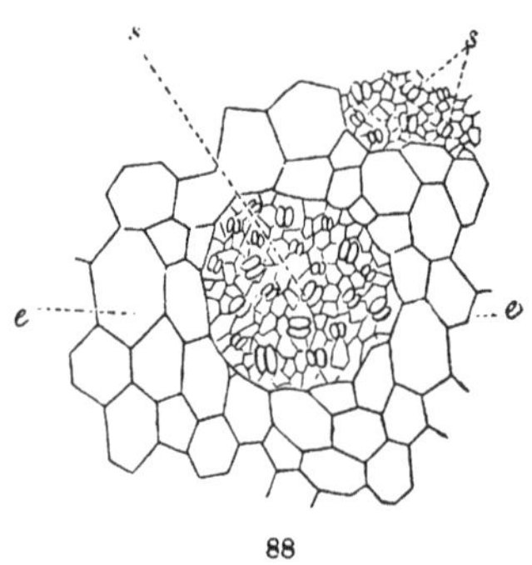

§ 43. The position of the stomates is not always the same. We cannot discover this by examining a flat piece of the epidermis; but we must, for this purpose, cut very thin vertical sections perpendicularly to the surface, and include a stomate in it; this is not very difficult when the stomata are large, and may easily be done by chance when we multiply the sections. It is in this way alone, that we can verify, in some cases, the existence of certain layers superposed to form the epidermis (*figs.* 86, 92 *e*), and determine the shape of cells, the outline of whose external walls could, otherwise, alone have been drawn.

We see, then, from such sections taken from different vegetables, that the cells which form the stomata are not always arranged

88. A fragment of the epidermis of a leaf of the SAXIFRAGA SARMENTOSA.—*ss* Stomata grouped on the surface of the epidermis, the cells of which are smaller when grouped around them than in the spaces *ee*, in which there are none.

in the same manner with regard to those of the rest of the epidermis. They frequently do not rise above the level of the surface; but they are also often buried more or less deeply under the surface, whether situated lower than the epidermal cells (*figs.* 89, 91 *s*), or, in spite of their juxtaposition, their size, generally much less than that of the latter, determines this difference of level (*fig.* 83 *s*). It also even happens, that this inequality may be still more enlarged by the projection, which forms the upper wall of the surrounding cells, by raising a sort of border round it (*figs.* 89, 91 *mm*), like the edges of the little well, at the bottom of which is the stomate. This edge may be found when the stomate is not buried, or may even belong to the stomate itself (*fig.* 90 *mm*). It is

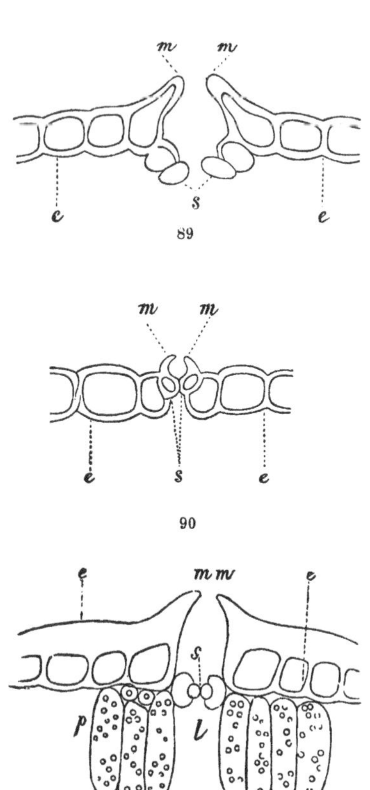

89. Epidermis of the Cycas revoluta.—*s* Stomate at the bottom of a cavity formed by the walls of the adjacent epidermal cells.—*m m* The margin rising above the surface of the epidermis *ee*, and formed by the projection of these very cells.

90. The epidermis of the Leucadendron decorum.—*e e* Epidermal cells.—*s* A stomate, the upper wall of which rises into a margin *m m*.

91. The epidermis of the Hakea pachyphylla.—*ee* Epidermal cells, with very thick upper walls. Those which border on the stomate *s*, project around it with the whole thickness, and thus form a kind of well at the bottom of which it is placed. They are produced into a free edge, which forms the margin *m m*.—*p* Subjacent parenchyma, the cells are filled with granules of chlorophyll.—*l* Lacuna.

clear, that mistakes may arise from observing not a vertical section, but a horizontal one, and from being no longer seen directly, but through the edge or the neighbouring cells; hence has arisen many erroneous opinions, and quite a different and false appearance has been ascribed to the stomata. This confusion of the neighbouring cells with the stomate has made botanists attribute to it more varied forms than those which really exist.

§ 44. When these stomates are arranged in groups, they commonly occupy the bottom of a cavity which was at first naturally taken for the stomate itself. This may be very plainly seen, for instance, in the leaf of the Oleander, (*Laurier-Rose*) (NERIUM OLEANDER), in which the upper side is quite bare, the lower one pierced here and there with little holes narrower at their mouth than at any other part, and bristled with hairs (*fig.* 92). It is at the bottom of these holes, hidden by these hairs, that very small stomata have been found, though presenting the usual structure and appearance.

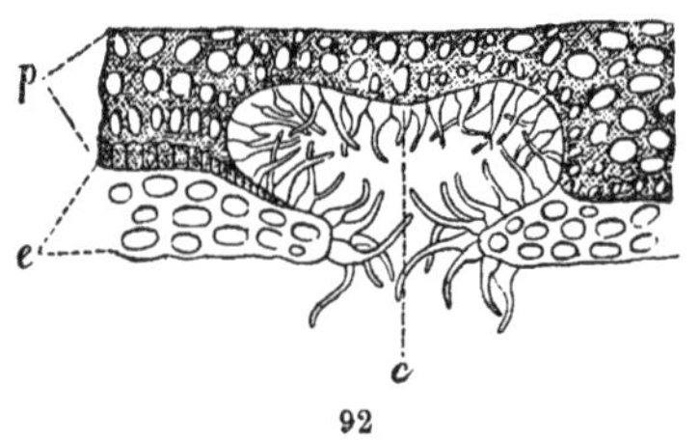

92

§ 45. What is the true nature of these stomatia?

There has been, and still is, a very great difference of opinion on this subject. The very names which they have received shew this. They have been called successively, *lengthened, exhaling or cortical pores; miliary*[f], *cortical or epidermoïdal glands.* Are they glands? This name is applied to the seat of the production of some peculiar matter, and the most minute observation has not yet discovered such a substance in the stomata. Sometimes, it is true, a certain substance has been found exuding from them; but, on looking for its origin, it has been discovered on a point more or less distant from the orifice, and not between the two bodies which

92. A portion of a vertical section of a leaf of the Oleander (*Laurier-rose*)(NERIUM OLEANDER), from the lower side.—*e* The epidermis composed of several rows of superposed cells.—*p* Parenchyma.—*c* Cavity, studded with hairs, at the bottom of which several botanists have recognised masses of stomata.

[f] This word is derived from *Millet*, indicating the smallness of the gland, as it were a grain of millet seed.—TRANS.

form the lips of the stomate. These bodies are hollow underneath and contain globules or granules (*figs.* 83, 84 *s*) of a different nature; sometimes colourless, often tinged green by chlorophylle. These are evidently two utricles, the produce of which is almost like that of the utricles placed immediately under the epidermis: and, if we did not always lift them along with it, we should conclude that they belong less to it than to the subjacent utricular mass. We will just state besides, in passing, that they resemble it in their outline; whilst in this point they differ remarkably from the epidermal cells, which are also commonly filled with a colourless fluid, and are, consequently, opaque or transparent according to the thickness of their walls.

§ 46. We have hitherto examined the epidermis in the parts of vegetables exposed to the air, excluding even the vegetables of an inferior order. Indeed, in Fungi, Mosses, &c., we cannot say that a true epidermis exists: the cellular tissue which forms the mass of the plant is not at all, or at most very slightly, modified at the surface. Some acotyledonous vegetables of a more complicated organization, such as the Lycopodia (*Lycopodes*) and Ferns (*Fougères*), are like the cotyledonous in their epidermis, which has an analogous structure as well as stomata. Aquatic vegetables have no stomata, and even no epidermis; and that not only in those, which constitute families, like the Algæ, placed on account of the simplicity of their organization amongst the lowest of the vegetable kingdom; but also in plants belonging, doubtless, to families of the most elevated ranks. It is the medium in which the plants live, that determines the presence or the absence of the epidermis. This is so true, that in the leaves which swim flat on the water, the upper side in contact with the air is furnished with an epidermis and stomata, whilst the lower one is entirely without them both.

§ 47. Roots, excluded, though less absolutely, from contact with the air, are equally void of stomata (*fig.* 87); and even although a layer of epidermal cells may be recognised, they differ much less from the subjacent tissue than those of the stem, and sometimes this difference is completely effaced.

§ 48. Epidermal Pellicle. (*Pellicule épidermique*).—Called also *cuticle.*—We have already stated (§ 38) that a prolonged maceration separated the epidermis into two parts, the one of which, the epidermis properly so called, which we have just considered, is the innermost, and is covered over its whole extent by the other thin pellicle which follows its surface throughout its whole outline, throughout the whole of its inequalities. This may be seen on a Cabbage leaf; and the epidermal pellicle, which is easily detached, is moulded exactly to the epidermis which it covers, even over its hairs, to which it forms sheaths, (*fig.* 93, *pppp*), and pierced with little button-holes in all the places corresponding to the stomata (*ff*). The presence of this continuous membrane, which is found on the surface of certain plants or of certain parts bare of the true epidermis, is more general than the existence of the latter: so it has been proposed by several authors to reserve the name of epidermis for the epidermal pellicle.

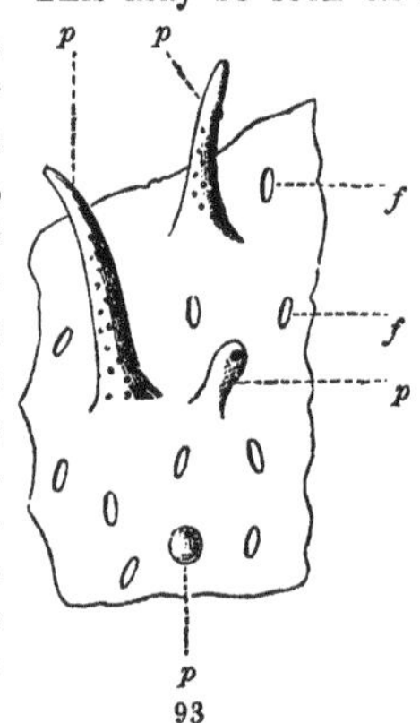

There are several opinions as to its origin. M. Ad. Brongniart[g], to whom we owe the most complete explanation, regards it as quite independent of the subjacent layers, and dubiously indicates the possibility of its being composed of granules, because in certain vegetables its inner surface presents a very granular texture: but in many others this is not the case. Some authors suppose that it arises from the deposit of a coagulating matter elaborated and furnished by the epidermal cells themselves. M. Mohl at first proposed an ingenious theory, which is connected with another previously explained (§ 15), and according to which it would be intercellular matter, which, spread between the cells of the epidermis and over their external covering, forms on the surface of the latter a continuous membrane exactly moulded to it. But, fresh observations made him abandon this opinion to adopt that of Meyen, according to whom this cuticle is nothing else than a portion of the wall of the epidermal cells; a wall which, like that of all the utricles, is thickened by the addition of several layers. Of these the external ones, at first distinct from those of the neighbouring cells, would

93. A shred of the epidermal pellicle detached by maceration from a leaf of the Cabbage (*Chou*) (BRASSICA OLERACEA). We here see the sheaths corresponding to hairs of different stages of developement (*pppp*) and the slits (*ff*) answering to the stomata.

g He was the first man to describe it, having discovered it on the Cabbage leaf.—TRANS.

afterwards be joined and united at their edges, so as to constitute a continuous membrane or cuticle. But, whilst we admit this origin, we should remark that, in order to form this organ, apparently so different, the outer wall of the epidermal cells is singularly modified; that it is thickened in a different way from the rest of the walls of these very cells, and, that at the same time its chemical composition is modified through the whole of this extent; this is very easy to verify: for this purpose we must employ certain reactives: for example, iodine, which colours this portion of the wall differently from the rest; secondly, sulphuric acid, which does not act upon it, whilst it dissolves the other parts. These new properties acquired by the external layers allow us to separate them from the subjacent ones: this origin explains the common appearance of a net of lines projecting from the lower surface of the pellicle, thus divided into small compartments. These answer to the primitive cells, and result from the enlargement of the upper wall being extended to the lateral walls to a greater or less depth.

Let us now return to the fundamental organs after having examined their common envelope.

§ 49. We have seen, that the axis of the young plant developes itself in two opposite directions, and we have called its upper part, which bears leaves, and commonly ascends, and is in contact with the air, the *stem* (*tige*) (CAULIS); its lower part, which has no leaves and most frequently is buried in the ground, *the root* (*racine*) (RADIX). The starting point, as it were, of these two parts, where they meet and touch, has been called *neck* (*collet*) (COLLUM) or *vital node* (*nœud vital*), because it was considered as the centre of the life of the vegetable, and has, consequently, had a greater importance ascribed to it than it has in reality; or again *coarcture* [h], on account of the contraction of the axis, which often indicates its place in the very young plant. As the plant increases these indications are generally weakened and effaced, and it becomes very difficult to discover the real place of the neck at the end of a few years of vegetation.

At first we will examine successively these two parts of the axis, the STEM and the ROOT; then the lateral organs, the LEAVES, which spring from the stem.

[h] So Grew. Turpin, led by a supposed resemblance to the animal kingdom calls it the *Median Horizontal Line* (Iconogr. pl. 4 *bis*, *fig.* 1, 2, *a a*).—TRANS.

THE STEM.

§ 50.—Fully developed stems present, according as the embryo is *acotyledonous*, *monocotyledonous* or *dicotyledonous*, such great differences, that it might involve a considerable degree of confusion to examine them together, and hence it has been thought advisable to treat these three classes separately. We will begin with dicotyledonous vegetables, as offering to us the best object of comparison with the others. Indeed, all the trees of our climates are dicotyledonous, so that we can observe them at all stages of their developement in various species, under a great number of different circumstances, and the student may refer to the objects of his study without trouble or inconvenience.

THE STEM OF DICOTYLEDONOUS PLANTS.

§ 51.—In the embryo the little stem was like all other parts, entirely composed of cellular tissue. During the germination, sooner or later, some cells begin to lengthen into fibres, to change into vessels, and, as they multiply, to be grouped into several bundles or fascicles (*fig.* 94 *f f*), which, when seen together, are ar-

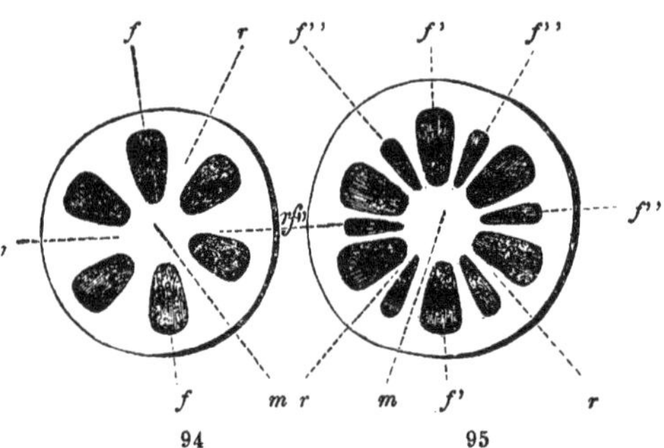

94 95

ranged in a regular circle. This surrounds a central circle entirely cellular, which is the *pith* (*moelle*) (MEDULLA) (*m*); it is surrounded itself by an external cellular zone which will belong to the bark: and the bundles are separated from one another by bands of cellular tissue (*r*), which establish the communication between the pith and the bark; these bands are the first indications of *medullary rays* (*rayons médullaires*) (RADII MEDULLARES).

§ 52. At the beginning, these rays, necessarily equal in number to that of the fascicles, were very large. A little later they are narrower and more numerous, because fresh fascicles (*fig.* 95 *f'' f''*) have been developed and interposed between the first ones (*f' f'*) After some time, the bundles are multiplied and near enough to form a continuous circle which the medullary rays traverse under the form of very fine lines. The stem is then composed from the interior to the exterior of:

1st. The parenchyma of the pith;
2nd. The fibro-vascular circle;
3rd. The cortical parenchyma;
4th. The epidermis.

§ 53. The stem of herbaceous plants, which only live for one year, generally stops at this or even at one of the preceding stages of developement. The proportion of the pith and the medullary rays to the fibro-vascular part is generally very large.

§ 54. The stem of ligneous plants, which exist several years, undergoes ulterior changes. But, up to this point, it was like that of annual plants; it was in an herbaceous state and presented the same proportions in its parts, with the exception of the fibro-vascular circle sooner becoming more complete and solid. A one year's branch taken from one of our trees is, therefore, a good subject upon which to make our observations of the successive changes mentioned above. This branch grows at the top, so that its base is the portion first formed, while its summit is only just springing into existence. Thus, its different degrees of height present the same difference which several stems of the same plant would present at different periods of their developement, from that of the embryonary state, which is represented by the summit of the branch, to that which has grown for one year, represented by its base. On comparing very thin horizontal sections taken at different heights of this branch, we shall then determine very easily and readily all the changes which take place in one year of the life of a stem; and the state of the base will serve in its turn as a starting point, by comparing it with the modifications which we may expect another year to bring, or, in other words, which we may observe on a second year's branch.

§ 55. Now, let us look more in detail into the elements which compose this one year's stem: and, since an example will render these researches still clearer and easier to be understood, let us take a branch of Common Maple (*Érable commun*), and describe all its parts, from the middle to the outside, magnifying them sufficiently to determine the nature of all the elementary organs which compose it.

Let us take a very thin horizontal section towards the top of the branch, where it measures about 1½ French millimetre or ·05899 English inch in diameter (*fig.* 96). Its outline is circular, or rather approaching to a hexagon. The pith (*fig.* 96 *m*), situated at the centre, is equal to one-half of the whole diameter, or even more. At the middle, it is composed of large cells, loosely joined, transparent, dodecahedrical or spheroïdal; towards the circumference, the cells diminish gradually (*fig.* 97 *m*), and are coloured green, so that the pith appears to be surrounded by a zone of rather a deep green colour and of a very fine close tissue, from which commence the medullary rays (*fig.* 97 *rm*) of the same colour, which divide into a great number of fascicles the fibro-vascular zone, which we find outside the pith and concentric with it. These fascicles (*fig.* 97 *fb*) are distinguishable from it by a much more compact tissue.

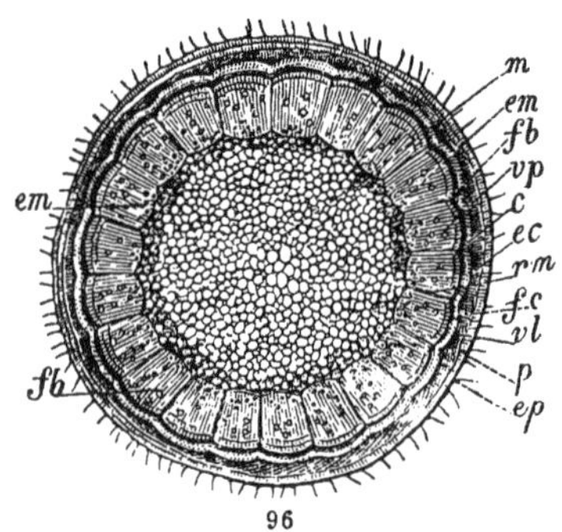

Each has the form of a blunted angle. On looking at the horizontal section, we even now see that they must be composed of different elements, since, in the midst of a compact tissue, perforated with little holes, we observe other openings very much larger. If we magnify these parts very highly, and then try to determine their nature, we shall find, when aided either by the transparence of an exceedingly thin vertical section, or by the separation of these parts by a very sharp needle, that the great openings belong to vessels (*fig.* 97 *vp*); whilst the rest of the tissue, which appears to

96. A horizontal section of a young branch of Common Maple (*Erable commun*) (ACER CAMPESTRE), magnified twenty-six times larger than nature.—*m* Pith.—*em* Medullary sheath.—*fb* Bundles of wood.—*vp* Dotted vessels.—*rm* Medullary rays. —*c* Cambium.—*fc* Cortical fibres.—*vl* Lacticiferous vessels—*ec* Cellular envelope.— *p* Suberous envelope.—*ep* Epidermis.

be almost full, is composed of fibres (*fig.* 97 *fl*) of a moderate length, the walls of which are rather thick, and, consequently, the in-

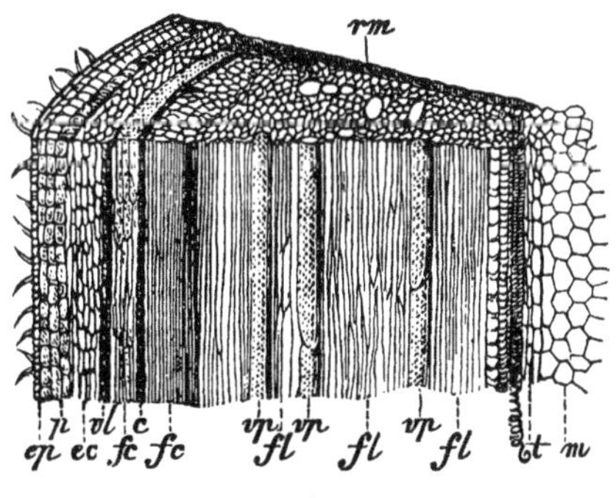

97

terior canals or ducts so small that their openings are only like little dots or points. They are, for the most part, arranged in series diverging like rays from the medullary centre. The vessels are not all of the same order: the largest and most external are dotted tubes (*vp*), whilst those immediately in contact with the pith are unrollable tracheæ (*t*). They are never found in any other place in the stem than this. They are formed before any of the other vessels. This union of tracheæ and fibres, the first parts of the wood which are formed along with those which immediately surround the pith, has received the name of the *Medullary sheath* (*Etui médullaire*) (VAGINA MEDULLARIS) (*fig.* 96 *em*).

§ 56. Outside of each of the fibro-vascular fascicles will be perceived in the horizontal section another collection of fibres (*figs.* 96, 97 *fc*) of a duller white, united in the form of a crescent with its convexity turned outwards. This crescent is separated from the rest of the fascicle by a zone of greenish cellular tissue (*c*). This zone merits our marked attention, since it separates the bark from the wood, and afterwards becomes the seat of the formation of fresh layers, which increase the thickness of the stem. Those fibres, which it separates from the wood, are cortical fibres, longer and more tenacious than the ligneous.

97. Vertical section of the same branch, cut parallel to the diameter through the middle of one of the ligneous fascicles, still more highly magnified than the preceding.—*t* Tracheæ.—All the rest of the letters refer to the same parts as in the preceding figure.

At the moment that the branch of the maple is cut we see a whitish milky liquid ooze out. It is from the bark that it flows, immediately outside the fascicle of cortical fibres; and, indeed, the microscope enables us to discover there a system of proper lacticiferous vessels (*figs.* 96, 97 *v l*).

Still nearer to the outside we only find cells, the entire mass of which forms the cortical parenchyma. It is covered by a reddish cuticle; it is the epidermis (*ep*), composed of a single row of cells, and covered over the whole of its surface with small fine whitish down. We have already pointed out the formation of these hairs, which are themselves only the cells of the epidermis modified in their form.

Under it we find several rows of cubical or lengthened cells, like those of the rays in the horizontal section (*fig.* 97 *p*), rather different in shape from those of the epidermis. The external ones are, however, slightly coloured with its reddish tinge; the interior ones are rather brown. The rows are arranged in regular layers over one another. Within this brownish zone is found a green one (*figs.* 96, 99 *e c*), formed by cells filled with chlorophyll. These may be distinguished not only by their colour, but also by their rounder or polyhedric form. They are also arranged less regularly, and not in close rows in concentric layers. We see, then, that in the cortical parenchyma may be recognised two systems: the outermost, with brownish rectangular cells, which M. Mohl has termed the *suberous layer* or *envelope* (*couche* ou *envelope subéreuse*), (*p*); the innermost, of a green colour, has received the name of *cellular envelope* (*ec*), (*envelope cellulaire*), under which name botanists have frequently confounded the whole mass of these two layers of such different natures.

It is in the green cellular envelope that the medullary rays (*r m*), which are themselves of the same colour, terminate. They are each formed by one or more rows of cells, which, compressed between the ligneous bundles or fascicles, are not long in being included and lengthened in the direction of the ray, and in forming by their union the thin laminæ. The name of *muriform* (*muriforme*) has sometimes been given to the tissue of these laminæ, since they may be compared to a wall; the flattened cells placed in layers the one above the other, resembling the stones.

§ 57. We have just examined a one-year's branch of Common Maple; but, we shall find in the greater part of our other dicotyledonous trees the same parts, only in different shapes and proportions. This branch or stem presents at first, then, two very distinct systems:

the one, the inner or ligneous; the other, the outer or cortical; these are separated from one another by a zone of cellular tissue.

The ligneous system is itself composed of the pith in the centre, and of an exterior zone of fibro-vascular bundles, which are formed, 1st, in the inside, by the first developed tracheæ immediately touching the pith, around which they constitute, together with some fibres intermixed with them, the medullary sheath; 2nd, nearer the outside, by masses of fibres mingled with spiral vessels of another kind.

The cortical system under the epidermis is composed of three distinct layers; the two exterior cellular ones, the epidermis and the cellular envelope properly so called, on the edge of which are scattered some lacticiferous vessels; the interior fibrous one, in which these same vessels are also intermingled.

§ 58. Let us compare all these parts which we have thus determined in a branch of the second year (*fig.* 98). In the spring, the

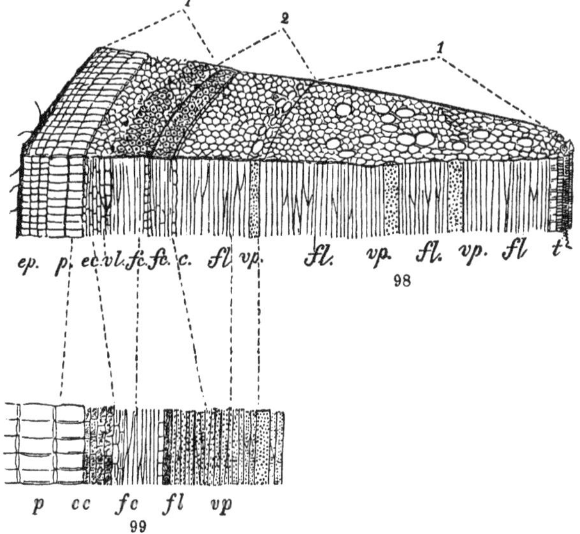

98. Vertical section of a branch of the second year, from the Common Maple, in which 1 indicates the portions formed during the first year; 2, the parts formed in the second.—The letters refer to the same parts as in *figs.* 96 and 97.

99. Some portions of the preceding figure more highly magnified, so as to shew their structure more plainly: for instance, the dots of the ligneous fibres.—The letters again refer to the same parts.

little cellular zone which we observed to be interposed between the ligneous and the cortical systems, really appears to be much narrower and as if filled with an almost liquid jelly. This jelly is gradually thickened, and skilful observers agree in recognising in it the organization of a nascent cellular tissue, although when they attempt to determine the mode of its formation they are involved in some differences, which may readily be supposed, when they are trying to examine the developement of parts of such excessive tenuity, and still almost fluid. Whatever may be its formation, it is the tissue which will present in a little time all the developements of elementary organs analogous to those which we have observed in a one-year's branch; and hence is derived the name of *cambium*, which was given to it when the property of transforming itself into all these different organs was assigned to it.

We now see, that, at the end of some time, in this interval are formed two fresh zones (*fig.* 98, 2), the one a cortical (*f c*) the other a ligneous one (*f l*), commonly like zones of the first year, to which they touch, and are, as it were, moulded: the cortical, composed of fibres, like the innermost layer of the bark to which it is joined: the ligneous, of fibres and spiral vessels similar to the exterior of the fibro-vascular fascicle, to which it is in close juxtaposition; for it does not share in the nature of its internal part or of the medullary sheath. We never find in it any tracheæ. The portion of the zone of cambium which corresponded to the cellular rays is also organized like the former tissue with which it is connected; and remains cellular; so that the ray is continued without interruption and modification across the fresh layers. What has passed during the second year is renewed during the third and following years. Each, in its turn, produces between the wood and bark already formed its layer of wood and its layer of bark; and, thus, at the end of several years, we have a certain number of concentric layers both of bark and of wood. The ligneous layers are easily observable and form the greater part of the thickness of the branch; whilst the cortical are very thin, and, forming a very small zone, are not easily distinguished from one another.

§ 59. These changes, which we have observed in a branch of *Common Maple*, would have been equally conspicuous in the greater part of our indigenous trees; and what we have said with regard to this example may be held as true with regard to dicotyledonous stems in general. We will, however, extend our researches a little further; and successively reconsidering the parts which we have learned to distinguish in one of our own trees, we will compare them with others to see in what way they differ. Those, which, growing in our temperate regions, are exposed to the regular alternations of the seasons, and have naturally served as the principal, and for a long time, the exclusive objects of the researches of botanists, will alone at first occupy our attention; then, after we shall have considered for some time the nature of these stems, which we may be permitted to call normal, we will cast our eye rapidly over the stems of vegetables, equally dicotyledonous, the structure of which presents certain notable differences, and which, inhabiting for the most part tropical climates, the vegetation having been studied only in a passing manner by travellers, and that too, only very recently, are still very imperfectly known.

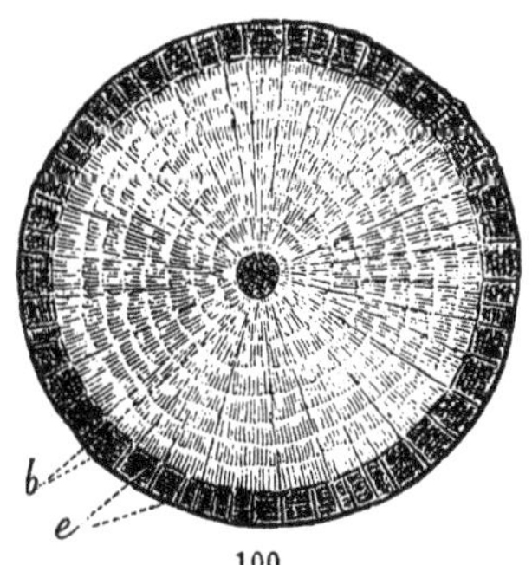

100

We will consider successively the ligneous and the cortical systems; the first consisting of the pith, the wood, and the medullary rays; the second, of the fibrous and the two cellular layers.

LIGNEOUS SYSTEM.

§ 60. PITH (*Moelle*) (MEDULLA)[i].—We have seen that the parenchyma, of which the stem was at first exclusively composed, is afterwards separated by the developement of the ligneous circle into two portions, one of which, the central, takes the name of pith; and the example we have chosen has shewn, that it is composed of cells which, from the centre to the circumference, diminish gradually in

100. Horizontal section of a branch of an Oak (*Chêne*) eight years old.—b Wood, in which we see eight concentric zones, separated from one another by lines of points answering to the openings of the large vessels.—*e* Bark, which also is composed of eight concentric zones, but much thinner, and less distinct.—The wood and bark are traversed by medullary rays, extending from the circumference, some to the centre, which forms the pith, others only to a circle formed during one of the subsequent years.

[i] Grew gave it the name of *Medullary cavity*, which was afterward changed into the *Medullary Canal* (*Canal médullaire*) (CANALIS MEDULLARIS).—TRANS.

size, whilst, at the same time, they take a deeper and a deeper green. The latter are filled with abundance of juices, which are wanting, on the contrary, in those in the centre; and, from these different characteristics, it is easy to perceive that they are younger, and consequently, their vital activity is more energetic. By degrees this energy is weakened, and, after the first year, the pith generally takes a uniform colour throughout, commonly white, though there are examples of other hues. Its cells, the size of which diminishes from the centre to the exterior, do not contain anything but air, and their vitality appears to be quite suspended: they often burst, also, and several lacunæ appear at the centre; this may often be observed before this time, especially in annuals which have a large pith, and are of a very rapid growth. In the first year, however, and especially towards the end, they exhibit a very active degree of vitality, and this action may be prolonged for some time. This is proved by the thickening and dots of the walls, which take place only by the formation of fresh layers in the interior of each cell, and leads us to suppose the existence of rather a long duration of action.

The pith forms at the centre of the stem, a column, the form of which is indicated by its horizontal section; this section at first presented the form of a star, on account of the spreading of the ligneous bundles leaving between them angles filled with the still green pith. It sometimes preserves its star-like form: it frequently however, changes into that of a circle, often also into a polygon of a greater or less number of sides. It is in some plants reduced as low as four, and then assumes the shape of a square or rectangle, (as for instance, in the Linden or Lime-tree (*Tilleul*), and also in several climbing Bignoniæ): sometimes even that of a triangle (as in the Oleander), (*Laurier-rose*); sometimes it takes very extraordinary shapes and forms, as that of a cross in several *Bauhiniæ.*

It has been supposed that the form of the pith corresponds to that of the stem, or perhaps, more correctly speaking, to the manner in which the leaves and, consequently, the branches are arranged. This theory, which appears likely enough, is supported by several examples, though contradicted by others. Thus, without mentioning the arrangement of the leaves, with which we are not now engaged, we sometimes find a cylindrical pith in a polygonal stem. Besides, the pith has not always the same form at different heights of the same stem.

Its diameter may also vary greatly, and that in two branches which themselves are equal in diameter. Does it vary at different times at the same height? Does it augment or diminish with age? Changes may also easily take place when the stem is still young. By the multiplication of cells, and the augmentation of each one, the pith would of course be enlarged; and afterwards, when the ligneous fascicles were developed in their turn, they might be swelled on every side, and by their inward extension press the pith a little closer. But a time must come when equilibrium is established, and then its size remains immutable; this we may confirm by examining old trunks and young stems of the Elder tree. For a long time it was the opinion of botanists, that being still compressed by the increasing quantity of wood, it was at last obliterated: but this was only an illusion arising from its relative smallness, when it was examined in a larger trunk. Correct measurements prove the contrary.

We have hitherto described it as composed of cellular tissue only. We sometimes find, however, fibro-vascular bundles or fascicles scattered in it, especially in certain herbaceous plants, the rapid and enormous developement of which is unusual in plants of this class: for example, in several great Umbelliferæ, especially in the genus *Ferula.* Such bundles are also observable in the pith of the Pepper Plant. The composition of these bundles, the vessels of which are unrollable tracheæ, seems to indicate that they are nothing else than detached portions of the medullary sheath approaching the centre. We sometimes, also, see along with these tracheæ masses of fibres, and frequently also lacticiferous vessels. The presence of the latter vessels in the pith is much more frequent than that of any other: for instance in the pith of the Dwarf Elder (*Yèble*) (SAMBUCUS EBULUS) and in that of the Fig-tree (*Figuier*), in which we may see, directly it is cut, the milky juice oozing out; but this is only in young branches, and, generally, lacticiferous vessels are only very few in number and much separated from one another.

§ 61. THE WOOD (*Bois*).—We have seen (§ 56) that the first layer of wood is composed of fibro-vascular bundles arranged in a circle around the pith; that they are separated from one another by rather large bands of cellular tissue extending like rays from the pith to the bark; that some little time afterwards fresh bundles develope themselves in the substance of these rays and increase in number at the expense of their thickness (*fig.* 95, *f'' f''*); that then these bundles, both by their multiplication and the augmentation of their whole volume, resulting from that of each of the elementary organs of which they are composed, at last approach and touch so closely, that the rays which separate them are reduced to extremely thin laminæ. They thus form a ligneous circle, (*fig.* 96 *f b*).

§ 62. We have seen, also, that this circle has a peculiar structure in its internal part which is in contact with the pith; that it presents in that part, and in no other, a mass of unrollable tracheæ (*fig.* 97 *t*) and that this internal part has received the name of the medullary sheath. The pith is moulded to it, or rather it is moulded to the pith; and the angles which it always presents during its earliest growth and retains in the stems of certain plants, correspond to so many angular projections, which form the internal boundary of each of the bundles.

The medullary sheath is that part of the wood which undergoes the fewest changes. Its tracheæ preserve the size which they acquired at first, and may be unrolled when taken from the stems of even rather old plants.

§ 63. The rest of the ligneous circle, and by far its greater part, is composed of fibres and of annular striped or dotted vessels, (*fig.* 97 *vp*, *vp*), generally of a large diameter.

Now, we know that the developement of herbaceous vegetables proceeds no further. But the parts developed at this stage may acquire a solid consistence in the first year by the increase in breadth of this first layer and the density attained by its elements. We know, in short, that in perennial vegetables, between the wood and the bark, the interval of which is filled with *cambium* (*fig.* 97 *c*), a matter at first almost fluid, but afterwards organized into cellular tissue, a fresh layer of wood moulded to the preceding one is formed during each year. It is very clear, then, that the number of layers represents that of the years which the tree has lived, so that its age is inscribed on its section; this is a truth which has long been known[k], and which several facts more or less convincing have con-

[k] The honour of having first made this discovery, is by some attributed to Malpighi; but it must have been a received opinion in his time, since this celebrated anatomist was born in 1628, and we may read in Montaigne's *Travels in Italy*, in 1581, the following passage; "The workman, an ingenious man, noted for making good mathematical instruments, told me that all trees had as many circles of wood as the number of years which they had lived amounted to, and shewed me this in all those which he had in his shop, whilst he was working in wood. And that the part which was turned towards the north was always narrower, and the circles closer and more compact than any other. Wherefore he boasted, that whatever piece of wood was brought to him, he could not only tell the age of the tree, but also in what direction it grew."

firmed. Thus, let us suppose that the layer of cambium is disorganized in certain places, (this may easily take place from the action of a very rigorous winter), no wood will be produced in these places, and consequently in the ligneous tissue there will be a series of gaps or breaks. During following years, when the cold has not exercised the same influence, as many layers of wood have been formed, and, consequently, have covered the gap. When we discover this, we may count as many circles outside of this gap, as years have passed since the severe winter which produced it. This experiment has been verified. On cutting down very old and very large Elms (*Ormes*), there were found in the interior some of these solutions of continuity. By counting the number of concentric layers by which they were covered, it was determined that the layer of wood, in which these solutions of continuity were found, was formed such a year; this year exactly corresponded to a very severe winter which occurred. Several sections of such trunks may be seen in the botanical galleries of the Museum of Paris. They preserve there also a trunk of a Beech-tree (*Hètre*) which bears the date 1750 inscribed on its bark, and the same concealed in the substance of its ligneous or woody layers; the two dates are separated from one another by a certain number of layers (55). This number was precisely that of years passed between the year indicated by the date and that in which the tree was cut down. The inscription was engraved on the bark when still young, and cut through the whole thickness of the bark, and a little way into the wood. Thence resulted a lacuna, which, like that resulting from frost, was covered by the successive layers of subsequent years. But, here the experiment is more complete, since it both comprehends the bark and the wood, and also bears an authentic date. Most botanical collections contain specimens of this kind, some owing to chance, others to experiments tried in order to discover the way our trees grow. It is easy to introduce between the wood and bark a foreign body, such as a metallic plate; and on cutting the branch at the end of a certain number of years we find this plate covered by an equal number of layers.

We have already said that each branch grows in the same manner as the stem. Those which were produced the first year present, therefore, the same number of layers as the stems; those which are produced one, two, or three years afterwards, offer, the third year, three layers less; the second, two; the first, one. When we cut a tree at its base, we may thus tell at what period of its life each one of the branches was formed by subtracting the number of the layers of the branch from that of the layers of the stem. If we count fifty concentric layers at the base of the trunk, thirty in such a branch, ten in such a one, the tree was twenty years old when it produced the first, and forty when it produced the second.

§ 64. But, how can these layers, composed of the same elements, be distinguished from one another? Because these elements are not scattered at random, but are distributed in a certain constant manner. Let us take for instance, a log of Oak (*Chêne*) (*fig.* 100, 101), or of Elm, and after having cut a thin piece from the surface of its section, let us examine its zones: we shall see that the inner edge of each is indicated by one or more lines of little holes which are wanting in the rest of the zone. These are the openings of so many large vessels which are only found towards this inner border, whilst the layer towards the outside is composed of close fibres, with walls so thick as to appear solid, and the canal which flows through them escapes the naked eye. In a piece of Horn-beam (*Charme*) (Carpinus Betulus), of Linden or Lime-tree, or of Common Maple, we do not observe these large vessels, the wide opening of which points out so well the inner border of the annual zone; but, it is almost entirely perforated by those of smaller and more equally sized vessels. They always cease, however, towards the outer margin, exclusively formed of several rows of fibres, which, as they approach this side, become more and more granulated, compact, and coloured; and hence results a line of separation between this layer and the following one, sometimes very strongly coloured, sometimes pale and rather indecisive.

In several species of wood, the line of demarcation is marked by a circular row of cells analogous to those of the medullary rays; more rarely this cellular tissue, interposed between the layers of wood, acquires greater thickness; in the Sumac, for instance, in which the cells, arranged in several concentric rows, are large and coloured like the pith. This observation gave rise to the following ingenious theory: namely, that each circle, produced after the first year, would itself have a pith represented by that cellular zone placed within its woody layer: and performing, relatively to it, the same functions as the central pith discharges, with respect to the ligneous circle which surrounds it. But besides the objection that these *medullary zones* (*zones médullaires*) are frequently wanting, the absence of the medullary sheath in every layer established posterior to the first year, would still establish an essential difference.

§ 65. We have already stated (§ 52) that the number of ligneous bundles is augmented, because fresh ones are developed in the cellular intervals, at first very large, which separated them, and served as foundations for the medullary rays (*figs.* 94, 95). These bundles are afterwards still more multiplied, but it is in another way, and, as it were inversely, since fresh medullary rays shoot (if we may be allowed the expression) into the ligneous elements (*fig.* 101).

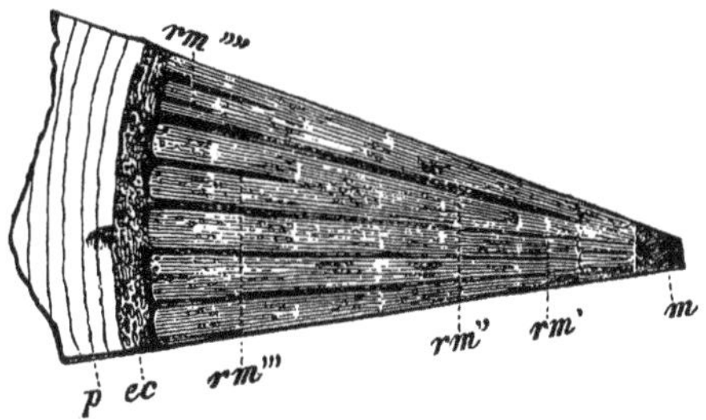

101

We know, that in each year's production, to each existing tissue, is moulded an analogous one: utricles on utricles, in order to continue the medullary rays; fibres and vessels on fibres and vessels, in order to continue the ligneous fascicles. But the fresh bundle, thus fitting to the old one, is not single like it, but is doubled or tripled, and thus divided into several pieces by rows of cellular tissue, which will be the commencement of new rays, (*rm*'' *rm*''' *rm*''''), differing from the first (*rm*') in that they do not begin from the centre. In the fresh zone, larger than those which preceded it, since it is exterior and concentric to them, a much larger number of bundles and of interposed rays is naturally formed.

§ 66. Each zone finishes growing in the course of a year: and when extended to a certain limit, it stops, and thus forms a fixed base to which the zone of the following year will be attached. The ulterior changes which it undergoes, only depend on those which are

101. Horizontal section of two ligneous fascicles of the Cork-tree (*Chêne-liège*), from a branch several years old. These fascicles, separated by the medullary ray *rm*', are each divided during the following years into several secondary fascicles, so much the more numerous and separated by shorter rays (*rm*'' *rm*''' *rm*'''') as these rays spring from a more external, and, consequently, later circle.—*m* Pith,—*ec* Cellular envelope.—*p* Suberous envelope, which, in this species, is largely developed.

passing in the interior of its elementary organs. When young their cavities, enclosed by very thin walls, were quite filled with liquid juices. As they grow old, the liquids diminish in relative proportion to the increase of the solids, as much because the walls of each organ are thickened by the addition of fresh layers (§ 6), as that the matter contained in these cavities on account of fresh chemical combinations, or of the evaporation of their fluids, is also thickened and gradually hardened. It is thus that the *ligneum* or *woody matter*, (*ligneux*) is formed, a matter which augments the density of the fibrous wall, by almost penetrating through it; whilst the fibre is of the same nature in every species of wood, this ligneum, varying in each, gives its particular qualities. But, there must come some time when the fibres thus solidified cease to be permeable to fluids.

§ 67. Since these modifications are brought on by age, they ought to be much further advanced in the inner, than in the outer circles: and, consequently, in the former the tissue will be fuller, harder, and drier than in the latter. In coloured woods, the centre will be the first coloured, and the colour will fade, as well as its hardness decrease, as the circumference is approached. Hence arises in several kinds of wood the distinction of two parts: 1st. the exterior one, which still retains the qualities of the young wood, i. e. which remains impregnated with the liquids and juices to which it is permeable, and is, consequently, more tender, and of a pale or white colour, whence has been derived its name of *Sap-wood*, (ALBURNUM), (*aubier*); 2nd. the interior one, dried, hardened, and coloured, which is commonly called the *heart* or *perfect wood*, (DURAMEN), (*cœur*, *bois-parfait*).

In deeply coloured wood, principally in those kinds used for cabinet-work, this distinction of the two parts is very apparent: we may thus conceive, without having seen it, how in ebony (*ébène*) violet-ebony (*Palissandre*) or mahogany (*acajou*) the heart, which is employed for our furniture, is united to an alburnum which is still white. It is useless to enumerate here those natural clouds and veins so different in different woods, which every one will imagine from those with which he is already acquainted, and of which we may find in collections several less common and much less known. Although this intensity of colour is most commonly remarkable in the trees of warm climates, it is found in some of our own, in rather an extraordinary degree. In the greater part, however, the change is slow, and the transition from the alburnum to the duramen is less sensibly marked. In several, as in the Poplar (*Peuplier*) or the

Willow (*Saule*), for instance, the wood is not coloured, and they are consequently called *White Woods*, (*Bois Blancs*); no difference is here perceptible to the eye. The durability is generally in accordance with the intensity of colour; the deepest coloured woods, ebony, and iron-wood, may be cited as examples of the most compact and durable. White woods, on the contrary, are the softest and are exceedingly perishable; for they still preserve in some degree the nature of alburnum. The bad quality of this is well known, and might be presupposed without any experiment, from what we have previously remarked on this subject, that its tissue contains the greatest proportion of liquids, and the least of solids. Then, besides the diminution resulting from this, with regard to the portion which alone ought to be preserved and assimilated, we can easily conceive that the super-abundance of liquids, brings, by their evaporation or by the new combinations which their condition favours, numerous alterations in the size and even in the composition of this imperfect wood, and especially invites the attention of insects, enemies always too ready to taste the sweets afforded by the mass of matter, which was destined for the nourishment of the vegetable tissue.

§ 68. The annual layers are very unequal in their thickness, much larger in soft woods, which, as is well known, are very rapid in their growth, much less in hard woods, which, on the contrary, grow very slowly. They vary also, in this respect, in trees of the same species, according to the circumstances in which they are placed. Thus, a tree will grow more slowly, if it be too closely surrounded by other trees, if it be in a less favourable soil, if it be in a colder climate, or the winter last longer. In the last trees which travellers have found as they approached the poles, the annual layers are still distinguishable, but are exceedingly thin. We shall see, for the same reason, a great irregularity in the successive layers of the very same tree; this answers to the differences presented by the seasons of the corresponding years.

Another cause of the inequality of the layers, which is difficult to appreciate, is the age of the tree. An old tree grows more regularly, but not so quick as it did when it was young; and during this latter period, there was a time when it increased more than at any other; in the Oak, for instance, from the age of twenty to thirty. On the other hand, a less breadth in the layer, arising from whatever cause, is commonly connected with a greater density in the wood. The business of a forester is to know these habits of each species of trees; so that he may know when to fell them, according as the quantity of wood is an object, when its yearly increase begins to fail; according as, on the contrary, the quality is to be preferred, a later period would be better, when it has reached its perfection.

The same zone is not always of an equal thickness throughout its whole extent, and, whenever there is any irregularity, it is commonly perceived always on the same side in a great number of successive layers, so that, it apparently arises from a permanent cause acting always in the same place. It was thought at first, that this arose from the diversity of aspect of the different sides of the tree: according as the tree stood with respect to the four points of the compass, it grew more towards the south, than towards the north. But, we may be convinced of the falsity of this opinion, since the influence ought to act generally and regularly; and it has been determined that the phenomenon is owing to purely local circumstances: for instance, the tree may be crowded and sheltered on one side, open on the other and exposed to the air and light: and the roots may find, and this is frequently the case, better soil on one side than on the other.

§ 69. We have already said that the alternation of the seasons determined the formation of the annual layers, by the regular interruption and the renewal of the work of Nature. What then occurs in the tropics, where the winters are too warm to interrupt vegetation? It would seem, that it is incessantly continued, and that the wood, forming every instant, will not be separated into distinct zones. They are, indeed, still marked in the greater part of the plants of these countries, although it is in a slighter degree. The reason of this is, that, for the most part, vegetation has here also a periodical repose, the dry seasons, which in several trees causes the fall of the leaf, and thus supplies in some degree our winter. Observations, made in these climates so unlike our own, cannot fail to tell and explain many facts very different from those to which we have been accustomed.

§ 70. We are led to suppose, that if in the same year, a vegetable was exposed to such conditions, that its vegetation was alternately excited or arrested, several partial zones would be formed in the general layer which is the produce of this year. This, indeed, may be sometimes seen in our plants, particularly in the herbaceous, the more rapid increase of which makes manifest the influence of the alternations to which a harder wood, more slowly and better formed, remains insensible. We shall see, when we come to examine the subject of leaves and buds, the reasons of such a multiplicity of concentric circles concurring in the formation of the commonly simple ligneous circle.

§ 71. The failure of distinct zones in wood of several years' growth is an inverse fact much more frequent and even constant in a pretty large number of ligneous plants. We should at first be inclined to explain this by the action of a climate where the alternation of the seasons is not apparent, especially since this is often observed in the trees cultivated in our green-houses and conservatories, where the changes of the season are, as far as we are able, artificially suppressed; the ligneous zones are not discernible in some, and in others, although distinct, cease to correspond exactly to the number of years. Since, however, the climate during the course of a whole year is very rarely the same, even under the tropics, as we have just mentioned; since, besides these vegetables, several others grow around them, which although necessarily submitted to the same influences, present a succession of concentric zones; since, besides, it is always in certain species, in certain genera, and even in certain whole families, that this peculiar organization is observable, we must suppose that it is inherent in the very nature of these plants. We will cite as examples of this the Cacti and the Peppers.

There are some, lastly, in which we observe several circles, each of which, however, is the produce of several years. The CYCAS is one of them, and enters into the class which we just now mentioned as comprising such trees and shrubs as are cultivated in a green-house.

§ 72. MEDULLARY RAYS (*Rayons Médullaires*).—We have frequently had occasion during the progress of the inquiries detailed in the preceding articles to speak of the *medullary rays;* we have shewn their construction, how they are formed and multiplied. Those which, being in existence from the origin of the stem, are continued without interruption from the pith to the bark, have been termed *great rays* (*grand rayons*) (*fig.* 101 *rm'*); those which appear only during succeeding years and commence from the layers corresponding to these years, have been termed *small rays* (*petits rayons*) (*fig.* 101 *rm''*, *rm'''*, *rm''''*). The latter are found even in those woods, where the distinction of the layers is not observable, and thus indicates, although very obscurely, successive formations, which the homogeneity of the whole ligneous mass does not allow us to perceive otherwise.

When the rays are examined, not only in an horizontal section, but in the stem cut longitudinally, it will be seen that the cells which compose it, placed one above the other in one or more rows, form thin plates (*fig.* 102 *r m*). If the progress of the fascicles is perfectly rectilineal, as in the Clematis, for instance, the laminæ formed by the rays extend uninterruptedly from one end of the stem to the other; when the wood splits, the cleft follows their direction and that too with great facility. But, most commonly the partial fascicles are more or less sinuous in their vertical direction: and then the laminæ are interrupted where they deviate from a straight line. This is very easily deter-

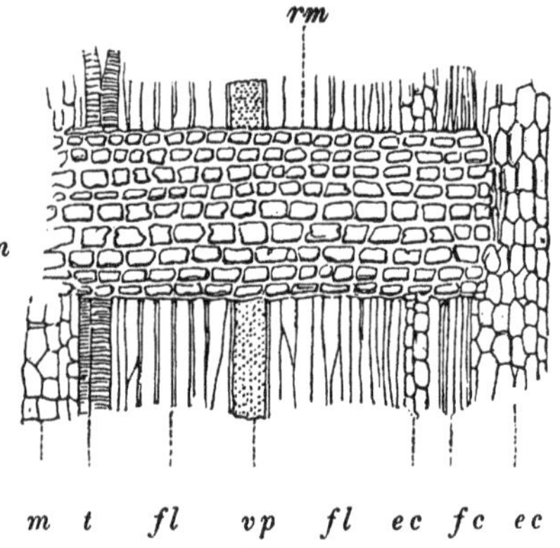

102

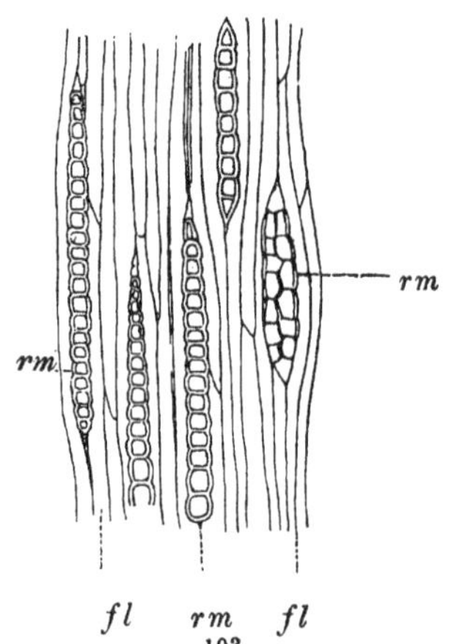

103

102. A section of a one-year's branch of the Common Maple, vertical to, and passing through the pith, and magnified very highly.—We see the lamina formed by the medullary ray *r m*, extending from the pith *m* to the cortical parenchyma *e c*, bounding in its progress, 1st, a ligneous fascicle composed from the interior to the exterior of tracheæ *t* and of ligneous fibres *fl*, in the midst of which is perceived a large dotted vessel *vp*; 2nd., a fascicle of cortical fibres *fc*.

103. Vertical section of the same branch, perpendicular to the medullary rays.—*fl*, *fl*, Ligneous fibres, forming small sinuous fascicles, which thus leave between them intervals traversed by the medullary rays *r m*, *r m*.

mined by examining both the surface of the barked wood, and, what is still better, very thin vertical sections perpendicular to the rays (*fig.* 103). The fascicles, at first united, separate a little, in order to unite afresh a little lower down, and thus leave between them an interval filled by the cells of the rays, the laminæ of which are moulded to these intervals, and are thus frequently rather thicker in the middle than at the top or the bottom.

It is towards the periphery of the wood, and consequently in the part which is connected with the cortical system, that the rays often present the greatest width; it is there also that their vitality appears to be most active. It appears to be gradually extinguished, as the ray approaches the heart of the wood, in the colour of which their cells participate, and to which they sometimes greatly contribute: whilst, in the alburnum, and especially near its circumference, they are filled with fecula or liquid juices according to the season, and often coloured green by the chlorophyll. We may hence consider them as allied more intimately to the cortical system than to the pith, and we consequently pass from the investigation of their properties to that of the BARK (*écorce*).

§ 73. THE BARK (*Écorce*).—We know that at first the cortical system is not distinguished from the ligneous: that a little later, in each of the fascicles developed in a circle around the pith, a thin layer of a demi-fluid tissue, the cambium (*figs.* 96, 97 *c*), describes a circle separating this bundle into two unequal parts, the exterior (*f c*), belonging to the bark, much narrower than the interior (*f b*), which belongs to the wood; that the whole of the cellular zone which is outside of these bundles forms the cortical parenchyma, in which may be discerned, besides the epidermis (*e p*), two very distinct modifications, the *suberous envelope* or *cortical layer*, (*envelope subéreuse*,) (*p*), and the *cellular envelope* (*e c*); that, lastly, in this envelope and mingled amongst the fascicles of the cortical fibres, generally circulate numerous lacticiferous vessels. The bark presents, as well as the ligneous system, two portions, one cellular, and one fibro-vascular. But, here the order is inverted both in the situation and the relative proportion of the parts; for the parenchyma, the pith, as it were, of the bark, occupies its circumference; and it is more largely developed, and presents forms more varied than the fibro-vascular fascicles, whilst, on the contrary, in the wood we have seen these bundles much more developed and complex than the pith.

In consequence of this inverse situation of the parts, we will pursue an equally inverse method in our investigation: we will at first consider the cellular and exterior part, which is the first formed: then the fascicles and the fibres which compose the interior of each layer; for we have seen (§ 58) that in our trees there is formed each year a layer of bark along with one of wood. But, from the inverse situation of the component parts of the former, a result, which it is easy to foresee, will occur. Whilst the zones of wood remain immovable, the fresh one fitting to the older one which it covers, the zones of bark are incessantly pushed to the outside, to make room for others which are still younger, and especially for the fresh ligneous layers which are formed within them. When they have once obtained the highest degree of developement of which they are susceptible, not being able to stretch to an infinite extent, they are necessarily rendered liable to alterations, the causes of which are augmented by their position on the outside; they break in different places and are detached in plates, &c., and that, in the order of their formation, the oldest and outermost bursting the first.

§ 74. The epidermis, which has already occupied our attention so long (§ 37), that it is useless to dwell on it any longer, is that part of the bark, which, from the distention caused by the progressive increase of the stem, and the action of external agents, disappears the first. Its existence, indeed, is only temporary: in a little time it breaks, peels, dries, and is lost.

§ 75. Under it were other cellular layers, which then filled its place in covering the stem, and which, sometimes disappearing in their turn, are themselves replaced by fresh subjacent layers. This cellular layer, stretched over the surface of the bark, is called its epidermis by several botanists, who thus assign to the word another meaning to that we have given it. M. Mohl proposes to call it *peridermis* (*périderme*), a name, which expresses its interior situation with regard to the epidermis properly so called, and its peripheric situation with regard to the bark (δέρμα). This peridermis does not always derive its origin from the same source, and may be formed at different depths. We shall understand this better, by examining more in detail the exterior and interior parts of the bark.

§ 76. 1st SUBEROUS ENVELOPE OR LAYER (*Envelope ou couche subéreuse*).—It has received this name, because it is that part, which, in some trees, constitutes the substance commonly known under the name of *cork* (*liége*) (SUBER). It has also been called *epiphlœum* (*ἐπὶ, on; φλοίος, bark*), on account of its superficial position. It is first perceived (*fig.* 97, 98 *p*) under the epidermis, forming one or more rows of cells sometimes cubical, more frequently tabular, closely joined together, without granules in the interior, with thin walls at first colourless, afterwards often of a brown colour. Sometimes its developement does not proceed any farther, sometimes, on the contrary, the rows multiply and the thickness increases, especially in that kind of oak so well known under the name of the Cork-tree (*Liége*) QUERCUS SUBER (*fig.* 101 *p*). All its cells have not the same identical form and colour; but, here and there, may be seen several which are more compressed or tabular, also arranged in rows, by which the whole layer seems to be subdivided into several secondary layers. This arrangement is not easily discernible in the Cork-tree or Common Maple, where these layers of tabular cells are much less clearly and regularly formed. In the GYMNOCLADUS CANADENSIS, the secondary layers alternately formed by the broad and by the narrow tabular cells are of an almost equal thickness. In the common Birch (*Bouleau commun*) (BETULA ALBA), the layers of tabular cells of a brown colour, have a much higher degree of developement than the others, which are very white and small, so that they are very easily broken by the growth of the stem: hence those small shreds brown on the inside, white on the outside, which are detached from the bark of the Birch. Lastly, in the Beech, (*Hètre*) (FAGUS SYLVATICA,) the tabular cells are the only ones which are developed.

§ 77. CELLULAR ENVELOPE (*Envelope cellulaire*).—This has also been called the *green layer* (*couche verte*), on account of its usual colour; *mesophlœum*, on account of its position in the middle of the cortical layer (*μεσὸς, middle; φλοίος, bark*). It is distinguished from the suberous layer which surrounds it by the chlorophyll, which fills and colours its polyhedric cells green, the walls of which are thicker, more loosely united, and consequently leaving between them several meati, and very frequently large lacunæ. In the midst of the green cells some are often found, quite colourless, and enclosing crystals.

§ 78. CORTICAL FIBRES OR LIBER (*Fibres corticales*).—These form fascicles placed opposite those of the wood, often separated from them by a thin lamina of the cellular envelope, and always, a little time after, by a layer of utricles belonging to the cambium.

These fibres of brilliant white are much longer and thinner than the ligneous. Their walls, as they grow old, become very thick and dotted by the formation of fresh layers in their interior. These have the greatest tenacity of any other part in the vegetable, and, consequently, in several plants, render man very important services, in furnishing him with materials for cordage, for thread, and for clothing, both of the strongest and finest description. Those most used are Hemp, Flax, and the Cocoa-nut tree. Even the method of preparing the first for use, shews how much stronger the cortical fibres are than any other part of the plant. It is first soaked, i. e. macerated in water, and then it is beaten; the fibres are thus obtained whole after this double manipulation, which successively destroys all the other parts.

Medullary rays generally continuing those of the ligneous system, but much larger, and naturally formed of cells less compressed and less closely united, separate the cortical fascicles, the whole constituting a zone concentric with the ligneous zone. In the same manner as in the wood, sometimes the fascicles follow a rectilineal direction (as in the Vine, and the Horse-chesnut, [*Marronier d' Inde*]), and then their rays, forming equally straight laminæ, are continuously interposed between two neighbouring fascicles from one end of the stem to the other; sometimes they are very sinuous, as in the Elm, the Lime, and the Oak, and then approaching the neighbouring rays alternately on the right and left, they touch one another, and are united, though only to separate again a little lower down, and thus interrupt the medullary rays, which, consequently, only form short plates, making by their numberless anastomoses, or openings, a net, the meshes of which are filled by these rays (*fig.* 104). Each layer of these cortical fibres represents a kind of cloth of a very loose texture. The *ensemble* of the layers of several years, each of which may be itself subdivided into

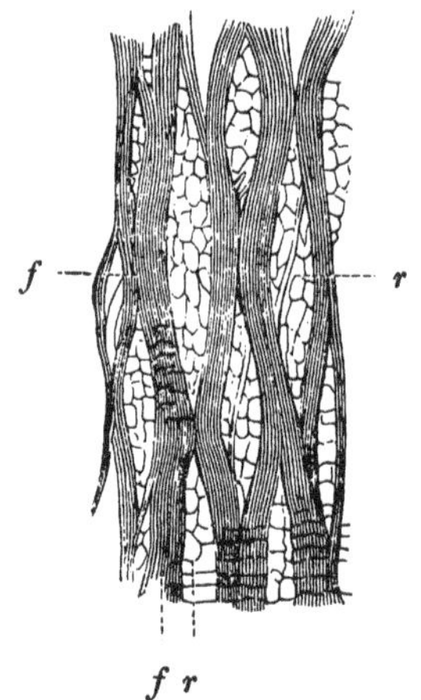

104. A net formed by the liber in the Spurge Laurel (*Lauréole*) (DAPHNE LAUREOLA).—*f* Fibrous bundle or fascicle.—*r* Medullary rays.

several others still thinner, (provided the fibres are in regular rows,) may be compared to a book, the layers of the several years forming the leaves: and, hence, the name of *liber*, which has been generally used to distinguish the cortical fibres. Some botanists have given them the appellation of *Endophlæum*, (ἐνδὸν, *within*; φλοίος, *bark*,) because it is the internal part of the bark.

These leaves, the produce of different years, are like the annual layers of the wood, sometimes separated from one another by utricular zones depending upon the cellular envelope, in the substance of which are formed the fibrous fascicles.

It is clear, that the progressive increase of the stem determines the proportional distention of the leaves of liber, the fascicles of which are thus progressively separating and consequently enlarging. The rays, by the multiplication of the cells which compose them, are extended in the same proportion, whilst the tissue remains alive, and thus continue to fill up the intervals resulting from their separation.

§ 79. In short, very active vitality is preserved in the parenchymatous system of the bark, and the production of fresh cells is unceasingly taking place, not at one point only, but at several at the same time, since, besides the annual formation of a layer of liber and of the utricles immediately surrounding it, there may also take place, as we have just seen, an increase of the suberous envelope, as much by the multiplication of the broad cells as by that of the tabular ones; and all these developements seem to take place independently of one another.

From the manner which these developements follow relatively to one another results that of the bark in general and of the peridermis. We have already (§ 76) examined the case in which the greatest activity of increase is observable in the suberous envelope. Then, that of the liber and of the cellular envelope is continued regularly, though in a less degree, and it still proceeds in the bark, the peridermis of which is formed by the layer of the tabular cells of the suberous envelope, whether this layer is developed alone, or, formed under a mass of cork which soon dies, it pushes it off, and makes it fall in large pieces.

But, in many trees, the suberous envelope, instead of being fully developed, quickly falls off along with the epidermis, and then the production of the peridermis takes place either on the surface, or else in the interior of the cellular envelope. In the first case, the peridermis previously formed on the cellular envelope is at length pushed off by that which is afterwards developed under it, and is detached in sheets, as in the Plane-tree, the smooth surface of which is thus always formed by fresh layers. In the second case, the peridermis pushes outwards the cellular and fibrous layers in the mass of which it is developed, and which are detached, sometimes in scales containing several zones of parenchyma and of liber, which remain for some time united (in the Oak, for instance, and the Lime); sometimes in large leaves, in appearance like to those of the Birch, but differing from them in their texture and origin, since instead of the suberous layer they comprehend the liber and the cellular envelope, as in the Juniper (*Genévrier*) (JUNIPERUS COMMUNIS), and several of the Protæaceæ. In the Vine and the Honey-suckle (*Chèvrefeuille*) (LONICERA), each year the layer of bark in its developement, causes that of the preceding year to fall: hence arises the thinness of the bark of these trees. Lastly it takes another form, (as in the Larch [*Mélèze*] and the common Pine,) in which the cellular envelope is more largely developed, and hus forms a *false cork* (*faux liége*), which in a little time falls in scales.

To resume, a continual destruction of the external parts of the bark is proceeding at the same time as a continual production of cells in the parenchyma, and according as this production takes place in the mass of the suberous envelope, at the surface or in the mass of the cellular envelope, the rejected portions comprehend a greater or a less number of the constituent parts of the bark; the superficial portion or peridermis is formed by a layer of those which constitute the primitive parts of the cortical system.

§ 80. We have passed over in silence the lacticiferous vessels which abound in the interior portion of the bark. Of their existence we may be easily convinced on examining the transverse section of a fresh twig or stem; we shall see oozing out of their orifices the juice or sap, commonly coloured, which they contain. Since it is only in the youngest parts, active and full of life, that their functions are performed, they soon die and are pushed outwards along with the layers through which they circulate.

§ 81. One consequence will be seen to result from all that has proceeded; this is the great activity which all the cellular system of the bark preserves in comparison with that of the wood, which ceases to grow, and even to live, after the first periods of the formation of the stem. It is then rather to the former than to the latter, that are attached the medullary rays of the wood, which also present towards the bark their maximum, both in number and size. They seem, moreover, to originate primarily in the layer of cambium.

§ 82. LENTICELS (*Lenticelles*) (LENTICELLÆ).—On the surface of several kinds of young bark, we may observe little spots of a variable form, commonly lengthened, following to the axis of the stem; and, by an attentive examination and delicate touch, may recognise that they form a slight abutment. They were at first called *lenticular glands* (*glandes lenticulaires*), afterwards, when it was discovered that they were not glands, *lenticels*. They grow along with the stem, but increase in thickness more than in length, so that they tend to enlarge and bulge out more and more. On examining them with a microscope, we discover a mass of utricles, and, on looking for their origin, that it is a little excrescence of the cellular envelope, which pushes the parts which cover it to the outside and causes a kind of external hernia. The suberous envelope, which it crosses, follows it and forms its outline. By means of the numerous lenticels scattered over its surface, the bark thus puts its interior layers in contact with the air, after that the stomata have ceased their functions and the epidermis has disappeared. De Candolle attributes to them another office. We know, that, when one puts a branch in the water or in the moist earth, in general it continues to live, and on its surface are developed numerous roots which are called adventitious. De Candolle having remarked that these adventitious roots often sprang from the centre of the lenticels, considered the latter to be predestined to produce roots, and to perform with regard to these roots the same functions that the buds do to the branches. But it has also been observed, that the roots spring from many other points in which there are no lenticels, and their ordinarily commencing from the centre of the lenticels has been explained by the cellular mass being here in direct contact with the medium which favours the production of the fresh parts.

ANORMAL STEMS OF DICOTYLEDONOUS VEGETABLES.

§ 83. We have hitherto examined the stems of dicotyledonous vegetables, as they are commonly presented to our notice in the

[1] First named by Guettard. They were also called *Cortical Pores*, but this confuses the botanist, as Stomata have sometimes been termed Cortical Pores.—TRANS.

trees of our climates. But we have already mentioned some exceptions to the rule; the formation, either of several ligneous layers in the same year (§ 70), or of only one, apparently resulting from the confusion of several produced during a series of years (§ 71). This last case is by no means rare; medullary rays most commonly exist in this case, and originating at a certain distance from the central medulla or pith, may still aid us in discovering several concentric formations. But the rays themselves may disappear, and the whole mass of the wood consequently present a homogeneous appearance, which can shew us nothing respecting its age and mode of formation; this is particularly observable in the PISONIA ACULEATA, a stem, of which twelve centimetres (*French*) or 4.7196 inches (*English*) in diameter, does not offer any appreciable difference between the pith and the bark, in whatever place we examine it, and it is entirely composed of very large and almost equal vessels, rather regularly arranged in an extremely fine utricular or fibrous tissue.

§ 84. At other times, it is the cortical layers which do not follow the general rules in their developement, although these rules are much more varied, as we have seen, than those relating to that of the wood. Thus, in several stems, those of the Aristolochiaceæ, or Birth-worts, for instance, the liber ceases to grow after the first year, and thus, instead of forming concentric zones, is reduced to little bundles placed in a circle around the wood.

§ 85. We find the greatest number of plants which differ from ours in their stem in the tropics; and it is especially among those which are called Creepers, and which, instead of supporting themselves, rest on other trees by clasping them with their tendrils, and thus climb to the top and hang down when there is no longer any support: our Vine and Clematis will convey the meaning of this most forcibly to our minds. Some are cylindrical like the branches of our trees; but many of them are flattened on one side or the other, and thus take very singular and extraordinary forms. One would be led to believe that this flattening is owing to some mechanical obstacle, which they find on coming in contact with the tree which they embrace, and this may, indeed, be true, when, flattened only on the side in contact with the tree, they bulge out on the other, and assume in the intervals, where they hang freely, a round or cylindrical form. But, in the greater part of these Creepers, the form remains the same, whether they are pressed against another body, or whether they are free; and we may, consequently, conclude that this irregularity of form depends upon their nature. In these compressed cells, the ligneous body is developed in two directions only and these diametrically opposite; whilst it is very small or almost disappears in every other. We find remarkable instances of this in climbing plants of a genus of the family of leguminous plants, the Bauhinia, and even in certain species (BAUHINIA SCANDENS, for example), the stems are not only flat, but, by their turning first to one side, then to another alternately, they present a very singular and remarkably zigzag form.

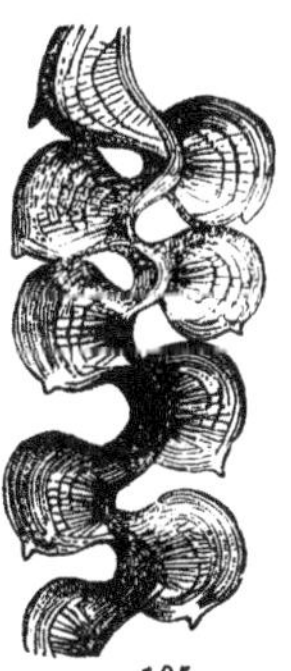

105

§ 86. The anormal form of several of the climbing plants may be assigned to an analogous reason: namely, the unequal developement of the ligneous body, which, instead of growing almost uniformly throughout the whole of its extent, and thus forming a cylinder or a cone, continues to thicken considerably in certain directions, whilst in others it hardly increases at all; so that it quite loses its cylindrical form, or else takes that of a column more or less deeply, more or less regularly fluted. Sometimes, the bark follows all the outline of the ligneous body, the form of which is reproduced externally by that of the bark (*fig.* 106); sometimes the flutings or channels are so very narrow and the bark so thick that the former are filled by the latter, so that the stem presents a smooth exterior, or at most only slightly fluted, whilst the ligneous body is very deeply channeled (*fig.* 107, 2). These intervals, thus filled by the cortical tissue, tend towards the centre in the same manner as rays, and thus, by meeting at the centre, cut the stem into as many fragments, the form of which is that of an angle, and each of which contains a portion of the medullary sheath. At other times, on account of a much stronger and more powerful convergence, these

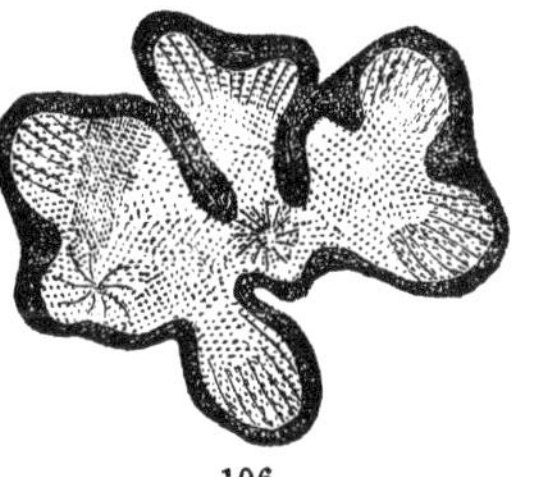

106

105. A piece of the stem of the BAUHINIA SCANDENS, much smaller than nature.
106. A horizontal section of the stem of the HETEROPTERYS ANOMALA.—In this and the following figures, the dots with which the wood is sprinkled indicate the orifices of large dotted vessels.

rays meet before reaching the centre, and there is, consequently, a central portion of the ligneous body which preserves the pith with its medullary sheath quite whole, surrounded by a circle

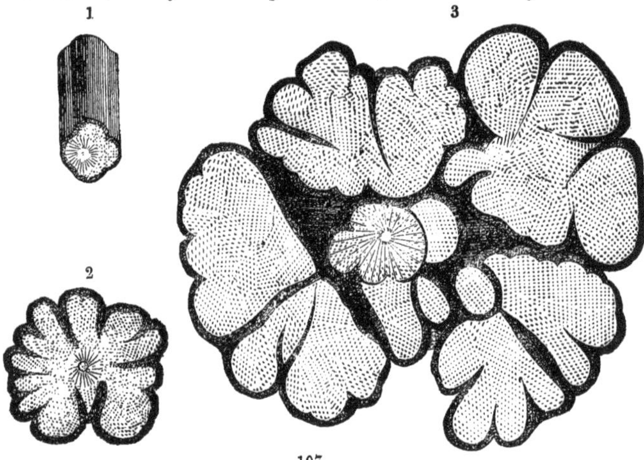

107

of other bundles which are quite deprived of it. The bark generally follows and covers all these divisions of the wood: and hence results the appearance of several stems united and twisted together when it is really only one stem (*fig.* 107, 3).

§ 87. These divisions may be either regular or irregular: and, in the former case, they sometimes offer very elegant figures, as for instance, in several of the Bignoniaceæ in which the ligneous mass represents a sort of Maltese cross by its tendency to develope itself in four opposite directions, following the direction of two diameters crossing at right angles (*fig.* 108).

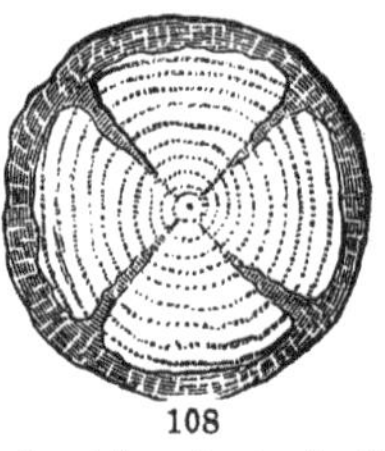

108

§ 88. It may also happen, that the ligneous fascicles, instead of

107. Three horizontal sections of the BANISTERIA NIGRESCENS of different ages. —1. When the outline only presents four superficial lobes.—2. When it presents several deeper ones, some of which are beginning to divide.—3. When they are completely separated, thus assuming the appearance of so many distinct branches simply placed close to one another. We may perceive that the middle one alone has a pith and a medullary sheath.

108. Horizontal section of the stem of the BIGNONIA CAPREOLATA.

being gradually detached from the central ligneous mass, are separated quickly from it, and seem to be so many branches growing parallel to the stem, but hidden by one covering of the bark. This is very easily seen in several of the Sapindaceæ (*fig.* 100), in which we may remark that each of these ligneous bodies, thus placed in a circle around a central ligneous body, itself presents a centre containing several unrollable tracheæ, which, consequently, will be part of the medullary sheath.

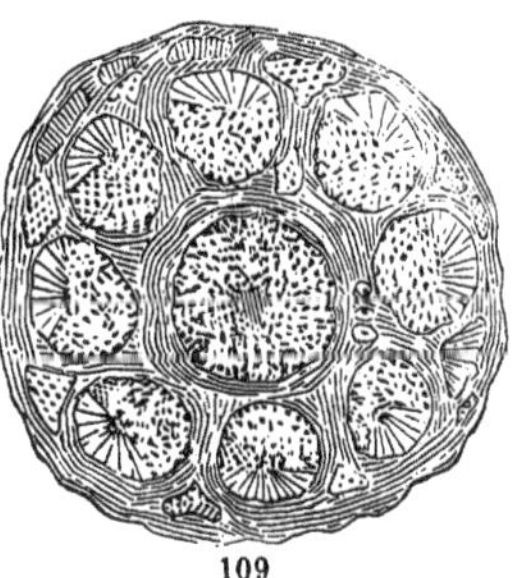

109

§ 89. The bundles thus arranged around a central cylinder, from which they are separated by a layer of the bark which envelopes them on all sides, may remain insulated from one another, or approach one another even as far as to touch, and thus form a circle concentric with the first; then fresh fascicles may be detached around the second, and by approaching, form a third circle, and so on. We shall thus have several concentric zones, but they can by no means be compared to those of the ordinary stems of dicotyledonous plants: for they are not generally the produce of the year, and are separated from one another by a layer of cortical tissue. These irregular alternations of cortical and cellular layers are presented to our notice in several of the Climbing Convolulaceæ, in the Mænispermeæ, in several Cissi, and especially in the Gnetum. In all this last series of plants, each ligneous zone is divided by medullary rays into so many bundles, and to each of

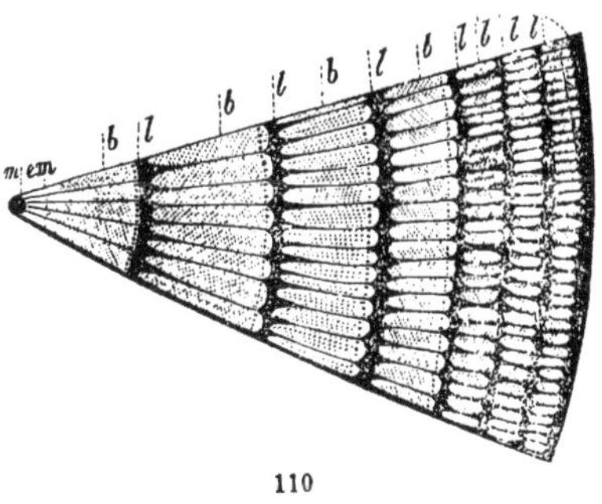

110

109. A horizontal section of the stem of one of the *Sapindaceæ* of Brazil.

110. A section of the horizontal section of the Gnetum.—*m* Pith.—*e m* Medullary sheath.—*b b b b b b b* Ligneous fascicles, forming seven concentric zones, each of which is the produce of several years.—*l l l l l l l* Small fascicles of the fibres of the liber corresponding to each of the preceding fascicles, and thus forming interposed circles equal in number to the ligneous zones.

these fascicles corresponds one composed of fibres of liber in the surrounding cortical zone. It is remarkable, that this liber is found in all the concentric zones of the GNETUM (*fig.* 110), but only in the innermost of the Mænispermeæ, in which plants all the rest are entirely cellular.

§ 90. It would take too much time to describe all the modifications of these anormal stems, which are still very imperfectly known; because, natives of distant countries, it is not possible to follow them throughout their developement so well as those of our own country. We must, then, be contented to sketch the principal differences, and to call particular attention to that mingling of the cortical and ligneous layers which we frequently observe in them. There can be no doubt, but that facts, better known and in greater number, will throw more light on the subject, and that, instead of several fresh laws, we shall discover only one which presides over the developement of all the stems of dicotyledonous plants, of which rule, that of the stems of our plants will be no more than a general case. Most likely the position of the leaves and of the branches, which spring immediately above them, causing on their own side the production of a much greater quantity of ligneous tissue, exercises a great influence on this division of the body of the wood, especially in the Creepers, the growth of which is extremely rapid, and the leaves often very far distant from one another. In our trees, the growth of the stem is rendered much more equal in the whole of its extent by its very slowness: but above all by the nearness of the leaves, and consequently of the branches. In herbaceous plants, which grow so very fast, and in which the leaves are often far distant, we often find also anomalies of exterior form and internal structure, sometimes comparable to those which we have just examined.

THE STEM OF MONOCOTYLEDONOUS VEGETABLES.

§ 91. We have followed the monocotyledonous, as well as the dicotyledonous embryo, from its first appearance through the first periods of its existence (*figs.* 74, 76, 79). Like the latter, it is entirely composed of cellular tissue (an exterior layer of which, of rather a different form than the rest, constitutes the epidermis), until its maturity, and generally even to its germination. It is then, that the fibres and vessels appear, and are in a little time grouped into bundles or fascicles. These are at first arranged in a circle, and up to this point nothing distinguishes in a clear and marked manner this little stem from that which springs from a dicotyledonous embryo.

But, in proportion as it increases and is covered with a greater number of leaves, and, consequently, the bundles or fascicles are multiplied in its interior, we shall have our attention directed to an arrangement very different from that which is observable in the dicotyledonous vegetable, in which, arranged in a circle, they approach, touch, and form a ligneous ring, interrupted only by the medullary rays. In monocotyledonous plants (*fig.* 111), the bundles are scattered without any apparent order, some near the centre, others, more numerous, near the circumference, in the midst of the cellular tissue. This tissue, interposed between them, does not, however, form straight lines extending from the centre to the circumference; it does not form medullary rays. The centre, which remains quite cellular, or at most contains but a very small number of bundles, gives a good representation of the pith or medulla, though only to a certain point; the medullary sheath, however, is wanting, and along with it the tracheæ which we have observed in dicotyledonous plants. The pith forms rather a large and regular cylinder, quite free from ligneous fascicles in several monocotyledonous plants, particularly in the Gramineæ or Grasses, as we may easily see in the *Maize* or *Indian Wheat* (*Maïs*), and in the ARUNDO or *Reed*. But most commonly it does not grow with the rapid increase of the stem, which in a little time becomes hollow or fistulous, on account of the perishing of the pith (*fig.* 112). Its remains may be seen on the sides of the pipe, under which form the stem then exists: this, also, happens in those dicotyledonous plants which have a very large medulla or

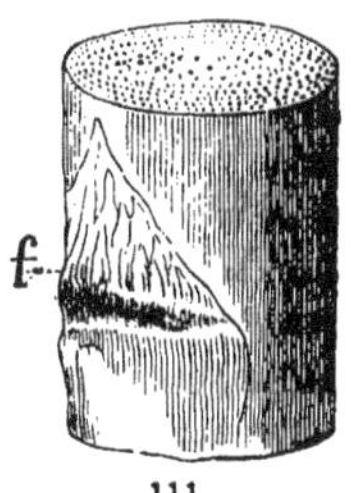

111

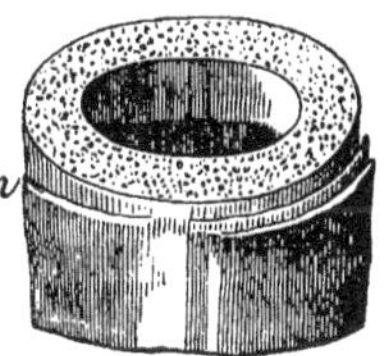

112

111. A fragment of the stem of the Asparagus (*Asperge*) (ASPARAGUS OFFICINALIS), the horizontal section of the upper end of which is plainly visible. In this figure, and all the following figures, the dots upon the section indicate the place of the ligneous fascicles.—*f* Leaves reduced to the state of a scale.

112. A fragment of the stem of the common Reed (*Roseau*) (ARUNDO PHRAGMITES) a little above a knot. It has become fistulous by the disappearance of the central medullary parenchyma, which is still seen at the surface of the knot *n*.

pith and grow very rapidly, the UMBELLIFERÆ (*Ombellifères*), for instance, of which the *Hemlock* is the commonest example.

§ 92. If we compare the anatomical structure of one of the fibro-vascular bundles of the stem of a monocotyledonous vegetable with that of one of the stem or a branch of a dicotyledonous plant, under a year's growth, we shall find it to be much the same. The former, in short, from the inside to the outside, presents (*fig.* 113):—1st, tracheæ (*t*), then larger striped or dotted vessels (*v p*), both surrounded by dotted cells (*u*), sometimes lengthened into fibres; 2nd, a mass of lacticiferous vessels (*l*), and of fibres with very thin single walls, surrounded by a crescent of other fibres (*f*) quite external, with thick walls, resulting from several layers fitted on one another. Now, do we not find in this combination all the elements of a fibro-vascular bundle of a dicotyledonous plant: the internal portion, which would correspond to the wood; the external, which would correspond to the bark? Thus, in the first year, the herbaceous stems of the Monocotyledonous and of several of the Dicotyledonous plants, especially among those in the latter class, in which, as we have mentioned, bundles of the medullary sheath are scattered in the bark, are very difficult to distinguish from one another: but, when we compare them more closely and prolong our observations, the resemblance fades away.

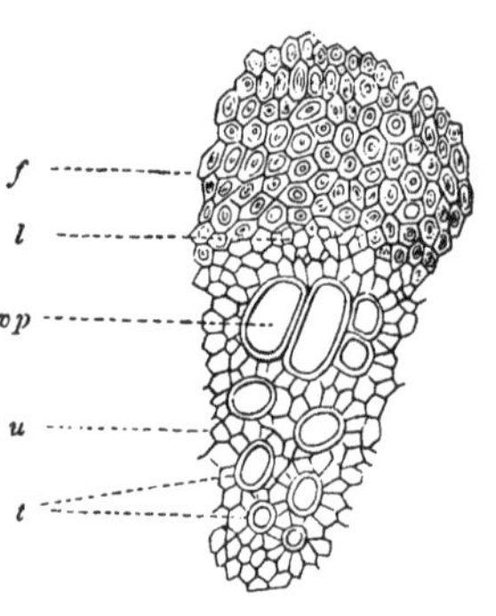

The fascicle of the Dicotyledonous plant presents the same structure throughout the whole of its length; that of the Monocotyledonous, examined at different heights, often changes in thickness and composition. The former, at a certain time, commonly after one year, is divided into two portions, the one remaining as the ligneous, the other becoming the cortical system; and between them is formed a fresh bundle, destined, at some future period, to undergo the same separation. The elements of the fascicle of the

113. A horizontal section of a fibro-vascular fascicle of a Palm (CORYPHA FRIGIDA).—*t* Tracheæ.—*v p* Large dotted vessels.—*u* Utricles accompanying the vessels, forming the parenchyma, or at other places lengthened into fibres.—*l* Proper or lacticiferous vessels.—*f* Thick fibres analogous to those of the liber.

Monocotyledonous plant are not disunited at another period; and, if the interior may be compared to the wood, and the exterior to the liber, we should then have a liber dispersed throughout the whole mass of the stem along with the ligneous bundles to which it would always remain united.

We now see clearly in what manner the growth of the Monocotyledonous stems differs from that of the Dicotyledonous, and that, consequently, we are not to expect in the former either concentric ligneous zones, one of which is formed every year, or leaves of liber.

§ 93. Unhappily for the prosecution of this branch of the investigation, ligneous Monocotyledonous plants are very rare in our climates; and we cannot, as we did during our researches among Dicotyledonous plants, bring before the notice of the student examples, which would be familiar to him, and, at the same time, be easily procurable. But he will see in the plates, which mostly accompany the narratives of travels, figures of Palms, which of all monocotyledonous trees perform the most important parts in nature; and on seeing them, he will instantly notice the difference between these trees and ours, arising from its lance-like trunk, almost uniform in thickness from the bottom to the top, and in the bareness of this trunk which does not separate into branches, and only bears on its summit a tuft consisting of a few large leaves (*fig.* 114, 1). The YUCCA ALOEFOLIA, which may be seen in most gardens, will give one a fair idea of the habit of Palms.

114. Two monocotyledonous trees belonging to two different families: the one, 1, to that of the Palmaceæ, is the Cocoa-nut tree (*Cocotier*) (COCOS NUCIFERA); the other, 2, to that of the Pandanaceæ, is the PANDANUS ODORATISSIMUS. The first gives us an example of a simple, the latter of a branched stem. The figures of two men have been placed at their foot to shew their height.

§ 94. If, from examining the exterior of a Palm, we carry our observations into the interior (*fig.* 115), we shall find this mass of fibrous fascicles scattered without order in the cellular tissue, which we pointed out from the first year of the plant's life. But these fascicles are very much multiplied; fewer and farther apart from one another in the middle of the stem (*m*), they become more numerous, more compact, and at the same time of a deeper colour as they approach nearer the circumference, towards which they form a close and blackish zone (*b*). Sometimes this is immediately covered by a zone of cellular tissue, which has been called bark (*e*); sometimes between these two zones is another one (*l*) composed of fascicles, more loosely united, thinner, less compact, and of a lighter colour, the situation and nature of which has often made botanists take it for a zone of liber.

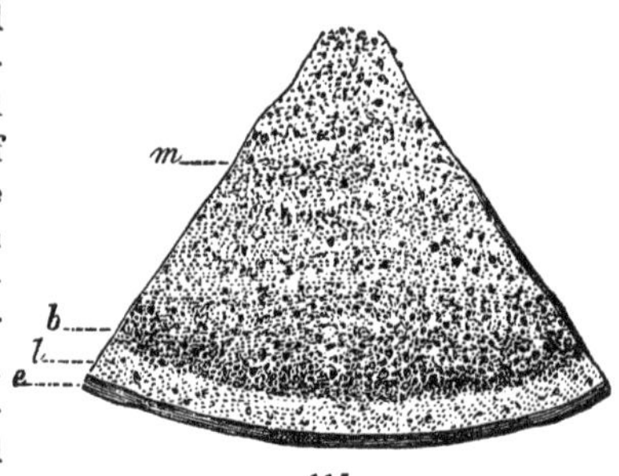

§ 95. This structure of the greater part of Palm-trees was known to the ancients. Desfontaines has the fame of discovering its generality in all Monocotyledonous plants, and of proclaiming to the world this most simple rule:—

"VEGETABLES *according to the internal structure of the stems are divided into* TWO GREAT CLASSES:

1st. THE MONOCOTYLEDONOUS,

Those, which have no distinct concentric layers; whose solidity decreases from the circumference towards the centre; in which the pith is interposed between the fibrous fascicles, without medullary elongations into diverging rays:

2nd. THE DICOTYLEDONOUS,

Those, which have distinct concentric layers; whose solidity decreases from the centre to the circumference; in which the pith is enclosed in a longitudinal canal with medullary elongations into diverging rays."

115. A segment of a horizontal section of the ASTROCARYUM MURUMURU.—*m* Central or medullary part, where the ligneous fascicles are fewer and more scattered. —*b* External or ligneous part, where the numerous and close fascicles form a compact and blackish zone.—*l* Zone of thinner fascicles, which have been compared to the liber. —*e* Cortical cellular layer.

This rule, thus established by Desfontaines, has never yet been overthrown.

It was not so, however, with the conclusions on the way in which the stems of monocotyledonous vegetables grow; he borrowed these very doubtfully from Daubenton, but, they were, a little later, generally adopted and recognised. All the leaves are commonly united at the top of the tree, and the youngest, the last formed, are placed in the centre; consequently, since all the fibro-vascular fascicles which constitute the solid part of the wood end with the leaves, the fascicles, which end with the youngest and were, therefore, formed the last, are situated in the middle of the others. Thus, the trunk is continually hardened by the addition of fresh fascicles formed in the middle, pushing to the outside the old ones, which approach and are more and more compressed, and at last form this external zone which is harder and more deeply coloured than the rest. This way of growing would be directly inverse to that of the dicotyledonous class, in which the newest and youngest layer is always the outermost, and each layer is older as we approach the centre. The name of EXOGENS (*Exogènes*) has been proposed for the stems of dicotyledonous plants which grow at the outside, the name of ENDOGENS (*Endogènes*) for those of monocotyledonous plants which grow in the inside.

§ 96. But, to make these conclusions true, the fascicles must necessarily preserve their relations to one another invariable, and, consequently, the same parallel direction, throughout the whole extent of their progress, whilst their *tout-ensemble* forms a kind of sheaf. But this does not occur: and in a stem of a Palm-tree cut lengthways, we see the fascicles bend and cross in all directions: this may be even seen, though not very clearly, in the short stem of the Leek (*poireau*) (ALLIUM PORRUM) or any other of our herbaceous monocotyledonous plants, in which the leaves are inserted on a very contracted stem.

If we follow one of these fascicles in its progress from top to bottom, i. e. from the point situated on the surface of the stem, where it enters into the leaf, we see that at first it is directed inwards more or less obliquely, and, when it has nearly arrived at the centre, it takes a downward course. It thus appears to come out of the centre; and, because botanists have not extended their examinations farther, they were deceived with regard to its origin, and admitted the theory of endogenous stems. But, if they had followed it a little lower, they would have seen it take a very oblique direction, quite inverse to the former, i. e. towards the circumference, and approach the surface more and more until it arrives under the bark, when its progress becomes almost rectilineal. It has, then, described an arc which turns its convexity towards the centre. In this course it would necessarily cross successively all the fascicles situated below it and formed before it, since they were connected with lower leaves and are consequently older, and it is at last placed on the outside of them. The most recent fascicles are, then, definitely the outermost, as they were in the plants of the dicotyledonous class; only the contemporary fascicles instead of remaining almost parallel in their progress, and thus forming a cylinder in the stem, converge towards one another at their top, and diverge towards the bottom. Let us add, also, that the arc which they describe is not contained in the same plane, and, therefore, a vertical section of the stem would not shew us the same bundles from one extremity to the other. Its tortuous course and the difficulty of following it in the midst of all this net-work complicate this investigation in a singular manner. But, two theoretic figures, indicating the course of four pairs of fascicles *a*, *b*, *c*, *d*, in the two systems, that of the endogenous stems,

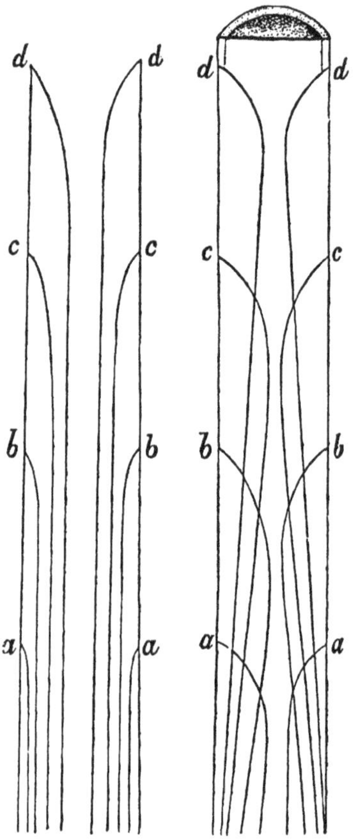

116

116. Relation of four pairs of fascicles, *a b c d*.—1, In the system of endogenous stems.—2. In M. Mohl's system.

and that which we have just explained, will shew very clearly the difference between these two opinions and explain the preceding remarks (*fig.* 116).

§ 97. We have already stated that the composition of the same bundle was by no means identical, when observed at different heights. At the top, the elements are those which we have compared to the wood: at the bottom, on the contrary, they are those which have been likened to the bark: the proportion of the one to the other thus changes gradually. In the upper part of the course of a bundle, that during which its arc is directed towards the centre, reckoning from the centre to the circumference, we find several tracheæ: then larger vessels of another kind surrounded by their cells: lastly, in an equal, greater, or less number, the proper vessels and thick fibres analogous to those of the liber. But, these last multiply more and more, and augment the size of the fascicle as it descends and approaches the circumference, so that, a little lower down, we find them in great numbers, still bordered on the inside by a little mass of ligneous cells surrounding one or two large vessels, and a little lower down still we only find them, i. e. the fibres. Quite at the bottom, when the fascicle touches the bark, it becomes completely fibrous, commonly very thin, and often even parted into several threads; and these, opening into those of the neighbouring fascicles, augment the confusion.

Thus, then, in a horizontal section of the stem, these threads form the exterior, and, dotted and loosely united by a parenchyma with very fine meshes, constitute that layer which has been sometimes taken for that of the liber, but which, as we here see, has quite a different origin from that in the dicotyledonous plant, and is even sometimes wanting. It is that part of the fascicle, composed entirely of a large mass of fibres with very thick walls, which forms the hard and coloured zone: it is their upper part, in which these fibres are united to vessels and ligneous cells, which forms the dots, which, few and far between, are scattered in the central parenchyma, and are found towards the insertion of the leaves. The merit of having formed this theory is wholly due to the learned labours and researches of M. Hugo Mohl.

§ 98. The stem has grown during the first stages of its existence, principally by the individual increase of each of the different elements which compose it. But, why should it present almost an equal diameter at the top as at the bottom, when it would seem that the continual addition of fresh bundles corresponding to fresh leaves would continually tend to thicken it? The number of these bundles can by no means be compared to that of those we find in dicotyledonous plants, because, for the most part, the stem, instead of being quite covered with branches and leaves, only presents these at its summit, and grows in height only by means of a terminal bud. We know, moreover, that these fascicles, instead of being equally thick from top to bottom, grow gradually thinner towards the bottom, and most probably disappear altogether. The base of the stem, then, does not present the sum of all the bundles, and the number of those which do traverse it, is compensated by their thinness and the thickness of the upper fascicles: the same thing takes place at each degree of height. Sometimes, however, this compensation is not quite exact at every height, and we see the stems swell towards the bottom, the middle, or the top, according to the period at which the tree has vegetated most actively[m].

§ 99. We have hitherto represented the monocotyledonous stem as free from ramifications and as increasing only by a terminal bud. This case, however, though by far the most general, has several exceptions. We see many among our monocotyledonous vegetables, as the *Asparagus* (*Asperge*), the *Asphodels* (*Asphodèles*), and a great number of the *Grasses*, which branch; but, their stem only lives for one year, and, consequently, we cannot calculate very correctly the influence which the developement of the branches exercises on their increase. This observation may be made in a more conclusive manner on the trees of warm climates, which also branch. They also augment in diameter and sometimes attain an enormous one. As an example of this, we need only cite the Dragon-tree of the Canaries, one of the greatest known trees of our globe. This is so large that a small chapel has been built in the inside of its trunk, hollowed like some of our willows[n]. When lateral buds are developed on a stem of a monocotyledonous plant already of a pretty good size, the fascicles which correspond to them, instead of piercing this stem by taking their direction towards the centre, creep between it and the

m De Candolle in his *Organographie Végétale*, vol. i. p. 218, mentions the case of a Cycas in one of the conservatories of the Jardin des Plantes, at Paris, which has in the middle of its length a very distinct contraction, corresponding to the time it was brought from the Isle of France. During the voyage it received very little nourishment, and the vital action being very sluggish, the external fibres were solidified before they attained their proper size.—TRANS.

n This destruction of the central part of monocotyledonous stems, which is very frequently observable, is an incontrovertible argument against the theory of endogeneity. The endogen, with its centre destroyed, could not continue to exist any more than the exogen deprived of its circumference for a certain depth.

bark; and there is then an increase in diameter, analogous to that of dicotyledonous plants: always differing from them in the relative situation and composition of these fascicles, which remain undivided like those of the central part.

§ 100. We have given the name of bark to the cellular layer, which, covered at first by the epidermis and commonly thickened by the base of the leaves, forms the external portion of the stem. Its composition is clearly distinguished from the fibrous portion which it covers, and from which it is sometimes detached so as to fall off. Sometimes, on the contrary, being extremely thin and adhesive, it is confounded with it; in some very rare cases it is very largely developed. Thus, the stem of the TAMNUS ELEPHANTIPES, now rather common in our conservatories, presents the appearance of a kind of dome, the surface of which is divided into numerous compartments separated by deep furrows, and these compartments are so many plates of a cortical substance analogous to cork; but, in spite of this seeming resemblance, its uniform cellular tissue never shews the distinct envelopes, the suberous and the cellular, which have been described in dicotyledonous vegetables. We know, besides, that the liber is not met with in the bark of monocotyledonous plants, since that which was supposed to be liber comes from a very different source, and is nothing else than the lower extremities of the ligneous fibres. A little higher, it really performs the functions of wood and may, perhaps, be called by that name. The bark, then, differs as much as the ligneous system in these two great classes of the Vegetable Kingdom, and even, rejecting as false the theory of endogens and exogens, they are no less distinct in their anatomical structure which presents important differences very easy to be appreciated.

The Stem of Acotyledonous Vegetables.

§ 101.—We have seen (§ 37) that the embryo or *spore* of an acotyledonous vegetable does not present any difference in the parts destined to be developed into roots, stems, and leaves: that it is commonly a single and simple utricle filled by a granular matter. If it happens to be placed in conditions favourable to its germination, that portion, touching the earth or any other surface which is sufficiently humid, is prolonged into a tube which performs the functions of the root: the other extremity is enlarged by the production of fresh cells placed in juxtaposition to the primitive cell into a lamina, commonly horizontal, and several of these cells throw out in their turn radical tubes, similar to the first. The vegetation of a great number of these plants does not proceed farther than this; they do not produce any stems. In several of those which live in the water, the Chara, for instance, at the same time that the roots grow in the water, on the opposite side there springs a cylinder, which we may either call a stem or a branch: it is only a continuance of tubes or of lengthened cells joined end to end. Others have a kind of stem much more complicated, since it results from a union of cells: the outer ones, preserving their primitive round or polyhedric shape, form the envelope of an axis consisting of cells of a different form, i. e. lengthened, or even of true fibres: this may be observed in the Mosses (*Mousses*), and the Liverworts (*Hépatiques*) (HEPATICÆ). But all these vegetables are entirely cellular: we never find any vessels in them.

§ 102. They appear, however, in the Marsiliaceæ, and in the Lycopodiaceæ, the stem of which, under a cellular envelope, presents a cellulo-vascular axis. This consists in a single fascicle, or in several united by a delicate parenchyma. These fascicles, instead of the cylindrical form we have observed in cotyledonous vegetables, are flat: they form a kind of plaited ribbon. If we examine by the aid of a microscope the nature of these vessels thus united into flattened fascicles, we shall only find annular vessels, or rather those we have distinguished under the name of scalariform; there are even long fibres more commonly independent of one another than united together in one long continuous tube by their ends. All these plants, such as we find at present on the globe, are herbaceous: but, it appears from the fossil remains of plants which have long since been extinct, that at a very early period, stems, which may be referred to the same families in the Vegetable Kingdom, were of much greater dimensions and of a ligneous consistence.

§ 103. There still exists a very large family of acotyledonous vegetables, scattered very widely over the face of the earth, (I refer to that of the Ferns,) which, with an analogous structure, may give us some idea of what these great antediluvian vegetables were. In our temperate climates, it is true, the Ferns (*Fougères*) (FILICES), are only found as herbaceous plants; or, even if their stems live for more than one year, they creep and are concealed in the earth. Like the Lycopodiaceæ, they have in their centre a single fascicle or else a small number of fascicles composed, for the most part, of scalariform vessels. We may see (p. 28, *fig.* 48), the figure of some fragments of these vessels taken from one of the largest Ferns of our country, the OSMUNDA REGALIS.

§ 104. Under the tropics and the warm climates which border upon them, the Ferns are often quite differently developed. They become large trees and are often as high as 15 to 20 metres (French) or 16·3875 to 21·85 yards (English): and it is only these which we can compare with the monocotyledonous and dicotyledonous trees which have previously been the subject of our examination. Externally, they bear the greatest resemblance to monocotyledonous trees: their trunks are thin, single and of an almost equal thickness from the base to the summit, and are crowned in the same way at their extremity with a tuft of large leaves, whilst they are entirely bare of them over the rest of their surface.

§ 105. For a long time it was thought that their internal structure was like that of monocotyledonous plants. But, if a horizontal section is made in one of their trunks (*fig.* 118), and we examine its elements, we shall see a very notable difference; for, instead of little ligneous fascicles placed in the midst of the parenchyma throughout the

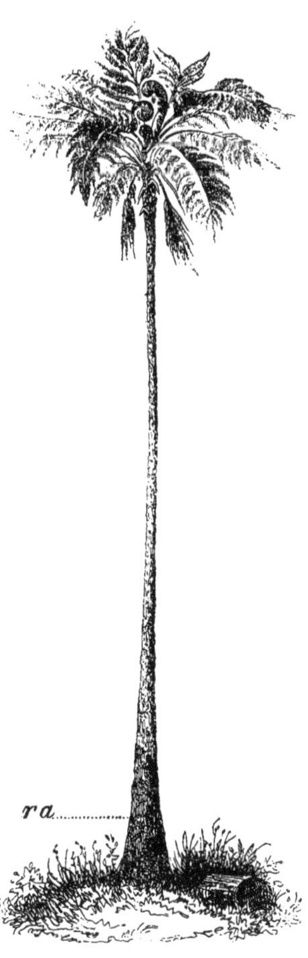

117

117. A Tree Fern (ALSOPHILA PERROTETIANA), a native of the East Indies.—The cylindrical stem presents at its base, *r a*, a conical enlargement resulting from a mass of adventitious roots, which spring from it and cover it in that part.

whole mass of the trunk, we find very large ones (*z l*) arranged in one circle towards its periphery. These fascicles are sometimes separated from one another by parenchyma, are sometimes united together by their edges, so as to constitute a continuous ring. They thus circumscribe a very large cellular central cylinder, which, by its position and nature, may receive the name of pith (*m*). Outside of the ring is another cellular zone (*p*), covered by the epidermis in the first stage of the vegetable, and afterwards by a hard envelope (*e*), formed by the bases of the leaves, which have fallen, in proportion as the trunk has grown and they have ceased to crown it.

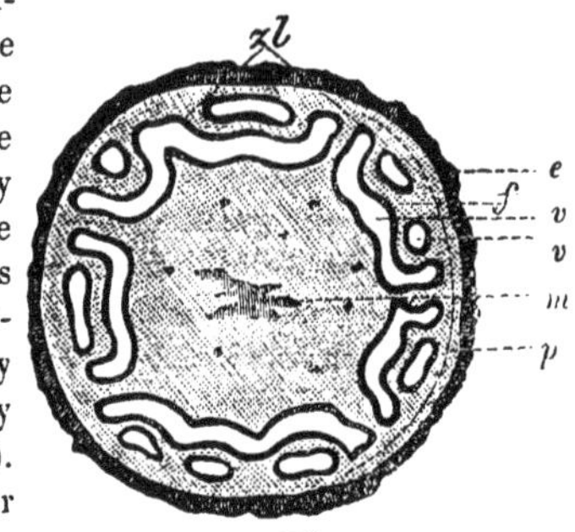

The fascicles are recognised in the horizontal section by the hardness and compactness of their tissue, and their colour, which is commonly blackish. This colour is owing to that of the prosenchyma (*f*), a zone of which round each bundle envelopes the mass of vessels (*v*), which all belong to those which we have pointed out under the name of annular, striped, and, especially, scalariform. The entire fascicles, and, consequently, the ring which is the result of their proximity and union, commonly present the form of a band, which bent or doubled on itself, thus forms very singular and sometimes exceedingly elegant patterns. Besides these elements, the whitish vessels which form the centre of these fascicles, the prosenchymatous and blackish cells which form their boundary, M. Schultz is said to have observed between the first and the second lactiferous vessels and elongated fibres, analogous to those of the liber. M. Mohl denies the existence of the liber and proper vessels.

Sometimes, in the central pith we find other little round fascicles, composed of vessels of the same order as those of the ring.

118. A horizontal section of the stem of a Tree Fern (CYATHEA).—*m* Pith occupying the whole of the middle.—*z l* Ligneous zone formed of large vessels arranged here in an interrupted circle (in others in a continuous ring).—*f* A mass of black parenchymatous fibres forming the border or edge of each of the fascicles.—*v* A mass of scalariform vessels occupying the middle of each of the fascicles, thus representing a whitish band differently bent and surrounded by the blackish border.—*p* External parenchymatous zone communicating directly or indirectly with the pith.—*e* Hard envelope taking the place of the bark.

If we examine the latter in a vertical and not in a horizontal section, we see that its great fascicles follow a course, not rectilineal, but undulating, so as to leave between them at certain distances, as they are alternately separated and reunited, intervals, occupied by cellular tissue, which thus forms a communication between that of the centre and that of the periphery. This arrangement of the parts may be easily seen by destroying the cellular tissue by maceration, which does not destroy the fibro-vascular tissue. It remains in the shape of a hollow cylinder, of a sheath pierced with a large number of rather regular openings, which may be easily compared to the ligneous cylinder of those of the dicotyledonous stems, in which also the fascicles follow an undulating course.

§ 106. This description will be sufficient to point out the difference between the stems of Arborescent Ferns and those of monocotyledonous and dicotyledonous plants; namely, the distribution of fascicles arranged in a circle, and not scattered without apparent order, as in the first class, only forming one circle, and not several concentric ones, with as many cortical circles as in the second, and in all cases the very different form and structure of these fascicles. No unrollable tracheæ have ever been found there, and we have seen that the elements are arranged otherwise than in cotyledonous vegetables. But, if the reader has paid attention to the description which we have given of these three classes, he will readily seize the points of difference which we cannot again detail.

§ 107. The trunk of the Tree, or Arborescent Ferns, acquires a certain diameter by the developement of the different elements which compose it; it then ceases to grow in width, and preserves the same size, whilst it is continually increasing in height. When it was hardly above the surface of the soil, it was almost as thick as it will be when it reaches the height of sixteen or twenty yards. The reason of this is that it only grows at the top, that its fascicles are lengthened without multiplying, that they continue the same at all ages, and at all heights.

This ligneous trunk has been represented as never ramifying; this, however, is not without an exception, and in the galleries of the *Jardin des Plantes* at Paris may be

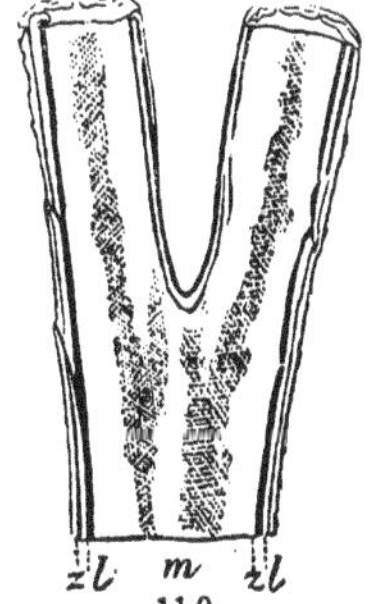

119. The vertical section of the ALSOPHILA PERROTETIANA, where it branches into a fork.—*m* Pith.—*z l* Ligneous zone or sheath.

seen a specimen of one of the Ferns of the Indies (ALSOPHILA PERROTETIANA) branching upwards. If we cut this vertically, following the axis (*fig.* 119), we see that it is not, as in the ramification of cotyledonous vegetables, a branch implanted on a trunk, but that the trunk is, as it were, doubled, that the ligneous sheath is equally continuous, and uninterrupted on both sides.

Several of the herbaceous acotyledonous plants which we have mentioned above, Ferns, Lycopodiaceæ, Marsiliaceæ, also appear to be ramified; but this is always, as in the preceding case, by the doubling of the extremity, and not by the implanting of a lateral branch. Each of these ramifications forms a fork; and when we follow it down the trunk throughout its formation, we see that it was owing to the existence of two terminal buds instead of one. They are then lengthened, sometimes equally, sometimes unequally; and are doubled in their turn, sometimes both, sometimes only one. The vegetable appears to have a larger or smaller number of branches in proportion to the number of times this division is repeated.

§ 108. It is, then, a general rule in the stems of acotyledonous plants, that they only grow at their summit and by the lengthening of the fascicles already formed; that they differ in this respect from those of cotyledonous plants, in which fresh fascicles are being continually formed over the old ones. It has been proposed therefore to call these stems ACROGENS (*acrogènes*), in opposition to those formerly admitted, *Endogens*, and *Exogens*. But we know that the last two ought to be suppressed, and, consequently, there is no need for the first. But, as it is rather inconvenient to use the words acotyledonous, monocotyledonous, and dicotyledonous, on account of their length, we may use these three words, if we forget their meaning and derivation, and apply to them a meaning in accordance with the notions which the actual theories of the science warrant.

§ 109. Let us mention, however, before we finish, one exception to the preceding law; this is met with in a singular family of acotyledonous plants, the Horse-Tails or Equisetaceæ (*Prêles* ou *Equisétacées*), the stems of which have nothing in common with those we have just described: they are hollow in their centre, which forms a large cylindrical lacuna, divided here and there by partitions, which answer to so many articulations; and in their solid part, which is almost quite cellular, are found much smaller lacunæ, arranged in one or two circles. Some annular vessels are found along these lacunæ. From the exterior of the stem, at the point of these articulations, a circle of branches grows. There is no trace of leaves, for we cannot term leaves the membranous sheaths which are also found at the articulations within the branches, whilst these ought to spring between the sheath and the stem, if the former was formed by a circle of united leaves; but, as a compensation, the epidermis is pierced by a great number of stomata arranged in rectilineal lines. We find in the structure of these Horse-tails nothing which can be compared to any of the stems of which we have spoken.

THE ROOT.

§ 110. (*Racine.*)—The root is that part of the plant which takes a direction contrary to that of the stem, i. e., towards the interior of the earth. We call, on this account, that part its base, which is the highest, and by which it joins the stem at a point we have termed the collum or neck; and its lowest extremity, the top. The long time and space which we have devoted to the consideration of the stem will allow us to abridge that of the root, since we have only to establish a comparison between them.

§ 111. We will follow it, as we have done in the stem, from its first appearance in the embryo. We have already shewn, that the whole of what we called the radicle, does not belong entirely to the root, but in its upper part rather to the stem, and very often, even through the greatest part of its length, its lowest extremity being excepted. It is only, then, by germination that the true radicle is developed, and here from the very commencement are distinguished the three great classes of vegetables. Putting to one side the acotyledonous embryos, in which there is no distinction of parts, and, consequently, of radicle, and in which the roots are only the tubular elongation of the cells touching the soil, the radicles of the dicotyledonous and of the monocotyledonous embryos, are developed in quite a different manner. In the former, the radicular extremity of the axis is lengthened, in the second (*fig.* 120), it is pierced with an

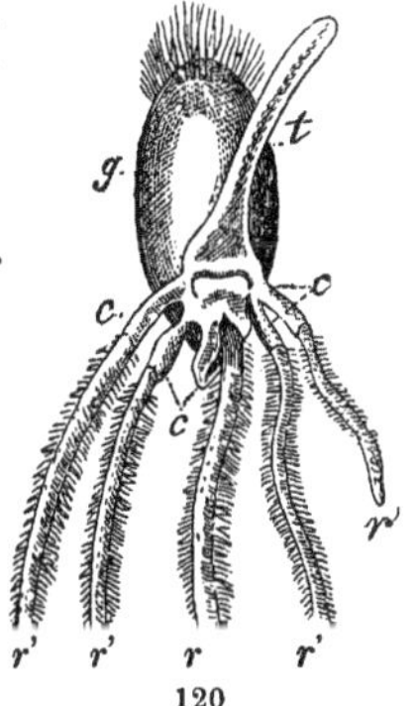

120. A germinating grain of Wheat (*Blé*).—*g* The mass of the grain.—*t* The young stem beginning to grow.—*r* Principal root.—*r' r' r' r'* Lateral roots covered like the preceding with small filaments.—*c c c* Coleorhiza or sheath, with which each root is enveloped at its base, when it pierces the superficial layer of the embryo.

opening to allow the radicle (*r*) to pass, which is covered by a superficial layer of the substance of the embryo adhering to it; this forms a sort of sheath (*c*) at the base of this the first root. This is the reason why dicotyledonous embryos have received the name of *Exorhiza* (*exorhize*), and the monocotyledonous that of *Endorhiza*, (*endorhize*), because the *root* (ʽΡίζα) is an external part (ἔξω, *without*) in the former, internal (ἔνδον, *within*) in the latter. Hence, has also arisen the name of *Coleorhiza* (*coleorhize*) (κολεός, *sheath*), given to the radicle of the latter.

§ 112. The radicle or lower continuation of the axis of the young plant, in its ulterior developement, presents two important modifications. Sometimes (*fig.* 121) it continues to be lengthened and thickened, to throw out more or less numerous ramifications, to perform, in short, with regard to the whole subterraneous systems of the secondary roots, the same functions that the stem performs with regard to the whole aerial system of secondary stems or branches: it forms, then, what is called the body of the root, or the tap-root (*pivot*); this is very frequently found in the roots of dicotyledonous plants. At other times, on the contrary, by the side of this first root are developed others almost equal to it, or

121

even much larger: they spring very near the base, and even appear to be already formed in the interior of several endorhize embryos, since they are pierced below with several openings around that which corresponds to the axis to give vent to so many lateral radicles (*fig.* 120, *r' r' r' r'*). These different roots, thus springing from nearly the same height, grow and are developed together, forming a tuft or bundle. Although, sometimes, each of them branches out lower down, it is not uncommon to find them undivided. Several botanists have, consequently, called these roots *compound* (*composées*), *fasciculated* (*fasciculées*), or *fibrous* (*fibreuse*) (*fig.* 122); and in opposition, the first have been termed *entire*, *simple*, or when the tap-root is very greatly developed in a vertical direction, *spindle-shaped*, or *fusiform* (*pivotantes*) (*fig.* 121).

121. Fusiform root of the Dwarf Mallow (*Petite Mauve*) (MALVA ROTUNDIFOLIA).

Of course between these two modifications, every intermediate degree may be observed, according to the variation of the relative proportions which the lateral roots may assume in connection with the axilary root. This, often alone, and always the most important in germination, may preserve its predominance, or lose it, or even die and totally waste away, the others supplying its place in the performance of its functions.

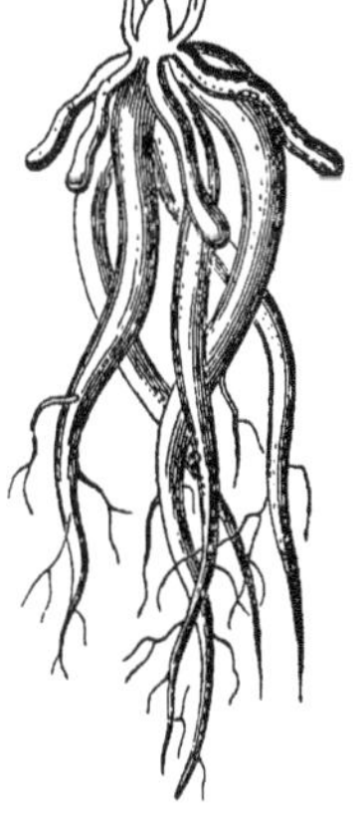

122

§ 113. The stem, moreover, placed in certain circumstances, throws out from its surface several roots which we call *accessory* (*accessoire*) or *adventitious*, (*adventitive*). Certain stems do this very readily, (the Willow or Poplar, for instance,) the lower extremity of which is planted in water or moist earth. These pieces of the stem are then called *slips* or *cuttings* (*boutures*). On different points of the surface of this extremity are quickly developed small threads which are gradually lengthened, taking a downward direction ; these are so many adventitious roots, relatively to which the lower part of the branch plays the same part as the tap-root of the true root would have done relatively to its ramifications. Certain vegetables need not, in order to throw out roots from the surface of their stem or branches, be put in contact with the earth or water. These roots, which are directed from the point where they originate to the soil, are suspended in the air, *aërial* (*aériennes*).

These adventitious roots do not spring indifferently at all points, but prefer those where there is a mass of juices and nourishment collected or a rupture of the epidermis, where the knots of the stem, accidental tumours and wounds, or, frequently, the lenticels are found.

§ 114. Whatever may be the origin of the roots, whether they result from the prolonging of the radicle or its ramifications, or they are formed, secondarily, on the stem or on the branches, they commence and are organized almost in the same manner. They appear at first under the form of a small mass of utricles, of which those, situated towards the centre of this kind of nucleus, are lengthened at

122. The compound root of a species of Asphodel (ASPHODELUS LUTEUS).

the same time as the whole of this little body: then, in a little time, some are organized into vessels which are entangled in a shorter or longer period with those of the fascicles of the stem. The definite structure of these roots is the same in the same vegetable, and these which we are going to examine briefly, by first explaining the common characteristics which they offer in the generality of plants, then the modifications which they present in the three great classes.

The roots have neither leaves nor analogous organs, nor buds arising in constant connection with them. Their ramification, when it does take place, is quite different from that of the stem, and obeys other laws which are not yet known, so irregular and various does it appear. The ramifications end with little threads or *fibrils* (*fibrilles*) (FIBRILLÆ), which have also been called the *chevelu:* in undivided roots, towards the end, the surface is often quite covered with these fibrils: sometimes, these only seem to constitute the root, at other times, on the contrary, they are completely wanting. The existence of the fibrils is temporary; they wither on the old parts of the root, and new ones are produced towards the younger extremities.

§ 115. It is, in short, at these extremities that one of the principal functions of the roots is going on, the passage of the liquids and juices of the surrounding earth into the plant. It was thought that this was done principally by means of cellular swellings, which, terminating the fibrils or the last and youngest roots, whatever might be the way in which they were divided, would swell like a sponge, as they imbibed the liquids in connection with them, and would thus come to be termed *spongioles* (small sponges). Microscopical research informs us, that this is an error arising from examining them with a lens of too weak a power, which led to the opinion that the mass of mucilaginous tufts or other small foreign molecules adhering to the surface belonged to the radicular extremity; that, in short, certain roots are terminated by a kind of swelling or cellular hood of a looser tissue (HYDROCHARIS), but in other cases, on the contrary, more compact (LEMNA); that very often there is no terminal swelling, and that the end of the fibril is covered with an epidermal layer, the same, indeed, as the rest of its surface.

As to the extremity of the larger divisions of the root, which do not wither away like the fibrils, but continue to grow, it, in general, presents a tissue in the state of birth, since it is the only seat of developement; and, hence, results necessarily a certain difference between this point and all the rest of the root nearer to the base, the tissue of which has already obtained every degree of formation of which it is susceptible.

§ 116. The epidermis of the roots (*fig.* 87) differs from that of the stems by the constant absence of stomata. It differs from it in its form also, which is much less distinct from that of the subjacent cellular tissue.

The cells which form it are very often lengthened into hairs or into papillæ. They are generally observed towards the base of the radicle, from the moment that it begins to lengthen by germination (*fig.* 120, *r r'*) on the last very young ramifications and on the fibrils. These extensions multiply the surface of the parts at a period, when it most probably assists, though in a less degree, the extremities in absorbing the surrounding fluids. These epidermal hairs have been termed by some authors *fibrils* (*fibrilles*) or *chevelus* (*chevelus*), and there may result some confusion from the same name being given to these two classes of organs, the former being simple, the latter compound and even presenting the former among their component parts.

§ 117. The vessels, with which one meets in the roots, even at their very ends, are analogous to those of the stems, excepting the tracheæ, which have only been found very rarely, and with great uncertainty.

The fibres are also the same.

The cellular tissue is generally filled with juices, and the presence of the fecula in great quantities in its cavities proves that the root, to the function of absorbing and conducting the nourishing fluid still crude, often joins another, that of serving as a depôt for the nourishment which is already formed. In this case, this portion of the tissue is often much extended, and large swellings are the result, sometimes in a portion, sometimes in the whole of the root. Sometimes it is the very tap-root which is thus thickened, the maximum of this increase may be seen near the base, as in the carrot (*carotte*), or towards the middle, as in the radish (*radis*); sometimes in a compound root all the branches, or only some, are swelled here and there like a string of beads, as in the PELARGONIUM TRISTE (*fig.* 123); in one part only, as in the *Drop-wort* (*Filipondulo*) (SPIRÆA FILIPENDULA) (*fig.* 124); or else the whole root, taking the globular or egg-shaped form, as in the *Orchis* (*fig.* 125). (This plant will shew us several singular modifications of this.) These feculiferous swellings take the name of *tubers* or *tubercles* (TUBER), and the roots the name of *tuberous*.

Let us now cast our eye over the differences between the roots of the three great classes of vegetables.

§ 118. THE ROOTS OF DICOTYLEDONOUS PLANTS.—It is in this class, and especially among trees, that we find tap-roots, and their ramifications represent in their number, their size, their extent, those of the stem. Sometimes the tap-root is not buried very deeply, and even withers away near the base, whilst the branches have a large lateral developement just in the same manner as in

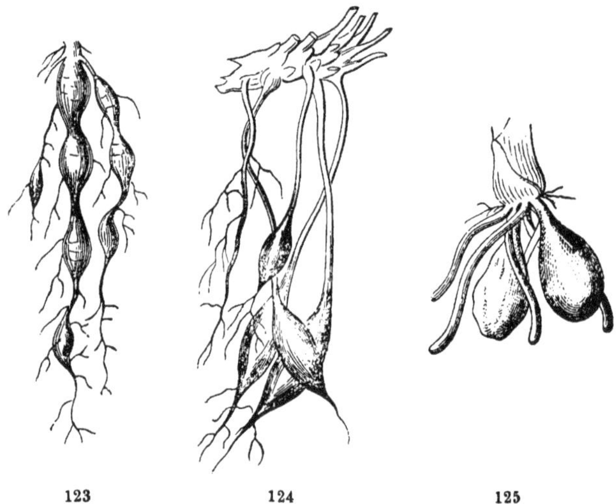

many stems. But, in spite of these frequent resemblances, the connection between the root and the stem is very far from being constant with regard to form and size. Thus, there are very voluminous roots for very small vegetables; and there are others which are very small for exceedingly large trees, and these are naturally pulled up by the roots very easily.

If we compare the internal structure of the stem and root of a dicotyledonous tree, we see that the latter differs from the first in the absence of pith and a medullary sheath. The wood, consequently, wanting tracheæ, forms the axis of the root. This characteristic of

123. Root of a species of Geranium (PELARGONIUM TRISTE).
124. Root of the Dropwort (*Filipendule*) (SPIRÆA FILIPENDULA).
125. Root of an Orchis, in which two fascicles only are swollen into tubercles, all the rest preserving their cylindrical form.

the root is rather exaggerated, when we say that the pith ceases altogether along with the medullary sheath at the collum. This is true in the greater part of herbaceous plants, but not in all the trees. The Walnut (*Noyer*) and the Horse-chesnut (*Marronier d'Inde*), for instance, continue the pith in a very largely developed state to a great distance into the root.

But the roots grow in thickness in the same way as the stem, forming each year a zone of wood, and another of bark; but their mode of increase in length is not quite the same. In stems and their branches, the shoots, up to the moment that they cease to be lengthened, grow throughout their whole length. The roots grow only at the extremities, as we have previously shewn. This fact is easily verified by marks traced here and there on a shoot of the stem and on one of the root: in the first these marks will be separated farther and farther from one another; they preserve the same intervals in the second, which will shew below the last the whole length which it has gained during the experiment. In stating the want of buds to be a characteristic which clearly distinguishes the roots from the stems, we referred only to normal buds, those which spring in a regular and common place, generally immediately above the leaves. We shall see that others may be produced on the stem at different points, in which they are not commonly developed, but which happen to be placed in peculiar circumstances favourable to their growth. Now, these buds, called *adventitious*, sometimes also appear on roots, especially when they are placed in the ordinary circumstances of the stem. This possibility of reciprocal production of adventitious buds on the roots, of adventitious roots on the stems is a very important connection between the two parts.

Let us add, to finish the comparison, that the anomalies of the stems are repeated exactly on the roots. Of this we may convince ourselves by casting our eye on that of one of the Menispermeæ, catalogued in our pharmaceutical collections under the name of Pareira brava, on that of the Climbing Bindweed, known under the name of *Turbith*, &c.

§ 119. The Root of Monocotyledonous Plants—Is most commonly compound (*figs*. 120, 122, 125), and its branches, though sometimes divided, often remain undivided. All these partial roots, which together form the compound root, do not continue to exist if the stem is perennial, but die in the order in which they were formed, so as to form circles extending farther and farther out, since those of the first year were formed around the radicle, which was the continuation of the axis itself. The aerial roots, extremely rare in dicotyledonous plants, are very frequently found here. They spring from a greater or less height in the stem. In several Palm-trees, they are produced in great abundance at the base of the trunk, which they quite cover, and thus essentially contribute to its thickness. The internal structure is that of the stem. In the large roots, we find fibro-vascular fascicles scattered in the parenchyma, fewer towards the centre, multiplied and more compressed towards the circumference, and a cellular cortical envelope often covering a fibrous layer. In the small roots, these fascicles are concentrated or often reduced to only one which forms the axis, surrounded by a cellular zone. One difference, however, may be remarked in the distribution of the elements of these fascicles compared with those of the stems. Their vessels, which are grouped in simple series, or often parted into a V (*fig.* 126), directed like rays with regard to the axis of the root, gradually decrease in size from the centre to the circumference, smaller (*vs*) and more anciently formed in proportion as they are more external, larger (*vp*), although of a more recent formation, in proportion as they are internal in the series: this appears to be quite contrary to the arrangement and developement of the vessels in the fascicles of the stems.

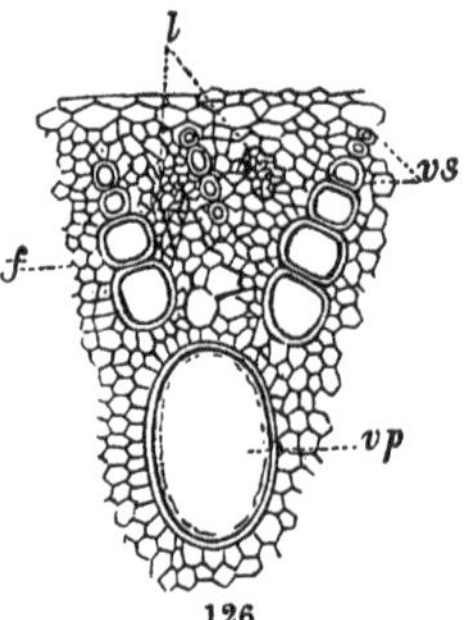

§ 120. THE ROOT OF ACOTYLEDONOUS PLANTS.—Here, there is no radicle developed by germination. Thus, as we have said several times (§ 110, 111), tubular lengthenings of cells, analogous only to those of the epidermis of other roots, fulfil their parts and pump up the nourishment for the young stem. This, when once developed, throws out adventitious roots, the only ones which are ever found in this class of plants. It is at the knots that they are produced, and this takes place all round, if the stem grows perpendicularly, and only on the side that touches the earth, if horizontally. On the trunks

126. Fascicles from a transversal section of the root of the DIPLOTHEMIUM MARITIMUM, shewing the relative arrangement of the vessels with regard to one another and the other elements.—*vp* Large dotted vessels situated within.—*vs* Scalariform vessels, nearer to the circumference, and smaller in proportion to their distance from the centre.—*f* Fibrous tissue, or one composed of lengthened utricles which accompany the vessels.—*l* Groups of proper vessels, large near to the centre, small as they approach the circumference.

of Arborescent Ferns, these roots are accumulated at the lower extremity in such a quantity, that they sometimes double or triple its thickness (*fig.* 117, *r a*); from that point, the conical form goes to a certain height, when the cylinder formed by their stem is bare and freed from this kind of fibrous mass produced at the bottom by the adventitious roots. These roots recal to our memory the organization of the plants to which they belong, purely utricular in those whose stem is so, shewing the association of vessels with cells in the acotyledonous vegetables, in which we have shewn it to exist equally in the stem. They then present the form of threads, more or less scattered, single, or branched, in which a fibro-vascular fascicle forms the axis, surrounded by a cellular layer, which is covered by a brownish envelope, becoming black as it grows old. The fascicle often takes the form of a column deeply channelled with acute angles, and, consequently, its horizontal section describes a very regular little star. Its fibres and vessels are of the same nature as those of the stem (§ 105). In several Ferns and Lycopodiaceæ, these fascicles, before being separated from the stem as adventitious roots, descend for a certain length through the parenchyma; and, even in certain old stems of the Club-Mosses, between this parenchyma and the central fibro-vascular fascicle, which are separated and thus leave an empty space between them.

LEAVES.

§ 121. (*Feuilles*).—We have examined the axis of the plant, 1st, in its ascending part or stem: 2nd, in its descending part or root. We have seen the latter throw out lateral branches, and in a certain number of cases these branches acquire a greater or less volume, relatively to the axis, which may stop sooner or later in the progress of its developement. It may even not be developed at all, and then these lateral productions form the whole of the roots. It may also happen, that there is no descending axis, and that all the roots spring from the bottom of the stem. This degradation, as it were, of the roots seems to be in connection with that of the vegetable, since we have observed by far the largest developement of the axis in Dicotyledonous plants: that it is hardly developed relatively to the lateral roots of Monocotyledonous plants, and that it is completely wanting in Acotyledonous plants.

§ 122 *a*. Let us now pass to the lateral productions of the stem, to the leaves and to the buds. We will consider the leaves at first isolated in their structure and in their form, then in their mutual connection with the stem.

§ 122 *b*. THE GENERAL STRUCTURE OF THE LEAVES.—The leaves are those expansions, generally flat and green, which spring from the circumference of the stem, and which every one knows under that form, which is the commonest. Their base is the extremity, generally the thinnest, by which they are united to the stem: their summit or point the opposite extremity.

The base is frequently contracted into a sort of tail, the length of which greatly exceeds its breadth, and takes the appearance of a thin branch: this is called the *petiole* (*pétiole*) (PETIOLUS).

It is not uncommon to see this petiole enlarged and widened at its bottom, by which it sometimes embraces the stem for a large part of its circumference: this is called the *sheath* (*gaîne*) (VAGINA). But this widening sometimes seems to be wholly or partly detached from the petiole; and it then forms on each side little appendages of various forms, very often presenting that of a little leaf: these appendages are called *stipules* (STIPULÆ).

We may then consider the leaf as formed of three distinct parts: 1st. the limbary, that which is formed by the terminal widening commonly flattened, or the *limb* (LIMBUS); 2nd. the petiolary; 3rd. the vaginal, formed by the sheath or the stipulæ. A leaf of Water Pepper (*Renouée*) (POLYGONUM HYDROPIPER) (*fig*. 127) presents these three parts very visibly at one and the same time. In other plants, the leaf is reduced to two of these parts or even to one. Since it is the limb which generally constitutes the greater part, the most visible and the first formed, since it is that part which is commonly known under the name of leaf, since it is that which performs the functions, which this fundamental organ is called upon to fulfil in the life of the vegetable, its

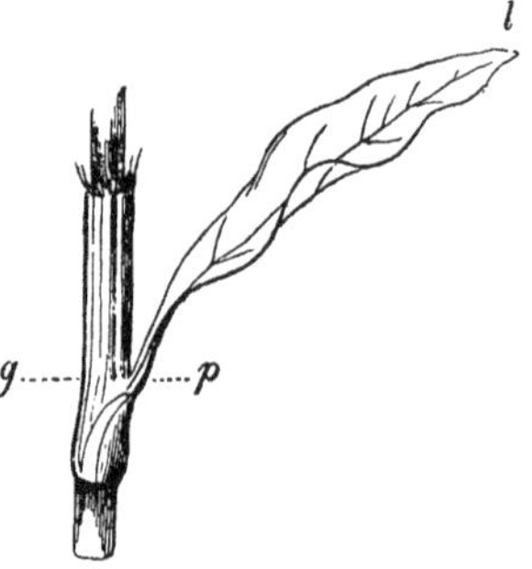

127. A leaf of the Water Plantain (*Rinouée*) (POLYGONUM HYDROPIPER), with a piece of the stem which bears it.—*l* Limb.—*p* Petiole.—*g* Sheath or vaginal part, embracing the stem and terminated at the top by hairs.

examination will principally occupy us. We will study it, 1st, in the leaves which live in the air; 2nd, in the leaves which live under the water.

§ 123. Aerial leaves (*Feuilles aériennes*).—Their structure.—It is in the course of the first year in which the stem or branch is formed, that the leaves appear and vanish. They are seen at first, under the form of little masses or laminæ, closely pressed against one another. They are separated from one another, in proportion as the stem is lengthened, and at the same time increase in size, gradually assuming the form and dimensions which they are at last to preserve. When they have attained them, if we examine their interior, we see that they are formed of the same elements as the stem, which seems to be continued into the leaf, of the same vessels, the same fibres and parenchyma. These vessels and fibres are in the stem united into a fascicle, and sometimes preserve this arrangement, when they detach themselves from the stem: it is then that we have the petiole. This fascicle is commonly not simple, but compound, on account of the juxtaposition of several: and when the lateral ones are separated a little from the others at the base of the leaf, we have a sheath or stipules. Sometimes near this base, sometimes at a certain distance from it, all these fascicles begin to separate: the commencement of the limb thus results from their expanding. The fibro-vascular fascicles form the most solid part of the limb, its frame-work or skeleton: their intervals are filled with the parenchyma. The whole is enveloped by the epidermis which is a continuation of that of the stem.

§ 124. The limb, formed by the flattened expansion, necessarily presents two surfaces (PAGINÆ) and two edges (MARGINES), which, parting at the base, are reunited at its summit. In the majority of vegetables, in almost the whole of those of our country°, its plane is perpendicular relatively to that of the stem, or more frequently rather oblique, so that it presents an upper and a lower face, and two margins, the one to the right and the other to the left.

° The aspect of the trees and forests of New Holland struck the first explorers with astonishment by the singular sensation the distribution of the lights and shadows imparted to the eye; this unwonted effect created great wonder in the minds of the beholders, a long time before the cause was known. Mr. R. Brown, on visiting this country, easily accounted for this extraordinary distribution of light and shade, by determining that the leaves of the greater part of the trees, instead of being placed on the branches in the normal manner, were situated in a contrary direction, so that the light shone between vertical, instead of falling on horizontal laminæ. In some species these are real leaves, but in others simple phyllodes (vid. § 141.)

§ 125. The limb, although flattened between the two laminæ of epidermis which cover it, has a certain thickness occupied by the fibro-vascular skeleton and the parenchyma. Do we observe there vessels and cells of a different kind, and, in this case, how are they distributed with regard to one another? We have said that the fascicles are a continuation of those of the stem: we know, on the other hand, that the latter, whether in the stems of monocotyledonous, or in those of dicotyledonous plants of the first year's growth are composed in the inside of unrollable tracheæ (*fig.* 128, *t*), a little more to the outside of annular, striped, or dotted vessels (*v*), with ligneous fibres (*f*), and quite to the outside of these of proper vessels and cortical fibres (*l*). The relations of these constituent parts are preserved in the leaf (*fig.* 128, *p*). The vertical fascicle in the stem, on becoming oblique or horizontal in the leaf, will turn to the top the innermost portion, and, consequently, the outermost downwards. herefore, a fibro-vascular fascicle of the leaf presents, in the half turned upwards, first tracheæ (*t*), then vessels of another order (*v*) accompanied by fibres (*f*); in the half turned downwards, proper vessels and fibres analogous to those of the liber (*l*): so that we may, up to a certain point, compare the upper to the wood, the lower to the bark.

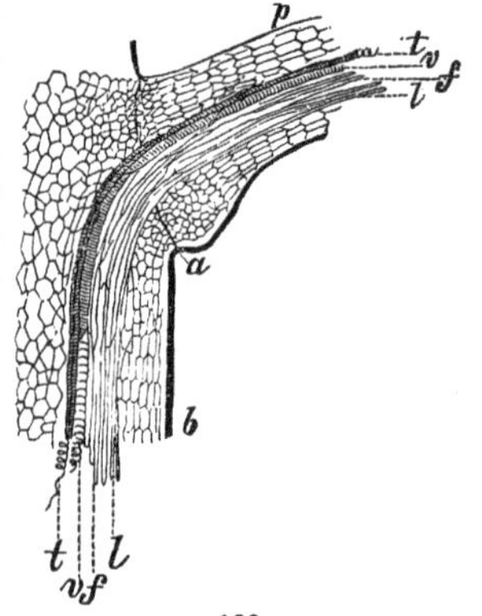

§ 126. The epidermis also presents a very remarkable difference between the two surfaces. We have already pointed out one (§ 41), that in which the stomata are so much more numerous on the lower than on the upper. The former very frequently has a number of hairs or scales which are either totally or partially wanting on the upper one, and from this circumstance it has a much greater resemblance to the external aspect of the epidermis of the young stem.

128. Passage of a fibro-vascular fascicle of a branch *b* into a petiole *p*. We see that the elements, vertical in the former, have a tendency to assume a horizontal direction in the latter, and to preserve, in spite of this change, the same relations with one another. We see, also, how they, as well as the surrounding cellular tissue, on passing them from one organ to another, are modified, whence results the articulation *a* between these two organs.—*t* Tracheæ.—*v* Spiral vessels of another kind; they are here annular.—*f* Ligneous fibres.—*l* Cortical fibres or liber.

In those leaves which float on the water, those of the Water Lilly, (Nymphæa Alba), for instance, it is the upper epidermis, on the contrary, which is pierced with stomata, whilst the lower one is entirely bare. In all leaves the stomata are only observed on the portion which corresponds to the cellular tissue, and never on that which answers to the fibro-vascular fascicles.

§ 127. As to the parenchyma, it merits a particular investigation as the seat of the special functions of the leaf.

Generally, in those which are rather thin or flattened (*figs.* 129, 130), we may easily distinguish two regions or layers of this parenchyma, the one the upper, the other the lower. In both the cells are filled in their normal state with granules coloured green by the chlorophyll; but they have generally neither the same form nor the same arrangement in both. For in the upper (*p s*), below the epidermis (*es*), we find one, two or three rows of oblong utricles, much narrower than those of this epidermis, obtuse at both ends, lying in a direction perpendicular to the surface of the leaf, pressed against one another, so as only to leave between them small meati (*m*), separating, however, sometimes, so as to leave between several of them a lacuna which is frequently found to correspond with a stomate, (*fig.* 83 *s*). The lower layer (*p i*) is composed of irregular utricles,

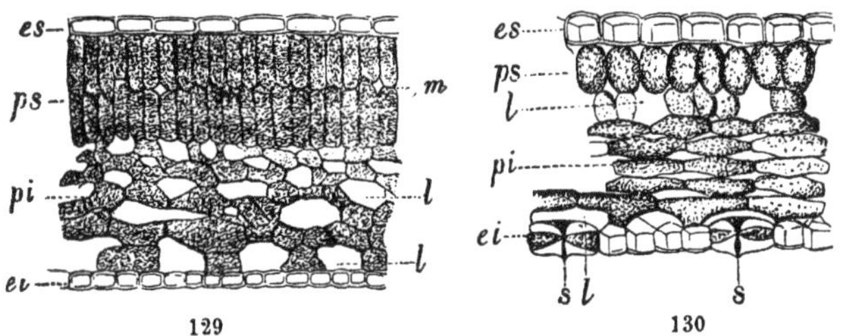

sometimes branched, united to one another only by the ends of their branches, sometimes simple and joined at the greater part of their surface, but in all these cases leaving between them numerous lacunæ (*l*), which communicate with one another and form a reticulated parenchyma, which might also be termed cavernous or spongy. Many of these lacunæ are situated immediately above the lower

129. A thin vertical section of a leaf of the Lily very highly magnified.—*es* Epidermis of the upper surface.—*ei* Epidermis of the lower.—*ps* Parenchyma of the upper region.—*p i* Parenchyma of the lower region.—*m* Meati.—*ll* Lacunæ.

130. A similar section of a leaf of the Balsam.—The letters refer to the same parts as in the preceding figure.—*ss* Stomata.

epidermis, perforated, as we know, with a much greater number of stomata than the other, and these stomata correspond precisely with the lacunæ. The parenchyma of these leaves is, then, more compact at the top (*ps*) and looser at the bottom (*pi*), hollowed out by a number of lacunæ, communicating with one another either immediately or by means of the meati, and with the outside by the openings of the stomata.

This arrangement is not quite the same in the thick leaves of those plants which are commonly known by the name of succulent (*grasse*), the cells of which being rather large, leave very few intervals between them, and enclose only a few green globules, especially towards the centre, in which their whitish mass is like a kind of pith.

It is useless to enlarge here on the different modifications which this parenchyma of the leaves may present, according to the vegetable which we are examining, according to the place the leaf under observation occupies on it, and even according to the age of the same leaf. But the existence in its mass of a certain number of meati or lacunæ, the outermost of which open under the stomata, and the constant relation existing between the frequency of these empty spaces and the intensity of the green colour are two general facts of which sight must not be lost.

This arrangement may be verified by means of very thin sections of the leaf, cut perpendicularly to its surface (*fig.* 130); they are especially instructive when they include some stomata. The correspondence of the latter with the lacunæ may also be studied in shreds of the epidermis (*fig.* 131), with which has been taken off a small layer of green cells (*pp*), which adhere to it, and which appear under the microscope as a great net, the meshes of which circumscribe colourless areolæ, in which is in general included a stomate.

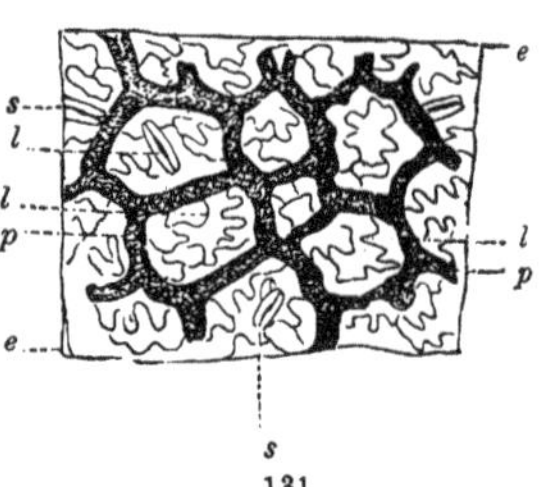

§ 127 *b*. Submerged Leaves.—The leaves which live under the water present a very different structure. They want the epidermis, and consequently the stomata also. The fibro-vascular skeleton is

131. A piece of the lower epidermis *e* of the leaf of the Balsam, on which is laid the net-work formed by the lower layer of the parenchyma *p*.—The meshes of this net-work are so many lacunæ *l*, frequently corresponding to so many stomata *s*.

also wanting, and if sometimes we think that we perceive it, when looking at the outside, a more attentive examination made with lenses of sufficient power, will prove that they are merely lengthened cells, where we first thought, by analogy, to find vessels. It is, then, the parenchyma alone of which the leaf consists; but its cells, for two or three rows only of thickness, (consequently, the greater part in immediate contact with the surrounding liquid,) are commonly regular, closely united together, without intervals, lengthened into meati or enlarged into lacunæ. They, however, all shew in their cavity green granules (*fig.* 132, *p*). It is true, that in those of the leaves which are thicker, we sometimes find lacunæ (*fig.* 132, *l*); but then these, commonly very regular in form and arrangement, have no connection with one another, nor with the outside, but are completely closed by their wall of surrounding cells. They appear destined to diminish the specific gravity of the leaf and thus to sustain it in the water and to perform functions analogous to the natatory bladder of fishes.

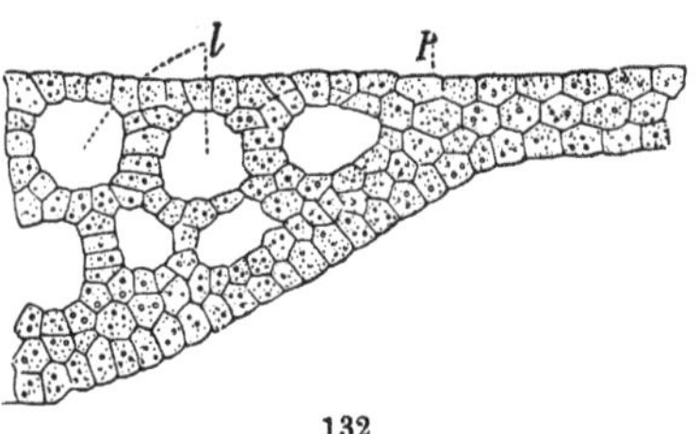

132

We may just observe in passing that these leaves when taken from the water are very rapidly dried and become crisped and deformed; this may be explained by the absence of an epidermis which retards the evaporation of liquids contained in the parenchyma, and of a solid skeleton which supports it.

§ 128. General form of the leaves.—We have just seen that these leaves are formed of parenchyma and compound fascicles, either of vessels and fibres, or in aquatic or inferior vegetables, of lengthened and thin cells. These fascicles which are commonly perceived on the outside, have received the name of *nerves* or *veins* (*nervures*) (NERVI), their arrangement that of *nervation* or *venation*. It is on them and on the extended part in which the intervals between the nerves are filled with the parenchyma, that the general shape of the leaf depends. The fascicles, destined to form the nerves, may remain for a longer or shorter time united, then be se-

132. A section perpendicular to the surface of a small portion of a submerged leaf of the Potamogeton perfoliatum.—*p* Parenchyma.—*l* Lacunæ.

parated from one another by a sort of expansion, and we know that there commences the distinction of petiole and limb.

§ 129. We will pass over the investigation of the properties of the former, and, for the present, confine ourselves to cite a case in which the fascicle is terminated without being divided, and consequently the whole leaf assumes the shape of a petiole. If it finishes in a point, it reminds one of that of a needle, and is called *acicular* (*aciculaire*) (FOLIUM ACEROSUM); this form is observable in several of our evergreens, Pines, Firs, Larches, (*fig.* 133).

§ 130. But more commonly these fascicles are separated into several others, and the latter are divided, either in the same plane, in which case, the leaf offers no other dimensions than length and breadth, or in different planes, in which case thickness must also be added to the other two.

133

§ 131. In the former case, the nerves on separating are carried forward either in a different plane with the petiole, and thus form an angle with it, or rather, which is the most common, remain in the same plane. Then sometimes the fascicle is divided into several almost equal ones, which are separated somewhat like the fingers of the human hand when open, whence this class of nervation has been termed *palmate* or *handshaped* (*palmée*) (PALMATA), and the leaf is then called *palminerved;* sometimes, it is continued in the direction of the petiole to the top of the limb, throwing out to the right and left secondary fascicles, which are arranged with regard to it in the same manner as the barb of a quill with regard to its barrel, whence this kind of nervation has been termed *pinnate* (*pennée*) (PINNATA), and the leaf is then called *pinninerved* (*fig.* 134). The large nerve (*n m*), continued from the petiole (*p*), is called the *median* (*médiane*) nerve or the *costa* (*côte*) of the leaf. The lateral ones (*ns*, *ns*, *ns*), springing from it at an angle more or less acute, are the secondary nerves.

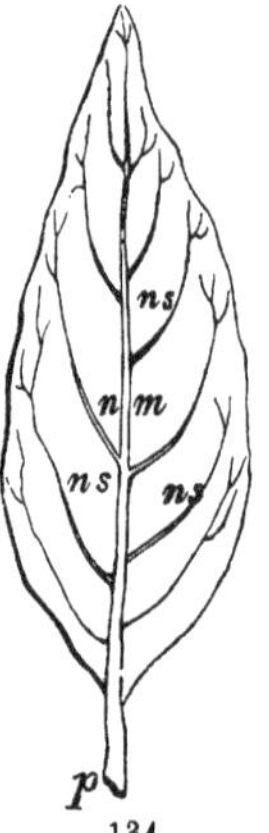

134

As to the case in which the secondary fascicles all diverge in the same plane on quitting that of the petiole at its summit, radiating

133. The under side of a leaf of the Pine.

134. A leaf of the Dwale (*Belladonne*) (ATROPA BELLADONNA).—*p* Petiole.—*n m* Median nerve.—*n s n s* Secondary nerves.

like the spokes of a wheel in connection with its axis, the nervation is termed *peltate* (*peltée*) (PELTATA), and the leaf PELTINERVED (*fig.* 135).

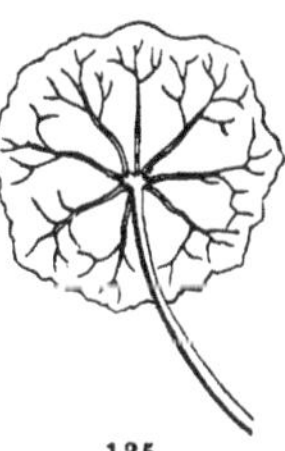
135

In penninerved leaves, the secondary nerves may be thrown out by the median one at every angle from the right angle to the straight line. They may be all equal, long or short, or unequal, by decreasing from the base to the summit, or *vice versâ*, or, on the contrary, increasing to a certain point and then decreasing, attaining their maximum length at the middle of the leaf. or above or below it.

§ 132. When the nerves on separating to form the limb are given off into different planes, there results a lamina variously folded on itself, or else a thick and solid mass. In the first condition, amongst the forms which the leaves take, we may mention the *fistular* (*fistuleuse*), i. e. of a cylindrical tube, as in several species of *Garlic* (*Ail*) (ALLIUM), as well as several other singular and curious forms, where the leaf puts on the figure of a horn, of a vase, of an urn, &c.; thus disguised the leaves are known under the name of *ascidiary* (*ascidies*) (*ἀσκίδιον*, *a leathern bag* or *wine skin*). In the second case, which takes place when the parenchyma fills up the interval between the nerves thus diverging into different planes, the leaf will acquire a solid form, bounded by a round surface or by several plane surfaces determining at their junction several points or prickles, or by a combination of both; this body will be sometimes so regular as to be represented by the different geometrical terms for solids, such as pyramid, prism, cylinder, cone; at other times its irregularity will defy all attempts at rigorous definition, and it will be easily described by names borrowed from well-known objects, such as, a sword, a cimetar, a tongue, a hump, &c.; whence we derive the epithets, *ensiform*, *acinaciform*, *linguiform*, *gibbous*, &c.

§ 133. Let us return to the flattened limbs, and examine the distribution of the parenchyma with regard to the nerves. It may completely fill the interstices, so that the line which passes through the extremities of the longest nerves may be continuous: it is then said that the leaf is *entire* (*entière*) (FOLIUM INTEGRUM, *fig.* 134). The parenchyma is often stopped before it reaches the termination of the nerves: the leaf is then, as it were, cut or slashed, its border

135. A leaf of the Marsh Pennywort, (*Ecuelle d'eau*) (HYDROCOTYLE VULGARIS).

formed by a continuation of broken lines. The outline will now present several degrees of regularity. These forms have received different names according to the distance from the median nerve to which the parenchyma is developed, and to the alternately projecting and retreating angles the margin presents. When the margin is only slightly cut, the projections are called *teeth* (*dents*); *saw-like teeth* (*dents en scie*), when they are very sharp like a saw (*fig.* 142); *scollops* (*crénelures*), when they are more obtuse. If the margin be more deeply cut, in which case the projections are generally much larger, they are called *lobes*. This depth varies much, and with its degrees the names by which the lobes are distinguished. If they only reach the middle of the limb they are termed *fissures*, and the termination *fid* is joined to the adjective denoting the number of fissures; if they penetrate deeper and nearer to the median nerve, they are called *partitions*; if to the nerve itself, they are *segments*. The form of the leaves is naturally described by epithets derived from these different denominations. Thus, we say that they are *dentate* or *toothed* (*dentées*), *serrate* or *saw-like* (*dentées én scie*), *crenate scalloped* (*crénelées*), *cleft* (*fendues*), *partite* (*partagées*), *sectate* (*coupées*), (FOLIA DENTATA, SERRATO-DENTATA [*fig.* 142], CRENATA, FIDA, [*fig.* 136], PARTITA, [*fig.* 138], SECTA, [*fig.* 139]), according to the depth, the figure, and the size of the incisions. But these words are not generally employed alone, they make part of other compound words and by that indicate several modifications at once: thus, when we say that a leaf is *trifid*, *quinquefid*, *multifid*, *palmatifid*, or *pinnatifid* (*fig.* 136), &c., we understand that its border is cut to a depth less than its half into fissures to the number of *three*, *five*, or *indefinitely*, arranged according as the nerves are *palmate* or *pinnate*, &c. If, instead of the termination *fid*,

136

137

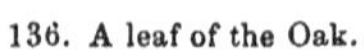

136. A leaf of the Oak.
137. A leaf of the Valerian (*Valériane*) (VALERIANA DIOICA).

we substitute that of *partite*, *multipartite*, *palmatipartite*, (FOLIA MULTIPARTITA, PALMATIPARTITA) (*fig.* 138), *pinnatipartite* (PINNATIPARTITA) (*fig.* 137), we understand that the incisions penetrate beyond the half of the demi-limb; if the termination *sect* (*sequée*) is employed, *palmatisect* (*palmatisequée*) (PALMATISECTA) (*fig.* 139),

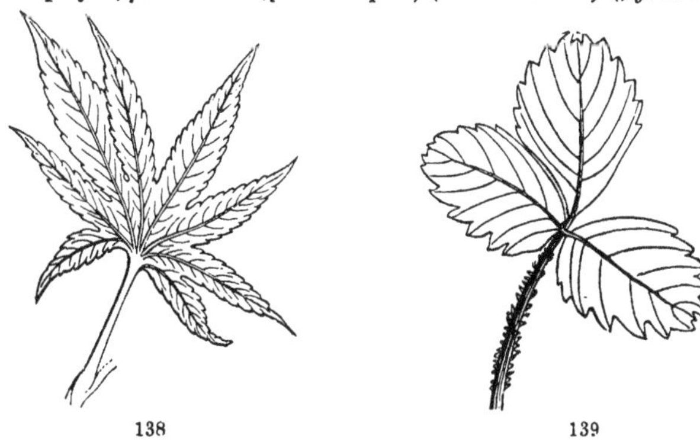

138 139

pinnatisect (*pinnatisequée*) (PINNATISECTA), &c., that they penetrate to the median nerve, and that each lobe is connected by it alone tc the neighbouring lobes, always with the modification which the beginning of the word indicates.

§ 134. In all the leaves of which we have hitherto spoken, the parenchyma disappears in the median nerve or axis, so that the lamina formed by the limb, interrupted towards the margin, forms in the interior a continuous whole. We have, nevertheless, arrived to the last degree of these incisions, when this continuity is only sustained by the median nerve; each segment, however, still holds to this nerve in rather a large portion of its extent, and often presents its maximum breadth at its base.

But another modification may also be found, namely, that the segment is only connected with the median nerve by the secondary fascicle or nerve, which is detached from it to form the segment, and that the separation of this fascicle and the interposition of the parenchyma takes place only at a certain distance from the nerve. It is clear, that, in this case, the secondary fascicle performs the same functions with regard to the median nerve, as the petiole does in relation to the branch from which it springs. This nerve has,

138. A leaf of the Castor-oil-plant (*Ricin*) (RICINUS COMMUNIS).
139. A leaf of the Strawberry (*Fraisier*).

then, the appearance of a branch; and these segments assume the form of so many independent leaves. But we still see that it is only one leaf, because all these segments of which it is formed are always in the same plane, and when the segments are detached from the tree they fall off in a single piece. The leaf then takes the name of *compound* (*composée*); its median nerve that of *rachis*, or *common petiole;* its segments take that of *leaflets* or *folioles* (FOLIOLA); and if the median fascicle of each of these remains undivided for some way beyond the base, this portion is called the *vetiolule.* This name *compound* is opposed to that of the *simple* leaf, by which we distinguish those of which we have previously spoken and all the parts of which are continuous.

Knowing the origin of the compound leaf, we know that it may present modifications analogous to those which we have pointed out in the simple leaf, depending on its palmate or pinnate nervation. It is to these leaflets, that we apply these epithets: for instance, that of the Horse Chesnut, (*Marronier d' Inde*) (*fig.* 140), are palmate; those of our Acacia (*fig.* 141) are pinnate. When we apply

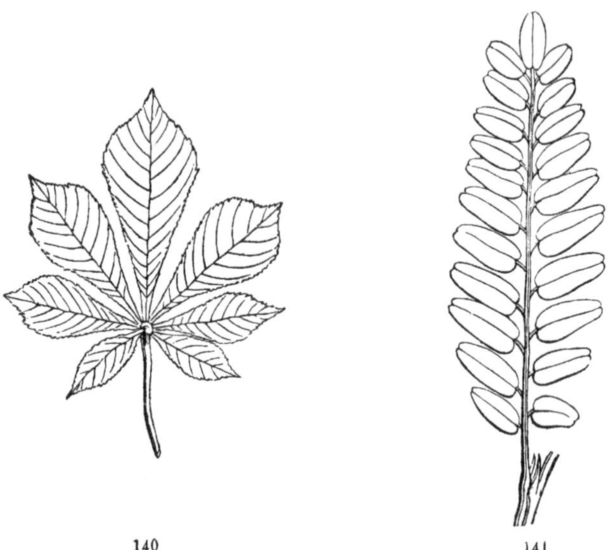

140. A leaf of the Horse Chesnut, (*Marronier d'Inde*) (ÆSCULUS HIPPOCASTANUM).
141. A leaf of the Acacia (ROBINIA PSEUDO-ACACIA).

to a leaf the name of pinnate (PINNATUM) alone, it is this last form which we mean. In the compound words by which we characterize the different modifications of these leaves, we employ the termination—*foliolate* (*foliolée*); thus we say, a *bifoliate*, *trifoliate*, *multifoliate* leaf (*fig.* 141), (FOLIUM BIFOLIATUM, TRIFOLIATUM, MULTIFOLIATUM), according to the number of folioles into which it is divided. They often grow by twos, one on each side of the leaf (*fig.* 148), and this is what is called a *pair* (JUGUM), whence are derived the epithets *bijugate*, *trijugate*, *multijugate*, according to the number of pairs. We say, when the leaf is pinnate without inequality, i. e., has no terminal leaflet, that it is *abruptly pinnate* (ABRUPTE PINNATUM), when pinnate with inequality, i. e., has a terminal leaflet, it is *impari pinnate* (IMPARI PINNATUM), (*fig.* 146) (*fig.* 141).

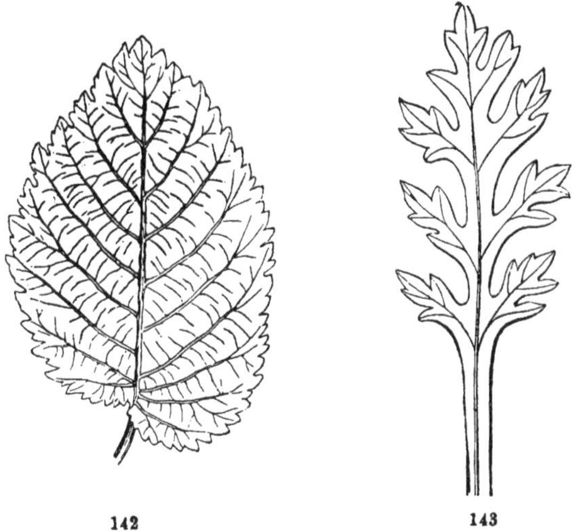

142 143

§ 135. We have hitherto spoken only of the median and of the secondary nerves. In a certain number of plants, the division does not extend any farther; in a much greater number, the secondary nerves are divided in their turn, and there thus exists a series of subdivisions, smaller and smaller, and more and more numerous. All, then, that we have said relatively to the secondary nerves, is re-

142. A leaf of the Wych-Hazel or Broad-leaved Elm (*Orme*) (ULMUS EFFUSA).
143. A leaf of the Poppy (*Pavot*) (PAPAVER ARGEMONE).

peated in the nerves of the 3rd, 4th, and 5th order, &c., each playing the same part with regard to that from which it springs, as the secondary nerves do to the median. In the simple leaves, the lobes may be in their turn either entire or divided, and these divisions themselves be subject to sub-divisions. In order to indicate this, the same epithets are employed, preceded by the syllables *bi* or *tri*, which point out the number of times that the leaf has been subdivided. For instance, a *biserrate* leaf will be a serrate leaf, the margins of the teeth of which are themselves similarly serrated (*fig.* 142); a *bi-pinnatifid* leaf will be a pinnatifid leaf, the lobes of which are themselves scolloped into smaller lobes or lobules, always following the same pinnate arrangement of the nerves (*fig.* 143).

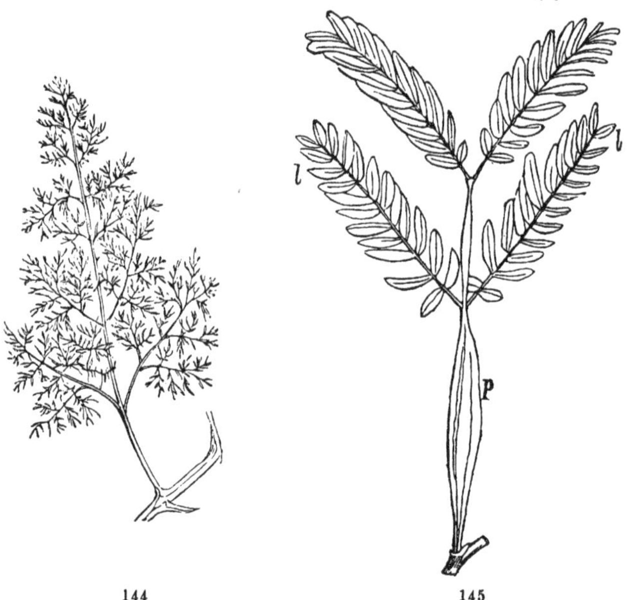

After this double division, the parts commonly become too small for us to take the trouble of examining their regular system, and then we group all the leaves indefinitely divided into a number of lobes (*fig.* 144) under the name of *laciniate* (*laciniées*).

144. Part of the leaf of the LASERPITIUM HIRSUTUM.

145. A leaf of the ACACIA HETEROPHYLLA. *p* Phyllode or enlarged petiole, which sometimes forms the leaf itself.—*l* Limbary part, composed of bi-pinnate folioles, which are entirely wanting in a great number of leaves.

In compound leaves, the folioles themselves may be either dentate or lobed. But, when they are divided, it is generally in the same manner as they were at first, each being divided itself into folioles (DECOMPOSITUM), which sometimes in their turn are again divided (SUPRA-DECOMPOSITUM). We then say that the leaf is twice (*fig.* 145 *l*) or three times pinnate or palmate (BI-TRI-PINNATUM, BI-TRI-PALMATUM). The secondary, tertiary, &c., nerves have become themselves so many *rachides* or *partial petioles* (*pétioles partiels*).

§ 136. We will not stop any longer to dilate on the shape of the leaves, the diversity of which is so great. The explanation of the principal terms which are employed in botany to describe all the known modifications which have not been set forth here, which are, besides, not applied only to leaves, but to every compound vegetable organ of a definite form, will be the subject of another chapter. It is necessary to know them in order to understand the books which describe and distinguish the different species of plants, but superfluous when we study plants in general. It is sufficient to learn how the distribution of these constituent elements of the leaf, the nerves which form its skeleton, the parenchyma which is the soft and essential part, determine these varied appearances: how, in reality, there are no differences except in the degree of developement of the one relatively to one another, and how the most compound leaf does not present a greater number of parts, but only the same part repeated a greater number of times.

§ 137. PETIOLE (*Pétiole*).—We have already pointed out the most common form of the petiole, that which results from the union of the partial fibro-vascular fascicles, which, on being detached from the stem in order to form the lateral expansion of the leaf, remain, in a greater or less extent, united in only one fascicle, and thus form a sort of little branch interposed between the stem and the limb. This fascicle is accompanied by the parenchyma, the greater part of which forms an envelope itself covered by the epidermis deprived of stomata, in the same way as it is on the surface of the nerves, which are a continuation of the petiole (§ 126).

§ 138. We have said in another part (§ 12) that the vessels, at the commencement of one organ springing from another at an angle which changes their primitive direction, present some changes in the form of their elements, cells or fibres, which are shortened and united end to end in a smaller extent. There results from this less intimate union a tendency to be more easily separated; a tendency increased by the parenchyma, the cells of which at the same places present analogous modifications. Thus, it frequently happens, that at a certain period of the developement the adherence of these parts is so weak, that they are separated, either spontaneously, or with the slightest effort. The point where this takes place is called an *articulation*, and such a structure is frequently observed at the point of juxtaposition of two organs, at the nodes of the stem, and at the origin of the branches and leaves. The leaves, which are often articulated with the stem, are, then, continuous with it at the narrowest part. Then they are detached after a certain period, when they have fulfilled the functions destined for them in the life of the vegetable and are beginning to wither. This time varies according to the plants: in a great number, it happens in the course of the year in which the leaf was produced; in others, it differs very much. But, when they are not articulated, they continue to remain united to the plant, although dying or dead. This is the reason why we see the Oak covered with their withered leaves during winter, whilst those of the Walnut and of the Horse Chesnut fall in the course of the autumn.

Those leaves which are persistent are generally found to be simple leaves, whilst compound leaves on the other hand are articulated; and in these not only is the petiole articulated with the stem, but the folioles are, also, with the partial petioles, and become detached on being disarticulated. Compound leaves have been defined by many botanists as possessing this characteristic, not accounting as such, those which remain all in one piece, although in their youth they presented all the appearances of being compound.

§ 139. When a petiole is disarticulated, we frequently observe on the stem at the point from which it grew a slight protuberance, which before appeared to make part of itself and served as a base for it. This little lateral excrescence has been termed PULVINUS (*coussinet*), the face of which turned upwards in an external direction, and which before had a face continuous with and similar to that of the petiole, is the CICATRIX (*cicatrice*), which is the result of their separation (*figs.* 162, 163 *c*). In general we clearly see on this face in the middle of the parenchyma several points which indicate the fascicles which concur in the formation of this petiole. These are diversely grouped in the different species of plants, and, from this grouping, as well as from the general form of the cicatrix and of the pulvinus, we may derive good authority for recognising the species of a tree in the state of nudity in which the loss of its foliage in winter leaves it.

§ 140. The petiole is commonly much shorter than the limb; sometimes it equals it in length, sometimes it surpasses it. It also varies in thickness; when it is very large with reference to the size of the limb, and, consequently, to its weight, it supports it without bending, especially when it is short, as any one would say, who had the least knowledge of the laws of mechanical force. When it is thin, long, or of a soft tissue in which the parenchyma predominates over the fibres and vessels, it bends, forming an arc, drawn down by the weight of the limb attached to the extremity of this flexible lever.

§ 141. Phyllode.—We have hitherto supposed that the fascicles in the petiole preserve their parallel direction as far as the limb, and this is, indeed, the general rule. It is not, however, without exception, and the fascicle sometimes begins to diverge in the petiole itself. If this divergence were to continue, it would be the commencement of the limb; but instead of this, a little higher up, the fascicles approach again as at their commencement, before they enter and expand in the true limb. In their progress, they have ceased to be parallel, but remain in the same plane and do not ramify. The result from this arrangement is, that the petiole thus dilated, itself presents the appearance of a limb (*fig.* 145 *p*), which we commonly consider as the leaf, and this is the reason why we have given it the name of *Phyllode*, as the diminutive of φύλλον, *a leaf.*

But the phyllode is distinguished from the limb, in as much as instead of pinnate secondary nerves thrown out from a median nerve which becomes gradually thinner as they are detached, it is traversed by a certain number of longitudinal nerves scattered over the whole of its surface, and very nearly equal both at the base and at the summit: it is also distinguished from them in being placed on the stem in the reverse way of true leaves, that is to say, that its plane is almost vertical instead of being horizontal.

§ 142. The Sheath (*Gaîne*).—The Stipulæ (*Stipules*).—The petiole, as we have said (§ 122 *b*), is sometimes widened at the base and thus embraces either the whole stem, or a portion of it, being continuous with it: this is what is called the vaginal part or sheath of the leaf. Now, the fascicles destined to form this part, instead of being concentrated and passing into the petiole in one body, are detached separately from the circumference of the stem, and form with the parenchyma which unites them a cylindrical surface, or rather a portion of a hollow cylinder instead of a small solid one. Sometimes the fascicles, at first separate, converge more or less as they advance in height, and the sheath is gradually narrowed into the petiole: it is a kind of phyllode commencing immediately on the stem; at other times (*fig.* 146 *s*) the lateral fascicles are stopped after a certain length, or are prolonged in another plane than that of the petiole, and we then have a clear distinction between the petiole and the sheath. In short, the parenchyma frequently does not bind these lateral fascicles to those of the middle which are continued in the petiole, and this is the probable origin of several of the stipulæ.

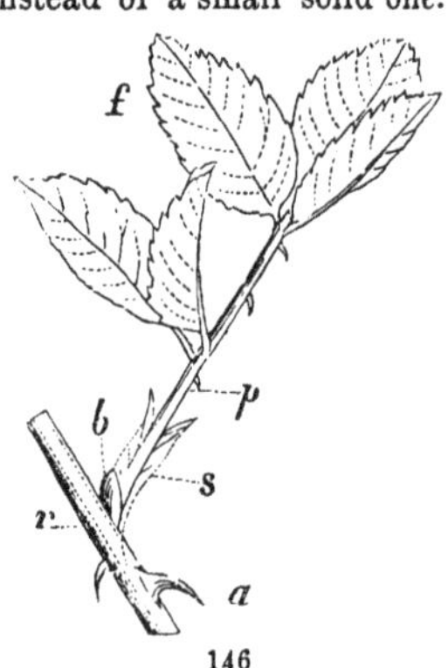

§ 143. These are generally defined as small foliaceous organs at the base of the petiole, and in the actual state of the science it is impossible to give a stricter definition. It is likely that their production, which has not yet been sufficiently explained by the anatomy of the vegetable, is analogous to that of the lateral lobes of simple or of the folioles of compound leaves; that they are the lateral expansions of the fascicles more or less distant from the base of the leaf, more or less connected with it by the intermediate parenchyma, more or less abruptly terminated. They may, although growing from the stem along with the proper fascicles of the leaf, remain independent of them, and, then, as they only seem to adhere to this stem, they are called *caulinary stipules* (*fig.* 147). They may be united to the

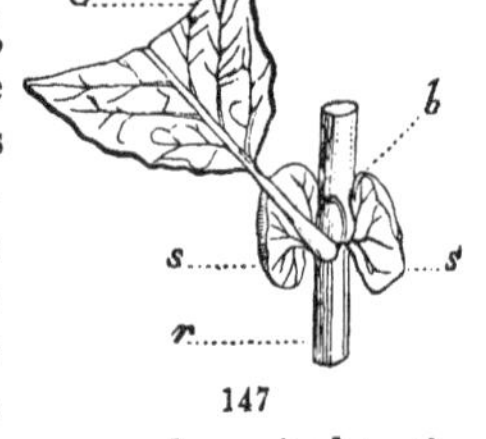

146. A fragment of a branch *r* of the Common Dog-Rose (*Eglantier commun*) (ROSA CANINA), bearing a leaf *f* with its petiole *p*, its petiolary stipules *s*, and its bud *b*.— *a* A thorn.

147. A fragment of a branch of the Trailing Sallow (SALIX AURITA), bearing a leaf *f* with its petiole, its caulinary stipules *s s*, and its bud *b*. The upper part of the limb has been cut off.

petiole (*fig.* 146) in a greater or less extent; and as they then seem to be attached to it, they are termed *petiolary stipules.*

The stipules are entirely wanting in a great number of cases, less frequently, however, than one would be led to suppose: for they are often concealed from the sight, either on account of their smallness, or of their short duration. But, on prosecuting our researches very carefully and diligently with strong lenses and in young leaves, they may be observed in a multitude of plants which were said to want them.

§ 144. Their appearance varies much. They are often reduced to one little point, a little thread, or a little scale. At other times they assume a greater developement, and they then assume a foliaceous consistence, the appearance of lobes or of folioles which themselves present nerves, teeth, lobules, and even a lower contraction in the form of a petiole. It is by no means rare, that their tissue, instead of becoming thick and green, remains in the state of a thin, colourless, and transparent membrane. But, if it varies so much in different plants, it is constant in the same plant, and frequently even in several plants related to one another, like those, which, resembling one another in a great number of their characters, form what are termed in Natural History, genera, families, &c. Entire families are characterized by the absence or presence of the stipules and, also, by some peculiar circumstance in their manner of existing.

§ 145. We have seen them either entirely free, or else united to the petiole. If they are very largely developed and thus embrace one half

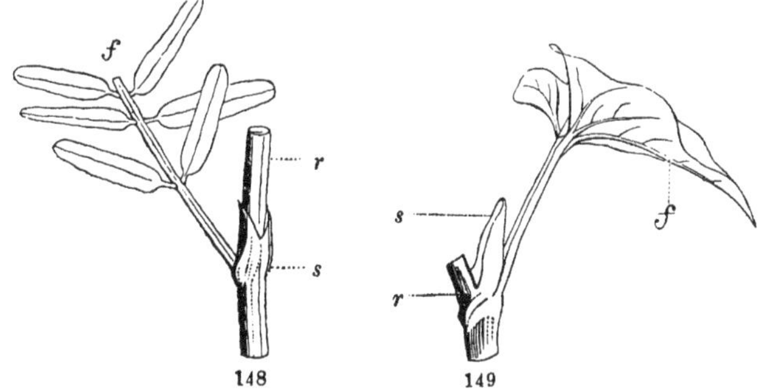

148. A piece of the leaf of the ASTRAGALUS ONOBRYCHIS.—*f* The lower part of the leaf composed of three pairs of folioles.—*s* Stipules running into a single one on the opposite side of the branch *r*.

149. A leaf *f* of the HOUTTUYNIA CORDATA, with a piece of the branch *r*, to which it is attached by its petiole, at the base of which we see the axillary stipule *s*.

of the stem, a stipule may meet the other on the side opposite to that from which the leaf grows, and their external edges on meeting may even unite either at the bottom or throughout their whole length. It is then that we have a divided (*fig.* 148) or an entire sheath (*fig.* 127) a vaginal stipula. If the stipules, as they extend, happen, on the contrary, to be united by their internal edges when they meet, they form one piece only, the middle of which is placed between the stem and the commencement of the leaf in the angle which we term the *axil* (*aisselle*) of the leaf: it is then called *an axillary stipule* (*fig.* 149). If two leaves begin at the same height, the one opposite to the other, each provided with its two stipules, and on each side the stipule of the right leaf meets that of the left, and is united to it, the two together seem to form only one intermediate between the two petioles, and are, therefore, called *inter-petiolar* (*fig.* 150). In all these different cases, it is very easy to find out the origin of the two stipules running into one, either because they are only partly joined, or because in the neighbouring plants of the family this union does not take place and the passages from one state to the other are found.

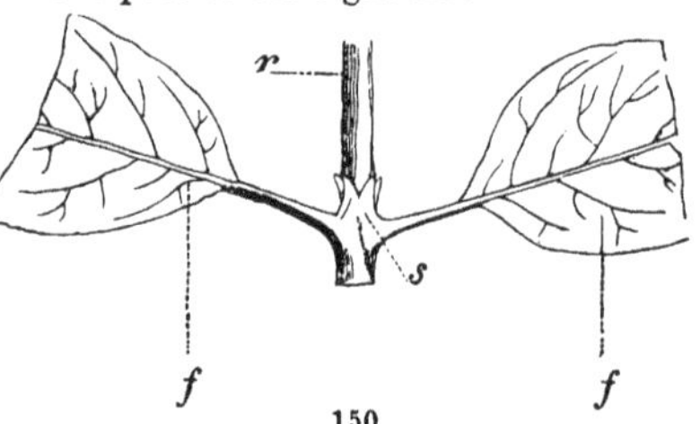

In all the preceding explanations, when we say that the stipules are enlarged, meet, and are united, we employ a figurative language, their enlargement and their union being likely to take place from the very beginning, and resulting from the arrangement of the fascicles in the stem and at the points at which they spring in order to form these new organs.

§ 146. The three parts of the leaf are not always seen together, and there may be two or even only one. We call that leaf which wants the petiole *sessile*, and that in which there are no stipulæ *exstipulate*. In short, the petiolary and vaginal parts are those which are most commonly wanting. Sometimes, however, it is the limb, and then the leaf has lost its common appearance, and is in some cases denoted by other names.

150. The two opposite leaves *ff* of the CEPHALANTUS OCCIDENTALIS, each shewing at their origin that of their two stipules turned towards the spectator. These two stipules are united on the median line into one single interpetiolar *s*. There is a similar one on the opposite side.—*r* Branch.

§ 147. What are the differences which are observed in a leaf at the different periods of its developement? It first appears as a little tubercle or lamina, without any distinction of parts, either in the inside which is entirely cellular, or else on the outside. It is in a little time afterwards that the cells situated on the median line are lengthened, and thus sketch out, as it were, the first nerve, which is then perfected by their organization into vessels of different kinds, of which the tracheæ appear the first. It is, then, as one would expect, a growth perfectly analogous to that of the stem.

As to the developement of the different parts of a complete leaf relatively to one another, the following observations have been made on a plant in which these distinctions are very clear; for instance, on a POLYGONUM, in which the vaginal part forms a complete sheath; on the MELIANTHUS MAJOR and MINOR, in which it is lengthened into a large axillary stipule, single in the one, and double in the other. This comparison of the leaf at the different stages of its growth is easily made on a branch which has not yet finished growing; so that it presents at the bottom several leaves arrived at their greatest dimensions, at the top leaves just coming into existence, and in the interval all the series of the intermediate ages. We thus see that the limb is the part which appears first, that the vagina or sheath is not slow in shewing itself, that the developement of both is nearly equal, and that it advances to a certain point, before we begin to perceive the petiolar. The sheath is the first that arrives at its full growth, and its size is almost always much less than that of the limb; and its growth is finished whilst that of the limb and of the petiole continues. In order to make these relations clear by tables, let us take a branch of the HOUTTUYINA CORDATA and measure six successive leaves from the uppermost, which is still young and only 3 millimetres (French) or .11799 inches (English) in length, to the lowest, which is 9.8 centimetres (French) or 3.85434 inches (English). Here is the relative length of the three parts in these six leaves:

FRENCH MEASURE.

	Limb.	Petiole.	Axillary Stipule.	Total.
1.	.2		.1	.3
2.	.6	.08	.25	.93
3.	3.0	.5	1.0	4.5
4.	4.0	1.5	1.0	6.5
5.	5.0	2.0	1.4	8.4
6.	6.0	2.3	1.5	9.8

ENGLISH INCHES.

	Limb.	Petiole.	Axillary Stipule.	Total.
1.	.07866		.03933	.11799
2.	.23598	.031464	.098325	.365769
3.	1.1799	.19665	.3933	1.76985
4.	1.5732	.58995	.3933	2.55645
5.	1.9665	.7866	.55062	3.30372
6.	2.3598	.90459	.58995	3.85434

The limb is at last nearly three times longer than the petiole and four times than the stipule. At first in the leaf 1 it was only double of it, rather more in leaf 2, triple in leaf 3, quadruple in all the others. The petiole, which was nothing in leaf 1, began to appear in leaf 2; it was the half of the stipule in leaf 3, much larger in leaf 4, and double in the two last. On choosing different examples, those of the POLYGONUM and MELIANTHUS, for example, which we mentioned above, we shall find the proportions a little different, but still analogous.

We may then conclude that it is at the part of the base situated below the sheath, that the leaf continues to lengthen, and not as in roots at its extremity. But is it not lengthened in the same manner as the stem in all its parts at the same time? It is easy to determine of the contrary by measuring the distances between several points marked, either designedly or naturally on the limb of a leaf during its developement: it has, thus, been determined that the upper marks do not separate from one another, whilst the lowest one separates gradually from the base, and hence we may of course deduce, that it is at this base and at the petiole continuous to it, that the leaf continues to grow, after it has ceased to do so in its upper part. An observation of the same kind, made, not in the longitudinal direction but in the transversal, would shew that the leaf sometimes continues to grow in breadth at its middle for a much longer time than in its length, and that the lengthening of the secondary nerves with regard to the median nerve is analogous to that of the median nerve with respect to the branch which bears the leaf.

What we have just said of the increase in length applies to simple leaves. The compound may, generally, be compared rather to a branch, because the developement seems to take place from the bottom to the top, and the folioles appear and are developed in proportion to their height, the lowest first, and the highest last. We will cite particularly one genus of the Meliaceæ (GUAREA), in which the whole of the upper part of the leaf is still in the state of a bud, whilst the lower has already acquired its perfect developement, so that these leaves seem to produce two distinct generations of leaflets.

§ 148. The leaves, when they have once arrived at their full growth, live for a greater or less length of time. We know, that, in the greater part of our trees, this does not exceed a few months. In some, however, especially in those of a warm climate, they remain two years or more, and these trees are called ever-green, because we see them constantly covered with a foliage which preserves its colour: but, of course, they are not always the same leaves. The first fall after a certain time; but the tree, already covered with fresh leaves, is not leafless, and, consequently, always preserves the same appearance. This is the case with our Pines (*Pins*) and Hollies (*Houx*), &c. We have seen that, amongst annual leaves, some remain withered on the tree, others are detached by disarticulation. It is useless to repeat that the green colour is gradually replaced by that which is known under the name of a dead leaf. But the leaves, especially those which are articulated, frequently pass, before they fall, through various shades, the variety of which, and in some the richness and brightness, gives them those boasted autumnal tints and hues, which in large masses impart such a pleasing effect to the landscape.

§ 149. A COMPARISON BETWEEN THE LEAVES OF THE THREE GRAND CLASSES OF VEGETABLES.—Up to this point, the comparison of the fundamental organs of the three great classes of vegetables has presented great and notable differences between them. Is it the same in their leaves? Let us recollect at first, that they are composed of nerves and parenchyma, and that the former are palmate or pinnate according as the petiolary fascicle is divided into several other nearly equal divergent ones, or is continued in the median line, and distributes smaller secondary fascicles to its right and left. But we have said nothing of its ulterior destination or connections with these secondary fascicles or nerves. Now, two things must happen: 1st, They must all continue in a straight line without throwing out side branches, or, at least, if they do throw out a few lateral divisions, it is without mingling with the neighbouring fascicles: 2nd, The secondary nerves are themselves branched, these ramifications are sub-divided in their turn, and passing from one nerve to the other, they are reunited, and, at last, the result is a sort of vascular net, the meshes of which are formed by the last divisions of the nerves, the areolæ by the parenchyma.

There appears to be a very constant connection between these two methods of the distribution of the nerves, and the two great classes of cotyledonous vegetables, the first being most frequently observable in monocotyledonous, and the second in dicotyledonous plants.

§ 150. THE LEAVES OF MONOCOTYLEDONOUS PLANTS.—Their limb does not present a net-work of nerves: sometimes their course is quite parallel (*fig.* 151), as in the leaves of the Iris, the Reed, &c.; at other times, there are secondary nerves which are detached from one or more principal ones and take another direction; but they are separated by a line, which is more or less bent, the convexity of which is turned towards the principal (*fig.* 152). From this equality and parallelism of the secondary nerves, it most frequently happens that the leaf is entire. The Palms, it is true, have leaves which are pinnatisect and palmatisect; but, we may, by following their developement, see that in their first stage they were really entire, and that it is not till they are older, that they are divided into several lobes parallel to their palmate or pinnate nerves.

The vaginal part is often very largely developed in the leaves of monocotyledonous plants, which sheath the stem to a considerable length. There are even several of these sheaths fitting one into the other, which strengthen the stem and seem to constitute the greater part of it in several of these plants, in the Banana trees for instance. The sheath at its termination is sometimes lengthened into a kind of a little collar, which is generally membranous and whitish (*fig.* 151 *gl*), sometimes whole, sometimes fringed or scolloped, and the most frequently divided symmetrically into two lateral parts. This is what is termed the *ligule*, with which we meet in almost all Graminales. They have

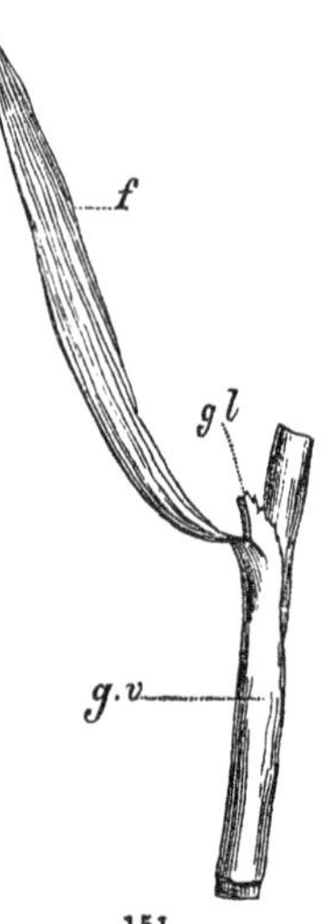

151. A portion of a leaf of PHALARIS ARUNDINACEA.—*f* Its limbary part.—*g* Its sheath.—*g v* The vaginal part of the sheath.—*g l* Its upper membranous part or ligule.

been compared to the stipules; for, if these are considered to be entirely distinct from the sheath, we shall not find any others in monocotyledonous plants.

Continuous with the stem in a great part of its length and following the direction of its axis, the sheath is not articulated, and the leaf does not fall before it dies.

If the nerves remain quite parallel from the bottom to the top, its form is most commonly that of a ribbon (the TYPHA or *Cat's Tail*, the *Reed*, &c.), and it is difficult to distinguish between the petiole and the limb. At other times they diverge a little as they proceed from the base, then converge again as they approach the summit, and we then have the appearance of a limb, as some orchidaceous plants, EPIPACTIS OVATA, LATIFOLIA, &c. If secondary nerves are separated from the principal ones in another direction, on being separated they determine a limb, quite distinct from the petiole, which they form when united lower down. The Banana-tree shews this on a very large scale (*fig.* 152). In the greater part of these cases, however, the whole leaf would be more to be compared with a phyllode, and a plant which is very common on the banks of our rivers, the Arrow-head (*flèchière*) (SAGITTARIA SAGITTIFOLIA,) justifies this comparison; for, one may see on the same plant, leaves bearing at the top of the long and straight petiole, a large limb in the form of an arrow; others, stretched along the water which laves them, are lengthened into long thin ribbons without any distinction between the limb and the petiole, and we may see the intermediate steps between forms which are so different from one another.

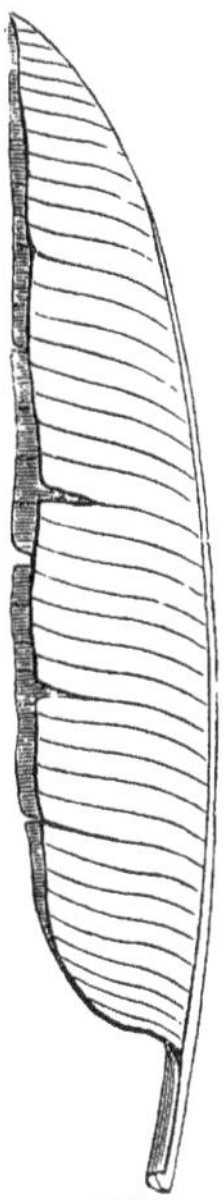

152

A small number of monocotyledonous families are an exception to the preceding rules, in the nerves branching and joining so as to form a net, in which also we find a true limb, frequently lobed in its outline. These are the Aroïdeæ, Smilaceæ, Dioscoreæ.

§ 151. THE LEAVES OF DICOTYLEDONOUS PLANTS.—It is amongst these that we find the articulated leaves, those which are truly com-

152. Leaf of the Banana-tree (very much reduced), shewing the parallel secondary nerves curvilineal to their origin.

pound (*figs.* 140, 141), those whose outline is dentate (*fig.* 142), crenate, divided into lobes by angles (*figs.* 136, 137, 138, 139), and not by right lines resulting from scalloping. The nerves, on springing from one another, form an angle properly so called most commonly acute (*figs.* 134, 136); they are divided and confounded by their last ramifications. Whilst the examination of the leaf in general occupied our attention, it was these and their constituent parts which we had almost constantly in view; it would, then, be superfluous to stop here to repeat all our observations.

Let us, however, remark that in some dicotyledonous plants, the leaves, on account of their nerves being parallel or converging without ramifications, would very well pass for those of monocotyledonous plants. Such are, for instance, some few of our Ranunculi (*Renoncules*) (RANUNCULUS GRAMINEUS, LINGUA, &c.) There are some which may be doubtlessly considered as phyllodes, as in the Acacias with entire leaves, in which we find to these false leaves the limb constantly added to the first which succeed to the germination. Several botanists have tried to explain in the same way all the leaves of the dicotyledonous plants, which are exceptions to the general rule in their shape, and the arrangement of their nerves.

§ 152. THE LEAVES OF ACOTYLEDONOUS PLANTS.—In this class, the leaves of the Ferns (*Fougères*) (FILICES) arrive at the greatest degree of developement, whether sessile or petiolate, entire or lobed. Their division may be carried to a very remarkable degree. Thus, in the PTERIS AQUILINA, which is a large Fern very common in woods, what one would think to be a stem loaded with leaves, is nothing else than one single leaf springing from a subterranean stem, and several times pinnatisect. The nerves present ramifications, and net-work even more varied than in the leaves of cotyledonous plants, and may furnish very clear and good characteristics wherewith to classify them. The petioles are traversed by fibro-vascular fascicles, similar in their composition to those of the stem, i. e. presenting a mass of vessels (for the most part scalariform), compressed together so as to form a band diversely bent and surrounded by a layer of blackish parenchyma. There results from the horizontal section of these petioles, very varied and singular figures which also serve to distinguish the several species from one another. We will content ourselves with mentioning here the rude resemblance to the eagle with two heads, on the arms of Austria, which has been observed in the fascicle of the petiole of the PTERIS

AQUILINA cut obliquely towards the base. It may serve as an example for studying this kind of vessels and of fibres which are so common in Ferns.

The leaves become very simple in all the other acotyledonous plants, the stems of which have offered to us a fibro-vascular system; still divided, (as into four folioles or leaflets), and marked with numerous nerves in the MARSILEA, they are reduced in the Lycopodiaceæ to a cellular plate transversed breadth-ways by one little fascicle. This is wanting, and is replaced by lengthened cells in those families which are deprived of vessels, like the Mosses (MUSCI) (*Mousses*), JUNGERMANNIA (*Jongermannes*); and lastly, this sketch, as it were, of leaves itself disappears along with the stem in the lowest families, as the Lichens, the Fungi, and the Algæ.

THE ARRANGEMENT OF THE LEAVES ON THE STEM, OR PHYLLOTAXY.

§ 153.—The leaves are arranged on the common axis which bears them in several ways. They are called *caulinary* (*caulinares*), and *ramal* (*rameales*), according as they are borne by the stem or the branches. Sometimes instead of appearing at different heights, they are collected together in a mass near the collum; they are then termed *radical* (*radicales*), although they by no means depend on the root, but are only in its neighbourhood[p] (the Primrose, Cowslip, Polyanthus, &c.)

More frequently, however, they are situated at different distances along the axis. The points of the stem situated at different distances from which spring the leaves, are called *knots* or *nodes* (*nœuds*) (NODI) (*fig.* 154, *n*); *internode* (*entre-nœud*) (INTERNODIUM) or *merithal* (*méritalle*) (*fig.* 154 *m*) is the name given to the bare interval which lies between one of these points and that which is situated either above or below it. Sometimes a node bears two or more leaves which consequently grow at the same height; sometimes each only produces one. It is the latter case which we will examine the first.

§ 154. ALTERNATE LEAVES, (*Feuilles alternes*).—This is the commonest arrangement, and the leaves are then said to be *alternate* (*alternes*). For a long time botanists were contented with this word; to this was added the epithet *scattered* (*éparses*), when the leaves seem to be arranged without any regard to order; for it had already been remarked that a certain regularity generally seemed to preside over the alternation of the leaves. Bonnet was the first

[p] The radical leaves arise from a stem below the surface of the ground.—TRANS.

to notice that by drawing a line from the bottom to the top through the successive points from which the leaves begin, this line would describe a spire round the stem; that this relation of the leaves to one another is almost always constant, each separated from the other by an equal part of the circumference of the stem, so that if we find one placed vertically above a first lower leaf from which it is separated by a certain number of intermediate ones, the next leaf will be above the second, the next above the third, the next above the fourth, and so on. He mentioned as the most common case that in which the leaves are thus superposed from 5 to 5, so that the 6th is in a straight line above the first, the 11th above the 6th, the 7th above the 2nd, the 12th above the 7th, &c. He saw, however, at the same time, that there were other more complicated combinations, in which instead of the 6th leaf, it was one still higher, the 9th, for instance, which was placed in a vertical line above the first.

§ 155. Modern researches on the same subject, pursued with much attention and sagacity and principally by M. M. Schimper and Al. Braun, have confirmed and extended these first theories and have led the way to the knowledge of a certain number of laws which preside over the relative positions of leaves and, consequently, of all the lateral organs of the vegetable. Let us first consider the combination of Bonnet, that in which the leaves are grouped in fives (*fig.* 153). We will draw a line through the successive points of their insertion, and we remark that before we have arrived at the sixth, this spiral line has passed twice

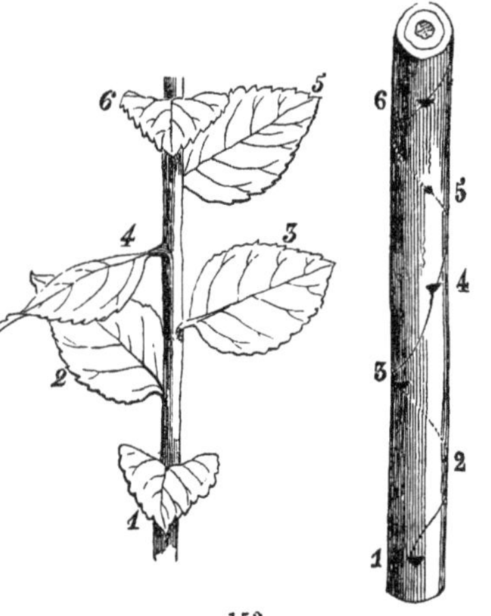

153. A fragment of a branch of a Cherry-tree (*Cerisier*) with six leaves, the sixth of which is placed vertically above the first after two revolutions of the spiral, and thus commences another cycle.—By the side of this figure, is represented the branch magnified, and stripped of its leaves, describing the spiral on which we find at certain distances the cicatrices marking the insertions of the leaves.

round the circumference. Since the five leaves are placed at regular distances on this line, which twice embraces the circumference of the stem, the distance from one of them to the following or to the preceding one will be expressed by the fraction $\frac{2}{5}$ of the circumference. This fraction which expresses the size of the arc interposed between the insertions of two successive leaves, measures that angle, called the *angle of divergence*; let us remark that the numerator is the number of times the spiral line winds round the stem before it brings a leaf directly above that at which it began, that the denominator is the total number of leaves contained in this interval. The sixth leaf will be the commencement of a new series of five arranged in the same way on two revolutions of the spire. Each of these systems of leaves which are found to be combined in the same way above the first, is called a *cycle*. All these terms once defined, it will be easy for us to take into consideration the different combinations which may be found in different vegetables.

§ 156. The most simple case is that in which the leaves are *distichous* (*distiques*) (δίστιχος, *consisting of two rows*), i. e. situated alternately on the two opposite sides of the stem (*fig.* 154). Each

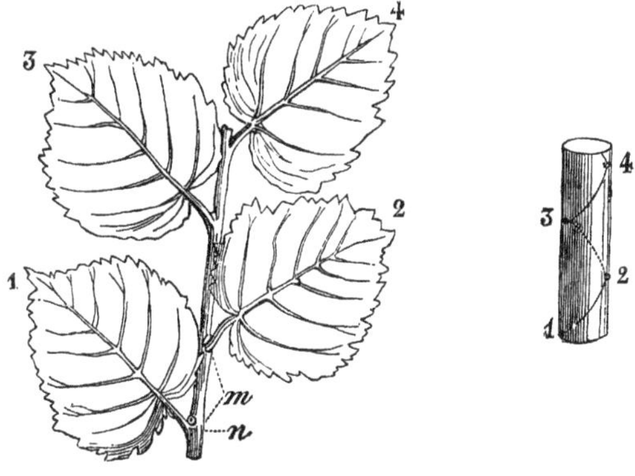

154

is then separated from the other by half a circumference; after a complete turn of the spire we find the third placed above the first

154. A fragment of a branch of the Lime-tree with four leaves, i. e. two cycles, the angle of divergence being $\frac{1}{2}$.—The magnified branch has been figured at the side with the spire and the cicatrix marking the insertion of each leaf.—*n* Node.—*m* Internode or merithal.

and commencing a fresh cycle. Let us note this angle of divergence $\frac{1}{2}$.

§ 157. A case which is much more uncommon is that which we observe in several Cyperaceæ (*fig.* 155), in which three leaves are

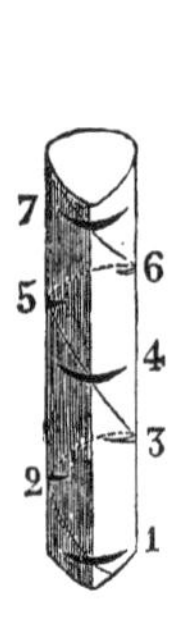

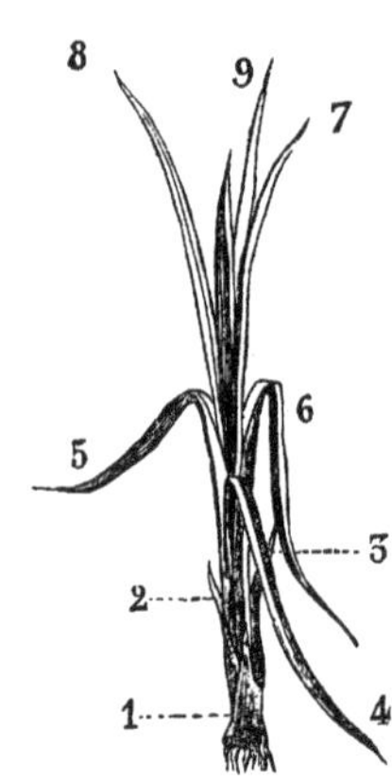

155

inserted in one revolution of the spire, and the fourth above the first. The angle of divergence of these *tristichous* (*tristiques*) (*τρίσ-τιχος, consisting of three rows*) leaves is $\frac{1}{3}$.

§ 158. All the cycles which we have mentioned are far from being the most common. We meet much more frequently a much greater number of leaves distributed on a much greater number of revolutions of the spire; for instance, 8 leaves on a spiral of 3 revolutions, 13 on one of 5, 21 on 8, or, in other terms, the angles of divergence, $\frac{3}{8}$, $\frac{5}{13}$, $\frac{8}{21}$, &c., between two successive leaves. If we write one after another these fractions, which the observation of a considerable number of vegetables has determined:—

$\frac{1}{2}$, $\frac{1}{3}$, $\frac{2}{5}$, $\frac{3}{8}$, $\frac{5}{13}$, $\frac{8}{21}$, &c., &c.,

and compare them with one another, we shall be struck with the connection which exists between them; for the numerator of each is composed of the sum of the numerators, the denominator of the sum of the denominators of the two preceding fractions: in the same way, this numerator and denominator may also be obtained

155. A young plant of the CYPERUS ESCULENTUS with tristichous leaves.—At its side, we see a fragment of the stem magnified with the spire and the cicatrix marking the insertion of each leaf.

by taking the difference of the two following fractions. Let us take the fraction $\frac{3}{8}$; $\frac{3}{8} = \frac{1+2}{3+5} = \frac{8-5}{21-13} = \frac{3}{8}$.

By applying this law to the determination of other possible combinations and thus adding the two terms of the last two fractions already known, we shall have $\frac{13}{34}$, which by a fresh operation will give us $\frac{21}{55}$, then $\frac{34}{89}$, then $\frac{55}{144}$.

These collected will be,

$$\frac{1}{2}, \frac{1}{3}, \frac{2}{5}, \frac{3}{8}, \frac{5}{13}, \frac{8}{21}, \frac{13}{34}, \frac{21}{55}, \frac{34}{89}, \frac{55}{144}, \text{\&c.}$$

Now, these numbers, which have been determined by such a simple calculation, are actually verified by observation; but it becomes more delicate and uncertain in proportion as the numbers increase; we may easily conceive, that if the internodes are rather long, and the leaves thus placed at some distance from one another, the superposition of a certain leaf, the 35th or 56th, for instance, above the first, would be very difficult to determine. If, on the contrary, by the shortening of the stem, the insertions are brought closer together and the leaves touch, as in the Artichoke (*Artichaut*), we may easily perceive that one leaf is placed directly above another; but it would be very difficult to follow the succession of the intermediate leaves.

§ 159. Nevertheless, even in this case, we may without much trouble number each leaf. The very simple method of doing it is owing to certain properties of this regular spiral arrangement, which we will now begin to explain. The demonstration may be simplified, by premising that the stem, which we have hitherto considered as a cylinder, really grows gradually thinner from the bottom to the top, and is, consequently, a very tapering cone; that, therefore, the revolutions of the spiral line around it have a constantly decreasing diameter; that they are not placed exactly one above the other like the spring of a brace, but rather like that of a watch, drawn out by its internal end; that if the axis be supposed to be very short and almost reduced to a plane, the spiral thread would take precisely the normal arrangement of the main-spring of a watch, which at each turn gradually diminishes in diameter, and thus forms a series of concentric revolutions, which gradually approach the centre as they reach the end of the axis (*fig.* 157). In this construction, the part situated at this central extremity represents that which would be the highest if the axis were pulled out; the part situated at the other extremity of the spire represents the lowest, the intermediate parts will, consequently, be nearer the outside as they are lower down on the elongated axis. Now, this is not a pure supposition; in nature, this arrangement is very often realized, when we see on a very short stem a very great number of leaves joined together in a tuft, which we call a *rosette* (*fig.* 156), as in the House-leek. Let us draw, then, (*fig.* 157) such a spire on paper, and in order to render this explanation at once both clear and complete, let us choose a cycle composed of a number of leaves and revolutions of a spire which will be neither too high nor yet too low: $\frac{5}{13}$, for instance, a combination which is most frequently found in nature.

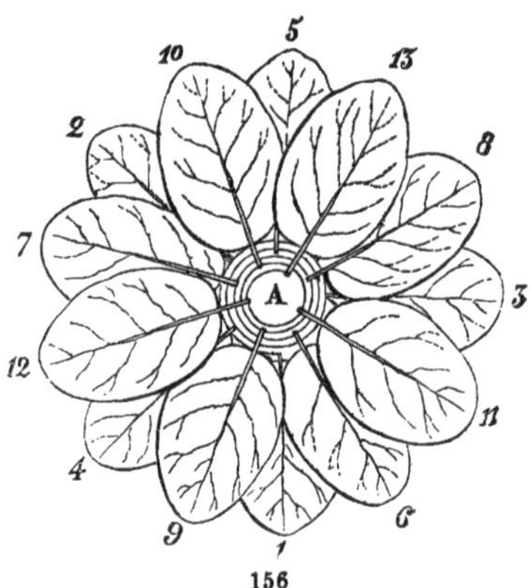

156

From the centre of the spire, at the distance of its opposite extremity where we place leaf 1, let us describe a circle and divide it into 13 equal parts by as many radii. Five of these parts form the angle of divergence. We count, then, 5 following the line of the spire, and we then mark the leaf 2; 5 other parts, and we then have found leaf 3; and so on for the rest; when we shall have determined the 13th, there will be one leaf on each radius, and the 14th will be placed on the same as the 1st, separated from it by 5 revolutions of the spire, and beginning another cycle. Let us suppose, then, that the spire describes 10 revolutions towards the centre, and let us continue to mark off the leaves from the 14th; we shall see, after five fresh revolutions the 27th placed on the same radius as the 1st, and beginning a third cycle which will terminate at the 40th. Three leaves will then be placed on each radius, separated from one another by 5 revolutions, and the difference between them will always be the number 13.

But each of them has at the same time certain relations with the leaves situated near it to the right and left of it; for instance, leaf 1 with leaf 6 to the right, and since each of the leaves may be

156. A cycle of thirteen leaves arranged in a rosette. The observer is supposed to be looking at their top. On a very short axis *A* which bears it, five revolutions of the spire have been described, and the origin of each leaf indicated.

taken indifferently as the point of departure, this leaf 6 is precisely situated with regard to leaf 11 placed on the radius to its right, as

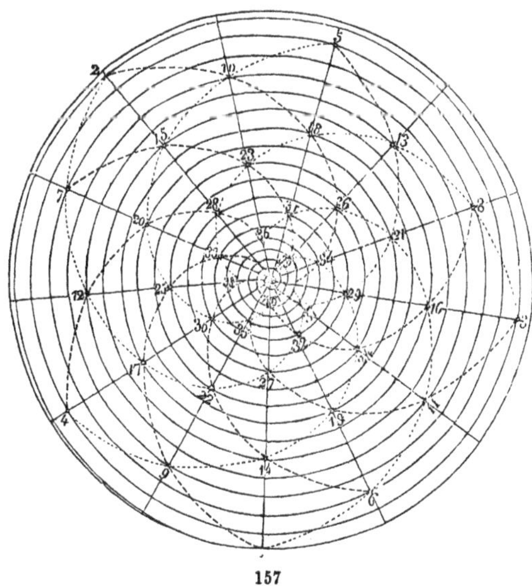

157

the leaf 1 was to the leaf 6; the leaf 11 in the same relation to leaf 16, 16 to 21, 21 to 26, and so on, so that on drawing a line through this series of points, 1, 6, 11, 16, 21, 26, 31, 36, it would thus be found to describe a portion of a spire. But the leaf 2 is found with regard to the leaf 7, 3 with regard to 8, 4 to 9, 5 to 10, to be in the same relation to one another, as 1 is to 6; we may then trace through the points 2, 7, 12, 17, 22, &c., 3, 8, 13, 18, 23, &c., 4, 9, 14, 19, 24, &c., 5, 10, 15, 20, 25, &c., 4 other spires like the first: these 5 spires will comprehend the insertion of the leaves. In each of them the number 5, that of the spires, will necessarily form the difference between the number of one leaf and that of the one which follows it.

If we find out the same connection of the leaf 1 with that nearest to it on the left, leaf 9, we shall construct in the same manner another spire comprehending the numbers, 1, 9, 17, 25, 33, &c.;

157. A plane projection of a spire from right to left bearing five cycles, each of thirteen leaves inserted on five revolutions of the spire.—The secondary spires formed on the right by the series of numbers from five to five, on the left by the series of numbers from eight to eight, are shewn by the dotted lines.

then 7 others successively beginning with the numbers 4, 7, 2, 5, 8, 3, 6, and shewing in the series of their numbers the constant difference of 8 between one number and that which follows.

The greater this difference is, the nearer the spire approaches the rectilineal direction of the radius, and the shorter is their course to the centre.

We have, then, one first spire which we will call *primitive* or *generative* (*primitive, générative*) passing through the leaves 1, 2, 3, 4, 5, 6, &c., that is, in the successive order of height on the stem; we have also several parallel secondary spires directed from right to left, i. e., in the same direction as the primitive one; the others parallel from left to right, i. e., in an inverse direction. Now, the latter refer to the number 5, precisely that which expresses the difference between the number of any leaf whatever, taken on one of these spires, and that of the following one, exactly the figure of the numerator of the fraction expressing the angle of difference ($\frac{}{13}$) in the generative fraction. On the other hand, the parallel secondary spires which have an inverse direction refer to the number 8, which added to the first 5 gives 13, or the denominator of this very fraction.

Now we see, that if we count the number of secondary spires parallel in one direction, and then in the other, we shall have the angle of divergence. The smaller of these numbers is the numerator; the sum of the two numbers is the denominator. We thus see that we may easily number all the leaves; for, (always keeping to the same case, that in which the angle of divergence is $\frac{5}{13}$), if we take some leaf for the point of departure and call it 1, we may then inscribe 6, 11, 16, &c., the consecutive numbers on the secondary spire to the right; 9, 17, 26, &c., the consecutive numbers on the spire to the left; and, when once a leaf is numbered on a spire, all the rest are obtained with the same ease, since it is only necessary to add the number 5 or 8 if you reckoned successively from the bottom to the top, or to subtract it if you reckon from the top to the bottom.

§ 160. But is it easier to determine the numbers on the secondary spires than on the primitive one? The latter, if the leaves are separated a little on an axis of some length, is determined the more clearly; it is not so, when the leaves are so near and on an axis so short, that it becomes difficult to appreciate their relative heights. But, in these cases, the secondary spires are very apparent: this will easily be seen, by casting our eyes on a rosette composed of a great number of leaves, such, for instance, as the House-leek before the developement of its stem, and, particularly that species cultivated in conservatories under the name of SEMPERVIVUM TABULARE. Our construction of figure 156 is perfectly realized in similar rosettes, at an angle of divergence which varies according to the species. We find, however, although on a longer axis, good examples in the cones of evergreens which are composed of a number of contiguous scales, representing so many leaves, and overlapping one another at their bases, so that it is impossible to determine on the outside the points of their insertion. Now, at the first glance we recognise several series of scales following the direction of very oblique parallel spires (*fig.* 158), some to the right, some to the left, and very easily counted in both directions: and we thus immediately obtain the angle of divergence of the primitive spire which we do not see. Thus the cone of a species of Pine (PINUS ALBA) presents precisely the combination $\frac{5}{13}$ which we employed for our demonstration (*fig.* 158). If we had taken that of the PINUS PICEA, we should have found 8 secondary spires on one side, and 13 on the other, and we should, therefore, have determined the angle of divergence to be $\frac{8}{13+8} = \frac{8}{21}$.

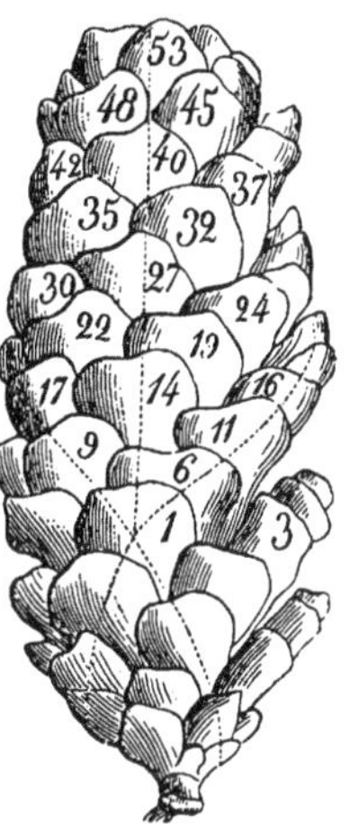

158

The problem may be resolved in several other ways: thus, if we can directly perceive the rectilineal series so as to count them, we have the denominator of the angle of divergence, and, this once known, we know the numerator which corresponds to it, in the series of the fractions which express the most common angles of divergence.

§ 161. We have spoken of two secondary spires only, the two which are the most visible, and the nearest to the vertical; but it is clear that there is a great number of others, since every line traversing a series of numbers and presenting between one another a constant difference would be spire: for instance, that which would pass through 1, 4, 7, &c.; or through 1, 11, 21, &c.

§ 162. The primitive spire may run from right to left or from left to right. It has been a question, whether this direction was

158. A cone of the PINUS ALBA, on which the scales are numbered according to the order of their relative position in height. The dots refer to a rectilineal series and two secondary spires, one from left to right, the other from right to left.

constant, either in a given vegetable or in certain branches of that vegetable with regard to others. This constancy has been determined in certain cases, though it has not been found in the majority. However it may be, two branches being given, *a* and *b*, the first of which *a* bears the second *b*, the first leaf of *b* is always arranged with regard to that of *a*, from the axil of which springs the branch *b*, so that the latter commences the spire of *b*. But, sometimes the spire of *b* is turned in the same direction as that of *a*, in which case it is said to be *homodrome* (ὁμὸς, *like*; δρόμος, *course*). Sometimes it revolves in the opposite direction, and is then termed *heterodrome* (*hétérodrome*) (ἕτερος, *different*).

In general, however, constant arrangement can be observed, whether in the direction of the spire, or in the angle of divergence, only when the terms of the fraction which expresses the latter are very small $\frac{1}{2}$, $\frac{1}{3}$, and at most $\frac{2}{5}$. Beyond that point, there is often a passage from one cycle to neighbouring cycles, and the direction of the spire from left to right or from right to left is met with indifferently on different twigs growing from the same branch, and is sometimes changed even on one and the same branch. The causes which determine these changes of direction have not yet been well appreciated. We can explain with ease the frequent substitution of one cycle for another, when we compare the different angles of divergence, which when reduced are found to differ but little from one another. For, if we express them in degrees and minutes, we shall see that to begin from $\frac{5}{13} = 138° 24'$, all the others equal 137°, plus a number of minutes which, though all vary a little, approaches nearer and nearer to 30′, and at last hardly differs from it. Now, we can understand what effect a few minutes have on so small a circumference as that of a branch: the slightest deviation, whether it results from a very small torsion of the stem which frequently takes place in nature, or from an imperceptible error often inevitable in the observation, changes the divergence these few minutes and thus substitutes one cycle for another. Thus, M. Bravais considers these different arrangements of the leaves on a continuous spire, as simple alterations of one arrangement, in which the angle of divergence would remain constant. This would be the angle 137° 30′ 28″, which is irrational to the circumference, that is, not capable of dividing it an exact number of times, and, consequently, not capable of bringing the leaf exactly in a right line above a preceding leaf. The leaves 6, 9, 14, 21, 35, 56, &c., which, in the series of the cycles hitherto observed, we have found to be placed above leaf 1, would not, in short, be exactly placed on the vertical straight line passing through the insertion of leaf 1, but alternately on both sides of this line, approaching it more and more without ever being able to reach it.

M. Bravais divides, therefore, leaves considered in relation to their arrangements on the stem into two great classes: 1st, The *curviserial* (*curvisériées*), those which, never being capable of being brought back one above the other in a straight line, thus describe an indefinite curve; 2nd, The *rectiserial* (*rectisériées*), those, the divergence of which is a component part of the circumference, and which, consequently, are necessarily brought one above the other in rectilineal series for the whole length of the stem: we have already seen several examples of this in the distichous and tristichous leaves, the divergence of which is one half or one third of the circumference.

§ 163. We have mentioned the angles of divergence which are usually presented in the spiral series and, as we have just seen, are all approaching to 137°. But we have sometimes found very different ones, as $\frac{1}{4}$, $\frac{1}{5}$, $\frac{2}{9}$, $\frac{3}{14}$, &c., which, as we may easily perceive, form a series analogous to that we have just examined, inasmuch as they may be obtained by the addition of the numerators and of the denominators. We will not stop to consider this new series, nor two others which have been discovered, for these are cases which are not very often found, and, in fact, may almost be considered as exceptions. The student ought, however, to be told of them in order to avoid confusion and doubt, if at any time he stumble on these rare combinations of leaves in his researches.

M. Bravais remarks, that these changes of divergence would take place, if all the leaves, growing on one of the secondary spires, were to disappear, and there would be such or such a change according to the spire which was suppressed. This is by no means a gratuitous supposition, for it sometimes takes place. Thus the stems of very young Cacti have quite a different form from that which in a little time they will present. Very round when still young, the leaves, or rather the small mass of little needle-like thorns, would be arranged according to a certain number of spires, of which we see several stopped a little higher, at the very time that the stem takes the form of a prism or of a fluted column, the projecting angles of which indicate the number of series of persistent leaves; and, if these series are reduced to two, the stem is quite flattened (Cactus phyllanthus).

§ 164. OPPOSITE LEAVES (*Feuilles opposées*).—Let us now examine the cases, in which each node bears several leaves. If there are only two growing at the same height, one on one side of the stem, the other on the other side, they are called *opposite* (*opposées*); if there is a larger number, they are *verticillate* or *whorled* (*verticillées*); and the group of leaves thus arranged in a circle around the stem is a *whorl* or *verticil* (*verticille*) (VERTICILLUS). In general, the leaves of the same whorl are separated from one another by equal intervals, and, consequently, the arc between two neighbouring leaves is equal to the circumference divided by the number of leaves of the whorl, a demi-circumference if there are two opposite leaves, a third if there are three, and so on. Another almost general law, is that in which the leaves of a whorl are not placed above those of the lower whorl, but in the interval, sometimes nearer to one side than to the other, sometimes exactly in the middle.

159 160

§ 165. In this latter case, it is clear that the whorls would be placed one above the other two and two; and if the leaves are simply opposite, the upper pair will cross the lower one at right angles. This arrangement is called decussation, and the leaves which presented it are *decussate* (FOLIA DECUSSATA) (*decussées* [*fig.* 159]). The whole of the leaves of the stem are then in four rectilineal series. If the whorls consist of three leaves (*fig.* 160), they

159. Decussate leaves of the PIMELEA DECUSSATA.
160. Leaves of the LYSIMACHIA VULGARIS, verticillate by threes. The whorls are placed above one another by twos.

will be in six; if of four, in eight. All these combinations are evidently included in the class of rectiserial leaves of M. Bravais.

Here, instead of one continuous spire we have series of circles placed above one another (*fig.* 161). Since a given leaf *f* has a certain relation to the two leaves *f' f'* which are immediately above it, since one of these, the right, for instance, has in its turn the same relation to its upper one on the right *f''*, and so on, it is evident, that on drawing a line through the series of these leaves, we shall have a spire revolving round the stem *a*. We may commence another from each of the other leaves of the whorl, and we shall have as many parallel spires, analogous to those which we have called secondary in the case of alternate leaves.

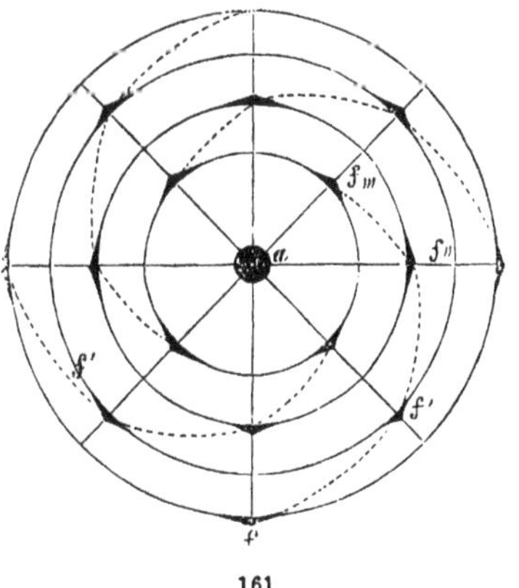

161

If it is allowable in this case to determine the divergence of two successive leaves by applying the rule by which we obtain it in the case of alternate leaves (§ 160), namely, according to a fraction which has for a numerator the number of parallel spires, and for denominator that of the rectilineal series, it is evident, that here the former, which is that of the leaves of a single whorl, is always the half of the latter, which is that of the leaves of two successive whorls; and, consequently, in every case in which the leaves are verticillate, whatever may be their number, the divergence of two successive ones would be $\frac{1}{2}$ or the half of the circumference. This numerical result seems to indicate, that every whorl is derived from the opposition of leaves by twos, and in that of more than two leaves, if we take one of them as the starting point, the second is not that which is nearest, but, on the contrary, that which is diametrically opposite. Indeed, it is not uncommon to find the leaves of a whorl separated, and then their binary association is clearly manifested

161. A plane projection of four whorls opposite by twos, and each composed of four leaves. The rectilineal series are marked by radii, the parallel spiral series by dotted lines. It is evident, that they would be the same from right to left, as we have represented them from left to right.

In one of these three leaves, for instance, the one, by being placed above or below the two others, indicates that it really belongs to an upper or lower circle.

§ 166. We have already stated the case, in which a whorl does not exactly cross that which is above or below it; several successive whorls must, therefore, be passed to find one situated directly above the first. In several Caryophyllaceæ, for instance, the pairs of the opposite leaves are placed upon one another by fives only, so that on taking one leaf as the starting point we shall count eight before finding one placed in the same vertical line as it is. When we study their relations carefully, we then see that this combination approaches rather to those of alternate leaves inserted on a continuous spire, and by shortening the axis so as to collect the leaves into a rosette, we shall obtain nearly the same proportion, as that of eight leaves having $\frac{3}{8}$ of divergence. In short, we shall frequently find two leaves of these Caryophyllaceæ, instead of being diametrically opposite to one another deviate a little from the same side of the stem, as if their divergence were really less than $\frac{1}{2}$ or $\frac{4}{8}$ of circumference.

§ 167. The transition of the opposition to the alternation of the leaves is by no means rare. We shall meet with it sometimes in the Myrtle (*Myrte*), Snap-dragon (*Muflier*), &c. Other oppositifoliate plants often shew this dissociation at the extremity of their young branches, when they grow very rapidly. It is possible, then, that there does not exist between these two classes of leaves so essential a difference as we should be led to believe. Nevertheless, the relative situation of the leaves is generally sufficiently constant in the greatest part of the species of plants to be noted as one of the characteristics which distinguish them. It is clear, that we ought to employ for this purpose only those modifications which are subject to very little variation; leaves, for instance, may be described as decussate, alternate, distichous, tristichous, &c., &c. We may sometimes go a little farther on, and the divergence $\frac{2}{5}$ of the leaves would still clearly characterize certain trees; but, we have said, that, when expressed by larger numbers, it often passes from one series to another in the same plant. We conceive what services the knowledge of these laws can render, when we shall have determined and studied them clearly in the greater part of plants. If we have only some branches with leaves, or even without leaves and only shewing by cicatrices the places in which they grew, if one wished to determine a fossil tree, we should be able to find some elements towards the explanation of a problem, which otherwise could not be solved.

§ 168. Monocotyledonous plants, the first leaves of which are necessarily alternate, preserve this method of arrangement much later. There are some few instances of opposite or verticillate leaves; but, even then, it is easily seen that they do not spring exactly at the same height.

In Dicotyledonous plants, the leaves often preserve the opposition which we have already observed in their cotyledons; they often also change this arrangement, and this takes place, either immediately after the first developed leaves of the gemmule, or else by degrees. All the plants of certain families present without exception opposite or alternate leaves, and sometimes even other secondary modifications. Thus, all the Labiates have decussate leaves; the greater part of the Tiliaceæ, distichous leaves.

We also find in Acotyledonous plants alternate and opposite leaves. Certain arborescent Ferns may be mentioned, as presenting the most regular verticils or whorls, perhaps, of the whole vegetable kingdom.

In short, the three great classes of plants offer the same combinations in the spiral arrangement of their leaves. There are some, doubtless, more rare in one than in another: thus, the angle of divergence $\frac{1}{3}$ is hardly ever met with in Dicotyledonous plants, whilst the leaves of a large number of Monocotyledonous plants are thus arranged.

§ 169. We have said, that the leaves are not always complete and may be reduced to one of their parts. As the limb attains the largest size and is usually considered as the leaf itself in common parlance, when it is not developed, the leaves have a very different appearance and we should be inclined not to give them the name of leaves. But their lateral position on the stem enables us to recognise them, and, when we have been able to perceive in the arrangement of these disguised organs the laws which preside over the relative arrangement of the leaves, we can no longer have any doubt as to their true nature. Thus, in the Asparagus, on observing the little scales (*fig.* 111 *f*) inserted on the stem and arranged in a spiral line, we shall not hesitate to pronounce them to be leaves reduced to the vaginal part. When they are thus represented by the sheath alone or by the petiole, or rather by a simple and short elongation of the fascicle which would have formed the median nerve, they most commonly take the form of little *appendages* compressed into thick scales, or extended into thin membranes, or contracted into threads. We shall in a little time see the form which they present when they come into the neighbourhood of the flowers.

BUDS.

§ 170. (*Bourgeons*).—The point, from which a leaf grows, has double importance in the life of a vegetable, since it is generally immediately under that from which the bud (GEMMA) springs (*fig.* 162, *b a*, *b a*, *b a*), in the angle contained by the stem and the leaf, and which is termed the *axil* (*aisselle*) (AXILLA) of the latter: whence we derive the epithet axillary (*axillaire*) (AXILLARIS). The bud is nothing else than the first stage in the developement of a branch, all the lateral parts of which, the leaves at the first stage of developement, are united on an exceedingly short axis. It has then been naturally compared to the embryo, from which it differs in these particulars, instead of being independent and of supporting and of nourishing itself by means of one or more fleshy leaves or cotyledons, it forms part of a vegetable already formed which supplies it with nourishment, and its first leaves, required to render it other services, by no means present the forms of cotyledons. Some botanists have consequently termed it a *fixed embryo* (*embryon fixe*).

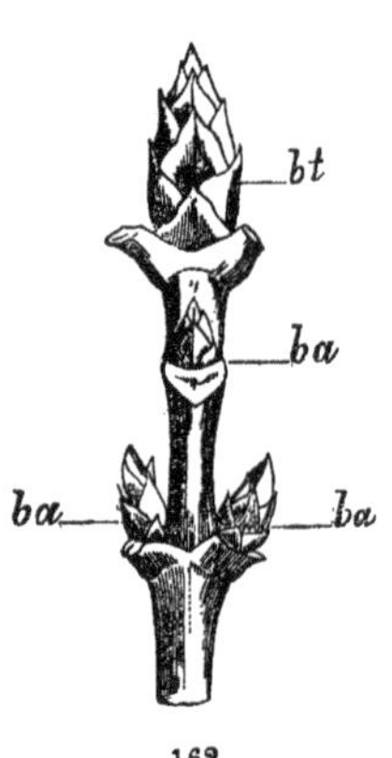

162

§ 171. It is, at first, a small mass or cellular nucleus in connection with the extremity of the medullary rays, and which, at first concealed in the inside pushes the bark before it and shews itself. In a little time the internal series of the cells of this little axis are organized into vessels, and its surface is covered with little cellular appendages, the first sketch, as it were, of the leaves, the developement of which will follow the laws which we have just laid down. We also know that the branch is but a reproduction of the developement and composition of the stem. The vessels and fibres of one are continued into the other; but there is not the same continuity of pith; the medullary sheath of the branch closes and ter-

162. The top of a branch of LONICERA NIGRA in the state of hybernation, i. e., after the fall of the leaves, and loaded with its buds; one terminal *b t*, and several axillary lateral ones *b a*, *b a*, *b a*.

minates at the point of its origin, in the same manner as that of the stem does at the commencement of the root.

§ 172. The bud, loaded with a generation of leaves which are to succeed to that from the axil of which it springs, naturally survives this leaf; and, when it falls or withers at the end of the year, the bud remains on the stem in a stationary condition until the spring, which, reanimating vegetation, will give it a fresh impulse and cause its developement into a branch. In warm climates, in which this period of repose is hardly appreciable, and where there is no danger to be apprehended for the young bud on account of the temperature, its first leaves are as complete and nearly the same as all the subsequent ones. But, in countries subject to the rigours of winter, during which such tender organs as the leaves could not resist the cold, the first leaves, the outermost, which, in the compact state in which they are like clusters, serve as envelopes to the rest, present remarkable modifications of form and of substance which render them capable of resisting the effects of the temperature and of protecting the inner parts from the inclemency of the weather. Their consistence is that which is commonly termed in botany *scaly* (*écailleuse*), i. e. hard and dry, like the envelope of a pip of a pear. They are often in addition to this impregnated with matter insoluble in water and a bad conductor of heat, such as resin, in several species of poplars; at other times covered with a thick down, as in several kinds of willows.

Sometimes these leaves or scales are large enough to envelope one another completely. More commonly, they are shorter than the whole of the bud, and then appear to be *imbricated* (*imbriquées*) in several rows, that is to say, the outer overlapping the inner like slates or tiles, (*figs.* 162, 163, 1). In this case although there are only a certain number and the bud is short, it is easy to recognise at the first glance the spiral arrangement analogous to that which we have pointed out in the cones of Pines. These buds are called *scaly*, when they are thus protected; *naked* (*nus*), when the outer leaves only present remarkable modifications as in the greater part of tropical trees. Some of ours, however, as the Buckthorn (*Bourgène*) (RHAMNUS FRANGULA), have naked buds; but this is a very rare case.

Several terms (TEGMENTA PERULÆ) have been proposed for these exterior modified leaves, which thus serve as protective organs. Linnæus ingeniously calls them HIBERNACULA, or *Winter-quarters*, De Candolle reserves for their whole the name of *bud*, giving to the

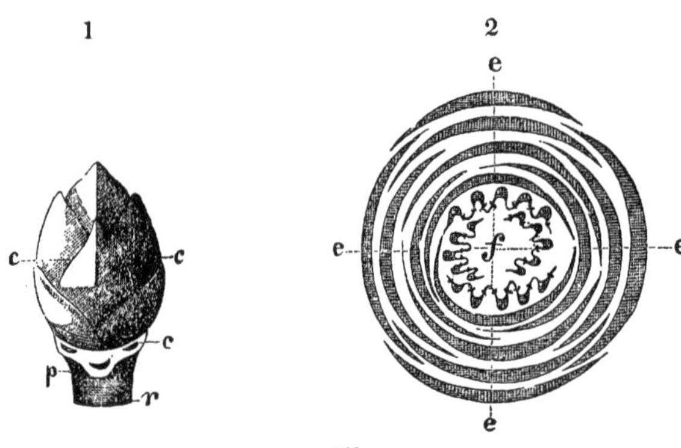

163

rest that of the *young shoot.* But to avoid the multiplicity of words, we will here call them *scales* (*écailles*) *:* warning the student at the same time that it is not the constant form, but only the most common.

§ 173. The leaf, in fulfilling this part of its duties, is reduced to one or other of its parts, and hence the different epithets by which these different organs are distinguished. The buds are called *foliaceous* (*foliacés*), if the scale is formed by the limb thus transformed alone; *petiolaceous* (*pétiolacés*), if it is formed by the lower enlargement of the petiole, which is termed the sheath; *stipulaceous* (*stipulacés*), if by its lateral expansions or stipules; *fulcraceous* (*fulcracés*), if by both the stipules and petiole at the same time. The determination of these parts is sometimes clearly indicated, and nature often confirms our theories by shewing us the gradual transitions from the innermost scales to the first true leaves: as in the PAVIA.

§ 174. The leaves properly so called, when their limb has acquired a certain size in the bud, are generally bent or rolled on one another, so as to adapt themselves to its round form and to occupy the least place possible. This stage has been called *præfoliation* (*préfoliation*), or more anciently *vernation* (VERNATIO), that is to say, the

163. 1. A scaly bud of the Sycamore-tree, (*Érable Sycomore*) (ACER PSEUDO-PLATANUS).—*r* Branch.—*p* Pulvinus bearing at its summit the cicatrix *c*, which remains after the fall of the leaf and in which we perceive the terminations of three fascicles.—*e* Imbricated scales of the bud.—2. Cross section of the same bud.—*e* Scales.—*f* Leaves.

spring stage. Each of these modifications is known by a particular name. If we consider the leaves at first independently of one another, we shall find that they may be: 1st, bent, which may happen in several ways, either in two halves, which admits of two modifications, the upper on the lower, the summit thus approaching the base; these are called *reclining leaves* (*feuilles réclinées*) (FOLIA RECLINATA), as the Tulip-tree (*Tulipier*) (*fig.* 164, 1); or the right half on the left, the ends and median nerve remaining unmoveable, *or conduplicate leaves* (*f. condupliquées*) (FOLIA CONDUPLICATA), as the Oak (*fig.* 164, 2); or folded up a certain number of times like a fan, *plaited leaves* (*f. plissées*) (FOLIA PLICATA), as the Sycamore, (*fig.* 163, *f*, 164, 3), and commonly along their principal nerves; 2nd, rolled, their axis either remaining straight, whether on themselves in

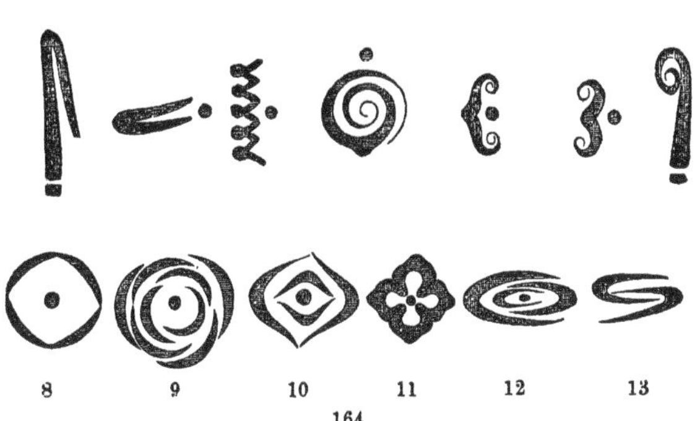

164

the form of a horn, *convolute leaves* (*f. convolutées*) (FOLIA CONVOLUTA), as the Apricot-tree (*Abricotier*) (*fig.* 164, 4); or on opposite sides, by their two edges, which are bent sometimes inwards, *involute leaves* (*f. involutées*) (F. INVOLUTA), as Violet (*Violette*) (*fig.* 164, 5); sometimes outwards, *revolute leaves* (*f. revolutées*) (FOLIA REVOLUTA), as the Rosemary (*Romarin*) (*fig.* 164, 6); or on their axis from top to bottom like a crook, *circinnate leaves* (*f. circinnées*) (FOLIA CIRCINNATA), as the Pill-wort (*Pilulaire*) (PILLULARIA PALUSTRIS) (*fig.* 164, 7). These modifications may sometimes be very much com-

164. 1—7. Leaves in the state of vernation, considered as isolated.—1 and 7. These are vertical sections.—2, 3, 4, 5, 6. Horizontal sections.—8—12. Several leaves in one bud; they are cut horizontally, so as to shew their relative position as well as their individual vernation. In all these and in the preceding figures, the median nerve is marked by a thicker piece of the section, and the axis which bears the leaves by a round dot at the side.

plicated, for instance, when a plaited limb is reclined on the petiole, or be compound, when the secondary nerves are bent relatively to the median as that is relatively to the axis which bears the leaf. This is frequently observed in deeply scolloped leaves (Ferns, for instance, the scollops of which are rolled in the form of a crook like the whole of the leaf), and especially in those which are truly compound.

If we now consider the leaves of the same bud relatively to one another, we see that;—1st, Plane or slightly convex, they touch at their edges without covering one another, *valvate vernation* (*vernation valvaire*) (FOLIA VALVATA) (*fig.* 164, 8), or only cover one another for a small part of their height, *imbricate vernation* (*v. imbriquée*) (FOLIA IMBRICATA), and then commonly at their edges also, following the spiral arrangement which they must in a little time assume *spiral vernation* (*v. spirale*) (VERNATIO SPIRALIS) (*fig.* 164, 9); 2nd, bent on themselves, they only touch at their opposite edges (*fig.* 164, 10) or at the neighbouring faces *induplicate vernation* (*v. indupliquée*) (VERNATIO INDUPLICATA) (*fig.* 164, 11); or else a conduplicate leaf completely embraces another and sits across it on a saddle, as it were, *equitant leaves* (*feuilles équitantes*) (FOLIA EQUITANTIA) (*fig.* 164, 12); or else it receives in its fold the half of another folded in the same manner, *demi-equitant leaves* (*f. demi-équitantes*) (FOLIA INVICEM EQUITANTIA SEU OBVOLUTA) (*fig.* 164, 13). In conclusion, we will just state, that these terms are by no means exclusively applied to the leaves in the bud; they serve to describe the analogous methods and relations of folding or rolling in all the plane parts of vegetables, in what organ or at whatever period we may find them. But it is principally in the young parts that we observe them; for instance, we frequently observe them in the flower-bud. We shall have need, therefore, to use them over and over again, and it is, therefore, useful to imprint them on the memory.

RAMIFICATION.

§ 175. We naturally come, now that we have completed our observations on the bud, to treat of the ramification of the vegetable, since it results from the developement of its buds which are lengthened into branches, each of which will, in its turn, be covered with fresh buds, which will develope fresh branches, and these last will form a third generation, to which will succeed a fourth, a fifth, &c. If we call the stem the primary axis, we may name those branches which immediately spring from it, secondary axes; those which grow from the secondary, tertiary; and so on. In practice the words *Branch* (Ramus) and *Twig* (Ramulus) are applied to these successive divisions; and, as they are often very numerous, these terms are modified, since their value is not at all fixed, by epithets or other means, so as to indicate approximately to what degree of division the branch of which we speak answers. Besides, it frequently happens that we give to these different terms a purely relative value, taking as our starting point, not the stem, but an axis at a certain distance from it. Thus, what we term ramus on a herbaceous plant would most commonly on a tree be called ramulus.

§ 176. It is clear, that if at the axil of each leaf a bud is developed into a branch, the relative situation of the branches will be no other than that of the leaves; it would permanently shew on a much larger scale these curvilineal and rectilineal series we have lately mentioned. In herbaceous plants, in which the number of leaves and of axes is necessarily much more limited, the greater part of the buds are very frequently developed. The arrangement of the leaves and the ramification reproduce one another and correspond pretty correctly, but it is by no means rare to find that a number of axillary buds are not developed at all. This takes place much more in ligneous vegetables, the prolonged life of which causes a more complicated ramification.

We have here the first cause which modifies the arrangement of the branches with regard to that of the leaves, namely, the suppression of a certain number of buds. A second cause is the addition of a certain number of other buds which are developed at certain points. Now let us examine successively these two causes and their effects.

§ 177. We have not hitherto spoken of a kind of bud, the existence of which is still more constant than that of the laterals placed at the axils of the leaves: it is the terminal bud, destined to continue the axis at the extremity of which it first grew (*fig.* 162, *bt*). The gemmule of the embryo was the first. When it has been developed as far as it was able, when the stem with the leaves, arrived at this the first stage of its life, stops in its growth, on its summit is formed a bud which is, as it were, the crown. After a certain period of torpidity, which in our climate corresponds to the winter, this bud begins to develope itself, then stops again in its turn, and prepares another bud for the following year. The stem is really composed, therefore, of a certain number of branches placed end to end: consequently, in our Dicotyledonous trees, we ought to find the number of ligneous layers gradually diminishing in proportion as they reach the top; and, if we could distinguish on the outside the shoot of one year from that of the preceding, we should have, so long as the tree continues to grow in length, a means of discovering the age of the tree on the outside.

There is rather a large number of vegetables, in which this terminal bud is the only one developed; and then there is no lateral ramification: the stem is simple. It is a very rare case among Dicotyledonous plants, which, nevertheless, sometimes shew it, such as the Cycadaceæ and the Papayaceæ, the trunk of which shoots up in the manner of a column crowned with a tuft of leaves; but it is very common among Monocotyledonous plants (*fig.* 114, 1), and we have seen that those which become trees habitually take this form; therefore, to obtain their age, it has been proposed to adopt the means which we have just explained. But, even if, towards the top of the tree, we find annular traces which indicate the successive shoots, these have been long ago effaced towards the bottom in old trees. And, besides, we do not yet know precisely, whether, in climates so different from ours and exempt from winter, the formation of each of these rings corresponds to a year or any other regular interval of time.

§ 178. Let us now take the case in which the greater part of the axillary buds are developed, but not all. The failure of the rest may be irregular and totally dependant on local or individual causes. Thus, on the side on which the plant is crowded, deprived of light, planted in a bad soil, or subject to any other unfavourable condition, its buds fall off, or shoot out weakly and badly, or soon perish. It is useless to consider the purely accidental circumstances, which, acting in all directions, impress so many apparent differences on vegetables of the same species. But the buds often fail with a wonderful degree of regularity. Thus, in Pine-trees, the leaves, very numerous and thick, are arranged in a spiral line, and yet the branches are arranged in rings or whorls at rather long distances. The reason of this is, that in the spire of the leaves, there are alternately long series without buds, then several buds at the axils of successive leaves, and then the revolutions of the spire which bears them are too near for us to appreciate the difference between the heights at which they are developed: so that these branches appear to the eye to grow in a circle. In the case in which the leaves are opposite, frequently one only of the two developes a bud at its axil; then, in the following pair, the leaf of the other side will develope one in its turn; we have, thus, from two rows of opposite leaves alternate distichous branches, as in the TRIBULUS; from the pairs which cross obliquely, branches arranged in a spire, as in several Caryophyllaceæ. It will suffice for us to state here, without examining in detail the diversity of the combinations which may be offered to us in nature, that sometimes, in a series of organs, several, situated with regard to the others in constant relation, have a tendency to fail in their developement, and that this law, which we shall find in every part of the vegetable, exercises a very great influence on the ramification.

§ 179. Let us now suppose, that the terminal bud fails whilst the lateral ones are developed: the stem will be short or almost nothing; the vegetable will grow at the sides, either all round, or else in some few directions, if any of those constant relations, such as we have just alluded to, occasion abortions.

Certain modifications, of which several are commonly ascribed to the stem, but all of which really depend on a particular method of ramification, may be placed here. In the cases of which we are treating, the stem, the produce of the embryo by its germination, ceases after a certain time to grow: and since it is not elongated by the production of a terminal bud, a lateral branch, springing generally near the base, performs its functions and takes the lead in the subsequent production of fresh plants. Besides, the stem does not always commence at the surface of the soil, it is often more or less deeply buried below it; and, thus, this branch which will replace it, may originate *in* the ground as well as above it.

§ 180. The plants commonly known under the name of *perennial* (PERENNES) belong to this case. The first year appeared a stem which underwent all the appearances of the annual plant, and which, like it, withered and died; but it was only the portion above the soil: below the surface, its root and the base of the stem charged with one or more buds still continue to live. When thus buried, they will brave the winter, and reviving the following spring, will be developed into as many stems, which, in their turn, will pass through the same circumstances of life. These buds frequently have a peculiar form; their axis thick and fleshy, is much lengthened before it produces leaves: to this has been given the name of TURIO (*Turion*). We find good examples of this in the Peony (*Pivoine*), or, to mention one more familiar to the majority of our readers, the Asparagus in the state in which it is eaten.

§ 181. Instead of remaining stationary until the following year, and coming to the light, they sometimes are lengthened under the ground forming subterranean branches. We have seen in another part (§ 113) that the stems, in this condition, commonly produce adventitious roots. This is what happens to our branches; and thus creeping obliquely or horizontally beneath the soil, loaded with elongations and radicular fibrils, they will take every appearance of a root. They are then called RHIZOMA (*Rhizome*). Sometimes the subterranean branch continues its course, throwing out from its upper face or from its sides buds, which shooting vertically, grow up and expand; sometimes it itself turns up and throws out a bud from its extremity; but this takes place, commonly after that a branch like to it, and springing from it, has usurped its subterranean place and course. The same plant may therefore cover a large space of ground, and grow to some distance from the place where it began to germinate. A series of cicatrices remaining on the upper face of the rhizoma often points out the place of the successive shoots: as (*fig.* 166, *cc*), in the Solomon's Seal (*Sceau-de-Salomon*).

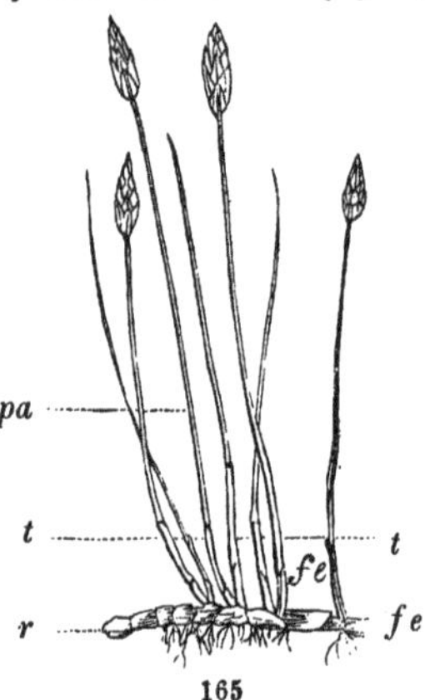

165

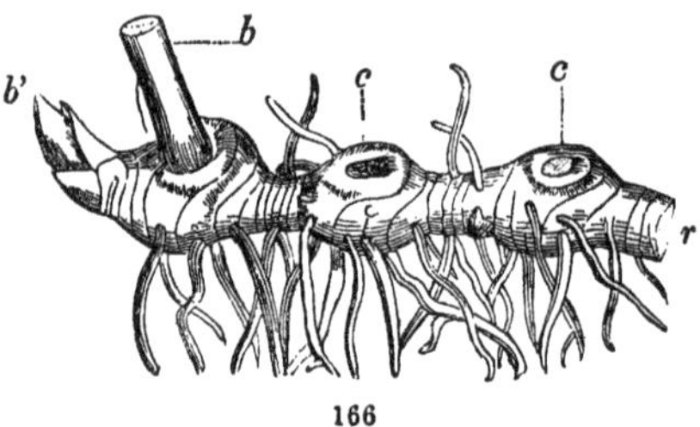

166

165. A piece of the rhizoma *r* of the SCIRPUS PALUSTRIS (much smaller than nature).—*fe fe* Leaves situated on the rhizoma, in the state of scales —*p a* Aërial part of the plant, its folioliferous or floriferous branches which rise above the base.—*t* Surface of the earth above the rhizoma.

166 A portion of the rhizoma *r* of the Solomon's Seal (*Sceau-de-Salomun*) (CONVALLARIA POLYGONATUM).—*b* A bud already developed into a branch at the extremity of the rhizoma.—*b'* A bud which is still to be developed.—*c c* Cicatrices marking the insertion of older branches which have withered and fallen off.

§ 182. The bulb, which was formerly classed wrongly amongst roots, is another modification of perennial plants, peculiar to Monocotyledonous vegetables. This stem, in its buried portion, produces laterally a thick bud, fleshy at its centre and covered with a greater or less number of leaves. Of these leaves, the outer ones which are necessarily inserted lower down, are reduced at their sheath to the state of scales, and represent what we have called by this name in aerial buds. Sometimes each of these thin sheaths completely envelopes the base of the stem (*fig.* 167, *e*), as may be seen in the Hyacinth (*Jacinthe*), the Crocus (*Safran*), the Onion (*Oignon*), wherefore, *onion-like plants* (*plantes-à-oignons*) has been

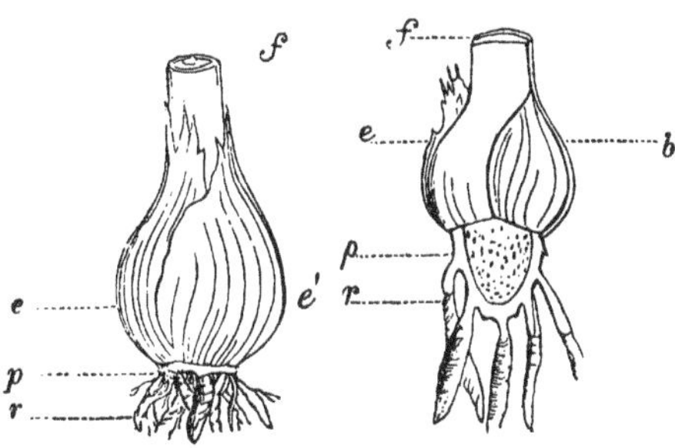

167

vulgarly applied to all those which present this characteristic: botanists call them *coated*, *tunicate* (*tuniqués*), bulbs. At other times, instead of these membranous tunicles covering the whole of the bud, we find several much smaller imbricate appendages over the whole bulb (*fig.* 168 *e*), which is therefore termed *scaly* (*écailleux*), because these appendages bear a great resemblance to scales, from which they differ only in having a fleshy substance; the White Lily (*Lis Blanc*) is a good example of this (*fig.* 168). At other times, we only find a little mass of tunicles; and, as then the mass of the bulb is almost entirely formed by its axis which is very much swelled, we give it the epithet of *solid* (*solide*) (*fig.* 169).[q] At the axil of these leaves thus metamorphosed, we observe much

167. A tunicate bulb of Garlic (*poireau*) (ALLIUM PORRUM), seen in its whole state 1, and in its vertical section 2.—*r* Roots.—*p* Plate between the roots and the bulbous swelling.—*e* Scales or lower modified leaves.—*f* Upper developed leaves which are cut near the base.—*b* Bud situated at the axil of a scale, which, when developed, forms a fresh bulb.

q This is more generally called CORMUS, a *solid bulb* being an anomaly.—TRANS.

smaller secondary buds, which are termed *young bulbs* or *cloves* (*cayeaux*), the number of which seems to be in connection with that of the leaves. Some of these buds may be developed on the bulb itself, and, in some plants, during several years: others may in their turn become bulbs, and since they adhere very feebly to the mother plant, which generally withers and dies, they are in general detached at a certain time, and all the plants thus formed, although apparently united at the beginning to the same stem, will be in a time so many different plants.

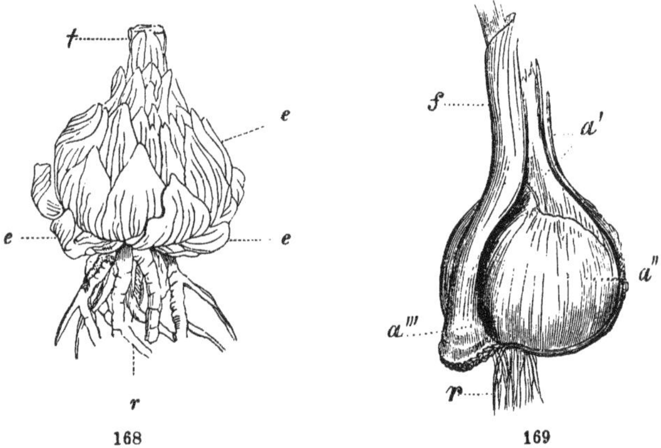

168 169

On one of the sides of the cormus, a single bud (*fig.* 169 *a''*), is often developed, which in its turn takes the form of that from which it emanates, and also in due time itself produces a lateral bud (*a'''*), most commonly situated on the opposite side, according to the law of the alternation of leaves and buds. In this manner a plant is produced every year; and if the second one springs from the right of the first, the third will spring from the left of the second, the fourth from the right of the third; so that we shall always find the plant at the same place, only changing alternately every year from the right to the left and from the left to the right. This is

168. A scaly bulb of the White Lily.—*r* Roots.—*e e* Scales.—*t* The stem is cut off here.

169. A cormus of the Meadow-Saffron (*Colchique*) (COLCHICUM AUTUMNALE). Roots.—*f* Leaf.—*a'* The withered primary axis belonging to the preceding year.—*a''* The secondary axis or bulbiform stem of the present year.—*a'''* The point where that of the following year will be developed.

very clearly seen in the Meadow-Saffron (*Colchique*) (COLCHICUM AUTUMNALE).

In each bulb, below the tunics is a short piece in the shape of an orbicular plate (*fig.* 167 *p*), on the lower surface of which are formed several fasciculated roots. Between the leaves and the roots, it was considered as the stem; it is rather the lower part of a branch, since it is nothing else than the base of a lateral bud; only this bud is detached; it has become, as it were, a natural branch of the mother plant.

§ 183. The lateral branches, assuming the office of the stem, whose vertical developement is stopped, and thus extending the mother-plant in a horizontal direction, may be seen also above the surface of the soil. They commonly give them the name of *creeping stem* (*tige rampante*). This false stem, for the most part slender and flexible, passes over a certain space without producing any leaves, or only a very small number of leaves at certain distances, or usually even

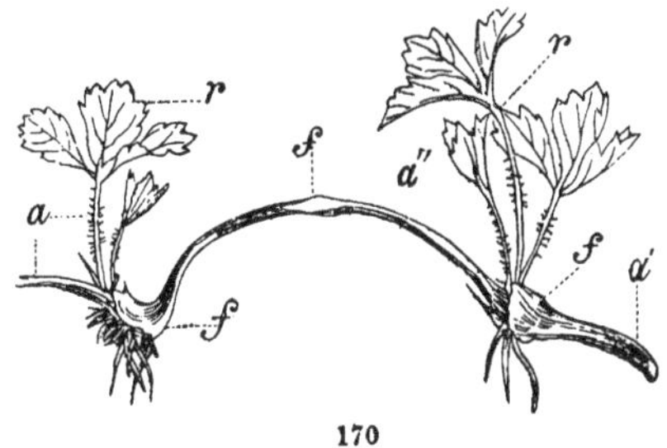

170

one at most (*fig.* 170 *a'' f*), the bud of which may develope itself, but is commonly abortive, the stem is then terminated by a kind of rosette naturally turned upwards (*fig.* 170 *r*); below this rosette roots are then produced which are buried in the ground, and at the axils of its lower leaves spring fresh branches (*a''*) which are terminated in their turn by a fresh rosette. Every one will recognise in this, the description of the formation of the *Strawberry plant* (*Fraisier commun*) (*fig.* 170), of the *Creeping Crowfoot* (*Renoncule rampante*).

These lateral shoots, which are commonly called *runners* (*coulants*)

170. A portion of the Strawberry plant (*Fraisier*).—*a'* The first axis which has produced a rosette of leaves *r*, the upper *r* green, the lower *f* rudimentary. From the axil of one of the latter a second axis or runner *a''* grows, bearing about its middle *f* a rudimentary leaf, and at its extremity a rosette *r*, from which arises a third axis *a'''* similar to the first.

(FLAGELLUM), most commonly wither away and are disarticulated; and the rooted tufts which they connect become so many distinct plants. Gardeners imitate this operation, when they *lay down* (*marcottent*) a plant, i. e. put a branch into the earth, which, buried up to a certain point, throws out leaves and roots, and thus developes a plant which soon begins to vegetate of itself, and may at last be detached from the mother plant.

In succulent plants, in which the leaves suffice for the nourishment of the plant for some time, there is no need of waiting until the rosette formed on the runner or layer shall have produced roots, in order to separate and plant it. This modification is called *cutting*, *slip* or *piping* (*propagule*) (PROPAGULUM).

§ 184. To all the preceding cases, in which we observe so great a tendency in the buds and their products to become at last independent of the mother plant and of one another, we must add the modification of the aërial bud, known under the name of *bulbel* (*bulbille*) (BULBILLUS), the diminutive of bulb, with which, indeed, it is most closely connected. It then takes that fleshy consistence peculiar to every organ or every group of organs which has to live for some time at the expense of its own substance. It has very few scales, which are thick, and, sometimes united together partly or wholly, form a little compact mass. It adheres very feebly to the axil of the leaf, is at length detached, and may even remain in that state for some length of time, and, at last, when planted, reproduces the plant which gave it birth. This is a true link between the bud and the embryo. The Orange Lily and Tooth-wort (*Dentaire bulbifère*) (DENTARIA BULBIFERUM) are very good examples of this method of reproduction.

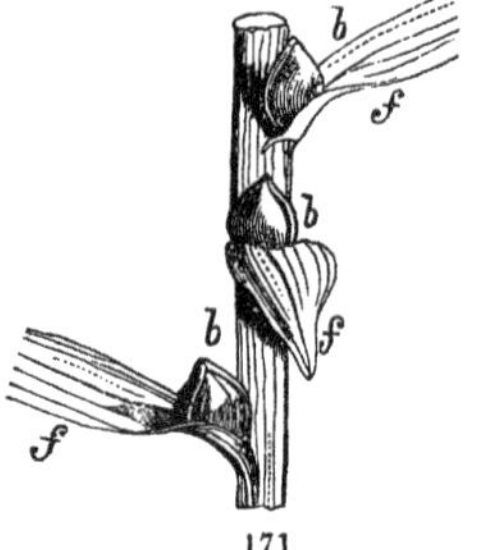

§ 185. In all the preceding cases, the branch, which represents the stem, preserves its lateral position with regard to it. But it may happen, that the branch, stronger than it, pushes it on one side, straightens itself, and takes its place. The relative position of the parts then determines their true nature. When, for instance, in the Vine (*Vigne*) (*fig.* 172) we see the stem produce here and there on

171. A piece of the stem of the Orange Lily (*Lis bulbifère*) (LILIUM BULBIFERUM), with three leaves *f*, and three axillary bulbels *b*.

one side a leaf without an axillary bud[r], on the other, without leaf, a little herbaceous branch, as well as branches commonly known under the name of *tendrils* (*vrille*), it is but reasonable to suppose, that the continuation of the stem situated between the leaf and the tendril, consequently at the axil of the former, is nothing else than the produce of the axillary bud, which in its vigorous developement, has pushed to the other side the extremity of the exhausted stem which becomes abortive under the form of a tendril. This arrange-

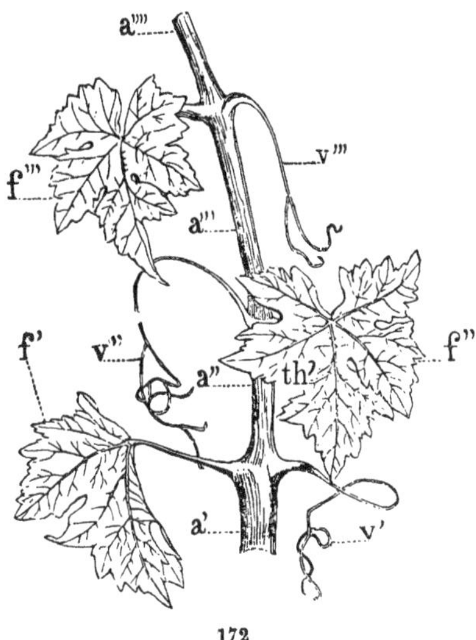

172

ment is more evident still in the RICINUS or the PHYTOLACCA, because the axis, of a fresh kind, whose purpose is to continue the stem, does not do so in the vertical line, but is thrown a little to one side, so that we cannot doubt but that it is the axillary branch. In these two plants, as in several others, we find a bunch of flowers in the place which is occupied by the tendril, and, besides, this

172. A portion of a branch of the Vine.—*a'* The first axis, terminated by a tendril *v'* thrown on one side, bears a leaf *f'*. From the axil of this arises a branch *a''* which seems to continue the axis *a'*, terminated in the same way by a tendril *v''*, and bearing a leaf *f''*.—*a'''* A branch growing from the axil *f''*, terminated by *v'''* and bearing *f'''* from which springs *a''''*.

[r] There frequently exists within this leaf a simple or double bud, but a little to one side, and not in the very middle of the axil.

tendril is only an abortive bunch of flowers, which appear in its place when the vine is in a flourishing condition.

§ 186. We are now going to examine in what way the ramification is modified by the irregular abortion of a certain number of terminal and axillary buds. This may also take place on account of the displacement of the latter, when, instead of being developed at the axil of the leaves, they appear to originate at a certain distance from it. The bud and the branch are then called *extra-axillary* (*extra-axillaires*). This may arise from several causes, from the complete abortion of certain leaves, or from the adherence of the stem either to their lower part, their petiole, for instance, or to the base of the axillary branch; so that the bud in the first case appears to be lower than the leaf, and higher in the second.

By these reasons and modifications we explain certain anomalies in the arrangement of the leaves of some plants, as in many Solanaceæ. We cannot enter more fully on these exceptions here, which, moreover, we shall meet with again on treating of the flowers.

§ 187. If the ramification varies on account of abortion, it will very often vary from quite an opposite cause, the multiplicity of buds. Thus, we sometimes find *accessory* (*accessoires*) buds, besides that which commonly exists alone at the axil of the leaf, from which we then see several branches arise. Sometimes they are placed one immediately above the other, commonly two in number, more in some

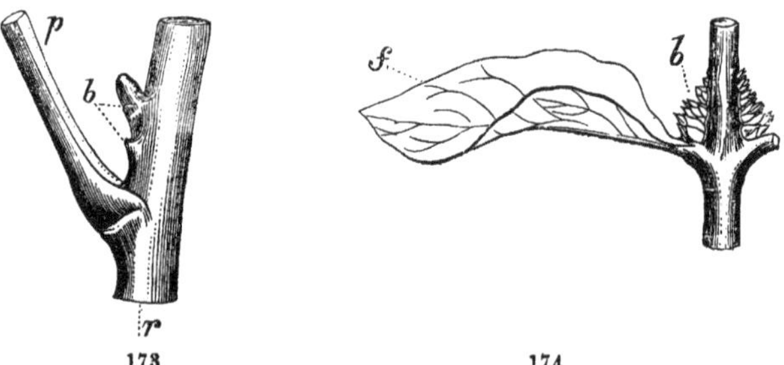

173. A piece of a branch *r* of the Walnut-tree (*Noyer*), bearing the petiole *p* of a leaf of which the rest has been cut off. At its axil are several buds placed above one another, so much the more developed as they are higher up the stem.

174. A piece of the branch *r* of the LONICERA TATARICA bearing two opposite leaves, one of which has been cut off, the other *f* preserved. At the axil is a series of buds placed above one another, which are more developed in proportion as they are situated nearer the axil.

vegetables: as, in the Honeysuckles and the Walnuts, where may be observed a series of three, four or five buds, which, thus placed above one another, decrease in the former from the bottom upwards (*fig.* 174), in the latter from the top downwards (*fig.* 173); sometimes they are placed in the same horizontal line, i. e. side by side, and then it appears, that besides the bud rising from the base of the petiole, one grows from each stipule. We may see this in the Willows and Poplars, where it is the origin of those little branches, which are continually pushed out two by two on the recently cut logs of these trees.

§ 188. The multiplication of the buds is much more frequently owing to those which we have already mentioned, and which we called *adventitious* or *latent*.

All the cellular parts adjacent to the surface of the stem appear to be disposed, when any cause excites vitality and accumulates materials there by the greater afflux of the juices, to be organized into buds. But, although more commonly on the stem, these productions may be found on other parts, namely, on roots exposed to the air; on leaves more or less fleshy, either on the edges, as in the Bryophyllum calycinum, the Malaxis paludosa (*fig.* 175), &c., or on the surface itself as in the Ornithogalum thyrsoideum (*fig* 176).

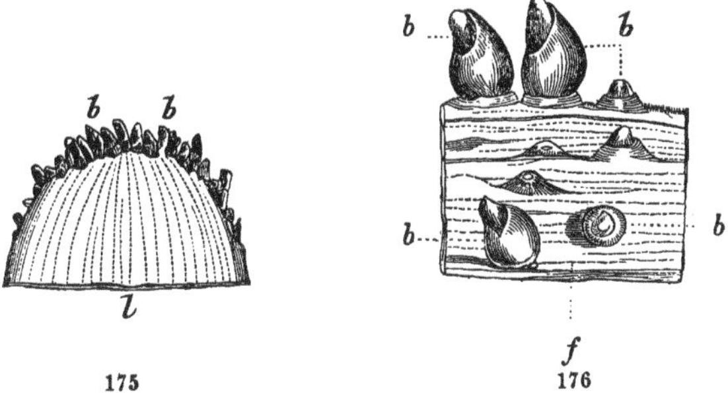

We may cause the formation of adventitious buds by artificial means, such as the use of ligatures, wounds, &c., which cause the afflux of the liquids and the swelling of the part which has been manipulated.

175. The end of a leaf *l* of the Malaxis paludosa, the edge of which is covered with bulbels *b b*.

176. A piece of the limb of the leaf *f* of the Ornithogalum thyrsoideum, on the surface of which are found adventitious buds or bulbels *b b b* at various stages of developement.

Their form naturally differs from that of normal buds. They have neither their dimensions nor the system of outer leaves modified into protective scales, since they are not prepared a year in advance, and fortified for passing the winter. Called to life in the midst of a vegetation in full activity, they are immediately and incessantly developed; they generally appear at first under the form of small excrescences, which, after being more or less elongated, are covered with leaves. The form of those which are sometimes seen on the leaves themselves may be rather compared to that of the bulbels (*fig.* 176).

§ 189. Some of these, instead of being developed on the outside, as is most common, grow enclosed in the substance of the cortical parenchyma. M. Dutrochet has fully investigated the formation of these singular bodies, which are particularly observable in the bark of certain Dicotyledonous trees (the Beech [*Hêtre*], the Horn-bean [*Charme*], the Cedar of Lebanon [*Cèdre du Liban*]), under the form of irregular spheroids of a ligneous consistence, and which at this first stage may be called *nodules*. Their centre is occupied by the pith surrounded on all sides by a ligneous layer traversed by medullary rays, and on which is moulded exactly every year a fresh layer (*fig.* 177). Of these as many as twenty-five have been counted. This ligneous mass, extending on all sides, often touches the wood of the tree, with which it previously communicated only by a thread-like prolongation, is united with it, and thus constitutes what is commonly known under the name of *knob* (*loupe*), which may then be complicated by the double growth of the stem and of the nodule. If, instead of one nodule, there are several, distinct at the summit, but meeting as they approach the surface and forming a knotty mass, we then call it an *excrescence* (*broussin*). The nodule may be compared to a branch which has to live without leaves or buds, not able to grow in length, but increasing in thickness, doubtless at the expense of the already elaborated nourishment furnished by the bark in which it was inserted, and, perhaps, also by its feeble communication with the ligneous body; shewing, in the single internode to which it may be said to be limited, the accumulation of annual

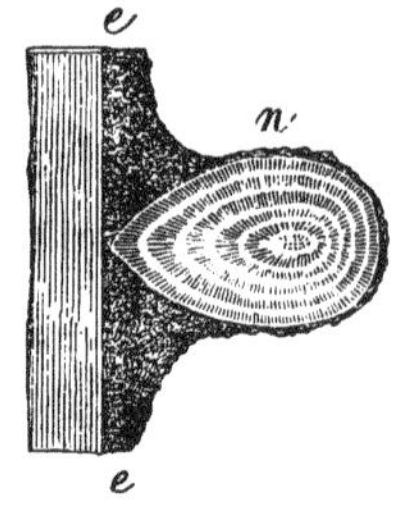

177

177. A vertical section of a nodule *n* of the Cedar enclosed in the bark *e*, from beneath the surface of which it projects.

layers, which would be found in the lower internode of every contemporaneous branch; being developed in every direction, and substituting the spheroidal for the cylindrical form in the arrangement of the layers and the direction of the rays, since it has no determinate base or summit. Sometimes, however, from the outside, springs a branch, which is seldom much developed, and is soon arrested in its progress, but suffices to demonstrate the analogy of the nodule to the adventitious buds.

§ 190. Although the parts dependent on the stem, the branches, often placed under ground and there covered with adventitious roots, may be easily taken for roots, a mistake which for a long time obtained; although, reciprocally, the roots, sometimes exposed above the earth, and then frequently covered, by means of adventitious buds, with foliage, appear to form part of the aërial system of the stem; we shall now know, how to determine in both cases, by means of all the external characteristics we have previously laid down, which is the root, and which is the stem and its dependencies. The latter is always characterized by buds produced at the axils of the leaves which are regularly arranged. These leaves, it is true, in the branches vegetating under the earth, are very much modified in their size, their shape, their consistence, their colour: in a word in their whole appearance; they are most commonly short, brownish scales or membranes; but, even when they are very small, the regular arrangement of the buds and their nature shew that it is not a true root; the absence of these that it is one. When a plant here and there throws up above the surface of the ground fresh stocks, we may know in this manner whether they come from secondary roots, running horizontally and throwing up adventitious buds, or from buried branches following a similar course and budding regularly in their progress.

The problem is sometimes complicated by the changes in the form and nature of the subterranean branches under the influence of the medium in which it vegetates: it is shortened, thickened and becomes fleshy on account of the extreme multiplicity of the feculiferous cells which constitute nearly the whole of its substance. Yet even then by means of the same rules the solution is easy. Let us mention an example familiar to all of our readers, the Potatoe (*pomme de terre*) (*fig.* 178, T). Its surface is covered with little eminences which we call eyes (*b*), at first concealed in the axils of small scales, which soon fall, and which are arranged with a certain degree of regularity, most commonly in a spire. These eyes are developed into branches if the tubercle is placed in favourable conditions o warmth and moisture, and then become green when they reach the light. We are thus compelled to pronounce that these are normal buds and that the potatoe is a true branch. This conclusion, which appears singular at the first sight, is, however, easily confirmed by the daily experience of gardeners, who, on earthing up the plant, i. e. burying its lower part, multiply the number of the tubercles by the conversion of the buried buds into potatoes. In rainy and gloomy seasons we see this change take place spontaneously and gradually in the open air, the axillary branches thicken and grow round as they shorten, and we may thus obtain all the transitions between the branch and the tubercle. At the first sight, the tubercles of the DAHLIA seem quite analogous to those of the potatoe: but they have neither scales nor eyes; they are the swellings of true roots.

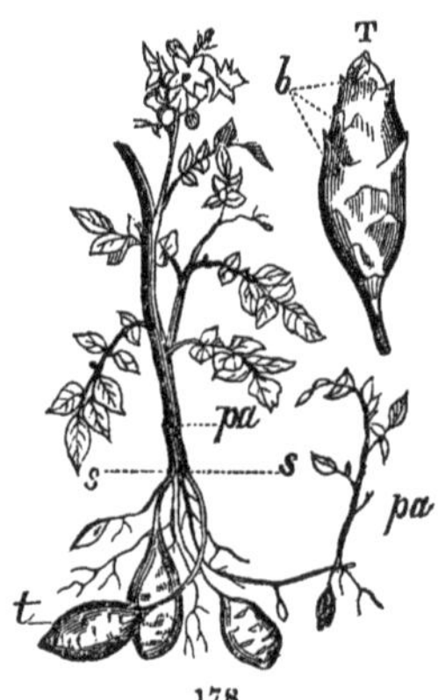

178

§ 191. We have examined the different methods of ramification; we have seen that the relative position of the branches would correspond to that of the leaves, if as many axillary buds were developed as there are leaves, but it is most commonly modified, either by the abortion of a certain number of these or of the terminal, or, on the contrary, by their being displaced and multiplied, to which changes adventitious buds contribute in no small degree. We have thus a large number of possible combinations, which would impart great variety to the shape and form of the plants. It is clear, that the subterranean part of the ramification only exerts an indirect influence on the external appearance; that, from this cause, we have a greater or less number of plants similar to and connected with the original, but which appear to be distinct and often actually become so; that the developement of these subterranean buds has exactly the same effect as the germination of the same number of seeds emanating from the same plant.

§ 192. Let us now lay this class on one side and direct our atten-

178. The lower part of a plant of the Potatoe (*Pomme-de-terre*).—*p a* Its aerial part or stems covered with leaves.—*t* Its subterranean part or tubercle.—One has been represented a little larger, T, in which we see the eyes or buds *b* still concealed by regularly arranged leaves in the state of scales.—*s s* Surface of the earth.

tion to those cases, in which the branches of the same plant are externally and visibly attached to one another. The stem may sustain itself in its vertical direction. When it has attained considerable dimensions we distinguish in it, the *trunk* (*tronc*), the lower part, which is bare of foliage : the *head* or *cyme* (*cime*) (CYMA), the upper part, which is covered with leaves. This nudity of the trunk is perfect and results from the failure of all the axillary buds, as we have remarked, for instance, in Palm-trees, for the stem of which some botanists have proposed the name of STIPES (*stipe*) ; or else it is only partial, resulting from the imperfect developement of the lower buds, and more commonly by the withering and falling of the branches which they have produced. Let us remark here, that the majority of our trees owe this appearance to the labour of man, who *lops* off these lower branches at an early stage of their growth: again, it is, on the other hand, the upper branches, those of the cyme, which are lopped, so that the natural appearance of the tree is quite changed. Elms, Willows, and Poplars may be cited as examples of the changes caused by these mutilations into *pollards* ; and we cannot then recognise them such as nature made them. It is well understood here, that on speaking of the external appearance of vegetables, we always refer to their natural state without the intervention of the pruning-knife and hatchet.

§ 193. A vegetable sometimes appears to have several stems, because its lowest branches, springing at the surface or only just below the soil, have been equally developed with the primary axis from which they spring, and have grown in nearly the same direction: it is then called MULTICAULIS (*multicaule*). Art often turns these lateral branches growing close to the earth (and, consequently, furnished with adventitive roots) to profit by taking them off directly they appear, and planting them separately to form so many distinct individuals: these are called *suckers* (*surgeons*, *drageons*) (SURCULI). Each of these stems, being in its turn capable of reproducing itself, without being detached from the common stock to which they belong, we may conceive to what a distance the circle may be extended. We are led to believe that this is the origin of some famous trunks, such as that of the Chesnut-tree of Ætna, called the Hundred Horses, because the cavity in its trunk would hold as many. Our readers may have observed, that it is generally the practice to cut down old Chesnut-trees at the surface of the soil, and that soon the base of the trunk, left in the ground with all its roots, throws out all around it vigorous shoots, which sometimes interlace and engraft with one another, and several of which, becoming in their turn great trees, will, perhaps, form, when united after a series of ages, one trunk analogous to the Chesnut-tree of the Hundred Horses. The form of this, with the great holes in the trunk, which form an entrance into the great area which occupies its centre, seems to indicate this method of formation. Similar forms in several of the largest trees known, would conduct us to the same conclusions; but, assuming this supposition, we shall have a cyme commencing in the earth rather than a real trunk.

§ 194. The size and the height of the trunk ought, in general, according to the theory, to be in constant relation with the age, and may serve to calculate it in all trees, the increase of whose parts is known during a certain given time. We know a certain number of trees whose origin goes back for several centuries or even remains hidden in the darkness of tradition. The greater part are Dicotyledonous; amongst these trees, Lime-trees, Sycamores, Chesnut-trees (*Châtaigniers*), Yews (*Ifs*); in the East, Plane-trees (*Platanes*), Figs (*Figuiers*) and Cedars (*Cèdres*); under the Tropics, Baobabs and several other species belonging to the family of the Bombaceæ: but amongst these also figure Monocotyledonous plants: the Dragon-tree (*Dragonier*) of Orotava, for instance, in the Canary Isles. Their circumference necessarily varies according to the individual and species; some have been mentioned, in which it exceeds thirty metres (French) or 32·775 yards (English); a great number are as large as the half or third of this; but these giants, since they are exceptions, ought to detain us no longer here.

§ 195. Amongst ligneous vegetables of ordinary size, we have, according to the limits at which the growth is stopped, formed several classes, which are denoted by names derived from their particular appearance: thus, we call that a *tree* (*arbre*) (ARBOR), which is several times the size of man, sometimes reserving the diminutive, ARBUSCULA, for that which is not more than five times the size: *shrub* (*arbuste* or *arbrisseau*) (FRUTEX), that which does not exceed it five times and branch out from the bottom, for less ones using the diminutive, FRUTICULUS: an *under-shrub* (*sous-arbrisseau*) (SUFFRUTEX), that which does not exceed the length of a man's arm. If the under-shrub is low and branches very much at the base, it is a *bush* (*buisson*) (DUMUS, DUMETUM). The adjectives, *arborescent* (ARBORESCENS), *frutescent* (FRUTICOSUS, FRUTICULOSUS), *suffrutescent* (*sous-frutescent*) (SUFFRUTICOSUS), *bushy* (*buissonant*) (DUMETOSUS), are derived from these different substantives and do not require definition.

§ 196. The stem, at other times, cannot sustain itself, and must, consequently, rest on some other body; if it is on the ground, it is said to be *trailing* or *procumbent* (*couchée*) (PROCUMBENS); if it is on an erect body, *climbing* (*grimpante*) (SCANDENS). In climbing, it sometimes preserves very nearly its rectilineal direction, as the Ivy (*Lierre*), which from its surface in contact emits small radiciform elongations by which it is fixed to whatever supports it; sometimes, it is twisted round its support and takes the name of *twining* (*volubile*) (VOLUBILIS), often describing spires, which are constant and turn from left to right (DEXTRORSÙM), as in the Hop (*Houblon*); or from right to left (SINISTRORSÙM), as in the Wild Convolvulus (*Liseron des haies*); or first in one direction, and then in another; irregularly and at intervals. In our cool, temperate climate, the greater part of the climbing stems are herbaceous, although some are ligneous, and may even acquire rather large dimensions, as the Honeysuckle (*Chèvrefeuille*), the Clematis, and especially the Vine; we then give to their branches the name of SARMENTUM (*sarment*). These are analogous to the LIANAS, so abundant in the tropics, of which we have already had occasion to speak (§ 85). These creepers, sometimes twined in spires round the highest trunks, sometimes falling in a straight line the whole of this height towards the ground, or from one branch to another, run from tree to tree, bind them to one another, and sometimes kill them: In this irregular progress, which baffles all attempt at description, they run through very long spaces without producing leaves, without branching; and travellers have frequently not perceived the foliage and flowers connected with these singular stems which surround them on all sides.

§ 197. The ramification exercises other influences on the general appearance of vegetables, than that which has hitherto occupied our attention, that of the developement of a certain number of buds situated in a certain manner. There is no need of stating that the external appearance of the plant is modified by the variety in the direction, the substance and the relative length of the branches and twigs. The branches spring from the stem, and the twigs from the branches, sometimes at a very acute angle, sometimes at a right angle, but most commonly at one somewhat less than a right angle: they are nearly *erect* in the first case (*dressées*) (RECTI), *spreading* or *patent* (*étalées*) (PATENTES) if they rise at an obtuse angle, and would, of course, form two perfectly different cymes, as those of a Cypress (*Cyprès*) and a Poplar compared with those of a Cedar or an Oak. In some trees, which are termed *weeping* or *pendulous* (*pleureurs*) (RAMI PENDULI), the branches assume a position quite the reverse of that which is most general, bending towards the earth, whether, that long and weak, they fall from their own weight as in the Weeping Willow (*Saule pleureur*); or preserving a great degree of stiffness, bend back from their origin (RAMI RETROVERSI), as in the Weeping Ash (*Frêne*) and Weeping Sophora. The patent branches, sometimes arising from the surface of the soil, without the stem taking a vertical developement, thus creep along the ground which they cover like a sort of turf, as they ramify. We may take advantage of this arrangement by grafting one of the species which present this peculiarity on one with a tall stem; the MESPILUS LINEARIS, for instance, on the OXYACANTHA (*Aubépine*) or Hawthorn. The former, growing from the top of the trunk of the latter, creeps in the air, as it would have done on the earth, and thus forms thick and elegant guards against rain and sun; examples of this may be seen in some of the walks of the Jardin des Plantes at Paris. Let us cite one more example, the COTONEASTER BUXIFOLIA, which presents this singular property, that on a declivity, it always follows it from the top to the bottom, instead of spreading on every side.

The relative length of the branches ought also to cause notable differences in the general aspect. If the lowest, formed the first, are continuously lengthened in the same proportion, the upper ones will be shorter and shorter in proportion as they approach the top, and the whole will have the shape of a cone or of a pyramid, as the Firs; if those in the middle surpass those of the base, the top will represent a ball or ovoid, as the Horse Chesnut; if those of the top are most developed, it will take the form of an umbrella, as the PINUS PINEA. We have only mentioned the extreme forms, between which may be found all the intermediate ones.

§ 198. Let us now recapitulate in a few lines all that has proceeded: we have considered as the stem properly so called, that which results, in Cotyledonous vegetables, from the act of germination by the evolution of the gemmule, that part of the primary axis always vertically directed towards the sky in a contrary direction to the other, the root, which descends into the ground. This vertical direction can be diverted only by means of mechanical obstacles, or the weakness of the stem, which causes it to fall to the ground from its own weight. Sometimes this stem is indefinitely lengthened in the same direction by means of terminal buds, which are formed and developed successively; sometimes it is stopped by the failure of one of these buds; and if the vegetable continues to enlarge itself, it is by means of the lateral buds, and, consequently, by a secondary axis which is substituted for the primary. The axis, thus substituted, may take the same vertical direction, or one more oblique, and even horizontal, sometimes on the surface of the soil, sometimes below; in the latter cases, the true stem is soon stopped in its course, and replaced by a bud situated near its base. We ought then to consider creeping or subterranean stems only as phenomena of the ramification.

Dicotyledons present us with the greatest number of true stems, attaining in trees all the developement of which they are capable, and much branched above the soil, arrested after their first shoot in annuals with the very life of the individual. Perennials present us with examples of the lateral and parallel substitutions.

These substitutions occur in the majority of Monocotyledonous vegetables which are for the most part perennial; this is the case which almost all the bulbs and rhizomas offer to us. When the stem, properly so called, attains a great length, it is most commonly unbranched.

We have in the whole of these observations thrown to one side the consideration of Acotyledons, the embryo of which has no gemmule; for the stem, when it exists, is not developed by the act of germination. Among them, are found the only *true subterranean stems*, as in the non-arborescent Ferns and the Marsileaceæ. When the stem is branched, whatever may be its direction, vertical or horizontal, it is by a double terminal bud, and consequently by a kind of bifurcation.

INFLORESCENCE.

§ 199.—The stem or its ramifications more or less multiplied bear flowers at their extremities. Each flower may be compared to a rosette of leaves, and, consequently, to a bud. These new leaves differ more or less from those we have hitherto examined in their form, their colour, and some points of their internal structure; they differ from them, moreover, that they never produce fresh buds at their axils; the part which constitutes a flower is essentially terminal, it is the limit of the ramification. We call the arrangement of the flowers on the branch, *inflorescence:* its examination ought, then, to succeed that of ramification, and to complete it, since it is the true end. In short, the vegetation of the branches loaded with leaves will be indefinitely continued by the production of fresh buds, if death, abortion, or some foreign cause does not happen to arrest it. The vegetation of a branch bearing a flower at its extremity is naturally stopped at this fresh kind of terminal bud which does not produce any others.

§ 200. Sometimes only one flower (*fig.* 179, *f'*) is developed immediately at the extremity of the stem or primary axis *a'*; sometimes at the extremity of the secondary *a''*, of the tertiary *a'''*, or of some other axis *a''''*. The leaves situated below them have preserved their nature; sometimes, also, they have begun to be modified both in their form and their colour, without producing at their axis any kind of bud, either foliaceous or floral. In all these cases, the flower is said to be *solitary* (*solitaire*).

179

§ 201. But, most commonly, from the axil of these leaves thus modified, spring branches sometimes naked, sometimes covered with other leaves, themselves modified in an analogous manner only to a still higher degree and terminated by a flower: this ramification may be repeated a greater or less number of times, and we have thus a group of flowers intermingled with their modified leaves; a group which is clearly distinguished from every part of the plant covered with true leaves, and which also is called *inflorescence.* This term has two acceptations in which we shall henceforth employ it: we have seen that it signifies the arrangement of the flowers; it also signifies a group of flowers which are not separated from one another by leaves properly so called.

179. A plant of the Butter-cup (*Renoncule bulbeuse*) (RANUNCULUS BULBOSUS). Its axis *a'* is swollen at its base into a bulb *b*, from the bottom of which grow roots, and from the top radical leaves, and terminated at its extremity by an expanded flower *f'*. Towards its middle it bears a leaf from the axil of which arises a secondary axis *a''*, terminated by the flower *f''* not quite so much developed as *f'*.—*a''* bears a leaf and a tertiary axis *a'''* terminated by a flower-bud *f'''*; and at *a'''* a fourth leaf hardly developed, at the axil of which we perceive another bud still less developed, the axis of which has not yet begun to grow in length.

§ 202. In this group of flowers, the different parts assume fresh names under fresh forms; the modified *leaves* which are often termed *floral* (*florales*), assume that of *bracts* (*bractées*) (BRACTEÆ) (*fig.* 181, *b*, *b*, *b*); the branches which only bear bracts and flowers, that of *peduncles* (*pédoncules*) (PEDUNCULI) (*a'*, *a''*). In the ramified groups, we distinguish the last branches, those which each immediately bear a flower, by the name of *pedicels* (*pedicelles*) (PEDICELLI) (*a'''*, *a'''*). The last bracts frequently present no branches at their axils, and the pedicels also may be furnished below the flower with several small bracts, which are described by the diminutive *bracteole* (*bractéole*) (BRACTEOLA). As their nature is evidently the same as the others, and as, in some cases, we see flowers thus developed at their axils, so that the pedicel when covered with flowers ceases to be one, it would be better to reserve this name for the last internode of the floriferous peduncle, and the pedicel will be characterized not only by the flower which terminates it, but by the absence of every bract.

§ 203. When we shall have to describe an inflorescence, it will be our first duty to indicate its connection with the rest of the plant, with the leaves properly so called: it will arise from their axil or will terminate a branch; it will be *axillary* (*axillaire*) or *terminal* (*terminale*), situated at a greater or less height in the vegetable, detached from its foliated part at a greater distance, &c. We shall then have to examine it independently of the rest of the plant, and its different parts relatively to one another. Treating it, then, as a separate organ, we shall first note as the primary axis, the common peduncle, which botanists have often termed spike-stalk or *rachis*, that from which all the others grow; and these, secondary, tertiary, &c., axes, according to the order in which they spring. In short, the greater part of the modifications which the ramification of leaf-bearing axes have brought under our notice, will be found again in those of the flower-bearing axes, but, with this difference, of which we must by no means lose sight; that is, that in the former, each axis is indefinitely continued by the production of terminal buds, and is only stopped by their failure; that, in the latter, on the contrary, it is the production of a flower which stops it, and that when one does not appear to terminate it, it truly enters into the class of the former.

§ 204. The primary axis of the inflorescence may, then:—1st, be terminated by a flower; in this case it stops there, and the inflorescence is only continued by means of secondary axes, stopped in their turn, by the production of terminal flowers, then by means of tertiary axes, and so on:—2nd, not be stopped in its growth by a flower, which will then only terminate either the secondary or the tertiary, &c., axes.

Hence two great classes of inflorescences: the *definite* (*définies*) or *determinate* (*terminés*), the *indefinite* (*indéfinies*) or *indeterminate* (*indéterminées*): the former, in which the primary axis bears a flower immediately; the second, in which it only bears flowers through the intervention of other axes.

We have here, therefore, the two grand modifications which the ramification of the stem has already laid before us, and we shall observe the same analogy in the secondary ramifications, which, in the same way, will depend on the number of times that the axis is branched before it is terminated, of the relative length of these different orders of axes, of the regular failure or abortion of a certain number among them, &c.

These observations, which we owe principally to M. Roeper, have enabled us to be a little more exact in the definition of inflorescences, which, previously, was too frequently very vague and confused. Always retaining the ancient terms and seeking to preserve, as much as possible, their original meaning, we have been forced to change it a little to render it more fixed and precise. Let us, then, define these words clearly, as we examine successively the two great classes of inflorescence which we have recognised.

§ 205. Indefinite inflorescence.—We have a primary axis lengthened without a flower.

The secondary axes may each be terminated by a flower, and in this case, the inflorescence stops there. If they are all nearly of the same length, they are called a *raceme* (*grappe*) (RACEMUS) (*fig.* 180).

At other times, all the secondary axes, or only some of them, the lower ones commonly, are not terminated by a flower, but throw out tertiary lateral branches: these may be branched in their turn. In all these cases, the raceme thus formed takes the name of *panicle* (*panicule*) (PANICULA) (*fig.* 181). Its commonest shape is that of the pyramid, on account of the unequal developement of its peduncles, which are largest at the bottom and grow less in proportion as they reach the top. But sometimes the peduncles of the middle are larger than those of the two extremities, and the panicle then receives the name of THYRSUS (*thyrse*).

§ 206. If, instead of peduncles almost equal in length or diminishing from bottom to top almost insensibly, we find the lower much longer than the upper, and erect, so that not one bears a flower till they have arrived at an almost equal height, and that the whole of the flowers thus forms a sort of umbrella, with unequal ribs, it is

180 181

called a *corymb* (*corymbe*) (CORYMBUS) *simple* (*fig.* 182, 1), or *compound* (*fig.* 182, 2), according as the flowers are on the secondary axes, or on peduncles which are ramified a greater number of times.

§ 207. Let us now suppose the secondary axes to be extremely short, so that the primary axis is thus covered on its sides with all the flowers, which in the raceme were separated from it by the length of the secondary axes, or, in a word, to be *sessile:* this is a

180. A raceme of the Barberry (*Épine-vinette*) (BERBERIS VULGARIS).—From the axil grows a leaf changed into a thorn *f*, which has preserved its two stipules *s*. —*a'* The primary axis from which grow small alternate bracts *b*, and at their axils secondary axes *a''* at the ends of which the flower is situated. We may follow the gradual evolution of these flowers from the base to the summit, the lower axes bearing fruit, the flowers of the top still buds, and the middle fully expanded.

181. A panicle of the YUCCA GLORIOSA.—*a'* The primary axis or rachis.—*a''* Secondary axes.—*a'''* Tertiary axes, each of which immediately bears a flower and takes the name of pedicel.—*b b b b* Bracts of different orders, from the axils of which these axes grow.—This inflorescence results from a series of racemes on a common axis *a'*, so much the more developed and precocious as they are lower on the axis, and in each raceme the flowers are so much the more advanced as they are lower down on its axis.

spike (*épi*) (SPICA) (*fig.* 183). It is said to be compound, if the primary axis bears secondary ones, themselves lengthened and not terminated, but bearing lateral, sessile flowers.

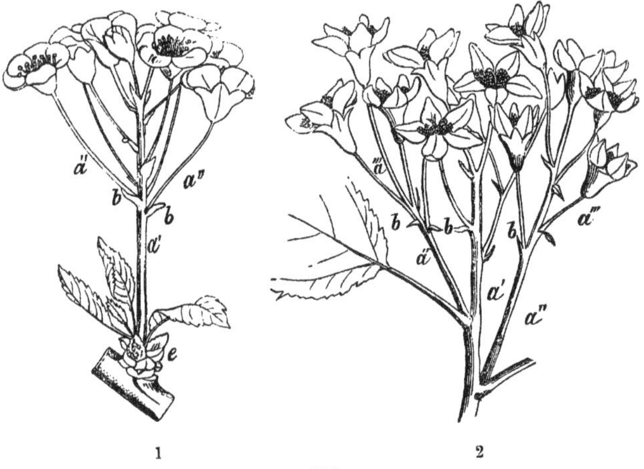

182

Different names have been given to different spikes which present certain characteristics, peculiar to certain classes or families of plants, as the *Catkin* (*Chaton*) (AMENTUM), a simple spike which falls as it disarticulates after blossoming, and is formed of flowers of the same sex, commonly male (*fig.* 184); the SPADIX (*Spadice*) (*fig.* 185), a spike of Monocotyledonous plants enveloped at its base with a large bract, which is called the *spathe*[s], and the flowers of which, extremely close, are most commonly, as it were, encrusted on a thickened axis, most commonly simple, sometimes branched, as in Palm-trees, and then taking the name of SPADIX RACEMOSUS (*régime*).

§ 208, *a*. Now let us suppose the contrary to all the preceding,

182. 1. A simple corymb of the CERASUS MAHALEB.—*a'* The primary axis, from which the alternate bracts *b* grow, throwing out from their axils the secondary axes *a'' a''* each of which bears a flower directly and is called a pedicel. We may follow their gradual evolution from the outside to the inside; the innermost being still in the state of a bud.—This corymb grows from the axil of a leaf which has already fallen and terminates an abortive branch, the lower leaves of which are in the state of scales *e*.

2. Compound corymb of the Wild Service tree (*Alisier des bois*) (CRATÆGUS TORMINALIS)—*a'* The primary axis.—*a''* The secondary axes.—*a'''* Tertiary axes bearing immediately the flowers or pedicels.—*b b b* Bracts.

[s] From σπάθη, *a broad blade, a ladle.*—TRANS.

namely, that the part of the primary whence the secondary axes spring is shortened, or at least is not lengthened, so as to appear

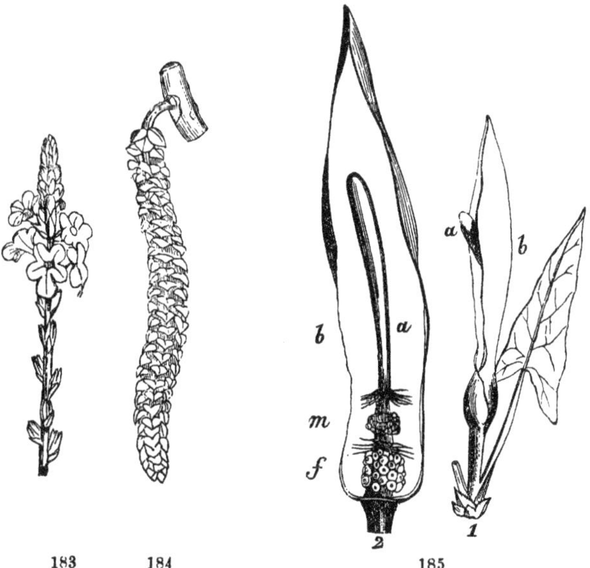

183 184 185

almost nothing (*fig.* 187 *a'*); whilst the secondary axes *a''*, which take the name of RADII (*rayons*), are largely developed. Then, commonly commencing together, they are almost equally lengthened, and if they are each terminated by a flower, these flowers, carried to the same height, form together a kind of umbrella with equal ribs, and are, therefore, called an *umbel* (*ombelle*) (UMBELLA) (*fig.* 186 *o' o' o' o'*). This is, therefore, a raceme with almost no primary axis, in which, by means of this shortening, the relation of the peduncles to one another is changed, the upper ones becoming the inner, the

183. The extremity of the spike of the Common Vervain (*Verveine commune*) (VERBENA OFFICINALIS).—The lower flowers have already passed to the state of fruit, those of the middle fully blown, those of the top still in bud.

184. The catkin of the Hazel-nut (*Noisetier*). The bracts, reduced to the state of scales, are so near that their spiral arrangement may be perceived. The end of the stamens of the male flowers may be seen projecting from them.

185. Spadix of the Lords and Ladies (*pied-de-veau*) (ARUM VULGARE).—1. Shews it enveloped with its spathe *b* rolled like a horn, which allows the extremity to be seen *a*.—2. The same shewing the spathe cut vertically so as to allow the whole axis *a* to be seen. It is loaded with a mass of female flowers *f* at the bottom, and with a mass of male flowers *m* a little above them.

lower ones the outer in the umbel. It is then said to be *stipitate* (*stipitée*) (STIPATUS), if the primary axis attains a certain length before giving birth to the secondary; *sessile* if the lower part is itself wanting. The secondary peduncles may, instead of bearing flowers, be ramified, in the same way (*fig.* 187), and we have thus several secondary umbels, or *umbellules* (*ombellules*) (UMBELLULÆ) (o''), arranged in a common or compound umbel. The compound is to the simple umbel, what the panicle is to the raceme.

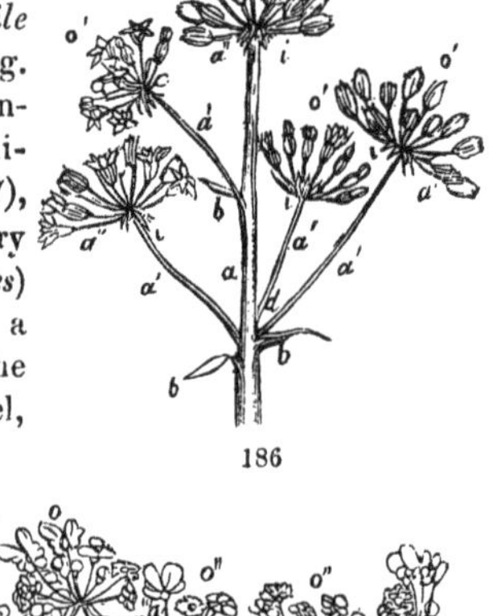

186

§ 208, *b*. When, from an extremely short primary axis there grow several secondary thin peduncles, which, instead of rising, diverging and thus spreading the flowers which terminate them into a kind of umbrella, approach and press against one another in the shape of a sheath, and often even lean all on one side when they are of a certain length, it is well to distinguish this arrangement from the umbel

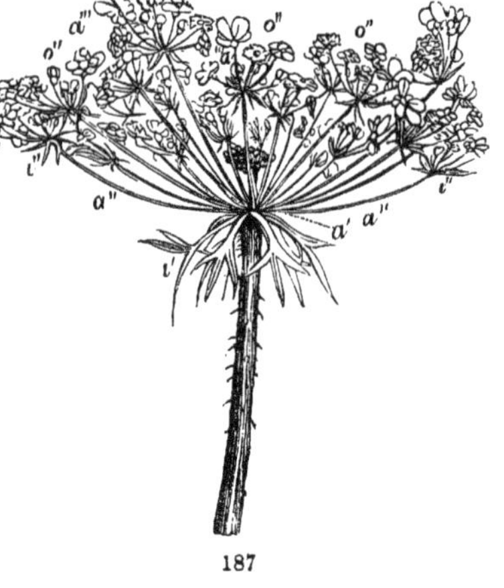

187

186. Several umbels of the ARALIA RACEMOSA, *o'*.—*a* A general axis or summit of the branch, itself terminated by an umbel more advanced than the rest.—*a' a' a'* Axes growing from the preceding, and secondary with regard to it, but, each bearing an inflorescence, and primary with regard to it.—*a'' a'' a''* Secondary axes or radii of the umbels.—*b b b* Alternate bracts on the general axis. We may remark at the axil of one of them *d* two axes of the same rank growing from it one above the other on account of a double bud.—*i i i* Verticillate bracts on the axis of each inflorescence, forming an involucre.

187. A compound umbel of the Carrot (*Carotte*).—*a'* The primary axis reduced in the inflorescence to a convex surface.—*a''* The secondary axes or radii each bearing an umbellule *o''*.—*a'''* Tertiary axes or radii of the umbellules.—*i'* Whorl of pinnatipart bracts forming the general involucre.—*i''* Simple bracts forming the involucelles.

by giving it the name of *fascicle* (*faisceau* or *fascicle*) (FASCICULUS) (*fig.* 202, *f*). These fascicles are most frequently found to be sessile at the axils of the leaves or bracts.

§ 209. Lastly, the secondary axes may not be lengthened, any more than the part of the primary from which they spring; and then all the flowers are necessarily brought together into a kind of disk or ball, in which the outer ones represent those which would be below on a lengthened axis; the inner ones, those which would be at the top. This arrangement is called the *head* (*capitule*) (CAPITULUM) (*fig.* 188). The capitulum as in the umbel may be either sessile or pedunculate.

188

A particular name, that of CALATHIS (*calathide*) or ANTHODIUM, has been given to a modification of the capitulum depending on that of the flower-bearing summit of the primary axis. This summit is thickened and enlarged so as to offer to the insertion of the flowers a large plane, concave or convex surface. We see a number

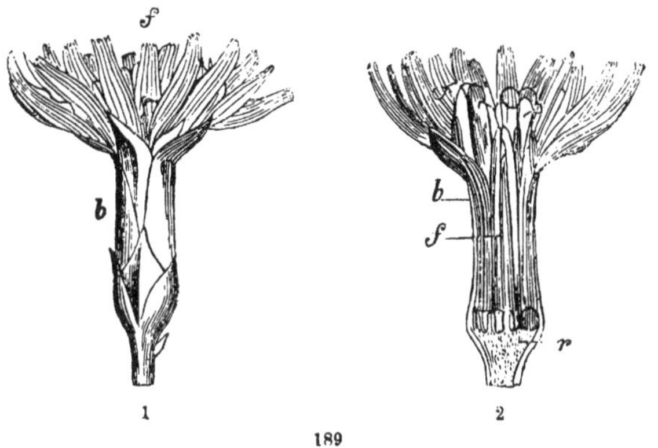

189

188. A capitulum of the SCABIOSA ATROPURPUREA.—The flowers are less developed in proportion as they approach the centre.

189. 1. A whole capitulum of the Scorsonera (*Scorsonère*) (SCORSONERA HISPANICA).—2. The same, cut vertically down the middle, so as to shew the peduncle enlarged into a receptacle *r* which bears the flowers *f* on its surface.—Those of the circumference have already been expanded, those of the middle are still in the state of a bud.—*b* Imbricate bracts forming the involucre.

of examples in flowers, termed in common parlance compound, and which are nothing else than many little flowers thus united into a mass, which, at first sight, presents the appearance of one flower, as those of the Chicory or Succory (*Chicorée*) or of the Scorsonera (*Scorsonère*) (*fig.* 189, 1), of the Thistle (*Chardon*) or of the Sunflower (*Soleil*). Every one will call to mind the stem of the last, dilated at its extremity into a round and fleshy plate, quite covered with yellow florets (*fleurettes*). This flower-bearing plate (*fig.* 189, 2, *r*) has received different names: formerly it was called the ***receptacle*** (*réceptacle*) more recently the CLINANTHIUM (*clinanthe*) or PHORANTHIUM (*phorante*), (according to the etymology, the bed or support of the flowers).

We have just said that its surface is sometimes concave. This concavity is commonly superficial and very wide; but, it is deeper at other times, in the shape of a cup or an urn; its edges may even be joined and touch most closely, so that it forms a completely close cavity. This is observed in the Fig (*figue*) (*fig.* 190); and this is the reason why persons unacquainted with botany do not discover its flowers, which are inserted on the whole interior surface of the receptacle which quite incloses them, and which is only perceived on the outside as a green pear-shaped fruit. In other plants of the same family, we find all the transitions from this close cavity, in which are concealed the flowers of the Fig, to the vase-like receptacle of the DORSTENIA (*fig.* 191), on which they are open to the day.

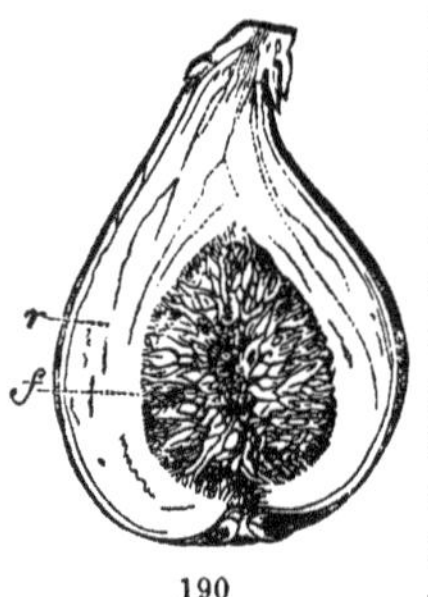

190

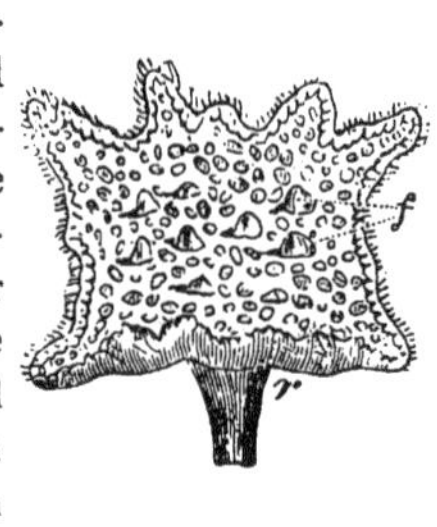

191

§ 210. After having compared all the kinds of inflorescence we have just explained with one another, we see that they differ, as we have said, only in the developement or the failure of certain axes and in their relative lengths. This is so true, that it is easy to give

190. A longitudinal section of a Fig (*Figue*), shewing its flowers *f* inserted over the whole of the inner surface of the receptacle *r*.

191. Inflorescence of the DORSTENIA CONTRAJERVA, in which the flowers *f* are half buried in a receptacle *r*, slightly concave, from the surface of which their extremities project.

each of these kinds of inflorescence a definition which is only a term of comparison with any other. Thus, we may say that the raceme is a spike with pedicellate flowers; the spike, a raceme with sessile flowers; the sessile umbel, a raceme without rachis or primary axis; the capitulum, a raceme in which the axes are suppressed, or a spike, in which the rachis is absent, or an umbel without radii: the umbel, a capitulum with pedicellate flowers, &c.

In all these inflorescences, the secondary axes or those of a lower order may be either opposite or alternate. Their relative position will always be the same as that of the leaves, since they spring from the axils of leaves modified only in their shape: indeed, this takes place very frequently. But we have said (§ 167), that in oppositifoliate plants it is not uncommon to see true leaves become alternate at the summit of the branches. We must not, therefore, be astonished if the bracts, always situated at the top of the branches, frequently become alternate, even when the leaves are opposite.

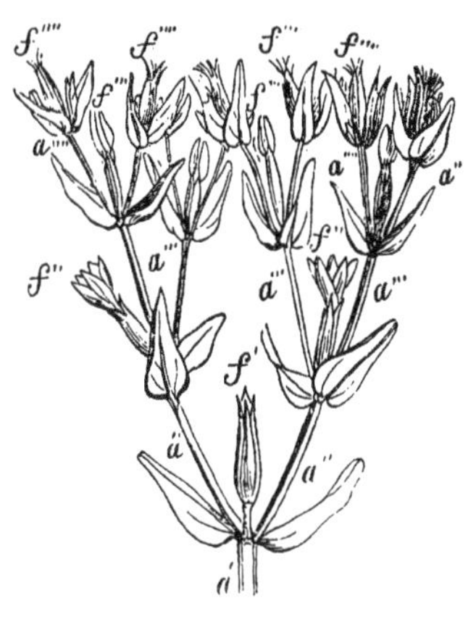

192

§ 211. DETERMINATE INFLORESCENCE.—We find determinate inflorescences most frequently and regularly in plants with opposite leaves. We will begin to examine them in one of these plants: in one of the Gentianeæ, the Common Centaury (*Petite Centavrée*) (*fig.* 192).

The primary axis *a'*[1] is terminated at a certain height by a

192. The top of a plant of the Centaury (*petite Centaurée*) (ERYTHRÆA CENTAURIUM).—*a'* The primary axis.—*a'' a''* The secondary axes, two in number.—*a''' a''' a''' a'''* The tertiary axes, four in number.—*a'''' a'''' a''''* The quarternary axes, eight in number.—*f* Flowers.—The flowers are more developed in proportion as they belong to an axis of a higher rank; *f'* in the state of fruit; *f'' f''* expanded; *f''' f'''* still in buds.

[1] The part of the primary axis situated above the birth of the two secondary axes thus seems to grow from the angle which they form with one another. The name of *axil* (*aisselle*) (AXILLA or ALA), which we have reserved for the angle formed by the branch with the leaf which grows from it, was formerly applied to the angle formed by every bifurcation of the vegetable. Hence, the name of alary flower or flowers, which we find in some authors, who, thus, designate the flower or the whole of the flowers, which thus terminate a primary above the fork formed by the secondary axes.

flower f'. Immediately, or, it may be, a little below it, grow a pair of leaves, from the axil of each of which arises a secondary axis a'' a'', the summit of which each equally bears a central flower f'' f'' and two leaves, one on each side ; each of these leaves throws out, in its turn, from its axil a tertiary axis a''' a''', terminated in the same manner. This ramification, termed *Dichotomy* (*dichotomie*), is thus repeated a greater or less number of times, and each time the number of axes and, consequently, of flowers, is doubled.

If, instead of two leaves we have three verticillate leaves beneath a central flower, and from the axil of each grows a secondary axis, which, in its turn, is divided into three, it is then termed *trichotomy* (*trichotomie*). We may have a still greater number of verticillate leaves and axes.

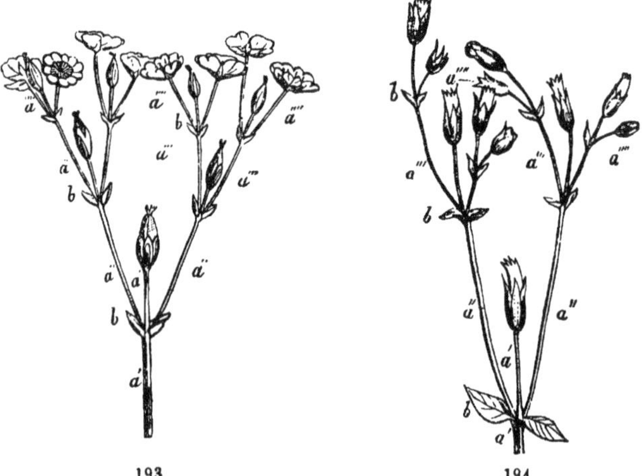

193 194

We have not described these methods of ramification before, since they appear with the flowers, and are, consequently, connected with the inflorescence. Sometimes, however, the axes, without being limited by a flower (which then fails), may be thus stopped at each production of fresh axes, and we have then a continuation of bifurca-

193. The inflorescence of the CERASTIUM GRANDIFLORUM.—The letters refer to the same parts as in the preceding figure.—*b b b* Bracts opposite to one another at each ramification.

194. The inflorescence of the CERASTIUM TETRANDUM.—The letters again refer to the same parts.—The quarternary axes a'''' are reduced to lateral ones by the abortion of the second.

tions, or, less commonly, of trifurcations, accompanied with leaves without flowers.

In the case in which the one class exists with the other, we have a series of solitary terminal flowers. But, if the leaves are modified and change into bracts, the whole system forms one single inflorescence, which is distinguished by the epithet di or tri-chotomous. The most modern botanists have called it by the name of *Cyme* (*Cime*) (Cyma). In this Cyme, the flowers may be more or less separated from one another, according as all the successive axes, lengthened and much surpassing the flower-bearing summit of that which precedes them, give to them the appearance of a panicle upside down, or that, shorter and shorter, they hardly surpass them, so that the flowers all grow at the same height like a corymb.

The division by dichotomy is not very frequently followed out so regularly from the bottom to the top of the inflorescence; but, at a certain height, whilst one of the axes gives birth, below its terminal flower, to two bracts, and to two fresh axes, that which is opposite to it bears its flower without bracts, or at least without developing any thing at their axils, and the ramification is thus arrested on one side, whilst it is continued on the other.

§ 212. Now, this lateral stoppage, frequent towards the top of the inflorescence, sometimes begins at the base, and we see at each node one of the axes, which ought to have been developed, fail. We have, then, the appearance of a raceme or spike: but, instead of only one axis without a terminal flower bearing a series of secondary flower-bearing axes, as in the indefinite inflorescence, it is in this case a series of axes of different ranks, which grow successively from one another, each terminated by its own flower; and this origin of the false raceme is also shewn by its axis, instead of being rectilineal, commonly presenting a series of bends or angles, and also by all the flowers being generally situated on the same side, most commonly in the inside; this shews, that the abortion generally holds on all the axes of the same side, the exterior.

§ 213. We may find, without abortion, the same arrangement of the flowers and of the axes which bear them, when the leaves are alternate instead of being opposite. The primary axis presents only one bract on the outside and is then terminated by its flower. From the axil of this bract arises a secondary axis, terminated in the same manner and furnished with a bract, from which springs a tertiary axis. This, also, in the same manner gives birth in its turn to a fourth; the fourth to a fifth, and so on: we have in this way a series of axes growing from one another, each immediately terminated by a flower, and from this combination results, as in the preceding case, a false raceme or a false spike. It is an arrangement just like that which we have described in the creeping stems (§ 183) or in certain stems (§ 185), in which each internode is apparently continued by an internode developed at the axil of the terminal leaf. The only difference is, that here each axis is terminated by a flower instead of a tendril or a rosette of leaves. In all these cases, it is easy to distinguish this inflorescence from the true raceme or the true spike, because the bracts are opposite to the flowers instead of being situated immediately below them.

The flowers, in this case, may occupy, in relation to the axis, two kinds of different positions, which depend upon the order in which the bracts are developed. Let us now call to mind, that the alternate leaves are arranged in spires; that these spires on two axes *a* and *b*, springing from one another, coil either the same way or in a contrary one; that the first leaf of *b* is always regulated by that of *a*, which bears *b* at its axil (§ 162). Now let us suppose, after *a* and *b*, a series of axes *c*, *d*, *e*, &c., growing successively from one another; although each may be reduced to a leaf or a bract, the position of the leaf of *b* is influenced by that of *a*, that of the leaf of *c* by that of *b*, and so on; and these two co-ordinate leaves are sufficient to determine the direction of the spire from right to left or from left to right. This series of axes is either homodrome or heterodrome. In the former case, if the spire of *a* was turned from left to right, the leaf of *b* will be placed at the right of that of *a*, that of *c* at the right of that of *b*, that of *d* at the right of that of *c*, &c., &c., and we shall thus have a series of bracts describing a spire from left to right. In the latter case, on the contrary, the leaf of *b* will be placed to the left of that of *a*, the leaf of *c* to the right of that of *b*, the leaf of *d* to the left of that of *c*, the leaf of *e* to the right of that of *d*, &c., &c., and we shall have a series of bracts situated alternately to the right and to the left, so that the line, which would pass through them successively, would describe a zig-zag.

Hence two very different arrangements in the false racemes or spikes which now engage our attention. Those which we find in the former combination bear a striking resemblance to the true ones, with which they have generally hitherto been confounded, because the flowers are distributed quite round an axis apparently continuous: this inflorescence has been called by M. Bravais *helicoid cyme* (*cime héliçoïde*) (*fig.* 195). Those, which we observe in the latter combination and are easily distinguished on account of the unilateral position of their flowers, have received the name of *raceme* (*grappe*) or *spike* (*Épi*), or more properly *scorpioid cyme* (*cime scorpioïde*) (*fig.* 196). The last epithet is owing to their extremity being rolled like a crook or a scorpion's tail; and it is easy to account for this rolling

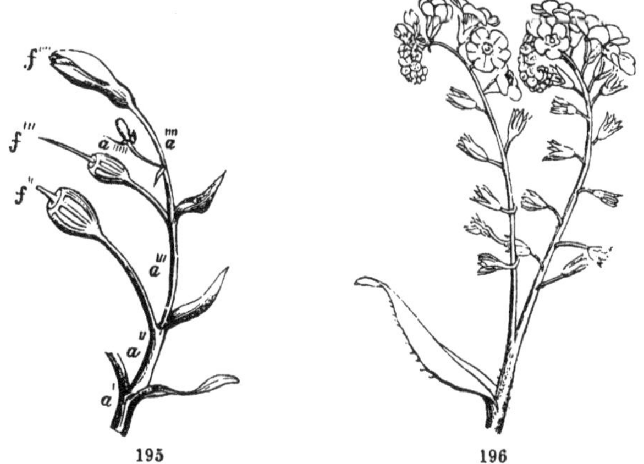

up (*fig.* 197). The axis *b* makes with *a* a certain angle to the outside; *c* nearly the same with *b*, *d* with *c*, *e* with *d* &c., &c., and the result is a broken line tending to describe a curve, which after a sémi-revolution returns upon itself. The axes are developed later, and are, consequently, shorter in proportion as they are separated from the point of departure; this curve ought, therefore, to be a spire with gradually decreasing revolutions, like a bishop's crozier. The slower or quicker approach to the centre of the crozier depends on the angle more or less acute, at which the axes spring from one another; the distribution of the flowers on the two rows is more or less manifest in proportion to the distance between two successive bracts. All these characteristics may be studied in a great number of plants: in the

197

195. The helicoïd cyme of a species of ALSTRŒMERIA.—The succession of axes a' a'' a''' a'''' is very apparent. They seem to form a single continuous one, of which they would be the internodes. Each of them grows from the axil of a leaf and is terminated by a flower *f* opposite to that leaf.

196. The scorpioïd cyme of the Scorpion grass (MYOSOTIS PALUSTRIS).—The flowers are arranged alternately in two rows and on the same side of a common peduncle.

Rue, in the *Stone-crop* or SEDUM, the inflorescence of which was in the opinion of Linnæus and, consequently, of several subsequent botanists the type of the true *cyme*, especially in the family of the *Borragineæ*, in which will be found numerous examples to clear up the truth of the demonstration.

§ 214. In all the definite inflorescences which we have just examined, in the first especially, it may happen that, in consequence of the extreme shortness of the axes, all the flowers are brought close together in masses, as in the Pink (*Œillet*). M. Roeper calls this arrangement *fascicle* (*fascicule*), when the axes preserve a certain length and a regular arrangement, as in the Sweet-William; *glomerule* (GLOMERULUS), when they are almost absent and the numerous abortions completely break through all regularity. We have pointed out, for the former of these words, another meaning (§ 202 *b*), and think with De Candolle, that that of *contracted cyme* is sufficient in both cases.

§ 215. De Candolle, under the name of *mixed inflorescences*, has pointed out several which belong at the same time to both definite and indefinite, because their different axes are not arranged in the same way. In the Labiates, for instance, the flowers form cymes arranged at the axils of opposite leaves on an indefinite common axis. If the leaves preserve their characteristic, there will be no embarrassment and they will be termed immediately axillary cymes; but, if the leaves have passed to the state of bracts, and all these same cymes are thus confounded in a single inflorescence, how shall we define them? The difficulty seems to us very easy to resolve: by describing them as cymes in spikes, racemes or panicles, we shall clearly indicate the double characteristics of this inflorescence.

§ 216. Again, it is not only the primary axis, but also several axes growing from it which do not immediately bear flowers; the inflorescence does not present below either dichotomy or a succession of different axes; in a word, does not appear definite, and yet becomes so at its extremities, its last ramifications only being divided by regular bifurcations or unilateral cymes. This case can only be presented in those inflorescences which are ramified a great number of times, such as the panicles (*fig.* 198) and the corymbs.

We may nevertheless find some analogy to the simplest inflorescences, the raceme, and the spike, when their primary axis is terminated by a flower, as in the greater part of the Campanulæ (*fig.* 199).

Ought we to invent fresh appellations or change the signification of the ancient ones to represent these modifications? It appears to us more simple and more clear to employ terms which are known, indicating by an adjective the modification which the object they express has undergone. Thus, we may say paniculate or corymbate cymes, or rather panicles or corymbs terminated in cymes, or else, still more briefly, definite panicles, corymbs, or racemes. The transitions cannot generally be expressed in a single word, since they suppose two terms of comparison.

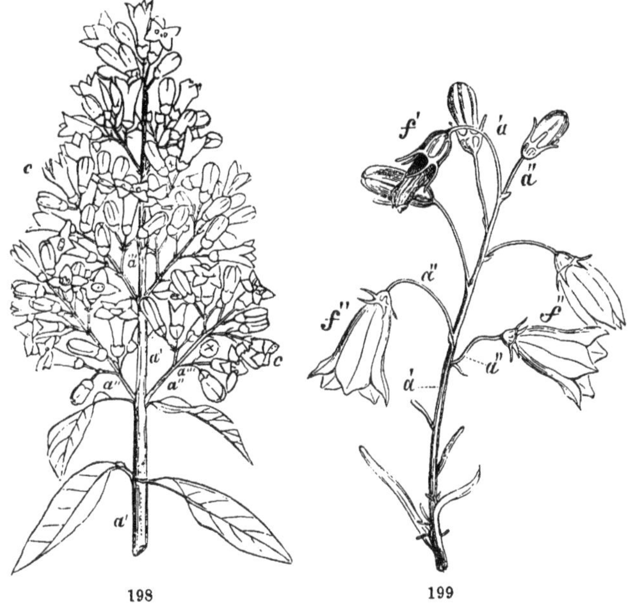

198 199

§ 217. Whatever nomenclature we may adopt, we must remark,

198. A definite terminal panicle of the Privet (*Troëne*) (LIGUSTRUM VULGARE).—The primary axis *a'* throws out secondary opposite axes *a'' a''*, whence the tertiary axes grow *a''' a'''*, terminated by dichotomy, and, consequently, by small trifloral cymes *c c*. In each of them, the central flower is more developed than the lateral ones.

199. A definite raceme of a CAMPANULA.—*a' a'* The primary axis, terminated by a flower *f'*, which has already withered and begun to pass into the state of fruit.—*a'' a'' a''* Secondary axes, each terminated by a flower *f''*, which is more advanced in proportion as it is situated lower on the raceme.

in passing, that these mixed inflorescences are extremely frequent, and that the definite, which once were considered much more uncommon than the indefinite, are being continually multiplied under the eyes of observers, who are now much more attentive and practised in their researches.

In the case in which the last peduncles alone are each terminated by one flower without being divided by dichotomy, and which seems at first to return altogether to indefinite inflorescences, we are frequently shewn the contrary by the presence of several small bracts or *bracteoles* (*bractéoles*) (BRACTEOLÆ) on this last peduncle.

We very frequently find two of them opposite to one another a little below the flower, and we here easily recognise a small trifloral cyme the two lateral axes of which are not developed. Again, these bracteoles are alternate; and, now that we have learned how to discover the definite spikes and racemes, we can no longer hesitate to recognise them in the peduncle, terminated by a well-developed flower and furnished lower down with bracts, at the axils of which would be developed other flowers of a secondary order with regard to the preceding, if they did not prove abortive. This is the reason, why the name of pedicel is reserved for the last internode alone of this peduncle, and not for its whole, every pedicel necessarily being unifloral.

The passage, therefore, of the definite to the indefinite inflorescences, is frequent and sometimes almost insensible. On account of this we must take care not to employ fresh words for varied cases separated by undecided limits. It is much better to preserve the ancient words and define them well, without, nevertheless, entirely changing them from their primitive acceptation, and explaining them, when needful, by epithets or even short descriptions: we have endeavoured to do this in general cases, and the others are only modifications of these, which practice alone will teach us.

Flowering.

§ 218. (*Floraison*).—In what order are the flowers of an inflorescence developed? It is a question to which our previous observations render an answer easy. We have followed into all its different details the investigation of branches bearing leaves and of branches bearing flowers; we have seen that their ramification follows the same laws, and that every peduncle terminated by a flower may be likened to a branch. Now, every branch is necessarily developed before those which spring from it; its evolution is, therefore, advanced to some degree, more or less, before that of its lateral buds commences. It will be the same with the flower-bearing axes; each of them will then be developed before any of the axes which are secondary with regard to it: *the flowers terminating different axes, will, therefore, expand in the order in which the axes which bear them succeed one another.*

Instead of a branch several times ramified, let us suppose one limited to its lateral buds, or, in other words, a primary axis with a certain number of secondary axes only. We know that this branch increases in height, so that its parts are formed so much the sooner as they are lower on the axis. Its buds follow this manner of evolution from bottom to top, and develope themselves in proportion as they are lower. The same rules ought to hold in a series of flower-bearing axes springing immediately from the same peduncle; the evolution of the flowers which terminate them ought to begin with the lowest axis, and thus to proceed gradually from bottom to top. We have a second law, that *flowers terminating axes of the same rank situated on a common axis, expand from the bottom upwards.*

The establishment of these two laws is the result of observation as well as theory. They preside over the evolution of all inflorescences, with the exception of some cases of irregularity from internal or external causes which require a separate examination. These known laws teach us to distinguish with ease different methods of inflorescence, to determine which would be difficult without their aid.

§ 219. We already know, that when all the flowers of the same inflorescence are borne at the same height by the lengthening of some axes and the shortening of others, the lower ones are naturally outside, and the upper ones inside. We may then make use of the terms external for lower, and internal for upper, outside to inside, for bottom to top, inside to outside, for top to bottom. If we look at an umbel or a capitulum, we shall see the expansion of the flowers gradually proceed from the circumference to the centre of the inflorescence, as we have seen it from the base to the summit in the raceme. Hence the name of *centripetal* (*centripète*) evolution, which has been given to that of the flowers of these kinds of inflorescences, and, consequently, to all indefinite inflorescences.

Let us now take a cyme composed of a primary axis and of two or more secondary axes. The flower which terminates the first, and necessarily occupies the central part of the inflorescence, will be the first to expand: then will come those which terminate the second, and which occupy the circumference.

In this case, the evolution follows a course quite inverse to the preceding; it starts from the centre towards the circumference: it is termed *centrifugal* (*centrifuge*), which is, consequently, applied to that of all definite inflorescences. We may understand, however, that it is far from being exact in every instance, since, if we suppose a continued series of floriferous dichotomies, tertiary axes will arise between the central flower and those which terminate the secondary axes, which will flower much later than the secondary, although nearer to the centre. We must then in accepting this word not take it in its rigorous sense.

§ 220. All this being stated, we may easily foresee, that, at the first inspection of a flower, we shall be assisted in the determination of its kind of inflorescence by the relation in the position of flowers developed at different stages. If we perceive in the centre or at the top, a flower more developed than those which are around or beneath it, we know that we have got a definite inflorescence: if the top or the centre presents us, on the contrary, with flowers less developed than the bottom or the circumference, we know that the inflorescence is indefinite. The degrees of developement to which the different flowers have arrived relatively to one another, clearly point out to us the ranks of the axes which bear them, or their relative position on a common peduncle.

§ 221. We have, for greater clearness, supposed the most simple cases, those in which the inflorescence is not much ramified. If it be much branched, the examination is complicated by the dispersion of axes of the same rank over several parts of the inflorescence. We have already pointed it out in the cymes; but, we may say as much for an indefinite inflorescence; for a panicle, for instance, in which each of the secondary axes situated at different heights bear through the medium of other axes flowers of several different ranks, we may, towards the bottom, find flowers less advanced than others which are above them, and which seems contrary to the centripetal evolution.

We ought to remark, in this instance, that the *ensemble* of the inflorescence is only the repetition of a certain number of groups of flowers nearly similar, arranged on a common axis: thus, a panicle is generally only a union of racemes on the same peduncle. We are thus led to admit *compound inflorescences*, in which in the *general inflorescence* (which alone we have hitherto studied) we may distinguish several *partial* ones, each of which in its flowering will obey the laws we have just laid down. So also, on comparing the inflorescence of one with that of others, we shall see that each of these groups of a compound inflorescence may be likened unto a flower of a simple inflorescence; that, if they are all lateral with respect to the axis, they will be less advanced in proportion as we examine them higher up, that they will be developed from bottom to top; that, if one terminates this axis, it will be developed before the others (*fig.* 186), and even then the flowering of these others will often proceed from top to bottom: it will become centrifugal. From these data, we can now give a third law: *In a compound inflorescence, the partial inflorescences follow, in their relative evolution, the same laws as the flowers in a simple inflorescence.*

We see that the ramification of an inflorescence may be sufficiently complicated to render it capable of being divided several times: a general panicle, for instance, into partial panicles, each of these into racemes, &c., &c.

§ 222. The flowering sometimes reveals in the inflorescences an arrangement which it is impossible to discover otherwise: as in contracted cymes (§ 214), which we could not, without this aid, easily distinguish from capitula. Thus, the Teazle or Fuller's Thistle (*Chardon à Foulons*) (DIPSACUS FULLONUM), has its flowers united in long elipsoidal heads, which, at first sight, we should call spikes; but, in a true spike, the flowering would take a regular course from bottom to top: but here it commences about the same time at several stages, and we are thus led to conclude that this spike, apparently only one, is formed by the junction of several, one of which, larger than the rest, is the terminal. Thus, to cite another example, the ECHINOPS SPHÆROCEPHALUS has its flowers united in a ball, in which the

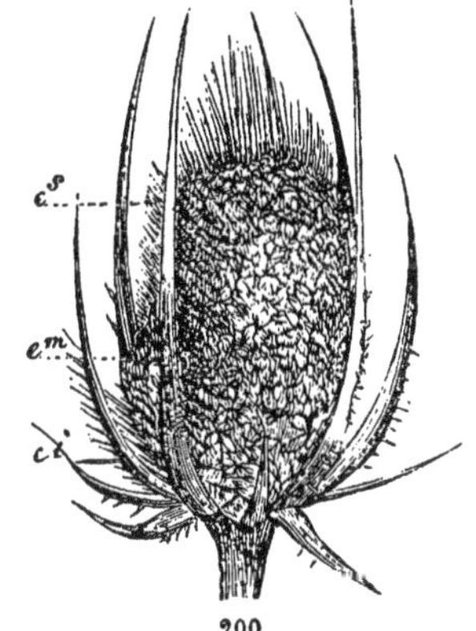

200. The inflorescence of the DIPSACUS SYLVESTRIS.—The flowers are separated by long pointed bracts, which bristle the whole of the capitulum, and are developed at several heights e^s e^m e^i, at each of which points the lower are fully blown, the upper still buds.

flowering proceeds from top to bottom, and not from bottom to top; and instead of a capitulum, we are obliged to call it a compound definite inflorescence.

§ 223. We have now to treat of some kinds of inflorescence, which in their point of departure, seem to be an exception to the general rule, since the peduncles, which are nothing else than the last branches of the plant, ought always to spring from the axils of the leaves, and instances exist in which the contrary appears to obtain.

Thus, a *radical* inflorescence was formerly admitted, as if the flowers sprang immediately from the root; but the very definition of them which was given indicates that the nature of this inflorescence was very well understood, that it was well known that the extreme shortness of the leafy part of the stem gave good reason for suspecting this illusion. In this case, in the lower part of this stem, the internodes are so near that all the leaves (called themselves *radical*) form at the surface of the soil a rosette from the middle of which, the flowers, which terminate the stem thus contracted, arise, or else grow from the axils of leaves thus united into a mass. But they frequently do not spring so near the ground, and the naked stem continues to grow to a certain height, at which it begins to bear bracts and flowers: it then takes the name of *scape* (*hampe*) (SCAPUS). In short, the inflorescence is ranked amongst those we have already explained. Examples are frequent in several bulbous plants, as the Hyacinths, in the Cowslip, the Polyanthus, the Daisy, the Primrose, &c., &c.

§ 224. Petiolary, or rather epiphyllary inflorescences were also admitted, the flowers being thus made to spring from the leaf. In this case, the branch which assumes the form of a leaf, was generally taken for a leaf, as we shall see a little farther on (§ 236, *fig.* 202). At other times, it is a true leaf, from which the flowers grow: then, the peduncle which bears them is partly joined to the leaf from the axil of which it springs, either to its petiole, as in the HELWINGIA, the CHAILLETIA, and several species of HIBISCUS, or for a greater distance to the limb itself, as in the ZOSTERA. The flower-bearing leaf may then be, as it were, in the state of a bract, as in the Lime-tree, in which it is easy to determine this partial junction of the peduncle.

§ 225. We may explain also by the junction of parts, which are commonly distinct, many extra-axillary inflorescences, i. e., inflorescences which seem to spring from another point than the axils of the leaves: the Solanaceæ furnish good examples of it. The insertion of the peduncle joined to the branch seems then to be carried to a certain height above the leaf, and if the part of the peduncle thus blended is much longer than the internode, we may find one or several leaves between that from the axil of which it ought to grow, and the point at which it really detaches itself: a straight line drawn from this point, ought on coming down, to meet this leaf leaving the others on one side; and we thus arrive at the true relation of the parts which this complicated structure renders more obscure.

As to the case in which the inflorescence is oppositifoliate, i. e. is directly opposite to the leaf, instead of springing from its axil, it has already been sufficiently explained (§ 185).

BRACTS.

§ 226. (*Bractées*) (BRACTEÆ).—We have already said, that bracts are modified leaves from the axils of which grow the axes laden with flowers. Sometimes the modification is not complete, and the bracts preserve, especially towards the base of the inflorescence, the green colour and all the appearance of leaves, although lessened and shortened, so that we hesitate as to what name we should give them: they are not the leaves of the stem or body of the plant, nor are they, on the other hand, bracts. We indicate this state of transition by adding to the inflorescence the epithet of *foliaceous* (*feuillée*) (FOLIOSA). Thus we say, foliaceous panicles, racemes, &c.

Again, on the contrary, the abortion of the leaves accompanying the flowers is complete; we do not find the least trace of them, whether at the origin of the general or partial inflorescences, or that of each flower singly. This failure of bracteæ is marked by calling these inflorescences FLORES EBRACTEATI. This is observed in the Cruciferæ.

§ 227. Between these extremes, the foliaceous developement or complete absence of bracts, we find all the intermediate ones, and then the reduction of the leaf may present different modifications, which we have pointed out in the envelopes of the bud (§ 173). It is most commonly the petiole which is wanting, and the sessile limb is more or less shortened. It frequently happens that the vaginal part is the only one which remains. It may be only one end of the petiole, under the form of a point: a piece of the central fibro-vascular fascicle, the prolongation of which would form the median nerve of the leaf. Sometimes we find the stipules alone, which may attain the same degree of developement as in the leaves properly so called. If the rest of the leaf is completely abortive, we have then the appearance of a double bract; if, with them we have the limb, the petiole, or the sheath, developed themselves to a certain point, we have the appearance of a trilobed or even a triple bract, according as the stipules are petiolar or caulinary. We must carefully distinguish this case from that in which the bract appears triple, from the immediate neighbourhood of two opposite bracteoles.

§ 228. In the majority of cases, the metamorphosis of a bracteate-leaf is complete in proportion as we observe it on an axis of a higher rank in the inflorescence; and in the same we may sometimes see, from its base to the summit, all the transitions we have just enumerated. This diversity of appearance may complicate the description which ought to mention them in a general manner.

The bract, when it is the limb which remains, may preserve, with a more enlarged form bearing a greater or a less resemblance to the leaf, its structure and its green colour, and it is then called foliaceous. At other times it is shortened and thickened into a scale, or else expanded into a coloured or transparent membrane, and is then generally formed by the vaginal part. When it is reduced to a thin fascicle, it takes the form of a thread; or, when very short, that of a prickle, or only of a small point, commonly stiff and blackish.

It frequently takes a hue somewhat similar to the flower, and the shades which are seen in the flower, are found, either weakened or quite as deep, in the bracts, which are then commonly rather wide: the scarlet bracts of the *Scarlet Sage* (*Sauge éclatante*), of some kinds of *Cow-wheat*, or MELAMPYRUM, furnish examples very easy to procure.

When the limb is reduced into the bracts, its shape is generally preserved; sometimes, however, it is dentate more or less deeply, as in these very species of MELAMPYRUM.

§ 229. The bracts may be persistent for a long time at the base of the peduncle: but most frequently they are articulated and fall down very early: this ought to be recollected, so as not to describe inflorescences as wanting bracts which have only lost them: and of this we can, of course, only be sure when they have not yet grown much.

§ 230. The bracts, laying aside the exceptions (§ 225) of the axillary inflorescences, will preserve with one another the relation in the position of the flower-bearing peduncles: when these by the reduction of their common axis spring from the same height or from neighbouring points, as in the umbels and the capitula, the bracts are then at the same height, and form around the axis a kind of whorl which is generally called a *fence* or *involucre* (INVOLUCRUM), and in which they each take the name of *foliole*, or sometimes from their substance, *scale*. If the inflorescence is compound, besides an involucrum at the base of the general inflorescence, there is another at the base of each of the partial ones: these are distinguished by the use of the diminutive *involucelle* (INVOLUCELLUM). Thus in Umbelliferous plants, the umbellules are frequently *involucellate* (*involucellées*) (*fig.* 187, *i''*), the general umbel *involucrate* (*involucrée*) (*fig.* 187, *i'*). The very characteristic French word *collerette* (*collar*) is also often applied to these whorls of bracts: it may be either *general* or *partial*.

The folioles of the involucrum may be arranged in only one circle (*uniserial*), as is most common among these very Umbelliferæ; or else at different stages or heights (*pluriserial*), as may be often seen in the flowers of the Compositæ. In the latter instance, pressed against one another, the outer ones cover the bottoms of the interior ones like the slates of a roof: they are then termed *imbricate* (*fig.* 189, 1, *b*). If they then are numerous, their general spiral arrangement is easily recognised by describing the secondary spires: the bracts of the Artichoke, commonly termed leaves, are a familiar example of this. But this is sometimes not shewn: this is when the bracts are not numerous, and especially when arranged on two rows, in which those of the external, the smaller, are not like those of the internal. Some botanists have termed this last arrangement a *Caliculate involucrum* (*involucre caliculé*.)

§ 231. Sometimes all the folioles of the involucrum are free, sometimes they are united together either totally or only at their base: the involucrum is then called *monophyllate* or *polyphyllate*, according to one or other of these cases. In the former, if the folioles are in only one row, they form a *collerette* either entire, or scolloped in its circumference, as in the species of BUPLEURUM or *Hare's-ear;* if they are on several rows, they form a sort of cup bristled on the outside with scales or points, which are the free extremities of these folioles united and blended with one another throughout the rest of their body. Such is the origin of the *cup* or *cupule* (*cupule*) (CUPULA) of

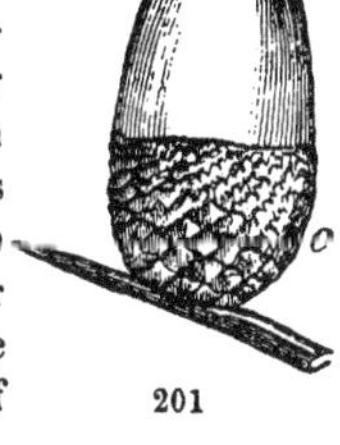

201

201. An acorn of the Oak.—*c* Cup or cupule formed by the union of a large number of bracts, the points of which are free and arranged spirally.

the Acorn (*fig.* 201 *c*). The spiny envelope of the Chesnut has an analogous origin: it is an involucrum, and its bark-like and brownish skin is an involucellum enclosing several flowers, as may be seen from the plurality of fruits which we frequently find in the inside. We here see that all resemblance to the leaf is completely effaced, on account of these changes and unions which are so often placed between our senses and the facts of the case.

§ 232. It is clear, that when the bracts are thus united for several rows, no flowers can be developed from the axil of any but the top ones: but the same thing frequently happens, even when they are free, and the imbrication most commonly brings sterility to the axils of all the exterior folioles of a capitulum. These are often developed so much the more, and those which bear flowers at their axils are very different and much less. Let us take the Artichoke, the receptacle of which, the fleshy part which is eaten, bordered by these steril, long, thick and green folioles, bears on the whole of its upper surface other short membranous, whitish bracts, mingled with its flowers: this is the part which we reject under the name of choke (*foin*).

There are several plants, in which beneath a single flower we find several of the sterile bracts arranged in a monophyllate or polyphyllate involucrum, which has then been called calicule or external calix: the HIBISCUS, the MALOPE, the *Mallows* and several other plants of the same species present this characteristic.

§ 233. Sometimes it is only one bract which envelopes the inflorescence either partly or altogether. We have already called this the *spathe* (SPATHA), which we meet with in a great number of Monocotyledons, around a spike of a peculiar nature, simple in the spadix, compound in the Palms. It is a kind of leafy sheath at the base, often rolled up like a horn, sometimes prolonged at the summit into a lateral tongue, sometimes green, as in the *Cuckoo-flower* (*pied-de-veau*) (ARUM VULGARE) (*fig.* 185, *b*), sometimes coloured, as in the CALLA ÆTHIOPICA. Its edges, which are bent underneath and united, often split and separate, when the inflorescence or the fruit, as it increases, pushes back the walls of the cavity, now become too confined. At other times, we see the spathe divided into two pieces or valves, because it was composed of two distinct and united pieces, but always, necessarily, the one exterior to the other, according to the law of constant alternation of the leaves in Monocotyledons. At the higher parts of the spadix, smaller bracts, called *spathelles* (SPATHELLÆ), are sometimes found at the base of separate flowers or little groups of flowers. The spathe appears destined to protect the inflorescence in its youth, for during this period it always envelopes it, although in several plants, as in the *Cat's Tail*, it is not developed in the same proportion, but remains cast down at the side of the base of the spike, or is even detached very early. A rather similar arrangement is observable in several Dicotyledonous plants, in which this envelope (which we then sometimes call a spathe by analogy, and which would be better called *spathiform involucrum*), generally results from the union or proximity of two great opposite bracts.

§ 234. In the Gramineæ, at the base of the little spikes or spikelets, which have for some time been considered as their flower, and which themselves are arranged either in compound spikes, or in panicles, are observed one or most commonly two green scales which have been named *glumes* and which are quite analogous to the spathes. But we shall have occasion to explain them in a little time, when we shall treat of the Grasses in the part describing the families of plants.

Transformed Organs.

§ 235. We have followed the plant through the whole of its developement, from the first appearance of the embryo to the production of the flower which terminates it; we have seen, how the extreme diversity of appearance in these organs really results from that of the forms which a very small number of fundamental organs may take, and we have examined those which most commonly present it. The study of these transformations, the determination of the organs under all the disguises which cause them thus to vary under our eyes, is one of the objects of botany, which, without it, would only be employed in comparing the numerous species, which it is called to distinguish and classify. Since botany has sprung from its infancy and is ranked amongst other sciences, its chief object is this comparative organography, which, especially in later times, has been simplified and perfected by ingenious theories and numerous applications, and which, thus taking a fresh flight, has also assumed a fresh name, that of Morphology (*Morphologie*) (μορφὴ, *form*; λόγος, *treatise*). The rule which guides us most surely in the prosecution of this study is drawn from the knowledge of the constant relation in the position of the parts to one another. We have then searched into the reason of all these relations, and we have verified them by following through all its successive steps each organ, the nature of which is not at all doubtful, and by seeing it always subject to these laws of position throughout all these changes of form. The most difficult cases, those which would almost necessarily lead the observer into an error, who did not use this thread to conduct his steps out of this labyrinth, are those in which one organ takes the exact form of another, so much so in short as to be mistaken for it. Thus we have seen the aerial roots, especially in climbing plants, often take the form of branches; we have also reciprocally seen branches sometimes take that of roots, as in the example of the Potatoe (§ 190).

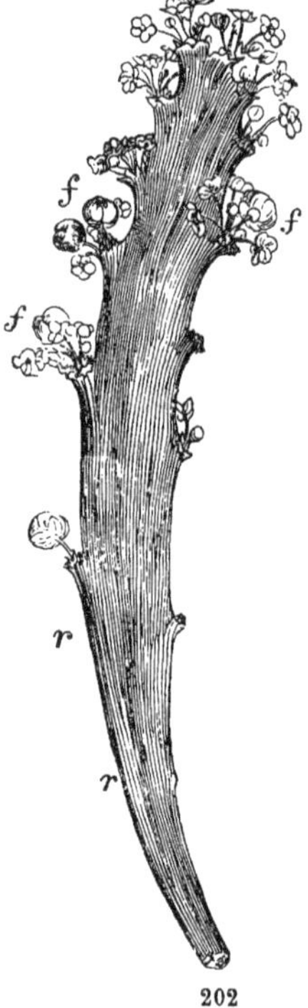

202

203

FASCIATION.

§ 236. It is not at all rare to find the branches assuming the appearance of a leaf by a change, analogous to that which we have described in the petiole enlarged into a phyllode (§ 141). Thus, in the species of XYLOPHYLLA, the cylindrical branches, at their last ramification, give birth to green leaf-like expansions (*fig.* 202 *r*), on the surface of which we are astonished to find here and there small bunches of flowers (*fff*). The flowers of the RUSCUS ANDROGYNUS

202. A leaf-like branch of the XYLOPHYLLA LONGIFOLIA.—*fff* Fascicles of flowers growing from it.

203. The tops of the stem of the CELOSIA CRISTATA or Cock's-comb, expanded into a kind of fleshy *Cock's-comb*, covered with sharp bracts, and flowers on the top.

are arranged in the same way, and in the Butcher's Broom (RUSCUS ACULEATUS), they also apparently grow from a similar leaf, but from its median nerve: this was formerly described as an epiphyllate inflorescence. But, since we have discovered that we can only find flowers growing from branches, we shall not hesitate to recognise them from this very fact in these pretended leaves, and this will be farther confirmed by finding the true leaf existing under the form of a small membranous foliole, from the axil of which springs the false one. Nature takes particular care to confirm our theory, by sometimes causing other similar branches to spring from these leaf-like organs, in the same way as the flat flower-bearing branches of the OPUNTIA and other CACTI grow from other flat branches, which may be compared to the fasciated ones, since leaves properly so called do not spring from one another. Besides, their stiff and ligneous structure is not that of leaves. *Fasciation* has been applied to this unusual arrangement of the ligneous fascicles of a branch, which, instead of remaining cylindrical, is thus flattened, as in a band (FASCIA). The large and thick tuft like a Cock's-comb (*fig.* 203), which we see bear flowers at the widened extremity of the stem of a plant commonly cultivated in our gardens, the *Cock's-comb* (CELOSIA CRISTATA), is a fasciated expansion. We often see them in the turio of Asparagus, in the branches of the Ash, of the Daphne; but here it is an exception, a case of monstrosity.

TENDRILS.

§ 237. (*Vrilles*) (CIRRHI).—We have already had occasion to speak of the tendrils of the Vine (§ 185, *fig.* 172 *v' v''*), and we have seen that it was a metamorphosis of the inflorescence, in which the flowers have proved abortive, and the peduncles, reduced in number and sometimes even to the primary, are lengthened into flexible herbaceous threads, capable of twining themselves round the bodies they meet. They are then the last ramifications of a climbing stem, quite similar to young shoots, but differing from them in that they never put out their leaves. We know, that in the Vine these tendrils, representing terminal inflorescences thrown out on one side, are opposite to the leaves: in the greater part of other plants furnished with them, they occupy their normal place, either at the extremity of the branches, or at the axils of the leaves, as in the Passion Flower (*Passiflore*). There are instances in which the conversion is not complete and the inflorescence presents, along with flower-bearing peduncles, others changed into tendrils. They are sometimes the result of the change of other organs than the branches or peduncles, of the different parts of the leaves themselves. In this case, the nerves are lengthened into this shape, sometimes the median alone at the extremity of the limb, either simple, as in the FLAGELLARIA INDICA, the METHONICA GLORIOSA, or most frequently compound, as in the *Pea* (*Pois*) the *Vetch* (*Vesce*) and the *Vetchling* (*Gesse*). In these pinnate leaves terminated in a tendril, it frequently produces lateral ones owing to an analogous metamorphosis of the upper folioles. It is not uncommon for the parenchyma to disappear completely in these leaves which are thus changed and reduced, either at their principal nerves, and then the tendril is branched; or at their median nerve, and then the tendril is single as in the *Yellow Vetchling* (LATHYRUS APHACA). As the median nerve and the petiole are the continuation of the same fascicle, we give these tendrils the epithet of *petiolar*.

Very rarely, the lateral fascicles of the vaginal part are prolonged into tendrils, and the leaf then has two growing at its base like stipules, one on each side: such appears to be the origin of those of the several species of SMILAX.

In all these instances, the point of departure of the tendrils points out to us the organ which is thus disguised. If it results from the metamorphosis of several axes of different ranks, as in the Vine, we frequently observe at the origin of each lateral thread, a small rudimentary leaf to which it is axillary.

SPINES.

§ 238. (*Piquants*) (SPINÆ).—Another form, quite different and almost inverse, under which the same organs are frequently disguised, is that of a *Spine*. Instead of a flexible and soft, thread-like tendril, we have a small stiff branch, shortened and terminated in a point, either simple or branched.

The branches most frequently present this change to our notice. Sometimes all the branches of a plant are changed, as in the Furze (*Ajonc*) (ULEX EUROPÆUS) and the COLLETIA; sometimes they are only the last, and sometimes even only the extremity, which instead of ending with a terminal bud, tapers into a spine and hardens. These spiny branches sometimes preserve a part of their characteristics in being covered with leaves and even with flowers, as in the Sloe-tree (*fig.* 204), or else naked throughout their whole length, lose these common marks completely on the outside, as in the GLEDITSCHIA. But an anatomical examination of their interior will always shew a structure identical with that of the branches.

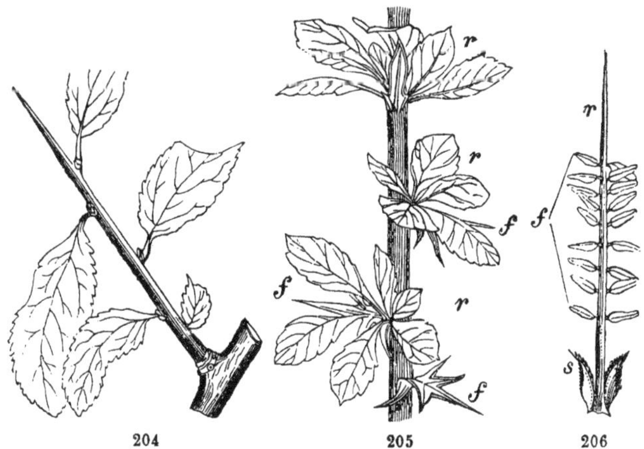

Peduncles very rarely terminate in spines, as they do in the ALYSSUM SPINOSUM.

In the leaf, the fascicles belonging to its different parts, may become spines: thus,—1st, The median or principal nerves, either with a portion of the parenchyma still uniting their base, and then we have a limb terminated or bordered with points as in the Thistles (*Chardons*); or without any parenchyma, as is most common in the Barberry (*fig.* 205 *f*). The spine is sometimes formed by the petiole alone. Sometimes it is only when it grows old that it takes this form; as the rachis of the pinnate leaf of the ASTRAGALUS TRAGACANTHA (*Astragal adragant*) and others, after the fall of the leaves which it bore when it was young (*fig.* 206). 2nd, The stipules hardened into two shorter thorns at the base of the leaf, as in our Acacia (ROBINIA PSEUDO-ACACIA), (*fig.* 207). The pulvinus itself, sometimes also spiniform (*fig.* 208), will be easily distinguished, if it only forms one point immediately below the leaf; but, when it

204. A branch of the Sloe (*Prunellier*) (PRUNUS SPINOSA) terminated in a spine.

205. A branch of the Common Barberry (*Epine-Vinette*) (BERBERIS VULGARIS), the leaves of which *f f f* have assumed the form of branched spines. From the axil of each a rosette *r r r* of regularly shaped leaves arises.

206. A compound leaf of the ASTRAGALUS MASSILIENSIS, the rachis of which *r* is terminated in a spine.—*s* Petiolary stipules.—*f* Folioles grouped in nine pairs.

is in two points, the distinction is not so clear. There is no need of explaining here, how the origin of the spikes is determined by the relations of their position with regard to the other parts of the plant, as in the tendril.

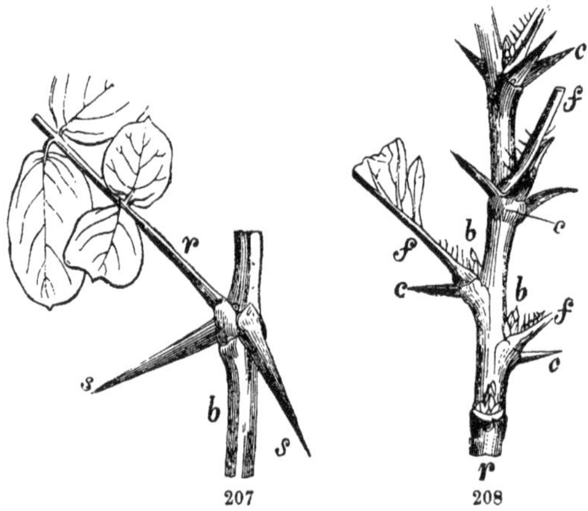

§ 239. We have now to examine some organs, which should have been done, perhaps, in the chapter treating of cellular tissue or of that of the bark, since they are only a form of this tissue, particularly of the cortical, but which it was better to leave till now, that we might not interrupt the general examination of these tissues, since their existence is far from being constant in them, and, when they are found, they are quite local.

PRICKLES.

§ 240. (*Aiguillons*) (ACULEI).—We shall begin with prickles, a natural transition from the spines, with which we have just been occupied, and with which they were confounded for a long time. Those of the Rose-tree (*Rosier*) will serve us for examples. If we look at them on the outside we shall see they occupy no fixed place on the branch, sometimes few and far between, sometimes close

207. The base of a compound leaf of the Acacia (ROBINIA PSEUDO-ACACIA): the stipules *s s* have assumed the form of spines.—*b* Branch.—*r* Rachis.

208. A branch of the Smooth Gooseberry (*Groseillier-à-muquereau*) (RIBES UVA-CRISPA), in which the pulvinus *cc c c* of each of the leaves is developed into a simple or triple spine.—*fff* The base of the leaves.—*bb* Buds growing from the axils of these leaves.

together without any order; we also see that they hold on very feebly, and are detached by a slight effort without causing a rupture. An examination by the microscope demonstrates, that they are composed of a cellular tissue analogous to that of the suberous envelope, dry like it is, and preserving its vitality only at the base, where it continues to increase whilst it is thick and hard over the rest of the superficies. These thorns of the Rose-tree cannot, then, be compared to those which result from the change of a fundamental organ or of one of its parts, and which, consequently, has a regular position and a fibro-vascular tissue. They are compared on better grounds to the hairs, from which they differ only in their greater size, and the agglomeration of a greater number of cells which compose them. The prickles are observable not only on the stem and its ramifications, but also on the leaves, and even on the parts of the flower which preserve most of the characteristics of leaves, but almost exclusively on the petioles and nerves. Their form is generally that of a cone, sometimes straight, more generally bent into a hook, commonly flat in one direction.

HAIRS.

§ 241. (*Poils*) (PILI).—We have already spoken several times of the hairs, but only in their most simple form, when each of them results from the prolongation of a single epidermal cell (*fig.* 87). The base of this cell being buried in the middle of the others, the rest of its body projects outwards, sometimes perpendicular to the surface of the epidermis (*fig.* 209, 1), sometimes obliquely, either from bottom upwards, which is most frequent, or in a contrary direction (PILI RETRORSI) (*fig.* 87), sometimes, in the last place, almost parallel (PILI ADPRESSI) (*fig.* 214). Its sur-

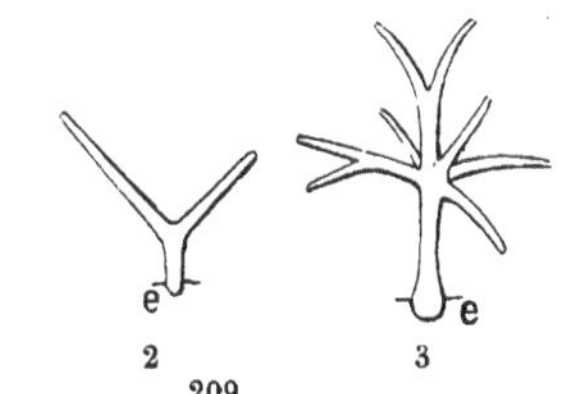

face is smooth, though sometimes covered with little asperities (*fig.* 210, 4); its most common form is that of a long thin cone (*fig.* 209, 1), of a needle; but it may be cylindrical, and even swell into a club-shaped mass at the top. Lastly, it may separate into two or more directions (*fig.* 209, 2), become branched (*fig.* 209, 3)

209. Hairs formed of a single cell growing from the epidermis *e*.—1. A simple hair.—2. A bifurcated hair of the SISYMBRIUM SOPHIA.—3. A branched hair found on the leaf of the ARABIS ALPINA.

although always presenting a continuous cavity in the inside. It may be ramified either at its base, or at a certain height.

Several hairs are formed, not by one single cell, but by several cells joined end to end; and the faces in contact seeming to interrupt the continuity of the hair by so many partitions, it is termed articulated. These different forms are almost the same as when they are formed by only one cell, those of a cone, in which the cells placed one above another diminish as they approach the top (*fig.* 210, 1), or of a cylinder, in which they are equal in diameter, or of a club-shaped mass, in which the upper ones are enlarged, or of a little tree more or less ramified (*fig.* 210, 2). When several hairs grow from a common centre, they form a pencil (PILI PENICILLATI), or a star (PILI STELLATI SEU RADIATI) (*fig.* 210, 3), according as they are directed obliquely or parallel to the face of the epidermis. These latter arrangements are characteristic in whole families of plants, as in the Malvaceæ. The cells united end to end may present, each in its length, not an equal or gradually decreasing diameter, but be contracted either at the middle or most commonly at their extremities, so that the hair takes the form of a small necklace, *moniliform hairs* (PILI MONILIFORMATI) (*fig.* 210, 4).

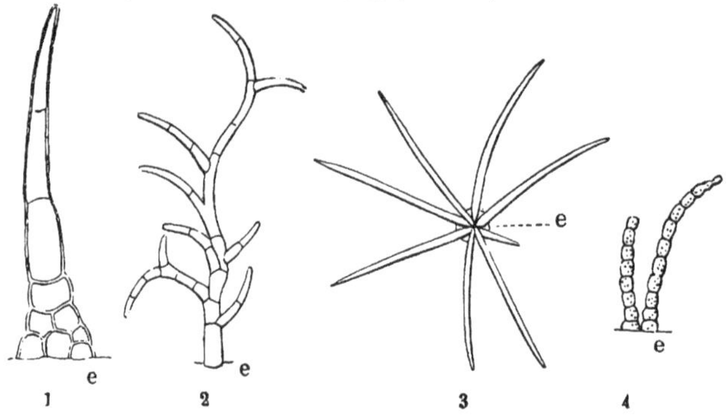

210

The compound hair is not always formed of a single row of cells: but we may find several of them placed together at the same height.

210. Compound hairs formed by the union of several cells.—*e* The epidermis from which the hair grows.—1. A simple articulated hair from the stem of the Bryony (*Bryone*) (BRYONIA ALBA).—2. A branched hair from the flower of the NICANDRA ANOMALA.—3. A stellate or star-formed hair from the leaf of the ALTHÆA ROSEA.—4. A moniliform hair from the LYCHNIS CHALCEDONICA; its surface is rough with small asperities.

This is the first transition to the state of prickles, which only differ from them in as much as they consist of a deeper layer.

Under the microscope the cells of the hairs will appear to be composed of a double membrane. We have seen this (§ 48, *fig.* 93), where the epidermal pellicle extends over the hairs as well as over the rest of the epidermis, and forms so many sheaths by which the proper membrane of each leaf is covered by another external membrane.

§ 242. The hairs which radiate from a common centre are sometimes united to one another, probably by means of this pellicle which envelopes their *ensemble*, and then, instead of a star, they represent a kind of membranous plate (*fig.* 211), adhering by their centre only to the surface which bears them, and are easily detached like little scales by desquamation. These are called, therefore, *scaly* or *shield-like hairs* (PILI SQUAMOSI SEU SCUTATI[a]), or a single word borrowed from the Greek, LEPIS (λεπὶς, a *scale*). They generally shine brightly and often with a metallic lustre, as in the leaves of the Elæagneæ.

211

We may mention besides these other squamiform or membranous expansions, which, instead of adhering to the surface by a central point, are fastened to it by their widest edge. It is like a fold of the epidermis, or, if it would simplify it, it may be described as a compound hair formed by the union of a great number of cells, and extended in breadth rather than in length. These are called *ramentaceous hairs* (PILI RAMENTACEI, or, in one word, RAMENTA SEU VAGINELLÆ). We find them particularly developed on the petioles and limbs of the greater part of the Ferns. Their colour is generally brown.

§ 243. We should suppose, that, if the hairs are only an elongated cell, projecting to the outside of the others, we shall meet with them wherever such a projection is possible; and, in short, we observe them in the internal cavities of some vegetables, in the lacunæ of

211. A scale or shield-like hair from the leaf of the Sea Buckthorn (HIPPOPHÆA RHAMNOIDES).

[a] They are sometimes also called *scurf*.—TRANS.

the stem or of the petiole of the Water-Lilly (Nymphæa) and other aquatic plants. But, this internal production is almost exceptional, and it is on the epidermis of the different parts of the vegetable that they are generally found, chiefly on the parts exposed to the air, although we also see several on those parts which are sheltered, as on the seeds or on the surface of the loculus of the fruit which contains them, and very commonly on young roots, as we have seen in a previous part (§ 116).

They often abound on the branches and on the leaves, and in the latter appear most frequently and abundantly on the lower face and on the nerves and petioles. Their existence and their functions appear to be in relation to the youth of these parts, to the afflux of the liquids with which they are filled, and the activity of the evaporation which is a consequence natural to them, and which they are probably destined to moderate. As these surfaces extend by the growth of the different parts as they grow old, there is not always a proportionate production of fresh hairs. Those which cover it with a thick down, separated by a space which is continually increasing, cover it at last very imperfectly. This is the reason, why hairs, often so abundant on a young shoot, seem to have disappeared, when it acquires a certain developement. Sometimes they are detached or are dried, and it is rare to find them on the bark of old branches in ligneous plants.

§ 244. We have pointed out the most general forms of the hairs when considered separately. Generally, in our descriptions, we stop at the point which the naked eye or lens can see, and say that the hairs are simple or branched in such or such a manner, without looking whether they are unicellular or multicellular ; this, in fact, can have no great importance since both are found by the side of one another.

But the most important part in the description is to represent the appearance, which results from the meeting of a greater or less number of hairs on a part of the vegetable, and we have only now to state what are the principal modifications and the names under which they are known. These names are as follows :

Smooth (*Glabre*) (Glaber); when there are no hairs on the plant.
Glabratus ; a surface which has lost its hair.

Hairy (*Poilu*) (Pilosus); covered with hairs.

Pubescent (Pubescens) ; covered with soft hairs, rather short and not many of them, with *down* (*duvet*) (pubes), like that on the chin of a youth.

Villose (*Velu*) (VILLOSUS); with long, soft hairs, which are rather oblique.

Silky (*Soyeux*) (SERICEUS); with recumbent, silky hairs, having a sublucid, silky appearance.

Hispid or *hairy* (*Hispide*) (HISPIDUS, HIRTUS); bristled with stiff, upright hairs. HIRSUTUS expresses a state between this condition and that expressed by VILLOSUS.

Velvety (*Velouté*) (VELUTINUS); covered with a short, soft, but rather rigid down, like velvet.

Cottony (*Cottoneux*) (TOMENTOSUS); covered with crisp hairs like cotton, entangled into a kind of felt (TOMENTUM). This state generally results from the accumulation of hairs into pencils.

Woolly (*Laineux*) (LANATUS, LANUGINOSUS); covered with long, soft, entangled hairs, like wool.

Scaly (LEPIDOTUS); covered with scaly shields.

Ramentaceous (RAMENTACEUS); covered with scales or RAMENTA.

When the hairs are long and form a fringe to the margin they take the name of CILIÆ (*cils*). [Adj. CILIATUS]. If they are arranged in tufts they are called *Beard* (*Houppes*) (BARBA). [Adj. *Bearded*, BARBATUS].

There is no need of explaining the different degrees which we express by the diminutives GLABRIUSCULUS, PILOSIUSCULUS, VILLOSULUS, TOMENTELLUS, HISPIDULUS, CILIOLUTATUS, to indicate the state of a surface on which the hairs are comparatively shorter, &c.

GLANDS.

§ 245. (*Glandes*).—A GLAND, in vegetables as in animals, is an apparatus containing some liquid of a peculiar nature differing from those which are to be found in the rest of the body; a liquid, which is *secreted*, that is to say, drawn from substances in connection with it by the action of the organs which compose this apparatus. In the vegetable, it is a cellular tissue, which is always entrusted with this function, and it is not distinguishable from that which we have hitherto described. It is only recognised by means of its contents; but it is impossible to judge of its action from its shape: thus, organs which are now regarded as glandular, were, for a long time, confounded with others which do not secrete any peculiar kind of fluid, with the hairs, for instance. These hairs have been quoted under the name of glandular.

§ 246. GLANDULAR HAIRS (*Poils glanduleux*).—These secretory hairs sometimes preserve, even without the slightest change, one of the forms we have just described. We do not see anything different, except the liquid which accumulates in their last cells and oozes out. But most commonly the secretory power presupposes a slight change of form, generally a terminal swelling. If the hair is formed by a single cell, it is widened either all the way down[v] or only at its summit, into a globe, an egg, or a club (*fig.* 212, 1); if it is formed of several cells, it is always the highest which secrete, sometimes a single one (*fig.* 212, 2), the last, more or less dilated into one of those forms we have just named: sometimes several terminal ones placed end to end (*fig.* 212, 4), or at the same height, two, one by the side of the other (*fig.* 212, 3), four in a cross, &c.; sometimes, lastly, several united into a single mass which constitutes the swelling. The other cells of the hair, placed underneath, present their ordinary conformation, and support, as well as attach to the epidermis, the simple or compound cell-gland, which is then called *pedicellate.*

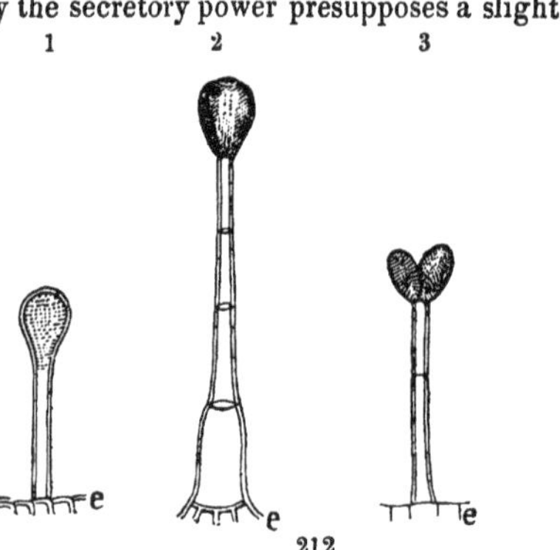

§ 247. The *stinging hairs* (*poils urticants*) (SETÆ URENTES) have been described as being formed quite differently, those whose prick causes a very painful itching; of the Nettle (*Ortie*), for instance.

212. Glandular hairs.—*e* Epidermis from which the hair grows.
1. A hair composed of a single cell from the leaf of the SISYMBRIUM CHILENSE.
2. A hair composed of several cells, terminated by a secretory cell, from the peduncle of the Snap-dragon (*Muflier*) (ANTIRRHINUM MAJUS).
3. A hair composed of several cells, terminated by two united secretory cells, from the peduncle of the Loose-strife (LYSIMACHIA VULGARIS).
4. A hair composed of several cells, terminated by several secretory cells united end to end, from the Avens or Herb-Bennet (*Benoite*) (GEUM URBANUM).

[v] We must rank among these glands reduced to one superficial and almost sessile utricle those yellowish granules known by the name of *Lupulin* (*Lupuline*), which are so abundantly scattered over the leaves, the bracts, and the flowers of the Hop (*Houblon*) (HUMULUS LUPULUS). These are so many simple vesicles filled with a fluid and bitter resinous principles, which harden and alone remain, and in which are the properties for which this plant is so much used.

It was supposed, indeed, that the liquid was secreted in a mass of glandular cells concealed under the epidermis ; from the middle of this mass sprang the hair, the tube of which would serve as a duct for the poison into the wound, exactly like the fang of a viper, pierced with a canal in communication with a little gland situated at the base of the tooth; but it is not so. The hairs of the Nettles (*fig.* 213), of the LOASA, of some JATROPHA, are all equally formed by a single conical cell, long and stiff, swelled into a bulb at its base (*b*) and terminated at its other extremity, either in a straight line, or a little to one side with a small button (*s*). It is in this cell that the poison is formed; and when it buries itself in the skin, it breaks and leaves its end there, which is retained by the small terminal button. Hence, there is a double reason for the irritation; the presence of a foreign body and the peculiar nature of its contents.

248. The hairs of some species of MALPIGHIA (MALPIGHIA URENS, FUCATA, &c.) (*fig.* 214) have been placed amongst the class of stinging ones; they are remarkable for their shape, that of a shuttle placed on the leaf to which it adheres by the middle. To this middle a circular opening corresponds which leads to a canal occupying the centre of the hair, and ended by a little gland (*s*) situated near the superficies of the leaf. It was once thought, that the hair, when it pricked the skin, forced out a liquid secreted by this gland. But this hair, very stiff and with very thick walls, especially at its two points, is not open at its extremities and is not broken in the wound as may be easily seen after it is extracted; it cannot, therefore, pour out any fluid, and only irritates in the same way as a common thorn; the irritation ceases directly it is extracted. The name of *shuttle-shaped hairs*

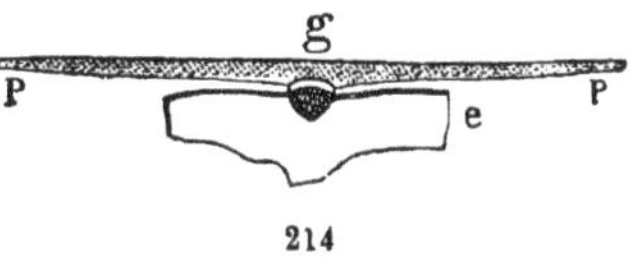

213. A hair of the Common Nettle (*Ortie Commune*) (URTICA DIOICA). It is conical and terminated at its summit *s* by a button-like swelling; at its base, by a large dilation into a bulb *b*. This base is surrounded by cells of the epidermis *u e* which are raised round it in order to support it. In the cavity of the hair we find currents of a granular substance.

214. A hair *p* of the MALPIGHIA FUCATA situated on a shred of the epidermis *e*.—*g* A gland, which, buried on one side in the epidermis, on the other in the opening situated in the middle of the hair, unites them.

(*poils malpighiacés*) (*poils en navette*) (PILI MALPIGHIACEI) has been given to these recumbent hairs, which are fixed and correspond to a gland at the middle: they have often been confounded with all recumbent hairs in the shape of a shuttle; the absence of glands in the latter will enable us with a little care to distinguish them.

§ 249. GLANDS PROPERLY SO CALLED.—The transition from glandular hairs to glands is almost insensible. When a mass of secretory cells is without pedicles, but attached to the epidermis by a kind of contraction, is it a compound sessile hair? Is it a superficial pedicellate gland? In short, the name is of no moment. We, however, remark two modifications of some moment; 1st, the gland is hollow in the interior, its cells forming the envelope in a single row (*fig.* 215); 2nd, the gland is solid, without a central lacuna. We thus pass gradually from this pedicellate gland (*fig.* 216) to that which expands and is fastened by a large surface, like a kind of wart. Thus, in Roses, Brambles (*Ronces*), we find their summit to be not much more swelled than the base.

§ 250. At other times, the glands are buried in the mass of the cortical parenchyma, but, generally, very superficially and immediately under the epidermis, and, even then, it is not unusual to see them project above the surface, covered by the epidermis, sometimes somewhat modified, which follows them and is moulded to their surface, or else to see the broken epidermis surround the upper part of the gland which is uncovered.

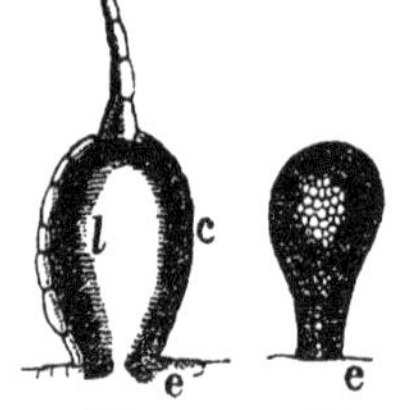

§ 251. Amongst these internal glands, we ought to mention those known under the name of *vesicular* (*vésiculaires*), and which, furnished with pellucid walls, secrete a volatile oil which is colourless or almost so, and appear like transparent points, on the green leaf which bears them, when the light strikes through it. The leaves of the different kinds of St. John's Wort (*Millepertius*), of the Orange (*Oranger*), of the Myrtle (*Myrtle*), of Rue (*fig.* 217), are examples which are familiar to every one and will serve for our researches on

215. A gland from the peduncle of DICTAMNUS ALBUS, cut vertically, so as to allow its central cavity *l*, to be seen: it is filled with a greenish oil, and its envelope is formed of a layer of cells *c* filled with a red juice.—*e* Epidermis.

216. A gland from the ROSA CENTIFOLIA. It appears under various shapes.—*e* Epidermis.

this point. We here see that these transparent points are formed by a small number of utricles (g) much larger than those of the surrounding tissue (ue) and loosely united to one another. They are even sometimes separated from one another, leaving between them a central lacuna (l) in which the liquid is collected. These are the glands which form nearly the whole of the skin of the orange. On its white flower we perceive them like so many little green spots, which proves this to be the colour of the liquid, although it is so transparent.

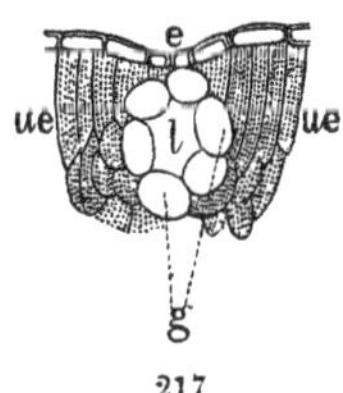

217

The reservoirs of proper juices, of gums and of resins, which we consider as quite distinct from the glands, are lacunæ with a wall of peculiar cells, in which their liquid is formed and from which it is spread, and, consequently, are very like the vesicular glands from which they differ in being situated more deeply.

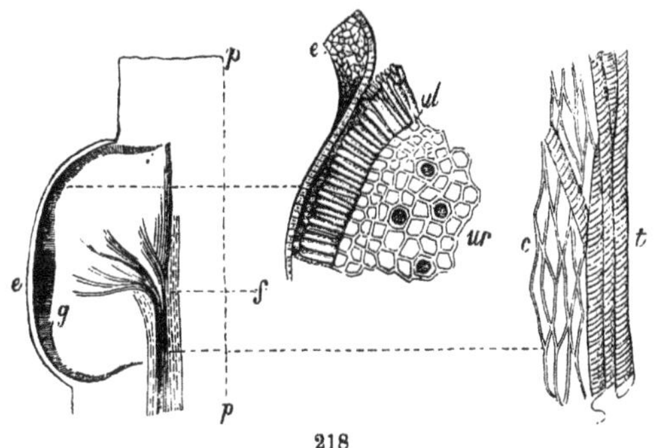

218

§ 252. The greater part of the internal glands is, unlike those

217. A vesicular gland from the leaf of the Common Rue (RUTA GRAVEOLENS).—g The gland formed by large transparent utricles, separated from one another so as to leave a central lacuna l.—e Epidermis of the upper surface of the leaf.—ue ue lengthened cells of various shapes filled with chlorophyll and forming the green tissue of the leaf.

218. A vertical section of a gland taken from the base of the petiole of the HETEROPTERYS CHRYSOPHYLLA.—p A part of the petiole.—g A gland joined to it and covered by the epidermis e.—f The vascular fascicle which borders on the gland and apparently throws out several branches into it.—On the side are represented small portions of the same parts much more magnified, namely: 1st, A portion of the glandular tissue with its epidermis e, its inner rounded utricles ur, some containing crystals, and its outer long and filiform utricles ul; 2nd, A portion of the fibro-vascular fascicle composed at its centre of tracheæ t, nearer the outside of lengthened cells c.

we have just examined, more opaque and composed of cells very closely united and smaller than the surrounding tissue, does not determine by separation a central reservoir, and forms, besides, in their mass some accidental lacunæ. Sometimes, at least we observe it in the Malpighiaceæ (*fig.* 218), all the surface of the gland presents a layer of cells (*u l*) quite different from those which compose the inner mass (*u r*). They are like so many very fine, small, obtuse hairs which cover this surface in great numbers and give it the appearance of velvet.

§ 253. The matter which is formed by the glands is sometimes limpid, sometimes thicker, of very different natures, according to the plants in which it is produced. We have seen it accumulate in the interior of the cells which form it or else in neighbouring reservoirs. It is frequently spread abroad, either because the external surface is itself secretory, or else transudation takes place through the cellular wall. Then, in contact with the air, it frequently changes its nature, is thickened or concentrated, and it is generally in this last state, when it is found on the outside.

§ 254. A certain relation has been discovered between the glands and the tracheæ, which are supposed to enter into the composition of the most perfect. We have lately said that the glandular tissue is exclusively utricular. But it is no less true that tracheæ are found in its neighbourhood, and penetrate, if not the very tissue, the surrounding one (*fig.* 218 *f t*). We see them come to the very foot of the pedicellate glands; in the Drosera or Sun-dew, for instance. Can their functions have anything in common, or is it not a natural consequence, that these two organs equally developed in the young parts, and only in those which are formed that year, will frequently be put in juxtaposition with one another?

FUNCTIONS OF THE ORGANS OF VEGETATION.

§ 255.—We have now examined the organs of the vegetable from its first appearance in the embryonary state to that of the flower. These organs, which have hitherto occupied our attention, all concurring in the support of the life of the plant of which they form part, are, under the collective name of organs of vegetation, distinguished from those of reproduction, which in the flower join in producing fresh embryoes, each destined in its turn to undergo the same progressive changes of its parent. Now, before we begin the examination of this new class of organs, in order to complete that of the former, which we have prosecuted in treating of their *Organography*, i. e., their structure, their shape, and their arrangement, we must say a few words on their *Physiology*, on their functions, their actions during their life.

§ 256. We have followed (§ 32—35) the first changes a young plant undergoes, when it commences to live by itself, or in a single word, to germinate. When its germination is accomplished, we find, in its lower part, its roots which are in connection with the ground; in its upper part, its stem and its leaves in connection with the air. Its roots pump the liquids from the earth or from any other humid medium in which they may happen to be placed; this function is called ABSORPTION. The liquid, when it has once entered into the plant, runs through every part of its tissue, for which function we have found (§ 17) the means of communication most wonderfully prepared: this is called CIRCULATION, a term borrowed from Zoology, although, in animals, the analogous function is exercised by different agents and in a different manner. The liquid, which takes the name of sap, is modified in its progress, especially over all the surface of that part in connection with the air: this action of the air on the sap is called RESPIRATION. The sap, thus perfected, has become fit for the nourishment of the tissues, that is, by means of particles like to them, for strengthening the organs which already exist, and also for producing others of the same nature: hence arises the term NUTRITION or ASSIMILATION. At some places, however, it furnishes substances more or less different, either destined for a special use, or set aside to undergo in a little time a fresh elaboration, or useless or even hurtful to the plant, which rejects them from the living tissue. These are SECRETIONS, which in the latter case are called *excremental* (*excrémentitielles*). Such are the principal functions which exist in common between the Animal and the Vegetable Kingdoms. After having studied them in detail, we will cast our eye over the differences which exist between both the kingdoms of organized beings.

ABSORPTION BY THE ROOTS.

§ 257.—We have seen (§ 115) that the roots are covered by a continuous layer of cells without any openings. How does the liquid in contact with them penetrate them, and how does it pass into all the other cavities which compose the vegetable tissue, separated from one another by thin membranes? These membranes are, it is true, permeable to the fluids; but, in order that these fluids may traverse them, a sufficient power is requisite to force them. That power, called ENDOSMOSE by M. Dutrochet, who has so well explained it, accounts very plainly not only for the absorption of the roots and also that which takes place consecutively from cell to cell, but also for a part of the circulation of vegetables, which, before this discovery, remained inexplicable.

If a small bladder, the wall of which is either an animal or a vegetable membrane (that of the Bladder-Senna) (*Baguenaudier*), is plunged into pure water, itself containing a denser liquid, such as a syrup or gum water, the two liquids will tend to restore equilibrium between themselves, and through the walls a double current will be formed, the one from the exterior to the interior, from the pure water to the syrup in the bladder, the other from the interior to the exterior, from the syrup to the pure water. But the two liquids do not filter through the membrane with the same facility or with the same rapidity; the less dense passes quicker than the other. The mass of enclosed water thus gains more than it loses, and the exterior loses more than it gains; hence will result a difference in the height of the two, and the ascension of the liquid contained in the bladder; an ascension which is only stopped, when the two liquids shall be, on account of this change, in equilibrium with one another. If we adapt to the bladder a graduated vertical tube (*fig.* 219), we may calculate the quickness and force of the ascension. If, instead of the straight tube we

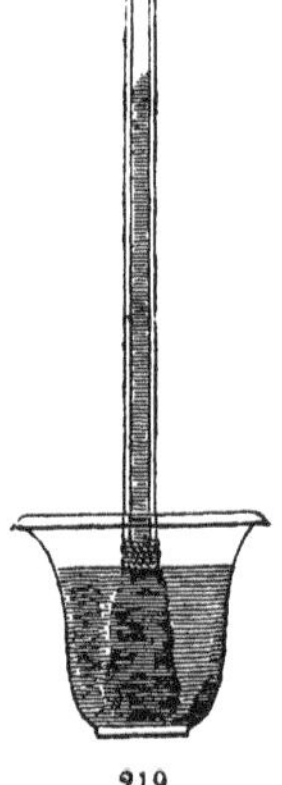

219

substitute one with a double bend, the lower one filled with mercury, which will indicate by the height of its column, as it mounts into the external graduated part of the tube, the resistance which the column of syrup has to surmount. By such experiments we have been enabled to determine that the speed and strength of the endosmose proceed together, that they are considerable, and that the action lasts for a long time. An aqueous solution of one part of sugar to two of water, in two days causes the column of mercury to rise more than a metre (French) or 39·33 (English inches), and at the end of this time it only contains three of water to one of sugar.

§ 258. The absorption of the roots is now very easy to explain. The cells, which form their tissue, are filled with denser juices than the water which the earth has imbibed: and this water, by means of endosmose, is infiltrated through their membranes, swells the cavities of the external cells, diminishing the density of the liquid which is there, and passes thence into the interior cells. If we attempted to assist the growth of the plant, by furnishing it with its nourishment already prepared, by putting its roots in contact, for instance, with a syrup of sugar and water, far from assisting it, we retard it by preventing endosmose, and, consequently, absorption.

§ 259. At which part of the roots is this process most actively carried on? Experiment tells us, that this takes place at its last and most recently formed ramifications, at their extremities, as well as at the fibrils or chevelu with which they are covered. We know that, to be sure of success in transplanting a plant, we must preserve as many fibrils as we can, and maintain them in the state of humidity and turgescence which is natural to them. We have seen (§ 116) that, at the first stage of their growth, they are bristled with soft hairs which we may suppose to be designed to multiply their surface, and, consequently, their parts of absorption. Observation, however, teaches us that the action of these two surfaces is very weak compared with that of the extremities themselves. We may, indeed, so arrange the roots of a plant at a little distance above some water, that their extremities alone are plunged into it, whilst all the other parts do not touch it; and we here see from the activity of the vegetation that that of the absorption is very great. If, on the contrary, we arrange these roots so that they are quite plunged into water, their extremities alone excepted, which must be outside; here we shall find, that the vegetation does not quite cease, but languishes: it is clear that absorption still takes place, but in a slight and insufficient degree.

We have already said (§ 114), that lengthening of the root and of all its ramifications takes place only at its end, which is, consequently, in the state of nascent tissue, during the whole time that the activity of vegetation is carried on. It is not, then, by a peculiar modification of the swelled tissue acting like a sponge, as it was supposed, that the radicellary extremities suck up the moisture which surrounds them ; it is, on the contrary, because their nascent cells already swelled with thick juices, happen to be placed in the most favourable conditions for the endosmose. Their epidermis is not yet formed, as it is higher up, when it opposes a drier and less permeable layer to the absorption.

§ 260. The surrounding liquid is better absorbed and in greater quantities in proportion as it is more fluid. In the earth, it is water holding in solution different soluble solids, which vary according to the soil. The solution of these matters must be complete; for, if they are in suspension only, they cannot pass, however fine they may be. On mixing with water the finest and most impalpable powder, but which is not directly dissolved, such as that of charcoal, and then, in this state offering it to the absorption of the roots, we see that the water alone passes into these roots, and that all the charcoal remains on the outside without its being possible to find a single atom within. We obtain the same result with almost all coloured solutions: the water on passing through the radicellary extremity, is deprived in its passage of colouring matter, which is deposited on its surface.

§ 261. It has been a question, whether this absorbent surface can shew a vital action, a sort of choice in the matters which are presented to it. The answer appears to be in the negative, since it easily admits a number of solutions which are poisonous to the plant, and soon cause it to die when they have penetrated into its interior. A contrary fact, however, is mentioned; the plant was fed with a solution of nitrate of strontian : here the water was absorbed, and the whole of the salt remained on the outside, as if it had only been in suspension. As to the experiments of Saussure, according to which, on presenting to the plant certain substances in solution, we find, that the water passes in a much greater proportion than the dissolved matter, or, that of two matters dissolved in the same water, the one penetrates in much greater quantities than the other; he himself draws a very wise and ingenious conclusion, that this is not owing to any affinity, but to the fluidity or viscosity of the various substances. He attributes to the fineness of the filter furnished by the vegetable membrane that effect of which the endosmose would be a still more powerful cause.

CIRCULATION.

§ 262. ASCENDING SAP.—The moisture of the earth has penetrated the extremities of the roots. From these it ought, by a similar operation to pass immediately into the cells situated directly above them, then into those situated still higher. Thus, from each cell to one still higher, it mounts in the root until it comes to the stem, in which its ascending movement ought to continue. For we are right to compare the plant to an endosmical apparatus, in which the earth performs the part of the receptacle of water; and this apparatus is more energetic, since it is not an empty and inert tube, but is itself a tissue filled with numerous depôts of substances analogous to those which have already provoked the action of the roots; so that this action instead of being spent, is renewed at each row of cells. The liquid has not, as in our experiment, become thinner in proportion as the mass is augmented, and, consequently, is raised in height; on the contrary, acting on those substances which it finds in its passage, it dissolves a portion of those which were in a solid state, and is thus more and more thickened. Being thus modified since its entrance into the vegetable, it has taken the name of *sap* (*sève*). If, at several places in a tree, we perforate the trunk very deeply, and introduce a tube into each hole, and collect in separate vessels the sap which flows out of each of these canals, we shall find that it is thicker and more viscid in proportion to the height of the hole; we shall see in a few minutes what changes its composition undergoes and by what means it may be determined.

§ 263. We have hitherto spoken as if the plant was only composed of cells: and such is, indeed, the structure of some vegetables. But, we know that most frequently, in Cotyledonous vegetables, several vessels are found in the middle of this cellular tissue, which follow the direction of the axes. We may easily conceive how much the progress of the sap, continually forced upwards, would be accelerated by these long ducts, in which it will find no obstacle, and how it may thus run rapidly through great distances, which would have been traversed from cell to cell very slowly and in doing so have occupied a very long time.

Let us now remark, that the centre of the roots is occupied by fascicles of vessels which extend lengthways almost to the extremities where the absorption commences. The absorbed liquid thus finds almost immediately this more speedy way of transport; and this is, doubtless, one of the reasons which explains the fact that the absorption by the extremities is much quicker and more useful than that by any other part of the plant.

§ 264. Physics tell us that in extremely fine tubes, which are termed capillary from being likened to hairs, the internal wall of the duct exercises on the contained liquid which moistens it, a sort of attraction, which destroys a portion of the weight, and thus determines the ascension of this liquid above the level which it would otherwise retain. The greater part of the vessels in the vegetable are, on account of their smallness, so many capillary tubes, and exercise on the contained liquid that action which makes it rise to a certain height and is thus added to the force of the endosmose. Before the latter fact was made known, the greatest part of the ascending motion of the sap was attributed to capillary attraction, without, however, being able by it alone to explain all the attendant phenomena.

When we plunge into water or any other liquid sufficiently limpid, the end of a branch which has been recently cut, this liquid penetrates through the open orifices of the vessels, and mounts immediately, by means of capillary attraction, up to a certain point. We may suppose that in this passage the action of the endosmose is exercised through the walls of the vessels and surrounding cells; so that the cut end of this branch thus supplies the absorbing action of the roots. We may now understand why gardeners, in planting a vegetable, the fibrils of which as well as the radicellary extremities, have dried and become incapable of absorption, as frequently happens during transplantation, refresh the roots, that is, cut them to the point, at which they see their freshness and vitality perfectly preserved. This is the very cause which permits the multiplication of plants by cuttings, putting into a medium sufficiently humid, the extremity of a branch which sucks up at its section the juices, by means of which its life will be prolonged sufficiently to cause it to produce adventitious roots and to take upon itself all the conditions of a rooted plant. The freshness preserved by bouquets, when we leave their stems in the water, is a phenomenon familiar to all our readers. The necessity of cutting the end touching the fluid cleanly and evenly is explained by the necessity of opening the vessels, which are either open or shut as the bud has been separated by being cut or twisted off. The capillary tubes of the vegetables present rather a large passage to the liquid, so that it penetrates them more easily than through the cellular walls. They may then admit liquids holding in suspension very fine pulverised substances, such as some colouring matter; and botanists have taken advantage of this property to study in their interior the progress of the sap, which we may follow without much difficulty when it is thus coloured. But we must be very reserved on the conclusions which we may draw from this, since things do not take place exactly in the same manner as in ordinary life, when the absorption takes place by means of roots and the vessels at the same time as from cell to cell.

§ 265. But the endosmose and the capillary attraction are not the only forces which cause the continual ascension of the sap. We may foresee, indeed, that a time will arrive when they shall have produced their effect, and then there will be established a kind of equilibrium in all the liquid parts of the vegetable. But, although this takes place to a certain point, and after a certain period of extreme activity, this movement is considerably slackened and ceases entirely in certain parts, yet it continues in others, and the absorbing action of the roots is maintained in the same proportion. We know, that if we pluck out of the earth a plant arrived at its perfect state, it lives for a longer or shorter time; and on plunging its entire roots into water, if they are fresh, or with the ends cut off if they are dry, we see it revive rapidly from one end to the other; there is therefore an absorption and transmission of a very considerable quantity of water from the lower extremity to the upper, and the liquids contained in the plant were not in a state of equilibrium from which would result their definite immobility.

§ 266. Let us now mention a very interesting observation which may throw some light on the subject. Under the tropics, a certain number of Creepers, especially of those of the genus of Cissus, a neighbour of the Vine, are filled with an abundance of sap, fresh and agreeable to the taste. The water which flows copiously from their cut branches may serve as a drink and men, in their travels through the deepest forests, employ it to quench their thirst: this has caused these plants to be termed Water or Hunter's Lianas. M. Gaudichaud, who has discovered one of these at Brazil, to which he gives the name of Cissus hydrophora, has remarked that if we content ourselves with cutting the Creeper across the stem at one place, there issues from the two surfaces of the section very little liquid. It continues to mount rapidly in the upper part, in which we may be assured that the vessels are being emptied from the bottom to the top. This ascension cannot be attributed to the roots, with which the upper part is no longer in connection, and the vessels are of much too large a diameter for capillary attraction to have any influence. But if we cut it at two different heights, so as to detach a fragment of the stem of a certain length, we immediately see a great abundance of sap flow from that extremity which is held the lowest down, consequently obeying the laws of gravity. Now, previously, the sap continued to mount very rapidly. This can be caused by no force which is placed beneath or at the sides; it can only, therefore, be from some force situated above the second section and drawing the liquid upwards. (See Appendix A).

§ 267. It is not difficult to discover the cause of this fresh force. The vegetable, at a certain height, is furnished with a greater or less number of buds. From the moment that they begin to be developed, they draw from the stem or the branch to which they are attached, materials destined to nourish them, and the quantity of these must be in proportion to the branch, which will result from this developement. The leaves appear at the same time, expand into the air, and become the seat of a considerable evaporation from their surface, which is perforated with pores. All, that evaporates in this manner from the leaves and at the same time from the young bark of the branch, all, that is employed to form and nourish these fresh parts, is so much taken from the liquid of the stem; and there results from it towards the surface and the origin of each branch, empty places, which are immediately filled by a proportional quantity of the sap raised by the stem, replaced itself at the same time by that of the neighbouring parts, and thus determining from each part an ascending flux beginning from the root, the absorption of which compensates this loss[x]. There is no need of explaining what in-

[x] The force of the suction of the buds and leaves may easily be determined by placing the lower extremity of a branch into a liquid, of which the portion absorbed in a given time (§ 301) is determined. Recent experiments conducted on a very large scale have proved that its power is very great, and even surpasses the limits which were given to it. M. Boucherie, indeed, proposed to communicate to wood certain qualities serviceable in their consumption, to render them indestructible, harder, softer, incombustible, or to colour them with various hues, and tried to make various solutions penetrate into the tissues, so that, on combining, either with the substance of the vegetable or with another solution afterwards introduced, the nature of these tissues might be modified conformably to the particular end he had in view. He was fortunate enough to think of employing this very force, and proved that it was sufficient to imbue the whole tree from bottom to top with the liquid imbibed by the cut trunk. It is not even necessary to retain all the leaves and branches; a tuft preserved at the top is sufficient to cause this upward motion; and this property remains, always gradually growing weaker, for a fortnight after the tree has been cut down. All the tissues are penetrated by the liquid except the heart-wood when it is hard.

fluence the state of the air, warm or cold, dry or humid, the presence or absence of the sun, its action direct or through the clouds, exercises over the evaporation of vegetables.

§ 268. Let us now explain the different phases in the ascending motion of the sap occasioned by the succession of the seasons which constitute our climate. In the spring, from the moment that the degree of heat, which appears necessary to the enjoyment of life in the greater part of the vegetables, is established, a little sooner in some, a little later in others, we see that the buds, which have remained stationary during the winter, swell a little, and at the same time the roots begin their functions. The excitement communicated to the organs by the renascent heat appears to lead the wakening, as it were, of vegetable life. In a Vine trained against a warm greenhouse, if a branch be introduced into the inside of the greenhouse, its buds with their leaves are soon developed, whilst the branches on the outside remain in a state of hibernation. It is the heat which has excited the bark and the buds, and their action has consecutively determined that of the roots for a long time before they react on the branches still exposed to the cold. That of the buds seems then to precede that of the roots, nay even to contribute towards its cause; this is not very astonishing, since after the winter has passed, the air is more quickly warmed than the earth.

The absorption of the roots, when it is once established, the motion, whether it commenced immediately at their extremities or was stimulated to commence at the buds, goes on with much activity, the effect of which is to be perceived on the buds only after a very long interval. In short, at this time, in the tree despoiled of its leaves, the young shoots of which are still enveloped with impermeable teguments, the endosmose is almost exclusively the acting force, even supposing that it has first been roused by an effort of the buds; and, before having forced up to them the juices destined for their support and their developement, it ought to have taken place at every interval to mix and prepare these juices which it puts in motion. The quantity of substances, more or less thick and solidified, formed by the work of the whole of the preceding year, and amassed together in a depôt, as it were, in the interior of the vegetable during winter, softened and dissolved when the current arrives at them, is a powerful cause of the excitement of the motion of the liquids, which the endosmose incessantly pushes from the earth into the plant. They ought, then, to mount in great quantities and with extreme force; and this we may see, by breaking off a piece of the stem, from which water will flow as from a fountain; the stem must be in that state called the *spring-sap* (*sève du printemps*); this forms the watery emission caused in the Vine by an incision, and known under the name of *the tears of the Vine.* Since these tears flow copiously from the end of the stem stripped of every leaf, and even cut to the surface of the soil, it is impossible to attribute in this phenomenon any influence on the excitement of the fluids to the suction of the buds and the evaporation of the leaves. On adapting a tube to the cut extremity, we see the sap mount in it to a certain height, which is thus determined and is found to be considerable. Hales, the Englishman, to whom we owe a series of experiments, as exact as ingenious, intended to determine the motion of the juices in the plants and explained in his *VegetableStatics*, has applied, to find out the value of the force and quickness of the ascending sap, the same apparatus, that M. Dutrochet has, to discover the force of the endosmose, that is, this

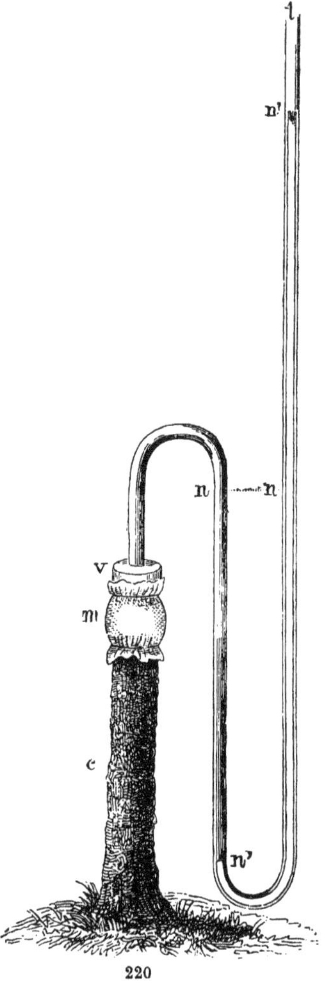

220. The stem of the Vine cut about two feet from the ground.—*t* A doubly bent glass tube adjusted to a brass ring *v*, which is fitted and cemented to the cut extremity of the stem, the apparatus being covered and supported by a piece of bladder *m*.—*n n* Surface of the column of mercury in the two branches of the lower curve of the tube at the beginning of the experiment.—*n'n'* Surface at the end of the experiment.

tube with a double bend, the ascending branch of which is fitted to the end of the cut stem and the lower bend of which is filled with mercury, which, pushed by the accumulated sap as it mounts into the interior branches, shews itself on the outside and indicates by the height of its column the values we want (*fig.* 220). Hales has seen the column raised 1 metre (French) 39·33 inches (English) which is equivalent to 14 metres (French) or 45·886 feet (English) of water; and he has calculated that the force which thus pushes the sap in the vine is five times as great as that which forces the blood in a great artery of a horse.

§ 269. The buds are developed, the leaves are expanded, and their action is added to that of the endosmose, which will be partly exhausted, since the ascension, which always continues actively, does not leave off slackening, and is gradually weakened. Thus, by experiments analogous to the preceding, we may determine the influence which the new force, acting in conjunction with the endosmose, exercises on the ascensional movement. Thus, if we adapt to the lower end of a branch a long tube full of water, which is itself plunged into a bath of mercury, the branch will suck up a certain quantity of water, which will be indicated by an equal ascension of the mercury in the tube. We may change the state of the branch by leaving on it a greater or less number of leaves, stripping off part of them, or even the whole so as to leave its buds alone; that of the atmosphere, which may be at different degrees of dryness or humidity; we may observe it at various times, at different hours of the day and night, and we shall see that all the causes, which exercise an influence on the different degrees of evaporation from the branch, exercise an analogous one on the quantity of water which it sucks up.

§ 270. The branches, however, are successively developed as well as their leaves; they have acquired little by little their full dimensions and the consistence which characterizes their tissues at this state, which we may call their adult age, at the same time that fresh tissues are organized in certain inner parts of the vegetable. It has thus arrived at a kind of equilibrium of which we have spoken, an equilibrium, which by no means presupposes the immobility of the sap, but only its movement moderated according to the wants of a state, in which nothing more is requisite than to compensate the continual losses attending the very possession of life, to complete whatever was wanting in certain parts, and to prepare for the following year the organs then to be developed and the materials destined for that purpose.

§ 271. If all this vital work begins early, if the year has been precocious, it may happen that these materials are ready too soon, in a season which is not yet sufficiently advanced, and thus present requisite conditions to provoke their developement and so anticipate the proper season; this frequently happens towards the end of summer, when we see some of the recently formed buds shoot forth, some of the phenomena of spring renewed, and necessarily with them the ascending motion of the sap brought back; this is called the *autumn sap, (sève d'août).*

§ 272. It then languishes afresh. During autumn, the evaporation from the surfaces has diminished more and more; the tissues are dried and solidified; the leaves die and fall by degrees, and the tree arrives at that state of almost complete repose, during which life appears to be suspended. The motion of the sap then ceases with its causes and stops more or less completely throughout the whole of the winter.

§ 273. In order to follow the different appearances of this motion of the ascending sap, we have chosen examples, in which they are shewn most clearly and completely, those, with which, at least, we are best acquainted, those of the trees of our own climates. That, which takes place in one of their branches, may, or very nearly so, take place in every herbaceous plant, with more activity, however, since it is most commonly ramified, and thus developes in the course of the same year several generations of buds. As to the vegetables of warmer climates, the times are changed; and, under the tropics, there seem to be hardly any periods of repose, and the motion, consequently, almost continual. But we may judge of it by the seasons and by the external phenomena of vegetation, rather than by direct observation which would offer so much interest to us.

§ 274. An important point still remains to be explained. What course, in the midst of the elementary organs combined in the stem, does the sap follow in its ascensional motion? That of the spring invades all the tissues, filling the cells, the fibres, the vessels, the meati. It mounts almost entirely by means of the ligneous body, as we may easily see by cutting the branch of a plant. We see the liquid flow from the surface of the section; from the whole of the ligneous body if it is young; if it is old, only from the exterior zone, which is in the state of alburnum. After the spring sap, several vessels appear to be empty, but, on examining them under water, we find that they are filled by gases which rise to the surface in little bubbles. The passage of the sap, for the most part, takes place by means of the cellular tissue, but by a slow and hardly sensible motion from the bottom towards the top, the vegetable being then, as it were, saturated with liquids and nearly in the condition of an apparatus full of water, which, pierced with small openings at its two extremities, permits a small quantity to flow out at one end, and receives an equivalent quantity at the other, without there being any apparent current. If some cause happens to disturb this equilibrium, as rain succeeding to a drought, or the developement of fresh buds, the ascension of the sap is revived and partly resumes the passage which it has momentarily abandoned.

§ 275. Descending or elaborated Sap (*Sève desendante ou élaborée*).—The sap, enriched by all the substances it has dissolved and incorporated during the course of its passage, has arrived at the young branches; then, traversing them, it has passed near to the surface of their bark, through the cellular tissue of the medullary rays and of the cortical parenchyma to that of the leaves, by the most rapid way of the vessels as well as by that of the parenchyma. These green surfaces are from the greater or less number of stomata which cover them in immediate connection with the atmosphere, which penetrates into these little openings and circulates in the network of the lacunæ of the subjacent tissue. The sap is now separated from the air only by the thin membranes of this tissue, through which the principles of both may act reciprocally, be exchanged and, consequently, modified. We shall see in detail, under the article of respiration and of nutrition, what these changes are. It is sufficient for us to mention now, that they take place; and that, consequently, the sap changes in its nature at the same time that it loses the greater part of its aqueous particles, which escape in the form of vapour.

It is easy to convince ourselves by the inspection of the parts, that the leaves and the young bark enclose juices different from the sap, which we have hitherto examined. In the interior of the cells, the chlorophyll (§ 24) renders them green in different degrees of intensity, and from the vessels or lacunæ of the bark, we see a thicker liquid, which is often coloured, ooze out. Its properties are as different from those of the sap as its appearance. The Euphorbia of the Canary Isles (Euphorbia Canariensis) furnishes a deadly poison; it is the milky juice of its bark; but, after the inhabitants have stripped the tree of its bark, they find a limpid and innocent drink in the ligneous body of the same plant, by extracting the ascending sap which flows through it.

Has this cortical sap a general motion like the other? If we cut a stem horizontally where it is coloured, we see that the lower surface of the section furnishes very little of this juice compared with the upper. If we take off the whole ring of bark, we see the juice ooze out and collect on the upper edge of the wound and not on the lower. If we bind a tight ligature round the stem, we see at the end of a certain time, the bark swell and form an excrescence above the ligature, and the stem below preserve its original diameter. There is, then, a flux of the cortical sap from the top towards the bottom, that is, in a direction inverse to that of the ascending sap. This is the reason why the name of *descending sap* has been given to it; it has also been sometimes called *elaborated sap*, on account of the organic process which it has undergone to acquire its new properties.

§ 276. We have seen (§ 73) that the bark is composed of parenchyma, of elongated fibres (those of the Liber, § 78) and of lacticiferous canals (§ 14). It is the juice contained in the last, the *latex* which is often coloured and which, in this case, is generally known under the name of *proper juice* (*suc propre*). At other times, the same vessels convey a colourless juice, but it appears to be of the same nature; and some observations have shewn that the same plant, which in our climate affords a colourless latex, in the tropics produces a milky one. In every case it is composed of very fine unequal granules swimming in some liquid. The presence of these granules and the transparency of the walls of the lacticiferous vessels allow us with the aid of a microscope to see the movements of the latex. Place, for instance, under a thin piece of glass upon the object-stand of the microscope, a young leaf of the Celandine (*Éclaire*) (CHELIDONIUM MAJUS), a plant which is so common on our walls and is well known on account of its orange-coloured acrid juice; let this leaf be chosen as thin and transparent as possible, fixed to the living plant, and, consequently, partaking of its life, and, sprinkled with water to prevent its drying, let it be examined with a strong light and a powerful magnifier, we shall perceive (*fig.* 221) in its substance several little series of a granular substance in motion; some of which take one direction, some another, and even contrary to the first, some of which remain insulated, others approach, are united and blended together. If we embrace a sufficiently large field, we shall find that these series are united to one another, and thus form a net; it is that of the lacticiferous vessels (*figs*. 55, 56). The latex descends in one branch to arise in another, and we thus observe a true circulation, quite comparable to that which exists in the capillary vessels of the animals. M. Schultz, who claims the honour of the discovery, has proposed to call it by the name of CYCLOSIS (*Cyclose*).

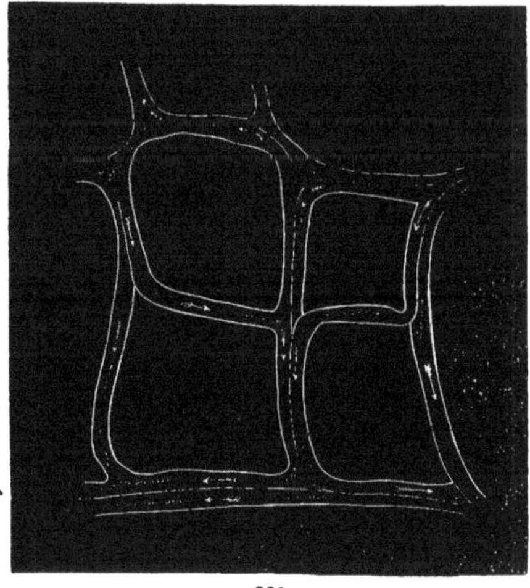

221

If the direction of these partial currents varies, it is probable that it is most commonly descending, since it is incontestable that the general motion is from top to bottom, as we have proved. But the cyclosis, prolonging and multiplying the connection of the latex with the tissues through which it passes, aids the effect which results from the presence of the nourishing juice.

§ 277. What is the force which gives the impulsion to the latex? Several different explanations have been proposed, merely because we have not yet found a satisfactory one. Some seem to consider it to be purely a physical phenomenon, such as M. Amici, who attributes it to heat acting on these tubes in the same way as on a thermometer, and demonstrates that the direction of the currents may be changed when they approach some substance which is rather warmer. But how would the influence of heat alone determine the motion in a constant direction, when it is necessarily so confusedly distributed over surfaces arranged with so little regularity as the assemblage of the branches and leaves of a great tree? Without altogether denying this influence, it is difficult to recognise it as the only cause. Other physiologists have supposed that

221. A small fragment of the leaf of the Celandine (*Eclaire*) (CHELIDONIUM MAJUS) very highly magnified and shewing several meshes of the net of the lacticiferous vessels.—The arrows indicate the direction of the currents.

there exists an action of the granules upon one another, or of their whole with respect to the walls of the vessels, alternatively attractive and repulsive; but to admit which suppositions, we want proofs, with which observation hitherto has not furnished us. Some authors have proposed as the reason a contraction of the walls; this contraction must not take place along the whole tube at the same time, since then it would, when one of these tubes is open at the two ends, drive the liquid out of both, whilst in this case it flows out of one end in the direction of the current. The contraction would commence at one end and would gradually proceed to the other: but we must state that these walls are frequently united very firmly to the surrounding tissues, and even so closely, as to lead several authors to deny their existence and to suspect that the latex circulates in the intercellular meati. In short, are we to think that the endosmose acts here; that the sap, thinner, arriving from the ligneous body to the bark, penetrates into the lacticiferous vessels filled with a juice which is more thick than it is; and that, commencing to act towards the upper ends where the latex is formed, it necessarily forces it in an opposite direction towards the lower ends? Whatever the cause may be, there reigns a great obscurity on the nature of the impulsive force of the latex, and we have contented ourselves with mentioning the different hypotheses which are proposed to explain it, without adopting any one definitely.

§ 278. The motion still continues for a pretty long time in the detached part of the vegetable. In this manner, it is observed with the greatest facility and convenience: in a thin layer of young bark from which the epidermis has been taken, like that of the Sycamore or of the greater part of the Figs; on the stipule which envelopes the terminal bud of the latter, particularly that of the FICUS ELASTICA; on the corolla of the Great White Convolvulus of the hedges, &c., &c. The lamina under observation ought always to be put in a drop of water, to prevent desiccation which necessarily stops the motion.

The lacticiferous vessels are found in the bark in the outermost layers of the liber, or, more frequently in the innermost, and we then follow them to the extremities of the roots. But we also find them in other parts and even in the pith, as we have previously (§ 60) said.

§ 279. We have now to examine in the bark the functions of the fibres of the Liber. M. Mirbel considers them as making part of the system of lacticiferous vessels, from which they differ externally by their simple unbranched tubes, as well as by the nature of their walls, very analogous to that of the ligneous fibres in the wood. The opinion of M. Mirbel is confirmed by the transitions of form, which we may sometimes observe between the fibres of the liber and lacticiferous vessels, as in a large number of species of the Apocyneæ and Asclepiadeæ; the second may entirely replace the first in several of these plants and in others, as the Euphorbias, in which the milky juice is found in such great abundance. It is true, that in the majority of the vegetables in which the liber exists in its most usual forms, the contained liquid is colourless and thus differs from that conveyed by the neighbouring lacticiferous vessels. The descending sap is then found in two rather different states in the two kinds of vessels which are generally found together, taking the place of one another when needful, analogous, but not identical. It does not appear in such an elevated state of organization in the fibres, the long tubes of which without ramifications, without circumvolutions, carry it more directly to the bottom.

§ 280. On examining the places in which the cambium (§ 58), the element or first appearance of every vegetable organization, is deposited, we remark that it is generally over the passage of lacticiferous vessels. In the stems of Dicotyledons which have alone been employed in every explanation we have hitherto given, the great depôt is formed between the bark and the wood, precisely along the cylinder which is made on the outside by the mass of lacticiferous vessels and fibres of the bark. The little cellular mass, prepared at the axil of the leaf to form the bud, is over the passage of the vessels, which, bringing back all the latex formed by this leaf, are compressed in the petiole or are expanded in the sheath. In Monocotyledons the fibres and the tubes, which are considered as the liber and the lacticiferous vessels, are contained in the fibro-vascular fascicles scattered throughout the stem, and it is by a mass equally dispersed that the cambium is deposited. Their terminal bud, often single, is the first to profit by the juice prepared by the leaves of that which has preceded; and this is, for a much stronger reason, true for the vascular Acotyledonous vegetables. In all, in short, the lacticiferous vessels end at the extremities of the roots, the seat of an almost incessant formation.

The reservoirs of several other very highly elaborated substances are also in relation with the numerous lacticiferous vessels around the apparatus which secretes them. It is, then, naturally in the bark that this kind of reservoir most commonly exists; for instance, those of the Resins. But we sometimes meet them also in other parts, the pith, for instance; and we ought not to be astonished at this, when we remember that the lacticiferous vessels may be found almost everywhere.

§ 281. We will now recapitulate in a few words what we know of the general motion of liquids in the most perfect vegetables; the water of the earth, holding several substances in solution, enters into the roots at their extremities; hence, under the name of sap, it rises by the roots, then by the stem through the ligneous body, as much by the direct and uninterrupted ducts which the vessels present, as by the fibres and cells through which it passes successively, dissolving and appropriating to itself fresh substances. This progress from bottom upwards, and from within to without, leads it into the leaves and to the surface of the bark, where it is in contact with the air; then, completely organized by this act of respiration, it takes a retrograde direction and descends for the most part through the bark, sometimes directly, sometimes by a series of circumvolutions; depositing in its passage, in the solutions of continuity ready prepared, masses of substances destined generally for the nourishment and formation of the tissues; and it at last arrives at the extremity of the roots where the absorption commenced.

§ 282. Rotation or intra-cellular circulation.—The vegetables, which have hitherto served as examples in our study of the general motion of the juices, are provided with various cavities and ducts in which this motion takes place. But we know, that there exist many other plants of a more uniform structure, composed only of cells without spiral or lacticiferous vessels. We may conceive that the liquids may arrive from the lower to the upper by the force of the endosmose alone; but observation teaches, at least in several instances, that something else takes place besides this physical phenomenon. Let us take as a well known example, and one in which this may be easily seen, the Chara. These are little plants common in our stagnant pools, and composed (§ 101) of a series of cells joined end to end: in several species, a single cell forms, as it were, a sort of internode; in several others, it is enveloped with other parallel and narrower cells which form a sheath to it; and to see the central cell, we must, by scratching lightly, take off those which surround it. Placing this in water under the microscope, we perceive in its inside a very sensible motion: it is that of a very great number of granules of different sizes swimming in its cavity in the midst of a transparent liquid which fills it, and moving together along the walls in two principal directions, the one ascending, the other descending. We soon see that this is the result of a single current, which, as it rises, follows one side of the tube, bends at its upper end, descends on the other side of the tube, and thus reaches the point of its departure to commence the same course again; it thus describes an ellipse, the length of which is in proportion to the length of the tube. The name of *rotation* has for this reason been given to this intra-cellular motion of the juice. After the first days of youth are passed, the cell appears to be slightly twisted on itself, and the current follows a direction which is rather oblique with regard to the axis, instead of being perfectly parallel; we may then remark that it is moved along a band of green granules, which, covering the wall, form part of it. If we interrupt the continuity of the cell by closing it in the middle with a thread, in each of the two cavities thus formed at the expense of one, the circulation continues between the ligature and the corresponding articulation. We thus prove that there is none in the interval between two membranes as several botanists have supposed; for then it would be necessarily stopped by means of the ligature.

§ 283. A little time elapsed, and then the same motion was shewn to exist in the cells of several other aquatic plants of a single organization, although not in so great a degree as in the CHARAS, such as the NAIAS, HYDROCHARIS, VALLISNERIA. The phenomenon is seen very clearly, especially in the cells which form the radicellary hairs; but we observe it also in the other parts of the same plants in the cells which occupy the interior of the stem or of the leaves, and which are, consequently, not in direct communication with the water. The current indicated by the progress of the granules also describes in these cells an ellipse in the direction of the axis of the plant and commonly parallel or a little oblique to that of the cell. As the cells are in this case not detached, we may study the motion in several neighbouring cells at the same time, and thus discover that the circulation of one is completely independent of that of another.

§ 284. Is this intra-cellular circulation peculiar to aquatic vegetables of a simple structure? Researches, extending to a multitude of plants belonging to every degree of organization when conducted with skill and attention, have almost always shewn an analogous motion in the interior of the cells, especially in the tissues abounding in sap and the actual seat of a rapid growth. The plants of the family Commelinaceæ, and among others the TRADESCANTIA VIR-

GINICA, are particularly mentioned as presenting this phenomenon in a remarkable manner in their articulate hairs, and also in several other parts of their flower and of their stem (*fig.* 222).

§ 285. The current is not always single as in the first examples we have mentioned. It is sometimes divided; and although even then its divisions only appear like ramification deviating from a principal body, we see the internal wall of the cell furrowed by small series moving in different directions, and thus forming a kind of very irregular net (*fig.* 222, *a*). It may be compared on a smaller scale to that of the lacticiferous vessels, and M. Schultz goes so far as to think that they are very fine ramifications of these vessels, penetrating into the interior of the cells. This in the vessels would be a phenomenon of the cyclosis, although in the CHARA and other cellular vegetables he admits the existence of rotation. But the phenomenon appears to be similar in all these different plants; amongst the modifications which it may present, we pass so insensibly from one to the other, and the penetration of vessels through the cellular wall appears to be so singular,

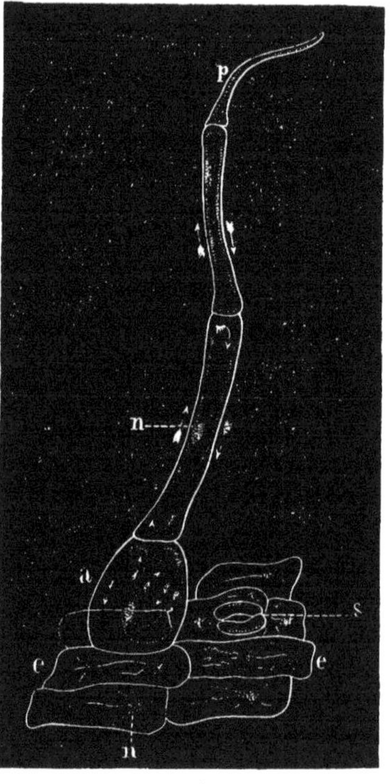

222

222. A hair *p* from the calyx of the flower of the TRADESCANTIA VIRGINICA, with a small portion of the epidermis *ee* in which we find a stomate *s*. In each of the cells which compose both the epidermis and the hair, we find a nucleus *n* and rotatory currents whose direction is indicated by the arrows. It appears obscurely in the cells of the epidermis and even in those which form the stomate; very clearly in the cell *a* which serves as the base of the hair: this only allows us to perceive one (although several exist) in the upper lengthened cells.

that it is generally agreed to attribute to the intra-cellular motion of the juices the same nature in vegetables of different degrees of organization. Besides, if we observe for some little time these cells with several currents, we shall see very marked and numerous changes. They do not follow a constant direction, as they would, if they were imprisoned in separate tubes.

M. Schleiden, in the young cells of the perisperm of the CERATOPHYLLUM, describes a current which traverses the axis of the cell from end to end, and is separated by a kind of expansion towards the walls. This is the only observation of the kind that has been made, and all the others have hitherto shewn the currents applied exclusively to the walls. They are, in their passage, covered with a mucilaginous fluid, the flakes of which are often drawn into the midst of the granules in motion. The existence of the nucleus is frequent in these very cells, and seems to exercise an influence on the currents, which generally, the principal at least, terminate at it (*fig.* 222). It sometimes happens, that they detach and drag it along with them; its nature also appears to be very like that of the substance which composes them.

§ 286. The rotation, which was at first considered as a method of circulation peculiar to the lower and aquatic vegetables, between which and vascular vegetables there is no analogy, is, from all that has preceded, an almost universal fact in the Vegetable Kingdom, and its very generality shews it to be a fact of some moment; thus its activity appears to be commonly in relation to that of life; the one is modified in the same way as the other and by the same circumstances. The physical or chemical agents, which, according to the experiments, augment or diminish, or stop the first, exercise the same influence on the second.

§ 287. Several plants, especially those whose thick and fleshy tissue has caused them to be called *succulent* plants, shew in their cells, instead of a well-defined rotation, a vague and uncertain motion of the juice from one point of the wall to the other; partial currents which begin without finishing, or which, at most, finish in a corner of the cavity. In fact, there are several vegetables in which we look in vain for any trace of the intra-cellular motion. But we can draw no conclusion from these negative facts in presence of numerous positive facts. Thus, under the apparent immobility of the vegetable is concealed a real, general, well-defined motion in each of its least parts as well as in all together.

Respiration.

§ 288.—Before explaining the phenomena of vegetable respiration, let us examine what organs are entrusted with this function, and the value of a theory, which was predominant for a long time, and is now once more revived in many books, and ascribes respiration to unrollable tracheæ as one of their functions. The external similarity between them and the tracheæ of insects has naturally caused this idea. We know that on the sides of the body of the insect are opened a number of pores, which form the orifice of so many vessels, each formed of two tubes fitting closely to one another, between which is a spiral fibre; that these fibres are distributed by a series of ramifications, which grow thinner and thinner as they proceed, throughout the whole of the interior of the body: that they are thus bathed, as it were, with the vital fluid which fills the body freely; that the fluid on one hand, and on the other the air, which, entering from the outside, circulates in the tubes, react upon one another communicating with each other through the thin walls; that the respiration is thus performed in every part. When vessels furnished with a spiral fibre were found in vegetables, which were distributed throughout the whole system of the branches and of the leaves of the first year's growth, i. e., throughout all the green parts, in which respiration takes place, botanists were led to conjecture that they contributed to this function. It was even formerly supposed, that the tracheæ were terminated immediately at the stomata, and then the analogy with that of insects would have been complete and almost have insured conviction. But it is not so, and we now know, that the tracheæ are separated from the stomata in the branches, by the whole thickness of the parts interposed between the medullary sheath and the epidermis (§ 62), in the leaves, by that of the parenchyma, and in these they also approach the upper faces in which there are very few stomates (§ 125). The air, then, instead of entering directly into the tracheæ through the stomata, only gets there after having traversed the layers of other parts and is introduced into their cavity only through their walls.

All that anatomy tells us of the entrance of the air into the vegetable tissue (§ 127), is, that below the surface which is exposed to it, we find a great number of lacunæ which communicate with one another, and that it may also circulate around the cells, the separation of which forms that series of interior cavities which are in communication with the air by stomata. It is possible for it to go still deeper, where it would find meati. Here the direct and open means of communication are limited.

What then is the physiological function of the tracheæ? Are they designed for the same purposes as other vessels, and, if so, why are they different in form and situation? Since they are always the first which are formed, and this in their tenderest parts which are still growing in length, since in proportion as the speed of this growth is slackened and tends to stop, the fresh vessels which are developed assume fresh forms which depart more and more from the spire, we may be permitted to suppose that their structure and method of developement have more relation to the wants of this lengthening, which takes place in the whole branch at once, and not at its extremity only, as in the root. During the ascent of the sap, the tracheæ, like all the other parts are completely filled with it, and even convey it more quickly than the other vessels. They may afterwards like them contain gases instead of liquids; but since they are not the only ones, this by no means seems a function peculiar to them. If they do serve as a means of respiration, it is neither exclusively nor constantly. This deduction is confirmed by some plants, the Ferns for instance, in which the respiration is perfect, and which, however, want true tracheæ.

§ 289. We know that the atmosphere is a mixture of two gases, Oxygen and Azote or Nitrogen. One volume of air presents nearly 79 of azote for 21 of oxygen in the 100; to this we must add a very small quantity of another gas, Carbonic acid gas. This is a combination of 8 parts by weight of oxygen to three of carbon, a body well known in the solid state under the name of charcoal. This small quantity of carbonic acid gas is respired by the plants; and, at first sight we may be astonished that such a small quantity should suffice, considering that carbonic acid gas only forms a millionth part of the weight of the air. But this astonishment will soon give way, when we reflect on the extent and height of the atmosphere pressing on our globe, and that this weight reduced to its millionth part would still represent an enormous weight, many times above that of all the vegetables of the earth together: for this calculation proves that the atmosphere contains about 1500 billions of kilogrammes (French) (which are equal to 13,8616,075,892 tons English) of carbon.

§ 290. Chemistry has determined the changes produced in air thus composed in two different ways. 1st, We permit a plant to vegetate under a bell-glass filled with air which cannot be renewed, and then after a certain time this air is analyzed. We may vary the experiment by composing an artificial atmosphere under the bell-glass, in which the elements of the air are not in their natural proportion or are replaced by others, and we thus see the change that takes place, both in the composition of this atmosphere and in the plant itself. 2nd, We cause a seed to germinate in pure sand, moistening it with water equally pure, and, when the plant is once raised, we continue its vegetation by giving it determinate quantities of water; then its chemical composition is found by analysis. We know that of the seed exactly, according to other similar seeds of the same weight; we know all that the plant could extract from the water, which has been its only nourishment. Every thing else, which the plant may shew beyond what composed the seed at first and what was drawn from the water, has been extracted from the air. Since, in this case the experiment takes a long time and the plant is in continually renewed air, we may find substances, which would have been wanting in one occupying a limited time and in a small body of air, and which, consequently, in the first method of experimenting would not have been found.

§ 291. By the first it has been determined that the atmosphere, which a plant has respired, has lost a certain quantity of carbon and gained a certain quantity of oxygen. Now, these two quantities are almost those which would have been required to form carbonic acid gas: there is only a little less oxygen. The plant, in respiring, decomposes the carbonic acid gas, retains its carbon and a little oxygen and gives out the remainder which is disengaged. But does this decomposed carbonic acid gas consist only of that portion which is in the air, or else partly of that in the interior of the vegetable, where it was already formed? This latter supposition is probable, because, if we place a plant in an atmosphere entirely deprived of carbonic acid gas, in pure azote, for instance, we find mixed with it, at the end of a certain time, a certain portion of oxygen, which is formed by the decomposition of the carbonic acid gas which the plant held in its own tissues.

§ 292. This is what takes place when the plant is exposed to the solar light. In complete obscurity, the case is quite different: for we find in the receiver more carbonic acid gas and less oxygen. The action is then inverted, and the green parts of the plant have taken and retained the second, whilst they released the first. Thus, the alternation of day and night brings on that of the respiratory phenomena: the retention of carbon and the release of oxygen during the day, the release of carbon and the retention of oxygen during the night. During the day even, the vegetables deprived of light are submitted to an analogous influence; when they are kept in the dark, they soon etiolate, that is, lose their colour, and grow in length, whilst they lose much of their solidity, thus shewing the loss of carbon, which renders their surfaces green and solidifies their tissues. It is clear, however, that all are not equally sensible to this influence and do not want the same degree of light, since several grow vigorously in the dark. Between the two extremes of intense solar light and complete obscurity, there is a proportionate graduation in the intensity of respiratory phenomena. A very strong artificial light will cause the etiolated parts to become slightly green.

§ 293. In this reciprocal action of the light and of the green parts of the vegetable can we discover anything beyond the apparent result? We know that a pencil of light is composed of several rays of different colours, of which each body absorbs some and reflects others, and that the result of this reflection gives the colour of these bodies, the impression of which is carried to our eye. We also know that the pencil of light warms the body at the same time that it lights it, but unequally by each of the differently coloured rays; that the heat of each is not invariably united to the light, but, that in certain media they separate and follow a different course, by which the ray is a second time decomposed into two, the one luminous, the other calorific. To this double property there is joined a third: light modifies the chemical nature of the body which it strikes; this is proved by certain substances which are preserved in the dark, but are altered in the light; whence we are led to admit chemical rays into the composition of the pencil of light. Now, in the pictures formed by the daguerrotype (the principle of which is founded on the latter property), the green parts of vegetables, as well as the representation of green bodies in general, are not reproduced. We must, therefore, conclude, that the chemical rays have disappeared in the vegetable, retained and absorbed by it, and that it is a property peculiar to its green parts. As to the calorific rays, they can exert no influence on the respiration of plants, which takes place in the light and not in the dark, although the temperature may be equal.

§ 294. The parts, the natural colour of which is not green, undergo the same changes in the light, as the green parts do in the dark: they are oxygenated and decarbonized. The roots and other subterranean parts are thus operated on; and the oxygen they imbibe seems to be necessary to their existence: for when they are plunged into a gas from which it is excluded, they die in a very little time. The free access of the atmosphere to them is a condition favourable to vegetation, and their being buried to a depth, to which it penetrates with difficulty, is adverse.

§ 295. It is the same with the seed. When it begins to germinate, during the first developement of its parts it disengages, even in the light of the sun, carbonic acid gas and absorbs oxygen. The former is the result of a certain portion of carbon contained in the tissue of the seed, and combined with the absorbed oxygen, the whole of which is employed for this purpose; for if the germination takes place in pure oxygen, the quantity of carbonic acid gas disengaged equals that of the oxygen absorbed. This continues until the germination, as it advances, brings forth to the light the green parts of the little plant; and, henceforth, the phenomenon is inverted, the inspiration of carbonic acid gas and the expiration of oxygen commence. This want of oxygen for forwarding the first process of germination, explains why seeds may remain so long unaltered at great depths.

§ 296. A curious experiment of M. M. Edwards and Collin shews, that in this action of the germinating seed on the medium which surrounds it, another decomposition besides that of the air may take place. For, when beans are put into water to germinate, several gases are evolved, amongst which carbonic acid gas is found in large quantities, and nearly eight times more than the small quantity of atmospheric air, which may happen to be in the water. The oxygen of this air would not suffice for the production of all this carbonic acid gas by combining with the carbon of the seed; and its excess can only be accounted for by the decomposition of the water, one of whose elements, oxygen, would enter into combination with the carbon; the other, hydrogen, not being found among the disengaged gases, will have been absorbed by the seed. It has been concluded, therefore, that water does not play purely a physical force in the respiration of plants, but that it may also be decomposed.

§ 297. On the other hand, M. Boussingault has demonstrated, by the second method of experimenting we have mentioned before (§ 290), that is, by the chemical analysis of a vegetable, which is developed only by pure water and atmospheric air, that at the end of some time it has gained a small quantity of azote, which could be furnished no other way than by the air. This quantity is very easily appreciated in certain Leguminous plants, such as Trefoil (*Trèfle*) and the different kinds of Peas, but in the Gramineæ there is none, as in Wheat and Oats. Is this gained at the expense of the ammoniacal vapours of which a very small quantity is, according to some authorities, always suspended in the atmosphere? We shall see, that ammonia or the volatile alkali, a combination of azote and of hydrogen, has, in the earth, a great influence on the nutrition of vegetables.

§ 298. It follows from all that has preceded, that the plant may extract from the atmosphere carbon, oxygen, hydrogen and azote; and it is remarkable, that it appears to look for these elements only in bodies allied, as it were, to the air, in variable proportions and most commonly infinitely small, as carbonic acid gas, and the vapours of ammonia and of water. But, let us pass over the absorption of hydrogen and azote from the atmosphere; since it is generally inappreciable, and in respiration its part seems quite secondary, relatively to that of oxygen and carbon. The inspiration and decomposition of carbonic acid gas by the green parts appears to be the essential phenomenon of the respiration of plants.

Some botanists, it is true, supposing that in the vegetable, when it is entire in the state of a seed, and afterwards, when it is developed, several parts during the day and all during the night inspire oxygen and expire carbonic acid gas, have considered this act to be that of vegetable respiration, which would then be perfectly similar to that of animals; and in the decomposition of carbonic acid gas in the light, they have only seen an act of nutrition. The truth of this depends upon the definition we adopt to explain the act of respiration. If it is the act, by which the fluid, destined for the nourishment of the organic body, comes into contact with the air, and thence derives those properties which render it at last proper to accomplish their destination, we shall remain faithful to the old theory, which considers the green surface of the vegetable in connection of the atmosphere, the bark and especially the leaves, as respiratory organs. We are attracted more towards this idea, because there is a parallelism between it and the respiration of animals. In inspiration, they take from the air oxygen, which the blood carries along with it into all the parts of the body, to bring in its retrograde motion the carbon, which it gives out into the air by expiration under the form of carbonic acid gas. Vegetables take from the air carbonic acid gas, which, carried into the interior of their tissue, leaves its carbon there and its oxygen is given back to the air. Thus, the respiration of vegetables is quite inverse to that of animals, it compensates the effects of their respiration on the atmosphere; and the air, after having run through this circle of the respiratory organs of organic beings belonging to the two different kingdoms, is then composed as it was at first. It is true, that with regard to vegetables, the effect produced on the air during the night seems to counterbalance the effect produced during the day; but when we reflect on the enormous quantity of carbon accumulated in vegetables, and when we think that it has been produced by the act of respiration, we see that it is not counterbalanced, that the diurnal gain of carbon considerably outpasses the nocturnal loss, and that all the carbonic acid gas expired by animals has employment given to it in them. The currents of the atmosphere unceasingly re-establish the equilibrium which may be disturbed at certain points by the accumulation of many individuals of either animals or vegetables.

§ 299. This comparison between the two kingdoms naturally leads us to the examination of the respiration of vegetables living under water, which M. Ad. Brongniart by an ingenious theory has shewn to be in connection with that of fish. We know that in these and in a great number of other aquatic animals, the respiratory organ is connected with the atmosphere only through the water which immediately surrounds them; that it borrows from this water the air contained in it, and decomposes it in the ordinary manner, retaining the oxygen and expiring the carbonic acid gas. We know the structure of submerged leaves (§ 127 *b*, [*fig.* 132]), which, without epidermis and, consequently, stomatia, present immediately to the water their parenchyma with very thin walls, pressed against one another without intercellular meati, commonly only a few in thickness. The water can, therefore, act very easily on this parenchyma by means of the air which it holds in solution, and which penetrates it and is decomposed in it. Carbon is left in the parts which grow green; oxygen is exhaled. Light has its usual influence on this phenomenon, and at a certain depth we see the plants grow pale and etiolate. Like the branchiæ of fishes, these leaves, once out of the water, are quickly dried and thus become incapable of continuing the respiration. This rapid drying is owing to the absence of the epidermis, which, in aerial vegetables, moderates the swiftness of the evaporation, protects the respiratory cavities against such a danger, and generally gives time to the liquids contained in the interior of the plant to come and replace what is lost by evaporation.

§ 300. Evaporation.—Evaporation or aqueous exhalation from

the parts of the vegetable exposed to the air, of which we have already spoken as one of the most powerful causes of the ascension of the sap, takes place almost entirely by means of the stomata, although it also takes place from all the rest of the surface, and, especially from the green surfaces, but so feebly that it is almost insensible. We may easily assure ourselves that the evaporation takes place by means of the stomata, when we see that there is scarcely any when they are absent, hardly observable when there are few, generally much more active on the lower than on the upper surface of the leaf, in a word always in proportion to their number. This evaporation, which has been compared to the perspiration of animals, merits then, rather, on account of its seat, which is precisely at the respiratory surface, to be assimilated to the pulmonary exhalation, that large emission of water in a state of vapour which escapes with the breath; and this is the reason that we have mentioned it in this place. Let us add, in order to strengthen this comparison, that its activity is influenced by the same cause as that of respiration, by exposure to the light. In the shade, an equal and even greater heat has had comparatively but little effect, whilst it has exercised a very remarkable one on the insensible transpiration. During night the exhalation is stopped.

§ 301. The quantity of water contained in the vegetable must necessarily regulate to a certain point that which it exhales. It is interesting to determine the proportion between the quantity of water exhaled and the quantity of water absorbed during the same time; which may be easily done by introducing the lower end of a branch into the neck of a flask which contains a known quantity of water, placing the whole under a receiver, and weighing, after the experiment, the water contained in the flask and that evaporated into the receiver. We can again, that the experiment may not disturb the natural conditions so much, permit the branch to remain in the open air, and plunge it into a known quantity of water, which we must cover with a thin layer of oil, lest it lose anything by evaporation from its surface. When the experiment is ended, we must weigh the water which remains, as well as the plant, which must have been weighed before it was put into the water. The total loss of the water indicates the portion which has been absorbed: the augmentation of the plant in weight that which has been retained; and the difference that which has been exhaled. The proportion necessarily varies in different plants: it may also vary even in the same. Thus, a plant of Mint (*Menthe*) loses by exhalation in one season, $\frac{2}{3}$ of the water which has been imbibed, in another $\frac{13}{18}$. This influence of the seasons is very much complicated on account of several causes, amongst which we may rank in the first place the age of the plants, since in an equal temperature and equally bright weather, they exhale in the summer less than in the spring, and more than in the autumn. The exhaled water is not always pure, but contains a small proportion of substances in solution, with which it has mingled in its passage through the vegetable.

§ 302. The state of the atmosphere, dry or humid, augments or diminishes the exhalation, which is almost stopped in a medium saturated with vapour. It was formerly thought, that the leaves sometimes absorbed instead of exhaling the vapour of water, and by this means it was explained, how certain plants could live for a long time without roots. But these are plants with very few stomata and very rich in juices, which, on account of this fact, are parted with very slowly, and nourish the plants for a long time. It is true, that a branch, whose leaves are covered with numerous stomata, plunged into the water, will also remain living for a very long time; but then it has been put into the same condition as the preceding, by impeding evaporation through the stomata which are stopped up by the water. If absorption takes place at all in the leaves, it must be exceedingly minute. Indeed, how can it take place at all? For it, in that case, would assume the functions of the roots, and the sap would then have to follow an opposite direction.

Nutrition and Secretion[y].

§ 303. Nutrition is that function by which the organized body extracts from the matters around it the elements, proper not only to nourish and strengthen the parts already formed, but also to form fresh parts, not only to preserve it, but to cause its growth. This organic operation is divided in vegetable life into three acts:

[y] These subjects are considered in one chapter on account of the difficulty of clearly distinguishing them in vegetables. If a secretory organ is a local apparatus, in which a special substance is elaborated and deposited, different from those which are generally scattered throughout the tissue, we cannot find any organ which would perfectly justify this definition. The glands (§ 246, 250) are frequently confounded with the surrounding tissue; the secretory walls of the gum-bearing or resin-bearing lacunæ cannot be distinguished from it; and the organ can only be recognised by its contents, and is frequently limited to a simple cell. Thus, in the opinion of the majority of authors, almost all substances stationary or circulating in the bark are secretions. We have preferred to rank them according to the common purpose in which they all seem to concur, and especially as we cannot discern them from what may be exclusively the descending sap.

1st, These substances, coming from the outside in a crude state are introduced into the body: 2nd, They undergo in its interior certain preparations, owing, for the most part, to fresh and more complicated associations of the elements which are introduced; they are organized: 3rd, Each part extracts from these substances thus prepared, that which suits its particular nature and destination, retains and communicates to it the properties which it wants, and with which itself is endued; it assimilates it to itself.

The first act, which has already come under our notice, seems to take place under the almost exclusive influence of physical forces. The second consists in a series of transformations, which chemistry will generally explain to us. The third is in a great measure the secret of life, and the unknown force which causes it has been termed *vital*. The vital force, in short, presides over all this succession and union of phenomena, which without it cease, either to be produced or to be connected with one another in their order; and we are always obliged to recognise it as a support of these mechanical, physical and chemical forces, which it employs and sets in motion.

§ 304. Whilst we were treating of the absorption of the roots and of respiration, we studied the introduction of the substances from the outside of the vegetable, some of which are furnished by the earth, others by the air. Some of them meet others, and where this occurs near the surface of the branches and leaves, a chemical operation takes place, which is exhibited by the different composition of the air at its ingress and egress. There has, therefore, been produced one of those acts which characterize nutrition, the transformation, in the inside of the plant, of the substances which have been collected from external objects. Respiration is thus closely connected with it.

§ 305. Chemical analysis always finds four elementary bodies only in every part of the vegetable: carbon, oxygen, hydrogen and azote. These have been furnished by the air to the plant, and, consequently, the earth can only convey the same. It is true, as we have already said, that different mineral substances, which the water may have dissolved in the earth, are introduced along with it into the roots, that they traverse the tissues and that some of them remain there. But their presence is variable, frequently accidental; their function, still very obscurely known, often appears to be next to nothing, although at others it seems to exercise an indirect, but useful, influence; they preserve their nature and often even their crystalline forms (§ 22). We will for the present lay the consideration of these on one side and resume it again (§ 320 to 325) in a few pages, and we will now direct our attention to the substances, essentially organic, which are formed by the elements we have just enumerated.

§ 306. Although only four in number, they may form several compound bodies. We know, indeed, that elementary bodies are combined in several different proportions. Let us suppose that a body A is combined with a body B: the single body C resulting from this combination may be composed of equal parts of A and of B; or else 2, 3, 4 &c. parts of A and one of B; or else 2, 3, 4 &c. parts of B and one of A; in a word, a certain number of the one for a certain number of the other; and from all these different proportions result as many bodies, which differ in their characteristics and properties. We admit that this combination takes place between infinitely small particles of both bodies, beyond which there is no farther division and which are called atoms: between two atoms of A, for instance, and three of B, to form one molecule of C. But we may conceive, that these atoms may be grouped, relatively to one another, in two, three or several different ways. The molecules of C may, therefore, be arranged and also be grouped in two or three methods: and from this there may result three bodies, C, C', C", differing in their characteristics and properties, although chemical analysis cannot discover any difference between them, for all three are composed of 2 parts of A and 3 of B. These are different bodies, although composed of the same elements in the same proportion, and have been termed isomeric. After these elementary ideas, which we have named to form a foundation for the details which follow, we easily understand that four elements susceptible of being combined in two, three or four different ways, and each time in different proportions, may form different bodies, especially if some of these combinations make several isomeric substances.

§ 307. Crude or mineral bodies may be formed by one single element or two or several combined together. But then the proportions of the latter in general are very simple and marked by rather low numbers. Let us cite as examples those, which will interest us the most as furnishing to the vegetable the elements, of which it forms by combination its organic substances. Water is composed of 1 of oxygen to 2 of hydrogen *by measure;* and, the first weighing 16 times heavier than the second, it is composed of 8 of oxygen and 1 of hydrogen *by weight;* carbonic acid gas, of 1 of carbon and 2 of oxygen *by measure;* 3 of the former and 8 of the latter *by weight;* ammonia, of 1 of azote or nitrogen and 3 of hydrogen *by measure;* 14 of the former and three of the latter *by weight*[1].

Vegetable substances, when compared with inorganic bodies, present a higher degree of composition. They result from the union of three elements at least, of carbon, hydrogen and oxygen, and sometimes of four by the addition of nitrogen, and their proportions are always more complex and marked by much higher numbers. The composition increases in complexity in animal substances.

We will not take into consideration every vegetable substance, but will confine ourselves to those which are found in the generality of plants, and will proceed from the most simple to the most compound. We will begin, then, with the principal substances which result from the combination of carbon with oxygen and hydrogen.

§ 308. Amongst these, the first which attracts our notice is that, which forms the skeleton of the vegetable, the walls of the cells, of the fibres and of the vessels; for M. Payen has determined, that it presents throughout the same composition, that the apparent differences which are observed in it are owing to other variable products, deposited on its surface, or filtered, as it were, into its mass (§ 20), and, that, after it has been separated from them and brought to its state of purity, this substance, which may be called *cellulose*, is thus composed; 24 molecules of carbon, 20 of hydrogen and 10 of oxygen; which in weight is equivalent to 70 parts of carbon, 10 of hydrogen and 80 of oxygen. The *fecula* (*fécule*) or

[1] As in some works the elements are mentioned by their symbols, those of the commonest substances are given.

	Elements.	Symbols.
	Carbon	C
	Hydrogen	H
	Oxygen	O
	Nitrogen or Azote	N
	Sulphur	S
	Phosphorus	P
	Iodine	I
	Bromine	Br
	Chlorine	Cl
	Silicon	Si
Generally as oxides, and all except last two are common.	Potassium	K
	Sodium	Na
	Calcium	Ca
	Magnesium	Mg
	Aluminium	Al
	Iron	Fe
	Manganese	Mn
	Copper	Cu

These are combined to indicate compound substances; as, Water is expressed by HO, the symbols of hydrogen and oxygen; Carbonic acid by CO^2, the symbols of 1 of carbon and 2 of oxygen; Ammonia by NH^3, the symbols of 1 of azote and 3 of hydrogen.—Trans.

starch (*amidon*) (§ 20), that substance, of which we have already had occasion to speak as being so abundant in vegetables, and so commonly found in the interior of the cells in the shape of solid granules insoluble in cold water, has exactly the same chemical composition; and we also find this to be the case in another substance equally common, but soluble in cold water, and not coloured blue or violet by a solution of iodine, this has been called *dextrine*[a]. Here are, then, three substances composed of the same elements and with different characteristics, they are, consequently, isomeric. It is easy to conceive, how they may all be changed into one another, whenever their molecular arrangement happens to be troubled, and how a substance may, in the tissues of vegetables, sometimes be preserved in the state of granules, which retain their solidity and insolubility; sometimes lose this property, and become a syrup[b], which the sap dilutes and carries with it into every part of the vegetable; sometimes, in the last place, be stretched out and solidified into membranes, which form the walls of fresh cells or double those of one already in existence.

§ 309. We have already spoken also of sugar, as a substance which is commonly found in the vegetable. Several species of sugar are easily distinguishable; those which will take our attention here, as the most common, are those which are called cane sugar, and grape sugar, after the plants in which they are most commonly found and have been longest known to exist. Cane sugar is thus composed: 24 molecules of carbon, 22 hydrogen, and 11 of oxygen. Grape sugar: 24 molecules of carbon, 28 of hydrogen, and 14 of oxygen. When we compare this compound with the cellulose, the starch and the dextrine, which we have just mentioned, we see that there is very little difference, since we have only to add to the last 2 molecules of hydrogen and 1 of oxygen, or, which

[a] When we look at a ray of polarised light through an aqueous solution of dextrine, it deviates to the right of the plane of the po'arisation in proportion to the strength of the solution; hence the name of *dextrine*. This property, however, is not peculiar to it, starch possesses it to the same extent; cane sugar also, but in a weaker degree. Grape sugar and gum, on the contrary, cause it to deviate to the left. M. Biot, who discovered this, employed it in studying the vegetable juices, the nature of which he determined by a simple optical experiment. He has found, that they change according to the height from which we take them, and the time at which we examine them; that the sap, which, ascending in the wood, made it turn to the right, makes it deviate to the left when it has been elaborated in the bark and in the leaves.

[b] In this state it has often been confounded with gum; in fact, it almost presents its appearance and composition. But real gums, distinguishable, also, by several chemical and physical characteristics, appear to be the produce of the elaborated juices in the bark where they are found, and, generally, like other juices, which are also found there (§ 312), are substances more or less compound.

is the same thing, 1 molecule of water, to have before us cane sugar, and to this 6 fresh molecules of hydrogen, and 3 of oxygen, that is to say, 3 molecules of water, to have grape sugar. It is curious to see, that the gradual loss of water, which marks the ascension of the sap, is reproduced in the very combinations which it conveys along with it, since we see grape sugar replaced by cane sugar, when we look a little higher in the plant, or, in other terms, when the grape sugar loses a little water which enters into its composition.

§ 310. Let us now pass to the consideration of more complex substances, but the existence of which seems, like that of the preceding, to be very common in vegetable tissues: these are the combinations of four elements, or quaternary compounds, in which nitrogen is associated with the three other elements. They constitute *fibrine*, which, according to the most recent researches, those of M. M. Dumas and Cahours, is composed of 52·7 of carbon, 6·9 of hydrogen, 16·6 of azote, 23·8 of oxygen, *by weight; albumen* (*albumine*) and *caseine*, compound, isomeric bodies[c], according to the same authorities, of 53·5 of carbon, 7 of hydrogen, 15·7 of nitrogen, 23·8 of oxygen. Fibrine is insoluble, caseine soluble in cold water, and albumen coagulable by heat. These azotised substances are under the same laws, as the ternary ones we have just examined, in moving and remaining stationary according to the wants of the vegetable (§ 311).

§ 311. But what force in the vegetable changes these substances into one another, modifying their first composition, either in the moleculary state, or by the addition of some drops of water? We may produce, even in our chemical laboratories, some of these reactions artificially; but it is most frequently by means of agents which we do not find in organized bodies, which otherwise could not support their quick and energetic action. The greater part of these phenomena seem to be produced by those slow forces, which are spread over a large extent, which can hardly be determined at an isolated point, but which, operating at a great number at the same time, cause by means of all these little local actions a general effect, by which we recognise their existence without being able clearly to appreciate their nature. Chemistry, however, has thrown a little light upon some of these problems. Let us give an example of one of the most interesting, the conversion of starch into dextrine, which renders it soluble in cold water, and thus permits its passage

[c] A fourth substance, *gluten* (*glutine*), found in Cereals, is also isomeric with these two.

through the tissues. M.M. Payen and Persoz have found that in the starch accumulated in certain Cereal grains, towards the point of insertion in the Potatoe, and even beneath the buds of certain trees, directly the seed begins to germinate, the tubercle or the bud to shoot, a part of the starch disappears to give place to a fresh substance which has been called *diastase*, and which has the singular property of separating the granules of fecula and of changing them into dextrine; and, if the action is prolonged, this is itself converted into sugar[d]. This takes place in winter, since even at the thawing point, 12 parts of diastase produce in four and twenty hours from 100 of starch, 11 of sugar; at 20°, they produce 77. We hence see that heat favours this action; and the effect increases up to a temperature of 70° to 80°, at which the diastase dissolves 5000 times its weight of fecula. It is a powerful agent, and science has borrowed it from nature to manufacture the gummy syrup of dextrine and sugar of starch, which is now in such general use.

§ 312. We have spent a good deal of time in considering the commonest substances of vegetables, which appear to serve as a base for others. This knowledge will be of use in giving us some idea of the processes, which nature employs to bring about that series of metamorphoses, from which organization results, and in which we have seen simple chemical combinations of elements, the origin of which we know, take place under the influence of causes sometimes appreciable, sometimes still unknown. We will now content ourselves with casting a rapid glance over the rest of the vegetable substances so numerous and so varied, which, by a still higher degree of elaboration, are formed at the expense of those we have examined.

The fresh substances have modified the proportions of the first, either by diminishing those of one or of two of their elements, or by augmenting them; whence they will be found, as the case may be, richer in carbon, in hydrogen, in azote, or in oxygen. *Lignine* (*ligneux*) will serve here as one of our examples; we know that this is a substance, which in the wood incrusts the membrane of the cells, and was, therefore, for a long time confounded with the cellulose. When we compare their composition, we see that the lignine contains more carbon and a little more hydrogen, the quantity of oxygen remaining the same; and from these two causes, it ought to be a better combustible; experience here comes to our

[d] The sugar of the fecula is, as we have said, analogous to grape sugar, now better known by the former name.

aid and verifies this, the best woods for burning being those in which the proportion of the lignine is greater than that of the cellulose. This proportion varies much, and thus contributes to the great variety of the different kinds of wood, even were the composition of the lignine not itself variable; thus, the wood of the Beech contains nearly equal parts of cellulose and of incrusting matter, whereas, in the Oak, the latter amounts to $\frac{2}{3}$, and only reaches $\frac{9}{10}$ in the Ebony.

The most manifest effect of respiration is to fix in the vegetable an additional quantity of carbon and to deprive it of oxygen; this deprivation of oxygen, which may partly arise from the decomposition of the water entering into the composition of the juices of the sap, may then also determine a larger proportion of hydrogen in the fresh substances formed at their expense. Indeed, we find all the substances formed in the bark under the influence of the solar light, impressed with this double characteristic, an augmentation in the proportion of hydrogen and especially of carbon: this is shewn by the chlorophyll and latex, as well as the resins, the essential oils and the wax [e].

There can be no doubt, that all these products result from the action of the light; for deprived of it, they gradually lessen and disappear. We have already spoken of the etiolation which the plant undergoes after a prolonged stay in the dark, and which supposes a change in the chlorophyll and the conditions, which are necessary to prevent its developement. Now, an analogous effect is produced by the same cause on the proper juices, the resins and the essential oils; and amongst the proofs of this truth, we need only cite the common practice of gardeners in the culture of certain culinary vegetables, which, when developed in full light, would have juices of too strong an odour, of too bitter a taste, and sometimes even dangerous to be used, such as several Umbellife-

[e] We do not speak of *fixed oils* (*huiles fixes*) here, because they are generally found in the fruit and kernels of the seed, scattered in drops in the interior of the cells in which they are formed. "These oils comprise bodies insoluble in water, fluid at the ordinary temperature and not susceptible of being volatilized without being decomposed. *Waxes* differ from the preceding merely by being solid at the ordinary temperature. *Volatile* or *essential oils*, which resemble fixed oils, are distinguished from them by an odour, a slight degree of solubility in water, and, lastly, by being volatilized without decomposition. *Resins* include dry bodies more or less fragile, slightly soluble in water and more or less altered by the action of heat." These definitions, borrowed from M. Chevreul, are all we can give here. As nature presents them more or less compound, it is necessarily extremely difficult, if at all possible, to determine their chemical characteristics precisely in general terms. A detailed examination would occupy more space than could be afforded in an elementary work like this.

rous plants and Lettuces. They cover the lower portion of the plant, which is to be eaten, with earth; this process is termed *blanching*, on account of the loss of the green colour. A familiar instance of this may be seen in the earthing up of Celery. But, at the same time, the plant loses the great strength of its juices, which are thus reduced to such a degree that they are innocent and agreeable.

§ 313. Does respiration exercise any influence over the compounds with regard to the proportions of nitrogen? We know that the air furnishes it sometimes, but in small quantities and not in a constant manner. On the contrary, Saussure says, that during the day a little is disengaged with the oxygen. Yet there exists a great proportion in the juices of the bark, whether the azotised combinations, which furnish the juices have been changed or have been simply concentrated. The latex extracted from the vegetable in a sufficient quantity and left to itself, is acted upon in the same manner as milk and blood, being divided into two separate portions, the one liquid and transparent, the other coagulated and opaque; and in some plants we may almost say that it is animalized. This is the only vegetable liquid in which we find this property.

§ 314. It is in the cells of the bark that those quarternary combinations also are formed, now known under the name of *Alkaloids* (*alcaloïdes*), because they have the property of combining with acids in the same way as alkalies. We meet with them when thus combined, and that only with a small number of vegetable acids, whilst the plant is alive. Modern researches have multiplied to an immense extent the number of these substances, which are generally designated by the termination *ine* (as *Morphine*, *Quinine*, *Strychnine*, &c.) These researches have been pursued on vegetables which have the most remarkable properties; each has furnished its alkaloids; more than one has furnished several different kinds. These alkaloids, extracts of the same vegetable, appear to have a sort of affinity for one another, which will be easily understood by reference to a well-known example. Peruvian bark contains three of these alkaloids, cinchonine, quinine and cusconine; when analyzed, the three give 20 atoms of carbon, 24 of hydrogen and 2 of azote; cinchonine has in addition 1 of oxygen, quinine 2 and cusconine 3; so that the first three elements seem to be united in a simple body, which, oxydised to three different degrees, would furnish the three known alcaloids. The most energetic properties of the vegetable appear to reside in these substances, and the small number of substances which we mentioned above, remind us of very active and noted medicines and poisons.

§ 315. We have just mentioned the principal substances which are formed by the extraction of a portion of oxygen. If the contrary takes place, and the proportion of oxygen augments, we shall have acids. The vegetable acids, the number of which modern chemistry has greatly augmented, are very rarely met with in a free state in living tissues, but commonly combined either with the alkaloids or with the inorganic alkaline substances conveyed in the sap. The most common is oxalic acid; this is a binary compound and is remarkable for differing from carbonic acid only by a small proportion of oxygen, of which it contains 3 parts for 2 of carbon. Several are ternary compounds, such as acetic, citric, pectic, malic, tartaric acids, &c., &c.; very few are quarternary compounds with rather a large proportion of nitrogen, such as aspartic acid, &c. As to hydrocyanic acid, formerly known under the name of prussic acid, instead of being ranked among super-oxygenated substances, it does not contain any oxygen at all, but an enormous proportion of nitrogen, rather more than half its weight. This nitrogen, united to carbon forms a base, called *cyanogen*, which is itself united to 3 parts of hydrogen to form the hydrocyanic acid, which we find in the ALMOND (*Amandier*) and in several trees of the same family.

§ 316. The production of the true acids, those which result from the augmentation of the proportion of oxygen, is favoured by nocturnal respiration, by means of which this element penetrates in great abundance throughout the whole of the vegetable; it is also in the parts deprived of solar light, or coloured otherwise than green, parts in which this is constantly the case, such as the roots and the fruits, that we meet with the greatest number and the greatest quantity of acids. Let us remark that these are the same parts which we often see become places of deposit for fecula and sugar, i. e., for those substances, which present the combination of almost equal parts of water and carbon, and which, under the influence of light, are modified by assuming a larger portion of hydrogen and especially of carbon. It is necessary that, where they are accumulated, they should not have been modified when once formed, or that when once they are changed they should be brought to their primitive composition by the subtraction of the excess of carbon fixed in the green parts by the diurnal respiration. Now, this effect is produced by the other process or manner of respiration, and explains, perhaps, the want of oxygen which is manifested by the subterranean parts and the favourable influence which night has over vegetation, in re-establishing the equilibrium after the energetic action of the day.

§ 317. Besides, this absorption of the oxygen gas by the vegetable parts is not a phenomenon which is peculiar to the living state. If, after death, these parts are put in connection with oxygen and with water, the former disappears and combines with the carbon of the vegetable substance, and forms carbonic acid. During this very slow combustion, the substance changes both in form and colour, and passes by degrees into the condition of a blackish dust, which is known under the name of *vegetable earth* (*terreau*) or *humus*, in which are found all the elements which it contained during life; but in different proportions. A portion, however, of the carbon is still combined with the elements of water and forms a compound very analogous to starch, from which it differs by the smallest possible quantity of water, 6 instead of 10 parts for 24 of carbon. This substance, which is called *ulmine*, is not soluble in water, but it becomes so when combined with alkalies, commonly mixed with the humus, the formation of which is accelerated by their presence. The majority of physiologists have assigned very high functions in vegetation to it; for, when it is combined, with lime, for instance, it is quickly dissolved in water; the plant, rooted in the humus, absorbs it; when it is absorbed, there only wants a few parts more water to convert it into starch and thus to become a most fertile source of organic matter. M. Liebig has, however, recently cast some doubts on this point; his objections are, that ulmine requires 2500 times its own weight of water to dissolve it, and, consequently, the whole of the water contained by the plant would only contain a very small proportion; that this water has also to carry along with it, and, after its decomposition, to leave in its tissues the alkaline substance to which it was united, and that, moreover, in the sum of the substances, which are found in the ashes of the plant, the ulmine is very insignificant. He, therefore, concludes that the greater part of the carbon found in the sap, does not spring from this source, but rather from the carbonic acid, formed during the decomposition of the vegetable *débris*, dissolved in the water of the earth and absorbed along with it into the roots. Whatever may be the explanation, the earth is the abundant source of carbon, and is so much the more fertile, as it contains more of it in a state to facilitate its solution in water.

§ 318. We have seen, however, that the plant, which draws a

great deal from the air, can dispense at an emergency with that from the earth, and that a seed reduced to pure water for its only support, may be developed into a very fair plant. Nevertheless, this developement is soon arrested in its progress, and we do not find that any fresh tissues are formed. Amongst the necessary causes of this stoppage, we find in the first rank one which we have already had to state several times. All the tissues, when they are springing into existence, are remarkable for the abundance of the azotised substances which they contain, and which may, consequently, be regarded as one of the conditions necessary for their developement. Now all the azote of the seed (in which its presence is constant), is soon extended and weakened by the growth of the plant, in spite of the supplementary provision which it may extract from the air. The earth, therefore, is necessary to supply that quantity of nitrogen which is necessary for a continuance of the growth of the plant.

§ 319. Whatever the water of the earth may contain, whether from the solution of several salts of humus, into the composition of which nitrogen for the most part enters, or from that of the ammoniacal vapours which rain brings down with it, would be far from being sufficient without the deposit, which the *débris* of vegetables and animals causes to accumulate on the surface; that of animals being exceedingly rich in azotised substances. Hence the utility of manures, when man wishes to cultivate vegetables crowded into a limited space, which, generally destined for the nourishment of man, ought themselves to imbibe in their tissues several of these principles; hence their necessity in the cases in which these vegetables, such as the Cereals, do not borrow nitrogen directly from the atmosphere. It is in the state of ammonia that it first mingles with the sap.

§ 320. It now remains for us to examine amongst the materials furnished by the earth, those which belong to the mineral kingdom, and the influence they exercise over vegetation. This influence may be of two kinds; the one exercised by those, which, insoluble in water, remain about the roots mingled with the vegetable and animal refuse, of which the earth, so called vegetable, is composed; the others, by those, which dissolved, are introduced and mingled with the sap.

§ 321. We shall understand, without entering into very long explanations, to what degree the power of the first extends, and how it varies according to the primitive constitution of the soil. Clay retentive of water, sand, which permits it to pass freely, present the two extreme conditions in which different vegetables only can live; and, if we remember the imperious want which most plants have for water in the inside, and of air in contact with the roots, we shall judge that a proper mixture of parts of a different nature, which hold a sufficient quantity of water, and permit the air to circulate freely, present the most favourable conditions to vegetation. Mineral substances, adapted to fix carbonic acid and ammonia, will thus be able to retain round about the vegetable a portion, which, otherwise, would be volatilized, and to preserve it as in a store-house: this will be added to the quantity furnished directly by the air and to that which the water already holds in solution. To an analogous reason M. Liebig attributes the magic influence of gypsum and ferruginous salts on vegetables, because gypsum, as well as the oxides of iron and alumina, attracts ammonia, and forms a solid compound with it, from which it is then separated by degrees at each shower of rain, to be conveyed by the water which is pumped up by the neighbouring plants.

§ 322. But what interests us most here, is the knowledge of the mineral substances, which, themselves soluble, penetrate and are incorporated in the vegetable. When they are once introduced, they may either preserve their liquid state, or be solidified, which may happen, either by the evaporation of the water which holds them in solution, or by forming an insoluble salt as they combine with acids in their progress; and where this is once formed it remains. We have already shewn (§ 20—22) the shapes, which these mineral bodies generally assume in the tissues of the plant, and the places where they are most frequently found; it is chiefly near the bark, which is the principal seat of the evaporation. The quantity of the mineral substances is generally in proportion to the activity of the vegetation, since it determines the passage of a larger quantity of water, and, consequently, of mineral substances dissolved by it. The quantity of these which remain soluble may vary according to the time; that of the insoluble ones can only augment with age.

This proportion may be easily found out by the combustion of the parts. Fire destroys without exception all vegetable substances, and this is one of their characteristics. It does not destroy the mineral substances, which the plant contained, and the residue of which forms the cinders. By weighing a vegetable body and then the cinders arising from its combustion, we can obtain the desired result.

§ 323. The mineral substances most commonly found in the vegetable are potash and soda, lime, magnesia, flint and rarely alumina, sometimes a little iron and manganese. These bodies may already be in the condition of salts, combined with certain mineral acids, sulphuric or phosphoric acid, for instance, and this explains in certain cases the presence of sulphur and phosphorus. The combination with carbonic acid may take place either inside or outside of the plant. The salts which are formed in the inside by the combination with vegetable acids and, therefore, are called vegeto-mineral substances, are most frequently composed of lime or potash and oxalic, malic, citric, &c. acids.

§ 324. It is clear that the nature of these compounds is always correlative to that of the soil in which the plants grow. The one can only receive what the other can give. But does it receive it indifferently? In other terms, is it because such ground contains such mineral substances, that the plant which grows there itself contains them, or else does it grow on this ground for this reason? In certain vegetables, the answer would not be doubtful. Thus, the greater part of the plants which grow on the border of the sea contain much soda, arising from the chloride of sodium or sea-salt; and they do not grow in any other part, except near salt lakes situated far in the interior of the country where they find this same salt. It is, therefore, necessary for them; they do not take it because they find it there, but they grow there because it is necessary to their existence. Certain families of plants, which are similar in all the principal points of their organization, present the same mineral substances in their tissues, and we may conclude that their presence is in connection with this organization. That of the Gramineæ presents a double example of it: its perfect fruits (in the Cereal plants) contain large quantities of the phosphates of magnesia and ammonia; their stems almost without exception, silica, which incrusts their epidermis and nodes (§ 20). It is the flint which gives straw the power of preserving its stiffness and durability for a long time without rotting, so that it often notches the sickle, and the stems of Rattans the property of striking fire like a flint and steel.

It happens, however, that the same plant, growing in different soils, will not present the same salts. The reason of this is, according to M. Liebig, because certain bases may supply the place of one another: those which are susceptible of entering into combination with the same vegetable acids. He even thinks, that the proportions of these acids, in the vegetable into which they enter as organic substances, are bounded at a certain degree, and, consequently, the bases which are united to them, although different according to the ground, are nevertheless nearly equivalent.

§ 325. We must then suppose, that all these substances, although inorganic, play an important part in the organization; that their quantity and their quality are in a certain relation to the wants of vegetables, and constant in a given plant, or in a class of plants. Mineral bodies would, then, have a double influence on the life of vegetables: the one, general, by retaining around them a much greater provision of their essentially nutritive principles; the other, special, by penetrating them and communicating to them materials, which are mingled with organic substances without being assimilated to them, exciting them by their presence, solidifying them or partly neutralizing them; which, in spite of our ignorance of the manner of their action, appear necessary to the exercise of the life of which they are themselves deprived, and which, lastly, are not always the same in different vegetables. Agriculture mixes with the earth as manures different inorganic substances (gypsum, marl, ashes, &c.), which are varied according to the nature of the ground and the kind of produce which is to be assisted.

§ 326. EXCRETIONS (*Excrétions*).—The organized body has received into its interior substances, which it has drawn from the outside; it has set apart, i. e. SECRETED, everything which can be employed in its nourishment. But there may remain a certain part which is improper for it, and the body tends to free itself, to throw it to the outside, to EXCRETE it, according to the scientific mode of expression. These excremental substances may have preserved the composition which they had, when they entered into the body, or else, by a series of combinations which take place in the interior, may be changed.

In animals (always excepting those whose organization is less perfect, which are placed at the lowest part of the scale) the excretions find ways prepared for their escape; these are most commonly ducts or canals, which are destined for this use and are called *excretory;* they may, therefore, be easily studied; but there are none of these in the vegetable. It is true, that little excretory canals have been observed by M. Ad. Brongniart in the glands which are found at the bottom of the flowers of certain Liliaceæ; but, we may assert, that these canals are for the most part wanting, and the substances which are to be rejected do not find any other open ways of egress, except those which serve for the transmission of the nutritive substances. When they do not escape by means of superficial glands directly open to the air, they will then escape either by transmission through the walls of the epidermis, or through the stomata or other natural solutions of continuity which may exist on the surface of the vegetable.

§ 327. The substances, thus rejected to the surface, may be distinguished into three classes. They have been wrongly confounded under the same name of secretions;

1st. Those, which extended over the surface, are preserved to protect them and thus continue to be employed in the life of the plant. These in general are resinous matters or wax, impermeable to water, and preventing by a kind of varnish, as it were, the effects of the humidity of the air on the tissues, and also moderating the evaporation. We see it employed in the first way on the scales of the buds of several of our trees, which are thus protected by this layer from the cold damp atmosphere of winter; the buds of the Poplar and Horse-Chesnut present this resinous exudation in a remarkable degree. This excretion may be supposed to be for the second purpose, when we see it on the greater part of plants which exude it during the summer, especially, if they grow in dry sandy districts, where they frequently want water and it is necessary to allow as little as possible to escape from their tissue by evaporation. In several parts of Chili as well as the Andes of Peru, where the soil and air are dry for a long time together, a great number of shrubs present in common this characteristic of resinous exudations on their surface. The abundance of glandular hairs or little glands on the surfaces which are thus covered is common. The SILENE VISCARIA and NUTANS, the LYCHNIS VISCARIA, and especially the DICTAMNUS ALBUS, are examples of this and may be readily found.

Waxy substances have analogous uses and characteristics. They form the white dust which covers several leaves, those of the Cabbage for instance, and several fruits, such as the Plum, the Grape, &c., which we call the *bloom* (*fleur*). If we plunge one of these leaves into water, we shall see that it is not wetted; and, on the other hand, since, in a great number, it is on the lower face, commonly the active seat of evaporation, that the waxy covering is exclusively or most abundantly found, we may infer from this, that its function is to moderate and retain the liquids in the interior. This is evident in those fruits in which a liquid and copious juice is amassed. No peculiar organ secreting wax has yet been discovered; it is formed in the external cells, and exudes thence to the outside. The stems of several Palms, especially of CEROXYLON ANDICOLA, are covered with a thick layer of this substance, which has probably flowed down from the base of the leaves.

The viscous matter, which covers the greater part of the plants submerged in sea-water as well as fresh water, is doubtless for the purpose of protecting the tissues from the action of the surrounding fluid which would otherwise macerate them. But is this substance which forms it produced in the tissues, and then exuded like resin or wax? This is more doubtful, and M. Mohl thinks that it is nothing else than the extravasated intercellular matter which has flowed to the outside (§ 15).

§ 328. 2nd. The substances, which are thrown to the outside, may be identified with those which are preserved within. It is only part of these extravasated; and, it is evident, that on this account they cannot be considered as true excretions, i. e., rejected as being improper to nutrition. They have become so momentarily, because each body can consume and retain for its nourishment only a given quantity, and the excess must necessarily find some other passage than its natural one. The bark, pressed on the inside by this exuberance of the substances, often splits and thus allows it egress. Through these fissures flow the gums, proper juices, resins, as we may see in the case of several of our trees: gums from the Cherry-tree, the Plum-tree, the Acacia, and resins from our Fir-trees. The superabundance, which determines these overflows, often corresponds to the healthy and vigorous vegetation of the plant and is even the result of it; but, at other times, it is a kind of disease, a sort of derangement in the functions. In this continuance of chemical transformations, which the materials of nutrition have to undergo, there is some condition wanting, and, consequently, there is some point stopped. The substance, which is formed, is not resumed by the organs to undergo its subsequent changes; it is accumulated and may even in some cases pass into fresh and unusual combinations.

The substances of which we have just spoken, exude in the liquid state and flow over the bark; but they are soon thickened by exposure to the air and are frequently even solidified into concrete masses. There are some which escape under another form, as the essential oils, which, when they are volatilized, are disengaged through the unbroken tissues.

§ 329. 3rd. Substances not adapted for nutrition, and then

ejected to the outside, only merit the name of excretions; but it is very difficult to determine those which are really so. Even the productions of the glands, which we have seen flowing out and evaporating on the outside, may cause some doubt as to this, since it is possible that they are partly reabsorbed in order to be reconveyed into the mass of nourishing fluid, and then the part which is evaporated would only be as in the preceding case, an excess from which the tissues wanted to be free. But, does there not exist some general way, by which the vegetable body after having drawn from the nutritive substances all the particles which are necessary for that purpose, excretes all those which are not adapted for it? Several authors have thought that this was one of the functions of the roots, and this theory appears to be justified by reason. The sap, entering into the roots, first runs through all the ligneous body, is then completely organized in the bark, by means of which it descends, furnishing in this passage all the elements of nutrition to every part, and thus returns to the ends of the roots, deprived of a part of all those elements, which it has distributed on its way. The question is, to know, if the residue is there thrown out as an excrement or is mingled in the still imperfect sap, like the venous blood in animals. It is certain, that around the extremities of several roots we may find small clotted lumps or flakes of a matter which has the appearance of a jelly or mucilage, and absorbs water as it swells. This is what holds the little lumps of earth or sand, which we find collected and attached at the ends of the roots, with whatever care we may pull them up. It is difficult to believe that this is not an excretion of the roots.

But are we to conclude that it is the residue formed by the particles of nourishing matter which were not adapted for nutrition? The experiments of M. Macaire seem to place this beyond all doubt, both by the analysis of some of the materials compared with the nutritive substances of the same plant, and also by the observation that different solutions of poisonous salts (acetate of lead, nitrate of silver, &c.) absorbed by a plant, whose roots are then placed in pure water, are soon found in this water; whence we should be inclined to think, that the plant rejected in this manner all substances which are improper for or opposed to the preservation of life. But similar experiments have given MM. Unger, Meyen and Walser negative results. It would, therefore, be difficult, if not dangerous, to affirm anything about it.

§ 330. This doctrine, when once admitted, will lead to very important consequences. If it be true, the plant would thus accumulate around itself in the earth, substances which would not be adapted to nourish it; and this would explain why these roots are always obliged to extend farther, in order to find subsistence, and why one tree languishes where another of the same species has previously grown. There are plants which are hurt by being too near; there are others, however, which prosper when close together. The reason of this is that the radiculary excrements differ according to the vegetable, the substances rejected by one being either hurtful to the other, or, on the contrary, suiting it, as we find the excrements of some animals becoming food for others; the plants which avoid one another rank among the first class; those which grow near one another among the second. The farmer in the culture of annual plants, cannot generally reap from the same ground several good successive crops of the same plant; he, therefore, takes care to vary them, and experience teaches him what plant should succeed another and thus rest the soil instead of exhausting it. This is called the art of *rotation of crops* (*assolements*), which is only the application of the preceding theory, and would have light thrown on it by all researches which may be made on the excrements of the roots.

Let us however add that it is not necessary to explain all the facts mentioned. It is clear, in short, that one plant growing in a piece of ground, will take from it a greater or less proportion of the substances necessary for its nutrition; that another plant of the same species with the same wants, which replaces it, can only find for itself a certain quantity; whilst a plant of another species with different wants will find substances suitable to it, both in nature and quantity. Recent experiments have thrown great light on this question. There are some vegetables, which, like the Cereals, extract all their azote from the earth; there are others which, as in the case of certain Leguminous plants, can draw it from the air. The second may, without any danger of being deprived of food, succeed the first. Now, this rotation of Cereal crops and of Leguminous plants is one of the commonest in practice. Between several crops thus varied, the earth is gradually regaining its first composition under the incessant action of the elements of the atmosphere.

§ 331. We ought, perhaps, to treat now about the colours of vegetables, as well as their own heat; for these different phenomena depend, like the greater part of those we have previously explained, on chemical composition and decomposition. Nevertheless, since the colours are observed with more variety and intensity, since it is in the developement of flowers that that of heat is most sensibly manifested, we shall postpone these questions till we shall have treated of the flower in detail (§ 633—661), which we have hitherto recognised only in its relations with the rest of the vegetable.

Growth of the Tissues.

§ 332.—The nourishing of the vegetable naturally causes its growth. Its elementary organs, increasing in size and number, determine a proportionate augmentation in its compound organs. It is the manner of growth in both that we must now consider successively.

We shall not repeat here what has been said in the first chapters concerning the manner in which the cells, the fibres and the vessels grow and are thickened: nor of the order in which these organs are generally developed with regard to one another; but their manner of multiplication has not yet occupied our attention, and this is the place to treat of this question. Now, since several vegetables are exclusively composed of cellular tissue, since all other forms at first begin with that of a cell, the problem will be simplified into researches on the manner of the multiplication of cells: this has been seen by the most skilful physiologists in directing their attention to this fundamental point, the origin of the tissues. Since they had to observe the smallest parts into which the plant may be divided, when they first appear, we shall not be astonished at the various differences in these observations and in the theories which are deduced from them.

§ 333. Growth of the cellular tissue.—The multiplication of the cells may take place in several different ways. In the most simple vegetables, as in several Confervæ (those green filaments which are frequently found on water or on moist surfaces, and are composed of cells placed end to end forming a necklace-like tube), we clearly and distinctly see these tubular cells, when they have acquired a certain length, present one or more rarely several strangulations, as if their wall was bent transversely and turned inwards; this projootion gradually increases and, at last, forms a complete partition, which is afterwards doubled, and we have then one or more cells resulting from the division of a single one; it is most commonly, but not constantly, that which forms the extremity of the general tube. Some of the middle cells often at their upper end throw out a small lateral lump which is gradually elongated, then, having nearly attained the length of the cell from which it springs, is separated from it, at the point at which it deviates, by the formation of a partition, such as we have just described. In this manner are formed the lateral branches of these Confervæ. The tubes of the Charas are divided in the same way, and sometimes not by a transversal, but a longitudinal partition; hence results from one single cell, two collateral cells which are not placed above one another. In these different cases, the inner cavity was filled with a granular substance, the mass of which is subdivided at the same time and in the same way as the cell. We may suppose that this method of multiplication may be also found in the cells of more elevated vegetables, especially, in those which are arranged in rectilineal series.

§ 334. Another manner, which many botanists for a long time considered, and which several still consider as most general, is that, which results from the formation of several cells in the cavity of a pre-existent cell. Sometimes, from the walls of this mother-cell slight abutments are formed inside, which, at last, meet and thus cut the single cavity into several, each of which encloses granular matter. Up to this time this mode of developement seems to be similar to the preceding; but in a little time each granular substance is covered with a proper envelope, and there are thus formed so many single utricles, contained under one general envelope, or, and it often happens, if this disappears, constituting so many separate utricles. The utricles are most commonly, from the beginning, free in the cavity of the mother-utricle, which, in this case also, either remains or is more commonly reabsorbed.

§ 335. The cells, in short, may be formed in an interior cavity of the vegetable, in the interval of the pre-existent cells, as we see, for instance, between the wood and the bark of Dicotyledonous plants, where the cambium is deposited.

§ 336. But, under what form are the cells, thus formed, whether in the cavity of the pre-existent cells, or in the lacunæ of the tissues, at first observed and by what gradations do they reach the form of those cells which have preceded them?

According to M. Schleiden, the first state of the cellular tissue is always that of a gummy solution, which is then thickened into a jelly. We next perceive in it a great number of extremely small opaque points. Some of these points become centres around each of which is formed a granular mass, which is the *nucleus* or *cytoblast* (§ 21), the germ of the cell. These cytoblasts are commonly lenticular bodies, rather rounder among Dicotyledons, more oval and generally larger among Monocotyledons. When they have acquired their definite dimensions, we see on one of their faces a blister rising, which covers it similarly to a watch-glass, then it swells more and more until it represents a vesicle, in which the cytoblast occupies only a small space, buried and even forced into a corner of its wall which is doubly thick at this place.

Most commonly it is in a little time absorbed and disappears, when the utricle has arrived at its perfect shape; it, however, sometimes remains, examples of which may be seen in several organs, such as the hairs, in which we observe the rotatory movement (*fig.* 222, *n n*); and in certain families, such as the Cactaceæ and the Orchidaceæ.

It has been objected against M. Schleiden's theory, that sometimes the cytoblast is first seen after the appearance of the utricle. Besides, although its existence has been verified in the young tissues of several organs, he himself has not recognised it in the two most active seats of this cellular formation, the radicellary extremities (in which he is not sure of having seen it) and the cambium. Wherever he has found it, the young cells were being produced in a mother-cell.

§ 337. M. Mirbel affirms, that wherever a tissue begins to be formed, cambium is found. It is at first a liquid of mucilaginous consistence, which is gradually thickened into a jelly. We then see it covered with little dots which resemble little points; but these are so many little cavities which are gradually increasing (cellular cambium); the cavities continue to increase, in proportion as the partitions, at first very thick and soft, grow thinner, and they take a regular form, which as yet they had not: it is the cellular tissue, still a continuous tissue, which may be compared in its form and appearance to the lather of soap-suds. At last, the partitions are doubled, either throughout the whole of their extent, or only at their angles, and the tissue is thus dislocated into as many distinct bodies as there are cavities; we have utricles and more tissue properly so called. The cambium is found nearly everywhere, not only in the great interstices in which we are accustomed to find it, but in the meati which it dilates, in the interior of the cells and of the vessels. If it is developed there, we shall see this interior obstructed by a cellular tissue; but, generally, a single utricle by its rapid growth outstrips and puts the others to one side, and, at last, fills the cavity and doubles the wall; and it may in its turn be doubled in the same way by a third: and in this manner do we explain the thickening of the cells, fibres and vessels. It often also happens, that the cambium deposited in the tissues is stopped at the first period of its developement, and is absorbed.

§ 338. M. Unger admits this series of apparent transformations in the cambium, but regards the fact of the tissue being at first continuous as a mistake. He says, that he has discovered that the utricles are distinct from the beginning, although their thick, soft and closely pressed walls do not permit the line of demarcation to be easily seen. This original separation of the cells was a fact universally admitted, till doubt was cast upon it by the researches of M. Mirbel.

§ 339. How is the cambium itself formed? We may suppose that it is at the expense of the most elaborated juices, as the numerous and ingenious experiments of Duhamel on the formation of the wood in Dicotyledons have proved. A thin plate of tin, introduced between the bark and the wood, shewed that all the cambium came from the bark. A shred of bark, attached by its upper surface to the tree, was raised, and the surface of the wood beneath it was destroyed; and, yet, this did not prevent the production of cambium when the bark was replaced. An annular decortication, prevented from becoming dry, shewed the cambium oozing out between the wood and the bark in great abundance from the upper edge of the wound, much less at the lower. We, therefore, conclude, that the juices of the bark, coming from the top to the bottom, furnish the cambium, and that it is not the ascending sap, as we should suppose from the identity of its chemical composition with that of the cellulose and, consequently, of the greater part of the organic substances found in this sap.

§ 340. The multiplication of cells sometimes takes place with extreme rapidity. The young shoots of some of our trees in a favourable spring may give us an example of it, and yet the speed of their vegetation is not to be compared to that which a more elevated temperature causes. In our very green-houses, we see the Aloe at the moment of blossoming, the Bamboo, &c., &c., grow at certain times in twenty-four hours more than 2 decimetres (French) or 7·866 feet (English). Certain plants entirely composed of cellular tissue are developed very quickly even in our climate; for instance, Mushrooms and other Fungi, the rapid growth of which has given rise to a proverb. There is one, Lycoperdon giganteum, which, in two or three days, acquires the form of a ball, 3 decimetres (French) or 11·799 feet (English) in diameter.

§ 341. Growth of the Stems and Roots.—As to what concerns the growth of the compound organs, we have already, under the articles *Stem*, *Root* and *Leaf*, explained what passes; we must now shew how these changes take place. We will prosecute our researches on Dicotyledons, which are better known, since they acquire a great size in our climates, and which, by the formation of lateral buds, furnish more numerous and clear data for the solution of the problem. Let us now mention in a few words, that the stem is lengthened from bottom upwards, the root in an inverse direction; that the former presents a pith and a medullary sheath partly composed of unrollable tracheæ, which are wanting in the latter; that, in a little time, between the sheath and the bark fresh fibres and vessels of another order are interposed, and that, from this interposition, which is repeated every year, results the growth in thickness.

The origin of these fibres and vessels is a subject on which botanists have not yet agreed. As we explain and discuss the different theories and the facts on which they are based, we shall treat at some length of those which are connected with the growth of compound organs.

§ 342. An ingenious theory, which is supported by a great number of botanists, was proposed by a French astronomer, M. Lahire, at the commencement of the eighteenth century; but his treatise was only in a few pages and wholly unsupported by facts. A century afterwards, another Frenchman, M. Dupetit-Thouars, proposed it again, having, doubtless, re-discovered it from his own observations, and, as he sustained his theory by works, which are rich in facts and reasoning, he had the honour of the discovery, and it is now known only by the name of the theory of Dupetit-Thouars.

The buds, as we have repeatedly stated, may be compared to so many embryoes; they are each developed into a branch like the stem, which is the result of the developement of the embryo. But this, fixed in the earth, in germinating produced from its lower part roots entrusted with the function of pumping up its food and nourishment. The buds, which, arrived at maturity, are detached from the stem, as we have seen in the bulbs (§ 182), the suckers or off-shoots, the bulbels (§ 184), the rosettes of creeping stems (§ 183), are so many imitations of true embryoes, and send their roots downwards. Do those buds only, which remain fixed on the stem, want the roots? Dupetit-Thouars does not think so; and, as this mass of fibro-vascular fascicles, formed between the bark and the medullary sheath, only appears after the buds have begun to grow, as, on one hand, we see them attached to the base of these buds, and, on the other, we may follow them to the extremity of the roots, he thinks that they are nothing else than the roots of the bud, running downwards between the bark and the sheath, until they escape to the outside in the form of roots, either normal or adventitious. The cambium itself is nothing else than a nourishing fluid which these roots extract in this passage through the vegetable. Each year a fresh set of buds or stationary embryoes thus produces a fresh series of corresponding radicular fascicles, the whole of which adds a layer of wood and fresh ramifications to the roots.

§ 343. This theory has recently received a fresh impulse in the hands of M. Gaudichaud, who does not restrain it to the buds, but extends it to its constituent parts, its axis and its leaves, the one performing with respect to the other exactly the same functions that M. Dupetit-Thouars attributes to the buds with regard to the stem. A Monocotyledonous embryo (apart from the gemmule) is composed of a tigelle, of a leaf or cotyledon and, in a little time after by the process of germination, of a root. M. Gaudichaud esteemed this the type of the vegetable individual or the *phyton*, thus formed of an ascending system (tigelle and leaf) and of a descending system (root). When the gemmule is developed, above the cotyledon the first internode terminated by a leaf is produced and bears the same relation to the leaf as the tigelle to the cotyledon. They form, therefore, the ascending part of a fresh phyton, the descending part of which can only reach the earth through the stem, which it traverses under the form of fibro-vascular threads, inside the cortical envelope. It is the same with all the successive leaves, each placed on its internode, each demitting its radicular fibro-vascular fascicles through those situated below it. Thus, the stem, which results from the evolution of the gemmule, is a series of tigelles united end to end, each of them enveloped by the radicular fascicles of all those above it; and it would exactly represent a twig, were it not that in the twig, the whole of the radicular fascicles reaching its lower end is implanted in the branch from which it arises, and in which they continue their internal descending course. The Dicotyledonous embryo or every internode bearing two leaves is only the union of two phytons.

§ 344. The fibres and vessels of the bark originate in the same way as the fibro-vascular fascicles of the wood, to which they are united at the beginning. They grow in the same way from the buds and belong to the descending system.

§ 345. As to the cellular tissue, its production is everywhere local and results from the multiplication of cells already in existence; consequently, in the wood, from the extension of the medullary rays. Thus, when it grows in thickness, the increase of the tissue takes place horizontally, whilst that of the fibres and vessels takes place vertically; it is a kind of cloth, in which one forms the warp and the other the woof. When the fascicles are separated on the outside in order to form roots, they borrow the layer which accompanies them from the neighbouring cellular tissue, and this, growing as they lengthen, forms a kind of sheath for them.

§ 346. The principal facts brought forward to support this theory are the following.

The wood of the roots, which without any dispute belongs to the descending system, never has any unrollable tracheæ; now the wood of the stem has none also and in its elements presents perfect similarity.

It is continuous with that of the root; and, since this is formed after the stem, we are authorized in our belief that the first formed is continued into the other, and that the fascicles of the stem follow a descending direction. But these fascicles are attached in the stem by their upper end to the base of the buds, and, in the bud to the origin of the leaves; they come, then, from buds, or, in the first place, from the leaves.

In the production of aerial roots, we may remark, that their origin and the situation of the buds and of the leaves regularly correspond; for they commonly grow immediately above the nodes, consequently, at the base of what M. Gaudichaud calls the ascendant system of the phyton. In several Monocotyledons, especially, we see one of these roots grow from the base of each internode directly beneath the leaf which terminates it. In a very small number of plants (Pourretia, Kingia, several Lycopodia), these roots, instead of appearing on the outside immediately, creep for some distance under the cortical envelope. The phytons produce, therefore, real roots, either free or concealed in the substance of the stem. We find all the transitions from this arrangement to that in which they are united and blended with one another in order to form the ligneous tissue.

§ 347. The descending course of the fascicles of the wood is especially manifested by the obstacles with which it may meet. If we cut off a ring of bark or bind a ligature tightly round the stem, at the upper edge of the ring the tissues thicken and form a collar-like excrescence; whilst the lower side is not at all expanded. On dissecting this excrescence, we find that it is composed of a network or plexus of interlaced fascicles, twisted and turned in every direction, but always continuous with those which terminate at the buds. If we take off a piece only of the ring of the bark, these fascicles are not long in turning round the edges of the wound in order to resume their vertical direction. If the stem be compressed by a spiral ligature (several climbing plants, as the Honeysuckle, are thus naturally pressed), an excrescence is formed in this case also, but in a spiral line following the spire of the ligature, and the fascicles also, when dissected, appear to be accumulated in this direction. If the stem, around which we make this annular decortication, have no buds beneath the ring, and all that existed have been destroyed, so that the only buds that are found are above the wound; and, if the wound be large enough for the two ends of the ring to meet, the whole portion of the vegetable above it continues to grow in thickness by the regular production of fresh ligneous layers, and the growth and formation of the wood below it are stopped. From these experiments we conclude, that the ligneous fascicles come from top to bottom, and are produced by the buds, whose roots they represent.

§ 348. But are there not some facts opposed to those, which have been mentioned in support of the theory of Dupetit-Thouars? And, besides, is this the only theory which will explain it? We will now treat of the objections to it.

If the ligneous fascicles are true roots, they must be lengthened at their lower extremity only; their organization ought, therefore, to be the more advanced, as they are nearer the bud from which they are demitted; we should see them stopped lower or higher, according as the time for the first evolution of the bud is more or less near. Now, it is true, that, in some cases, we find beneath the bud a mass, like a hank, as it were, of ligneous filaments, which are arrested at a little distance; but most frequently, it is impossible to follow the developement of the ligneous filaments, which are formed at almost the same instant from one end of the stem to the other: examination by the microscope tells us, that they are generally softer as they are higher, their fibrous and vascular elements more incompletely and, consequently, more recently formed. It would seem from this that the fascicles are formed almost simultaneously throughout their whole length, and that, in some cases, at least, it rather begins at the bottom and goes towards the top, than the reverse.

§ 349. If we examine the wood of two stems or branches of different species which have been cleft-grafted, one *A* on the other *B*, we remark that each has preserved the nature of its wood, although, according to the theory of Dupetit-Thouars, all the fascicles formed by *A* after grafting ought to have been continued and to have descended into *B* and formed its ligneous layers. If a young stock *B* has been grafted, as yet furnished with few roots, at the end of a certain number of years all the fresh roots, which have grown, should have come from the buds of *A*, and the suckers we should take off should reproduce this species *A* : but experience proves that it is *B* which is reproduced. To this double experiment, from which it would result, that neither the wood nor the roots could come from the radicular fascicles belonging to the buds, the partisans of the theory answer, that it is the cellular tissue which communicates the colour to the wood and imparts to it its nutriment, and that, consequently, in the midst of the cellular tissue of *B*, which, formed in its own place, has preserved all its characteristics, the wood of *A* borrowed its colour from it; that it has modified its grain on account of the change in its food. Their answer to the second experiment is, that it is the cellular tissue of *B* which, accompanying the ligneous fascicles of *A* in the roots, has given the special characteristics to the suckers taken from the roots, since these suckers could only spring from the developement of the adventitious buds growing from this tissue. These questions, therefore, border too near on the mystery of assimilation, to enable us to speak of them in a decisive manner.

§ 350. If the lateral ramifications of the roots begin, like the majority of the other organs, as a small cellular nucleus, in which we find in a little time vessels organized (§ 114), they are not simple prolongations of descending fascicles previously formed. This is, therefore, a very important point of organogeny, which we must verify, when we consider the theory of Dupetit-Thouars.

§ 351. The facts, on which it leans for the greatest support, are those which shew us the stoppage of the descending fascicle above every natural or artificial obstacle, above ligatures and decortications; and the accumulation of the ligneous tissue there, which on the contrary, ceases to be produced underneath it (§ 347). But, is it not as clearly and naturally explained by the course of the nourishing juices which furnish the materials of the Cambium? These juices, descending in the bark, will accumulate above every obstacle they meet with in their course, coming to the outside, if it be impassable, twisting round it, if there be a lateral passage open, stopping if there be not; and in all these cases, the afflux of the materials determines a more abundant production of the tissues, their want produces atrophy, according to the general rules of all organized beings. The afflux of the juices precedes the appearance of the fibro-vascular fascicles; they are formed there instead of coming there already formed.

The developement of the aerial roots close to the nodes, where there is frequently a slight stoppage in the circulation, is easily understood, since they mostly grow (§ 113) from all the points where the juices have been amassed and, consequently, where there is a great deal of cellular tissue. Since the evolution of the buds is one of the determining causes of the ascension of the sap (§ 267), since the sap, once risen, is elaborated in the young bark and, especially, in the leaves (§ 275), it is clear that the suppression of the buds and leaves will put a stop to this ascension, elaboration and, necessarily, from the want of materials to the consecutive formation of the ligneous fibres; it is clear, that it cannot be formed in any part of the vegetable, from which we cut off communication with that covered with leaves and buds.

We cannot very well conceive by the theory of Dupetit-Thouars, how a tree, which has undergone an annular decortication so as to prevent the formation of ligneous layers beneath it, can continue to live and grow; for this growth presupposes that the roots continue to grow in the same proportion. How can the radicular ligneous fascicles, stopped in their course, produce roots?

§ 352. The arrangement of the lacticiferous vessels into a net allows their canals to supply one another and the juices to overcome with greater ease the obstacles they meet in their descent; and, as these lacticiferous vessels, although accumulated inside the bark, are frequently extended to all the other parts of the vegetable, we may conceive that the suppression of the bark for a certain extent around the stem does not necessarily and completely retard the arrival of the latex to the lower parts. Could it not also pass along something analogous to what we observe in animals, when a member, in which the circulation of the blood arrested in a large artery is carried on only by some weak lateral branch, continues to exhibit a small degree of vital activity? This hypothesis would explain (it must, however, be verified) the reason, why, below an annular decortication, the stem is kept alive, produces roots and even in some cases thin ligneous layers. The production of the last has been explained by the existence beneath the decortication of adventitious buds, from which a certain quantity of ligneous filaments would be demitted. This result, however, would only arise from the developement of a certain number of buds and those of a certain size, and, then, they would be clearly seen. Observation has yet to decide the value of this hypothesis.

§ 353. M. Dutrochet has mentioned this singular fact, that the trunks of some species of Firs (ABIES EXCELSA and, especially, A. PECTINATA), cut down a few feet from the earth, continue to live and thicken with a succession of layers during a long series of years. There can here be no descending fascicles. The action takes place incontestably from bottom upwards, and this direction is proved by the existence of rather a thick excrescence which surmounts these layers, formed after the trunk has been cut down. The ascending sap, therefore, in this tree without the assistance of the leaves acquires a degree of organization sufficient to form these annual layers, which are extremely thin, not attaining a millimetre (French) or ·03933 feet (English) or even a demi-millimetre or ·019666 feet (English) in thickness.

This fact, which does not seem to be included in any of the preceding theories, has recently been explained by a very probable theory. It has been observed that these stumps, the growth of which continues in spite of being deprived of their top, are in the neighbourhood of other trees of the same species in full vegetation, and, on digging under into the ground, some of the roots have been found grafted on one another. The roots, therefore, of the untouched trees will nourish the others, by conveying to them juices already organized; this is confirmed by the larger developement of the ligneous layers on that side of the stump corresponding to the engrafted roots. This union of the roots of neighbouring trees of the same species, or even of species a little different, is by no means rare between those of the family of the Coniferæ, nor probably also between some other trees, the Beech, for instance. It may account for the vital force preserved by the roots of trees in the earth, a long time after all their aerial system has been suppressed.

§ 354. To recapitulate what we have said on the growth of the stems and roots in thickness, botanists have agreed on the formation of those parts which are purely cellular: but they are still at variance on the formation of the fibro-vascular fascicles of the wood and the bark. Two theories have been proposed: the one considers these fascicles as the roots of the buds and, consequently, as developed from top downwards; the other considers their elements as spread at once in the state of a half fluid jelly (the Cambium) over the whole internal surface of the bark and as being developed there. We think, however, that these two theories are not so much at variance as they would appear at first sight. They would be so, doubtless, were we to admit that the ascending fascicles produced buds at their extremity. But is this theory received now? We have admitted that the leaves have at the beginning no continuous vascular connection with the branch (§ 147), nor the bud (§ 171) with the stem which bears them; that the juices, elaborated in these leaves and these branches, descend from them to the extremity of the roots through the bark, on the internal surface of which is deposited a demi-fluid substance in which the tissues are organized. M. Gaudichaud, on his side, admits that "elaborated and partly organized juices (the Cambium) and tissues, still fluid, are formed and solidified as they descend from the buds to the branches, from the branches to the stems, and from the stem to the roots, by an act of lengthening analogous to that of the roots, if it be not perfectly the same." Between these *descending* tissues in a *demi-fluid state* and the tissues *formed* in a *demi-fluid state* and furnished by *descending* juices, we cannot establish a fundamental distinction clear enough to place these two hypotheses in direct opposition.

§ 355. We were right in stopping with the examination of Dicotyledons, both on account of the greater facility in studying them, and because in Monocotyledons the growth of the stems in thickness is generally arrested very soon on account of the want of lateral buds. Besides, the absence of true bark, as well as the permanent union of the lacticiferous vessels and fibres analogous to the liber with the vessels and fibres of the wood in the same fascicle, and the serpentine course of these fascicles in the interior of the stem, would have rendered the explanation more complicated and obscure. We will, therefore, simply refer the reader to what we have already said on this subject (§ 96), as well as on the growth of Acotyledonous trees (§ 107).

SUMMARY.

§ 356. On recapitulating what we have previously explained concerning the functions of vegetation, we shall compare summarily the Vegetable and the Animal Kingdoms. In this comparison, we shall reject from both those, which present the most imperfect organization, and in which the functions of life are incomplete and most generally obscure.

§ 357. 1st, The Vegetable absorbs by the extremity of its roots bodies on the outside in the state of liquid; bodies, perfectly inorganic, namely, oxygen, hydrogen, carbon and azote, under the form of water, carbonic acid, ammonia, as well as some others of a metallic nature; 2nd, The latter preserve their nature in the inside of the tissues; but the former are differently combined, whence result other more compound bodies called organic, suitable for becoming the constituent parts of the vegetable; as, the fecula; 3rd, These constituent parts are in their turn modified, under the influence of respiration and secretions, so as to give birth to all the products of organism. It is very necessary to conceive all these combinations as taking place, not between the four elementary bodies, but between the compound organic substances, which, when introduced into the Vegetable, they have formed under the influence of the vital force. These compound bodies are called the *immediate principles of Vegetables* (*principes immédiats des Végétaux*).

§ 358. In the nutrition of Animals, these first preparatory operations on the aliments do not take place, since they always feed on animal or vegetable substances, which, consequently, are already more or less organized. The fresh degree of elaboration, which they undergo in the animal body, continues the vegetable act commenced in the plants, but at the same time traverses, as it were, a circle; for the Animal destroys these substances by employing them, and those which they reject by its respiration and its excretions are precisely those crude substances, which are the first food of the Vegetable.

§ 359. The Vegetable absorbs its aliments in the earth by the extremity of the last radicular ramifications; the Animal absorbs its nutriment by the last ramifications of the venous and lymphatic vessels in the intestinal tube. But they have, ere this, undergone in this tube a previous preparation by the act of digestion, which act, as well as its organs, is completely wanting in vegetable life[f],

[f] Some authors have considered the greater part of the phenomena of nutrition and respiration to be those of digestion. According to them, the stomata and radicular extremities would represent so many mouths giving entrance to the aliment, and the elaboration, which they then undergo in the interior of the vegetable, especially the fixation of carbon, would represent a kind of digestion; the respiration would be limited to the fixation of the oxygen derived from the atmosphere, and would be quite similar to that of animals. As we have already said (§ 298), the truth of these theories is totally dependant on the definitions which we adopt for these functions. But, we should be obliged to give rather a different meaning to digestion to what it has in animals, i. e., the first preparation of the aliment in the intestinal tube.

since all the operations, by which the substances are found in a state of solution in the earth ready to be absorbed by the roots, are performed quite apart from any vital function.

§ 360. But, if we pass to the subsequent acts of nutrition, we shall find a kind of parallelism between the two kingdoms. In the Animal, these substances, absorbed in the liquid state by the vessels in contact with the intestinal tube, are conveyed into the mass of the blood, and along with it into the respiratory organs, where, submitted to the influence of the air, they undergo certain modifications, by means of which the fluid becomes adapted for the purposes of nutrition and is carried in a retrograde course to all the parts of the body. In the Vegetable, the fluids absorbed by the roots are carried[g] through the central parts to the circumference, the seat of respiration, and, in contact with the air, undergo fresh modifications, after which, endowed with fresh and essentially nutritious properties, they are in a retrograde course distributed throughout every part. Thus, in both cases, external substances are conveyed towards the respiratory organ, then from this to all the parts of the body to be nourished. The liquids brought back by the bark and, especially, those circulating in the lacticiferous vessels may, therefore, to a certain point be compared to arterial blood; those, which pass from the roots along the ligneous centre, to the liquid furnished by the chyle. Does that part of the nourishing fluid, deprived of the materials which were employed in nutrition, return in Vegetables again to mix with the circulating mass, as it does in Animals, in which it forms venous blood and probably lymph? This is very probable, and seems deducible from the frequent absorption of the vegetable tissues which has been discovered. But we must not affirm any thing decisively, so great is the confusion of opinion concerning the ways open to the circulation of the juices in the vegetable.

Indeed, if from the examination of the function in general and of its products, we pass to that of the organs by which it is performed, all similarity ceases. Vegetables have not the same regularly ramified systems of vessels as Animals, which clearly indicate the course of the nourishing fluid; nor any centre from which it receives the impulse which determines its motion. The absence of all contractile tissue may be esteemed an essential difference in the cause and nature of this circulatory movement.

[g] They undergo during this progress a certain degree of elaboration which begins to organize them, and we need not look for an analogous process in animals, whose alimentary particles are previously organized.

§ 361. We have already (§ 298) instituted a comparison between the respiration of the two Kingdoms, and shewn how it is performed in both. We have explained how its organs are modified, according as it takes place under the surface of the water (§ 299), or in the air, and pointed out under these relations a certain analogy between Vegetables and Animals, according as they respire in either medium, likening submerged leaves to branchiæ, aerial leaves to lungs or rather to the bodies of insects (§ 288) traversed by their tracheæ, which communicate with the outside by means of pores. In the stomata we have found something analogous to these pores; but, in the rest of the respiratory apparatus there is an essential difference, since in Animals the air traverses a series of branched vessels (branchiæ or tracheæ), and acts through their walls on the blood contained in capillary vessels creeping along their surface, whilst in Vegetables it passes through a series of cavities formed by the interval of the cells more or less separated from one another (§ 127), and acts on the fluid contained in its cells. This arrangement may be compared to that portion of the respiratory apparatus of birds, which is extended from the lungs to every part of the body, and puts them in connection with the air circulating through a quantity of large lacunæ placed in the cellular tissue. We have also called attention to another fundamental difference between the respiration of Animals and Vegetables: animal respiration is performed without intermission during life, in the dark as well as in the light, whilst Vegetables only respire in the light (§ 292), the chemical rays of which seem to take part in the fresh combinations formed under its influence (§ 293).

§ 362. As to the exhalation from the superficies of the body we have seen that it generally takes place over the whole of the Vegetable, but it is by no means large and depends for the most part on external circumstances, the insensible perspiration, which is naturally compared to the exhalation from the surface of an animal body: then a more active one, which accompanies respiration, and takes place by the same passages, and which we compared to the pulmonary exhalation (§ 300).

§ 363. As to the secretions, we find this very difference so frequent in the tissues of the organs in Animals and Vegetables, composed in the former of a plexus of vessels, in the latter of a mass of cells. This last structure is seen in the glands of Vegetables, which are only comparable to the simplest among Animals, the majority of which are, besides, provided with excretory ducts, not observable in the former. We have not found in Plants certain excretions constantly existing as such, though there are such in Animals; these, the existence of which is general in the greater part of these Animals, seem to have some special connection with digestion, a function wanting in Vegetables. And, besides, on account of the great uniformity of the vegetable tissue, it is so difficult to distinguish all those which may secret a particular fluid, and to follow the subsequent course of these fluids in the midst of others, that there is still a great deal of uncertainty with regard to vegetable secretions and excretions (§ 326).

§ 364. Let us conclude, therefore, that if there is a certain analogy in the series of the great actions which constitute the functions of nutrition in Animals and Vegetables, it disappears almost entirely in the organic apparatus which performs these functions, an apparatus clearly definite in the former, but obscure in the latter; that, in short, in the products which result from it we find differences so combined, that the two Kingdoms may make continual interchanges between one another; a counterbalance, always ready to maintain an admirable equilibrium in the midst of the disorder, into which the business of the life of this large crowd of organized beings unceasingly tends to throw Nature.

THE FLOWER.

§ 365. Although led to speak of the flower, during our examination of the inflorescence and blossoming, yet it has only been in a general manner, as one whole, which we have not hitherto separated into its component parts; and we have compared it to a bud or a rosette of leaves (§ 199), premising that these fresh leaves commonly differ more or less from those of the stem in their shape, their colour, their sizes, in short, in every property. They differ no less in their functions, and these differences added together, have naturally led ancient botanists to consider them as distinct organs. But how is it, it may be asked, that modern botanists have recognised leaves in the different parts of the flower? To this it may be answered, that the different stages in the passage from a leaf to a flower have been observed, and this has been strengthened by the application of the rules laid down (§ 235) for determining organs, which are frequently disguised under forms very dissimilar to their normal ones. A few examples will clearly explain this to the reader.

§ 366. Let us take the most beautiful of our wild flowers, the White Water-Lily (*Nénuphar blanc*) (NYMPHÆA ALBA) (*fig.* 223), which spreads on the surface of stagnant pools its rosettes of green folioles around the edge of the flower, yellow ones in the centre and white ones in the interval. The green ones present this colour only on their outside, the inside being white; there are four (*c c c c*), of a very long oval. The numerous folioles which follow (*p p p p*) are white on both surfaces, the outer ones of the same shape, and as long as, if not longer, than the green ones, the inner ones growing gradually shorter. Nearer the centre (*e*), they become yellow; and they grow narrower, passing gradually from the oval shape to that of a narrow ribband. At the same time, a more apparent modification is visible at their upper extremity, which presents two swellings, like two longitudinal folds, which are lengthened in proportion as they are found farther in the interior, among the folioles of that part, and finish by becoming rather more than half as long, (*e*, 4, 5), and quite as thick as they are, whilst at first they could hardly be distinguished at the swollen top of the outer leaves (*e* 1). Lastly, the middle is occupied by a circle of yellow bodies (*s*) much shorter than the preceding, also formed by a swelling, but a single instead of a double one at the top of each; they form the apex of a much larger body,

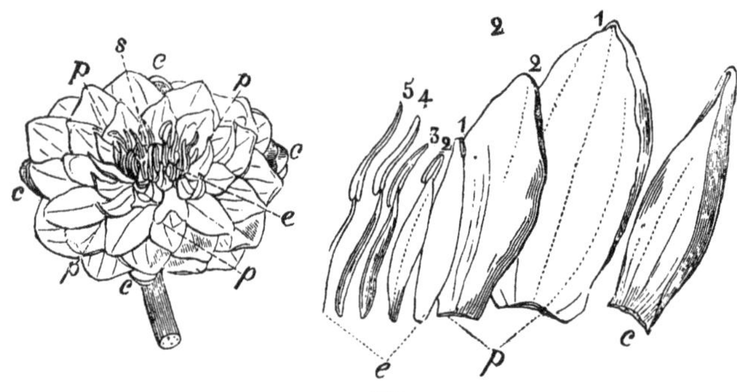

223

which, cut horizontally, presents in the inside a cavity divided by partitions arranged like so many radii, and equal in number to the yellow bodies of the summit. The central body is called the *pistil* (PISTILLUM); we cannot recognise any apparent resemblance between it and the bodies we have just described; between the latter, however, there must exist an undeniable resemblance, because the passage from the outer to the inner shews all the modifications in succession. Now, botanists have assigned names to the principal modifications which we have pointed out; the collection of green folioles is called a *calyx* (*calice*) (*κάλυξ*; *a covering*); that of the white leaves, each of which has received the name of *petal*, (*pétales*) (*πέταλον*; *a leaf*), is termed COROLLA; the *stamens* (*étamiines*) are all the yellow parts, which are thickened at the top, for different lengths, by a double fold. In the greater part of other flowers, the constant and well-defined distinctions between these various parts warrants their being distinguished by different names. In the NYMPHÆA ALBA they are not so clear; and in the series of the intermediate forms from the folioles of the calyx to the innermost stamen, it would be difficult to settle a point in which one kind of organ ends or another commences, so that this example in a great measure allows us to recognise in the folioles of the calyx, in the petals and in the stamens, one and the same organ, more or less modified.

223. A flower of the White Water-Lily (*Nénuphar*) (NYMPHÆA ALBA), taking a bird's eye view, as it were; it is several times smaller than nature.—*cccc* The four folioles of the calyx.—*pppp* Petals.—*e* Stamens.—*s* Pistil.—We may follow the degradation of the parts in form from the circumference to the centre. By the side of the flower are placed a series of modified folioles from the green one of the calyx and the white one of the corolla *p* 1 to the stamens, becoming more and more distinguishable by the difference of their form *e* 4, 5.

§ 367. But is this organ a leaf? The bracts are doubtless modified leaves (§ 226), for the passage from one to the other is frequent and is manifested by insensible gradations. Then, that of the bracts to the calycinal folioles is not less so, and in several cases it is impossible to distinguish them from one another, the Peony (*Pivoine*), the Broom-rape (*Orobanche*), &c. In other cases (the Rose for instance [*figs.* 285, 369, *cf*]), the calycinal parts present the shape of true leaves, and the name of folioles long since applied to them, shews that this analogy did not escape the notice of the first botanists.

§ 368. The parts of the pistil, then, are the only ones in which we cannot yet recognise the leaf. But if in the NYMPHÆA, their transformation has so completely disguised them, other examples, on the contrary, will shew us, that they are not always in such great perfection, and that most commonly they are less so than the stamens.

Let us take a flower of the MAGNOLIA, that of the YULAN (MAGNOLIA CONSPICUA), now so commonly cultivated in our gardens. It has first a spathiform involucrum composed of two green velvety bracts, then shews a rosette of nine large lily-white folioles; inside of these on a lengthened axis, a mass of narrow bodies terminated in a point; the lower ones of a yellow colour, drawn into a thread at their bottom: the higher ones green and swollen, on the contrary, at the base, which is empty within and the swelling of which thus corresponds to a closed cavity. If we compare this flower with the NYMPHÆA ALBA, we shall see in the white folioles the calyx and the corolla, here only distinguishable by their respective places; in the yellow bodies, we see the stamens, the form of which is here always the same, and clearly differs from the petals. The green bodies, which cover the top of the axis and occupy the centre of the flower, correspond by their position to what we have just pointed out as the pistil. But in this

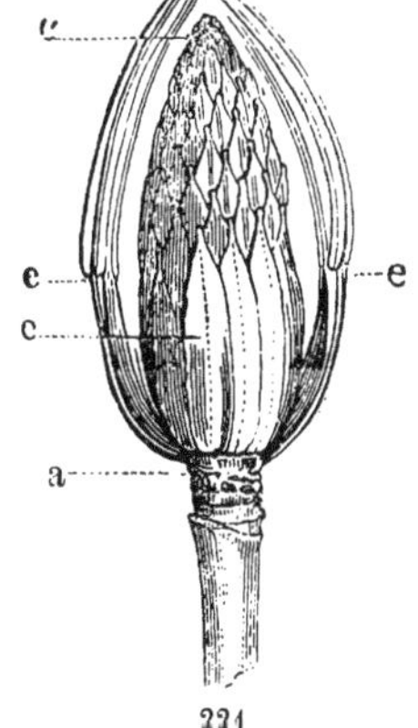

224. The central part of the flower of the Tulip-tree (*Tulipier*) (LIRIODENDRON TULIPIFERA), composed of carpels *cc*, the whole of which, when united, forms the pistil. They cover the upper part of an axis *a*, and beneath it are implanted numerous stamens of which some *ec* have been left, and others have been taken away leaving small cicatrices on the axis *a*. These stamens are hypogynous and extrorse.

flower they are composed of a great number of separate parts, of so many little rolled leaves, as it were. In the flower of a genus bordering on the Magnolias, the Tulip-tree of Virginia (*Tulipier de Virginie*), (LIRIODENDRON TULIPIFERA), we shall find first on the outside three green calycinal folioles; then, in two rows, six petals also green but spotted with red; nearer to the centre, a great number of stamens drawn at the bottom into a thread (*fig.* 224 *e*) and occupying the base of a central axis (*a*), the rest of which is covered by little green leaves (*cc*), flat, thick at their point, swollen and hollow at their base, where they are confounded with one another, till, when they have arrived at maturity, they are completely detached from one another. These are the component parts of the pistil, each of which has received the name of *carpel* (*carpelle*) (*καρπὸς; a fruit*); we shall call them also *carpellary leaves* (*feuilles carpellaires*) in the pursuit of this investigation.

§ 369. In the different examples which we have here taken, the spiral arrangement of this succession of parts composing the flower (calycinal folioles, petals, stamens, carpels) is quite evident, and thence results in the case of the White Water-Lily, in which the axis which bears all these parts is extremely short, a rosette analogous to that which we have represented (*fig.* 156); in those of the MAGNOLIA and Tulip-tree, in which this axis is very long, an arrangement similar to that delineated (*fig.* 158). This collocation of the parts of the flower seems to establish a certain analogy between it and the true leaves, provided that the rules which we have before laid down be true.

§ 370. In most flowers, this spiral arrangement of the parts is less manifest, and this may arise from several causes which it is easy to explain. The surface, which bears them, does not form a lengthened axis, as in the MAGNOLIA and Tulip-tree, nor does it extend in breadth, as in the Water-Lily; then, the parts, compressed into a small space, are inserted on points, which are too close for their relative positions to be clearly defined, or their direction to be strictly preserved, whilst their developement is frequently so irregular. It is the same here as in a plantation; if the trees be placed a good distance from one another it is easy to determine the order in which they are planted at the first glance; if, on the contrary, they are crowded as in a nursery or a wood, it will be difficult to perceive their arrangement, even supposing that they have been planted in regular order; and in time, this regularity even would be totally lost, because, among the trees, especially if not all of the same nature, some will have outgrown, or even killed the others.

The surface, also, presented for the insertion of the folioles of the flower is not always perfectly regular, like that of a cylinder, cone or circular plane; and this defect may necessarily cause a similar one in the position of the various parts of the flower.

In the examples which we have chosen, the great number of these parts placed on an extended and regularly developed surface, allows us to ascertain with ease the multiple and secondary spires, from which we may conclude that one primitive spire exists (§ 159). But, let us suppose for an instant, that, even in one of our examples, the Tulip tree, we had been contented to have compared the five stamens, or the five carpels situated lower down, together, it would then have been very difficult for us to have appreciated the small difference in the degrees of height, which exists between their insertions, and we should have the whole five in both cases arranged as if in one circle. This is what happens in most flowers, in which the number of parts is much more limited than in those from which we have borrowed our examples. The carpels, thus reduced in number, appear to rise from the same height; the stamens, reduced in the same way, seem to be arranged in a circle around them; the petals and calycinal folioles in two other concentric circles. Sometimes we may still recognise by means of certain signs, which we shall learn to appreciate in a little time, slight inequalities in the relative heights of the folioles of the same kind with respect to one another; sometimes it does not really exist, and these parts of the flower enter exactly into the class of verticillate leaves. Since they present almost the same appearance in both cases, they have been considered, indeed, as arranged in *whorls* (*verticilles*) (VERTICILLI), and it has, consequently, been agreed to designate by this name the groups of the different organs which we have already pointed out in the flower. If it be complete, it will be formed of four whorls, that of the calyx, that of the petals, that of the stamens and that of the carpels, which united form the pistil.

§ 371. If the flower is perfectly regular as well as complete, in each of these different whorls, the parts will be equal in number, and then we shall again find the law to prevail, which we have mentioned as general in the arrangement of the leaves of two whorls one above the other, i. e., the alternation of those of one with those of another (§ 164). Let us shew this by an example. The flowers of CRASSULA LUCIDA and of C. RUBENS, (*figs.* 225, 235,) &c.

present, 1st, A calyx composed of five green, equal, tongue-shaped folioles, arranged in a circle (*fig.* 225, *cc*); 2nd. A corolla of five rosy petals (*p p*) longer than the folioles, which grow on an inner row, exactly in the five intervals which separate the five tongue-shaped folioles; 3rd, The five stamens (*eee*) in the intervals of the petals and, consequently, placed in front of the divisions of the calyx; 4th, Five carpels (*o o*) arranged like a star, alternating with the five stamens and, therefore, placed in front of the petals. Let us add before we proceed, that these carpels will give, better than any of our previous examples, a just idea of the foliaceous nature of the pistil. They have each the shape of a small folded leaf, turning its median nerve outwards and its two edges inwards, which touch and unite during the blossoming, only to separate in a little time afterwards. Thus, in the case of whorls composed of the same number of parts, the alternation of these parts of one whorl with those of one nearest to it, and the opposition, on the contrary, of these parts in one pair of whorls to those in another pair, is a law common both to true leaves and also to the modified ones, which enter into the composition of the flower.

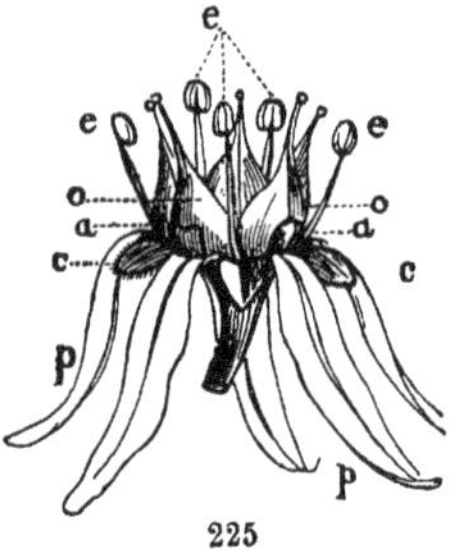

225

§ 372. In the midst of this prodigious variety, which enables us to distinguish so many millions of plants by their flowers, we might expect to meet with a great diversity in the number of the parts of which the floral whorls are formed; and this is what really takes place. Nevertheless, among these numbers, there are two which occur most frequently, viz., five and three; and it is a fact worthy of notice, that the former is found in

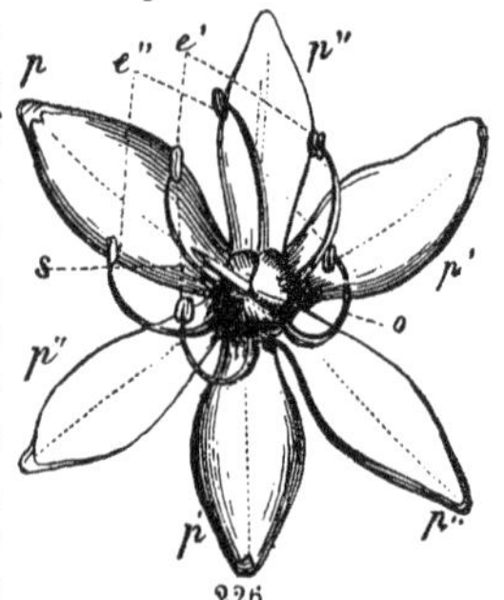

226

225. A flower of the CRASSULA RUBENS.—*c c* Folioles of the calyx.—*p p* Petals.—*e e* Stamens.—*o o* Carpels, to each of which a small appendix *a* in the form of a scale corresponds.—The horizontal section or diagram of this flower is represented in figure 234.

226. A flower of the SCILLA ITALICA; a bird's-eye view.—*p' p' p'* The three external folioles of the perianth.—*p'' p'' p''* The three inner folioles.—*e'* Outer stamens opposite to the former.—*e''* Inner stamens opposite to the latter.—*o* Ovaries united into one.—*s* Three styles united into one.—The diagram of a similar flower is represented in figure 233.

most Dicotyledons, and the latter is still more common among the Monocotyledons. The flower of the CRASSULA, which we have just described, may be cited as a type of the first; and that of the Lily (*fig.* 267), of the Tulip, of the SCILLA (*fig.* 226) and of the greater part of the Liliaceæ, as types of the second. This is composed of a whorl of three folioles (*fig.* 226, $p'p'p'$), of three others ($p''p''p''$) on an inner circle alternating with the first, to which they are more or less similar; of three stamens (e') opposite to the first folioles, then of three others (e'') opposite to the second, and, consequently, on a circle nearer the centre of the flower; lastly of three carpels (o) united in the centre of the flower, alternating with the inner folioles and stamens. This type may, therefore, be considered as formed of five ternary whorls, two of calycinal folioles, two of stamens, and one of carpels.

§ 373. ADHERENCE OF THE PARTS OF THE FLOWER.—Two flowers, in which the number of the whorls is equal, as well as that of the parts which compose each of them, may, however, be distinguished by many characteristics, by differences of size, of shape, of colour. One of those, which contribute most to determine the various combinations, is the union or joining of the neighbouring parts with one another; so that they seem to present one piece only instead of several distinct ones. In the flowers we have before mentioned, in spite of the care which has been taken to choose examples, in which all the parts of the flower were as independent as the leaves of a branch, we have found some of these unions; that of the Water-Lily and of the Scilla, in which the pistil apparently constitutes a simple body; that of the calycinal folioles of the Crassula, which are lost in a kind of cup at their base. These cases frequently exist, sometimes in one part, sometimes in another, and sometimes in several at the same time. Let us in a cursory manner examine the principal modifications which may result from it.

§ 374. This junction may take place between the parts of one or of several whorls, and, as we clearly understand, there are several degrees between complete union and complete separation. The pieces of the calyx may be united with one another by their edges at different parts of their height, or, it may be, the petals. In this case the calyx is said to be *monophyllous*, (*monophylle*) (a term we have already seen employed in a parallel instance, when the bracts form the involucrum [§ 231]), the corolla is termed *monopetalous* (*monopétale*), in opposition to *polyphyllous* (*polyphylle*), *polypetalous* (*polypétale*), by which is designated the contrary state, in which the folioles and petals, many in number and composing the calyx and corolla, are all independent and quite distinct. The former of each two terms has been justly criticised, for, according to the etymology (*μόνος, single*), it would seem to indicate that there was only one foliole, one petal. But, they have been so long and so generally adopted, that it is well to preserve them, always bearing in mind that the calyx or the corolla thus named are composed, not of one single part, but of several parts united together into one piece. It has been proposed to substitute for them the words *gamophyllous* (*gamophylle*), *gamopetalous* (*gamopétale*), (*γάμος, marriage, union*); but, besides the inconvenience of substituting fresh names for others which have been constantly employed, they are not even themselves free from objection, which will be clearly seen when we come to study the developement of these parts (§ 431). Let us, as a general principle, preserve as much as possible all the ancient names, at the same time taking care to define them clearly and so guard against any misconception.

§ 375. This coherence may take place between the stamens. If they spread like petals, they may be joined in the same manner by their edges (*fig.* 322); but most frequently they appear like little narrow threads which only touch and are confounded when they are rather numerous; then we often see them unite, not in one cylindrical tube, but in several fascicles or bundles, each of which is known by the name of *adelphia* (*adelphie*) (ἀδελφεῖος, [Ep.] *brothers*) (*figs.* 238, 322), as *monadelphia*, *polyadelphia*.

§ 376. Lastly, this union may exist between the innermost of the whorls, the carpels; and, as they then present themselves to one another by their faces and not by their edges, and as also they occupy the centre of the flower, the body, which results from their adherence, is a solid much more simple in appearance than the parts we have seen result from the junction of the other whorls.

§ 377. It is clear, that all these adherences have a tendency to disguise the foliaceous nature of these parts. In those parts which remain quite distinct, it is very easy to recognise the leaves; especially, if, placed at different heights, they manifest by the manner in which they mutually cover one another their spiral arrangement, as the calycinal folioles of the Hellebore or the CAMELIA (*fig.* 258, *c*). When, united at their base, they are separate at their upper part, we may still recognise, although with less facility, so many leaves, as in the calyx of the Borage (*Bourrache*) (BORAGO OFFICINALIS). We can only recognise it by analogy, when they are joined in a great part or even in the whole length of their edges, so as to form a tube, (the calyx of the Pink [*Œillet*, *fig*. 286, 1, *c*]; of the RHINANTHUS, or a kind of cup (calyx of the Orange blossom).

§ 378. These junctions will be so much the more frequent as the parts of one whorl are more crowded, whether they are larger, or the space allotted to them is narrower. We may conceive, therefore, that the stamens with wide filaments are more frequently joined than those with thread-like ones; that the stamens are generally less frequently joined than the petals, which are always so much larger; that, on the contrary, the carpels, commonly much thicker than the other parts and also concentrated in a much smaller circle in the middle of the flower, are more commonly united, when the axis is not long or broad enough to separate them; that, the shorter and thinner this axis is, the greater tendency the whorls, which grow on it, have to unite with one another.

§ 379. But, this union may not only take place between parts of the same whorl, but also between those of two different ones, and under the influence of causes analogous to those we have just indicated. It is of course in the lower part, in which these parts have less play in their developement, that this juncture most frequently takes place. The floral whorls may thus be united in pairs (the corolla with the calyx or with the stamens), in threes (the calyx, the corolla and the stamens), in fours. This last case will always happen when the calyx is joined to the pistil, since the bottom of the stamens and petals, situated between them, are necessarily comprised in this union. But it is extremely rare that the pistil enters into a union, the calyx remaining independent, with the stamens (NYMPHÆA ALBA), or even with the petals (RASPAILIA), although we may see by these very examples that this combination may be met with.

§ 380. When several different whorls are thus united to one another, the parts of one and the same whorl are themselves also united; this is almost a necessary consequence of the law of the alternation of one whorl with another. If the parts of the two whorls, A and B, are alternate, some one part of B, thus situated between two parts of A, cannot be united to these parts without joining them one with the other, if they are not so already of their own accord. We may, however, conceive possible exceptions, in the case, for instance, in which this part of B would be joined by one of its sides with one of the parts of A, whilst it remains independent of the other; this happens sometimes, though very rarely, (as in the Olacineæ). It is more common to find certain portions of the flower opposite those of a neighbouring whorl, whether from an apparent reversion of the laws of position which we have just stated, or from the deduplication of the parts of one of the two whorls; and here two pieces placed one above the other may easily adhere together, and yet remain independent of those which are on the right and left. This is observed pretty frequently in the petals and stamens that are opposite to one another (STATICE ARMERIA, AGROSTEMMA GITHAGO, and several other Caryophylleæ).

§ 381. The traces of the juncture are frequently very manifest. These parts remain distinct, although they adhere together; and, even, in some cases, a slight effort will overcome this adherence. Thus, in several monopetalous corollas, we may see the filaments of the stamens adhering to the tube which is formed by the lower part of the united petals: the stamens are then distinguishable by their projecting and their colour, which is frequently different, and we may follow these even to the commencement of the tube (*figs.* 227, *f*; 326, *i*). In other cases, the traces of the union disappear; of two parts united, the inner appears to spring from the other at the very point, at which it becomes free, and below which the two tissues are confounded in one.

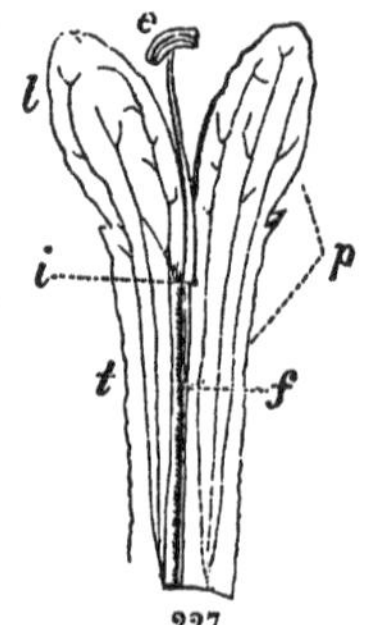

§ 382. But, very often, in the whole space where two whorls are thus united, a singular tissue may be remarked, differing from that of the parts which compose them; this tissue is most commonly glandular, that is, presenting in its structure the mass of little close, dense cells, which are characteristic of several glands; it is even frequently prolonged beyond it, in the shape of a wen-like excrescence or of a projecting ring. On a careful examination of the surface between the calyx and the pistil, which surface has hitherto received the name of the *receptacle* (*réceptacle*) (RECEPTACULUM) of the flower, more recently that of TORUS, and bears the parts of this flower, we frequently find it, where they originate, tapestried, as it were, with their tissue, which is sometimes extended in one superficial layer, sometimes takes the form of concentric projections

227. A detached portion of a monopetalous corolla of a COLLOMIA, shewing a strap-shaped piece of the tube *t* terminated by two lobes of the limb *l*, and to which a stamen *e* is inserted, the filament of which, free when it leaves the point of insertion *i*, is also perceived below *f*, till at the base of the tube it is confounded with its tissue.

similar to whorls. This projection, designated by various terms, but most frequently by that of *disk* (*disque*), generally gives birth to the parts of the corresponding whorl; it may, then, be compared to the pulvinus of the leaves. The parts may grow from the very edge of the disk, or from its inner or outer face. It may be more or less lengthened, and thus bear them at a greater or less distance from the torus. Of different degrees of thickness, it may fill the interval, frequently very narrow, which separates two whorls, and thus become the commonest means of union between them. It is on account of this, that we meet with its tissue so constantly in the union of several whorls, of the calyx with those which are nearer the centre, of the pistil with those which are nearer the circumference. This junction takes place, therefore, not at the lower part of the petal or of the stamen, but on the disk, which raises it and serves as a base for it.

§ 383. THE INSERTIONS OF THE PARTS OF THE FLOWER.—From the preceding facts, from the variety in the apparent point of departure of the whorls of the flower in connection with one another result those differences, so easy to appreciate, and so important in distinguishing different plants and flowers. Since each whorl seems to commence from the very point in which it is disengaged from the neighbouring whorls; since, examined on the outside it appears to be inserted at the corresponding height on the general axis which bears the flower, the term of characteristics of *insertion* has been applied to those, which result from these different relations of the whorls of flower, not united, or differently united with one another at their origin and throughout a greater or less extent. It is, principally, the relations of the stamens to the pistils (the essential parts of the flower, as we shall see in a little time), that botanists have endeavoured to express by terms invented to designate these different methods of insertion. If the stamens are united with the corolla, they are then said to be *epipetalous*, (*épipétales*), and, in this case, the insertion of both is considered as the same, as it is with regard to the remainder of the flower. If the stamens, whether thus joined to the corolla or independent of it, are equally so both of the calyx and of the pistil, it is clear that they are inserted on the torus below this pistil (*fig.* 228), and they are called *hypogynous* (*hypogyne*) (from ὑπὸ, *under*, γυνὴ, *a woman*). If they are inserted on the calyx (*fig.* 229), they are raised on it to a certain height above the base of the pistil; their position will appear in relation to it to be not beneath but on the side, and they are, consequently, termed *perigynous* (*périgyne*) (from περὶ, *around*). Lastly, if they are inserted on the ovary

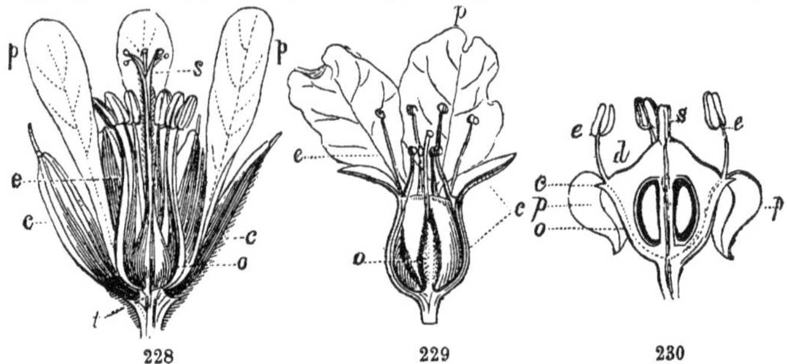

itself (*fig.* 230), they are *epigynous* (*épigynes*) (from ἐπὶ, *on*). We have already seen (§ 378), that, in this last case, the four whorls are most commonly joined together and, consequently, the stamens are at the same time inserted on the calyx and on the pistil, which sometimes leads observers to hesitate between these two manners of insertion and to confound them together. This is especially the case with De Candolle, who has termed *calyciflorous* (*calyciflores*), those plants, the flowers of which are in this last condition or else present stamens freely inserted in the calyx; *corolliflorous* (*corolliflores*), those in which the corolla bears the stamens; *thalamiflorous* (*thalamiflores*), those in which the whorls, independent of one another, are immediately inserted on the torus, otherwise sometimes called THALAMUS.

§ 383 *b*. We have just seen, that the several whorls of the flower may be separated from one another as a consequence of the adherences, which they contract with one another, and which carries them above the place that they ought naturally to occupy on the axis; but they may also be separated, whilst they preserve their relation

228—230. The vertical sections of three flowers, shewing the three principal ways in which the stamen is inserted.—*c* Calyx.—*p* Petals.—*e* Stamens.—The pistil is composed of an ovary *o*, of a style and of stigmata *s*.—*t* Torus.

228. A section of the flower of the GERANIUM ROBERTIANUM. The petals are hypogynous, and the stamens both hypogynous and monadelphous.

229. A section of the flower of the Almond-tree. The petals and stamens are perigynous. The pistil is free as well as in the preceding figure.

230. A section of the flower of the ARALIA SPINOSA. The petals and stamens are epigynous, inserted on the circumference of a large disk *d* which covers the whole of the ovary. This, adhering to the calyx, is open so as to shew its loculi and the suspended orules which they contain.

with this axis, and this takes place when it is lengthened, although it only bears a very limited number of parts. The whorls are by that separated from one another, and so much the higher, as, in a common flower with a plane torus or very slightly projecting, they would be nearer the centre. The Capparideæ (*fig.* 231) present very remarkable examples of this lengthening; the petals *p* remain at nearly the same height as the calyx *c*, but the pistil *o* is situated

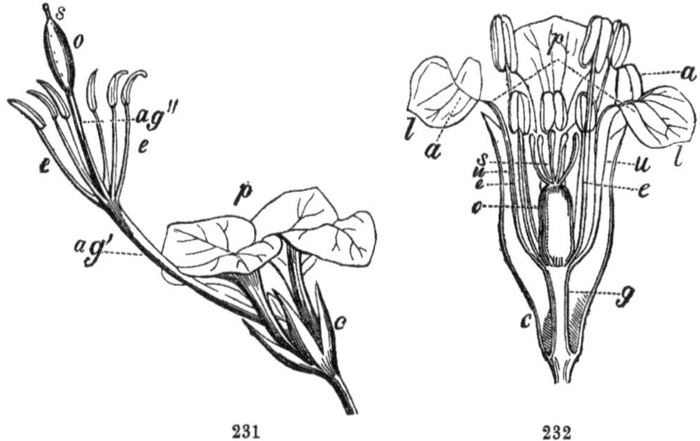

at the end of a long cylinder *ag'*, which rises above the flower and is nothing else than the axis, thus developed, on which the whorl of the stamens *e* may itself be placed at a very great height. In the Caryophyllaceæ (*fig.* 232), it is very common to see the axis, after having produced the calycinal whorl *c*, continue its evolution sometime before producing the following ones, which are thus raised on a column *g*, sometimes of great length. It is very clear, that this arrangement of the whorls in stages, as it were, does not change the true relations of insertion of the parts; it only, on the contrary, exaggerates the hypogyny (*hypogynie*) in the examples we have mentioned.

231. A flower of the GYNANDROPSIS PALMIPES.—*c* Calyx.—*p* Petals.—*e* Stamens. —*ag'* Gonophore or internode of the axis bearing the stamens.—*ag''* Gynophore or internode bearing the pistil.—*o s* The pistil, composed of an ovary *o* and of a style and a stigma *s*.

232. A flower of the LYCHNIS VISCARIA, cut longitudinally, so as to shew the relations of the parts.—*c* Calyx.—*p* Petals with their lengthened claw *u*, their limb *l* and the appendix *a* which is found at the junction of both.—*e e* Stamens.—*o* Ovary surmounted by five styles *s*, forming along with them the pistil.—*g* A lengthening of the axis bearing the petals, stamens and pistil, (it has been proposed to call this the *Anthophore*).

Several words have been proposed as names for these internodes of the flower, according as they bear the petals, the stamens, the carpels or several of these whorls at the same time. The general name of *stipes*, which formerly was employed solely for all these cases, still seems sufficient, as well as that of axis, which we must modify by a suitable epithet, according to the length, thickness, shape and direction of the internode. That, which raises the pistil a certain distance from the other whorls, exists most frequently and perhaps may deserve a particular name (§ 493).

§ 384. THE NUMBER OF THE PARTS OF THE FLOWER.—We have already in all these combinations and in the different modifications, which each of them may present, a certain number of characteristics by which we may distinguish a large number of flowers from one another. We have, however, hitherto supposed the number of the whorls and the parts which compose them to be constant; we have only admitted the marked difference between the flowers of the Monocotyledons in which there are five whorls, of three parts each (*figs.* 226 and 233), and those of the Dicotyledons, which have four whorls, of five parts each (*figs.* 225 and 234). But, beginning from these two types as the point of departure, we shall find innumerable instances of variation, which remain for us to examine. They may be distributed in two grand classes. The numbers we have just stated either augment by the addition of fresh parts, or else diminish by the substraction of some parts. We will, therefore, study these important modifications separately.

§ 385. THEIR AUGMENTATION.—The number of whorls may remain the same, whilst that of the parts is augmented by an equal quantity in each whorl. Thus, let us compare the example we have chosen as type of the flower of Dicotyledons, the CRASSULA (*fig.* 234), with that of a neighbouring genus, the House-leek (*Joubarbe*), SEMPERVIVUM TECTORUM, a species which commonly grows on our walls; we shall see in each whorl, that to the five parts which compose that of the *Crassula*, there is added from one to four, which will raise the number to nine. In other species of the same genus this number is again increased, and there are instances in which it is carried as high as twenty, in which it is, consequently, quadrupled in each whorl in particular and in the flower in general.

§ 386. More commonly the numerical increase of the parts results from that of the whorls themselves. The calycinal folioles, as well as the petals, may be thus found double in number and arranged in two concentric rows. But, it is principally among the stamens that this doubling is most frequent, and it generally takes place without the two external whorls participating in it, so that they are double in number to the folioles and petals: it is then said that the flower is *diplostemonous* (*diplostémone*), (διπλοῦς, *double;* στήμων, *stamen*): it would be said to be *isostemonous* (*isostémone*) (ἴσος, *equal*), if the stamens were equal in number to the petals.

Now, the diplostemony may take place without the number of whorls being really increased. This seems an enigma; let us explain it by examples. The flower of the CORIARIA MYRTIFOLIA (*fig.* 236), presents five calycinal folioles, five small, short, thick petals, alternating with them, then, ten stamens in two rows, the outermost opposite to the folioles, the innermost to the petals, and lastly, five carpels alternating with the last; we have, therefore, the addition of a whorl of stamens, which, intercalated between the first five and the carpels, occupies the normal position of the latter opposite the

233—236. Diagrams of different flowers, i. e., the relative position of their different parts, such as the horizontal section of the flower, not yet or hardly expanded, presents. In these and all the following diagrams, the same symbols have aways been employed to represent the same parts; namely, 1st, a double line *c* for the folioles or divisions, either of the calyx of Dicotyledons (*fig.* 234), or of the perianth of Monocotyledons (*fig.* 233); 2nd, a single line *p* for the petals or divisions of the corolla; 3rd, a small circle for the stamen with a unilocular anther; two circles united together for the stamen *e* with a bilocular anther, or else by a kidney-shaped figure: 4th, an egg-shaped figure, the small end of which is turned towards the centre of the figure for the carpel *o*, a large circle for the ovary composed of several united carpels (*fig.* 250).—Small accessory bodies *a* may be found, and are indicated by a small point or scratch.

233. The diagram of the flower of the ORNITHOGALUM PYRENAICUM.
234. ———————————— CRASSULA RUBENS.
235. ———————————— SEDUM TELEPHIUM.
236. ———————————— CORIARIA MYRTIFOLIA.

petals; the general rule is maintained, the successive whorls alternate with one another. Let us compare it with a flower of a SEDUM (*fig.* 235) almost similar to that of the CRASSULA (*fig.* 234); it only differs by the addition of one circle of stamens and, consequently, presents apparently the very same number of whorls and parts as the flower of the CORIARIA. Nevertheless, if we observe more attentively the relative situation of its parts, we find that of the ten stamens, the five outer ones are precisely placed before the petals and even united with them at their base. We should, thus, have two successive opposite whorls, contrary to the rule. We find ourselves led then to ask, if this is really a double whorl, or, if we ought not to esteem it as a whole composed of double parts, so that this flower would be brought under our primitive type, that which is composed of a whorl of five calycinal folioles, one of five petals, one of five stamens, and one of five carpels; only each petal is doubled by a stamen. This conclusion is justified not only by a consideration which we have already had occasion to repeat several times; namely, that the surest guide in determining the true nature of the parts of a vegetable, so various as their form is, is in the determination of their constant relations of position; but also by the frequency of a phenomenon, which we shall examine before long, that of the deduplication of the vegetable organs.

§ 386, *b*. The multiplication of the parts of the flower by the augmentation of the number of the whorls is not always limited to that case, in which one or several of them become double; they may be tripled or quadrupled, &c., &c. This is frequently observed in the stamens, more rarely in the calyx and corolla, and more rarely still in the pistil. But frequently, when this number is very large, the parts are not grouped in whorls regularly alternating with one another; the general arrangement in the insertion of verticillate or spiral leaves re-appears on a torus either extended in breadth, or lengthened into an axis. This is what we have seen in the petals and stamens of the NYMPHÆA, in the carpels of the MAGNOLIA, and what may also be observed in the flowers of a great number of Ranunculaceæ, in those of the CACTI, of the CAMELLIA, &c. &c.

§ 387. DEDUPLICATION (*dédoublement*).—The parts of the flower may be multiplied in another way. In a flower of the Pile-wort let us look at the base of each petal on the inside, and we shall see a small body of the same colour and of an analogous tissue, something similar to a fold (*fig.* 237 *a*). In those of the CRASSULA, of the SEDUM and of the SEMPERVIVUM which we have cited, on the outside and at the base of each carpel may be observed a small greenish scale (*fig.* 225 *a*), inserted at the same point and appearing to depend upon it. It seems that in these two cases, among the vascular fascicles which pass into these modified leaves, destined to form the petals or the carpels, several are detached in order to form, on an anterior or posterior plane, these little accessory bodies. It may be supposed, that these bodies do not stop at this stage, but are sufficiently developed to be nearly equal to the part of the flower to which they are united, and then it will appear to be double, as in the petals of the ERYTHROXYLON. Those of several of the Sapindaceæ, of several Caryophylleæ (SILENE, LYCHNIS [*fig.* 265 *a*], CUCUBALUS) present something analogous in the fold which doubles a part of their inner surface. This kind of production, which has been named *deduplication* (*dédoublement*) or *chorize* (from χωρίζειν, to *separate*), is most probably the cause, to which in a large number of cases may be ascribed the multiplication of the parts of the flower independently of that of the whorls.

This deduplication, which substitutes two parts for one, may substitute a greater number. Thus, in the flowers of the LUHEA (*fig.* 238) the five stamens alternating with the petals are replaced by five fascicles each composed of numerous stamens; in the flowers of certain Myrtaceæ there are only five stamens; in those of certain others, of the MELALEUCA, for instance, we find in their place five

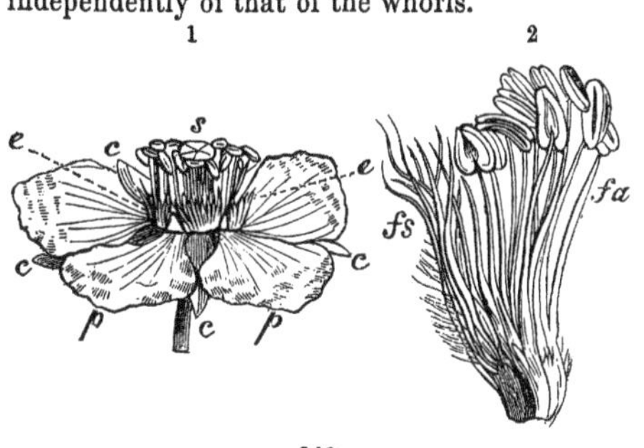

237. A petal of the Pilewort (*Ficaire*) (FICARIA RANUNCULOIDES); the inside is here represented.—*l* Limb.—*a* A small appendix at its base.

238. 1. The flower of the LUHEA PANICULATA.—*cccc* Calyx.—*pp* Petals.—*ee* Stamens grouped into fascicles alternating with the petals.—*s* Stigma composed of five pistils.

2. One of the preceding fascicles magnified. All the filaments are united into a single mass at the base, and are then separated in the upper part; the inner ones *fa*, longer than the rest, are each terminated by an anther; the outer ones *fs*, shorter and sterile, do not bear anything.

groups of stamens pressed against one another and united together at the bottom.

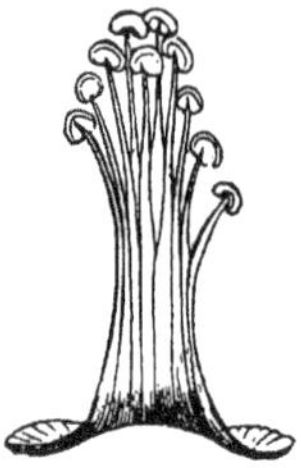

If this multiplication resulted from that of the whorls or from a series of parts arranged in a spire, these parts would, in both cases, be distributed over the whole intermediate zone between the pistils and petals, and not collected into five points having a constant relation with these petals. We conclude, therefore, that each of these groups answers to one of the solitary stamens which we found in the first case, and that it is from the deduplication that we have several. Certain Hypericaceæ and Malvaceæ (*fig.* 239) would present analogous examples and are more easily procured.

239

We can now understand how a petal and a stamen growing immediately before it, frequently joined at their bases, may be considered as resulting from the deduplication of the same kind of organ. It is true, that the parts thus substituted for one another ought naturally to be of the same nature. But the intimate connection which exists between the petals and the stamens will soon be clearly shewn, when we come to a more detailed examination of them, and we have already presented it to our reader's notice in the almost insensible passage of one to the other in the flower of the Water-Lily (§ 366).

In all the preceding examples, the deduplicated parts are situated on several planes, one before the other; but they may also grow on the same plane, one by the side of the other. The flower of the Flowering Rush (*Jonc fleuri*) (Butomus umbellatus [*fig.* 240]) presents from the outside to the inside a whorl of three external calycinal folioles; one of three others, which are situated nearer the centre and coloured; a circle of six stamens opposed in pairs to the outermost calycinal folioles; a second concentric circle of three stamens alternating, on the contrary, with these very folioles; and, lastly, six carpels; it is evident, that, in the circle of six stamens, each pair occupies the place in which we generally find one single stamen. We have, then, here, instead of only one, two situated by the side of one another formed by a deduplication, which may be

240

239. One of the five fascicles of stamens of the flower of the Malva miniata.
240. The flower of the Flowering Rush (*Jonc fleuri*) (Butomus umbellatus).

termed collateral and which we may recognise, when we thus meet an exact whorl, in which the number of the parts is a multiple of that of the others. In the BUTOMUS, the number of the carpels is six instead of three, most common among Monocotyledonous flowers; but of these six, three are alternately on an inner row. We have here then multiplication by the addition of the whorl, and not by deduplication.

We must avow that this faculty of deduplication in the parts of the flower is not often met with in the true leaves to which we have assimilated these organs. Doubtless, the composition of the leaves, which seems to substitute several for one, presents something analogous; then, the folioles of the same leaf are found in the same plane and, consequently, may be compared to collateral deduplications. But, it is in vain to look for several leaves growing in tufts from the place of one, as we have seen in the case of deduplicated petals or stamens. We should have found these tufts in the leaves which we formerly termed fasciculated, if modern observation had not proved that they are those of an entire branch brought close together by the extreme shortness of the axis. Some stipules, those we have called axillary (§ 145), and which appear on a plane anterior to that of the leaf from which they arise, might pass for a deduplication, and we should be confirmed in this opinion by an examination of the ERYTHROXYLON, in which the base of each leaf is united to one of these stipules, in the same way as each petal is united to a petaloid expansion. These facts, however, are far from being numerous: on following the developement of the axillary stipule, we shall be convinced that it results from the union of two lateral parts of the edges, and, consequently, in the same plane as the petiole.

The frequency of deduplication is, then, one more characteristic which distinguishes the parts of the flower from true leaves. Thus, the nearer the parts of the flower approach the nature of leaves (as the calycinal folioles and carpels), the rarer is it to see them deduplicated; the farther they depart (as the petals and stamens) the more frequently, on the contrary, is this method of multiplication observed.

§ 388. REDUCTION IN THE NUMBER OF THE PARTS OF THE FLOWER. —After having examined the differences which the multiplication of the parts, taking place in several ways, may cause in a certain type of the flower, chosen as a point of general comparison, let us now consider those, which result from a contrary reason, the diminution in number of these same parts.

The number of the whorls remaining the same, that of the parts of which each is composed may be equally diminished. Thus the common Rue (RUTA GRAVEOLENS) has at the bottom of its unilateral cymes flowers of five parts, whilst all the rest are reduced to four, namely, a whorl of four calycinal folioles, one of four petals, each joined to a stamen; one of four stamens, one of four carpels (*fig.* 242). This number four is observed in all the flowers of another genus of the same family, the ZIERIA (*fig.* 241), in which also there are only four stamens alternating with the petals; it is reduced to three in the flowers of the CNEORUM TRICOCCUM (*fig.* 243), in which three calycinal folioles alternate with three petals, three carpels with three stamens; to two in those of the CIRCÆA LUTETIANA (*fig.* 244), in which may be observed two calycinal folioles, two petals, two stamens, two carpels.

§ 389. The number of the whorls being always the same, that of the parts which compose one or several of them may be diminished. Thus, the flowers of the STAPHYLEA (*fig.* 245), which have five calycinal folioles, five petals, five stamens, have only two or three carpels; in those of several Caryophylleæ (POLYCARPON, HOLOSTEUM, [*fig.* 246] &c.), we find the stamens reduced to three or four with five calycinal folioles and five petals; in the Balsamineæ (IMPATIENS, [*fig.* 247]), although there are five carpels, five stamens

241—244.—Diagrams of regular flowers, in which each whorl is diminished by one or more parts.
241. Diagram of the flower of the ZIERIA.
242. ——————————— RUTA GRAVEOLENS.
243. ——————————— CNEORUM TRICOCCUM.
244. ——————————— Enchanter's nightshade (*herbe à la sorcière*) (CIRCÆA LUTETIANA).

and five petals, yet the calycinal folioles are reduced in number to three. On the contrary, with five folioles there are only two petals in certain Indian-cresses (TROPÆOLUM PENTAPHYLLUM [*fig.* 248]), and only one in the AMORPHA. Several whorls may be diminished in the same flower. Thus, in the same genus (*fig.* 248), there are only three carpels; there are two circles of stamens, the outer one opposite the petals; but in each of these rows there is one stamen less, so that their total is only eight instead of ten.

§ 390. Is this numerical inequality of the parts composing the different whorls under any fixed laws? There is one which we may deduce from the very position of these parts. The nearer the whorl is to the centre, the narrower is the circle on which it is inserted, and, consequently, so much less field do its parts find to develope themselves. It is, therefore, natural that there should be a tendency for some of them to be suppressed proportionate to the distance of the whorl from the centre; and this is what really takes place. In a complete flower, in which the parts are equally whorled, it is extremely rare to find the calycinal folioles less in number than the petals, the contrary however is less rare; it more frequently happens that there are fewer stamens than petals, and, lastly, it is very common for the carpels not to be equal to the number of the parts of the outer whorls.

§ 391. The reduction may take place not only in some parts of one whorl, but in an entire whorl. Of the external whorls, when

245—248. Diagrams of flowers, in which certain whorls only are diminished by one or more parts, consequently, more or less irregular.
245. Diagram of the flower of the STAPHYLEA PINNATA.
246. ———————— HOLOSTEUM UMBELLATUM.
247. ———————— IMPATIENS PARVIFLORA.
248. ———————— TROPÆOLUM PENTAPHYLLUM.

one remains, it is always the calyx; the complete disappearance of the corolla is pretty frequent, and the plant is then said to be *apetalous* (*apétale*). Thus, the little flower of the GLAUX MARITIMA (*fig.* 249) is composed of a calyx of five parts, of five stamens alternating with them and of one pistil, which is at last divided into five pieces thus representing so many carpels. It is much more common in these apetalous flowers, to find the stamens before the calycinal folioles, placed just as they would have been, had the intermediate whorl of petals been in existence; (in the CHENOPODIUM ALBUM [*fig.* 250], and several other Atripliceæ, &c.); then, indeed, we often find some vestiges, or else we see them re-appear in plants which are incontestably neighbours. Some Caryophylleæ, also, shew this suppression of the petals, which, however, remain in the greater part; amongst the Paronychieæ, which have so many things in common with the preceding, the half of the genera is furnished with petals, whilst the other half is destitute of them.

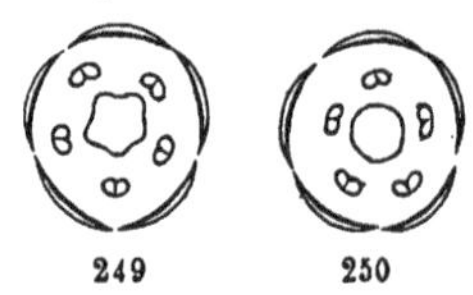

§ 392. In other flowers the stamens or the pistils are wanting. Thus, among the flowers of the JATROPHAS, within a calyx of five folioles and a corolla of five petals, some (*fig.* 251, 2) present a pistil without stamens, others (*fig.* 251, 1) ten stamens without a pistil. We shall see a little farther on, that the pistil, which afterwards becomes the fruit, in which are contained and nourished the seeds or eggs of the vegetable, performs the part of the female, also destined to the production of eggs in animals; and that the stamens, which fecundate the eggs, answer to the male. Hence the pistils are commonly designated by the term *female organs*, the stamens by that of *male organs*, and their whole by that of *organs of fecundation*. Hence also the name of *hermaphrodite flowers* given to those which contain both these organs; that of *male flowers*, to those which are only staminiferous; that of *andræcium* (*androcée*) (ANDROCEUM, from ἀνὴρ, *man*; οἰκία, *habitation*) to the union of stamens; that of *female* flowers to those which are only pistiliferous. We have previously described the flower of the CORIARIA (§ 386) as furnished with both stamens and pistils; but we commonly find on the same plant other flowers, in which either the pistil or the stamens are suppressed.

249, 250. Diagrams of two flowers, in which the whorl of the corolla is suppressed and the ovary compound with central placentation.
249. Diagram of the flower of the GLAUX MARITIMA.
250. ———————— CHENOPODIUM ALBUM.

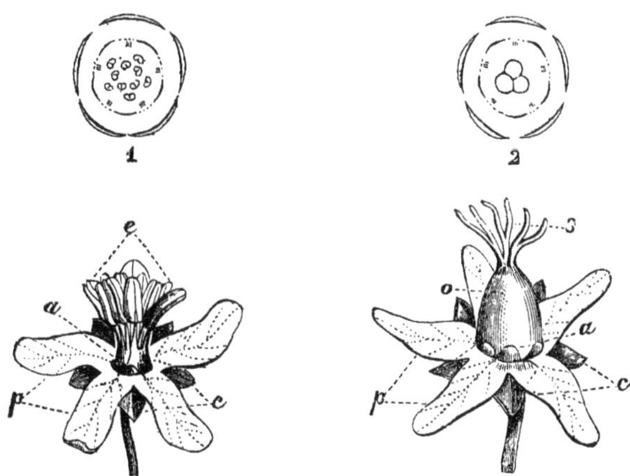

251

When a plant presents this mixture of hermaphrodite flowers, of male and of female flowers, the flowers are then said to be *polygamous* (*fleurs polygames*). If the hermaphrodite flowers are completely wanting in a plant, the flowers, provided either wholly with stamens, or wholly with pistils, have received the name of *diclinous* (*diclines*); then the male flowers may be found on the same plant as the female (as in the Castor-oil plant, the Arrow-head, &c., &c.); they inhabit, as it were, a common domicile, and the plant is then said to be *monœcious* (*monoïque*) (*μόνος, one; οἰκία, house*). At other times, in the Mercury, for instance, certain roots of the plants bear male flowers only, others female flowers only; the flowers occupy two separate domiciles and are called *diœcious* (*dioïque*) (διοικέω, to *live apart*).

§ 393. The flowers are destined to propagate the plant by means of the seed, which is the limit of their developement. The pistils, in which these seeds are contained, are, therefore, essential organs; but it has long since been shewn, that if there be only pistils, the seeds are abortive and the plant does not reproduce itself; that the neighbourhood and action of the stamens are necessary to their

251. The male (1) and female (2) flowers of the JATROPHA CURCAS.—*c* Calyx.—*p* Corolla.—*e* Stamens which occupy the centre in flower 1, on account of the suppression of the pistil, and which are completely wanting in flower 2.—The compound pistil of one ovary *o*, surmounted by three bifid styles.—*a* Small glandular appendices alternating with the divisions of the corolla.—The diagram of each of these flowers is placed above it.

fecundation and to the production of the embryo which served us as a starting point in the history of plants (§ 27); the stamens are then equally essential. As to the calyx and corolla, they only perform a purely secondary part in the flower and are destined to serve the stamens and pistils as *envelopes*, under the protection of which they are developed and perfected. We can conceive that these envelopes may be totally wanting without the flower becoming unadapted to fulfil its functions, whilst it would be, if the stamens and pistils were wanting, a sterile ornament, totally useless in the reproduction of the plant. Thus, we give the appellation of *neuters* to some flowers thus limited to the whorls of the calyx and of the corolla, which are, then, often remarkably developed. The flowers, limited, on the contrary, to the pistil and the stamens, but completely destitute of *envelopes* are called *achlamyds* (*achlamydées*) (ACHLAMYDEÆ) (*α, not; χλαμὺς, chlamyd, clothing*), or more commonly *naked* flowers (*nues*) (FLORES NUDI).

§ 394. We have seen that the parts of the flower may be reduced, 1st, By the suppression of some parts in each whorl; 2nd, By the suppression of one or several entire whorls. Now, let us combine these methods of reduction, and we shall arrive by successive suppressions, of which Nature herself presents us with all the examples, to a much greater degree of simplicity, in which the limit will be an isolated stamen or carpel. It is to this point, that the flowers of the genus NAIAS are reduced, two species of which (N. MAJOR and N. MINOR) grow in our rivers. The family of the Euphorbiaceæ alone would shew us in a series of instructive examples (*figs.* 252 to 256), the progressive degradation in the number of the stamens which constitute its

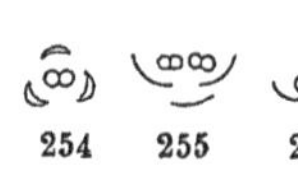

252—257. Diagrams of flowers gradually becoming more and more simple, in which we see: 1st, The calyx, the only envelope, reduced to three parts (252, 253, 254), itself completely disappearing (255, 256, 257), and replaced by a bract, from the axil of which the flower grows, sometimes accompanied by two inner bracteoles (255, 256); 2nd, The male flowers reduced to three stamens (252), to two (253, 255), to one (254, 256), and, lastly, this single stamen to a single loculus (257, 1); and the female flowers (257, 2), reduced to a carpel.

252. Diagram of the male flower of the TRAGIA CANNABINA.
253. ——————————— TRAGIA VOLUBILIS.
254. ——————————— ANTHOSTEMA SENEGALENSA.
255. ——————————— ANDENOPELTIS COLLIGUAYA.
256. ——————————— EUPHORBIA (*Euphorbe*).
257, 1. ——————————— NAIAS MINOR.
— 2. ——— female flower of the NAIAS MAJOR.

male flowers, and which we shall at last find reduced to three, to two, to one (EUPHORBIA) (*Euphorbe*).

§ 395. When the flowers thus reduced to one organ are solitary, there will be no difficulty in recognising them; but, some doubt may arise, when they are grouped in a common inflorescence. Thus, for a long time botanists considered as one flower the inflorescence of the EUPHORBIA, in which several male flowers, each formed by a stamen, surround a female formed by one pistil, the whole enveloped by an involucrum, which was called a calyx. Thus, at first sight, the fruit of the Mulberry (*Mûrier*) appears to be the same as that of the Bramble (*Ronce*), although the former represents the pistils of several flowers crowded on a short spike, and the latter those of one single flower arranged on a slightly lengthened torus. This is what M. Rœper has so well shewn, that there exists a great analogy between the inflorescence and the flowers which are its component parts, and that their difference disappears almost completely, if the parts of the inflorescence become as simple as those of the flower; this necessarily happens, when we find the degree of reduction we have just described. How, then, are we to distinguish between one flower and an inflorescence formed of very simple flowers? If these (stamens or pistil) are mixed with small bracts, if they do not follow one another in the accustomed order, from the outside to the inside, we may be sure, from the presence of these fresh parts or from this uncommon combination of the usual parts, that we have a compound part under our eyes. The comparison of neighbouring plants, especially, tends to throw light on these doubtful cases. The genus EUPHORBIA, already mentioned, whose examination is very easy on account of the frequency of several of its species, will again serve us as an example under these different phases. Seeing that its stamens are articulated near the middle, and that at the base of the lower articulation are some small strap-shaped bodies, we should already have suspected that these articulations were so many pedicles accompanied with their bracts; but, we can no longer entertain any doubt, on finding in the neighbouring genera of the Euphorbiaceæ, the flowers constantly diclinous and extremely simple, reduced, on one part, to a very small number of stamens, on the other, to one pistil, and, on reflecting that some of the former, grouped around one of the second, would exactly reproduce a flower of the EUPHORBIA. It must be avowed, however, that in certain cases, these distinctive signs may be wanting, and analogy only can then throw a doubtful light on the subject. Thus, the LILEA has a spike quite covered with bracts arranged in a spire, and at the axil of each of them is a stamen, then a carpel, and each of these small combinations is a flower. Let us take six similar ones and group them into a whorl, and we shall have a flower of the TRIGLOCHIN, with its calyx of six folioles, six stamens and as many carpels. It is very clear, that we call here parts of the flower what in the preceding case we called the flower itself, calycinal foliole what we called bracts. What shall we conclude from this? The gradual transition of the parts of a Vegetable into one another, compound as well as simple. We follow here imperceptibly the passage from the inflorescence into the flower, as we have before followed that from the branches to the inflorescence, as we have seen the leaf pass successively to the bract, to the calycinal folioles and the other floral organs. But, if, by tracing these transitions, we are prevented from exaggerating the difference of these parts as was formerly done; on the other hand, their differences are most frequently too clear to allow us to confound them in one whole, which would destroy the whole science by means of simplifying it.

§ 396. We have already found these parts of the flower to be susceptible of a considerable number of different combinations by multiplication or by diminution, which may act as much on the whorls themselves as on the elements of each. These two principal causes of modification may act together. Thus, in the MAGNOLIA and the Tulip-tree, the calycinal whorl, limited to three folioles only, is less in number than most Dicotyledons; the petals were also arranged in ternary whorls, having, consequently, undergone the same reduction; but there were several of these whorls, and from this multiplication followed necessarily that of the petals. In the genera of the neighbouring family of Anonaceæ (HEMISTEMMA, PLEURANDA) the stamens are completely wanting on one of the sides of the flower, but are, as a kind of compensation, multiplied on the other. In the St. John's Wort (*Millepertuis commun*), the stamens are multiplied; but they are arranged in three fascicles resulting from the deduplication, and their whorl is thus reduced to three; whilst it rises to five in some others.

The law of alternation of the successive whorls remaining in force, we may conceive how their augmented number in the flower may alter the apparent relation of the parts. It was singular to see the stamens opposite to the petals, and these to the calycinal folioles, in the flower of the Barberry; but all is explained by observing that the whorls are reduced to three parts, and at the same time each is doubled, so that the parts ought to be opposed, if we take them six by six, as we have done; the alternation of six by six would have been an exception to the rule.

§ 367. Degeneration and Transformation of the parts of the Flower.—Now that we have examined the manner in which the flower is modified by its various combinations of number and the situation of its constituent parts, let us examine the differences which depend on quite another class of causes, of the modifications of form of these parts. These modifications will be explained in a fuller manner a little farther on, and we shall content ourselves with announcing here in general terms, that they may occur either in every part of a whorl or only in some of them; that these parts may be modified not only in their size and shape, but also in their structure and, consequently, in their functions. The stamens of the great genus Diosma, the species of which are rather common in conservatories, will furnish a good example of these kinds of modification, and this very genus has been divided into several from this very consideration. The type of the flowers of this family is the most common among Dicotyledonous plants; five calycinal folioles, five petals, each doubled with a stamen, five stamens and five carpels. Now, in the different genera formed at the expense of the genus Diosma, the stamens opposite to the petals have quite changed their shape and structure; sometimes, they have those of the petal itself, but smaller (Agathosma); sometimes, of a short petaloid tongue (Barosma); sometimes, those of a thread either extremely short (Acmadenia), or long (Adenandra) and bearing a gland at its summit; sometimes, even of a simple glandular fold.

The place, which the organs thus metamorphosed occupy in the flower, will always point out their origin and the part they represent. Thus, when in the Clavija, a genus of Myrsineæ, we find five small palettes of the consistence of the petals, intermediate between them and the stamens and alternating with both, we see that they occupy the normal place of a whorl of stamens, and we, consequently, pronounce them to be transformed stamens. We can then easily explain, how it is that the stamens, which have preserved their true form, are found opposite the petals; and, although we do not find these modified stamens in all the other genera of the family of Myrsineæ, remarkable for this constant opposition of the petals to the stamens, we understand, that these do not form the whorl immediately following, but, that between them was another, which has sometimes changed its form, sometimes completely disappeared; and, generally, whenever two following whorls are opposite to one another instead of alternating according to the rule, it is well to look for the traces of the whorl which is wanting, and we shall mostly find some. These transformed parts, which are commonly called the accessories of the flower and are most frequently described according to their form and nature, were mostly confounded by Linnæus and several of his successors among the bodies to which has been given the name of nectary (*nectaire*).

§ 398. When the parts of the same whorl are unequally developed, so that they are not all equal and similar in shape or size, it is then said to be irregular. It is, therefore, regular in proportion to the perfection of this equality and similitude; and when it is so, it is clear, that, when the whorl is divided in half, the segments are similar, in whatever direction the division be made. An irregular flower is that which has one or more irregular whorls; but, generally, this name is only given to it, when the irregularity is noticeable in the external whorls, forming the envelopes, and much more apparent than in the internal ones.

§ 399. Can any of the causes, which act upon this unequal developement of the homologous parts of the flower and, consequently, upon its irregularity, be determined? Whenever the parts of the same whorl are not situated in almost similar situations, their inequality will be produced naturally; and this is so true, that, even in flowers constantly regular, if one of the sides is accidently deranged by some obstacle or deprived of the light which illumines the other, it is opposed, modified, arrested in its developement. Now, the position, which the flowers assume in connection either with one another or with the different axes of the inflorescence, differs according to the plant, and is constant in the same; so that in a great number of cases it would create, either in all the flowers of certain plants or only in several of them, some few of these obstacles, which, not being transitory or accidental, but resulting from the same state of things, would produce a constant effect upon it. Thus in a Scabious (*Scabieuse*) (*fig.* 188), we see the flowers crowded together on a capitulum, where all those, that form the outer circle and are not incommoded during their developement, have become much larger than those in the centre; and since they are freer at the outside than at the inside, their corolla is less developed towards the centre than towards the circumference; all the other flowers of the capitulum inside of this circle, equally pressed on all sides, have remained less, but regular. We have, then, here a double example, that of the flowers of the same inflorescence dissimilar to one another, that of the parts of the same flower unequally developed, the whole from their relative situation.

In the umbels (*fig.* 187), we frequently find the same effects arising from the same causes. In spikes, if the flower is not quite perpendicular to the axis, but more or less oblique to it, on the side nearest to the axis, it finds an obstacle to its free developement, which tends to stop on that side rather than on the other, and hence flowers thus arranged are often irregular (*fig.* 183). There is no need of explaining, that analogous phenomena may be observed in several other inflorescences on account of combinations of the same kind.

§ 400. If we now consider one flower as isolated and look for the causes of irregularity which it may contain in itself, we shall see that they are not wanting. Among the leaves of the branches, those which are arranged in whorls are generally developed concurrently in each and present the same forms and the same dimensions; those, situated at different heights, so much the later as they are more elevated on the branch, present from the bottom to the top decreasing dimensions and frequently also different forms. It is the same with the modified leaves which constitute the parts of the flower. We have already seen (§ 390), that these parts, crowded on a circle narrower in proportion to its proximity to the centre, tend so much the more to abortion. We know, moreover, that, if they are often perfectly whorled, i. e. on the same plane, these parts are also frequently not arranged in a circular, but rather in a spiral line, some, consequently, being a little lower or nearer the outside than the rest. Then, in this case, they are not placed under identical conditions, and those which are a little higher or nearer the centre have not so free a field for their developement, which takes place a little later; they will more readily prove abortions or remain of smaller dimensions. We shall see, not only in very irregular corollas, as those of the Balsamineæ or of the Papilionaceæ, but also in almost regular flowers, as those of the Malpighiaceæ, the petals so much larger as they are nearer the outside.

§ 401. There is still an arrangement which tends to place the parts of the same whorl under different conditions with respect to one another; it is the obliquity of the torus with relation to the pedicel, an obliquity which tends to raise some by debasing others.

Now, on examining a great number of irregular flowers, we shall be convinced, that the axis of the flower does not continue that of the pedicel in a straight line, but is more or less inclined; that the flower is situated more or less obliquely at the summit of the pedicel. In very regular flowers, on the contrary, the plane of the table forming the torus is perpendicular to this summit.

§ 402. There are, then, inherent causes of irregularity in the very relations of the parts of the flower to one another, in their relations to the pedicel, in their relations either to the axes, to the other flowers of the inflorescence or, lastly, to every other part of the plant to which they belong. These relations, on account of which the parts are predisposed for unequal developement, may act in another manner by sometimes forcing them to unions, which join and blend several of them together more or less completely. Yet, all these rules ought to be admitted as a general thesis only; they may be modified or inverted according to many secondary circumstances, which it would be too long to detail here, and the knowledge of which, besides, has not yet attained such a degree of precision as to enable us to lay down clear and constant laws. It is apparent, that in the small space in which the parts of the flower are accumulated and crowded together, the relations are difficult to appreciate and are frequently altered. It is sufficient for us to mention, that this point of vegetable organization does not entirely escape the observation and calculation of botanists, and opens a fresh field of research to them.

§ 403. We must not mistake regular flowers for symmetrical flowers: the former may be divided in every direction into similar halves; the latter only on one plane, and this plane is perpendicular to that of the axis which bears the flower. It may be verified on the flowers of the Vervain and Scabious (*figs.* 183, 188) which we have already mentioned; and we shall see that by a plane in this direction they are divided into two parts perfectly similar, one on the right, the other on the left. Divided by any other plane the halves would cease to resemble one another: if the conditions with respect to the parts of the corolla, were different from the outside to the inside, from the top to the bottom, they would be precisely similar on the right and on the left.

There may be, therefore, symmetrical flowers, although they be irregular, and this is most frequently the case: that in which there is at the same time a want of symmetry as well as of regularity is much more uncommon.

§ 404. PREFLORATION OR ÆSTIVATION, (*Préfloraison*).—There is one period, at which all these relations of position of the parts of the flower which have just occupied our attention, are most manifest and are most easily determined; this is in the bud (*bouton*), the first state of the flower, which is with regard to it what the bud (*bourgeon*) is with regard to the branch. Then, the real situation of the parts is perceived not only from their higher or lower point of situation on the torus nearer the centre or the circumference, but also from the order in which they are superposed or are enveloped by one another, since every enveloping part is almost necessarily external to the part enveloped. Linnæus has called *æstivation* (ÆSTIVATIO, whence was derived the verb, ÆSTIVARE) or *state of summer* this arrangement of the parts in the bud (*bouton*), as he called *vernation* that of the leaves in the bud (*bourgeon*) (§ 174). This name has been preserved, but we often substitute for it almost indifferently that of *prefloration* (*préfloraison*) (PRÆFLORATIO).

We see represented, in the different modes of arrangement of the envelopes of the flower at this first stage, the two principal modifications which we have recognised in that of the leaves as well as of the parts of the flower; their arrangement in a spire or at different heights, in a circle or at the same height.

§ 405. The spiral prefloration is also termed *imbricate* (*imbriquée*). This last epithet which is very significant, when the parts only cover one another in a part of their height, like the tiles of a roof (*fig.* 258, *c*), ceases to be so, when they completely envelope it, and then some have substituted the name of *enveloping* or *convolute* (CONVOLUTIVA) (*fig.* 260). The parts are frequently long enough to permit the first to be placed over the following one at its top, but not broad enough for it to reach it at its edges. On numbering the parts, after the order in which they are placed above one another from the exterior to the interior, we often find the arrangement of the leaves on one continuous spire, that in which the angle of divergence of two successive ones approaches to 137° (§ 162), and in which, consequently, the parts alternate by twos, by threes, by

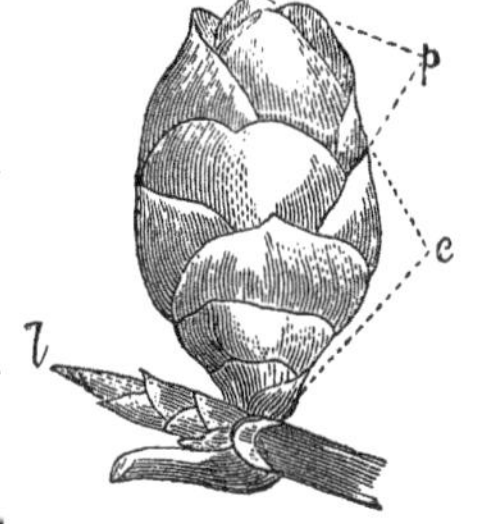

258. Flower-bud (*bouton*) of the CAMELIA JAPONICA.—*c* Imbricate folioles of the calyx.—*p* Petals with a convolute prefloration.

fives. In the flower of the MAGNOLIA this spiral envelopement of the parts may be observed (*fig.* 260).

We know that most frequently the parts of the same whorl do not exceed five. We see, then, arranging them according to the spire (*fig.* 259), that if they are not broad enough for two successive ones to join at the edges, there are two placed nearer the circumference in relation to the others and covering their neighbours by the two edges, two placed more internally and covered at the two sides, the fifth always placed between one of the two first which covers it by the corresponding border, and one of the two second, which it itself covers similarly. This *ensemble* of five parts thus arranged is termed *quincunx* (*quinconce*), and this manner and the manner of prefloration *quincuncial* (*quinconcial*) (QUINCUNCIALIS).

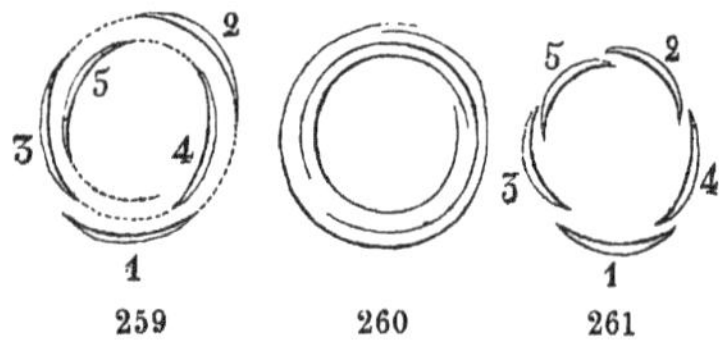

But, it is by no means very rare, that, from one of the causes of irregularity, several of which we have pointed out, the insertion of some one of the five parts is carried a little within or a little higher, which necessarily alters their relation. Thus, we frequently (*fig.* 261) find that of the two folioles inverted, which in the regular quincunx would have been numbered 2 and 4; the foliole 2 is placed nearer the centre, and is then covered by the corresponding border of the foliole 4, which it would properly cover. This arrangement, which is found in Papilionaceous flowers, has sometimes received the name of *vexillary* (*vexillaire*).

§ 406. There are several other combinations, according to which the parts of the same whorl are in the same relation to one another; we may believe, then, that they are all placed in the same conditions, regularly in a circle and at the same height. They may touch at their contiguous edges throughout their whole length, like the leaves of folding doors, they are then said to be *valvate* (*valvaire*) (PRÆFLO-

259. The horizontal section of the calyx of the flower-bud of the White Convolvulus (*Liseron des haies*) (CONVOLVULUS SEPIUM). The line of dots shews the direction of the spire which passes through the successive insertions of the five folioles.

260 Shews the way in which three external folioles (corresponding to the calyx) of the flower-bud of the MAGNOLIA GRANDIFLORA are arranged, cut transversely and very much smaller than nature.

261 Shews the way in which the three folioles of the calyx of the Snap-dragon (*Muflier*) (ANTIRRHINUM MAJUS) are arranged. They have been numbered to correspond with figure 259.

RATIO VALVATA [*fig.* 263, *c*]). At other times, wider, they are bent back on the sides either inwards or outwards; and those, which correspond to one another in two neighbouring parts, are applied one to another by a certain portion of their external face in the former case (*induplicate prefloration* [*fig.* 263, *p*]) (*préfloraison indupli-cative*), in which the bud presents all the external appearance of the valvate arrangement; of their internal surface in the second case, in which so many angles project externally from the bud, as there are parts thus united (*reduplicate* [*fig.* 262, 1; 264, *c*]). In these cases, which ought to be considered as simple and slight modifications of the valvate prefloration, the part of the folioles thus bent back either outwards or inwards, at the same time generally becomes thinner and frequently almost membranous. The folioles of the same whorl, instead of forming arcs of a circle or the sides of a polygon having for its centre that of the flower, may take a direction at a certain obliquity to it, as if each foliole underwent a kind of torsion on its axis; by that, one of these sides, the same with regard to all the folioles, is carried more inwards, the other outwards, and, in this case, the summits commonly enlarged, would be imbricated in a circle, each covering with one side one of its neighbours and itself covered at the other; this is *contorted* (*tordue*) prefloration (PRÆFL. CONTORTA [*fig.* 262, 2, *p*; and 264, *p*]). Sometimes, then, a slight

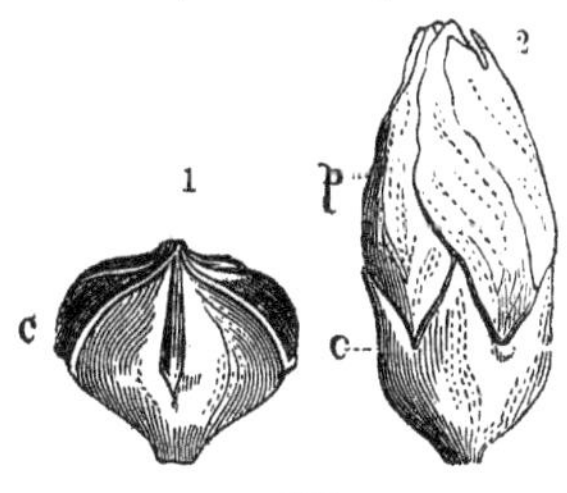

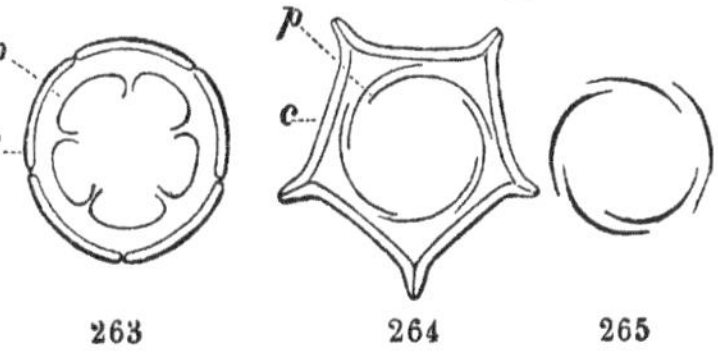

262. The flower-bud of the ALTHÆA ROSEA.—1. Not much developed; the calyx still completely envelopes the other parts and the edges of the divisions still touch one another.—2. More developed; the edges of the calycinal divisions *c* have been separated from one another in order to allow the corolla to pass through them; the petals *p* are contort. The diagram is represented in figure 264.

263. The diagram of the calyx *c* and of the corolla *p* in the flower-bud of the GUAZUMA ULMIFOLIA. The prefloration of the folioles of the former is valvate; that of the petals, induplicate.

264. The diagram of the calyx *c* and of the corolla *p* in the flower-bud of the ALTHÆA ROSEA. The prefloration of the calyx *c* is reduplicate; that of the petals *p* is contorted.

265. Deviation from the contorted prefloration.

deviation of one of the five folioles, placing it quite on the outside, brings back the spiral arrangement, but, in such a manner, that they are all on one coil of a single spire (*fig.* 265), which very nearly approaches the form of a circle.

§ 407. In the flower-bud, each foliole, considered independently of the others, may sometimes, in the same way as the leaf in the leaf-bud (§ 174), besides the modification which results from the mutual arrangement of the parts, present one which is peculiar to itself, i. e., may be bent on its axis into two halves which are turned inwards or outwards, (in which case the flower-bud will be raised by so many projecting angles, separated by so many hollows answering to the median nerve or to the interval of the folioles,) be wrinkled (*chiffonnée*) (CORRUGATA) and frequently then, as it were, wound on itself, as are the petals in the flower-bud of the Poppy (*Pavot*); &c., &c.

§ 408. In studying, in the bud, the possible relations of the parts, we have hitherto examined them only between those which belong to the same whorl: let us now look for them in the succession of several. The spiral arrangement may be continued without interruption from one to the other; this is to be expected in the flowers, in which the transition of one whorl to the following is gradual, as in those of the White Water-Lily, in the calyx and corolla of the MAGNOLIA, so that we frequently are puzzled to determine the limits between them. But, when the different whorls are distinguished in a very well-defined manner by perfectly fresh forms and colours, in spite of this quick passage, the spiral series may be regularly continued, the first foliole of the second whorl nearly occupying its regular place after the last of the first, and the following ones in their turn arranging themselves in their proper places with regard to it. An interruption, however, sometimes takes place, as if there were wanting several intermediate folioles between the outermost of the fresh whorl and the innermost of the preceding. Sometimes even the direction of the spire is inversed: that of the calyx proceeds from right to left, that of the corolla from left to right.

§ 409. We very frequently observe in two successive whorls a different manner of prefloration: this change is constant and characteristic in several families. Thus, for instance, in the Malvaceæ (*figs.* 262, 264), the Convolvulaceæ, the greater part of the Caryophylleæ (as in the AGROSTEMMA GITHAGO) the prefloration of the corolla is contorted: that of the calyx is, nevertheless, valvate in the first (*fig.* 264, *c*), imbricate in the others. This last example suffices to shew us, that in the same flower, the parts of one whorl may be arranged in a spire, whereas those of the neighbouring one are in a circle.

§ 410. In short, in these parts accumulated on such a limited space, where the successive insertions are generally separated only by very small intervals, very frequently quite unappreciable, we cannot expect to find the same regularity of relations as on an axis extended in length and in breadth, on which all the parts can grow in their own place and also choose their own time. This is the reason, why the arrangement of the parts of the flower in the same whorl, or of one whorl with regard to the following, is far from being invariable, and observation informs us in what limits it varies. The spire, often interrupted, inverted, even completely destroyed between the parts of two successive whorls, undergoes frequent and slight modifications between those of the same whorl. The valvate or contorted arrangement is much more settled, and, as it indicates parts placed in a circle and all in the same condition, it is almost necessarily in habitual alliance with the regularity of the flower. Indeed, with the exception of very few instances, the calyx and the corollas with valvate or contorted prefloration are regular, whilst we meet with almost as many irregular ones as regular ones in those, in which the prefloration is arranged in a spire.

§ 411. The prefloration only points out more clearly the relations of position between the parts of the flower and allows us to determine them more easily; from their importance it borrows all its own. In several flowers, the expansion separates these parts, which cease to cover, to touch one another, and these relations so manifest in the flower-bud are then more or less effaced. But, there is also a large number of flowers in which they remain to a certain degree. Thus, the quincuncial arrangement may be observed in several of the corollas of the Rosaceæ; those of the Apocyneæ remain always strongly contorted, and it is not rare to find that those of the Malvaceæ also preserve some traces of this previous arrangement. It is clear that the closeness of the valves could not remain without preventing the bud from opening; we see the calyces which are in this state burst, either laterally by the separation of two of their edges only, leaning on one side after the manner of a spathe (in the Hibiscus esculentus, for instance), or circularly at their base separating from the rest of the flower, which bends them back or throws them off as it lengthens ; we find instances of this in certain Myrtaceæ, (CALYPTRANTHES, EUCALYPTUS). Most commonly the contiguous edges in the valvate prefloration are thick enough to assist us in recognising them, even after their separation, whilst the sides which are covered are commonly thinner. Of this we may be convinced by comparing the calyx of a RHAMNUS with that of an ALSINE.

§ 412. We have not spoken of the organs of fecundation, the stamens and pistil, because they are not extended like those of the envelopes in laminæ of a certain breadth, and are not, from their form, susceptible of these different ways of mutually covering one another, by means of which we may discover those delicate modifications in the relative position of the calycinal folioles or of the petals. It is not impossible, however, to deduce, on this point, a few facts as to the state of these organs in the bud. It sometimes happens, that all the carpels or all the stamens of the same whorl, although destined definitely to acquire equal dimensions, do not gain them simultaneously, but, that among them some are a little more advanced in their developement than others: and we may, perhaps, be led to believe that they are in more favourable conditions than the later ones, as are the outer leaves in a rosette with regard to the inner ones. If the comparison holds good, no longer in one whorl of stamens but in several at the same time, then the relative position of these whorls is shewn in the bud much more clearly than in the expanded flower; their concentric circles are clearly distinguished, instead of being afterwards confounded in one. It is in this manner, that we frequently observe the whorl of the stamens opposite the petals surrounding that of the alternate stamens, as we have observed it in the flower of the SEDUM (§ 386); and on this point some buds may shew us a peculiar fact, which will confirm the conclusion that we drew from the exterior position of the stamens opposite to the petals. The five petals of the diplostemonous flowers of the Malpighiaceæ successively envelope one another in the prefloration, like so many hoods fitting to one another. Now, in some species, on taking away the outermost petal, we see immediately before it the stamen opposite to it, placed between it and the following petals; then, on raising these successively, we see the stamens opposite to each of them interposed in the same manner between it and the rest of the flower (*fig.* 266). How could this entanglement and this position of certain stamens with regard to that of certain petals be explained, if these stamens formed a whorl really distinct from and nearer the centre than that of the petals? Is it not rather a fact of the same kind as the frequent union between the base of the petals (§ 385), which leads us to conclude that in this case both are the doubled parts of one and the same whorl?

§ 413. We have learnt to determine as far as the actual state of science will permit, the relative position of the parts of the flower with regard to one another; it now remains for us to determine the place of the flower with regard to the rest of the plant. To discover this, we look how it is placed relatively to the axis from which its pedicel starts. On taking some part of this flower for an example, its outermost foliole for instance, we may suppose this foliole to be turned to the axis, or diametrically opposite, or to the right, or to the left. Now, we must remark, that one of these positions, whatever it may be, when it is found in one flower, is generally found in every other flower of the same plant; and it has even been determined that this uniformity is sometimes extended to all those of the same family. Thus, in the Scrophularineæ and in other neighbouring groups, there are two carpels turned one to the side of the axis, the other to the opposite side; if we find a flower similar in appearance to those of the Scrophularineæ, but the two carpels turned, one to the right, the other to the left, we may then conclude that the plant does not belong to one of these groups. Thus, the single stamen which we see developed in the Canneæ and in the Marantaceæ, turning in the one upwards, in the others on the side, is sufficient to distinguish at the first sight these two neighbouring families.

In general, the folioles of the calyx are arranged on the bract which accompanies the flower, or, when it is wanting, on the point of the axis where it would have developed itself, so that the series of the leaves of one branch is co-ordinate with the leaf, from the axil of which springs this branch (§ 162). When the pedicel is twisted on itself, or when it is long, thin or flexible, the primitive position of the flower with regard to the axis, whence this pedicel springs, may be more or less disguised. The study of the bud (*bouton*) will enlighten us in this case also, because the pedicel

266. The diagram of the flower-bud of the TRIOPTERYS OVATA. We see the position of the different parts of the flower with regard to the bract *b*, from the axil of which it grows and which, consequently, corresponds to the outer side in the general inflorescence, the other side being turned towards the axis.

is so much the less twisted, so much the less lengthened and reduced in size as the flower is younger.

§ 414. This assemblage of characteristics which results from the position of the parts of the flower relatively to the branch which bears them, and some relatively to others, is what is termed its *symmetry* (*symétrie*): a word taken here in quite another acceptation from that in which we have precedingly spoken (§ 403) of symmetrical flowers.

Envelopes of the Flower.

§ 415. We know that two whorls of parts commonly different from one another in their form and their colour, the calyx and the corolla, compose the envelopes of the flower when they are complete. We know also, that it is not uncommon to find only one, and that, in this case, it is almost always the corolla that is wanting. In one genus of Rutaceæ, the Diplolæna, in which the flowers accumulated on a capitulum are crowded and mutually incommode each other, the outermost parts are not developed, the calyx is a total abortion and the petals are reduced to the form of short thin scales, some even are quite wanting. Here, then, is an example of a corolla without a calyx; we could, perhaps, mention a few others, but they are so rare that it would be very useless to designate by a particular name such an exception to the general arrangement. It is not so with respect to the calyx without petals, which is rather frequent, and we have seen (§ 391) that the flowers which present it are termed apetalous.

§ 416. This term has given rise to no objection in Dicotyledonous plants, in which, when the floral envelopes are limited to one single whorl, they generally present the appearance and all the other characteristics of a calyx. But, in the flowers of Monocotyledons it is not always so. We have said (§ 372) that their envelopes are most commonly formed of six parts arranged by threes in two concentric circles. Very frequently, all the six are similar to one another and then they may be green (as in the flower of the Asparagus) (*Asperge*); but more frequently they are painted in different colours, sometimes very lively, as in the Lily (*fig.* 267), the Hyacinth, the Tulip, &c. &c. At other times the three outer ones differ from the three inner ones; the first are green and similar to the calyx, the second coloured and similar to the petals, as in the Spiderworts, the Water-plantain or Alisma, &c. In this case, we should be

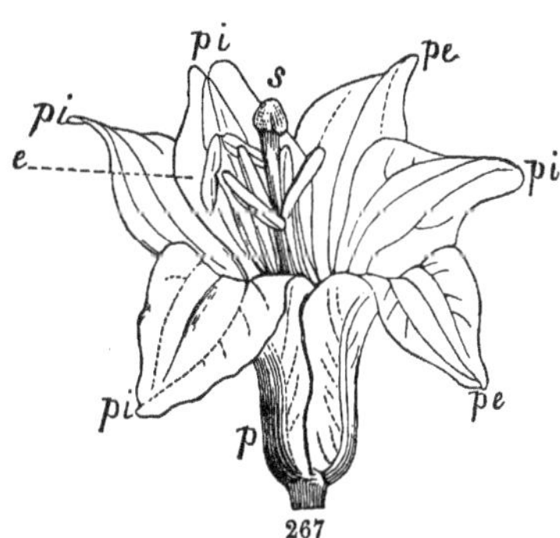

tempted to call the outer whorl the calyx, and the inner one the corolla; but, we must necessarily apply the same names to the same parts in all the flowers of other Monocotyledons, in which, however, the parts do not present any difference between them. This is what several botanists have done. Others, more anciently, only taking the characteristic of colour as a guide, admitted in these flowers, sometimes a calyx and a corolla, sometimes a calyx alone, sometimes a corolla only, although, evidently, the six parts in their constant relations ought always to represent the same part. Others, indeed, calling them in every case a calyx, which they defined to be the system of the outermost envelope of the flower, could not recognise two different systems in those of the greater part of Monocotyledons. It is necessary to warn the reader of this want of agreement in the terminology of different botanists, to avoid the confusion to which it might lead.

§ 417. De Candolle tried to do away with it. Struck by this diversity of appearances, which in the flower of such Monocotyledons, presents to our notice a calyx, a corolla in that of such others; seeing even, that some seem to join this double nature in their folioles, green on the outside and coloured within, he supposed, that each of the latter arose from the union of a calycinal foliole to an opposite petal, and is thus formed of two laminæ representing its double nature. It is vain to discuss this hypothesis, which is rejected by the considerations drawn from both the anatomy and the position of the parts. It becomes untenable, especially, when applied to the flowers of Dicotyledons, to which its author extended it, and the opposition of five petals to five folioles of the calyx would be contrary to nature. However it may be, De Candolle rejected, according to this idea, the epithet of apetalous for flowers

267. The flower of the White Lily (*Lis blanc*) (LILIUM CANDIDUM).—*p* The Perianth *p*, the three outer parts of which *p e* alternate with three inner *p i*.—*e* Stamens, the filaments are surmounted by their oscillating anthers.—*s* The Stigma terminating the upper part of the style.

furnished with a single envelope, and called them *Monochlamyds* (*Monochlamydées*) (μόνος, *single;* and χλαμὺς, garment). He called, with Erhart, the assembly of floral envelopes *perigone* (PERIGONIUM), and employed this word instead of calyx and corolla, whenever he saw them confounded in one single system, and, consequently, in the description of the flowers of all Monocotyledons.

§ 418. Several authors follow his example, without admitting the hypothesis which served him as his foundation, and not deciding on the nature of the single envelope, described it under this name of perigone, or more commonly under that of *perianth* (*perianthe*) (PERIANTHIUM; from περὶ, *around;* and ἄνθος, *flower*), which Linnæus proposed for the calyx, whenever it is in immediate contact with the organs of fecundation. This name may be admitted advantageously in the description of Monocotyledons; but it occasions real inconvenience in that of the Dicotyledons, in which we frequently find one plant after another, some furnished with, others wanting petals (in the Caryophylleæ and the Paronychieæ, for instance). Now, in two flowers, very similar in other points, we cannot call in one the perianth, what in another we call the calyx. It appears then more reasonable to apply this latter name constantly to the whorl of envelopes, either external or single, of every Dicotyledon, and to employ for Monocotyledons either the same word, which is modified by various epithets according to the case, or that of perianth. We shall use both indifferently in the following examination.

§ 419. CALYX (*Calice*) (*κάλυξ, a covering*).—We have said that the calyx is the outermost whorl of the envelopes of the flowers, that it is composed of several pieces representing so many leaves, and which have been, consequently, named calycinal folioles. M. Link has proposed to designate them by the single name of PHYLLA (*phylles*) (*φύλλον, leaf*), which has been already employed in the composition of the adjectives *Monophyllous* (*monophylle*) and *polyphyllous* (*polyphylle*); De Candolle has generally adopted that of *sepals* (*sépales*) (SEPALA): hence the epithets *polysepalous* (*polysépale*) or *monosepalous* (*monosépale*) given to the calyx, according as its folioles remain entirely independent of one another, or else are united together (§ 374). In future, therefore, we shall employ indifferently either of these two words, calycinal folioles or sepals.

§ 420. We have considered these parts to be true leaves, and their structure justifies this supposition; they are, indeed, formed in the same way of a parenchyma in the inside, which follows from bottom to top, the general direction, of the fibro-vascular fascicles composed of unrollable tracheæ and of thin fibres, and are externally clothed by an epidermis covered with stomata, much more abundant on the external face of the sepal, which, on account of its upright position, corresponds to the lower side of the leaf. The epidermis is frequently clothed with hairs similar to those, which are frequently found on the leaves themselves and the young shoots, consequently, much more frequent and abundant on the outer than on the inner face. To express the absence and the presence of hairs, and the different ways in which they modify the surface of the calyx, we employ the terms already explained (§ 244). We sometimes also find on this external surface, glands analogous to those which the leaves of the same plant bear on the lower surface, as in the Malpighiaceæ. These are so many common characteristics contributing to confirm the connection between leaves and the calycinal folioles.

§ 421. The fibro-vascular fascicles cause the nerves to be apparent on the outside (of which the median alone frequently projects) and follow, although in a manner much less visible on account of the smallness of the parts, the same laws as in the leaves of Dicotyledons and Monocotyledons, uniting with one another by ramifications in the calyces of the former, proceeding parallelly and without division in those of the latter. When the calycinal folioles are blended together into one single body at the bottom, the median nerves, which are found on the surface of this body, may indicate the middle of each (*fig.* 271). We frequently find as many other nerves, placed precisely in the intervals of the first on the line of junction of the united folioles, and each resulting from the union of fascicles belonging to the two neighbouring folioles; for, we see them, at the height at which they separate, double in two branches which follow the two corresponding edges (*fig.* 273).

§ 422. The sepals, like the leaves and generally like all vegetable organs, at first appear under the form of small cellular nipples; and it is remarkable that these nipples are always at first quite distinct, even though they are intended to form a Monophyllous calyx, and are quite equal, even when they are developed unequally afterwards. M. Schleiden has shewn this in the very young bud of the Lupin. But this is only at the very beginning; the adhesion and inequality are soon established, if either of these is one of the characteristics of the calyx of the flower. The vessels and fibres appear progressively as in leaves (§ 147).

§ 423. The form of the sepals may in general be compared to that of the bracts rather than of the leaves; it is commonly that of a lamina which grows narrower by degrees as it reaches the summit, and which, consequently, represents either the reduced limb, or the vaginal part of the leaf. We see them sometimes narrower also at their lower part, but it is extremely rare to find this narrowing lengthened into a petiole. It is rare to find the edge serrated or lobed (RUMEX MARITIMUS, and other species of the same genus [*fig.* 268], Rose [*fig.* 369]); is commonly entire. We shall not describe here all the possible forms of the sepals; the most frequent is that of an oval, obtuse or acute at its summit. When we describe them, besides their number and their form, we ought to mention their direction, sometimes upwards, *erect sepals* (*sépales dressés*) (SEPALA ERECTA), sometimes inwards, *converging* or *connivent* (*s. connivents*) (S. CONNIVENTIA), sometimes and most frequently *diverging* (*s. divergents*) (S. DIVERGENTIA), *patulous* (*s. étalés*) (S. PATULA), *reflexed* (*s. réfléchis*) (S. REFLEXA), according as they are more or less inclined, their apex turned upwards, horizontally or downwards.

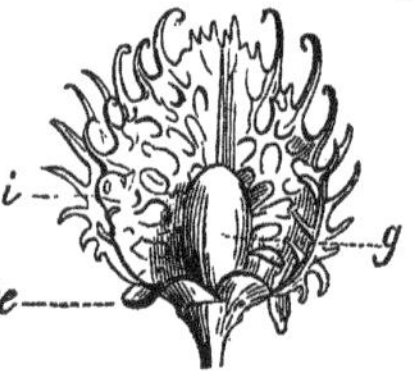

268

§ 424. When the calyx is monophyllous, the union of the parts may take place for a greater or less extent. If it only takes place at the base, this short lower portion is called the bottom of the calyx; if it takes place to a pretty considerable height, the united portion bears the name of tube. In both cases, the upper portion where the sepals remain free is the *limb*, and according as they remain more or less completely separated, as the limb, consequently, is composed of parts (LACINIÆ) of a greater or less length relatively to the bottom or to the tube, we give them names analogous to those we have given (§ 133) for the divisions of the edge of the leaf of different depths. Thus, they are segments or partitions, if the sepals remain distinct till they nearly reach their base; fissures, if they are united above the middle (*fig.* 270); or lobes, if they are at the same time widened; and teeth (*fig.* 271) or scollops (*fig.* 288, *c*), if they are free at their summit only, acute or obtuse. These words are frequently employed in the compound epithet, by which

268. The calyx of the RUMEX UNCATUS. It is composed of two whorls, the outer one *ce* with short, entire divisions, the inner one *ci* with much larger divisions, cut at their edge into narrow ribands or hooks, reticulated on the external surface, at the bottom and in the middle of which we remark a glandular swelling *g* in the shape of a seed.

the calyx is characterized and which indicates, at the same time, the number of these divisions. Thus, we say that the calyx is quinque-part, or quadrifid, or trilobed, or sexdentate, &c., &c. If the form and the union of the parts is such that there is no sensible degree of division, and that the whole of the calyx only forms a tube bordered on the top by a circle, it is said to be *entire* (*entier*) (INTEGER) or *truncated* (*tronqué*) (TRUNCATUS). Let us mention, that all those words, which apply to the parts of a single leaf, are applied in the calyx to the union of several leaves considered themselves as the parts of another whole, so that there is, therefore, analogy only and not identity in the use we here make of them.

269

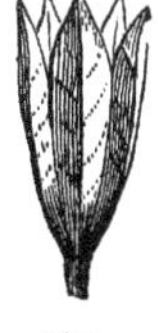

270

271

Besides these general forms owing to the different degrees of union between the several pieces of the calyx, it presents many secondary modifications by the lengthening of the tube and its swellings at different heights, by the varied directions of the limb with regard to it, &c. We shall give the terms by which these are designated when we treat of the corolla, in which these same modifications are more pronounced on account of the greater size which it generally attains (§ 439).

We have supposed, in all the preceding cases, the calyx to be regular; but it may not be so; and the irregularity is found, either in the tube, which may then be bent or dinted (in the SCUTELLARIA, for instance,) at certain places, or in the limb, of which certain parts are more developed than the rest. It is not very rare to see the sepals, either united, or free, prolonged below their point of insertion, either into a flat lamina (as in the Violet), or into a sac, which

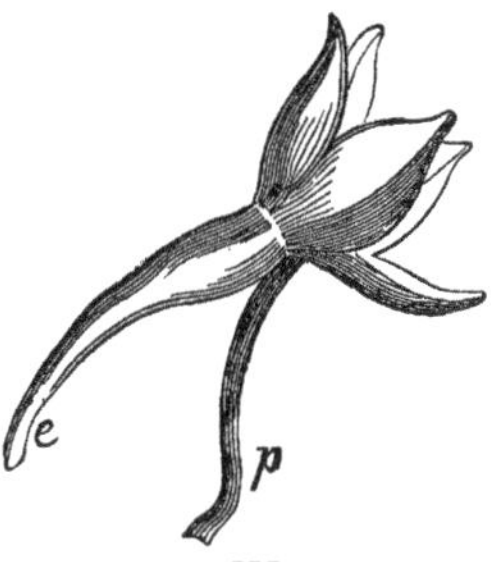

272

269. The pentaphyllous calyx of the Greater Stitchwort (*Stellaire*) (STELLARIA HOLOSTEA).

270. The calyx of the Oxlip (PRIMULA ELATIOR).

271. The calyx of the Bladder Campion (*Behen blanc*) (SILENE INFLATA).

272. The calyx *c* of the Indian-cress (*Capucine*).—*e* Spur.—*p* Pedicel.

then turns its opening towards the inside of the flower. If it be much lengthened it takes the name of *spur* (*éperon*) (CALCAR) and the calyx is said to be *spurred* (*éperonné*). This modification may affect either a single sepal (as in the Indian-cress (*Capucine*) [*fig.* 272]), or each of them (as in the Columbine [*Ancolie*]). In the PELARGONIUM this spur is in close union beneath the flower with the pedicel, which bears it and of which it seems to make part.

§ 425. The flower of some plants appears to be surrounded with a double calyx. We should suppose from the general notions which we have given on the multiplication of the parts, that the existence of this accessory calyx, sometimes called CALYCULUS (*calicule*), is then owing either to the deduplication of the sepals, or to the addition of an external whorl; the folioles of the two calyces ought to be opposite to one another in the first case, alternate in the second. Correct observations, however, tend to make us reject both these hypotheses. Thus, the existence of a calyculus is frequent in an extremely natural family, that of the Malvaceæ; but when these parts are equal in number to those of the calyx, they alternate with them: it is not, then, a deduplication. On the other hand, they are frequently less in number (as in the Mallow) or more, multiple or not (as in the HIBISCUS) (*fig.* 273), and they vary extremely, even in the species of the same genus: it is not then a regular whorl, like all the rest of the flowers; it is rather an assemblage of bracts crowded into an involucrum immediately beneath the flowers (§ 230). It is one transition more to note between the leaves and the floral envelopes. The folioles of this involucrum may be united and thus form a monophyllous calyculus (*fig.* 276, *i*, 277).

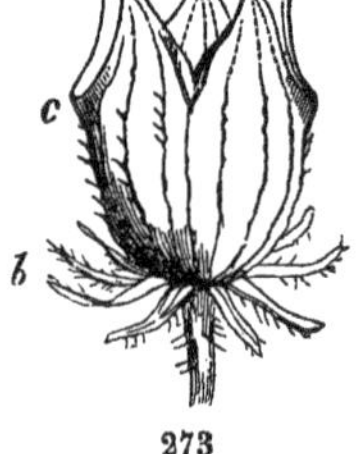

273

But, the calyculus traces its origin at other times to quite a different source. Thus, the leaves of the Rosaceæ are furnished at their base with two stipules thrown out on a plane a little exterior relatively to the limb. Let us suppose five of these leaves to be grouped in a whorl and united together at their base; the stipules will form a circle of parts joined two by two, alternating with the limb, placed on the same base, but in a plane a little nearer the outside. These stipules, lastly, instead of being in simple juxtaposition, may be united at their edges and be thus reduced from ten to five. This is precisely what takes place in the calyces of several species

273. The calyx *c* of a HIBISCUS with its calyculus *b*.

of Rosaceæ, as in the POTENTILLA (*fig.* 274), the Strawberry, &c., in which between the five divisions *c* of the quinque-part calyx, are found externally so many tongue-shaped bodies *b* forming by their whole a calyculus. Nature aids us to divine, that these are the stipules of the calycinal folioles placed in pairs, by sometimes shewing us some of these very tongue-shaped bodies bifid or bipartite, and thus betraying their binary origin. If this explanation be true, the epithet of *bracteolate* (BRACTEOLUS), which has been applied to these calyces, is quite improper, and the double whorl would really constitute only one.

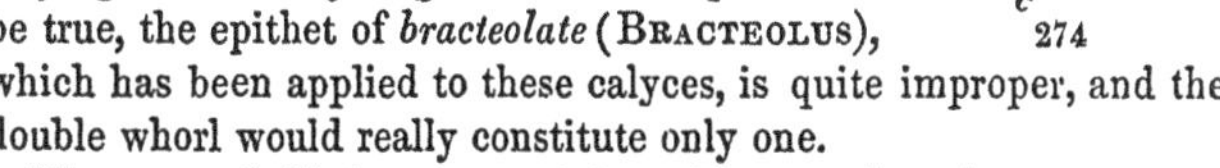

We cannot hold the same opinion with regard to the two rows of the outermost divisions opposite to the petals, which terminate the tube of the calyx in the Loosestrife (*Salicaire des marais*) (LYTHRUM SALICARIA). As the same arrangement is met with in all those of the same family, and as the leaves have stipules in none of these plants, we here admit the deduplication of the calyx by the addition of an entire whorl. We see from these examples, that different causes may bring similar results, and that appearances ought not to mislead us in the determination of the parts.

§ 426. The consistence of the calyx is most commonly that of the leaves, which we designate by the term *foliaceous* (*foliacée*) (FOLIACEUS) or *herbaceous* (*herbacée*) (HERBACEUS). The colour is commonly, therefore, green, but in some plants it passes into tints analogous to those of the innermost parts; to red in the FUCHSIA, the Pomegranate (*Grenadier*), &c.; to orange in the Indian-cress (*Capucine*); to rose in EPILOBIUM SPICATUM. Sometimes along with these other colours, which properly belong to the corolla, it also borrows from it its thinner, more delicate and softer tissue, and takes its external appearance, which then causes it to be named *petaloid*: the Columbine and the HORTENSIA present examples among Dicotyledonous plants. They abound among Monocotyledons, in which it is also the usual consistence of the calyx or entire perianth (White and Martagon Lilies, Jonquille, Hyacinths, &c., &c.), sometimes only of its inner row. The consistence in other Monocotyledons is, on the contrary, quite different, that is, dry, hard, of dimensions very much reduced and reminding one rather of bracts, of a green or brownish colour, for instance, in the Reeds (*Joncs*). The calyx, thus modified, is called *scaly* (*écailleux*) (SQUAMOSUS),

274. The calyx of the POTENTILLA VERNA, its under side with its calyculus *b*.

because its sepals are similar to the scales of the bud; and often also *glumose* (*glumacé*) (GLUMACEUS), from the name of *glume* (GLUMA), which is given to the envelopes of the flower of the Gramineæ, remarkable, in fact, for this consistence.

§ 427. The limb of the calyx is sometimes quite lost, under the form of a circle or of a tuft of bristles or hairs, which takes the name of PAPPUS (*aigrette*), whence the epithet of *pappose* (*aigretté*) (PAPPOSUS). Several families of plants, the Valerianeæ, the Dipsaceæ, the Compositæ, exhibit the transitions from the ordinary form to this. The last family, especially, presents us with all the possible modifications. Thus, in some species of the Valerianeæ (VALERIANELLA CORONATA) we see the teeth of the calyx already lengthened at their apex into a stiff bristle; in the VALERIANA this lengthening is much longer, softer and quite covered with a fine down; it has become a pappus. In the same way, in the SCABIOSA ATROPURPUREA (*fig.* 276 *pe*) the five lobes of the calyx, very short, but very distinct, are terminated by five long bristles; in those flowers composing the genus PTEROCEPHALUS, we find in its place a pappus; and its branches or rays, in the same way as in the Valerians, are either more or less than the number five, as if, not only the five median nerves, but at the same time several of the others were developed in this singular way (*fig.* 277 *l*). In one genus of Compositæ (CATANANCHE [*fig.* 275 *l*]) the limb is composed of five divisions widened at their base, gradually narrowing from bottom

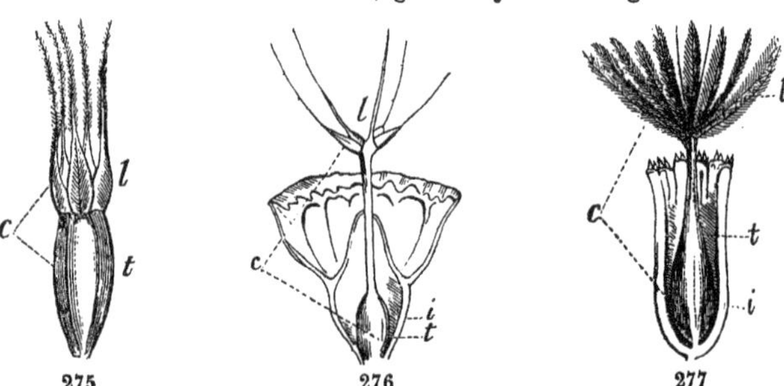

275—277. Examples of calyces, the limb *l* of which is gradually passing to the state of pappus.—*c* The calyx, the tube *t* of which makes one body with the ovary and grows narrower above it into a thin column in figures 276 and 277, the limb *l* of which consists of several divisions, tapering into a thread, either from their base or at their summit.—*i* The involucrum or calyculus divided longitudinally.

275. The calyx of the CATANANCHE CÆRULEA.

276. ———————— SCABIOSA ATROPURPUREA.

277. ———————— PTEROCEPHALUS PALESTINUS.

to top into a thin velvety thread, a real ray of a pappus. This is developed in its full form in a great number of other genera of this family; and its rays, often very much multiplied, no longer form simple whorls, but sometimes tufts, as if they sprang from several concentric circles; this really takes place and would also happen in the Dipsaceæ, if the involucellum or external calyx, with which each of the flowers is enveloped, had, like the interior, its limb tapering into a pappus. The pappus is said to be *plumose* (*plumeuse*) (PLUMOSUS), when each of its rays is covered with small hairs visible to the naked eye (*fig.* 275, 277), as in the SCORSONERA, in the CIRSIUM, &c.; *simple* (SIMPLEX SEU PILOSUS), when each ray, deprived of this down, has itself the appearance of a long smooth hair on its surface (*fig.* 276 *l*), as in the Dandelion (*Pissenlit*) (TARAXACUM). But even, on looking at it through a lens, we perceive this surface generally quite bristled with small asperities; when they are sufficiently large to represent so many small teeth easily visible, the pappus is said to be *dentate*.

Whilst representing the rays of the pappus as extensions of the nerves, we did not mean to imply that they were formed by the fibro-vascular fascicles separated from the parenchyma; they continue its direction, but not its tissue; reduced themselves to the cellular, they are analogous to the hairs both in their structure and appearance.

§ 428. The duration of the calyx is variable in different flowers. In some, it is detached from the torus by disarticulation (like the leaf from the branch which bears it [§ 138]), either in several pieces or one; it is *deciduous* (DECIDUUS) when it falls with the corolla, or soon after fecundation; when it falls earlier, as soon as the flower commences to expand, it is *caducous* or *fugacious* (*c. caduc, fugace*) (CALYX CADUCUS, FUGAX), for instance, in the Poppies. In other flowers, the calyx remains in its place even after the blossoming is over; it is then *persistent* (*persistant*) (PERSISTENS), in the Labiatæ, the Personeæ, Borragineæ, &c., &c. But, sometimes it ceases to live, fades and shrivels; sometimes, on the contrary, it continues to vegetate and sometimes even to grow, as in PHYSALIS ALKEKENGI. In the former case it is termed *marcescent* (*marcescent*) (MARCESCENS), in the latter *accrescent* (*accrescent*) (ACCRESCENS).

§ 429. COROLLA (*Corolle*).—The corolla is the coloured envelope of the flower, interior with respect to the calyx, composed of parts, which sometimes continue the spiral series commenced by the sepals

(§ 369), which at others, and most frequently, are arranged in a whorl and alternate regularly with these same folioles. We already know that those of the corolla are called *petals* (*pétales*) (PETALA; from πέταλον, *a leaf*). This etymology and the name of leaves, which is given in common parlance to the petals of the Rose and several other flowers, prove that the idea of comparing them with leaves is far from being recent. We have endeavoured to shew, that, in several cases, the passage from sepals (the foliaceous nature of which is incontestable) to petals takes place almost insensibly, and that the rules which may be deduced from the relations of position are applicable to the latter as well as to the former. Let us see if their anatomical structure will equally bear out this comparison.

§ 430. A petal, considered as isolated from all others, is a lamina of variable form, most commonly broadest at the upper and narrowest at the lower end; pretty frequently this narrow part is lengthened, as in the petal of the Pink (*Œillet*), and then takes the name of *claw* (*onglet*) (UNGUIS), (*fig.* 286, 2, *o*), whilst the upper expansion receives that of LAMINA (*lame*) or *limb* (*limbe*) (LIMBUS), (*fig.* 286, 2, *l*). The claw appears to be with respect to the lamina, what in the leaf the petiole is to the limb; the fibro-vascular fascicles are crowded and united in the one, are separated and expanded in the other. These fascicles are formed of unrollable tracheæ and of lengthened cells; their interval is occupied by cellular tissue, which sometimes completely fills it (in which case the edge of the entire petal is surrounded by a continuous curved line); at other times, is interrupted near the edge, so as to allow the extremities of the fascicles to project under the form of teeth, of fringes (FIMBRIÆ) (*fig.* 286, *p*), of lobes of various depths. These different modifications are generally indicated by the same terms as the analogous ones of the leaves. Much thinner than these, the petal does not present in its internal tissue, formed by a small number of rows of cells, those different layers we have described in the leaf. The epidermis, which clothes it, is also much less distinct from the rest; it is more so on the external face, where it is sometimes pierced with stomata, but much more rarely and less constantly. They are almost wanting on the inner surface and the superficial cells are frequently rounded on the outside, so as to shew under the microscope a delicately wrinkled surface, sometimes even as fine as velvet.

§ 431. When the petals commence their growth, it is under the form of small cellular projections, similar, therefore, to the first appearance of the calycinal folioles; then, they are widened into a small disk slightly concave or marked with a longitudinal wrinkle on the inner side; they are now true leaves and continue to grow from their base and to follow the same laws, so that the lower portion always appears the last, and the claw is not formed till some time after the limb. Several points are worthy of notice in this developement of the corolla; 1st, The parts of the monopetalous corolla are already confluent at this early period (*fig.* 278, *p*), and the small nipples, which are the first indications of the appearance of the petals, are, as soon as we can perceive them, united by a kind of circular excrescence on which they form so many slight projections. We cannot, therefore, say that they are united; they grow already united. 2ndly, The petals, although they represent lower or more external leaves on the axis with respect to the stamens and, consequently, ought to surpass them in their advancement, are, on the contrary, generally behind them, and we see, in very young buds, the stamens more developed than the petals still in the state of very small scales (*fig.* 279). 3rd, They are at this early age, of a green colour, frequently pale, sometimes dark, whatever may be their colour afterwards.

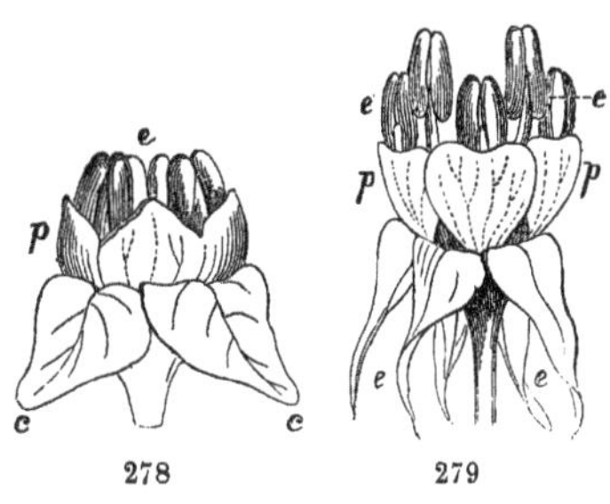

§ 432. Now, green is a very rare colour to find in the corolla, although we meet with it pure in some plants as in those of certain COBŒAS, of some Asclepiadeæ (HOYA VIRIDIFLORA, GONOLOBUS VIRIDIFLORUS, PENTATROPIS SPIRALIS), &c., &c. When it exists, it is most commonly pale and weakened by other tints, or striped and dotted by lines and spots of quite a different colour. The presence, therefore, of chlorophyll is rare in the cells, which are usually filled with granules or with a liquid of another colour (§ 24), or are empty. We shall enter a little further on into more ample details on this subject.

§ 433. To say, that the chlorophyll is wanting in the petals, is to

278, 279. Very young flower-buds from which the divisions of the calyx *c* have been bent down, in order to shew the comparative developement of the corolla *p* and of the stamens *e*.

278. The flower-bud of a monopetalous flower, the Fox-glove (*Digitale pourprée*) (DIGITALIS PURPUREA).

279. The flower-bud of a polypetalous flower, the GERANIUM STRIATUM.

state that the chemical phenomena of respiration do not take place as in the leaves (§ 294). The corolla and all other parts of the flower not of a green colour, under the influence of the light, absorb oxygen and exhale carbonic acid. The presence of a great mass of flowers, ornamented with brilliant tints, has, therefore, during the day, an effect on the atmosphere inverse to the salutary action of a mass of green leaves. But, this effect is not the only one, and is frequently complicated with the exhalation of the essential oils and other odoriferous principles so often concentrated in this very part of the vegetable.

§ 434. The consistence of the petals is variable, most frequently soft and delicate, sometimes thick and fleshy (STAPELIA), sometimes dry, like paper or a membrane, (Heaths) (*Bruyères*) sometimes hard and stiff (XYLOPIA).

§ 435. Since the petals properly so called belong to the flowers of Dicotyledons, their nerves ought naturally to be ramified and terminated by a net, which their last ramifications form by their union. The secondary veins are detached from the median, either at different heights, as in a penninervate leaf; or frequently from the base of the limb, as in a palmatinervate leaf; and this last arrangement, suggesting the idea of an open fan, is expressed by the epithet which is then applied to the petal (FLABELLATO-VENOSUM). The median nerve is sometimes prolonged to the apex of the petal, and even beyond it to a small free point (CUSPIS, whence *cuspidate petal* [PETALUM CUSPIDATUM]); but it has more frequently a tendency to double itself, by breaking into two parts, one of which turns to the right, and the other to the left. There frequently results from it at the summit a notch which causes the petal to be named *emarginate* (*p. échancré*) (P. EMARGINATUM); and, if it gradually grows wider, from its narrow base to its bilobed summit, it is said to be *obcordate* (*obcordé*) (P. OBCORDATUM), on account of its assuming the shape of an inverted heart. The division of the fascicles of the median nerve may be unequal, so that one half of the petal receives more than the other, which causes one side to extend at the expense of the other, throwing the axis a little to one side; the petal is then *unequally-sided* (*inéquilatéral*) (INÆQUILATERUM), or *oblique* (OBLIQUUM, OBLIQUÈ OBCORDATUM, or any other epithet which

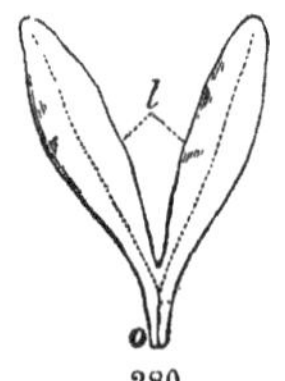

280

280. A petal of the Chickweed (*Mouron des oiseaux*) (ALSINE MEDIA).—*l* Limb.—*o* Claw.

best paints its general form). The division of the median nerve and, consequently, of the limb, may begin higher or lower, sometimes quite from the base, and the petal is then bifid or bipartite, and may almost appear, when there is no claw, to be composed of two equal (as in the ALSINE MEDIA [*fig.* 280]) or unequal collateral ones.

Let us remark that the irregularity of the oblique petal does not necessarily cause that of the corolla of which it is a part, since the different petals which compose it may, in this case, be perfectly similar to one another, and from their union there may result a regular whole; this is observed in several of the corollas of plants with a contorted prefloration like that of the Malvaceæ.

The petals are commonly inserted on a narrow base; but this narrow part is frequently not prolonged and they are said to be sessile. Sometimes, the base is wide; it may even be as broad as the limb; in the flower of the Orange, for instance. If, although narrow at its insertion, it does not gradually enlarge, it takes the form of a small riband and is said to be *linear* (*linéaire*). Between this last form and that of a circle, we may observe all the intermediate ones as in leaves. It is rather common to see the two sides of the limb prolonged at the bottom into two obtuse lobes or two angles parallel or oblique to the claw; it is then said to be *heart-shaped* (*en cœur*) (CORDATUM) [*fig.* 281]), or *arrow-shaped* or *sagittate* (*sagittée*) (SAGITTATUM), or *spear-shaped* or *hastate* (*hasté*) (HASTATUM).

The limb may be flat; but very often also it presents a curved surface, commonly turning its concavity towards the centre of the flower. Sometimes, the median nerve then projects sharply outwards like the keel of a boat, and the petal takes its name from its shape, *keeled* (*naviculaire*) (NAVICULARE), *cymbiform* or *boat-shaped* (CYMBIFORME). Sometimes, also, it is bent so as to bring its point to its base, as in several of the Umbelliferæ (*fig.* 282).

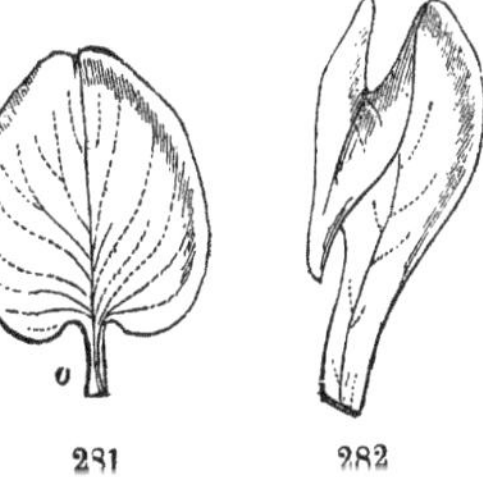

In the greater part of flowers, it is smooth; yet, in several it is covered with a down, commonly very short, fine and far apart, some-

281. A petal of the GENISTA CANDICANS.—*l* Limb.—*o* Claw.
282. ———— ERYNGIUM CAMPESTRE.

times thicker, which we observe more frequently and abundantly, perhaps, exclusively, on the internal face, the same as in both leaves and sepals. Although it appears upon the petals much more rarely and more widely dispersed than on the other parts of the vegetable, it is of the same nature: thus, in the plants characterized by stellate hairs, the Bombaceæ, for instance, those of the corolla are equally in the form of a star.

In botanical descriptions, the epithet, by which the form of the petal is indicated, is applied to the limb. When we mention petals, as *orbiculate*, *dentate*, *concave*, we mean to express petals with a claw and an orbiculate, dentate, concave limb.

§ 436. The corolla is said to be dipetalous, tripetalous, tetrapetalous, pentapetalous, &c., &c., according as it is composed of two, three, four, five distinct petals. We have already stated, that their number is generally equal to that of the divisions of the calyx with which they alternate; but that some exceptions may be found to this rule (§ 389), from the suppression of one or of several petals in the whorl of the corolla compared with that of the calyx, and *vice versâ*. Thus in the flower of the Horse-Chesnut, the calyx has five teeth; but we only find four petals alternating with four of them, the place of the fifth being vacant: in the Indian-cress (*fig.* 248), there are only two petals leaving three empty places. This circumstance is expressed by describing the corolla as *tetrapetalous* or *dipetalous by abortion*, (*tétrapétale* ou *dipétale par avortement*): an expression which is really true; for we see in other species of the Chesnut, and even in some flowers of the same species, the fifth petal re-appear; we constantly count five in several other species of Indian-cress. The number of the petals, which is five in almost all the Leguminosæ, is in the Amorpha reduced to a single one, placed between two of the five divisions of the calyx, and, in this case, the corolla is said to be *unipetalous* (*unipétale*), a word which we must by no means confound with *monopetalous* (§ 374).

§ 437. In descriptions, we ought to indicate, besides the number, the direction of the petals (erect, divergent, patent, reflexed, [§ 423]) with respect to the axis of the flower; that of the limb with regard to the claw, with which it sometimes makes an angle; their length with respect to the calyx; their shape, on the modifications of which we shall give a few details, and which, as well as their size, may be similar or very different, in those of the same flower. In the latter case, in which the polypetalous corolla is irregular, the dissimilar petals must be described, at the same time designating their place with regard to the axis of the inflorescence. When the irregularity is the same in the flowers of a great number of plants, one word is sufficient to indicate the principal characteristics. Such is the PAPILIONACEÆ, applied to the corollas of all the Leguminosæ of our country. Of the five petals (*fig.* 283) one, *e*, upper, that is, turned towards the axis, greater and ordinarily bent back upon itself, embraces the four others: it is called the *Standard* (*étendard*) (VEXILLUM); two lateral ones, *a*, called the *Wings* (*ailes*) (ALÆ), cover the two lower ones, *b*, which, close together and frequently joined at their edges, form by their union one piece in the form of a boat, the *Keel* (*carène*) (CARINA).

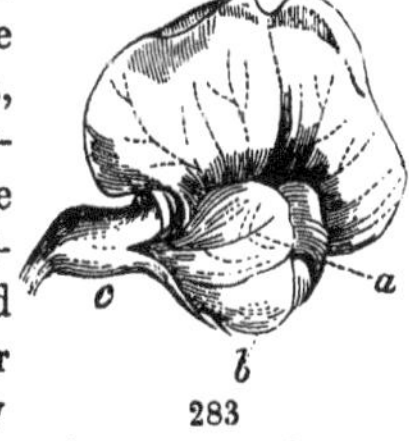

Certain modifications of regular polypetalous corollas, which we find in a great number of flowers, generally in those of the same family, have also received particular names. Thus, those, which have four petals opposite in pairs, in the form of a cross, are called *cruciform* (*cruciforme*) (*fig.* 284); those, which have five open petals without claws arranged as in the Rose, are termed *rosaceous* (*rosacées*) (*fig.* 285); those, which have five long petals furnished with claws, *caryophyllous* (*caryophyllées*) (*fig.* 286).

§ 438. The greater part of the remarks made on petals in general will apply equally to those, which by their union form the *monopetalous corolla* (*corolle monopétale*). We must, however, see that there is here no distinction between the claw and limb, since the bases are blended together. Frequently, however, these bases appear to represent the claws and the summit the limbs. Thus, we call by the same name, *limb* (*fig.* 287, *l*), upper parts free in their circumference, and describe their form by the same terms as isolated petals; the lower part in which the petals are closely united at their edges is called the *tube* (*fig.* 287—294, *t*), and commonly is of that form; the entrance to the tube, the inner circle, where the petals are detached from one another is known by the name of the *throat*, (*gorge*) (FAUX).

These names are equally applicable to the calyx or to every monophyllous perianth, in the same way as, on the other hand, the words, by which we have designated (§ 424) the different degrees of height, at which the pieces of the calyx or perianth are united, (or

283. The flower of the Sweet Pea (*Pois de senteur*) (LATHYRUS ODORATUS).—*c* Calyx.—*e* Standard.—*a* Wings.—*b* Keel.

if it pleases the reader better, the different degrees of depth in their divisions,) are equally employed in describing the monopetalous corolla.

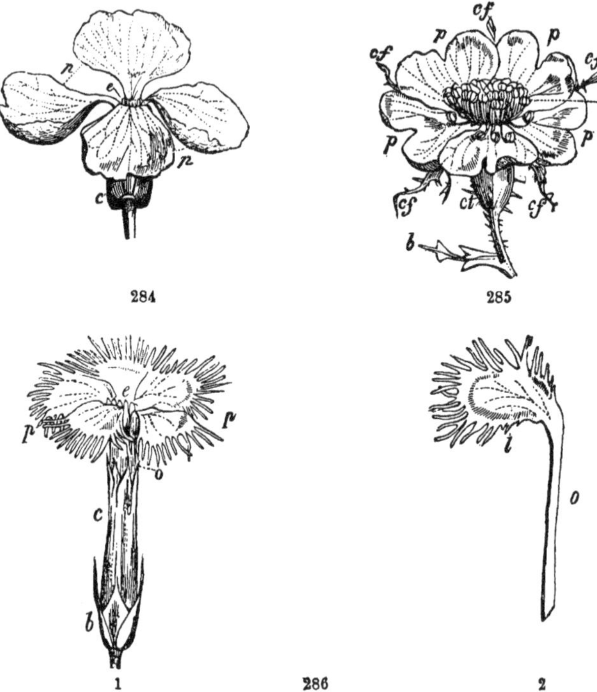

284 285

1 286 2

§ 439. But certain particular words have been invented to designate certain forms of monopetalous corollas common to a large number of flowers. We will mention among the regular, the

Tubular (*tubuleuse*) (TUBULOSA), the tube of which, long and cylin-

284. The flower of the Common Wall-flower (*Giroflée commune*) (CHEIRANTHUS CHEIRI).—*c* The lobes of the folioles of the calyx, two of which, the outer ones, are prolonged beneath into a boss.—*pp* Petals.—*e* The larger stamens, of which we only see the summit of the anthers.

285. The flower of the ROSA RUBIGINOSA.—*b* Bract.—*ct* The tube of the calyx. —*cf cf* Folioles of the calyx.—*p p p p p* Petals.—*e* Stamens.

286, 1. The flower of a Pink (*Œillet*) (DIANTHUS MONSPESSULANUS).—*b* Bracts.—*c* Calyx.—*p p* Petals with their claws *o* united into a tube.—*e* Stamens.—2. A separate petal of the same flower.—*o* Claw.—*l* Limb.

drical, seems to be continued by the limb, which follows the same direction, as in the SPIGELIA (*fig.* 287), in the Comfrey (*fig.* 288).

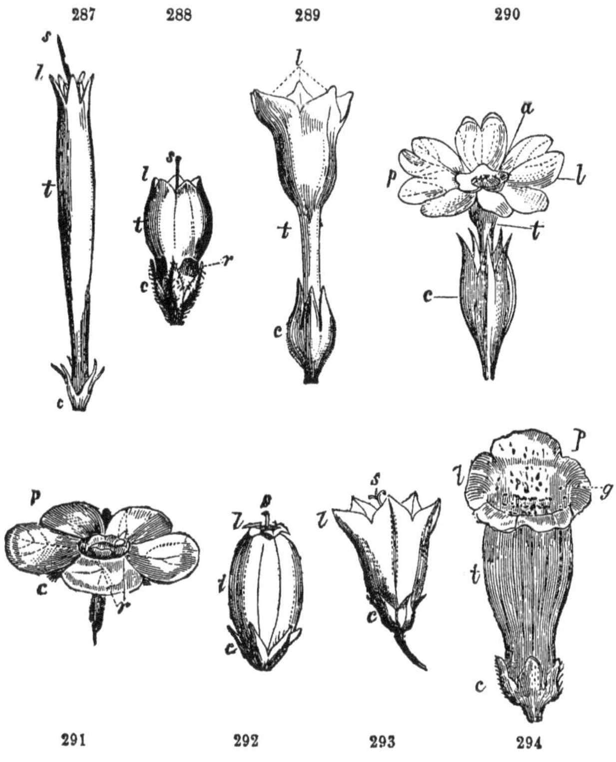

287—294. Regular monopetalous corollas.—*c* Calyx.—*p* Corolla.—*t* Tube.—*l* Limb.—*s* Summit of the style and stigmata.

287. The flower of the SPIGELIA MARYLANDICA.

288. The flower of the Comfrey (*Consoude*) (SYMPHYTUM OFFICINALE).—*r* External opening of the folds which project into the tube.

289. The flower of the Tobacco (*Tabac*) (NICOTIANA TABACUM).

290. The flower of the Common Primrose (*Primevère*) (PRIMULA ELATIOR).—*a* Anthers in the throat of the corolla and opposite to its lobes.

291. The flower of the MYOSOTIS PALUSTRIS.—*r* Folds of the corolla projecting beyond the entrance of the tube and opposite to the lobes of its limb.

292. The flower of the ERICA CINEREA.

293. The flower of the Heath-bell (*Campanule commune*) (CAMPANULA ROTUNDIFOLIA).

294. The flower of the Fox-glove (*Digitale pourprée*) (DIGITALIS PURPUREA.) This last corolla is not quite regular.—*g* Throat.

Funnel-shaped (*infundibuliforme*) (INFUNDIBULIFORMIS), that, which suggests the shape by its limb, separating at the summit of the tube into an inverted cone like a funnel, as in the Tobacco plant (*fig.* 289).

Salver-shaped (*hypocratériforme*) (HYPOCRATERIFORMIS), that, the limb of which, placed like a very broad saucer, surmounts a straight tube, as in the Primrose (*fig.* 290).

Wheel-shaped (*rotacée*) (ROTACEA), that, in which the limb presents open divisions like the spokes of a wheel and the extremely short tube represents the nave, as in the MYOSOTIS (*fig.* 291).

Stellate (*étiolée*) (STELLATA), the same shape, but with much smaller divisions, as in the GALIUM.

Pitcher-shaped (*urcéolée*) (URCEOLATA), that, in which the limb is almost wanting, the tube swelled in the middle, narrowed at the two points, as in the ERICA CINEREA (*fig.* 292).

Bell-shaped, (*campanulée*) (CAMPANULATA) or *Campanulate*, in which the flower represents the form of a bell by its tube gradually becoming vase-shaped, broader towards the limb, as in the CAMPANULA (*fig.* 293).

Glove or *Thimble-shaped* (*digitaliforme*) (DIGITALIFORMIS), longer than *Bell-shaped* and irregular, or in the form of a thimble or long bell (*fig.* 294).

Calathiform (*Calathiforme*) (CALATHIFORMIS), that, which is hemispherical and concave. This form is most frequently found among the calyces.

Cup-shaped (*cyathiforme*) (CYATHIFORMIS), that, which is in the shape of a wine-glass.

Amongst the irregular, the corolla:

Strap-shaped (*ligulée*) (LIGULARIS) (*fig.* 295), that, in which the tube at a certain height is split on one side and is thrown to the other in the shape of *strap* (LIGULA), which is terminated with small teeth (*l*). We may consider also the ligulæ as formed by the linear divisions of the limb, which remain coherent, either all of them, as in the SCORSONERA, the Dandelion and all other Chicoraceæ, or only several, as in the *Honeysuckle* (*Chèvrefeuille*). This last modification approaches the following.

Labiate (*labiée*) (LABIATA) (*fig.* 296), that, in which the divisions are arranged so as to form two kinds of open lips; the one, the upper, commonly composed of two; the other, the lower, of three, as in the Sages and all other plants of the same family. The calyx is then generally bilabiate, but in an inverse direction; that is to say, its upper lip consisting of three divisions and its lower of two.

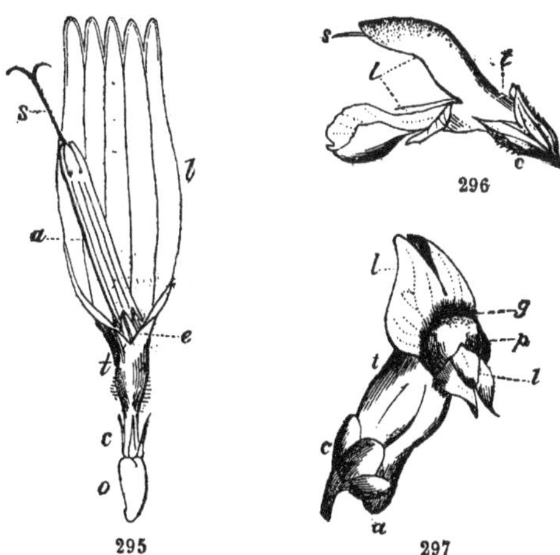

Personate (*personée*) (PERSONATA), like a mask (*fig.* 297), that, which has two lips like the preceding one, but nearer together, and closed by a swelling of the upper *p*, which is called the *palate* (*palais*) (PALATUM), as in the Snap-dragon (*Muflier*).

The tube may itself present to our notice irregularities, independently of those of the limb, for instance, in the LYCOPSIS, in which a regular limb is supported by a bent tube.

§ 440. We have yet to mention some strange and uncommon forms among the petals. In certain flowers, the limb, instead of remaining flat or slightly concave, is bent so as to represent a helmet, whence *helmet-shaped* (PETALUM GALEATUM), or *hooded* (P. CUCULLIFORME). The name is borrowed in these cases, as we may see, from the object whose shape it recals. When it is prolonged on the outside or at the bottom into a kind of long bag or spur, it is said to be *spurred* (*éperonné*) (CALCARATUM), as in the Violet. Instead of a bag we find at other times a simple fold, more or less short, and

295—297. Irregular monopetalous corollas.—*c* Calyx.—*p* Corolla.—*t* Tube.—*l* Limb.—*g* Throat.—*s* Stigmata and summits of the style.

295. The flower of the CATANANCHE CÆRULEA. The calyx, with a quinquefid limb *c* is united beneath to the ovary *o*. The anthers of the stamens *e* are united into a tube *a*, through which the style terminated by a bifid stigma *s* passes.

296. The flower of the Wild Sage (*Sauge des prés*) (SALVIA PRATENSIS).

297. The flower of the common Snap-dragon (*Muflier commun*) (ANTIRRHINUM MAJUS). The tube of the corolla is prolonged at the base into a boss *a* and is closed at its throat by a swelling *p*.

more or less compressed, the cavity of which may open either inside or outside of the flower, as in the Borage, the Myosotis (*fig.* 291) and several other Boragineæ (*fig.* 292). Instead of a hollow projection, we may have a solid one, formed by the thickening and extension of the tissue of the petal, as in several Asclepiadeæ (*fig.* 708, *a*). In these last cases, in which the corolla is monopetalous and regular, these projections opposite to the lobes form an internal circle, a kind of crown, and have received different names according to the different appearances which they present.

We have already seen (§ 387) that rather frequently a lamina of a greater or less size doubles, as it were, the limb, either on the outside, as in some Resedas, or in the inside, as in several Caryophylleæ, the Lychnis (*fig.* 298), the Cucubalus, &c., and that it may be considered as owing to a deduplication. The petal is then said to be *appendiculate* (*appendiculé*) (APPENDICULATUM).

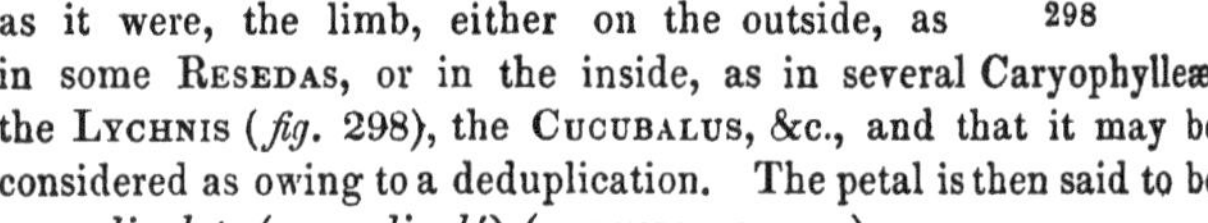

§ 441. The duration of the corolla varies like that of the calyx (§ 428), but it is always much more evanescent. It sometimes falls at the moment of expansion, almost always after fecundation, and when it remains any longer, it is only when dried and shrivelled, or in other terms *marcescent* (*marcescente*), as in the Heaths, the Campanulas. The monopetalous corolla is always detached in one single piece.

ORGANS OF FECUNDATION.

Stamens (*étamines*) (Stamina).

§ 442.—Up to this point we have only examined the stamens with respect to their position as to the other parts of the flower. To their shapes and peculiar structure we have paid little or no attention, and we have contented ourselves with representing them as narrow folioles, swollen at the top into two bodies, each of which borders one of the sides to a certain length (§ 366), or more fre-

298. A petal of the Lychnis fulgens, the inner side is turned towards us.—*o* Claw.—*l* Limb.—*a* Appendix.

quently even reduced to a thin cylinder which bears at its summit these same two bodies (§ 368). The upper swelling of the stamen is termed the *anther* (*anthère*), the lower part is called *filament* (*filet*) as it most commonly presents the form of a thread. The anther is the essential part of the stamen, and, if it be wanting or be incompletely developed, the stamen, incapable of discharging its functions, takes the epithet of *barren* (*avortive*) (ABORTIVUM, EFFÆTUM); but it is not so, if the filament only be wanting, when the anther is said to be *sessile*. We shall resume at the end of this chapter the examination of the anatomical structure, of the developement and of the functions of the anther, which are so closely allied to those of the pistil, that it would be inconvenient to prosecute the examination of the one without its being immediately followed by that of the other and we shall commence by examining the exterior and general characteristics of the stamens at first considered as isolated, then as a whole belonging to the same flower.

§ 443. FILAMENT (*Filet*) (FILAMENTUM).—This name indicates the most common form presented by this organ, which is most generally found in the shape of a body lengthened into a thin cylinder, or insensibly tapering from the base to the summit, *thread-shaped* or *filiform filament* (*f. filiforme*) (F. FILIFORMIS); much more rarely, it gradually thickens from bottom to top into the form of a club (F. CLAVATUM). It has often rather a great degree of solidity and sustains itself; but, at other times, as in the Gramineæ, the Plantains, &c., it is only as thick as a hair and is then said to be *capillary* (*capillaire*). It is not rare to see it flat or linear at its base, and tapering at its upper extremity, *awl-shaped* (*f. subulé*) (F. SUBULATUM). Flat throughout the whole of its extent, it may represent a lengthened riband, commonly entire on its edges, more rarely scolloped, as in the Elder; it may lastly be enlarged into a lamina, which acquires in certain flowers (CANNA and other Marantaceæ, NYMPHÆA ALBA) the developement and appearances of a true petal. Its direction is commonly continuous from one end to the other; we find, nevertheless, some examples, in which it changes quickly at an angle more or less obtuse, which has been compared to a knee, and hence the filament is then called *knee-jointed* (*genouillé*) (F. GENICULATUM).

§ 444. We have just seen that it frequently presents at its base a widened part; then instead of growing gradually narrower from bottom to top, it may, at a certain height, pass suddenly from the form of a lamina to that of a thread, as in the PEGANUM HARMALA, the TAMARIX GALLICA (*fig.* 324). This lower widening, which is often prolonged on both sides into a lobe or free point, reminds us of that forming the sheath of the leaves at the base of the petiole, which may be compared to the narrow part of the filament.

§ 445. But it sometimes happens that this portion widened at the bottom seems rather an accessory part united to the filament, with respect to which it occupies a plane either internal, as in the ZYGOPHYLLUM FABAGO (*fig.* 299), and several other Zygophylleæ, the Simarubeæ, &c. &c., or external as in the Borage (*fig.* 300), the TRICHILLIA and other Meliaceæ. These two cases, in which the filament is said to be appendiculate, evidently corresponds to those in which the petal receives the same term (§ 440); and in the second the stamen, thus united to a lamina placed on the outside, is, relatively to it, precisely as it is relatively to the petal, when it is fastened to its base, making part of a whorl immediately opposite (§ 386, 412). The appendix at the base of the filament receives different names according to its different appearances: such as glands, scales, &c., to which is added the epithet *staminiferous* (*stamnifères*).

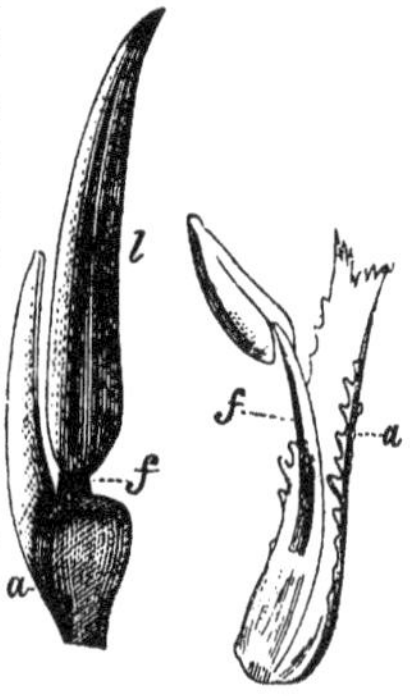

§ 446. ANTHER.—When we make a transverse section of the *anther* (*anthère*) (ANTHERA), that is, the swelling by which the stamen is terminated at the upper end, we see that it is not a solid body, but a hollow one (*figs.* 315, 318, 2), and filled with very fine dust. In all the examples which we have mentioned, the swelling was double and, consequently, the cavity also. Each cavity of the anther is called the LOCULUS or THECA (*loge*); and whenever there are two placed at the end of the same filament (and this is generally the case), the anther is said to be *bilocular* (*biloculaire*) (ANTHERA BILOCULARIS *seu* DITHECA). It sometimes happens that it is *unilocular* (*uniloculaire*) (UNILOCULARIS *seu* MONOTHECA [*figs.* 310, 311]), but much more rarely. Lastly, it is extremely rare to find it *quadrilocular* (*quadriloculaire*) (QUADRILOCULARIS *seu* TETRATHECA)

299. A stamen of the ZYGOPHYLLUM FABAGO.—*f* The filament borne on the outer surface of an appendix *a*.

300. A stamen of the Borage (*Bourrache*) (BORAGO OFFICINALIS).—*f* Filament borne on the inner surface of an appendix *a* lengthened externally into a horn.—*l* Loculi of the anther.

after it has attained its perfect state (*figs.* 314, 315). It is not absolutely necessary to divide the anther in order to determine the number of its loculi. It is clearly seen on the outside, because they each form a distinct projection, and, besides, when they have attained maturity, they each naturally open by a hole or more commonly by a split, thus allowing an escape to the dust, which fills them and which we call POLLEN.

The loculi of the anther appears then to be a kind of bag at first perfectly closed, the shape of which varies a great deal in different plants. We observe all the intermediate shapes between that of a globule (*fig.* 301), and that of a long and thin cylinder, either rectilinear, *linear loculus* (*loge linéaire*) (*fig.* 302, 1), or bent, *vermiform loculus* (*loge vermiforme*) (*figs.* 302, 2, 312); the most common is that of an oval, more or less elongated (*figs.* 303, 304 *l*); sometimes the loculus is reduced to a point at its extremity, the anther is then *pointed* (*aiguë*), in the Borage, for instance (*fig.* 299), if the two are united, *bicornate* (*bicorne*) (BICORNIS), if they are separated (*fig.* 319, 307 *l*); each of these horns may itself be bifurcated and the anther become *quadricornate* (*quadricornée*) (QUADRICORNIS [*fig.* 321]).

§ 447. The two loculi of a bilocular anther sometimes touch immediately and are united by their faces in contact. They may be joined at the summit of the filament, resting then on its internal or external side, or separated from one another by the whole of its substance: in all these cases, the anther is said to be *adnate* (*adnée*) (ADNATA) to the filament (*fig.* 304); but most frequently, it is not the filament which is interposed between the two loculi, it is a body which continues it, but with a change of structure, called the *connective* (*connectif*) (CONNECTIVUM), because it is thus the means of connection between the two loculi. Its proportions relatively to the loculi, are extremely variable; sometimes, equal to them in length, it is the means of union between them from one end to the other; sometimes it is shorter than they are, and may be even reduced to a point (*figs.* 301, 302) or to a shor line; sometimes, on the contrary, it is largely developed, and in this case it commonly follows the direction of the filament and is prolonged beyond the loculi in the shape of a bristle (*fig.* 306), or of a thick lump, reminding us of the form of a club or of a tongue (*fig.* 307), of a cone (*fig.* 308), &c., &c., or of a membranous expansion (*fig.* 317 *c*); more rarely, extended perpendicularly to the filament, thus resembling the beam of a balance bearing a loculus at each extremity (*fig.* 309 *c*).

We shall see in a little time, that the connective is distinguished from the loculi by its structure. But, the difference between them is also discernible at the first glance, on account of the dissimilarity of colour, which is a yellow, deeper or paler, the most common tint of these loculi.

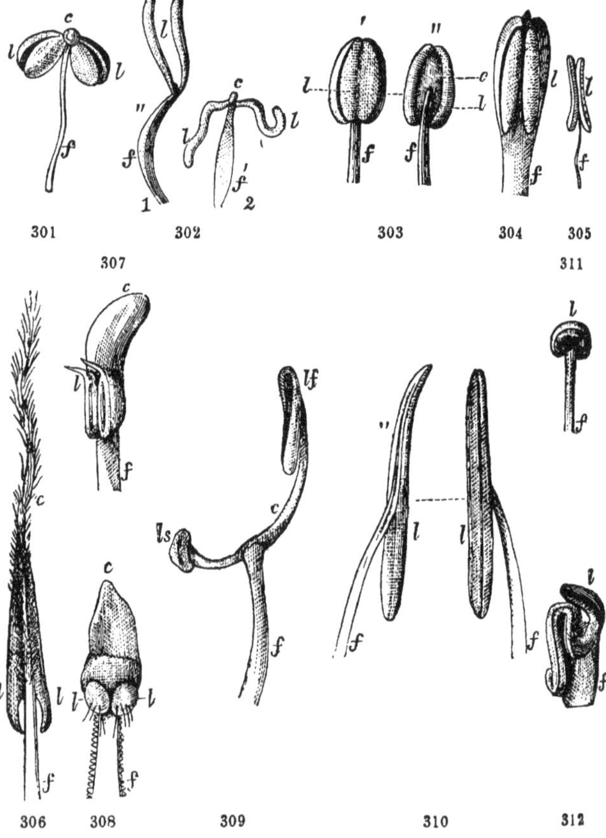

301—312. Different anthers with the summit of the filament *f*.—*l* Loculi.—*c* connective.

301. Anther of the Mercury (*Mercuriale*) (MERCURIALIS ANNUA).

302. Anther of the ACALYPHA ALOPECUROIDEA.—1. In the bud.—2. In the expanded flower.

303. Anther of the Almond.—1. Seen in front.—2. Seen behind.

§ 448. When the loculi cling to the connective by the greater part of their length, they are said to be adnate to it; when they are united by only a very short piece, they are said to be free. The point of union may be situated, either towards the middle of the loculi or towards the bottom, and then they are erect; or at the top, and then they are called pendulous. If, united throughout the whole of their middle part, they are free at the two extremities, they represent a long *x* (*fig.* 305); if, united throughout the whole of their upper part, they are not so at their lower ends, which are more or less separated, according as they are sharp or blunt, they are said to be ***arrow-shaped*** (*sagitée*) (*fig.* 306), or *cordiform* (*cordiforme*) (*fig.* 300, 303, 325): this shape is not by any means uncommon.

§ 449. The connective and the filament may continue together preserving the same direction and nearly the same thickness; then, in the case in which the loculi are adnate, the anther cannot change its position with regard to the filament; it is immovable (*fig.* 304, 307, 315). But, most commonly, the summit of the filament is thin and terminates under a very acute angle in a point of the connective, towards its middle (*fig.* 267 *e*) or nearer one of its two extremities. The anther is then balanced on the filament and takes different positions according to the different motions communicated to the flower; it is then *versatile* (*oscillante*) (VERSATILIS [*figs.* 310, 267]).

§ 450. When the anther is unilocular, the filament is directly attached to one point of the single loculus (*fig.* 310). We may conceive that there is no connective here: it may, nevertheless, be represented by a body different from the rest of the filament immediately between it and the loculus; and it is to be presumed in this case, that if this body does not bear a second loculus placed symmetrically, it is because it has not been developed. Indeed, we sometimes find traces of the second one; in the Sages, for instance,

304. Anther of the BEGONIA MANICATA.

305. Anther of the POA COMPRESSA.

306. Anther of the Oleander (*Laurier-Rose*) (NERIUM OLEANDER).

307. Anther of the BYRSONIMA BICORNICULATA. The loculi, empty at the summit, are detached from the connective under the form of two small horns.

308. Anther of the HUMIRIA BALSAMIFERA. This is an example of filament *f* ciliated with glandular hairs.

309. Anther of the Sage (*Sauge*) (SALVIA OFFICINALIS).—*lf* Fertile loculus, full of pollen.—*ls* Sterile loculus, empty.

310. Unilocular anther of the STYPHELIA LÆTA; the front is seen opened ', the back ''.

311. Anther of the Marsh Mallow (*Guimauve*) (ALTHÆA OFFICINALIS) before the dehiscence.

312. Anther of the Common Bryony (*Bryone*) (BRYONIA DIOICA).

in which the beam, which forms the connective, bears at one of its extremities a well-formed loculus full of pollen, at the other a malformed loculus without pollen (*fig.* 309): in such a case, the anther is only unilocular by abortion. We must also take care to regard it as such in two quite opposite cases, in which a mistake is easily made; that in which the two loculi separated from one another may each be taken for a distinct anther, that in which, on the contrary, they are continuous and are confounded at their bases and thus seem to form only one.

§ 451. The act, by which the loculi of the anther open to empty themselves, is called *dehiscence* (*déhiscence*) (DEHISCENTIA). We have stated that this most commonly takes place by a slit following the direction of their length. This slit, the place and direction of which are previously indicated by a line or stripe (*fig.* 303, 304), is naturally found on the side opposite to that, by which the loculus is attached either to the filament or to the connective. In the majority of cases, the loculi were parallel or inclined rather obliquely with regard to the filament or the connective; but, if they incline more and take a position which approaches the perpendicular (*fig.* 326, *ag*), the line of dehiscence will take the same direction: we shall say that the anther is opened longitudinally (LONGITRORSÙM) in the first case (*fig.* 310), transversally (TRANSVERSÈ) in the second (*fig.* 311); and it is in the latter that the false appearance of a single loculus may result from the circumstance that the two transverse splits sometimes seem to be continuous.

The loculus does not always split throughout the whole of its length at the same time; but the lips of the split, which separate from bottom to top, remain united for a longer time throughout the rest of their extent, and the dehiscence then seems to take place at an upper or lower opening (*figs.* 317, 319).

Again, at other times, there is neither split nor line which indicates it. Each loculus at its summit, by the separation of the walls which form it, is pierced with a hole or *pore* (*pore*), by which it is emptied; as in the SOLANUM, in the PORANTHERA (*fig.* 314). At other times, as in the TETRATHECA JUNCEA (*fig.* 315), these pores are united into one, the common issue of all the loculi of the anther.

Lastly, in a very small number of plants, a certain portion of the walls is circumscribed and is then raised after the manner of a casement, which is only fixed by one of its edges and completely detached from the rest. The anther of several Laurels (*fig.*

316) presents two windows of this kind on each side above the other; that of the HAMAMELIS only one.

§ 452. When the loculus is opened not by a pore at the summit, but

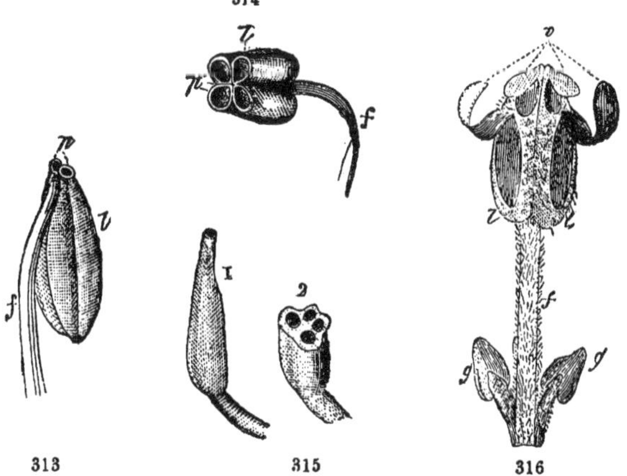

by a split, as is most commonly the case, or by other openings placed on one of its faces, this face may be turned either towards the interior of the flower (INTRORSÙM) or towards the exterior (EXTRORSÙM); these positions are indicated by the epithets of *introrse* (INTRORSA *seu* ANTICA) or of *extrorse* (EXTRORSA *seu* POSTICA) given to the anther. If the splits are turned towards the sides, which often happens when the *loculi* are fastened to the sides of the filament or of the connective, we must express this direction of the dehiscence intermediate between the preceding by ANTHERA LATERE *seu* RIMÂ LATERALI DEHISCENS). But how shall we determine these different directions, when the anther is versatile or when it opens at the summit? We may, in the first case, study it in the bud in which, still straight, it is not inclined on the filament; and in the other cases, if the filament is attached to the middle or the top

313. Anther of the PYROLA ROTUNDIFOLIA, suspended from the extremity of the filament and opening by two pores *p*.

314. Quadrilocular anther of the PORANTHERA, opening at the summit by four pores *p*.

315. Quadrilocular anther of the TETRATHECA JUNCEA, opening at the summit by a single pore.—1. Whole.—2. Divided horizontally.

316. Anther of the LAURUS PERSEA with four loculi, placed above one another by twos and each opening by a valve *v*. Two glands *g*, which seem two abortive anthers, are joined to the filament *f* at the bottom.

of the anther, it is on its internal or on its external face, and we may thus state its extrorse or introrse position.

§ 453. In the same way as the other organs of the flower which we have previously examined, the anther may have appendices. These are most commonly simple elongations of the parts which compose it. Thus the loculi may at one of the extremities taper to a point (*fig.* 321), be flattened into a lamina (*fig.* 319, *a*), &c., and at the extremity, thus modified, the internal cavity is interrupted. Sometimes, unwonted excrescences are seen on their faces in the shape of small points (*fig.* 320, *a*), or of warts, or of tufts (*fig.* 318 *a*). We have already stated, that the connective may be developed beyond the lobes to different sizes and into diverse forms. At other times, although more rarely, it is prolonged under-

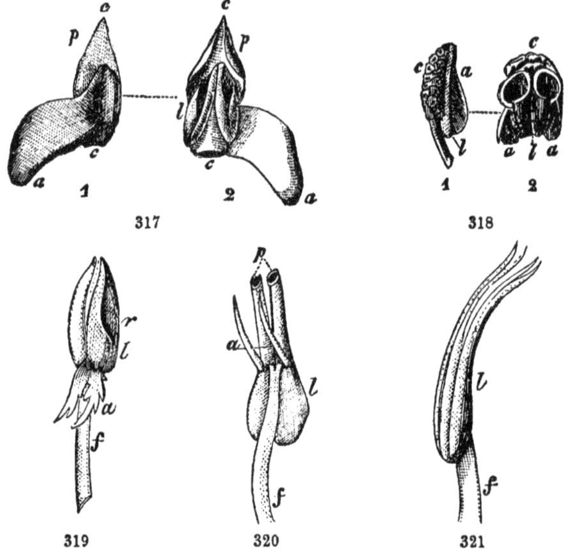

317—321. Appendiculate anthers.—*a* Appendix.—*l*, *p*, *c*, *f* have the same signification as in the preceding figures.—*r* slit.

317. Sessile anther of the Garden Violet (*Violette*) (VIOLA ODORATA).—Front, 1. —Back, 2.

318. Anther of the PTERANDRA PYROIDEA—1, Undivided, with a side turned towards us.—2. The lower half, after it has been divided horizontally.

319. Anther of the ERICA CINEREA.

320. Anther of the VACCINIUM ULIGINOSUM.

321. Anther of the GAULTHERIA PROCUMBENS.

neath or on the outside into a spur, as in two of the five stamens of the Violet, which is buried in that of the corolla (*fig.* 317, *a*).

§ 454. This relation of the petal and of the stamen is important and claims our notice by tending to prove their common nature, which we have already mentioned and shewn by several arguments and examples. It must, however, be avowed, that of the parts of the flower the stamen is that, in which the resemblance to the leaf is most effaced, especially in the anther, compared with the limb which it represents, as we have seen the filament (§ 443) with its base frequently widened represent the petiole with its sheath. It is in the developed organs that the difference is more and more pronounced, although even then we have found examples of the passage from one to the other, examples, which it would be very easy to multiply, if the limits of this work would allow us. But, on examining them at a more advanced period, this difference is much less sensible; as we shall see in a little time, when we come to study the formation of the stamen, and as we should be convinced on the other hand on following that of a great number of leaves.

§ 455. If, in the stamen, the anther is the essential part in the fecundation, the pollen is so in the anther itself, as we shall see in a little time. We call these stamens, therefore, in which this dust is wanting, sterile or barren. The loculi may exist, but famished and withered. At other times, they completely disappear, and it is often the connective alone which remains to develope itself. It is not rare to see in these cases the anther transformed into a petaloïd limb, sometimes compressed and wrinkled, sometimes expanded like a true petal; and this last transformation may become complete; it is from this circumstance that we owe our possession of double flowers. Lastly, the sterile stamen may be reduced to a filament; and this filament may itself be more or less diminished: it is then termed rudimentary.

§ 456. After having considered the isolated stamen, let us examine the stamens grouped and united in the same flower in their relations with the other whorls of this flower or with one another.

We have already shewn some of these relations: 1st, Those which depend on number, that of the stamens being equal to that of the sepals and of the petals (*isostemonous flower*, § 386), or unequal (*anisostemonous* [*anisostémone*] *flower*) (from ἄνισος, *unequal*, and στήμων, *stamen*); either double (*diplostemonous flower*, § 386) or less (*meiostemonous* [*méiostémone*] *flower*), (from μεῖον, *less*); or, on the contrary, more than double (*polystemonous* [*polystémone*] *flower*) (from πολὺς, *many*). We have seen that this last circumstance may result, sometimes from the addition of fresh whorls of stamens (§ 386, *b*), sometimes from the deduplication of some of them or of all (§ 387). 2nd, Those which depend on their position relatively to the parts of the neighbouring whorls opposite or alternate, or in an intermediate situation; 3rd, Those which depend on the different degrees of union that they may contract with these whorls, and according to which their insertion may be varied, i. e., their apparent point of departure, relatively to them and especially to the pistil, according to which we divide them into three great classes, *hypogynous* (*hypogynes*), *perigynous* (*périgynes*), *epigynous* (*épigynes*) (§ 383).

§ 457. As to their mutual relations, the stamens of the same flower may be completely independent of one another, *free* or *distinct stamens* (*étamines libres* ou *distinctes*) (STAMINA LIBERA *seu* DISTINCTA), or else contract adherences together, *united* or *connected stamens* (*étamines soudées* ou *connées*) (STAMINA COALITA *seu* CONNATA). This adherence takes place between the anthers, as we see in all the Compositæ, the LOBELIA, the JASIONE, and, in this case, the stamens are said to be *syngenesious* (*syngénèses*) or rather *synantherous* (*synanthérées*) (SYNGENESA *seu* SYNANTHERA, from σὺν, *with* [which indicates union in compound words]; γένεσις, *origin*, or ἀνθηρὰ, *anther*). But, it is most commonly the filaments, which are

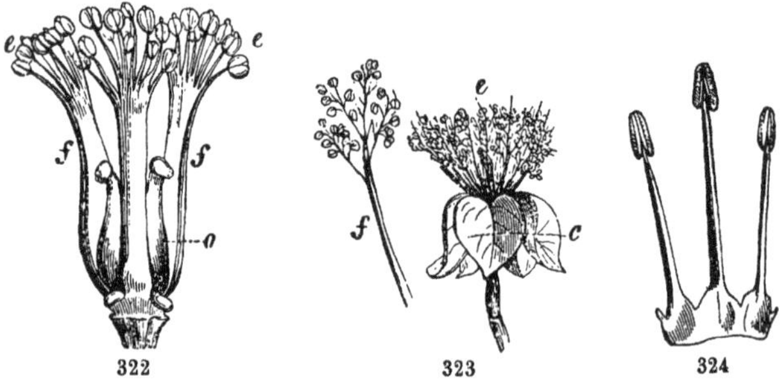

322. Triadelphous stamens *e e* of the HYPERICUM ÆGYPTIACUM surrounding the pistil *o*. The envelopes of the flower have been removed.

323. Male flower of the Castor oil plant, consisting of a calyx *c* of five reflexed folioles and polyadelphous stamens *e*. One of the branched fascicles *f* has been magnified and is represented by the side of the figure.

324. Three of the ten stamens of the TAMARIX GALLICA.—The filaments are united with one another only at their dilated base, so as to form a kind of ring, a part only of which is here seen.

thus united, whether they are thus blended together in one single body or are joined together in several groups to which we have applied the name of ADELPHIA (*adelphie*) (§ 373), so that the stamens are *monadelphous* (*monadelphes*), *diadelphous* (*diadelphes*), *triadelphous* (*triadelphies*) (*fig.* 322), *polyadelphous* (*polyadelphes*) (*fig.* 323), according as, by the union of their filaments, they form only one of these groups, or two, or three, or more. In the case of Monadelphia, if the pistil has not been suppressed, it is clear that the united filaments ought to leave it a free space at the centre of the flower and form around it a tube or ring (*fig.* 324); it is only when there is no pistil, when the flower is male, that these filaments are united into a fascicle which is itself central (*fig.* 251, 1). In the cases in which there are several groups of stamens, they form either so many segments of a circle (*fig.* 239) or so many fascicles (*fig.* 322). Sometimes, the filaments are united throughout the whole of their length; more frequently, united at the bottom, they are separated at their upper part (*figs.* 239, 322). In the former case, the fascicle takes the form of a column; in the latter, it is branched and bears a great resemblance to a small trunk divided into branches each terminated by an anther, especially, when all the filaments are not separated at the same height, but some are united to a still higher point than others (*fig.* 323, *f*). We have seen (§ 442) that the isolated filaments may be appendiculate; it is sometimes the same with the groups of filaments, with those which result from deduplication; in the LOASA, for instance, petaloid appendices alternate with petals and are covered with a small number of stamens; thus, also, in a genus of Tiliaceæ, the LUHEA (*fig.* 238), the stamens several in number are found to be united in five groups placed in the intervals of the five petals, and each of these groups to be joined so as to form a kind of lengthened amina, itself divided at its summit into a crowd of thread-shaped straps, which prove its tendency to deduplication into sterile filaments, as the anterior part of the group undergoes deduplication into fertile stamens.

§ 458. The stamens of the same flower, compared with one another, are equal or unequal in size, and, when they are unequal, they may be so with a certain degree of regularity. When they are numerous, they may be so much the longer as they are nearer the centre (*fig.* 239, 2) or, on the contrary, as they are farther from it, as in several of the Rosaceæ (*fig.* 369). In diplostemonous flowers, the stamens opposite to the petals are almost always shorter than the alternate ones. We term those of the Cruciferæ *tetradynamous* (*tétradynames*) (from τέτρα, *four*, and δύναμις, *power*, *rule*), four great ones of which, arranged in pairs, alternate with two smaller isolated ones (*fig.* 325); *didynamous* (*didynames*), (from δὶς, *twice*), those of the Labiatæ and other plants, in which the five stamens, alternating with the five lobes of the corolla, are, by the more or less complete abortion of the fifth, reduced to four. Of these the two greater ones answer to the upper part of the flower, two smaller ones correspond with its sides (*fig.* 326). It would detain us too long to mention all the possible combinations in the relative proportion of unequal stamens.

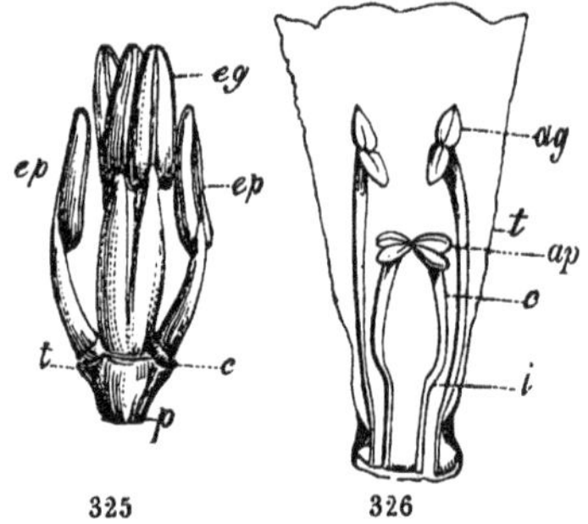

§ 459. Their proportion to the corolla ought to be noted in the description of a plant. When the stamens are longer than it is and protrude beyond it, they are said to be *exserted* or *protruding* (*saillantes*) (EXSERTA); when, on the contrary, shorter than it, they are concealed by it, they are said to be *included* (*incluses*) (INCLUSA [*fig.* 287 and following ones, 326]).

§ 460. They follow different directions, either turned upwards, *erect stamens* (*étamines dressées*) (S. ERECTA), or towards the centre of the flower *incurved stamens* (*étamines infléchies*) (S. INFLEXA), or, outwards, and are either simply divergent, or are extended horizontally (PATULA), or are quite bent (REFLEXA), or are pendent or even approach the vertical, *pendulous*, (PENDULA). Sometimes they are all inclined towards the same side of the flower, towards the top, or towards the base, *declinate* (DECLINATA), as in the Horse-Chesnut.

§ 461. STRUCTURE OF THE STAMEN.—After having examined the external forms of the stamens in several species of plants and the

325. The tetradynamous stamens of the Common Wall-flower (*Giroflée commune*) (CHEIRANTHUS CHEIRI).—*p* Summit of the pedicel.—*c* Wounds left by the fallen folioles of the calyx.—*eg* Two pairs of large stamens.—*ep* Small stamens.—*c* Glandular torus on which all these stamens are inserted.

326. Corolla of the Foxglove (DIGITALIS PURPUREA), cut open and expanded so as to shew the apparatus of the didynamous stamens which it bears.—*t* Tube.—*f* Filaments; beneath their insertion *i* we may see their prolongation into the substance of the corolla as far as its base.—*ag* Anthers of the great stamens;—*ap* of the small.

relations which exist between them with respect to the other parts of the same flower, let us consider the anatomical structure of the stamen.

Of the Filament.—The filament is composed: 1st, of a central fascicle of tracheæ, which extends from the base to the summit, without being branched throughout the whole of its course; 2nd, of a layer of cellular tissue enveloping this vascular fascicle; 3rd, of a thin epidermis, on which we sometimes, though very rarely, observe stomata. The fascicle of the tracheæ, is continuous and terminates in the connective; sometimes before it reaches it. This connective is formed of a mass of cells, rather different from those of the filament both in their colour and in their shape. Their consistence is frequently that of a glandular tissue.

§ 462. Of the Anther.—The loculi of the anther in their perfect state present internally a cavity filled with pollen, externally an epidermal membrane (*fig.* 327, *ce*) frequently dotted with stomata; in the interval, a layer of a peculiar tissue (*cf*), the nature and form of which we shall easily conceive, if we say, that it originates in a union of spiral cells (*fig.* 25) or annular (*fig.* 26), or, still more frequently, reticulated (*fig.* 27), arranged in one or several rows in thickness. But, commonly the membrane of these cells disappears as the anther approaches maturity, and there only remains the threads or bands, arranged consequently, in a spire, or more frequently in rings or in nets (*fig.* 327, *cf*). The name of *fibrous cells* has been given to the open cells thus reduced to laminæ, which primitively doubled them, to their fibres, attaching to this word not the idea of a lengthened utricle, as we have already done in the course of this work, but that of a thread or thick riband. This fibrous layer gradually diminishes in thickness in proportion as it approaches the line, where the dehiscence of the anther takes place, and at this line it is completely interrupted. These small, very elastic, hygrometric laminæ stretch, contract, lengthen and bend in different ways, according as the anther is dry or moist; and these variations are influenced, on the one hand, by the develope-

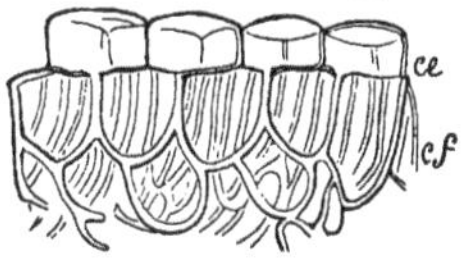

327

327. A part of the horizontal section of the wall of an anther of the Cobœa scandens, at the period of dehiscence.—*ce* The external layer consisting of the cells of the epidermis.—*cf* Fibrous cells forming the internal layer.

ment of the anther, the juices of which, at first abundant, are absorbed and evaporated by degrees; on the other hand, by the variable state of the atmosphere. The tissue, which forms the wall of the anther, thus subjected to a series of forces acting in different directions, naturally breaks where it offers but little resistance, that is, at the line or the point where the fibrous layer is interrupted; and it is thus that the loculus splits and communicates with the exterior, so as to allow the free egress of the pollen enclosed in the cavity, an egress, which the continual contractions of the elastic tissue favour and then complete.

Let us now examine what successive changes the stamen has undergone, since its first appearance in the flower to that perfect state which immediately precedes or accompanies the dehiscence of the anther.

It is the first to appear in the flower, and we find it in the flower under the form which all the foliaceous organs at first present to our notice, that of a solid cellular nipple-like projection. This projection then extends itself, and, it is remarkable, that it is then of a greenish hue, although afterwards it may take a perfectly different colour, most commonly yellow; it is then lengthened, but not differing in its shape from the other organs of the flower (§ 431). At its middle is commonly observed a superficial longitudinal furrow, indicative of its future separation into two loculi: a mark, which will correspond to the summit of the filament or to the connective, and which preserves its greenish tinge for a longer time than the rest. When the filament afterwards appears, the anther has taken its characteristic form, and sooner or later, on the sides are found two fresh furrows, generally parallel to the middle one, being the first indications of the lines of dehiscence.

§ 462, *b*. Developement of the Stamen considered generally.—The filament, when it has once appeared, continues to grow in length. Sometimes, whether it is to remain short, or has not gained the whole of its length in the flower-bud (*bouton*), we find it quite straight; at other times, when it attains its whole length, (which exceeds that of the bud,) it is twisted, compressed or folded up; and these modifications are constant in certain plants, and even in certain families. The filament was at the first completely cellular; it is not until after a certain time that the cells of the centre begin to be organized and lengthened into tracheæ.

It is easy to see in this developement an incontestable analogy to that of the leaf; that of the anther which represents the limb, preceding that of the filament which represents the petiole; so that formed first at its summit, the stamen continues to lengthen from its base. A few observations would tend to complete this resemblance, shewing us, that in certain cases, in which the filament presents at its insertion a wider part analogous to the vaginal part of the leaf, this part seems to be developed sooner than the thin or petiolary part of the filament; and we have seen this taking place in the foliary sheath (§ 147).

We have said (§ 431), that we frequently find in the flower-bud the stamens comparatively much more developed than the petals. These, however, appeared previously or at the same time; but it may happen that their slower evolution may be outstripped by the more rapid one of the stamens, especially of those, which alternate with them and are thus not only larger, but also more precocious than the oppositipetalous stamens. This is a fresh proof of the connexion of the latter with the petals.

§ 463. Developement of the Anther considered separately, and principally that of the Pollen.—But the most important part of the history of the developement of the stamen, is that of the proper tissue of the anther and the formation of the pollen, which constitutes its essential part, the agent of the functions which it is destined to perform.

We have stated that the tissue of the anther is homogeneous; the cells, which composed it, present nearly the same shape and the same dimensions (*fig.* 328). A little later, this tissue seems to be destroyed in several places situated at a certain distance from the periphery, and from its destruction result so many lacunæ, at first straight and linear, then gradually widened. These lacunæ are generally four in number, two in each half of the whole mass of the anther, which half at last constitutes the loculus. A mucilaginous fluid, formed, doubtless, at the expense of the destroyed tissue, fills the lacunæ, and is soon organized into cells (*figs.* 329, 330); the external, smaller (*c p*), the layer of which is extended as a wall over the whole surface of the lacunæ which we may name locellus (*logette*); the internal ones (*u p*), much larger, not only than those which are formed at the same time as they are, but also than those which were already in existence. We give them the name of *pollen-cells* (*utricules polliniques*), or *mother cells of the pollen*, because the pollen is formed within their cavities.

These utricles, indeed, are soon obscured by the presence of nume-

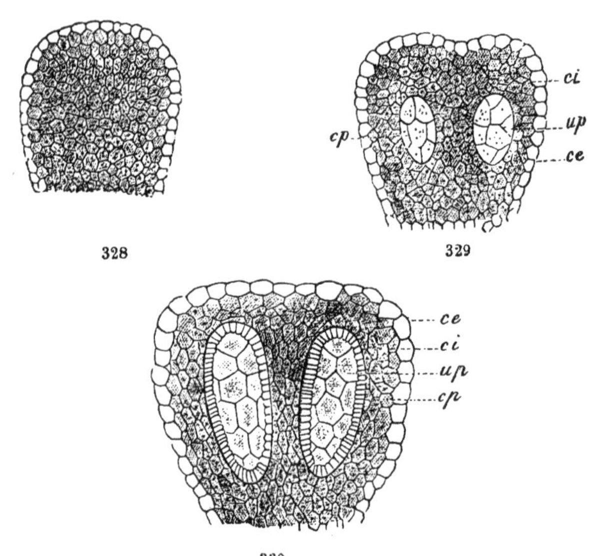

rous granules, which are gradually brought together into a mass (*fig.* 331, 1), which is then divided into four nuclei separated by a liquid matter, which fills the interior of the utricle and is gradually solidified (*fig.* 331, 2). This solidification is generally established from the periphery towards the centre of the pollen-cell; and we see, consequently, partitions gradually advance from the exterior towards the interior, until they meet in the centre and thus divide the cavity of the cell into four (*fig.* 331, 3). Each of the granular nuclei thus isolated is covered with a proper membrane and continues to grow (*fig.* 331, 4); in proportion as they augment, the walls and the partitions of the cell, which before were thick and

328. Horizontal section of an anther of the CUCURBITA PEPO, taken from a flower-bud about two millimetres long.

329. Horizontal section of the same in a bud a little more advanced.—*ce* External layer of the cells which form the epidermis.—*ci* Intermediate layer of several rows of cells, the greater part of which will be absorbed. Locelli filled by a tissue consisting of larger cells *up*, the first appearance of the pollen-cells, and covered by a layer of smaller cells *cp*.

330. Horizontal section of the same, still farther advanced. The letters again refer to the same parts.

succulent, on the contrary, grow gradually thinner, and so much so that they disappear, and the nuclei of the different utricles of the same locellus are free in its cavity; now, these are nothing else than the pollen-grains (*fig.* 331, 6). We here observe what we have already stated concerning the multiplication of the cellular tissue (§ 334), by the formation of several fresh utricles in the cavity of a mother-cell. The grains of pollen are themselves only so many utricles, remarkable for a peculiar form and structure and for becoming independent of one another, so as at last to form a kind of dust, instead of remaining a continuous tissue.

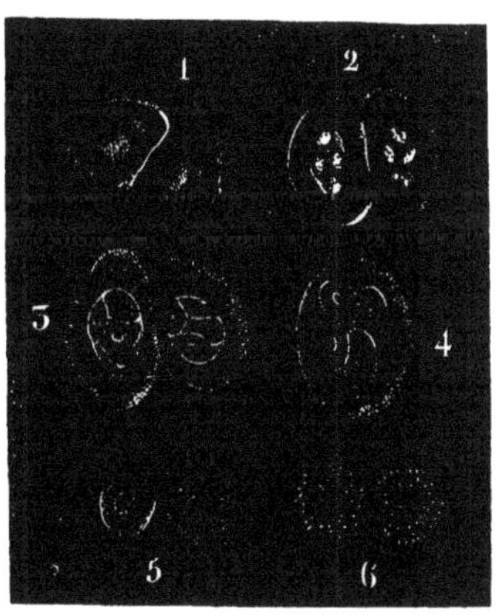

331

The increase of the pollen-grains in proportion as they are thus developed, seems to be made not only at the expense of the pollen cells, the substance of which is gradually absorbed, but also at the expense of other cells, the layers of which, more numerous at the beginning (*figs.* 329, 330 *c i*), end, in consequence of the absorption of the inner ones, by being reduced to a very small number, two, three or four; the most superficial (*c e*) constituting the epidermis of the anther, the innermost its envelope of fibrous cells. The metamorphosis, which gives their definite form to the latter, takes place very rapidly, almost suddenly, about the moment that the pollen-grains arrive at their perfect state.

In this gradual destruction of the cellular tissue of the walls of the anther, the part interposed between the two locelli grows itself progressively thinner and leaves nothing else between them than a weak partition, the external edge of which just touches upon the line of dehiscence. At the moment this takes place, the two locelli

331. The developement of the pollen in the Miseltoe (*Gui*) (VISCUM ALBUM).—1. Two pollen-cells filled with a granular mass.—2. The appearance of four nuclei in this mass.—3. Separation into four masses, each corresponding to a nucleus or to a fresh utricle.—4. Pollen-cell, in which these inner cells are already disunited.—5. Two of the latter or young pollen-grains escaped from the mother-cell.—6. Perfect pollen-grains.

communicate with one another and thus form a loculus of the anther, at the bottom of which the primitive partition is only seen as a short fold more or less apparent (*fig.* 310). We may presume, that, if the anther does not open by a split throughout the whole of its length, but only by a pore at its summit, this partition will not break and continues to separate the locelli: the anther is then said to be *quadrilocular* (*quadriloculaire*) (*fig.* 315). The greater part are so at first, and each loculus really results from the confluence of two originally distinct. The quaternary number remaining in the loculi is, therefore, only a slight modification of the most ordinary case.

§ 464. POLLEN[f].—We have said that the substance of the pollen-cells completely disappears by absorption and that, consequently, the grains of the pollen are free in the cavities of the anther: this is most commonly the case: yet, we sometimes meet more or less evident traces of the state which has preceded. Thus, in the anthers of the Onagraceæ, we find the ripe grains still imperfectly united by a crowd of long viscous filaments, which are nothing but the remains of the substance of the pollen-cells imperfectly absorbed. An analogous arrangement is observed in the pollen of several of the Orchideæ, the grains of which are united into several masses by a matter which has the same origin, presents the consistence of glue and by a slight force is drawn into elastic threads. By separating these masses, we arrive at agglomerations of granules united together in fours: these, formed in the same utricle, have preserved their primitive coherence. We could mention a great number of pollens, the grains of which are presented to our notice thus united in fours (*fig.* 332), or in eights (*fig.* 333), or even in sixes, either that those of two or four utricles are grouped together, or that in the same utricle a multiple number is developed. In the Asclepiadeæ, all the grains of the same loculus are united by their walls into a single mass, and thus form a continuous cellular tissue (*fig.* 709).

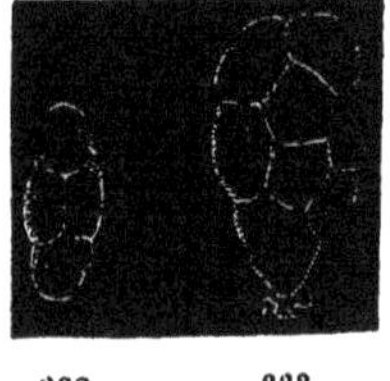

332 333

§ 465. But, let us lay aside these different kinds of exceptional

332. Pollen of PERIPLOCA GRÆCA.
333. Pollen of INGA ANOMALA.

f From παλή, *fine sifted flour* (Lat. POLLEN).—TRANS.

structure and return to that with which we usually meet, that in which the grains, having at last become free in a common cavity, fill it like a kind of dust and are scattered when it bursts. These grains, as we have already said, are themselves cells: we have, therefore, to study two parts: the one, the containing, or the envelope; the other, the contents.

§ 466. When the grain of pollen is ripe, its envelope is generally double, composed of an external and internal membrane[5]. The former is the first formed and is afterwards doubled by the latter. In some rare cases, we find a third intermediate membrane. In a few cases, much more uncommon still, we find only one, and then it is analogous to the internal one in its texture (*fig.* 337).

The external membrane gives to the pollen-grain its form and colour, constant in the same species of plant. It is, indeed, commonly rather hard and firm, sometimes smooth, sometimes sprinkled with little dots (*fig.* 334), or even granulations (*fig.* 335), which give it under the microscope the appearance of shagreen; sometimes covered with small eminences, which, when magnified

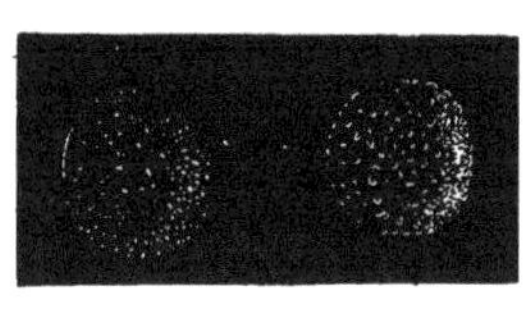

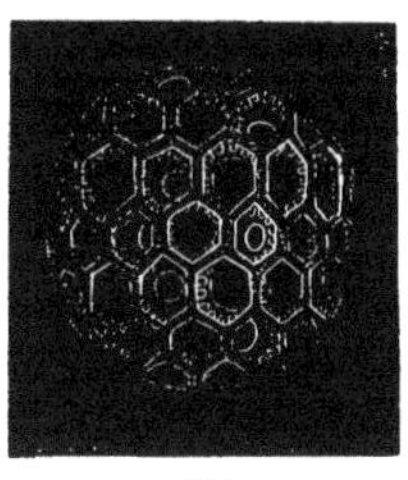

334 335 336

represent so many hairs or prickles (*fig.* 346). It sometimes happens, that these eminences, distributed with great regularity and united by an analogous substance, almost gelatinous, thus describe on the surface of the grains a net with projecting meshes (*fig.* 336). It is remarkable, that, whenever the external surface is thus covered with granulations or with other larger projections, there generally oozes out a coloured oily liquid: this gives it its colour; whilst it is commonly colourless, when the grain is perfectly smooth; and then we can perceive the inside through its trans-

336. Pollen-grain of IPOMÆA.

[5] The external membrane is called *extine*, the internal *intine*.—TRANS.

parent envelopes. In the other cases, we can only obtain this transparency by dissolving the oily covering by means of proper reactives, such as an essential oil.

§ 467. On the nature of this external envelope M. Mohl has published an opinion, which the majority of botanists do not support. He thinks, that, in rather a large number of cases it is formed by a kind of epidermis, a layer of cells in juxtaposition, which secrete in their interior the oily covering; that, in certain pollens they have walls evidently reticulate (*fig.* 345), but that they also exist in the greater part of the others, and that the granulations are nothing but very small cells, a mere sketch, as it were, united with one another by the intercellular matter spread into a membrane over the whole surface of the pollen. This substance alone would, therefore, form the external envelope, when it is single.

§ 468. As to the internal membrane, it is always identical in all the different kinds of pollen, smooth, very thin, transparent and capable of great extension. In some plants, the Gramineæ, for instance, it adheres throughout the whole of its extent to the external membrane; in others, at certain places only; in the greater part, wholly detached from it.

§ 469. Fovilla.—Within this inner envelope we find a substance, to which the name of fovilla has been given. It is formed of a thick fluid and of a crowd of small granular corpuscules, with which are frequently found small drops of oil, and more rarely granules of fecula substituted for them. The corpuscules are generally of two kinds (*fig.* 348, *f*), the greater part extremely small and spherical; some (*fig.* 349) much larger, globular or ellipsoïdal, or lengthened into short cylinders, rather tapering at their extremities. These corpuscules have attracted great attention from physiologists, who thought that they recognised in them the immediate agents of the fecundation; they have, indeed, discovered certain very remarkable motions in them. But is this power of moving really a vital faculty? Mr. R. Brown has determined that a very active tremulous motion, which agitates all these corpuscules, alternately approaching to and receding from one another, and thus susceptible of a very evident locomotion, is not a property peculiar to them, but found equally in the molecules of all bodies, not only organized, but also inorganic. This motion, called the Brownian, cannot be questioned here and appears to be a physical and general property of any substance very much divided; but, it has been thought that in the corpuscules of the fovilla some more characteristic phenomena of locomotion may be recognised, which soon stop in liquids improper to life, as alcohol, or some time after their egress from the pollen-grain, and reminding us to a certain degree of that of the infusorial animalculæ, especially, those which are larger or longer, and in which motions of contraction or of flexion (*fig.* 353) have been perceived. These delicate observations, the subject of numerous controversies, require verification, whether this phenomenon may be explained by an illusion or a purely physical cause. However this may be, whether the active principle is resident in the corpuscules or in the fluid in which they swim, or in both at the same time, the fovilla is indisputably the essential element of the pollen.

§ 470. Envelopes and external form of the Pollen.—It now remains to shew how its action takes place through the membranes which envelope it: we shall see this by examining the different forms of the pollen and the different modes of dehiscence. The pollen-grains are most frequently presented under the form of an ellipsoïd (*fig.* 339, 340), more or less tapering at its two ends (*pp*), which we call its *poles;* so that we may term the circular line (*e*) *equator*, which, equally distant from these two extremities, divides the grain into two equal parts. This line, most commonly ideal, is sometimes marked by the presence of certain peculiar points, as we shall see in a little time. When the grain is ellipsoïdal, as well as when it is spheroïdal, the surface presents a continuous curve. In a very small number of plants (Zostera marina and several other Zosteraceæ), the grain is lengthened into a tube or cylinder, a kind of hollow thread (*fig.* 337). At other times, the surface does not present this regularity, but seems to be formed by the meeting of several curved surfaces. Rather a common form is that which results from the meeting of three of these segments, and the pollen is then said to be *trigonal* (*trigone*) (*figs.* 348, 350).

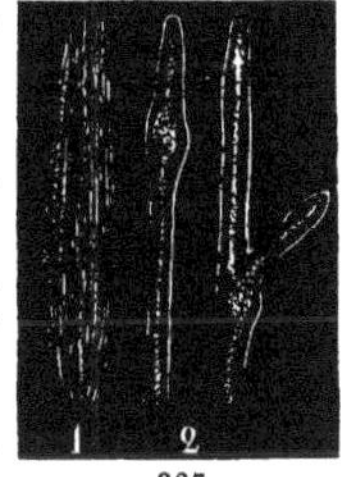

337

Lastly, it is not rare for the pollen-grains to affect the shape of a polyhedron. Then, the plane or scarcely curved faces are separated by solid angles, sometimes even projecting into tuft-like excres-

337. Pollen of the Zostera marina.—1. A mass of grains contained in an anther, and representing a skein of thread.—2. Two ends of these threads very highly magnified.

cences. These faces may all be similar to one another; in the majority of cases, however, they are not all so and, for instance, we find those, which answer to the poles (*p*), different from those which correspond to the equator (*fig.* 338 *e*).

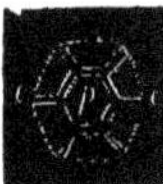

338

§ 471. We must remark, that the shape of the pollen is modified according to the degree of humidity with which it is penetrated. If we leave it exposed to the air for some time, it dries and contracts; its poles or its angles have a tendency to become more and more acute (*fig.* 347, 1). If, on the contrary, it be placed in water, it swells (*fig.* 347, 2); its angles are effaced, and it is not long in assuming the appearance of globule. Its true form must be something between these two extremes; it is that which it has in the interior of the anther, still closed, in a humid, but not liquid, medium.

§ 472. DEHISCENCE OF THE POLLEN.—The dehiscence of the pollen results from the difference in the aptitude of its two membranes to extend, when they are placed in connection with a liquid. The outer, which presents it in a less degree than the inner, will, at last, when too much pressed by it, open a passage for it. This passage takes place through openings, which are either accidental or previously marked out on the surface of the grain.

The former takes place, when its face is perfectly homogeneous throughout the whole of its extent, as it is, in truth, in a certain number of plants. Then, if the humid substance be applied to a certain part of the grain, the corresponding part of the internal membrane has a tendency to extend itself more than the rest, whilst that of the softened external membrane makes less opposition to it and, pressed from the inside outwards, at last bursts.

§ 473. But, in the majority of pollens, this is not the case, because we previously find on the surface of the external membrane some places which are weaker than others, whether it only appear to be thinner, or that there are real solutions of continuity. These thinner parts are generally found under the form of folds projecting towards the interior of the grain; these solutions of continuity under that of small circular openings, to which we have already given the name of *pores* (*pores*), but which, like that of the cells which are so termed (§ 17), are, perhaps, most commonly

338. Pollen-grain of the Succory (*Chicorée*) (CICHORIUM INTYBUS).

only small spaces, themselves extremely thin and, consequently, susceptible of being broken much more rapidly. Sometimes the same pollen-grains have only folds without pores, sometimes only pores without folds, sometimes both.

§ 474. The thin part of the membrane corresponding to the folds generally differs in its aspect from the rest of the surface, although in certain cases, it still has the same characteristics but weaker; thus, it may be partly covered by granulations or dots. But, it is generally smooth and transparent. The folds sometimes occupy the whole length of the grain, stretching from one pole to another; this is their usual direction. At other times, they are much shorter and equally distant from the two poles. Their number varies:

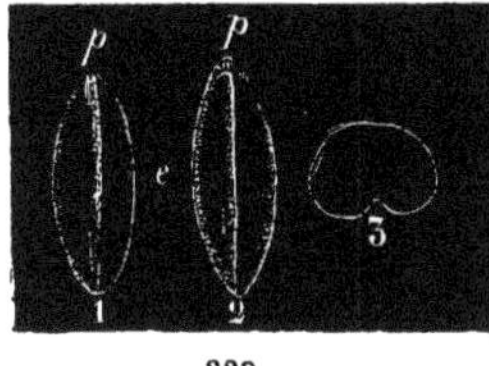

339

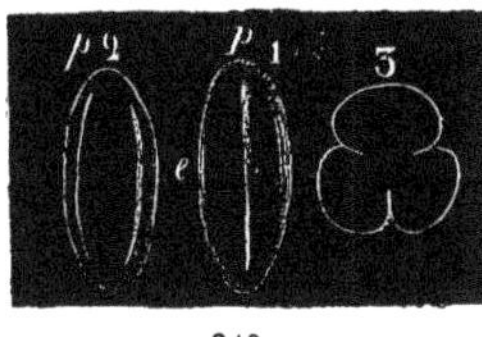

340

most commonly, there is only one in the greater part of Monocotyledons (*fig.* 339), or, on the contrary, three in several Dicotyledons (*fig.* 340). The existence of only two folds has only been observed in a very small number of examples, that of four is also very rare, that of six is much less so. We may find more, as high as to twelve or even beyond that.

These folds are almost always straight; there are only a very few cases in which they have a curved or even spiral direction, thus separating two spiral zones, as in the MIMULUS MOSCHATUS (*fig.* 341).

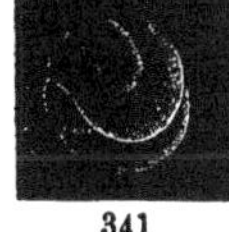

341

When the grain is swollen by nature, the fold disappears, and its membrane, by being unfolded, has almost the appearance of a spherical spindle, that is, of a segment of the surface contained between two arcs which converge towards the poles. In a small number of pollens this extension of the fold appears to be the normal state, and we observe in it thin, but not folded zones.

339. Pollen of the ALLIUM FISTULOSUM.—*p* Pole.—*e* Equator.—1. One face of the grain turned towards us.—2. The other face.—3. Transverse section, following the direction of the equator.

340. Pollen of the CONVOLVULUS TRICOLOR. The letters and numbers refer to the same parts as in the preceding figure.

341. Pollen-grain of the MIMULUS MOSCHATUS.

Then, they are frequently uninterrupted at the poles, but confounded together.

§ 475. The pores vary, in the same way as the folds, in their number and present in this respect the same combinations, that is, we often find only one, and that most commonly in Monocotyledons, as in the Gramineæ (*fig.* 342), often three, and that in Dicotyledons; sometimes two, at others four or more. When there are several, they may be arranged in a circle, and this circle is that of the equator (*fig.* 343); or regularly distributed over the whole surface, or else scattered without any very apparent order (*fig.* 344).

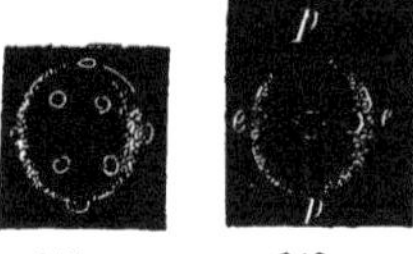

342 343

344

The pores are seen on the outside in different ways, but much better after the grain has been caused to swell by moistening it. We then find the pore under the shape of a small round body formed by a transparent membrane, either the outer one, extremely thin; or the inner one, presenting itself at the opening. The first opinion appears to be the most probable; at least, in some cases it is evident that the pore is covered by the external membrane, which has preserved its consistence and all its characteristics, except in a part circumscribed by a very fine line (*fig.* 345, *o*); the circle thus described, pushed towards the outside like a kind of cover, is at last detached (*fig.* 346, *o*): the pollens, which present this manner of dehiscence, have been termed *operculate* (*operculés*). The pore, at other times, occupies the extremity of a projection, which is larger in proportion as the pollen is moister: this is particularly observable in the trigonal grains of the Onagrariæ (*figs.* 350, 351), in which the three angles are lengthened in water to a remarkable degree.

§ 476. Lastly, the same grains, in a large number of plants all belonging to the Dicotyledons, may present at the same time both folds and pores; sometimes one corresponding to the other, or a single pore at the middle of each fold, or two pores at the two extremities of the same fold; sometimes every other or every third &c. fold only presents pores, so that we find, for instance, only three of the former for six or nine of the latter (*fig.* 347); sometimes, lastly, there are separate and alternate folds and pores.

342. Pollen-grain of the DACTYLIS GLOMERATA.

343. Pollen-grain of the Hemp (*Chanvre*) (CANNABIS SATIVA).—*e* Equator.—*p p* Poles.

344. Pollen-grain of the CORYDALIS CAPREOLATA.

In the polyhedric grains, as those of several of the Compositæ, the pores are situated either on the angles or on the middle of the faces.

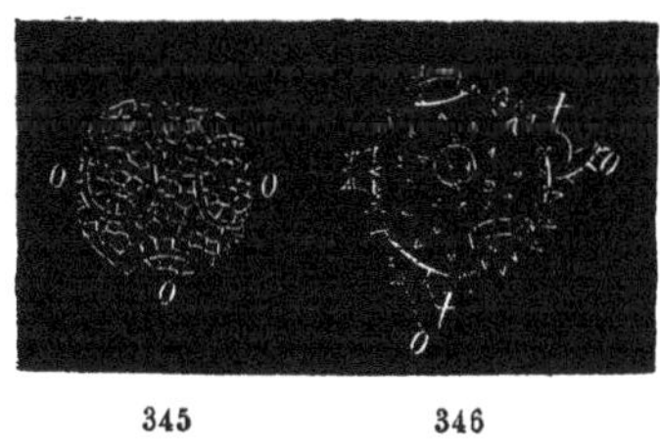

345 346

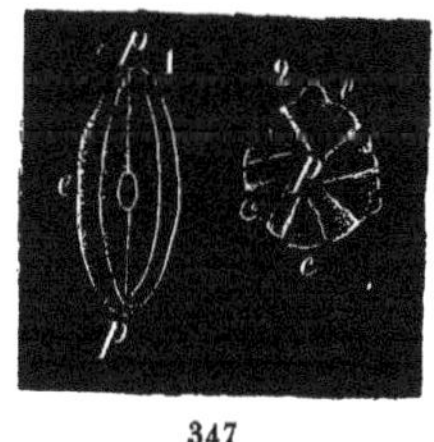

347

§ 477. If the grain of pollen is kept some time in water, it continues to swell, doubtless, from the effect of the endosme, because the water, less dense than the fovilla, will infiltrate in large quantities into the cavity which encloses the latter. The membranes being thus

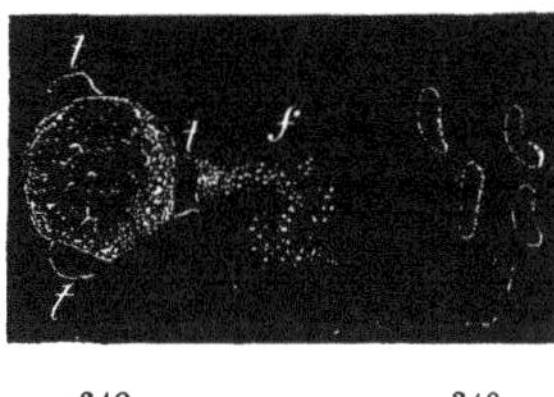

348 349

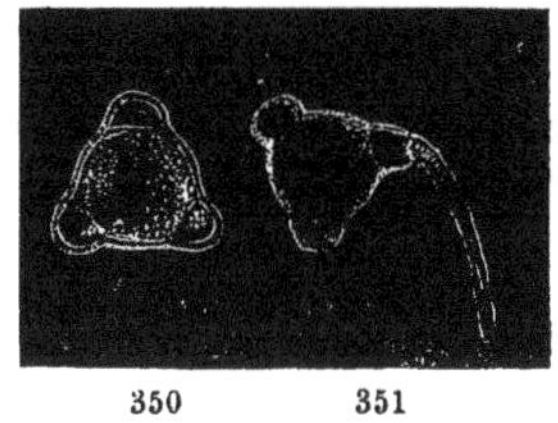

350 351

distended, if the external one is everywhere homogeneous, it breaks in some part or other; if it has folds, this portion, thinner and capable

345. Pollen-grain of a Passion-flower (PASSIFLORA KERMESINA) before dehiscence. —*o o* Opercules.

346. Pollen-grain of the Gourd (*Courge*) (CUCURBITA PEPO) at the moment of dehiscence.—*o o* Opercules already separated from the external membrane by so many projections *t* of the internal.

347. Pollen-grain of the Spiked purple Loosestrife (*Salicaire*) (LYTHRUM SALICARIA), in which we find six folds, three of which, pierced with a pore in the middle, alternate with three others without a pore.—*p p* Poles.—*e e* Equator.—1. Dry grain. —2. The same swollen with water, so that it has assumed the globular form and its folds are opened. The internal membrane is beginning to project through the pores.

348. Pollen-grain of the dwarf Almond tree (*Amandier nain*) (AMYGDALUS NANA), the internal membrane of which has begun to project through the three pores under the form of so many blisters *t* and is broken at the end of one of them, giving vent to a jet of fovilla *f*, in which may be seen granules of various sizes.

349. Large granules of fovilla of the HIBISCUS PALUSTRIS.

350. Pollen-grain of the ŒNOTHERA BIENNIS, entire.

351. The same, emitting through one of its angles opened, a prolongation of its internal membrane under the form of a tube *t*.

of greater extension, assists for a little longer time this augmentation of volume and forms a projection before it breaks itself. The internal membrane, which has this property in a much higher degree, projects through these ruptures of the external one, or rather protrudes through its pores, if they pre-existed. In the latter case, we see it come through all these pores like so many little blisters (*figs.* 346, 347, 348), and this furnishes the best means of clearly determining their number and position over the surface of the grain: this action may be aided by adding to the water some energetic acid, nitric or sulphuric acid, for instance. Thus pressed at a great number of points, the internal membrane itself is not long in yielding, bursts in a number of these points and allows the fovilla to escape under the form of a jet of a certain length (*fig.* 349). Early botanists, having always observed the dehiscence of the pollen in water, discovered this latter phenomenon, the eruption of the jet, which, as the most apparent, arrested their attention; and they thence naturally concluded, that it was in this manner that the pollen emptied itself of its fovilla in its normal condition, when it was placed on the humid surface of the stigma.

§ 478. But it is clear, that, in this last case, the grain, in contact with the liquid at a very small portion of its surface, is no longer in the same conditions, as when surrounded on all sides by water; that it swells more gradually; that the membranes thus gradually distended and only on one side may be much more lengthened without being broken. This may be easily observed, when the pollen is in contact either with the stigma itself, or with a slightly humid surface. Then, it is no longer at all its folds, at all its pores, that the internal membrane tends to burst outwards, it is only at one of them, rarely at two; but the blister which appears at first, is afterwards gradually lengthened into a kind of tube of a certain length, through the walls of which we may perceive the granules of the fovilla, which have partly followed to the outside the membrane immediately enclosing them. In some cases, even, they have been seen in these tubes moving in currents, with that motion to which we have given the name of *rotatory* (§ 282). This *pollen-tube* (*tube ou boyau pollinique*) is, as we have said, formed by the internal membrane; but, at its base it may be doubled by the external one, which it will have drawn with it some time before breaking. If there exist a third intermediate one, bearing greater analogy to the second, it follows it still farther.

§ 479. In the pollens, which have only one single membrane, we may presume that it is lengthened in this manner at some point or other of its surface, thus subjected to the action of the moisture, of which the curious pollen of the Asclepiadeæ will furnish an example, if, with several authors, we do not consider as an external membrane the cellular tissue which encloses them (§ 464); but, in the other cases in which the existence of a single membrane is incontestable, it is remarkable that the primitive form of the pollen is precisely that of a tube (*fig.* 337).

§ 480. Antheridia of Acotyledons (*Anthéridies*).—Do Acotyledonous plants present organs analogous to those which we have just described, to the anther or to the pollen? Some have denied these vegetables the organs of reproduction and have named them, consequently, *agamic* (*agames*), (*α*, *not*); others, by giving them the name of *cryptogamic* (*cryptogames*), κρυπτὸς, *hidden*) indicated this fact only, that these concealed organs have hitherto escaped their observation, but not denying on that account the absolute possibility of their existence. A little time afterwards, Hedwig, in a large number of these cryptogamic plants distinguished two kinds of organs, one of which, unknown before his time, has been compared to the male organ of phanerogamic plants. It is generally a small bladder, the form and situation of which vary according to the plant; at first perfectly close, then opening at a certain time at a point in its surface, allowing the matter which it enclosed to issue through this opening as a mass of corpuscules commonly united by a mucilaginous liquid. If these corpuscules are immediately contained in the bladder, and if this is formed by a simple membrane, it is evident, that it presents all the characteristics of a pollen-grain with its fovilla; but, in whole families the membrane is formed by a net of distinct cells, and the preceding comparison becomes false, unless, indeed, we admit along with M. Mohl that the external tegument of the pollen is an epidermis composed of several cells. The opinion that, in cryptogamic plants, the male organ exists, but reduced to a pollen-grain, has also been put forth and sustained. We have, however, now returned pretty generally to the idea that it represents an anther, imperfect it is true, and to which for this reason it has been proposed to alter the name to that of antheridium (*antheridée*). We shall better understand the reasons on which this rests by briefly describing the best known antheridia, those of the Mosses and the Liverworts.

These are bladders, sometimes immersed in a mass of cellular tissue which surrounds them on all sides (as in the MARCHANTIA and other Hepaticæ); sometimes fixed at their lower end and free throughout the rest of their surface (as in the Mosses [*fig.* 352]); sometimes collapsed at their upper extremity into a kind of neck, which gives to the whole the shape of a bottle; sometimes terminated without any lengthening in a blunt end, enclosed by a transparent membrane, by the rupture of which the tearing of the bladder takes place (*fig.* 352). The rest of the envelope is formed by a single layer of cells with simple and continuous walls (*a*). We do not find here the layer of fibrous cells, nearer the centre, which we have pointed out in true anthers. The cavity is filled with a demi-fluid substance, in which by microscopical aid we can recognise a cellular texture (*fig.* 352, 1 *f*), and, when it is fresh, we discern an active motion in the interior of the cells. It results from the rotation of a small body in the shape of a hoop, enclosed in each of these cells (*fig.* 352, 2). When this substance is freed from its

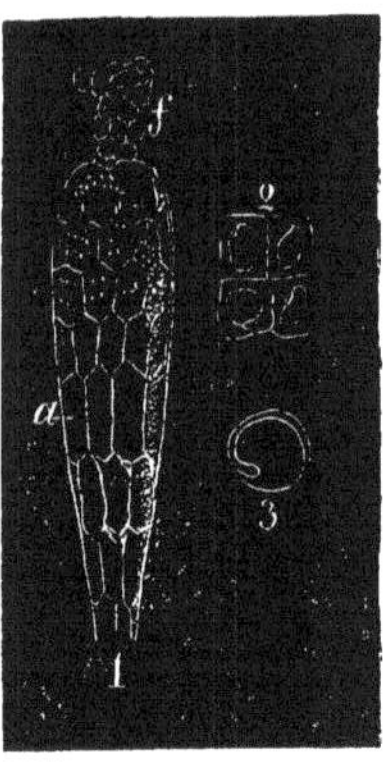

352

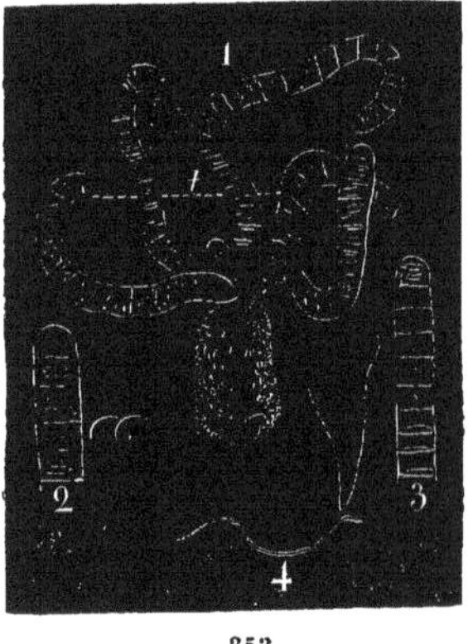

353

352. 1. The antheridium *a* of a Moss (HYPNUM TRIQUETRUM), at the time that its contained matter *f* is coming out of its open summit.—2. Four utricles of this substance each containing a circular moveable corpuscle or animalcula.—3. One of these animalculæ separated.

353. 1. Portion of the contents of an antheridium of the CHARA VULGARIS. Several articulate tubes *t*, attached to a utricle *b*.—A small mass of similar utricles, serving as bases to a much larger number of these tubes, fills the greater part of the cavity of the antheridium.—2. Extremity of one of these tubes, composed of several cells, in each of which is an animalcula. One of them is already more than half disengaged from its cell.—3. Extremity of a tube from which the animalculæ have gone, with the exception of the last cell.—4. An isolated animalculæ.

envelope and placed in water this motion gains fresh activity; the cells are separated from one another; their envelope very thin and soft, is not long in dissolving, and we may then more clearly see the circular corpuscules. They present the form of filaments thus rolled together, either in a single coil, whence results a circle, or in several crowded coils of a spire, swollen at one part and gradually tapering from this point to the other extremity which finishes the circle (*fig.* 352, 3). These filaments, when they have become free, frequently unroll themselves into a curved or undulating line, and they appear to be some of those little animalculæ, termed Infusoria, because we frequently meet with them in water, in which an organized substance has been infused. The resemblance is so complete that some naturalists have not hesitated to pronounce them to be true animalculæ. These infusoria have a kind of head corresponding to the swelling of which we have spoken, and a tail of a certain length and gradually tapering.

The antheridia of the Chara present similar ones; but, instead of being contained in the cavities of a cellular mass, they are enclosed in a heap in cells placed end to end, so as to constitute articulate tubes (*fig.* 353 *t*).

May each of these cells be compared to a pollen-grain, and each of these animalculæ to its fovilla? There still reigns much obscurity over the true nature of these parts, the discovery of which is quite recent: it presents, however, an analogy too striking to animal organization for us to pass it over in silence, in spite of the uncertainty in which we are plunged with regard to the part they play in vegetation. If these are the anthers of cryptogamic plants, it is evident, that the contents as well as container are perfectly different from what we described in phanerogamic plants.

PISTIL.

§ 481. We have already spoken several times of the pistil which occupies the centre of the flower, which is surrounded with the envelopes and stamens in the hermaphrodite and complete flower (§ 365), with the envelopes alone in the female flower (§ 391), and which is the only part when it is naked (§ 393). We have seen that this pistil is composed of modified leaves or carpels, the number of which varies according to the plant and may be reduced to unity; that these carpels sometimes remain distinct from one another (§ 368, 371), and sometimes are united into a single body (§ 366.

377). It now remains for us to examine the structure and the different modifications of this simple or compound body, which we have considered only with regard to its relative position. To render our explanation clearer, we shall first examine an isolated carpel, and shall then consider the case, in which several of these carpels are united in the same flower, and the different relations which they may then present with the other parts of this flower.

§ 482. Let us commence by following one of these carpels throughout its developement. This is easily done in a common plant that grows in our rivers, the Flowering Rush (*Jonc fleuri*) (BUTOMUS UMBELLATUS). If we open a very young bud of this plant (*fig.* 355), we shall find its centre occupied by a whorl of six small bodies (*c*), or rather by two whorls of three; these are small greenish substances, rather concave on the inside, and which do not differ from a true leaf, when observed in the first stage of its developement. Each of these small leaves becomes more and more concave by the gradual approaching of its edges, which then touch (*fig.* 356), and lastly unite. The leaf then forms the walls of a perfectly close cavity; and, if we carefully observe the internal surface of this cavity, corresponding to the upper face of the leaf, we shall see it quite

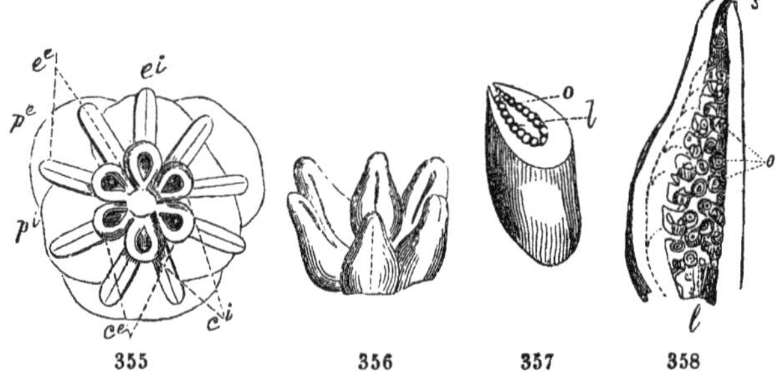

355. Very young flower-bud of the BUTOMUS UMBELLATUS, opened so as to shew the different parts of the flower. 1st, its perianth, with six folioles, three outer ones p^e, three inner ones p^i; 2nd, its nine stamens, three e^i opposite to the internal perianth, six e^e opposite to the external perianth; 3rd, its six carpels, three c^e opposite to the external perianth, three c^i to the internal perianth situated on an inner row. These carpels are still in the state of small leaves, concave on the inside.

356. These same carpels, a little further advanced, when the two edges of the small leaf which forms them touch one another, and the cavity, thus formed by the carpellary leaf, only communicates with the outside by a narrow slit.

357. The lower part of one of these carpels, cut across, so as to shew its loculus *l* and its ovules *o*.

358. Carpel, much more advanced, when it is completely closed, cut vertically so as to shew its loculus *l* and its ovules *o*.—*s* Papillæ of the stigma.

covered with small ovoïd excrescences which are attached to it (*figs.* 357, 358, *o*). This body, hollow in the inside, is termed *ovary* (*ovaire*) (OVARIUM), anciently *germ* (GERMEN); its cavity LOCULUS (*loge*) (*figs.* 357, 358 *l*); those small bodies adhering to its wall, *ovules* (*ovules*) (OVULA) (*figs.* 357, 358, *o*): these will afterwards become the seeds.

§ 483. The Cherry-tree affords in another manner the proof of the passage of the leaf to the carpel. If we take, for instance, a double flower of the Cherry-tree (*fig.* 359), we shall find its centre occupied by small leaves perfectly shaped and hardly folded, widened at the lower part into a green limb, (*l*), contracted at the upper part into an elongation, which seems to be a continuation of the median nerve (*s*). But, in a single flower, instead of these two central leaves, we shall find a single body (*fig.* 361, *o*), swollen

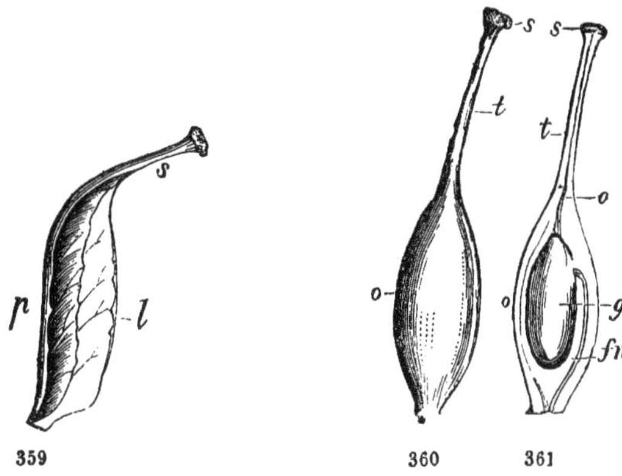

and hollow at the bottom, with a much smaller body (*g*) enclosed in its cavity, to the wall of which it is attached; we shall here recognise an ovary with a single ovule contained in its loculus. Above this cavity the ovary collapses into a cylindrical continuation (*t*), which is terminated by a wider part at its upper end (*s*). This

359. Carpels in the state of leaves, such as are found in the double flower of the Cherry-tree.—*l* Limb.—*s* Elongation of the median nerve *n* which becomes free at he top represents the style, and is terminated by a swelling which represents the stigma.

360. Carpel of the Cherry-tree, such as are found in the single flower.—*o* Ovary. —*t* Style.—*s* Stigma.

361. The same, cut vertically, so as to shew in its ovary *o* a central cavity filled by the ovule *g*, suspended from its wall at a point where a fascicle *fn* of nutritive vessels terminates; and in its style *t*, the small canal *c* which traverses it from the stigma *s* to the cavity of the ovary.

narrow elongation is termed *style* (STYLUS), and the wide terminal part *stigma* (*stigmate*) (STIGMA). We again find, therefore, the leaf which we have seen at the centre of the double flower, with this difference, that its limb is thickened and, by the meeting and union of its edges, has formed a closed cavity or loculus, in which an ovule has been developed.

§ 484. A complete carpel is composed of these three parts; the ovary or closed cavity, which contains one or several ovules; the style, the narrow, solid, upper elongation; the stigma, that, which terminates the style and is mostly distinguished from it by a swelling, always by a difference of tissue. Sometimes this tissue, instead of being on a style, which separates it more or less from the ovary, is placed immediately or almost immediately on the external surface of this ovary; the style is then wanting, or is so short, that it is considered to be so, and the *stigma* (*stigmate*) is said to be *sessile* (*figs.* 358, 397).

§ 485. What is the anatomical structure of these different parts? The ovary, like the limb of the leaf which it represents, is composed of a parenchyma traversed by fibro-vascular fascicles and covered with an epidermis. The parenchyma, sometimes very thin, is frequently rather thick, more fleshy and richer in juices than that of the leaf. The fascicles, formed of unrollable tracheæ, take their direction from bottom upwards and converge at the origin of the style; they are sometimes few, sometimes numerous, sometimes simple, sometimes ramified, and form by their ramifications a more or less complicated net. The cellular tissue, in the midst of which they lie, without presenting those layers of a different structure, which we have pointed out in the substance of several leaves (§ 127), is nevertheless modified a little from the exterior to the interior: this modification will be more clearly seen in proportion to the advancement of the ovary in its developement. The external epidermis, which corresponds to that of the under side of the leaf, is like it covered with stomata. The internal epidermis, which covers the cavity of the loculus, preserved from the action of light, is generally of a paler or whitish hue. Stomata are never found upon it.

§ 486. The ovary does not always represent the limb of the leaf, but sometimes also, and, according to some authors, most generally, its vaginal part. The style would then correspond to the petiole and the limb would be suppressed.

§ 487. The style, in its structure, appears rather to represent the upper portion of the leaf, collapsed and rolled up, rather than the continuation of the median nerve alone. It is formed, indeed, of a parenchymatous cylinder with small vascular fascicles, not united in its centre, but, on the contrary, dispersed through the whole of its circumference as a kind of sheath; they proceed directly from bottom upwards and terminate at a certain distance from the summit. An epidermis, continuous with that of the ovary, covers the whole system.

The centre of the cylinder formed by the style, which most commonly at first sight appears to be solid, when viewed more attentively and with a microscope of sufficient power, is occupied by a very narrow canal (*fig.* 361, *c*), terminated at one end by the internal wall of the ovary; at the other by the stigma. This canal is evidently empty in certain cases (*fig.* 262); in others it is obstructed by cellular tissue, but frequently loose and dislocated, as it were, (*fig.* 263, *p p*), and, therefore, thus leaving between the utricles which compose it several empty spaces: in every case, even when more compact, it differs notably from the proper tissue of the style. Its walls are generally rough with small projecting cells (*fig.* 362, *p*), or papillæ. At a certain time, we find, moreover, other cells which are lengthened in the direction of the canal, soft and humid; apparently, numerous filaments (*fig.* 363 *f f*)

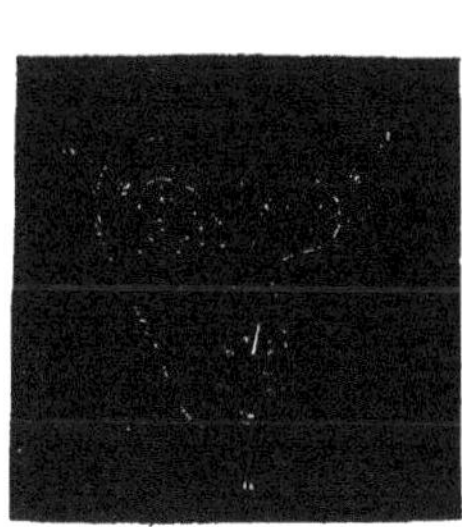

362 363

which cover it and at the same time partly fill it. The tissue,

362. Horizontal section of the style of the FRITILLARIA IMPERIALIS, composed of three united into one.—*v v* Three vascular fascicles, each corresponding to one of the three styles.—*p p* Papillæ projecting into the cavity of the canal.

363. The structure of the canal which occupies the centre of the style of a CAMPANULA.—*c c* Cellular tissue, which forms its walls, traversed by fascicles of tracheæ *v v*.—*p p* Utricles of another form, dislocated, as it were, which cover this wall and with other lengthened and filamentose cells *f f* partly obstruct the canal.

which thus covers or obstructs the canal of the style, has been termed *conducting tissue* (*tissu conducteur*) and we shall soon see the origin and applicability of this term.

§ 488. It seems to form the stigma, which is, as it were, a continuation and expansion of it; sometimes terminal, when the canal of the style is opened in a vase-like form at its summit only (*fig.* 360 *s*, 366 *t*); sometimes lateral, when this same canal, split to a certain length, is thus opened, either on one side alone (*fig.* 364), or on two sides (*fig.* 365, *s*). There is no demarcation between the conducting tissue and that of the stigma; the one passes insensibly into the other. The stigma is therefore composed of a cellular tissue of different degrees of compactness, of which the outermost utricles are most frequently lengthened into papillæ (*fig.* 366, 2), or even into true hairs (*figs.* 367, 3; 392, *s*). At other times, it is smoother and more compact on the exterior; but, in every case, at the time of fecundation, all its cells, as well as those of the conducting tissue, are filled with a liquid juice, commonly more or less viscid, oozing out of the surface of the stigma, which is, consequently, quite humid and sticky.

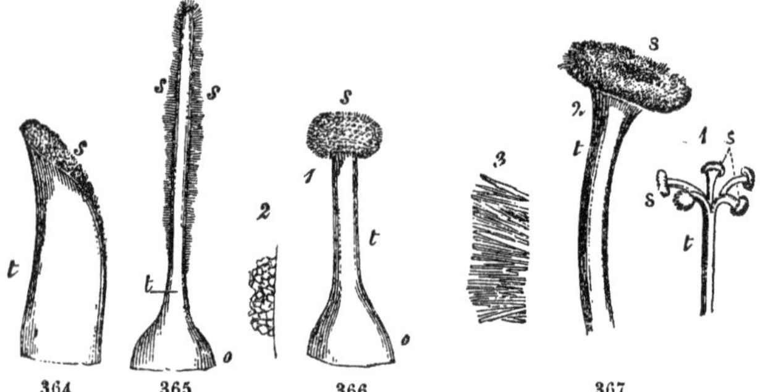

§ 489. When the anther opening elastically, ejects the pollen which fills it, these pollen-grains are naturally thrown on the stigma,

364. Unilateral stigma *s* of the ASIMINA TRILOBA—*t* Style.

365. Bilateral stigma *s* of the PLANTAGO SAXATILIS.—*o* Ovary.—*t* Style.

366. 1. Stigma *s* of the DAPHNE LAUREOLA, terminating its style *t*.—*o* Summit of the ovary.—2. A small portion of the surface of the stigma, very highly magnified in order to shew its papillose nature.

367. 1. Summit of the style *t* of the HIBISCUS PALUSTRIS, divided into five branches, each of which is terminated by a stigma *s*.—2. One of its branches more highly magnified.—3. Portion of the surface of the stigma very highly magnified to shew its parts lengthened like hairs.

either on account of the immediate proximity of these two organs in the flowers, or of the pollen being carried to a more distant stigma by the wind or by insects conveying it along with them from one part of one flower to that of another. The pollen, when once it has touched the stigma, is retained by its viscous fluid; and there commences an action which we may easily foresee, since we have seen what takes place, when the grain is in contact with a humid surface (§ 478). It swells slowly, as it absorbs this humidity; its internal membrane extends, projects through the external one, at a point of the surface in contact, is lengthened into a tube (*fig.* 368, *t p*), which enters between the inequalities of the surface of the stigma (*t c*) and between the interstices which it presents. Thus, it gradually passes through the substance of the stigma and enters the canal of the style in the midst of the conducting tissue, which continues to afford it a passage through parts greatly impregnated with liquids. It thus proceeds, at the same time continually growing longer, to the lower extremity of the canal, and arrives at the cavity of the ovary. Now, on the walls of the ovary, the conducting tissue is continued to the ovules, which at this period are so many small bags open at that end corresponding to this tissue. The pollen-tube at last traverses this opening, into which it is introduced, and an immediate connection is thus established between the pollen and the ovule, between the essential produce of the stamen and that of the pistil. We shall leave it here for the present and resume our observations on what passes afterwards under the head of OVULE.

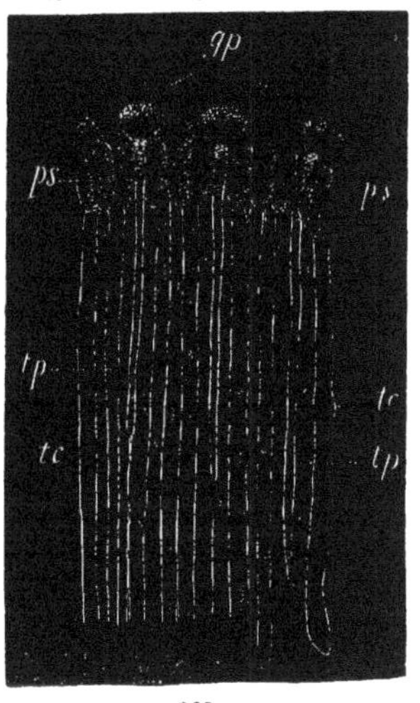

368

§ 490. We may now clearly conceive the structure and the functions of the carpel, 1st, One portion, that which corresponds to

368. Portion of the stigma of the ANTIRRHINUM MAJUS, at the period of fecundation.—*p s* Superficial cells forming the papillæ.—*t c* Long, cylindrical cells forming the conducting tissue.—*g p* Pollen-grains attached to the surface.—*t p* Tubes thrown out by each of the pollen-grains burying themselves in the interstices of this conducting tissue.

a leaf, is formed by the ovary and the style, and constitutes the nutritive system: it is joined, indeed, to the vegetable and communicates with the rest of the flower and plant by its vessels, which extract the juices necessary for its subsistence and increase: 2nd, Another part is formed by the stigma and the conducting tissue, and constitutes the fecundating system. It conducts a body from the outside to the bottom of the inside of the ovary. There is now no need of explaining why this name of *conducting tissue* has been proposed and adopted.

§ 491. In practice, the rigorous distinction of these parts is very difficult. It would be less so, if we could always call the microscope to our assistance. But, in the greater part of botanical descriptions, for the verification of which we are contented with a lens, and for which the study of the last tissues would demand too much time and, besides, would present great difficulties, since we frequently have only dried and dead parts at our disposition, we might experience much hesitation in determining the really stigmatical part in the style; and we generally term that, stigma which, from its situation, its aspect, its form, is easily distinguished from the rest of the style. In nicer observations, we take advantage of the presence of the pollen, the grains of which we frequently find after fecundation attached to points, which must belong to the stigma, although this diagnostic is by no means infallible. If we wish to specify in our description, the rigorous limitation of these organs, we must look at the elongation which gives rise to the embarrassment, whether it be pierced with a canal, in which case it is a style; or if it be solid and entirely cellular, in which case it is a stigma.

§ 492. After having shewn the organization and functions of the carpel generally considered, let us examine the pistil, composed of several carpels united in the same flower.

These carpels may not be all exactly similar to one another. Thus, in those which, three in number, form the pistil of certain Malpighiaceæ (ACRIDOCARPUS, HIPTAGE), one or two only are furnished with a long style which is wanting in the rest, or they may even differ from one another in their shape (GAUDICHAUDIA CONGESTIFLORA). These dissimilarities between the carpels of the same pistil are extremely rare; the most frequent is that which results from the more or less complete abortion of some of these carpels; but the most common case is that, in which all the carpels, at least in youth, are perfectly similar to one another, and we shall suppose them to be so in the following explanations.

§ 493. They grow sometimes at the same height, on the same plane and are then arranged in a whorl (*figs.* 374, 389); sometimes at different heights, and are then arranged in a spire. In this latter case, the cone or receptacle, which is quite covered with them, is lengthened into a cylindrical, as in the MAGNOLIA and the Tulip-tree (§ 368, *fig.* 224), or conical, as in the Raspberry (*Framboisier*), or expanded axis, as in the Strawberry (*Fraisier*); or else its widened surface, instead of remaining flat, is extended into a cup, or bent into an urn, as in the Rose (*fig.* 369). Sometimes, although the part of the axis which bears the carpels is rather largely developed in length, they only occupy its summit, thus crowded together and whorled on a small surface. It is one of those cases that we mentioned (§ 382, *b*), in which we observe between the different whorls of the flower, internodes of certain lengths. That, which thus appears below the pistil (*figs.* 374, 375, *g*), has received different names according to its different appearances, its different degrees of height or thickness, which vary much according to the plant. It is now pretty generally agreed to give it that of *gynophore* (GYNOPHORUM). Linnæus then gave to the pistil the epithet of *stipitate* (*stipité*), calling every similar elongation on which an organ is thus raised *stipes* or support; and if this term, taken in a general sense, may by its very generality give rise to some uncertainty, there is no inconvenience when used in descriptions, when we always know to what organ it is applied.

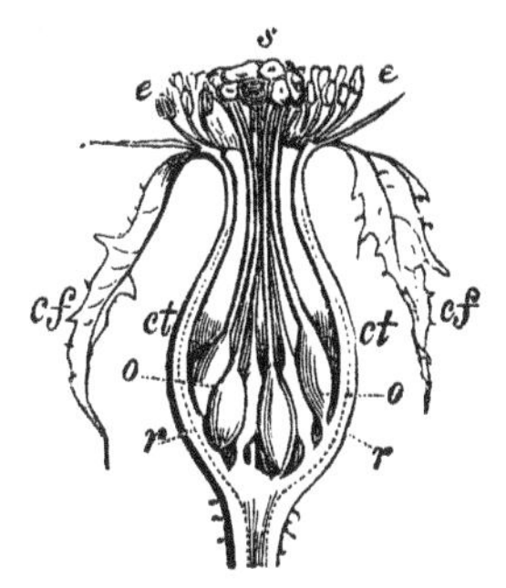

369

§ 494. We must not pass over in silence a remarkable modification, in which the torus bears not only the ovary but also the style which seems independent of it. To understand it clearly we must return for a moment to the relative position of the style and

369. Vertical section of the flower of the Rose (*Rosier*), shewing the position of the carpels at the bottom of the calyx on the concave surface of the torus *r*.—*c t* Tube of the calyx.—*cf* Its limb divided into folioles.—*e* Stamens.—*s* Ovaries each surmounted by its style which projects beyond the calycinal tube and is terminated by a vase-shaped stigma.

ovary, and look at the various places, which they may occupy with regard to one another. We have hitherto supposed, as is most commonly the case, the style to be *apiculate* (*apicilaire*), that is to say continuing the ovary at its summit (*fig.* 360): the leaf, which constitutes the carpel, has preserved throughout the whole of its length the same ascending direction; but we also suppose its limb to be bent back in a manner analogous to that which the reclinate vernation of certain leaves presents (§ 174, *fig.* 164, 1): then, the extremity which corresponds to the origin of the style will be found to be lower or higher on the side, the style will be *lateral* (*latéral*) (*fig.* 375). It will be almost at the bottom (*fig.* 370) or quite so (*fig.* 371), and the style will be *basilar* (*basilaire*), if the inflexion is such that the upper half of the limb is thus bent back on the lower one. The ovary presents to our notice examples of all these degrees of inflexion, all the intermediate ones between the apiculate and the basilar positions of the style. The latter is observed in the pistil of the Strawberry (*fig.* 370) and of several other Rosaceæ (*fig.* 371), a family, which would also furnish us with good examples of its lateral position.

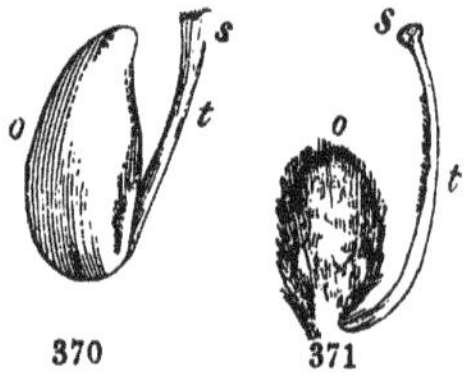

370 371

§ 495. It is clear that the basilar style is close to the torus at the point of departure; it touches it, if the ovary is sessile, and, if the base of the ovary is buried a little in the torus, it attracts the origin of the style, which then seems to spring rather from the torus than from the surface of the ovary. Such is the remarkable modification which we wished to explain and which has received the name of *gynobase* (GYNOBASIUM): the ovary is then said to be *gynobasic* (*gynobasique*). Generally, the styles of several gynobasic whorled ovaries are united together, and seem to form only one, a kind of central column, around which the ovaries without any other apparent style are arranged in a circle. This is observable in the Ochnaceæ, in all the Labiatæ (*fig.* 372), in the greater part of the Borragineæ. In the last the style is frequently rather lateral than basilar; but the ovary is placed on its anterior face and on an oblique plane which the torus presents to it, so that the origin

370. A carpel of the Strawberry.—*o* Ovary.—*t* Style.—*s* Stigma.
371. A carpel of the CHRYSOBALANUS ICACO. The letters refer to the same parts as in the previous figure.

of the style is planted in the ovary, although, on the other hand, it is higher than some parts of it (*fig.* 373).

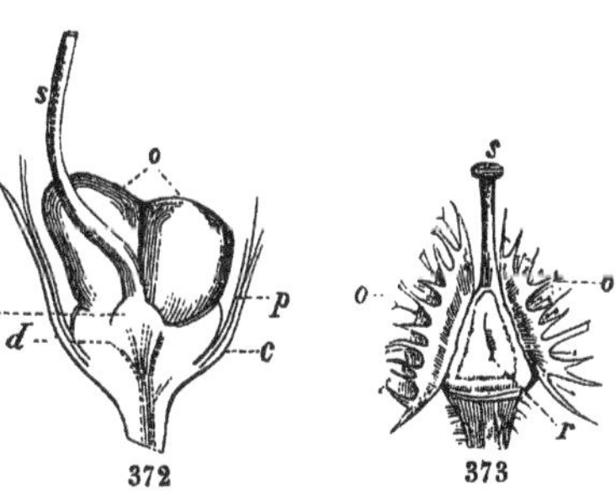

§ 496. We have hitherto considered the carpels as free, that is, as independent of one another. We know, however, that they are not always so, and that more frequently than any other of the floral parts, they are united together (§ 378), either partly, or altogether. This union may take place from top to bottom. Thus, we sometimes see several carpels united by their stigmas only, as in the Apocyneæ and the Asclepiadeæ, in the XANTHOXYLON FRAXINEUM[h] (*fig.* 374), or by the top of their styles (*fig.* 375), or by the whole of them. We shall in speaking of the gynobase, point out several styles closely united, although corresponding to distinct ovaries.

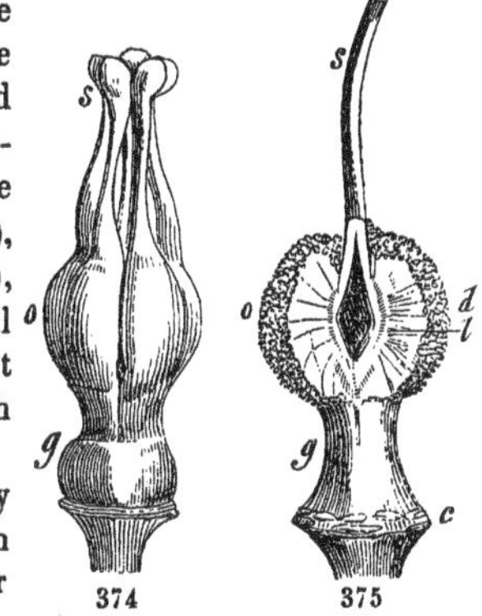

§ 497. But, much more commonly the union takes place from bottom upwards, the ovaries joining sooner

372. The pistil of the LAMIUM ALBUM, a part of the flower having been taken from it by a vertical section. Two of the four ovaries also have been taken off so as to shew the insertion of the style *s* on the torus *r*.—*o* The two remaining ovaries.—*d* Glandular disk placed beneath the pistil.—*c* Portion of the Calyx.—*p* The corolla.

373. Pistil of one of the Borragineæ (ERITRICHIUM JACQUEMONTIANA), from which the ovary opposite the spectator has been taken off, so as to shew how the ovaries *o o* are obliquely inserted on a pyramidal torus *r*, whence the style *s* with a vase-shaped stigma grows.

374. Pistil of the XANTHOXYLON FRAXINEUM, consisting of five distinct carpels elevated on a gynophore *g*. Each of the ovaries *o* leaves a terminal style expanded at its extremity into a stigma *s*, and the five stigmas are united for a long way by their sides.

375. A portion of the pistil of the DICTAMNUS FRAXINELLA.—Two of the five carpels have been taken away so as to shew how the five styles *s*, growing from the inner edge of these carpels and at first distinct, soon unite into one.—*o* Ovaries, those in the front shew their dorsal *d* and one of their lateral *l* faces. At the base of the gynophore *g* are the cicatrices *c* marking the insertions of the calyx, the petals and the stamens.

[h] The bark and capsules of this tree have a hot, acrid taste, and are used for the toothache: the name *Toothache tree* has, consequently, been given to it.—TRANS.

than the styles, the styles sooner than the stigmas. The ovaries may, therefore, cohere by their lower part only and remain distinct at their summit, as in the Rue; the description indicates this, *several ovaries joined at the base only* (OVARIA PLURA BASI TANTÙM COALITA), otherwise called an ovary with several lobes. When several ovaries are joined into a single body, this body then takes the name of ovary.

Formerly it was considered as a single organ differently divided in the inside, and then the simple or single ovary (that which results either from the existence of a single carpel or the union of several) was put in opposition to the compound or multiple ovary, that is, to the case of several free carpels in the same flower. Now, we generally continue to use the same terms, although we attach a different idea to them, considering that the term, *simple ovary*, ought, in reality, to belong to a *single free carpel;* the *compound ovary*, that which is formed by the *union of several carpels into a single body*. We must not lose sight of this usage in books on botany written at various periods.

It would be easy to prove by numerous examples, that this union of several carpels or leaves modified to form apparently one single ovary, which hitherto we have only recognised theoretically, is verified by practical observation and agrees with nature. We shall content ourselves here with mentioning a few instances, with which the plants common to our gardens will furnish us. The Lark's-Spur (*Pied d'Alouette*) (DELPHINIUM AJACIS) presents a single carpel, the ovary of which, with thin green walls, manifestly represents a folded leaf. Other species of the same genus (the DELPHINIUM JUNCEUM, for instance,) presents three similar ones entirely separated in each flower, some even have as many as five. Five analogous carpels form the pistil in a neighbouring genus, the Columbine. In a third genus of the same family, the NIGELLA, we also observe a whorl of five carpels; but here they begin to be united to one another only by the base in certain species (as in the NIGELLA ORIENTALIS); in others, much higher, and in others again, as far as the summit. We find in the flower of the NIGELLA DAMASCENA the ovaries thus completely fused, as it were, into an ovoïd body, which is surmounted by the five styles remaining quite distinct. We cannot admit that the whole of five ovaries of the Columbine or of the NIGELLA ORIENTALIS is a single organ; now the transition of these five separate ovaries to the single ovary of the NIGELLA DAMASCENA is too evident to allow us to hesitate in recognising the same composition in the latter, the presence of the five organs, the union of which we have thus been able to trace throughout all the different degrees.

Each of these isolated carpels presents an external or dorsal face and two lateral faces converging towards one another, and united at an angle of the side which is turned towards the centre of the flower. At these angles and lateral faces the carpels are united together to form an ovary more or less simple in appearance. Hence, if we cut this across we shall find it divided into five cavities separated by the lateral sides, which, united in pairs, thus form so many internal partitions, the plane of which is necessarily parallel to the axis of the flower, and which alternate with the styles, since they answer to the sides of the carpellary leaf, whilst the style corresponds to its middle. Each of these cavities is the loculus of the corresponding carpel and bears the same name LOCULUS (*loge*)*:* hence the epithet *multilocular* (*multiloculaire*) (MULTILOCULARIS) which is given to such an ovary; of *bi-*, *tri-*, *quadri-quinque-locular*, according as the number of the loculi is 2, 3, 4, 5, &c. The number of the *dissepiments* (*cloisons*) (DISSEPIMENTA) or SEPTA is equal to that of the loculi, and they are formed of true laminæ more or less closely united. The number of the styles, when they remain distinct, is also the same and will, consequently, indicate on the outside that of the loculi which we shall find in the inside.

§ 498. There is not, therefore, much difficulty in determining the number of the carpels which unite in the formation of an ovary, either by means of the styles, when they remain simple and distinct; or by means of the septa, when they preserve their integrity. But one of these resources may be wanting. Thus, in the majority of the Caryophylleæ, in which the septa disappear very soon, we are nevertheless informed, by the presence of several styles, that the ovary is really composed of several carpellary leaves, of two, for instance, in the Pink, of three in the Chickweed (*Mouron des oiseaux*) (ALSINE MEDIA), of five in the Corn-Cockle (*Nielle*) or the CERASTIUM (*fig.* 383, *s*). In several cases, on the contrary, the styles cease to indicate the number of the loculi, because they are united into one, or on account of their ramification they seem to represent a much greater number; then we are obliged to cut the ovary, and the number of the dissepiments or of the loculi determines that of the carpels.

But how shall we determine it, if both these aids are wanting at the same time? This, in a very great number of cases, is done from the position of the ovules. This is, consequently, the best place to examine their distribution with respect to the carpels.

§ 499. The ovules, after their connection with the pollen-tube has been established, or in other terms after their fecundation, are largely developed and at the same time are transformed into seeds. It is necessary, therefore, on the one hand, that the conducting tissue should direct towards them this fecundating principle; on the other hand, that they receive the nutritive principles necessary for their ulterior growth. They must draw this nutriment from the juices which come to them and are elaborated by the rest of the plant, and principally by the parts situated under them. Fibro-vascular fascicles, which have traversed these parts, are, consequently, distributed in the carpels and convey the juices through a branch peculiar to each of the ovules, which are thus united to the general system. To these fascicles coming from bottom upwards is joined a series of conducting tissue directed from the top downwards. This union of the two tissues determines, on some point or another of the walls of the loculus, a projection more or less marked, to which are attached the ovules they enclose and which has been termed PLACENTA. Some authors, reserving this name for the projection which corresponds to the attachment of a single ovule, give that of PLACENTARIUM (*placentaire*) to the body formed by the union of several placentas bearing several ovules. From this word comes also that of *placentation* (PLACENTATIO), by which is designated the distribution of the ovules and, consequently, of the placentas in a simple or compound ovary.

We have seen in the carpel of the BUTOMUS (§ 482, *fig.* 358) the numerous ovules and, consequently, the placentas covering the whole of the wall of the loculus. But this diffused state is rare, and most commonly the ovules are grouped on these walls in longitudinal and rectilinear series, and the nourishing fascicles follow a line, generally simple, in each carpel. We may readily conclude from this that in the absence of dissepiments or of loculi, the number of these series, of these lines, which the transverse section of the ovary lays open to our view, will indicate to the observer the real number of carpels that concur in composing the ovary, which is single in appearance.

§ 500. In the majority of cases, the line of the placentas follows the edges of the carpellary leaf, and, consequently, when this leaf is completely bent back so that its edges touch and are united, thus enclosing the carpel or the loculus, and forming by this union an angle which corresponds to the axis of the flower, this angle will be occupied by the placentas: and the *placentation* is qualified by the epithet of *axilary*. If the ovary be multilocular, this angle will be found in each loculus at the internal union of two neighbouring dissepiments (*figs.* 376, 379) which may even, when once arrived at the axis, be more or less bent from the inside to the outside towards the centre of the loculus (*fig.* 377).

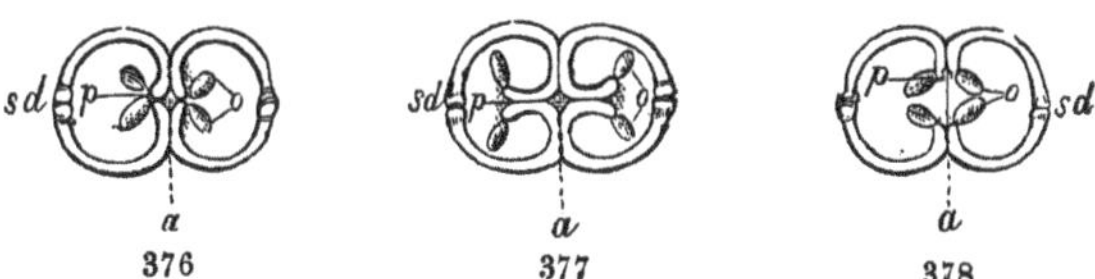

§ 501. But, suppose that the edges of the bent carpellary leaves do not reach the axis and, therefore, form in the interior of the ovary only incomplete dissepiments (*figs.* 378, 380), or even that they are not bent at all, and are united not by a lateral face, but only by their edges (*fig.* 381, 2), and that, of course, there will be no dissepiment: the placentary lines, which follow these edges, will be placed

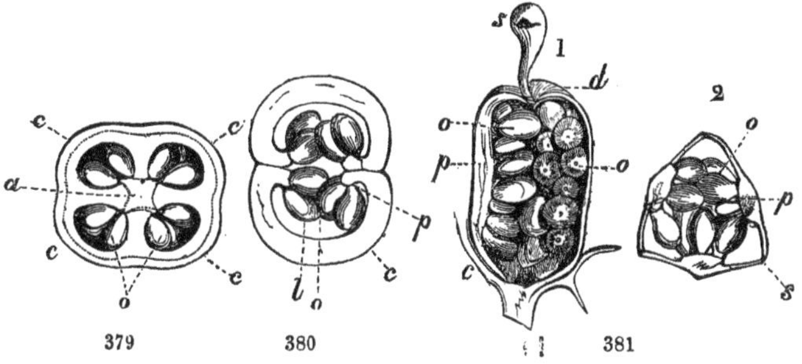

376, 377, and 378. Horizontal sections of ovaries, composed of two carpellary leaves, the folded edges of which meet one another at the axis *a*, in 376; bend within the loculus after having met one another at the axis, in 377; do not get so far as the axis, in 378.

379. Horizontal section of the ovary of the FUCHSIA COCCINEA.—*c c c c* Wall of the ovary or union of four carpellary leaves which constitute it.—*a* Quadrangular axis united to the dissepiments and joining one another.—*o* Ovules attached to the internal edge of the dissepiments.

380. Horizontal section of the ovary of the Centaury (*Centaurée*) (ERYTHRÆA CENTAURIUM).—*c* Wall of the ovary or carpellary leaf.—*p* Its edge, which forms the placenta and the ovules *o*—*l* Cavity or loculus.

381, 1. Vertical section of the pistil of the Pansy (*Pensée*) (VIOLA TRICOLOR) shewing how the ovules *o* are attached to the walls.—We see two rows, one in front, the other at the side, and the latter corresponds to a thickened line on the wall or to the placenta *p*.—*c* Calyx.—*d* Ovary.—*s* Stigma surmounting a short style.—2. Horizontal section of the same.—*p* Placenta.—*o* Ovules.—*s* Suture.

at a greater or less distance from the axis, and will be seen along the incomplete dissepiments or septa in the first case (*fig.* 380), on the very walls of the loculus in the second (*fig.* 381, 2): this is what is termed *parietal placentation* (*placentation pariétale*).

In this case, each placentary line corresponds to the edges of two different carpels, whilst in the preceding case, it corresponded to the two edges of the same carpel. The axilary placentas alternate, therefore, with respect to the parietal placentas; and this theoretic truth is frequently found to be verified in practice. In ovaries with axilary placentation (as in those of several Meliaceæ) the dissepiments are sometimes drawn back to a certain distance from the axis, and each series of ovules, which in well formed ovaries occupies the inner angle of the loculus and alternates with the dissepiments, is separated into two longitudinal series, each of which is united to a similar series of the neighbouring loculus, in order to form with it a placentary line on the free edge of the dissepiment or septum, which has become incomplete. In all cases, it is clear, that every placentary line is essentially a binary association.

§ 502. Let us suppose in the third place, that with the axilary placentas, as in the first case, that part of the septa, situated between them and the walls of the ovary, is arrested at a very early period of their developement, does not follow that of the other parts, and is not long before it breaks and disappears; the placentas with their ovules will then form a mass without any apparent lateral connection with the walls (*fig.* 382, 383); the different loculi, which are no longer separated by dissepiments, will be blended into a single cavity, in the midst of which will be elevated the placentary body (*p*) loaded with its ovules (*g*); this has been termed *central placentation* (*placentation centrale*).

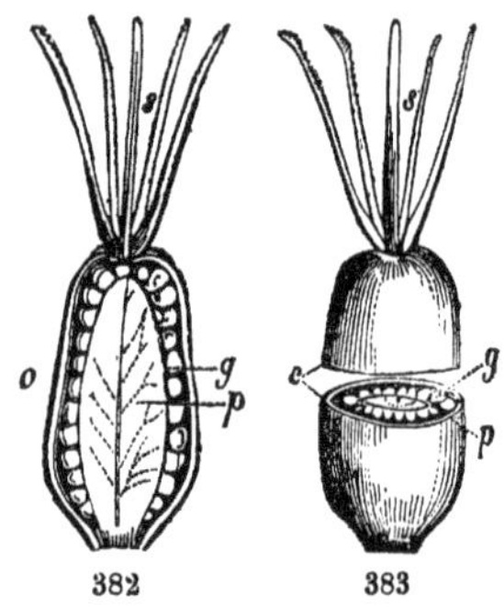

We have, therefore, three principal modes of placentation: the axilary, the central and the parietal; the last two differing

382. Vertical section of the CERASTIUM HIRSUTUM.—*o* Ovary.—*p* Placentarium—*g* Ovules.—*s* Styles.

383. Horizontal section of the same.—The pieces have been separated so as to shew the interior of the loculus with its central placenta *p* covered with ovules *g*.

from the first, the one by the destruction of the dissepiments or septa, the other by their incomplete formation.

§ 503. Yet these two last methods do not invariably shew the origin, which we have ascribed to them, and according to which the placentas would always follow the two edges of the carpellary leaf. In some examples, not common it is true, they seem to correspond to the median nerve and not to the edges; and in the BUTOMUS (*fig.* 358), we have already seen them scattered over the whole surface of the loculus. Here are two modifications of the parietal placentation, to which the rules previously enunciated are not applicable.

We may also conceive another origin of the central placentation than that which we have mentioned. Let us admit, in short, that the placenta is developed quite independently of the carpellary leaf, with which we have hitherto found it always associated, that several of these leaves whorled round the placentary body, which continues and terminates the axis of the flower, are bent around it, at the same time joining one another, and envelope it without touching it. We shall here have a placentation more essentially central than that which we have already explained: for, 1st, It will have been so from the beginning, whilst the other has become so on account of the unequal developement of the parts, whence resulted the disappearance of the dissepiments, of which we frequently still find vestiges in the lower part of the ovary, as in several of the Caryophylleæ: 2nd, It may even exist in a simple carpel, whilst the other requires for its formation the union of several of them.

§ 504. M. Schleiden admits that, in all cases the placenta is only the extremity of the floral axis, the ovules of which are the last buds modified; an axis, which varies in its form and divisions like that of the inflorescence, sometimes simple, sometimes diversely ramified; that the carpellary leaves arranged around this axis are sometimes expanded about it, either without adhering to it (central placentation), or affixing themselves to its divergent ramifications (parietal placentation), sometimes bending back on it, embracing a certain extent of the simple axis or one of its ramifications with the ovules, which are found there and frequently seem to arise from the internal axis (axilary placentation). This theory may be true in a certain number of cases and explains in a very satisfactory manner several facts, which are otherwise very difficult to understand. Yet there are many others, in which, following the developement of the carpels and of the ovules from their first appearance, we see the latter so clearly formed on the edges of the former, that we can hardly refuse to give our consent to the consequences of direct observation.

§ 505. Whatever may be the manner of its first formation, the placentation, taking it in the ovary when it has reached its perfect state, furnishes a very good characteristic in distinguishing plants; and if it varies in some families, there is a very much larger number, in which it is constantly the same: for instance, axilary in the Malvaceæ, Euphorbiaceæ, Campanulaceæ; parietal in the Violarieæ, Papaveraceæ, Capparideæ, Grossularieæ, Orobancheæ, &c., &c.; central in the Caryophylleæ, the Portulaceæ, &c. &c. It seems to be essentially so in the Primulaceæ, the Santalaceæ, the Olacineæ, &c.

§ 506. We have stated that the union of several carpels into one single ovary is observed only between those, which are whorled on the same plane, and, consequently, the axis of the ovary and its dissepiments are parallel. We may, however, also conceive the union of several carpels situated at different heights, but crowded together; in this case, the faces in contact by which the junction takes place are no longer the lateral ones, but one carpel will be joined by its upper face to the lower one of that, which happens to be above it, and the dissepiments will be horizontal or oblique. This case, an extremely rare one, appears to be found in the ovary of the Pomegranate (*Grenadier*), rather irregularly divided into several stages of loculi. Most commonly, when these different kinds of union take place between carpels arranged in a spire on a lengthened axis, they are blended only at their base and remain distinct throughout the greater part of their extent, so as to leave no doubt of their plurality, as we may see in several Anonaceæ.

§ 507. We have already seen, (§ 379), that the carpels may be united not only to one another, but also to the other whorls of the flower and, generally, to the calyx; so that the intermediate whorls are comprised in this union, and all the parts of the flower are thus fused at the bottom into one single body. The terms of *adherent calyx* (*calice adhérent*) and *adherent ovary* (*ovaire adhérent*) both equally indicate this circumstance: it was formerly designated by those of *superior calyx* (*calice supère*) and of *inferior ovary* (*ovaire infère*), when the limb (*fig.* 384, *l*), which constitutes the distinct portion of the calyx, appears to grow above the ovary (*o*), with which its lower portion or tube is blended. The tissue of the ovary and that of the calyx are in this case continuous, although

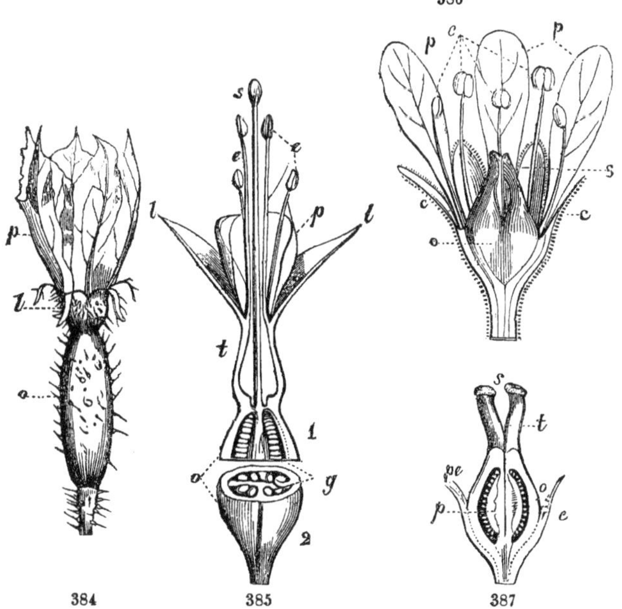

some perceptible differences frequently point out the line of demarcation between them; but we do not hesitate to describe them as

384. The flower of the Melon (CUCUMIS MELO).—*o* Lower swelling corresponding to the ovary adherent to the calyx.—*l* Upper part of the calyx projecting beyond the ovary, or the limb.—*p* Corolla.

385. The flower of the FUCHSIA COCCINEA. It has been cut horizontally through the middle of its ovary *o* into two pieces.—The lower part, 2, has not been touched and exhibits the four loculi with their ovules attached to their internal angle: figure 379 shews this section much more plainly.—The upper part, 1, has been cut vertically to display the ovules *g* arranged in a series in each loculus; the calyx joined at the bottom to the ovary, projecting above it as a tube *t* and divided at the top into several segments *l*; the petals *p* inserted on this tube, just where it is divided; the stamens *e* inserted in the same way alternately larger and smaller; the style rising from the summit of the ovary and terminated by an ovoid stigma *s*.

386. Vertical section of the flower of the SAXIFRAGA GEUM shewing its ovary *o* adherent for the half of its height to the calyx *c*.—*p* Petals.—*e* Stamens.—*s* Styles and stigmas.

387. Vertical section of a pistil of another of the Saxifragaceæ HOTEIA JAPONICA, exhibiting the interior of its two loculi.—*o* Two ovaries united into one adherent for the half of its height to the calyx *c*.—*t* Styles.—*s* Stigmata.—*p* Axilary, projecting placentas, covered with ovules.—*p e* Base of the petals.

the ovary, although the epidermis and the subjacent layer really belong to the calyx. Sometimes their union only takes place at their lower portion, and they are separated from one another at their upper, which is indicated by the expression *semi-adherent calyx* or *ovary* (*calice* ou *ovaire semi-adhérent* [*figs.* 386, 387]). In opposition to this, when they remain completely independent of one another, they are said to be *free* (*libre*) (LIBERUS); formerly the terms employed were *inferior calyx* (*calice infère*) (CALYX INFERUS) and *superior ovary* (*ovaire supère*) (OVARIUM SUPERUM). It is generally an important characteristic, that this relation of the calyx to the ovary, much more than the adherence, necessarily presupposes the perigyny or the epigyny of the stamens; this must then be carefully determined on commencing the examination of any flower. We frequently recognise with facility the ovary adhering to the swelling, which is found beneath the calycinal divisions (*figs.* 384, 385, *o*). The transverse section of this swelling determines whether we have a single body hollowed into one or several perfectly close loculi, as in the flower of the Apple-tree (*Pommier*). On cutting that of the Rose in the same way, in which there is a considerable swelling, we see, on the contrary, a cavity open at its summit and quite covered with distinct carpels (*fig.* 369). We shall, therefore, come to the conclusion, that there is an adherent ovary in the Apple-tree, several free ovaries in the Rose.

§ 508. The form of the ovary, whether free or united to the calyx, varies much. The most common form is that of a spheroïd or, still more frequently, of an ovoïd. When there are several loculi, their existence is frequently manifested on the outside by that of so many furrows, of various depths, extending from the base of the ovary to the origin of the style and indicating the lines along which the different united carpels are joined, consequently, alternating with the loculi. The middle of the dorsal face of each of these carpels or loculi is sometimes marked by another more superficial furrow, or, on the contrary, by a projecting angle. At other times, the whole surface of the ovary, perfectly smooth, does not shew any of its internal divisions. When the dorsal faces are very round and are separated by very deep furrows, the ovary is said to be lobed, (OVARIUM UNI- BI- TRI- QUADRI- QUINQUE-LOBATUM, &c.).

This surface is smooth or covered with hairs in different manners. The terms by which we designate the different degrees and manner of villosity have already been defined (§ 205). We frequently remark in the same plant a very great analogy in the nature and arrangement of the hairs between those which cover the ovary and those which clothe the leaves and young shoots.

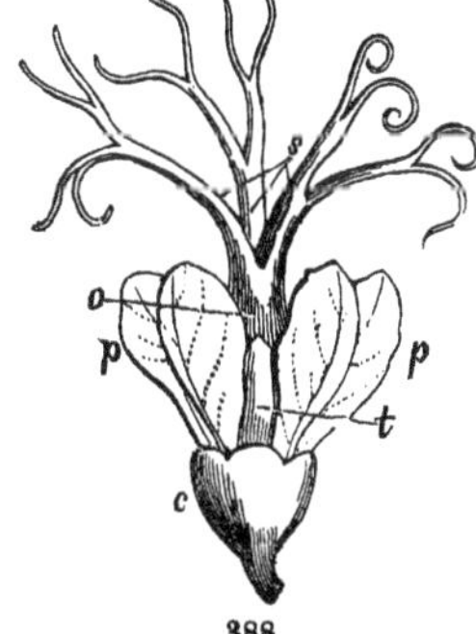

388

§ 509. The style takes its name from the Greek word στῦ-λος, *a column*, because we frequently find it in the shape of a cylinder of various lengths, frequently gradually tapering, either, as is most commonly the case, from bottom to top, or, on the contrary, from the top to the bottom. The style belonging to one carpel is often undivided, often also tends to bifurcate (*fig.* 251, 2, *s*), and sometimes each branch of this fork is itself divided in its turn (*fig.* 388, *s*).[i]

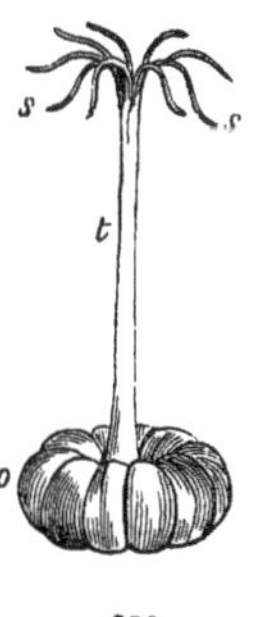

389

When the ovary has several loculi, the styles, which correspond to them, may be joined into one single one throughout the whole of their length; and, in this case, as well as in that of the undivided style for a single carpel, we make use of the term *simple style* (STYLUS SIMPLEX [*fig.* 385]). Again, they may be only partly joined together, generally at the bottom, and then we describe it as a multipart or multifid style (*fig.* 389) according to the greater or less height to which the styles are united. We indicate their number by joining the word or figure to the termination *partite* (*parti*) (PARTITUS) or *fid* (*fide*) (FIDUS) as *bi-fid*, *tri-partite*, *quadri-fid*, *sexpartite*, (BI-FIDUS, TRI-PARTITUS, QUADRI-FIDUS, SEX-PARTITUS, &c.); these are the expressions employed in the most ancient works; in more modern ones we often find the same fact denoted by 2—3—4, &c., styles united to the middle, or above it, or below it (STYLI USQUE MEDIUM, SUPRA MEDIUM, INFRA MEDIUM COALITI). Lastly, although the carpels may be completely united, the styles may remain quite independent (*figs.* 383, 387, 388), and we then say

388. Female flower of the EMBLICA OFFICINALIS.—*c* Calyx.—*pp* Petals.—*t* Membranous tube surrounding the ovary.—*o* Ovary surmounted by three styles *s* each twice bifurcate.

389. Pistil of the MALVA ALCEA.—*o* Ovaries, nine in number, united into a single one, on which are described as many furrows.—*t* Column formed by the nine styles united with one another till the top, where they separate, diverge and bend down, each terminated by a stigma *s*.

[i] Appendix B.

two—three—four—five, &c., free styles, or rather an ovary with two—three, &c., styles (OVARIUM DUOBUS—TRIBUS—MULTIS—STYLIS). These styles of a compound ovary, whether they remain distinct, or are united at the base, may themselves be simple (*fig.* 383) or divided (*fig.* 388). We have already said that their number will generally indicate on the outside that of the carpels or of the loculi, and that they correspond to the internal angle of the latter, consequently, alternating with the dissepiments.

The styles vary in their form, which is frequently different from that we have described as the most general: in the IRIS they take that of a petal. They also vary in their length and direction (which we habitually compare with that of the other parts of the flower, but more particularly with the stamens), in the state of their smooth or hairy surface. They are sometimes bristled with hairs quite different from those of the other surfaces of the plant, which have been termed *collectors* (*collecteurs*), because they appear to be destined to collect the pollen. In the large family of the Compositæ these hairs are rather stiff and cover the surface of the style to a certain height and to a certain extent (*fig.* 390 *p c*); and, since this style, by being later in its developement than the stamens, rises through the midst of the anthers which immediately surround it, these hairs, as they pass, act on the loculi of the anthers like small brushes, and thus become loaded with the dust of the pollen. In the Lobeliaceæ and the Goodenieæ they are arranged immediately beneath the stigma in a kind of a circle or little collar which has been termed INDUSIUM (*fig.* 391, *i*).

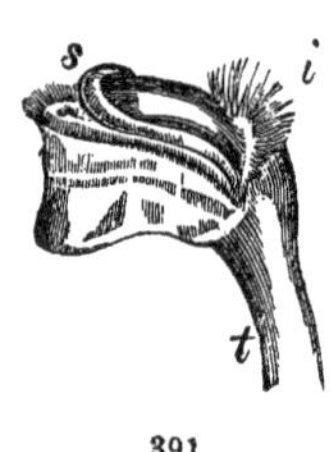

§ 510. STIGMA[k] (*Stigmate*).—We have seen, that in a single carpel the stigma may be sessile, that is, situated immediately upon the ovary (§ 484), or else placed on the style (§ 488), either at its upper extremity (*figs.* 360, 366, 367), or on its sides (*fig.* 365), or on one of its sides only (*fig.* 364), in which case it may be turned

390. Summit of the style *t* of an ASTER divided into two branches, each terminating in a cone covered with collector hairs *p c*.—The stigma *s* is seen below on the inner face of the branches under the form of a small band.

391. Summit of the style of the LESCHENAULTIA FORMOSA.—*t* Part of the style.—*s* Stigma.—*i* Indusium.

k From στίγμα, *a brand, mark or dot.*—TRANS.

either to the inside or the outside of the flower. We have seen, moreover, that the utricles, which compose it, sometimes form a smooth surface, are sometimes lengthened into projections more or less enlarged, into simple papillæ or into true hairs. These last are sometimes collected into a kind of paint-brush or bottle-brush, or dispersed so as to imitate the feathery part of a quill, *plumose stigma* (*stigmate plumeux*), as in a large number of the Gramineæ (*fig.* 392, *s*). When the style is divided, the stigma is also separated to form the termination of each one of the divisions, and probably the stigma alone constitutes the whole division. It tends, indeed, to become lobed by bifurcation, as may be easily seen in the Gramineæ and the Compositæ, in which it is double, although there is only one loculus.

But most frequently, its divisions, in the same way as those of the style, indicate that we are examining a pistil composed of several carpels united into one, as well as their styles. In this case, it may happen that the stigmas are the only parts, which do not participate in this union; they form at the extremity of the simple style a compound body composed of as many lobes as there are loculi in the ovary. Thus, the tri-lobed or quinquefid stigma of the Campanulaceæ (*fig.* 393), corresponds to three or five loculi; the bilobed stigma of the Scrophularineæ, of the Acanthaceæ, of the Bignoniaceæ, to two loculi, &c. These lobes assume different forms; they preserve this name when they are thick and obtuse; they take the form of a strap, *bifid stigma* (*s. bifide*), as in the Labiatæ, the Compositæ (*fig.* 295, *s*); *trifid* [*trifide*], as in the POLEMONIUM; *multifid*, &c. (*multifide*), when they are longer and sharper; of small blades, *bi-lamellate stigma*, ([*s. bilamellé*], as in the MIMULUS, the BIGNONIA LACTIFLORA, PANDOREA, &c. [*fig.* 396]), when they are flattened into plates. At other times, the stigmas are themselves united into one single body, either perfectly smooth on its surface, or frequently marked with as many superficial and radiating furrows as there are partial stigmas in its composition. It is said to be *capitate* (*en tête*) (S. CAPITATUM), when it is obtuse and larger than the style which it surmounts; it may be globular, as in the MIRABILIS JALAPA (*fig.* 395), hemispherical, ovoïd

392. Pistil of the CYNODON DACTYLON.—*o* Ovary.—*s* Stigmas.

(*fig.* 385, *s*), polyhedric, club-shaped, &c.; frequently flat at its summit, as in the Barberry, or even widened into a disk, which is supported at its centre on the top of the style, *peltate stigma* (S. PELTATUM), as in the SARRACENIA , ARBUTUS (*fig.* 394, *s*). The

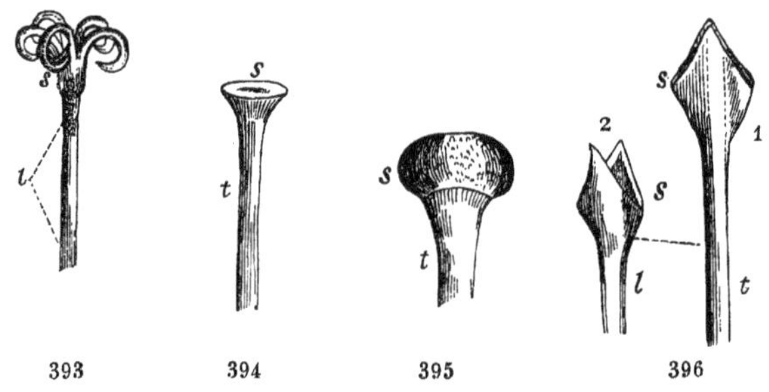

peltate, sessile stigma of the Poppies (*Pavots*) (*fig.* 397, *s*) is composed of two parts: the one formed of rays of a papillose tissue, which is really the stigmatical part; the other of a kind of shield hollowed at its circumference and smooth on its upper face, where these rays are, which seem, consequently, to represent a union of enlarged stigmatiferous styles over the whole length of one of their faces.

397

The stigmas, terminating really simple styles, those which correspond to a single carpel or a single loculus, will, if they are themselves simple, be opposite to the loculi with the dissepiments; if they are bilobed, their lobes will, on the contrary, be opposite to these.

393—396. Stigmas *s* of different flowers, with summit of the style *t* which supports them.

393. Stigma *s* of the CAMPANULA ROTUNDIFOLIA.

394. Stigma *s* of an Arbutus (*Arbousier*) (ARBUTUS ANDRACHNE).

395. Stigma *s* of the MIRABILIS JALAPA.

396. Stigma *s* of the BIGNONIA PANDOREA. The two plates are represented as naturally applied to one another in 1. They are artificially separated in 2.

397. Pistil of a Poppy (*Pavot*) PAPAVER SOMNIFERUM).—*o* Ovary.—*s* Shield laden with radiating stigmas.

[1] Known also by the English name Side-saddle flower.—TRANS.

FRUIT.

§ 511.—When the fecundation has once taken place, the organs, which have jointly performed it, die and disappear at various periods. Now, these organs, we have seen, are of two kinds; 1st, The essential ones: on one side the anther, on the other the stigma and the conducting tissue; 2nd, The accessory ones: the filaments, which bear the anthers; the styles, which bear the stigmas and through which the conducting tissue is insinuated; lastly, the envelopes, which protect all this apparatus; the petals, whose analogy to the stamens has been frequently pointed out; and the calyx, which differs from them still more, inasmuch as they represent leaves that are less modified. The more direct a part these organs take in the work of fecundation, the more transient is their duration. Thus, after fecundation, the stigma, the conducting tissue, the anthers, are not long before they wither and disappear. The styles, the filaments, the petals, may remain a little longer; but, they generally soon die, and then they fall or else remain attached to their place. The calyx itself, although a little later, and, if it be not these few cases in which it continues to vegetate and sometimes even to grow (§ 428), is stopped in its developement and ceases to live, whether it be detached, or remains after the manner of marcescent leaves. The name of INDUVIÆ has been given to these remains of the calyx, of the corolla, of the filaments, which may be attached for a greater or less length of time to the fruit, and which furnish some characteristics, either by their very permanence, or by enabling us to recognise the parts of the flower and their relations, when we have not been able to observe them in its previous perfect state. The style is sometimes persistent, and it is generally under the form of a point situated towards the summit of the fruit, which is then said to be *apiculate* (*apiculé*).

§ 512. At this period the vital principle is concentrated in the ovule, to which the fecundating principle has been directed, and in the ovary, which protects and encloses it. Both now continue to grow at the same time, assuming fresh appearances, fresh characteristics, and also even fresh names: the ovule becomes the *seed* or *grain* (*graine*), the ovary becomes the *pericarp* (*péricarpe*) (PERICAPRIUM: [from περὶ, *around*, and καρπὸς, *fruit*], i. e., the part, which forms the envelope of the fruit), and their whole constitutes the *fruit* (*fruit*) (FRUCTUS). Their vital principle and their developements are generally closely connected, so that, if the seed prove an abortion, the pericarp will not be developed; if the pericarp be abortive, the seeds will wither. We may, however, mention a few exceptional cases, in which either the seeds ripen without pericarp, or, on the contrary, the abortion of the seeds, far from stopping the developement of the pericarp, seems to favour it, as in the BANANAS, the Bread-fruit-tree, &c. Their varieties, which we eat and the fruit of which becomes so fleshy and so succulent, do not produce fecundated seeds; and when these are developed, the flesh of the fruit loses much of its substance and sapidity. We observe, moreover, something analogous in the fruits of our orchards; the wild stocks generally present a much larger developement of the seed with regard to that of the pericarp than the cultivated varieties.

§ 513. But, let us take the ordinary and normal case, that in which both the developements take place, and let us examine at first the changes that occur in the ovary. Those of the ovule and its structure will occupy us afterwards.

Let us at first recall the structure of the carpel, which is that of a leaf bent or turned on itself, the edges of which are united together, so that it presents an internal surface corresponding to a cavity and an external surface, each covered by its epidermis, and between the two layers of the epidermis, a parenchyma traversed from bottom to top by fibro-vascular fascicles. We may, therefore, reckon here three layers; the external epidermis (*fig.* 398, *e*) or *epicarp* (*épicarpe*) (EPICARPIUM [from ἐπὶ, *on*]); the intermediate parenchyma (*fig.* 398 *m*), or *mesocarp* (*mésocarpe*) (MESOCARPIUM [from μέσος, *middle*]); the internal parenchyma (*fig.* 398 *n*) or *endocarp* (*endocarpe*) (ENDOCARPIUM [from ἔνδον, *within*]). The use of these names is necessary from the various sizes to which these parts frequently grow in the developement of the fruit.

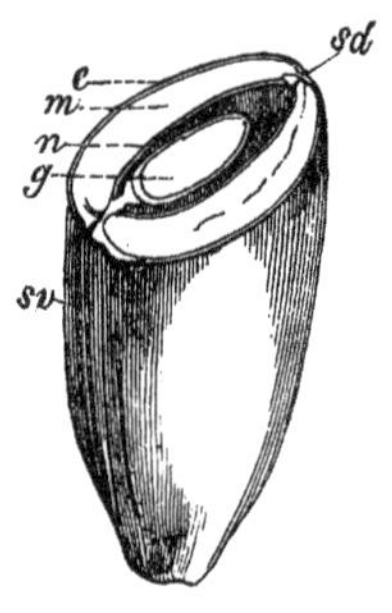

398

§ 514. The pericarp in its developement, preserves its resem-

398. Lower part of the carpel or pod of the FABA SATIVA cut across so as to display the composition of the pericarp.—*e* Epicarp or external epidermis.—*m* Mesocarp.—*n* Endocarp.—*s d* Dorsal suture.—*s v* Ventral suture.—*g* A seed at the place of section cut across in the same way.

blance to the leaf, as in the well-known fruit of the Bladder-nut; it is then said to be foliaceous or herbaceous. Sometimes this resemblance is more or less effaced by the different colour and consistence, which one or more of the three layers takes. The external one, (the epicarp), which forms what is more commonly termed the skin of the fruit, generally preserves its epidermal appearance, although frequently thickened by the addition of a certain number of cellular rows. The mesocarp is frequently developed in quite a different manner from the parenchyma of the leaf and is changed into a flesh of various degrees of succulence and thickness; this caused Richard to propose for this middle layer the name of *sarcocarp* (*sarcocarpe*) (SARCOCARPIUM [from σὰρξ, σαοκός, *flesh*, *pulp*]), a name, which from its etymology cannot clearly be applied to the herbaceous fruits, and which, consequently, it would be better either to abandon altogether or to apply only to fleshy fruits. The endocarp sometimes remains in the form of a fine membrane covering the surface of the loculus; but, at other times its cells are incrusted with a ligneous matter, and then the neighbouring cells of the mesocarp undergo an analogous modification, so that we have round the cavity of the pericarp an envelope of a certain thickness, of a certain degree of hardness; in several fruits this forms what is termed the *shell of the nut* (*noyau*) (PUTAMEN).

§ 515. Let us illustrate the preceding remarks by a few well-known examples. In a Cherry, an Apricot or a Peach, the skin is the epicarp; the edible part is the mesocarp or sarcocarp; the kernel, the endocarp. We find on opening the shell of the kernel a body, which is the seed. In the fruit of the Almond, outside the kernel, is the endocarp under the form of a thin and fragile shell, which is covered by a mesocarp with thin, green, coriaceous flesh. In that of the Walnut, the nut is the seed enveloped with its endocarp; the greenish fibrous envelope, which we take off when it is gathered and which is known under the name of the husk, is the mesocarp with its epidermis. It is the seed, then, of the last two fruits which we eat throwing aside the pericarps; whilst in the first we eat a part of the pericarp and throw away the endocarp and the seed. The Pear and the Apple result, on the contrary, from a compound adherent ovary; their skin or epicarp is the epidermis of the calyx joined to the ovary; their fleshy part is the mesocarp, and their centre is occupied by five small cavities enclosing the pippins, pips or seeds, and covered with a scaly covering, which is the endocarp. The last, in the Medlar (*Nèfle*), is much more largely developed into the form of a nut; we find here, indeed, five nuts corresponding to the five loculi. In other fruits, the demarcation is far from being so clear: in the Melon, for instance, the mesocarp varies from the outside, where it has a green colour and sharp taste, to the inside, where it takes another colour with a sweet sugary flavour, whilst the traces of the epicarp and endocarp are hardly visible. The skin of the Orange is the union of its epicarp and of its mesocarp; the thin membrane which divides the quarters is the endocarp, and these quarters themselves are so many loculi filled with additional tissue, which is the part we eat throwing aside the true pericarp. The different examples, which we shall have occasion to mention in the course of the work, will aid the preceding in shewing the diversity of parts, which give to the fruits their flavours, their properties and their different uses.

§ 516. The union of the two edges of the carpellary leaf is frequently indicated by an external line, by a furrow, when these edges are a little reflected towards the cavity of the loculus. We may observe it in several fruits produced by a single carpel, as in that of the Bladder-nut, of the Apricot, the Plum, &c.; and not only on their external surface, but even to the very kernel, the whole of the corresponding edge of which is hollowed by fluting more or less deep. The name of *suture*, which was applied to this fluting, proves that its true origin was known some time ago, since this word indicates that two separate surfaces have been united (sewn, as it were,) together. But, the leaf folded into a carpel may, besides this line corresponding to the junction of its edges and, consequently, like them always turned towards the axis of the flower, present another corresponding to its median nerve and turned, on the contrary, outwards. To this line the name of *suture* has been equally given; and, as in the carpel and the seed, we call that the dorsal face which is turned outwards, ventral face, that which is turned inwards, we have distinguished them by using in the first case the expression *dorsal suture*, in the second *ventral suture*, thus denoting the difference between the kinds of suture.

§ 517. It is clear that the dorsal sutures alone can be found on the surface of the multilocular fruits with axilary placentation, since the ventral ones are concealed and modified in the very substance of the fruit. But, if the placentation is parietal (§ 501) or central (§ 502, 503), the edges of the carpels always being carried back towards the circumference, their ventral sutures are equally so and may be seen on the outside.

§ 518. The suture, when carefully examined, seems to be formed by the union of two fascicles touching one another, which are easily separated by introducing the blade of a knife and working it backwards and forwards between them. This separation takes place spontaneously in several fruits at a certain time, either at the ventral or the dorsal suture, or at both at the same time. Hence there results, that the pericarp is separated into several pieces, the number of which will be in regular cases, either equal to or double that of the loculi. These pieces are called *valves* (VALVÆ), and according to the numbers the fruit is said to be *univalvate*, *bivalvate*, *trivalvate*, *multivalvate*, (UNI-BI-TRI-MULTI-VALVATUS, &c).

§ 519. We have just seen several of the changes, which the parts of the ovary may undergo in passing to the state of pericarp; but we have hitherto supposed that all its parts are regularly developed, an event which does not constantly occur. The different parts of the ovary may be so modified, as to render it difficult for us to recognise them in the fruit, when arrived at its maturity. The loculi, the seeds which they contain and their placentation, the dissepiments which separate them frequently present modifications, which it is important to study.

Several of the carpels, which, whether free or united, composed the pistil, often prove abortions, so that we no longer find the same number of them in the fruit. The failures sometimes occur with great regularity and are mostly connected with that of the ovules. Thus the ovary of the Common Ash (*Frêne*) presents two loculi, each enclosing two ovules with axilary placentation; but the two ovules in one loculus and one ovule in the other are generally not developed; the only one, which ripens, then repels the dissepiment or septum (*fig.* 414), which is forced against the walls, so that the second loculus is effaced and we only find at last one single cavity enclosing one single seed, attached to its side and no longer to the axis. The Horse Chesnut has an ovary with three loculi, each containing two ovules fixed to the axis, and by analogous abortions its fruit has apparently only one single loculus with a single large seed. We shall content ourselves with these two examples, which it would be easy to multiply. At other times, the abortions do not proceed so regularly, and among the fruits of the same plant, all do not exactly present the same number of loculi or of seeds, according as such or such an ovule has escaped fecundation. We must, therefore, study in the ovary the arrangement of the carpels and of the ovules, which may be disguised afterwards by these unequal or irregular developements and thus mask the true symmetry of the parts of the flower.

§ 520. The septa or dissepiments are also more or less modified during the maturation of the fruit. According to their origin, they ought to be formed of two laminæ placed close to one another, and each of these laminæ, of three layers representing those of the pericarp, such as we observe on the sides of a free carpel. But these laminæ in the multilocular fruit, pressed on one side against one another, on the other by the seeds which are filling the loculi, have not the opportunity of freely developing their layers, of which one or two partly fail. The innermost, the endocarp is most frequently only developed and is even closely united in the two adjacent laminæ, which are blended together into a single one. Sometimes they remain distinct, and even a small portion of the mesocarp is interposed between them ; but the epicarp disappears, only remaining on the free dorsal face of the carpel and thus merely covering the outside of the fruit; this may be clearly seen in the fruit of the Castor-oil plant, the EUPHORBIA or the Mallow. The septa, sometimes reduced to the state of a thin membrane, may in a few fruits be destroyed altogether, or only partly so, before its complete maturity; and we have already seen that this destruction (§ 502), happening at a very early period in the ovary still very young, determines the central placentation in several fruits, as in those of the Caryophylleæ.

§ 521. In a small number of fruits, we observe quite contrary changes, on account of the developement which the dissepiments or septa undergo. The ovary of the TRIBULUS has five loculi, and in the interior of each, we already see the wall formed of small folds (*fig.* 399, *c*), which project a little between the three or four ovules which are contained there. They continue to advance more and more from front backwards in proportion as the fruit ripens, and at last gain the opposite side of the loculus and are interposed between the seeds like so many transversal septa, so much so that each loculus is definitely divided into so many secondary loculi placed one above the other (*fig.* 400). In the fruit of several Leguminous plants we observe a continuation of analogous divisions ; these are what we term *false loculi* (*fausses loges*) or *false dissepiments* or *septa* (*fausses cloisons*),

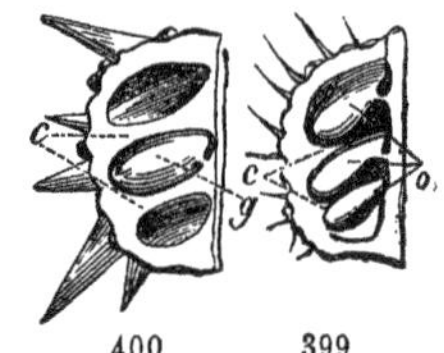

399. A vertical section of the loculus of the ovary of the TRIBULUS TERRESTRIS. The projections *c* of the wall are beginning to separate the ovules *o*.

400. A vertical section of a loculus of a ripe carpel of the same divided by transverse dissepiments into locelli, in one of which a seed *g* has been left.

and they are easily recognised by the horizontal position of these dissepiments and their formation subsequent to fecundation. But we may also conceive that these elongations or folds of the endocarp may also be parallel to the true septa; which is really the case, as in the ASTRAGALUS, in which each carpel is thus divided into two. These false vertical dissepiments, recognised with more difficulty, may, nevertheless, be determined by their relations of position to the styles, from their never bearing the seeds, and, especially, from the study of the young pistil.

§ 522. The loculi are sometimes filled with a pulpy substance, by which the seeds are enveloped, or, as it were, nestled in its substance (SEMINA NIDULANTIA); they then appear solid, and their cavity is effaced as well as their dissepiments, so that it is no longer so easy to discover the situation of the parts. We must look for it, therefore, in the ovary, and we may, moreover, also follow the formation of the pulp. Thus, in the Aroïdeæ we see that it is the conducting tissue itself which grows beyond its canal into the interior of the loculus. In the ovary of the Orange we observe in each loculus the ovules attached to the internal angle; whilst on the opposite face, the wall is quite covered with small vesicles or lengthened greenish cells, which, as they multiply, gradually fill the entire cavity, change in colour, distend with juices, and thus constitute the tissue which we eat in the Orange. In all the fruits called pulpy, these cells, thus filled with juices, are always those, which fill the loculus; but sometimes they depend on the pericarp, as in the preceding case; sometimes on the seed, as in the Gooseberry and the Pomegranate.

§ 523. Lastly, the placentas present also different changes in the developement of the fruit; these naturally occur in that of the vessels and cellular tissue which constitute the nutritive system of the seeds. One portion remains fixed to the walls of the loculus, on which it sometimes forms rather a large projection; another portion is detached from this wall to constitute as many elongations as there are seeds, being destined to bear these. They have often the form of a small cord, which has given rise to the name of FUNICULUS (*funicule*[m]). That of *podosperm* (*podosperme*) (from πούς, ποδός, *foot*, and σπέρμα, *seed*) is employed by some authors, who change that of placenta into *trophosperm* (*trophosperme*) (τροφός, *nourishing*, and σπέρμα, *seed*).

§ 524. We have cast our eye over the principal changes which

m Called also *Umbilical chord* (FUNICULUS UMBILICARIS) —TRANS.

take place in the ovary from the fecundation to the maturity of the fruit. When we reflect on the diversity of the modifications which the ovary presents in the immense variety of vegetables, and when we see them combined with those still more numerous mutations, which its developement into fruit causes in it; when we see it preserve in some almost the same size and consistence, in others acquire a shape, a size, a consistence perfectly dissimilar from its primitive state; when we remember, for instance, that the Gooseberry and the Pumpkin originate from ovaries which are nearly equal in size, we may readily conceive the numerous and decided characteristics, which different fruits will present to our notice both in appearance and structure; several kinds have, consequently, been distinguished, and several names invented to designate them. But, even admitting all these, numerous modifications still escape these names and their definitions, and we are unceasingly obliged to add explanations, descriptive phrases, to state clearly of what fruit we are talking. Now, since names are only adopted to avoid these descriptions by means of a single word previously well defined, and since, even with the assistance of these names, these descriptions cannot very often be dispensed with, it appears to be preferable not to multiply the names so much and to limit ourselves to those, which denote the most general and constant modifications of the fruit. This is at least what we intend to do in the classification we shall hereafter lay before our readers.

§ 525. We know already that the fruits, like the ovaries, are formed of carpels either independent of one another or united into one single body. Hence we gain the first division of the fruits into *apocarpous fruits* (*apocarpés*) (FRUCTUS APOCARPI [from ἀπό, indicating *separation*]), *syncarpous fruits* (*syncarpés*) (SYNCARPI [from σύν, indicating *union*]). We know, moreover, that the pericarp may preserve its thin foliaceous consistence, or swell into a fleshy mass of various degrees of thickness. In the latter case, the envelope thus thickened does not divide when it reaches maturity; it is only by being destroyed, by splitting irregularly, by decaying or by withering, that egress is given to the contained seeds. In the case, even, in which it is foliaceous, it may remain closed; but frequently, on the contrary, either by the disunion of the sutures of which we have previously spoken (§ 518), or by the rupture, much less frequent and less regular, of some other point of its surface, the pericarp when ripe is naturally opened, and the seeds thus find a passage allowing them egress from the fruit. We may here again distinguish two kinds of fruits, those which do not open, or *indehiscent* (*indéhiscents*), either fleshy or dry; those which open of their own accord, when they come to maturity, or *dehiscent* (*déhiscents*). This dehiscence, when it takes place at the sutures, may happen either at both at once or only at one of them, and thus separate each carpel into two or into a single valve. Lastly, each carpel or loculus may be *monospermous* (*monosperme*), i. e., enclosing only one seed; *oligospermous* (*oligosperme*), i. e., enclosing a small number; *polyspermous* (*polysperme*), i. e., enclosing a larger number. These are the different characteristics, the combination of which has served to define the different classes of fruits proposed by certain authorities, who have employed them, some in one order, some in another. We shall here adopt that in which we have just enumerated them.

A. Apocarpous Fruits.

A. Indehiscent.

§ 526. Some have a fleshy pericarp and an endocarp hardened into a kernel, and are commonly monospermous, either from the ovary only containing one ovule or from the abortion of one of the two This is what is called a *drupe* (DRUPA), of which the Cherry, the Plum, &c., are familiar examples. The fruits of the Almond and of the Walnut are only slight modifications of this, indicating the transition to the following fruits.

These have a much thinner and drier pericarp, in which the consistence of the endocarp and that of the mesocarp do not present much difference. A single seed fills the loculus, with the walls of which it may be found in different positions. Indeed, it most frequently adheres to it only by its point of attachment, its funiculus; and then we have an ACHÆNIUM (*achaine*) (from *α*, *privative*, and *χαίνω*, *yawn*) (*fig.* 401). But at other times the seed in its developement is united to the walls of the ovary which envelopes it, so that the pericarp, seeming to make a part of its proper teguments, apparently disappears. This fruit, which has been termed CARIOPSIS (*cariopse*)[n], for a long time bore the name of *naked seed* (*graine nue*), which has been extended even to a number of achænia, from the

[n] From *κάρα*, *head*, *ὄψις*, *appearance*.—Lindley places the cariopsis under *syncarpous* fruits, Intr. to Bot., p. 234.—TRANS.

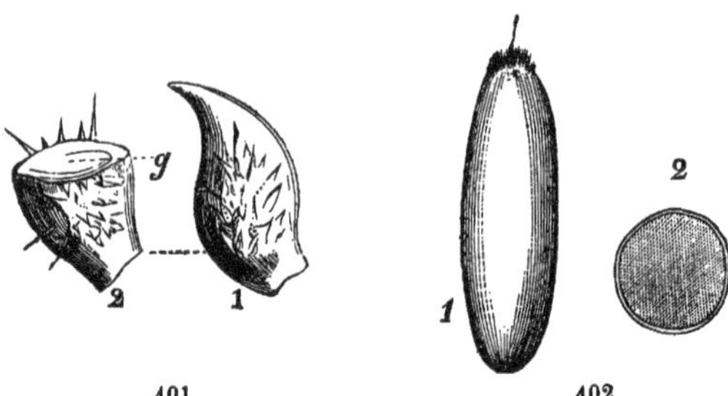

fact that botanists supposed the tegument of the fruit to belong to the seed which it immediately envelopes. But we are shewn the real state of the case, either by the presence of the style, which grows from this tegument, and can only grow from an ovary, or by the study of the latter, in which the separation of the ovules from walls of the loculus is still very manifest. Let us mention as examples of cariopses the fruit of the Gramineæ (as Wheat, Oats, Rye [*fig.* 402], Maize), which is commonly known by the name of

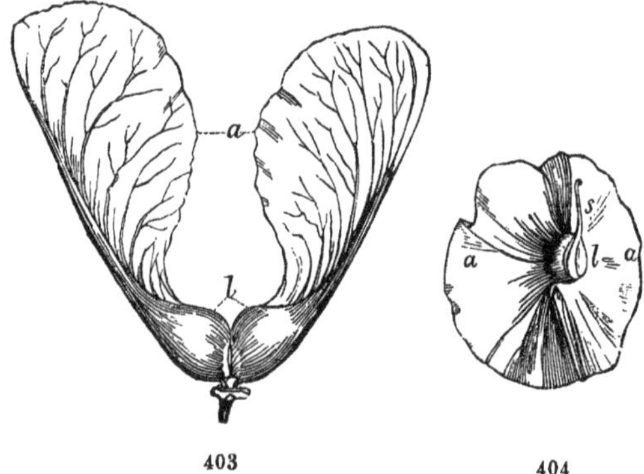

401. An achænium taken from the fruit of the RANUNCULUS MURICATUS. (The fruit is composed of several of these).—1. Undivided.—2. A horizontal section shewing the seed *g* not adhering to the walls.

402. A capriopsis of the Rye (*Seigle*) (SECALE CEREALE).—1. Undivided.—2. A horizontal section shewing the seed adhering to the walls.

403. The fruit of the Sycamore (*Erable*) (ACER PSEUDOPLATANUS), composed of two samaras.—*a* The upper part forming a dorsal ring.—*l* The lower part corresponding to the loculus.

404. A separate samara of the fruit of the HIRÆA.—*s* Persistent style.—*l* The part answering to the loculus.—*a a* Marginal wing.

corn or grain. The very thin pericarp and the membrane of the seed, closely united, form for it an envelope apparently single, which constitutes the chaff when it is separated by the thrashing. The carpels of the Bugloss and other Borragineæ, those of the RANUNCULI, of the Roses, are achæniæ differently grouped in these different plants. Those of the Compositæ are so also, but somewhat different in their pericarp adhering to the calyx and not being free. Some may serve as transitions to the cariopses, because their seed is united in places to the wall of the loculus. We sometimes term UTRICULUS (*utricule*) an achænium with a very thin and, as it were, membranous wall. Let us suppose that the pericarp grows thin beyond the loculus into a membranous lamina, when it is almost reduced to a fold of its epicarp; we shall have a SAMARA (*samare*). This fold seems to be an elongation sometimes of the median nerve of the carpellary leaf, sometimes of its lateral nerves, and thus to form a wing, sometimes dorsal (*fig.* 403), sometimes marginal (*fig.* 404).

B. DEHISCENT.

§ 527. When the carpel only opens at its ventral suture, it best justifies by its appearance the origin which we have assigned it, that of a leaf folded back; its name FOLLICULUS (*follicule*) recalls it to our mind, and yet it was adopted a long time before any one ever thought of such a theory. We find numerous examples of it in the fruits of the Ranunculaceæ (as the Hellebore [*fig.* 405], the Columbine, the DELPHINIUM, &c.), of the Asclepiadeæ, of the Apocyneæ (as the Periwinkle [*Pervenche*] [VINCA], &c.) The carpel, which, opening at its ventral and dorsal sutures, is separated into two valves, is, if it contains a very small number of seeds (commonly one or two), a COCCUM (*Coque*), with an endocarp commonly ligneous or crustaceous, as in the DICTAMNUS. If it contains a much larger number of seeds attached along its internal suture, it is a *pod* (*gousse*) or *legume* (*légume*) (LEGUMEN), which has given its name to the large family of Leguminosæ; (examples: the fruit of the Haricot-bean, of the Windsor-bean [*fig.* 398], of the Pea [*fig.* 406], &c.;) this family, however, presents some exceptions in which the pericarp remains closed instead of being separated into two valves. Others present the following singular structure, that the pod, instead of being open throughout the whole of its length, is contracted in several places, and at last is separated into so many

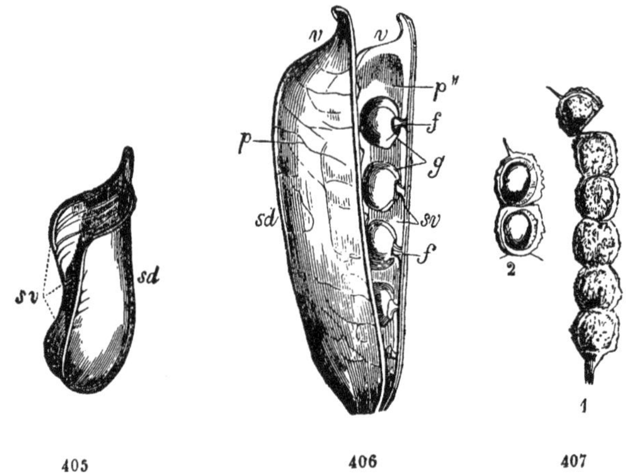

articulations each of which encloses a seed. This carpel thus cut into transversal partitions, which are doubled as they are disarticulated, enters into the number of those to which we have applied the epithet of false loculi (§ 521), and is called *lomentaceous* (*lomentacé*) (LOMENTACEUS), or substantively, LOMENTUM; (examples: that of the Saintfoin [*fig.* 407], of the CORONILLA, &c.)

§ 528. Let us now recall to mind, that in an apocarpous fruit as well as in the flower, where it existed under the form of the ovary, there may be there one single carpel, as in the Leguminosæ, the Plum, the Cherry, &c., or there may be several, and that these may then be arranged either in a circle or a whorl on the same plane (as in the DICTAMNUS, the SPIRÆA, the Hellebore, &c.), or at different heights, on a torus enlarged or hollowed into a vase, as in the Rose, the CALYCANTHUS, &c., or, on the contrary, lengthened into an axis, as in the MYOSURUS, the RANUNCULUS, the Strawberry,

405. A separate carpel of the Hellebore (*Hellébore*) (HELLEBORUS FŒTIDUS), after the dehiscence.—*s d* Dorsal suture.—*s v* Ventral suture.

406. The pod of the Pea (*Petit Pois*) (PISUM SATIVUM), opened.—Valvulæ formed of two pieces of the pericarp, composed of the epicarp *p* and endocarp *p'*.—*g* The seeds placed one above another, attached by means of short funiculi *f f* to a placenta, which follows, under the form of a longtitudinal cord, the inner edge of the valves corresponding to their ventral suture *s v*.—*s d* Their outer edge corresponding to their dorsal suture.

407. Lomentaceous fruit of the HEDYSARUM CORONARIUM.—1. Undivided. The upper articulation almost separated from the others.—2. A vertical section of two articulations, shewing two false loculi, each with its seed.

the MAGNOLIA, &c. In all these last cases, the spiral arrangement of these carpels is clearly perceived, and reminds us of that of the flowers in a spike or a capitulum. We may then describe it briefly according to this appearance, saying, for instance, drupes or achænia or cocca, or more generally carpels capitulate or in a spike (CARPELLA CAPITULATA, SPICATA). These appellations, by means of a small number of words appropriate to each particular case are preferable to the single words which have been proposed for some of these cases.

B. SYNCARPOUS FRUITS.

§ 529. In these fruits, formed by the union of several carpels united together, we ought carefully to note the placentation, which is susceptible of different modifications already described in the ovary (§ 501), that is, axilary, central or parietal.

The lateral faces of the loculi or carpels, which form the dissepiments, advancing from the outside to the inside, may change their direction and be bent on one side, or from within to without. They thus form a projection into the interior of the loculus, and the placenta which bounds them is said to be *projecting* (*saillante*) (PROMINENS), the more so in this case, as it is most commonly presented under the form of a thick mass adhering to the walls of the loculus by a lamina of a certain width. The dissepiment, when it is bent back, is doubled; of the two carpellary faces or laminæ, by the touching and union of which it was formed, each is bent back in the loculus to which it primitively belonged; so that each placenta frequently appears to be double or *bilamellate* (BILAMELLATA). If the dissepiments have become so, before they have arrived at the axis of the fruit (*fig.* 378), the placentation is necessarily parietal; but they frequently are advanced as far as the axis, and, bending back thence in a contrary direction, thus carry back the placenta which bounds them to a greater or less distance from the axis (*fig.* 420); but, since, in separating, it always adheres by that part to the rest of the pericarp, we must consider it as axilary. The two sides of the same carpel, on being thus folded, necessarily converge towards one another, meet, and most frequently are united. If the parts thus bent back are completely united, the placenta will appear to be simple; if they are only united for a little way and diverge afresh, it will appear to be double or bilamellate.

§ 530. The axis is frequently an ideal line, along which the

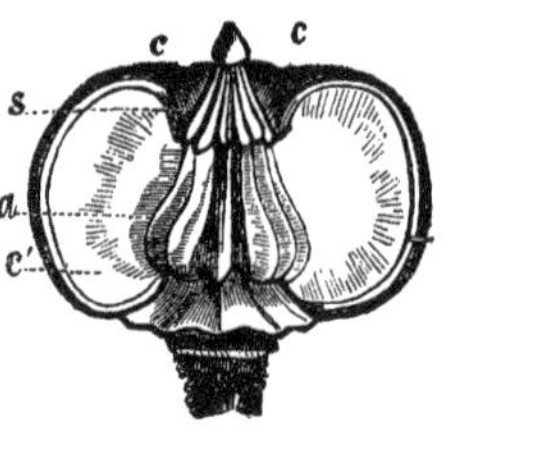

408

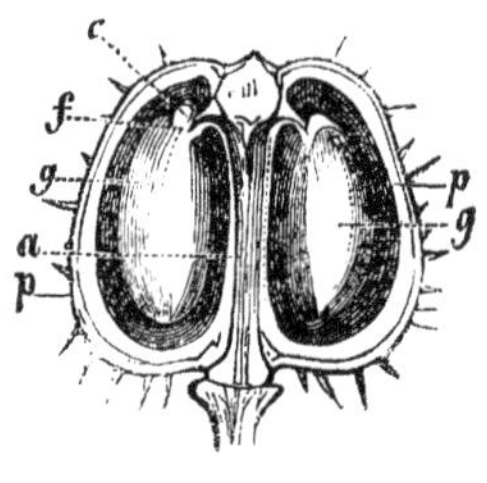

409

internal angles of the carpels meet and touch one another. But, in other cases, it really exists, continuing and terminating the axis of the flower beyond the insertion of the carpels, at the angles of which they are interposed, thus connecting them with one another. It is then formed by the cellular tissue, traversed by vascular fascicles which are distributed as much in the pericarp as in the placentas. It thus disappears as it rises, and ceases in general beneath the insertion of the styles; but, in some rare cases, we see it even prolonged beyond this point and interposed between the styles, as it is among the carpels: this is observable, for instance, in the GERANIUM, the fruit of which (*fig.* 410) when mature, shews its five carpels and their styles detached from the bottom to the top of a long pyramidal axis, to which they were thus united. The Malvaceæ (*fig.* 408), the Euphorbiaceæ (*fig.* 409), &c., present examples of axes very much developed, but terminated towards the origin of the styles.

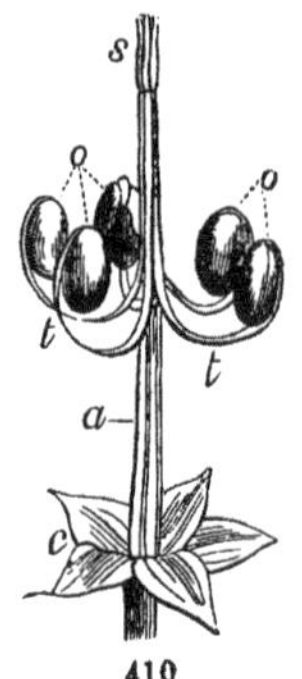

410

408. The fruit of a Mallow (*Mauve*) (MALVA ROTUNDIFOLIA). Half of the carpels have been taken away to shew the axis *a* interposed between them and terminated at the point whence the styles *s* grow.—*c c* The remainder of the carpels which have been left attached to the axis, around which they are arranged in a whorl. The lateral face of the nearest two *c' c'* is seen.

409. The capsule of the RICINUS COMMUNIS. It has been divided vertically, so as to display the axis *a* prolonged between the carpels and terminated towards the top in each of them by a small cord *f*, which forms the funiculus.—*g g* The seeds in their loculus are laid open by the division, each of them is surmounted by a fleshy carunculus *c*.—*p p* Pericarp.

410. The fruit of the GERANIUM SANGUINEUM.—*c* The persistent calyx.—*a* Axis.—*t* The styles, which at first were united to it, and which are at last detached, carrying the ovaries *o* along with them.—*s* Stigmas.

§ 531. Let us now pass to the enumeration of the different kinds of the commonest syncarpous fruits, separating them into two divisions like the apocarpous, according as they do or do not spontaneously open into several pieces when they are ripe. The former may be either fleshy or dry.

A. INDEHISCENT.

They are generally designated by the name of *berry* (*baie*) (BACCA): we are content with this word if the envelope be fleshy; we add the epithet of *dry* (*sèche*), if it be of a foliaceous or ligneous consistence. The berry may come from a free ovary (in the SOLANUM, for instance) or adherent (as in the Grossulaceæ); with axilary placentation, as in the former; or parietal, as in the second; or central, as in the ARDISIA. Several modifications have received particular names. We have already spoken (§ 515) of the POMUM (*pomme*), the fruit of the Apple, of the Pear, and other Rosaceæ, with thick flesh, covered with the adherent calyx, and crowned by its dried limb, at the part which is commonly known as the eye; of the HESPERIDIUM (*hespéridie*), the fruit of the Orange, of the Citron, of the Lemon, and others of the same family, free, with loculi filled with succulent vesicles, clothed, as it were, by a membranous endocarp, the whole surrounded by a bark or skin of a certain degree of thickness. We term PEPO (*peponide*) that of Melons, Pumpkins, Gourds and other Cucurbitaceæ with thick flesh, leaving in the

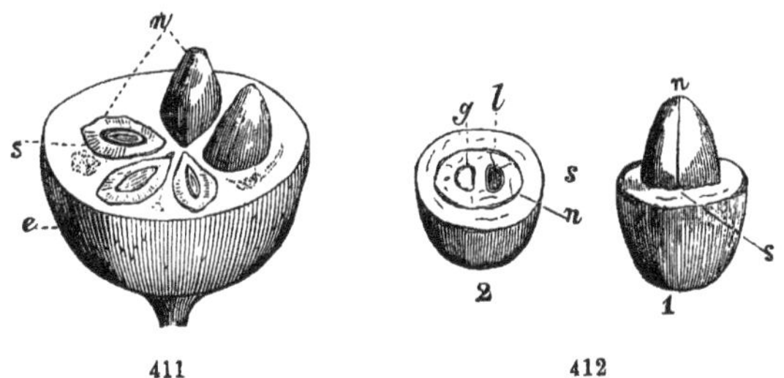

411. Fruit of the Common Medlar (*Nèflier*) (MESPILUS GERMANICUS). The upper part of the flesh has been removed so as to shew the kernels arranged in a circle around the centre.—*e* Epicarp.—*s* Sarcocarp.

412, 1. Fruit of the Cornel (*Cornouiller commun*) (CORNUS MAS). The upper part of the flesh *s* has been removed so as to shew the central kernel *n*.—2. The upper part of the kernel itself has here been removed, allowing us to see where it is hollowed by two loculi. One of them *l* is represented as empty, the other as filled by the seed *g*.

centre a cavity, on the walls of which are imbedded the seeds; NUCULANIUM (*nuculaine*), the fruit formed by the union of several drupes, consequently, presenting towards the middle of its substance so many seeds (*noyaux*) (PYRENÆ), which may grow from a free ovary, as in the Holly; or adherent as in the Medlar (*fig.* 411). Some authors call the latter modification the *pomum with kernels* (*pomme à noyaux*) and the former *pomum with pippins* (*pomme à pepins*). We may say also, instead of nuculanium, drupe with several seeds, indicating their number. We may conceive that the pyrenæ of a nuculanium may be united to one another, so that we only find one at the centre, and the fruit does not apparently differ from the drupe, such as we have already defined it. It ought, however, to be carefully distinguished, inasmuch as it comes from a compound ovary, and not from a simple carpel; and this is denoted by describing it as a drupe with a multilocular kernel, as in the Cornel (*Coroniller*) (*fig.* 412).

B. DEHISCENT.

§ 532. We may distinguish two degrees in the dehiscence of syncarpous fruits: 1st, The separation of the carpels from one another; 2ndly, The division of each particular carpel.

§ 533. The first degree, in which the carpels after being more or less united finish, in detaching themselves from one another, by becoming independent (CARPELLA AB INVICEM SOLUBILIA), evidently establishes the transition from apocarpous to syncarpous fruits, so that it is frequently difficult to determine to which of these two classes they belong: a fresh proof, that we must take care not to attach too great importance to these names in practice. The carpels, thus separated, may each remain indehiscent, as takes place in the Mallows, the Tropæolum or Indian Cress, the Umbelliferæ, &c. In these last (*fig.* 413), the carpels, instead of being completely detached, remain suspended to the axis, which is separated into as many threads as there are loculi, a particular arrangement, which has caused us to propose the name of *Cremocarp* (*Cremocarpe*) (CREMOCARPIUM) (from κρεμάννυμι, *hang*) for these kinds of fruits. In all these cases, when the loculus is monospermous, we may say that it represents an achænium, that it represents a samara when it is prolonged into a wing; its two loculi thus winged are separated when they are ripe, in the Sycamore (*fig.* 403), but remain united in the Ash (*fig.* 414) and the Elm, and all these fruits have been con

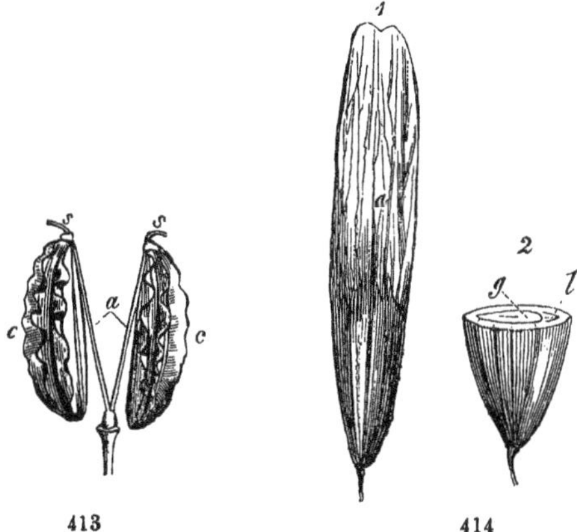

founded under one appellation, samara, which it would be preferable to confine to the simple carpel which presents this kind of characteristic, describing in these different cases the fruit as composed of several samara separating or not separating when ripe.

§ 534. The fruits, considered as truly dehiscent and known under the general name of *capsule* (CAPSULA), are those of which the carpels are themselves opened. But sometimes the sutures do not yield, and the pericarp is broken on the outside at a single constant point, either towards the top, as in the Snap-dragon (*fig.* 415, *t*), or towards the bottom, or in an intermediate point, as in the CAMPANULA (*fig.* 416, *t*). This opening, more or less irregular in its circumference, has the shape of a hole or pore, from which the pericarp is called *gaping* (*baillant*) (HIANS). In some fruits (those of the Red Pimpernel [*fig.* 417], of the Henbane, &c.), to which has been given the name of PYXIDIUM (*pyxides*) or CAPSULA CIRCUMSCISSA, or vulgarly Soap-box, the pericarp is cut transversely so as to be divided into two halves: the lower, which remains with the placenta attached to the torus; the upper, which is detached like a moveable covering (OPERCULUM). Is this singular dehiscence (CIRCUMSCISSIO) caused

413. Fruit of the PRANGOS ULOPTERA, after the dehiscence, which separated the two carpels *c c* and divided the axis *a* into two threads, to which these carpels remain suspended.—*s s* Persistent styles.

414. Fruit of the FRAXINUS OXYPHYLLA.—1. Undivided with its wing *a*.—2. The lower portion of the same divided horizontally, thereby shewing that it is occupied by two loculi, one of which *l* being abortive is reduced to a very narrow cavity; the other is very largely developed and filled by the seed *g*.

in a kind of transversal articulation analogous to that of the lomentaceous fruits? Does this transverse line correspond to a great effort or a less resistance at this height? Thus, in the fruit of the

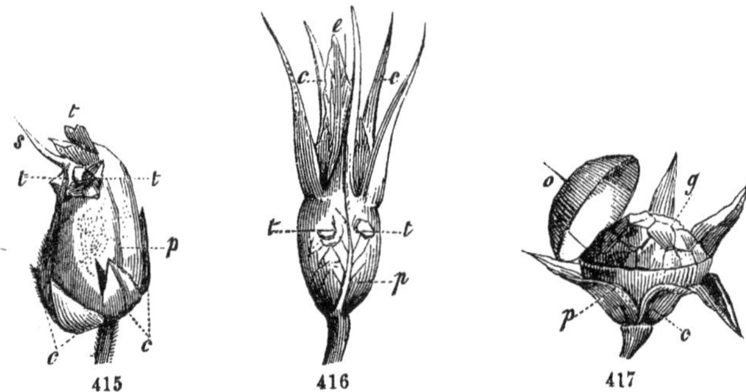

Lecythideæ, it is found precisely where the pericarp ceases to be doubled by the adherent calyx which is split in a circular direction.

§ 535. At other times, it is at the sutures that the dehiscence occurs; but they generally only yield incompletely at their upper part, and there is thus established at the top of the fruit an opening limited by the summits of valves which represent so many teeth, as in the Cerastium (*fig.* 418), Alsine and other Caryophylleæ.

§ 536. We now come to the most common case, that in which the sutures are completely disunited, so that the pericarp, throughout the whole or almost the whole of its extent, is separated from the summit to the base, or more rarely from the base to the summit, into several pieces or valves.

415. The capsule of the Snap-dragon (*Muflier*) (Antirrhinum majus) after the dehiscence.—*c c* Persistent calyx.—*p* Pericarp pierced near the summit, with three holes *t t*, which correspond two to one loculus and one to the other; the top surmounted by the remains of the persistent style.

416. The capsule of the Campanula persicæfolia, opening by holes *t t* beneath the middle.—*c* The persistent calyx blended at the bottom with the pericarp *p*, separating at the top into five tongue-shaped segments, in the midst of which we see the withered and crumpled corolla forming part of the induviæ *e*.

417. Pyxidium of the Pimpernel (*Mouron rouge*) (Anagallis arvensis).—*c* Persistent calyx.—*p* Pericarp, which has been separated into two halves, the upper of which is detached as an operculum *o*. We perceive on both three lines extended from the base to the summit of the fruit marking the sutures and, consequently, the true valves.—*g* Seeds forming a globular agglomeration around a central placenta.

418. Capsule of the Cerastium viscosum after the dehiscence.—*p* Pericarp, divided at the top into ten teeth, being the summits of as many valves *v* united together at the bottom of the capsule.—*c* Persistent calyx.

It may happen this disunion of the sutures may be preceded by that of the carpels themselves, i. e., of the loculi which they represent, and that these loculi begin to separate from one another by the deduplication of the dissepiments which kept them united (*fig.* 419). The dehiscence is then said to be *septicidal* (*septicide*): the

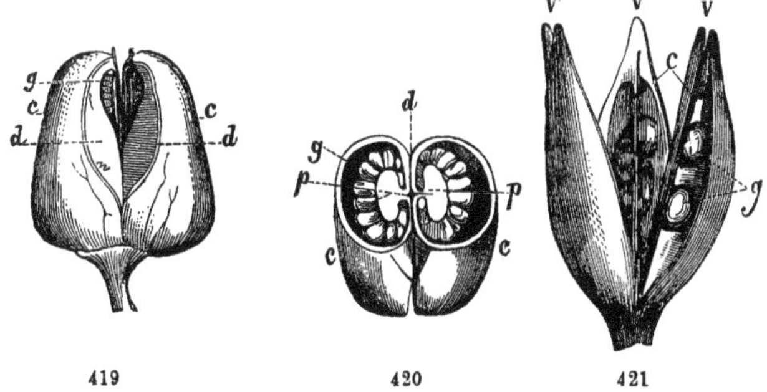

dissepiments form the sides of the valve, since this last corresponds to the carpel itself (VALVÆ SEPTIS CONTRARIÆ). At other times the dissepiments resist the separation, and the dorsal suture yields, thus opening at its middle the loculus which remains closed at the sides (*fig.* 421). This is the *loculicidal* (*loculicide*) dehiscence, by which the pericarp is divided into a certain number of pieces, each composed of two halves of the neighbouring united carpels, so that the dissepiments are placed on the middle of each of these pieces or valves (VALVÆ SEPTIS OPPOSITÆ). Sometimes the dissepiments yield along the whole length of their external edge and are thus separated from the valves (*fig.* 423): this is the *septifragal* (*septifrage*) dehiscence.

§ 537. In this last case, the dissepiments remain united to one another and to the axis, which in the centre of the fruit remains more or less developed, loaded with so many vertical laminæ as

419. The capsule of the Foxglove (*Digitale*) (DIGITALIS PURPUREA) at the moment of the dehiscence, which doubles by deduplication the partition *d* between two loculi *c c*, which thus reassume the appearance of distinct carpels. At the summit we see the inside of the loculi with the seeds *g*.

420. The lower part of the same, cut across so as to exhibit the composition of the partition *d* formed by the inner surfaces of the carpels *c* joined together.—*p* Placentas bent back and projecting into the inside of the loculi.—*g* Seeds.

421. The capsule of the HIBISCUS ESCULENTUS at the moment of the dehiscence.—*v v v* Valves.—*c* Dissepiment.—*g* Seeds.

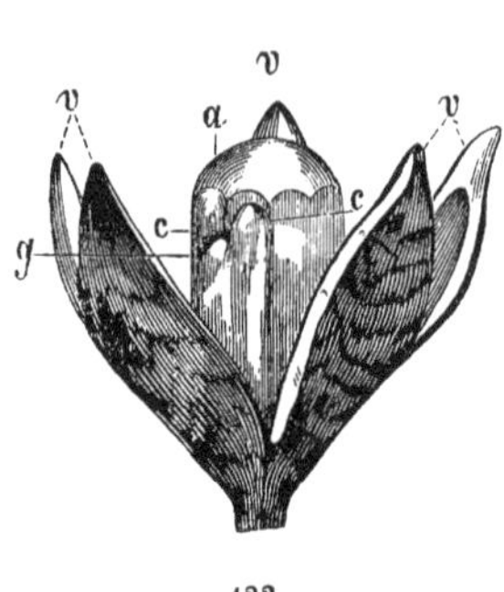

422

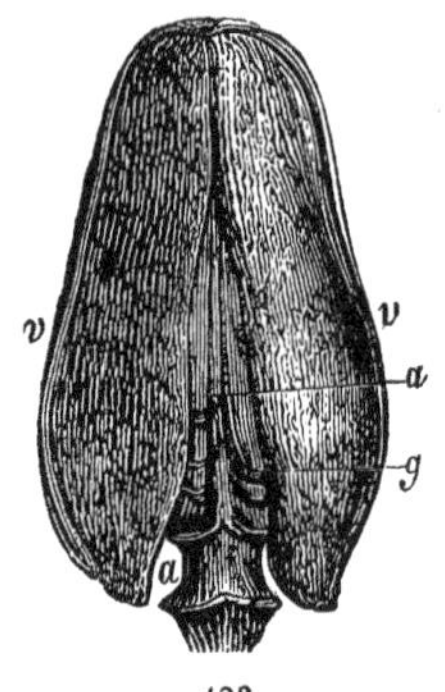

423

there are dissepiments, and in the angle, which is formed by their intervals, covered by the placentas to which the seeds are attached. In the capsules with central placentation, the body loaded with seeds, which occupies the middle of the loculus, is formed by the axis, quite similar to that which we have just described, except the dissepiments, whether they have disappeared on account of a premature rupture (§ 302), or have never even existed.

When the dissepiments are not separated from the valves, in the loculicidal and, especially, in the septicidal dehiscence, they are separated from the axis, and, if it be largely developed, we see it remain in the perpendicular direction under the form of a pyramid, of a cone, of a prism or of a cylinder, thus comparable to a kind of small column, and for that reason frequently termed COLUMELLA (*columelle*). Sometimes the placentas remain on this

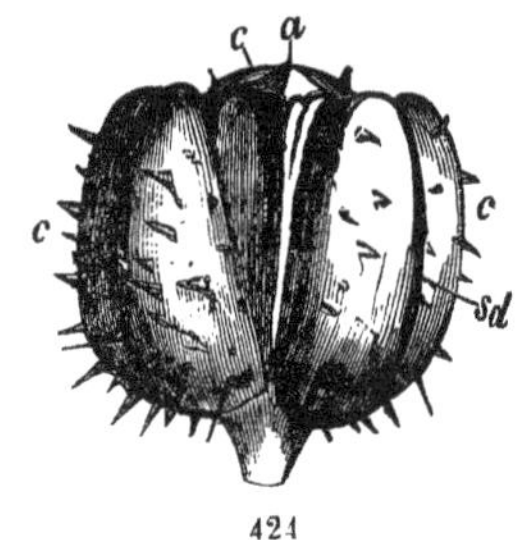

424

422. The capsule of the CEDRELA ANGUSTIFOLIA.—The valves *v v v* of which are separated by partitions *c c c* from top to bottom, so that the axis *a* remains in the centre, with five projecting angles corresponding to the partitions and separated by as many retreating angles which answer to the loculi and bear the seeds *g*.

423. The capsule of the Mahogany-tree (*Acajou*) (SWIETENIA MAHOGONI), which opens in a direction inverse to the preceding figure, that is from bottom upwards — The letters refer to the same as in the preceding figures.

424. The capsule of the RICINUS COMMUNIS at the moment of the dehiscence. The three carpels or cocca *c c c* are separated from the axis *a* which at first united them (see *fig.* 409), and which remains under the form of a small upright column. These cocca begin to open at their dorsal suture *s d*.

columella which is thus loaded with seeds, as in the EUPHORBIAS and other Euphorbiaceæ (*fig.* 424, *a*); sometimes they follow the edges of the carpels with the seeds and the columella does not bear them, as in several of the Malvaceæ.

It is evident, that the axis cannot be seen, when the placentation is parietal, since the vascular and cellular elements, which compose it, are then divided from the bottom of the loculus in order to form the placentas which follow the wall.

§ 538. We have said (§ 518) that the regular dehiscence generally takes place in the middle of the sutures formed by two united fascicles, which are disjoined when they reach maturity. But, sometimes the force of the union of these fascicles with one another is much stronger than with the rest of the walls, in which there happens what we frequently see in the material of our clothing, which tears along the side of the hem rather than come unsewn. In the same way, the pericarp may be broken on both sides of the placentiferous suture, which then forms a fillet of various degrees of thickness covered with seeds: the name REPLUM is employed by some authors to denote it. Although we have a few examples of this dehiscence in the capsules with axilary, it is more commonly observed in those with parietal placentation. Thus, in the fruits of Orchideous plants (*fig.* 425), in which the seeds are arranged in three longitudinal rows on the walls, we see, when ripe, the pericarp divided into six parts; three larger and thinner segments (*v*), which are detached at all parts and fall like valves; three segments (*p*), alternating with these valves, thicker and straighter, which continue to remain united at the top and at the bottom, and thus form a pericarp somewhat like a lantern. We find the inside of these three segments quite covered with thin seeds, and they correspond to the placentiferous suture.

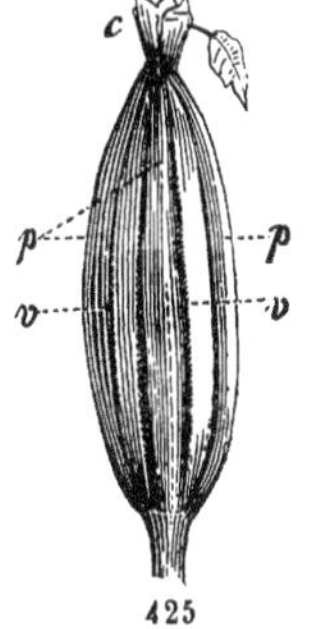

The fruit, so generally known under the name of SILIQUA (*silique*), is a capsule (*fig.* 426) analogous to the preceding one, except that it has two placentary lines only instead of three; so that, when it is

425. The capsule of the ORCHIS MACULATA at the moment of dehiscence.—*c* The remains of the calycinal foliole which crowns the fruit.—*v v* The segments of the pericarp which are detached in valves.—*p p* Segments which remain and bear the seeds.

ripe, after the two valves (*v*) are detached, the replum (*r*) remains under the appearance of a frame more or less elongated, studded along the edges with seeds (*g*) on the internal side of its boundaries. A thin lamina commonly fills the interior of the frame, thus forming a membranous dissepiment, which separates the cavity of the fruit into two loculi, contrary to the common custom, which is, that the dissepiments are stopped at the placentas and, consequently, that the parietal placentation necessarily causes the unity of the loculus. The siliquæ are frequently narrow and very long; when their length does not greatly exceed their breadth, we apply to them the diminutive, SILICULA (*silicule*). We may find every modification in the different plants of the large family of the Cruciferæ.

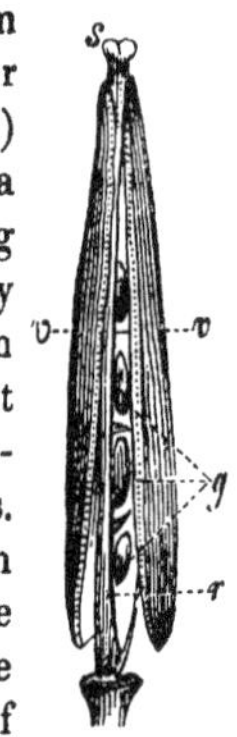

426

§ 539. In the most common case, in which the dehiscence takes place by the disunion of the dissepiments or of the sutures, it may occur by both at the same time; that is to say, it is at once both septicidal and loculicidal. If we look at the capsules of the small species of Flax, so common in our pastures (LINUM CATHARTICUM), we shall see the dorsal sutures separating at first, and each loculus thus open by the middle, so that then the dehiscence would be described as loculicidal. But a little afterwards, the dissepiments undergo deduplication in their turn and determine the separation of the loculi into so many distinct carpels or bivalved cocca, and the dehiscence then becomes septicidal.

After that the capsule is separated by the deduplication of the dissepiments into several carpels, these last represent so many folliculi, if they open by their ventral sutures only; if they open by both their sutures at once and are thus divided into two valves, they represent pods or legumes, containing a vertical row of seeds, or cocca, which only enclose a very small number. This last word is employed indifferently for apocarpous (§ 527) and syncarpous fruits; we say a bi- tri- multi-coccine (*bi- tri- multi-coque*) capsule.

§ 540. ANTHOCARPOUS FRUITS.—The fruit, besides its envelope formed of the pericarp, may have accessory ones furnished by some other part of the flower (ἄνθος) than the ovary. We have already seen, it is true, in several cases, the calyx accompanying the fruit;

426. The siliqua of the common Wall-flower (*Giroflée commune*) (CHEIRANTHUS CHEIRI).— *v v* Valves.—*r* Replum.—*g* Seeds.

but it was from the earliest period of its existence adherent to the ovary and partly confounded with it. It is otherwise in those fruits of which we speak. It is a whorl primitively independent of the

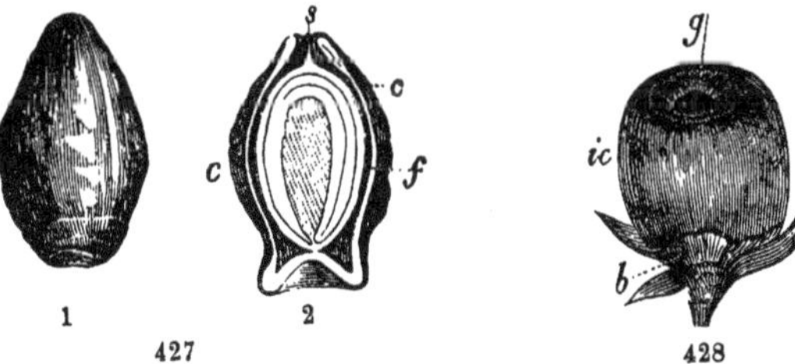

427 428

ovary, commonly a free calyx or involucrum, which, remaining around it, thickening or hardening in the same way as the pericarp, ends by forming a second external envelope over it. We may see it dry and representing a true achænium in the fruit of the MIRABILIS JALAPA (*fig.* 427), fleshy in the Sea Buckthorn (HIPPOPHAE RHAMNOIDES), in the Yew (*If*) (*fig.* 428), &c.

§ 541. AGGREGATE FRUITS.—The fruit, in all the modifications which we have hitherto examined, was the product of the pistil of the same flower. There are, however, some, although forming a single body, which grow from several different flowers. Thus, in the different species of Honey-suckle (LONICERA) we see two flowers grow from the same point, and their ovaries are thus brought into contact and unite, sometimes even so far as to be blended with one another, so that we have one single fruit really composed of two or more. In certain capitula or certain spikes, if the flowers are very close, the fruits, which will succeed to them, will not present any apparent difference from those which would follow a single flower, the carpels of which would cover an axis more developed in breadth or in length. Thus, at the first sight, the fruits of the Mulberry and those of the Bramble or of the Raspberry would appear to be of the same nature; and even the little succulent carpels of the Mulberry, united at their

427. The fruit of the MIRABILIS JALAPA.—1. Whole.—2. A longitudinal section shewing the parts which compose it.—*c c* The lower part of the hardened calyx forming the external envelope.—*f* The real fruit, hidden by the preceding. Its teguments are blended with those of the seed, which has also been divided. But it is easily recognised by the remains of the style *s* which surmounts the top.

428. The fruit of the Yew (*If*) (TAXUS BACCATA).—*b* Bracts imbricate at the base.—*ic* Fleshy envelope occupying the place of the pericarp and permitting us to see the top of the naked seed *g*, which it partly envelopes.

base into one single mass, would seem at first sight to be less independent of one another than those of the Raspberry, which are so clearly separated: and, yet, this latter fruit is the fructified pistil of a single flower; that of the Mulberry the union of the fructified pistils of a whole spike of flowers. Thus, we find at the base of the first a calyx for which we should search in vain in the second, in which the calyces are numerous and united to the bases of the pericarps.

The Pine-apple (*Ananas*) (*fig.* 429) represents the Mulberry on a large scale, and the Bread-fruit (*Arbre-à-pain*) on a still greater. In all they are spikes with compressed flowers, the pistils of which are united with one another; and the calyces, the bracts, the axis itself, distended with the same juices, contribute to augment this mass, in which they are fused into one another. The Fig presents something analogous, with this difference, that in it the dilated axis is bent around the mass of small fruits and thus forms the general envelope of the fruit (§ 209 [*fig.* 190]). In all these fruits we see the pericarp enriched by the association of some accessory parts, and, under these circumstances, they enter into the number of anthocarpous fruits.

§ 542. The *Cone* (*cône*) (STROBILUS), the fruit of the Pines, Firs, Cedars, &c., which has given the name of *coniferous* (*conifères*) (CONIFERÆ) to the family of evergreens, of which these form a portion, results from an analogous aggregation. It is a true spike more or less lengthened and loaded with scales, each of which bearing two ovules may well be compared to a carpellary leaf not bent back. They are very clearly independent in the cone of the PICEA; but in others they are sufficiently coherent to form by their union an apparently simple body. This body, which, in spite of its name, is far from presenting the conical form in all the plants of this family, assumes rather that of a spheroïd, when its scales are few in number, as in the Cypress (*fig.* 431); and even in the Juniper-tree (*fig.* 432), grouped in a globe, more fleshy and, consequently, united together, they represent a berry, which name the fruit of this tree wrongly, but commonly, bears.

429. Pine-apple (*Ananas*).—*a* Axis covered with fruits *c* crowded and united together into a single mass, and crowned by a tuft of leaves *f*.

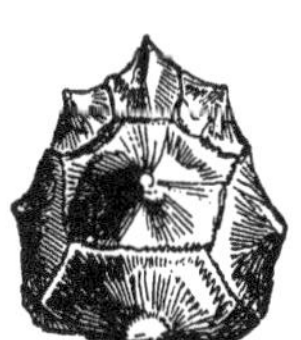

430

431

432

§ 543. Ripening of the Pericarp.—It now remains to enquire what changes take place in the substance which forms the pericarp, from the moment that it has passed from the state of the ovary to that of the fruit, when perfectly ripe. In this research, we ought to examine separately the pericarps, which preserve their foliaceous consistence, and those, which lose it and become fleshy.

The analogy of the former to leaves appears in their nutrition, as well as in their external characteristics. Like leaves (§ 292, 293), although to a weaker degree, under the action of light, they absorb from the surrounding air carbonic acid gas, discharging oxygen; in the night, they absorb oxygen and discharge carbonic acid gas. Their existence passes through the same phases; their tissues at first soft and rich in juices, are gradually solidified and, arrived at a certain point, begin to dry, to lose their green colour, to assume another, either that of a dead leaf or different tints analogous to those, which certain leaves assume in the autumn; and the withered pericarp continues to remain attached to the tree or falls, as it disarticulates. The phenomenon of disarticulation takes place at the union of the carpellary faces lying adjacent to one another and forming septa, and at that of the fascicles united into sutures thus determining the dehiscence. In a few thicker and indehiscent

430. The cone of the Pinus sylvestris.
431. The cone of the Cypress (*Cyprès*) (Cupressus sempervivens).
432. The cone of a Juniper (*Genévrier*) (Juniperus macrocarpa).

pericarps, maturity brings phenomena rather like those we observe in the bark; their external layers are detached as they split irregularly by a kind of decortication.

§ 544. During the existence of the fleshy pericarps, we distinguish two phases; one, where they grow like the greater part of the preceding, are of a green colour, disengaging oxygen during the day and carbonic acid gas during the night; the other, where they cease to disengage oxygen: this is the period of maturity and that which borders on it. The fruit increases by a large cellular developement; the vascular fascicles are not multiplied, or if they are multiplied, the stringy flesh does not acquire the desirable quality. The water, which comes with the sap relatively to the size of the fruit, is in proportion to the ripening of the fruit, although the evaporation is gradually diminished. The reason of this is, that a part of this water is fixed by combination with other principles. If it remains in the aqueous state, and continues to arrive in great quantities, the fruit, it is true, increases much more in size, but acquires much less flavour, as we find in moist summers, as well in the fruits growing on young trees as on those which grow in too moist a soil. The proportion of the lignine diminishes also, relatively to size, towards maturity; that of the sugar, on the contrary, is gradually augmented. Besides the water, the lignine and the sugar, we find in the flesh, gum, malic, citric and tartaric acids (this always, and the others sometimes combined with organic matters, as lime and potassa), vegetable albumen, and, lastly, an aromatic substance peculiar to each fruit. Such are the elements, which we meet in the generality of our fruits (those which we are naturally led to study), but mixed in different proportions in different kinds of fruits.

§ 545. The lignine, which is sometimes accumulated in such a remarkable degree in the cells of the endocarp, appears also to be very much developed in the sarcocarp of certain fruits, for instance, of the Pears, especially in certain varieties (the St. Germain, &c.), in which every one must have remarked the flesh quite sprinkled with small hard and stony-like granulations. They are so many cells encrusted with lignine, disseminated in small masses in the midst of the other cells, which are filled with juices more or less liquid. The lignine is generally in excess in young fruits, and its proportion gradually diminishes: its formation has ceased, although that of the flesh is still going on, and a part may, perhaps, have changed its nature. If we remember that starch combined with one or three atoms of water, becomes sugar (§ 309); that a small proportion of carbon and hydrogen added to the sugar causes it to become lignine (§ 312), we shall readily conceive how the lignine may change into sugar. Gum, which is identical in composition with starch, may easily undergo a similar alteration. We conclude, therefore, that it is by analogous changes of a part of these substances, that the fruit becomes so rich in sugar; for that which it contains has not been conveyed by the sap, since a fruit detached from the tree, and, consequently, receiving no sap, continues to ripen and to form sugar, and even gains more during its insulation.

Now, chemistry teaches us that these conversions take place under the influence of acids, and we have found acids to be more or less abundant in the fruit; that this action is aided by that of heat, and we know that heat greatly increases the ripening process, which is proved by the practice of gardeners and particularly by the use of walls. This effect is continued even after they are plucked from the tree, since the quantity of sugar in fruits increases as they are cooked. The acids appear, therefore, to contribute to the flavour of the fruit in two different ways: indirectly, by favouring the formation of the saccharine principles; directly, by mingling with them in a certain proportion, which is gradually weakened by the afflux of alkaline principles partly neutralizing the acids, when the fruit is ripe. We may mention as an example, the Grape; in proportion as it ripens, the tartaric acid, which abounds in it, attracts the potassa from the combinations, in which this alkali arrives in the fruit, and the increasing formation of the tartrate of potassa coincides with the diminution of the acidity, which at last almost entirely disappears. The purgative properties of certain fruits are owing to this presence of vegeto-mineral salts accumulated in their tissue.

The changes, which we have seen established during the ripening process in the relations of the fruit to the atmosphere, and which are resumed in a gradual loss of carbon and a gain of oxygen, to the ceasing of the evaporation of the water, which, stagnating in the pericarp, may take part in the fresh combinations of which it is the seat, appear to agree with those we previously discovered in the interior of the tissues.

§ 546. There is one principle of which we have not yet spoken, and which seems, however, to play an important part in several acts of the ripening process; it is that which forms what is termed vegetable jelly, the pectic acid. M. Fremy, who has made it his particular study, remarks, that in the fruit still green, in which the vital phenomena have all the activity, and, consequently, all the mobility of youth, it is very difficult to define the exact composition of the pulpy mass, which is modified without the least intermission. If we insulate this mass and treat it with acids, we obtain a substance soluble in water (which was before insoluble), composed of 24 atoms of carbon, 34 of hydrogen, 22 of oxygen, together with 1 of water: this is *pectine*, the gummo-gelatinous substance, which certain fruits, such as Pears, Apples, Gooseberries, Strawberries, &c., contain in large quantities. This pectine, in contact with the albumen, changes its characteristics without changing its composition and becomes *pectic acid*, insoluble in water, but possessing the property of absorbing it and of being changed into jelly. It is a body isomerical with the pectine, from which it differs only by the addition of an atom of water (2 instead of 1). We understand that these transformations take place spontaneously in the fruit; that the pulpy mass becomes pectine under the action of the acids which are there developed; the pectine becomes pectic acid under that of the albumen which is also found there. The housewife's processes of preserving fruits and making jellies from fruits are quite in accordance with these different notions.

§ 547. The presence of fecula in such abundance in fruits is almost always owing to its large developement in the seeds. But, if we separate them from the pericarp, the fecula will disappear, or, at least, will be more rarely found. We find it, however, in very large quantities in the Banana, and especially in the Bread-fruit, and mostly in those varieties, in which the flesh is developed at the expense of the ovules which are abortive.

§ 548. In a small number of fruits, the pericarp contains a large proportion of fixed oil. We need not adduce the Olive as an example. This is formed in the interior of special cells. The volatile oils also are formed in these cells, and will be found most frequently in the pericarps of those plants the leaves of which also contain it. But, in both, this oil is secreted and accumulated (§ 251) in utricles of a peculiar form differently grouped, in the vesicular glands. We may under this relation compare the pericarps of the Orange, of the Lemon, of the Rue, the Dittany, &c., with their leaves. We remark that, in these fruits, the pericarp is not very fleshy but sometimes even foliaceous.

§ 549. What is the precise time of the ripening of the pericarp?

The time is pretty clearly determined by that, which immediately determines the dehiscence in those pericarps, which are foliaceous or dehiscent; but in those which are fleshy, it is much more uncertain, since each day brings fresh changes in the composition of the fruit, and since it is not fixed in a certain state of equilibrium, in which the combinations that take place are maintained without alteration during some time. In common parlance, in those fruits which we eat, we have agreed to call them ripe, when the combination of the different sugared, acid and other principles has imparted to them the highest degree of flavour, which from that moment begins to deteriorate. Then, in the different fruits, this maximum does not evidently refer to the same period, since in taking, for instance, the half-decayed state of a soft pear, this pear is still eatable, although it has lost the greater part of its qualities; an apple, at the same point, is in a state of rottenness; a medlar is, on the contrary, only eatable when perfectly decayed.

§ 550. However this may be, the same thing occurs in fruits that we have shewn in other tissues, once deprived of their vital principle (§ 317): a more or less gradual combustion, resulting from the combination of the oxygen of the air with the carbon of the vegetable, induces the disengagement of carbonic acid and sometimes of other carbonised gases and of water, the phenomena of fermentation or decay. The pericarp is thus softened and separated into pieces; and the seed, which, far from participating in this decomposition, has grown in this atmosphere of carbonic acid and water, is at last perfectly free, disengaged from the envelopes which imprisoned it in the fruit.

OVULE AND SEED.

§ 551. Whilst we were occupied with the ovary, we had occasion to speak frequently of the bodies enclosed in its cavity, called *ovules* (OVULA), on account of their analogy to the eggs of birds; since like them they are developed to a certain point attached to the mother-plant, are then detached and continue to develope themselves into a being similar to that from which they have sprung. We have seen, that they are found at certain parts of the walls of the cavity or loculus of the ovary, and that at these parts there is a peculiar modification of the tissue of the walls, so that the nourishment is transmitted from the base of the ovary to the interior of the ovule. This nutritive system generally consists of a small fascicle of tracheæ surrounded by lengthened cells, the whole commonly covered with shorter cells, bearing greater similarity to the rest of the parenchyma of the walls of the ovary. A swelling arises from this, and it is called PLACENTA, if it corresponds to a single ovule, *placentary* (*placentaire*) if it corresponds to a row of several ovules. Sometimes the ovule springs immediately from the placenta; it is sessile; sometimes they are fastened to one another by an elongation, most commonly narrow, which presents the same structure and is called the *funicular* or *umbilical chord* (*funicule*) (FUNICULUS) (§ 523). The point, by which the funiculus adheres to the ovule, has received the name of HILUM (*hile*) or more anciently UMBILICUS (*Ombilic*). We shall soon learn to distinguish on the surface several other external points in relation to its internal parts, and which it is, consequently, necessary to understand well.

§ 552. We must at first determine the position of the ovules relatively to the loculus which encloses them. Let us commence with the most simple cases, in which the loculus encloses only one, *uniovulate loculus* (*loge uniovulée*), and let us suppose that the ovule has its most common form, that of an ovoïd attached by a rather short funiculus, which affects nearly the same direction as it does. The placenta may be situated at the very base of

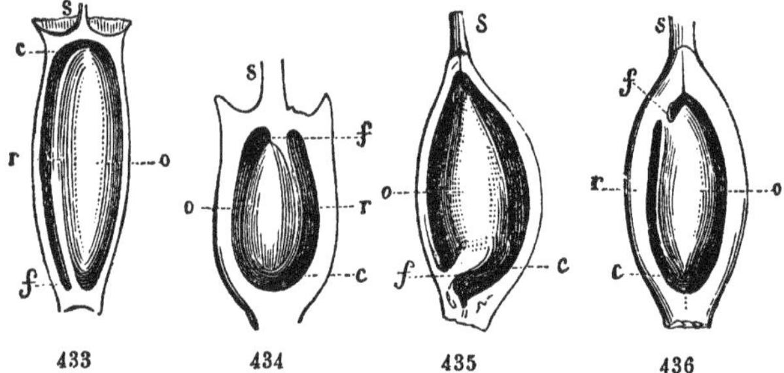

433—436. Carpels belonging to several flowers cut longitudinally so as to shew the various directions of the single ovule *o* which is enclosed in it.—*f* Funiculus.—*r* Raphe.—*c* Chalaza.—*s* Base of the style.

433. A carpel of the Groundsel (*Seneçon*) with erect ovule, anatropous.

434. A carpel of the Mare's-tail (*Pesse*) (HIPPURIS VULGARIS) with reversed, anatropous ovules.

435. A carpel of the Pellitory (*Pariétaire*) (PARIETARIA OFFICINALIS) with ascending, orthotropous ovule.

436. A carpel of the DAPHNE MEZEREUM with a pendulous, anatropous ovule.

the loculus, and the funiculus, as well as the ovule, be elevated in almost a vertical direction (*fig.* 433); we then call it *erect* (*dressé*) (ERECTUM). It may, on the contrary, be situated at the upper end of the loculus, from which is suspended into the interior the funiculus with its ovule, which is then said to be *inverse* (*renversé*) (INVERSUM [*fig.* 434]). It is usually found, as we have said, on the side of the loculus where the placenta corresponds to its dorsal, or more frequently, to its ventral suture; if it is attached near the top and its apex is looking downwards, it is *pendulous* (*pendu*) (APPENSUM [*fig.* 436], PENDULUM); if it is attached near the base and the apex is directed upwards, it is *ascending* (*ascendant*) (ASCENDENS [*fig.* 435]); if it is towards the middle, the ovule may direct its point, either towards the base or towards the top of the loculus, and we apply to it, according to these cases, the two preceding epithets. In some cases it assumes a direction nearly horizontal and we then use this adjective.

§ 553. Sometimes, although rather rarely, the funiculus, very much lengthened, follows a direction precisely inverse to that of the ovule; it rises vertically from bottom to top, and the ovule, attached to its extremity, falls from top to bottom, as in the STATICE (*fig.* 437), or, on the contrary, the funiculus hangs in the direction of the base, whilst the point of the ovule looks towards the top, as in the majority

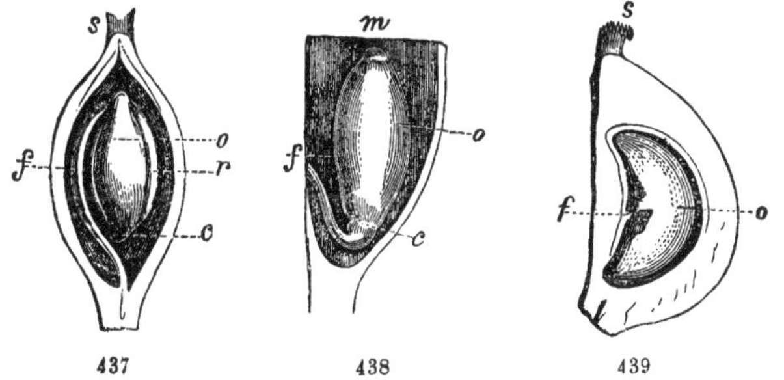

437, 438. Carpels, the solitary ovules of which are directed inversely to the ovary. —The letters have the same signification as in the preceding figures.

437. The carpel of the STATICE ARMERIA with a pendulous ovule from the end of an erect funiculus.

438. The carpel of the ZYGOPHYLLUM ALBUM with an ascending ovule from the end of a pendulous funiculus. The hilum is confounded with the chalaza *c*; the micropyle *m* at the opposite extremity

439. The carpel of the MENISPERMUM CANADENSE with a curved or campylotropous ovule.

of the genus ZYGOPHYLLUM (*fig.* 438). Care must be taken to indicate this double circumstance by a short phrase, saying a suspended ovule from an erect funiculus, an ascending ovule from a suspended funiculus (OVULUM FUNICULO ERECTO APPENSUM, E PENDULO ASCENDENS).

§ 554. Some embarrassment might arise, when the ovule, instead of being straight, is bent on itself. If this bend is not very large, we take no notice of it and describe the direction of the ovule as if it were straight. If it be very much pronounced, so that the two extremities of the ovule are very close to one another and are turned towards the same point of the loculus (*fig.* 439), we indicate this conformation in adding the epithet of *campylotropous* (*campulitrope*) (from καμπύλος, *crooked*; τρόπος, *direction*).

§ 555. Let us now suppose a case a little more complicated, that in which there are two ovules in one and the same loculus, *biovulate loculus* (*loge biovulée*). They may, by being inserted at the side of one another, follow the same direction, and we call them *collateral* (*collatéraux*) (COLLATERALIA [*fig.* 440]); or, more rarely, follow an inverse direction, so that, for instance, one is suspended and the other ascending, as in certain SPIRÆAS, in the Horse Chesnut (*fig.* 441). They may also be inserted at unequal heights, so that they are placed one above another, *superposed ovules* (*ovules superposés*) (OVULA SUPERPOSITA) and, in this case, they generally have the same direction.

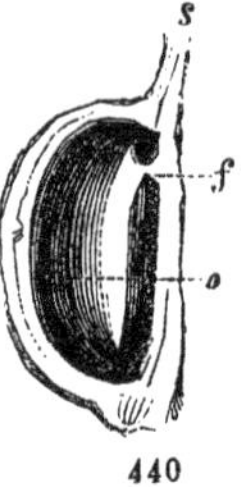

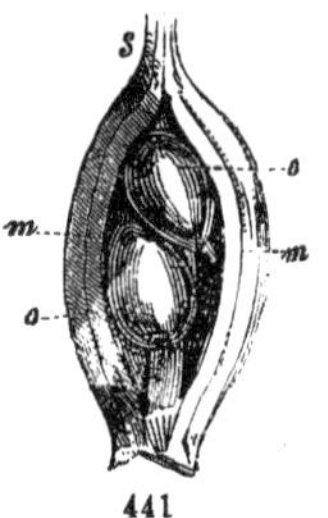

The same rules are observed, when there is in each loculus three ovules attached either at unequal heights or at the same height. In the latter case, they generally take different directions; one at the top, the other at the bottom, the other intermediate; the first ascending, the second pendulous, the third horizontal. It is

440. The carpel of the NUTTALLIA CERASOIDES with two collateral pendulous ovules.—The letters refer to the same parts.

441. A loculus of the ovary of the ÆSCULUS HYBRIDA, opened so as to shew the two ovules inserted at the same height, but directed inversely.—*m* Micropyle which indicates their summit.—The letters still refer to the same parts.

almost a necessary result from the field given to their developement, when the placenta is at about the half of the height of the loculus

§ 556. The direction of the ovules becomes less and less constant in proportion as we find a much greater number in the same loculus, *multiovulate* (*multiovulée*) and inserted on a much smaller space; for it is evident, that, as in the preceding case and with much greater reason, they will be developed according to the space which is presented to them, that is, the lower ones from the top downwards, the upper ones from the bottom upwards, those of the middle in intermediate directions (*fig.* 442); then, mutually crowding and pressing against one another, as they are developed, they will become angular at their surface and the polyhedric will be substituted for the ovoïd form. But, if the loculus is lengthened and the ovules are superposed (as in the Leguminosæ [*fig.* 443] or the Cruciferæ), they will not crowd one another and all will be in the same direction.

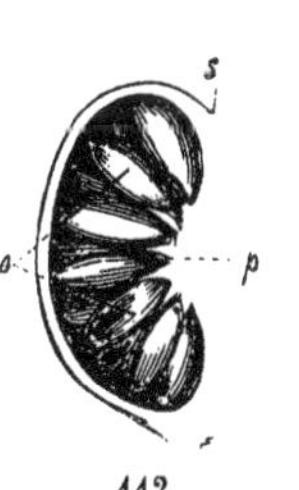

442

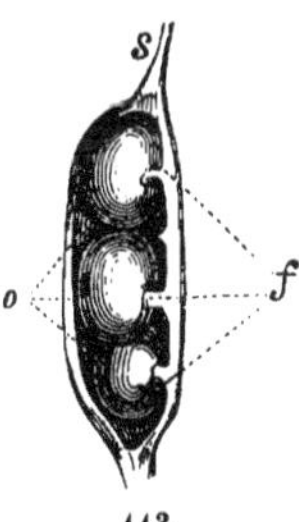

443

§ 557. In all these cases we employ the terms mentioned above to denote these directions, which, as we see, greatly depend on the shape of the loculus and the situation of the placentas. The position of the hilum, either at the top or the base of the ovule, determines its ascending or pendulous state.

But in this manner we have only hitherto learned to recognise the situation of the ovule relatively to the loculus which encloses it, and we may meet with some difficulties; for instance, if the hilum is placed at the middle of the ovule and no longer at one of its extremities. We should proceed with more certainty, if we could in every case recognise by constant characteristics a base and an apex in the ovule, and by determining these two points, discover its absolute direction. Now, observation will teach us these points. We shall learn to recognise them by studying more intimately the structure of the ovule, which we have considered up to this moment

442. A loculus of the ovary of the PEGANUM HARMALA, with numerous ovules inserted on a projecting placenta *p* in several different directions.

443. A carpel of the ONONIS ROTUNDIFOLIA with several campylotropous ovules placed above one another.

only in a very general manner, in its relations with the other parts, and not with regard to those which constitute it. The best manner of studying the ovule, is to follow it in its developement from the moment when it begins to appear to that when it attains its perfect state.

§ 558. We shall find the most simple example of an ovule in the Miseltoe. It first appears at the bottom of the loculus under the form of a small nipple composed of uniform cells; it is then lengthened into an ovoïd mass which gradually thickens, but always formed of a homogeneous tissue. At a certain time, this mass grows hollow (*fig.* 445, *c*) towards its centre; and then, after the fecundation has taken place, towards the top of this cavity there begins to appear a fresh body suspended by a delicate thread resulting from the union of several cells. This body, the forms of which will be more and more determined, is the outline, as it were, of the little fresh plant, the *embryo* (*embryon*). The name of NUCLEUS (*nucelle*) has been given to the cellular mass (*fig.* 446, *n*), which in these cases constitutes the ovule; of SUSPENSOR (*suspenseur*), to the delicate thread by which the embryo is attached to its summit. We may use the term of *embryonary cavity* (*cavité embryonaire*) for the cavity which is made at the centre of the nucleus.

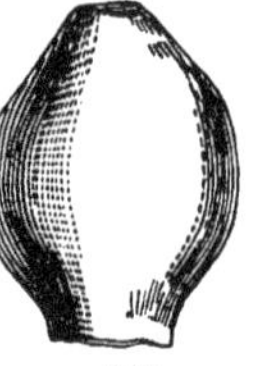

444

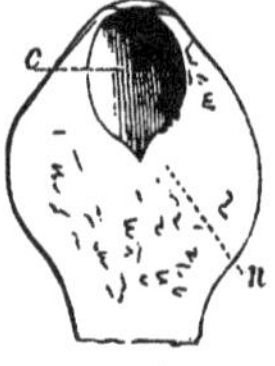

445

§ 559. In other plants, before the appearance of the embryo, the internal cavity is covered by a membrane which is commonly single; with a kind of bladder or sac, which is gradually extended from the apex to the base, adhering to the surrounding tissue by its two ends, but scarcely touching it at any other part of its surface. This is the *embryonary sac* (*sac embryonaire*). Sometimes its continuity with the base of the nucleus is broken, or rather merely exists on account of a series of some accessory utricles united end to end.

§ 560. But we have yet to state the most common case, when the nucleus instead of being thus naked in the loculus of the ovary, is covered with an external envelope. This appears later than the nucleus under the shape of a small circular enlargement at the base (*fig.* 446, 1, *t*). This enlargement is gradually lengthened into a

444. Ovule of the Miseltoe, undivided.

445. The same, divided so as to shew the embryonary cavity *c* and the rest of the substance *n* formed of a uniform tissue and thus constituting a nucleus without teguments.

sheath, beyond which we still see the summit of the nucleus project (*fig.* 446, 2, *t*), but which at last covers it completely like a sac. The upper opening of this sac is contracted in the same proportion and is at last reduced to a very small hole always corresponding to the point of the nucleus; this hole has been called the *micropyle* (*micropyle*) (MICROPYLUM, from μικρός, *small*, and πύλη, *gate*). All these changes may be very easily followed in the ovule of the Walnut.

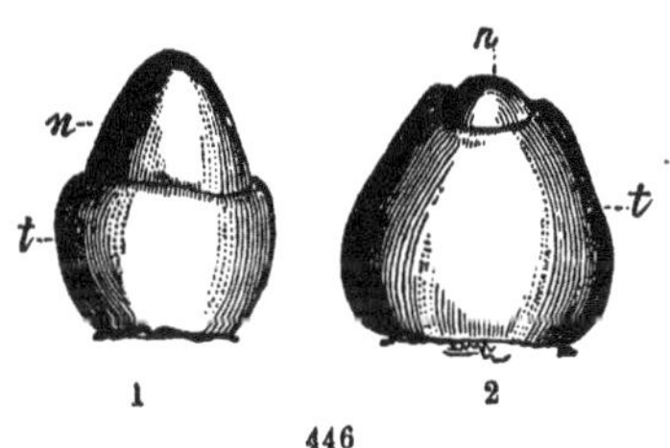

§ 561. But much more commonly a second envelope is formed, and then, above the first enlargement we find a second which grows in the same way and at the same time as the first; so that the nucleus is surrounded by two sheaths or coats fitting one over the other, the inner one at first longer than the outer one (*fig.* 447, 3), which,

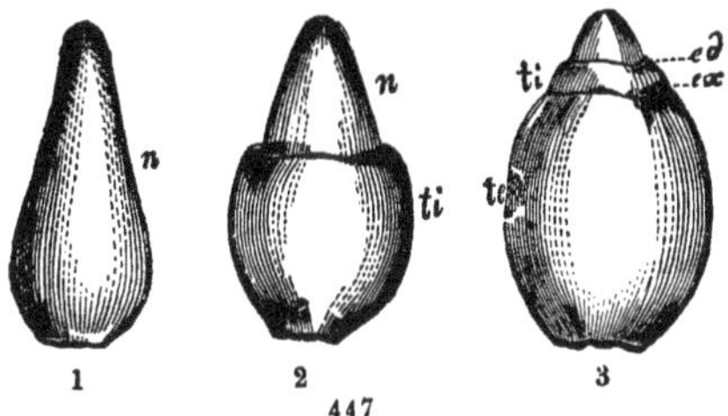

however, frequently overtakes it and passes over it in its turn. When both completely envelope the nucleus, we still observe above its point a micropyle, which in this case is composed of two openings, the one (*ex*) corresponding to the outer tegument, and which M. Mirbel calls *exostome* (ἔξω, *without*, and στόμα, *mouth*); the other (*ed*), corresponding to the inner tegument, which he terms

446. Ovule of the Walnut (*Noyer*) (JUGLANS REGIA).—*t* Simple tegument.—*n* Nucleus.—1. The first stage, in which the tegument covers only the base of the nucleus.—2. The second stage, in which the nucleus is almost completely covered.

447. The ovule of the POLYGONUM CYMOSUM at different ages.—*n* Nucleus.—*te* External tegument.—*ti* Internal tegument.—*ex* Exostome.—*ed* Endostome.—1. The first stage, when the nucleus is still naked.—2. The second stage, when the nucleus is covered at its base by the internal tegument, as yet single.—3. The third stage, when the two teguments form a double sheath, beyond the summit of which the nucleus projects.

endostome (ἔνδον, *within*). These two openings may exactly correspond and thus form a small canal, or else not correspond, if one of the two teguments passes much beyond the other.

§ 562. The complete ovule is, therefore, composed of a cellular nucleus, within which we find a cavity covered by the embryonary sac; enveloped on the outside with two other sacs or teguments, the one external, the other internal, adhering to it at the base and open at the opposite extremity. Their texture is cellular; it has been remarked that their cells are often two rows thick, and that those of the internal tegument commonly present the same appearance as those, which form the external layer of the nucleus, something like an epidermis: whence some botanists have concluded that this tegument is formed by a fold of the nucleus.

§ 563. These different parts have received different names. M. R. Brown, who claims the priority of pointing out the real structure of the ovule, calls the teguments TESTA and *internal membrane;* the pulpy conical mass the NUCLEUS; the embryonary sac AMNIOS° M. A. Brongniart calls this nucleus the *amande* surrounded by a TESTA and by a TEGMEN. Amongst the authors who preceded them, some few of them had paid some attention to the ovule, since we already find some good remarks on its organization in the writings of Malpighi and Grew, but they always considered the two external envelopes as a single one. M. Mirbel, to whom we owe the most complete account of the gradual developement and different modifications of the ovule, proposes to name all these sacs fitting one over the other, according to the order in which they are placed above one another from the outside to the inside, *primine*, *secundine* (*secondine*), *tercine*, *quartine*, and *quintine*. This last is the embryonary sac. The quartine is a layer sometimes formed at a later time around the sac, and appears to be rarely found and very transitory, so that it has escaped the observation of most authors. Other names also have been proposed. But we shall content ourselves with employing here those of which we have made use in the preceding explanation, those of the simple or double tegument, the one external, the other internal, of nucleus and of embryonary sac.

§ 564. In the ovule, such as we have described it, the base by which the nucleus is continuous at the middle with the placenta, at the outside with its own teguments, is filled in the inside with a peculiar tissue, denser and generally of deeper colour than the other parts,

° From ἀμνίον, *the membrana fœtus* or *caul*, (ἀμνός, *a lamb*).—TRANS.

frequently formed of lengthened cells pressed parallelly against one another, and in which the fibro-vascular fascicles, coming from the placenta and destined to nourish and support the ovule, expand and are terminated. This tissue forms an areola rather clearly defined to which has been given the name of CHALAZA[p] (*chalaze*). It is clear that it here precisely corresponds to the hilum, that is, to the point where the fascicle from the walls of the ovary are attached to those of the ovules. If the ovule be developed uniformly throughout the whole of its extent, all these points, the hilum with the chalaza and the micropyle, situated at the two opposite extremities of the ovules, preserve their primitive relations : this ovule is *straight* (*droit*), or, according to the nomenclature of M. Mirbel, *orthotropous* (*orthotrope*) (from ὀρθός, *straight*).

But, it very frequently happens that the developement does not proceed so equally on all sides; that on one it is very great, whilst it is almost stationary on the opposite side. Hence, the point of the ovule, with its micropyle at first turned upwards, is directed on one side (*fig.* 448, 3 *n*) then in a little time outwards, lastly downwards (4 *n*) after having thus made half a revolution. The chalaza, conveyed in the same way along with the extended teguments and preserving its relations with the micropyle, makes an analogous revolution, but in an inverse direction, and proceeds from bottom upwards; so that it departs farther and farther from the hilum, the

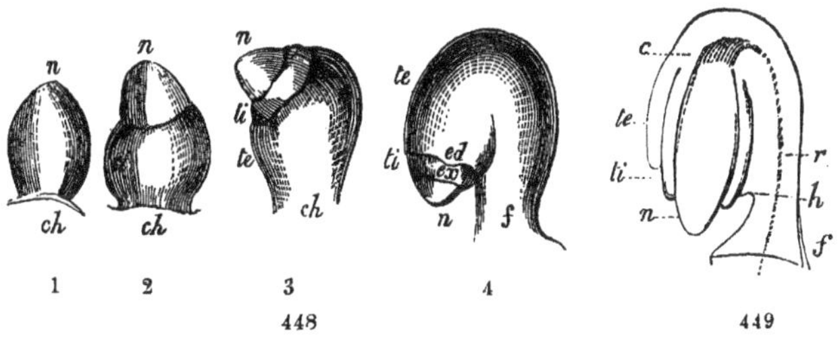

448. The different ages of the ovule of the CHELIDONIUM MAJUS.—*h* Hilum.—*c* Chalaza.—*f* Funiculus.—*r* Raphe.—*n* Nucleus.—*t i* Internal tegument.—*t e* External tegument.—*e d* Endostome.—*e x* Exostome.—1. The first stage. The nucleus is still naked.—2. The second stage. The base of the nucleus is covered by the internal tegument.—3. The third stage. The external tegument in its developement has covered the base of the internal. The ovule on account of the developement of one of its sides, has begun to bend back and turn its point on one side.—4. The fourth stage. The ovule has bent itself quite round and has turned its point downwards.

449. The same figure, cut longitudinally so as to shew the relations of the different parts.

p From χάλαζα, literally, *anything let loose*, thence *a pimple* or *tubercle*.—TRANS.

micropyle, on the contrary, approaching nearer and nearer. We may say, that the ovule is then *reflexed* (*réfléchi*), or, according to M. Mirbel, *anatropous* (*anatrope*), (from ἀνατροπὴ, *overturn*). The vascular fascicle, terminating at the chalaza, follows it in its revolution, at the same time, of course, being lengthened, and this elongation forms, in the substance of the teguments (of the external, when there are two), a small cord or riband, which, coming from the hilum, is terminated at the chalaza. We have called this cord Raphe (*raphé*) (from ῥαφὴ, *a seam*).

§ 565. At other times the ovule in its developement is bent or folded on itself, so that its upper half is in a direction nearly inverse to the lower, and that its micropyle approaches, as in the preceding case, the hilum. In this *bent* (*recourbé*) ovule, sometimes the two sides are almost equally developed, *camptotropous ovule* (*o. camptotrope* of Schleiden, from καμπτὸς, *flexible*); sometimes the outer side is much more developed than the internal one, *campylotropous ovule*[q] (*o. campulitrope* [*fig.* 450]), and then the chalaza (*c*) has been carried back a little to the outside of the hilum, which is found between it and the micropyle, these three points being very close and turned in the same direction. It frequently happens that the two faces correspond to the concavity of the curve and are even united together.

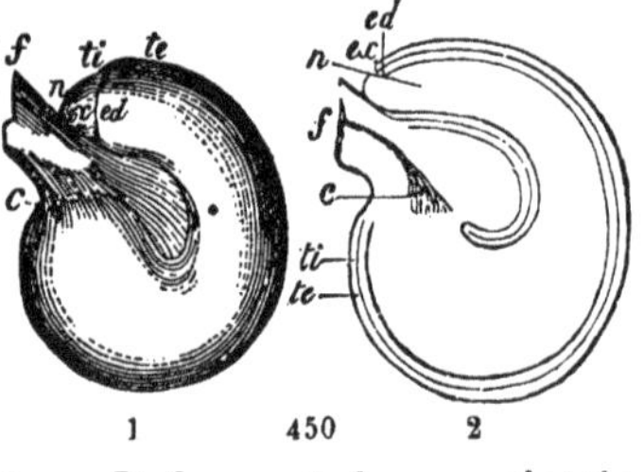

§ 566. The cavity of the ovule is bent, when the ovule is curved backwards, straight, when it is straight. The point of the nucleus generally continues to correspond with the micropyle, because its developement and that of its envelopes proceed at an equal pace. But, if they are unequally developed, it is clear, that they no longer exactly correspond; and this sometimes, though rarely, takes place, but only after the process of fecundation.

§ 567. We see that, to determine the absolute direction of the ovule, we must find out three points: the hilum; the chalaza, which we may consider as its *organic base*; the micropyle, which

450. The campylitropous ovule of the Wall-flower.—1. Undivided.—2. A longitudinal section.—The letters refer to the same parts as in the preceding figure.

q From καμπύλος, *crooked*.—Trans.

we may consider as its apex. The first two are generally seen so much the more clearly as the ovule is more advanced; the last, on the contrary, has a tendency to be gradually effaced. Its position, from which we shall be able to deduce that of the embryo, is no less necessary to be stated, and its physiological functions are extremely important, since the pollen-tube, having passed through the conducting tissue of the style to the cavity of the ovary, insinuates itself through this opening into the ovule and is thus placed in direct connection with the nucleus.

§ 568. Sometimes on the walls of the loculus we find above the ovule a small fleshy swelling, which at a certain time covers, as it were, its summit, and a small part of it is even inserted into the canal of the micropyle, connected doubtless, to the axis of the fecundation. This is the origin of certain *caruncles* (CARUNCULÆ) (*caroncules*) which we afterwards observe on certain seeds.

§ 569. At other times its origin is different, and it is the funiculus itself which is expanded close to the seed, thus developing quite another part, and by extending itself over the surface of the ovule, envelopes it more or less completely and forms what we call an ARILLUS (*arille*). This commenced, as in the preceding cases, with a swelling of the funiculus, which is gradually expanded into a

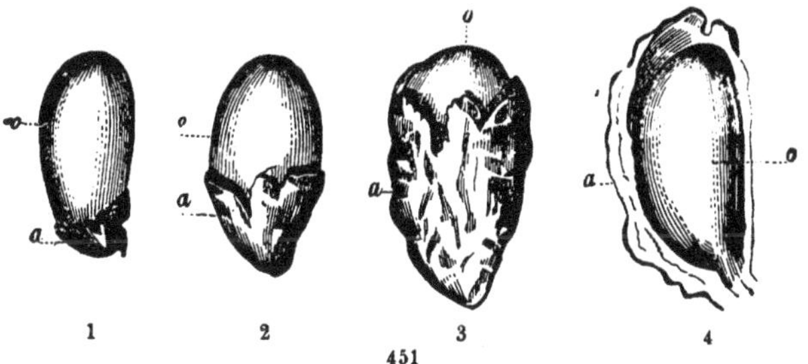

451

kind of cap (*fig.* 451, 1, *a*), then into a sac surrounding more or less loosely a part or the whole of the ovule (2, 3, 4, *a*), open at its other extremity, sometimes, however, as in the NYMPHÆA, completely closed. Its developement, which may be easily followed in a seed of the Spindle-tree (*Fusain*) (*fig.* 451), is, therefore, analogous to

451. The process of the developement of the arillus *a* round the ovule *o* of the Spindle-tree (*Fusain*) (EUONYMUS EUROPÆUS) during four successive stages 1, 2, 3, 4. In 4, the arillus has been divided longitudinally and permits us to see its relations to the ovule, which it completely envelopes.

that of the other teguments; but, it may be easily distinguished, not only because it is afterwards formed, because it constantly begins from the hilum, and, consequently, takes a direction inverse to the others which begin from the chalaza, but also by its consistence and its whole appearance. It is frequently fleshy, painted with brilliant colours, elegantly fringed at its edge, as in the URANIA and HEDYCHIUM, and sometimes in the form of network, as in the Nutmeg (*noix de Muscade*) (MYRISTICA MOSCHATA), in which it constitutes what is called Mace (*macis*).

§ 569, *b*. We have seen that the tubes emitted by the pollen-grains, when placed on the stigma, are lengthened through the interstices of the conducting tissue which is in the canal of the style, and thus come to the inside of the loculus into the neighbourhood of the placentas; that they there meet the ovules, which the wide mouth of the micropyles presents to them; that they enter into them, and that after that the connection is thus established between the extremity of the pollen-tube on one hand and the extremity of the nucleus on the other, we see a fresh body, the embryo, appear at the summit of the cavity formed in it. Now, it may rather frequently happen, that this connection is not established, that the ovules do not receive the pollen-tube: these are then arrested in their developement, they are abortive; and this is frequently the reason, why, among the ovules of the same loculus, some of them only ripen. When they are numerous, the abortion of a part among them is rather usual. It is not even rare for all those of the same loculus to escape fecundation, and, in this case, we see it gradually waste away and more or less completely disappear. The other loculi and the fecundated ovules continue, on the contrary, to grow, and even with so much the more strength as they have profited by the juices, which would have been absorbed and employed by those that have proved sterile.

§ 570. GRAIN OR SEED (*Graine*).—Let us examine the successive changes that we observe in these fecundated ovules which assume the name of *grain* or *seed* (*graine*) (SEMEN). We suppose the ovule to be as complete as possible, that is, a nucleus covered internally with an embryonary sac, externally with a double tegument. Sometimes all these sacs, thus fitting one another, remain and grow together, some parts more, some less, so that we find them in the ripe seed (*fig.* 452). More frequently, some are gradually effaced and at last disappear, whilst the others, on the contrary, are remarkably developed in several of their dimensions. Thus, the two teguments are most commonly blended into one, either because they are closely united together, or because one of the two, the inner one, most frequently becomes thin and wastes away. We frequently find the nucleus disappearing, forced outwards by the embryonary sac and the fresh body which fills it, as it increases. Thus pushed back, the nucleus may be expanded and become thin under the form of a membrane; it may even, whether it joins and is thus blended with the teguments, or is completely absorbed, leave at a certain period only feeble traces, or even no traces at all of its anterior existence. As to the embryonary sac it is preserved more generally, but it changes its nature; for a cellular net is organized over its internal face as over a mould, and we then have a sac formed no longer of one single cell, but of a layer of cells united together. This is the reason, why we find in the ripe seed the envelopes of the embryo most commonly reduced to two instead of four: the one, external, comprising the two blended teguments of the ovule; the other, internal, the origin of which varies, since it may result either from the reduction of the nucleus, or from the embryonary sac, or from both united together, or, lastly, in a few cases from the internal tegument, which has not been united to the external one. In the seeds, in which the whole of this developement has not been observed with strictest attention, it is almost impossible to pronounce to what part of the ovules the modified envelopes correspond, to discover what has been reabsorbed or wasted away, what has been united and blended together. We ought, consequently, to be contented to describe the actual state of things: the most common state is the existence of two envelopes; we generally give to the external one the name of TESTA, to the internal one that of *internal membrane* (*membrane interne*) (MEMBRANA INTERNA).

§ 571. But other changes have at the same time taken place in the interior of the growing ovule. After the appearance of the embryo, the embryonary sac is filled with a mucilaginous fluid, which is soon organized into a cellular tissue, at first soft and loose. This organization proceeds from the outside to the inside, and the utricles, at first soft and floating, are in a little time deposited on the walls of the sac, then others are applied to this layer and thus gradually thicken it. There may be established a nearly similar formation outside of the embryonary sac, consequently, in the nucleus itself, which is thickened by a cellular developement. This case is precisely the reverse of that we have shewn in the preceding paragraph, of that in which the nucleus, crowded and gradually absorbed, disappears.

§ 572. These juices, at first demi-fluid, then organized into a continuous tissue, are destined for the nourishment of the young embryo, which itself continues to grow (*fig.* 474); sometimes it absorbs them before this tissue is solidified, and, always advancing, gradually invades the whole interior of the seed, and at last fills it, covered immediately by the envelopes which we have just described.

§ 573. At other times, it takes up much less room and the remainder is occupied by this last formed tissue, either in the nucleus, or more commonly in the embryonary sac, or in both at once (*fig.* 452); a tissue, then forming a solid mass, to which has been given the name of *perisperm* (*perisperme*) (PERISPERMUM) (from *περὶ, around*; *σπέρμα, seed*). Richard called it *endosperm* (*endosperme*), and Gærtner before him ALBUMEN. This last name, which is that of the white of an egg, was borrowed from the comparison of our vegetable egg with that of birds; a comparison, which, although false in certain points, is so true as to enable us to form a good conception of the structure. We know, indeed, that in the egg the young animal developed at a point on the surface of the yolk or vitellus, absorbs for its nourishment this yolk, then the white surrounding it placed under the shell doubled by a membrane. It was natural to compare to this egg the embryo or young vegetable situated in the same way within these two concentric depôts of different substances collected together, the outer one in the nucleus, the inner in the embryonary sac, thus comparable in their relation to the albumen and to the yolk or vitellus; and Gærtner has finished the comparison between the two by giving the latter term to the internal perisperm in those rare cases, in which we meet two of them in the ripe seed.

This is what we see, for instance, in that of the NYMPHÆA (*fig.* 452), in which the developement of every pre-existent part of the ovule is observed with great clearness. Under a thin arillus (*a*) which covers this seed, under rather a thick testa (*t*) and a fine membrane (*mi*) representing the two teguments of the ovule, we find a large farinaceous body (*n*) filling almost the whole of the seed; but the axis of which is occupied by a kind of long tube, attached beneath to the chalaza and expanded above into a small sac (*se*) with thick walls, within which is the embryo (*e*). It is difficult not to recognise in this the embryonary sac, thickened by a cellular developement at its extremity where that of the embryo is stopped; not to recognise in the farinaceous body the nucleus developed to a much more remarkable degree.

§ 574. It has been proposed to distinguish these two deposits originating in different ways, by different names: to call endosperm, that which is formed in the embryonary sac, perisperm or albumen that which is formed in the nucleus. This would indeed be a valuable distinction, if it could be constantly established. But, the developement of the seed in a large majority of known plants has not been followed, and even in those which grow under our eyes, these researches, requiring a long and constant practice and minute observation, has only been prosecuted in a small number. We ought, therefore, in the actual state of science to be contented with a single term, that of perisperm, which alone we should employ in the majority of these doubtful cases: and we might, in those in which its origin has been often determined, qualify it by the addition of an epithet, that of internal or endospermous, of external or albuminous.

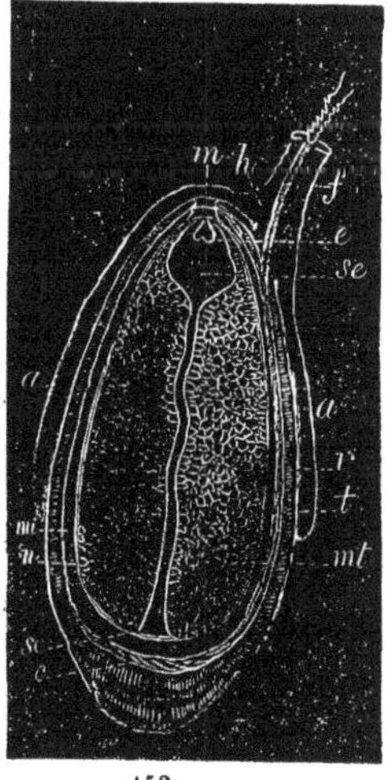

452

§ 575. According to M. Schleiden, certain perisperms would have a different origin from any of the preceding. Thus, in the CANNAS, the ovoïd body of the ovule shews on its upper half the distinction of the nucleus enveloped with its tegument. The whole of the lower half is occupied by a continuous mass which seems to belong wholly to the chalaza. The embryonary sac of the apex of the nucleus gradually extends from the bottom to the top and is buried in this mass, which continues to grow and at last surrounds the greater part of the sac and of the embryo, which has been developed in the interior. This part thus forms the perisperm, which may in this case be termed chalaza. It is composed of cells, the greater part lengthened into

452. A young seed of the NYMPHÆA ALBA. A horizontal section.—*f* Funiculus.—*a* Arillus.—*r* Raphe.—*c* Chalaza.—*h* Hilum.—*m* Micropyle.—*t* Testa.—*mi* Internal membrane.—*n* Farinaceous perisperm formed by the nucleus.—*se* Fleshy sac or inner perisperm formed by the embryonary sac.—*e* Embryo.

small cylinders and directed from the teguments towards the surface of the embryo.

§ 575 *b*. However it may be, the perisperm varies in its nature and its consistence and thus furnishes useful characteristics for the determination of the seeds. 1st, Its cells are rather frequently filled with granules of fecula, and it is then said to be *farinaceous* (*farineux*) (*fig.* 453). To this property of the perisperm several seeds,

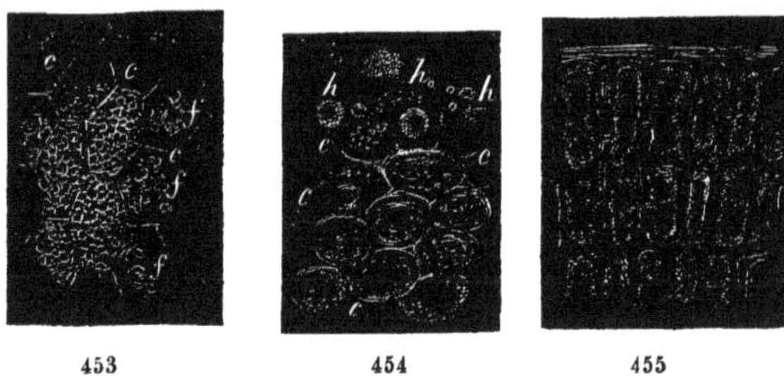

453 454 455

those of the CEREALES, for instance, owe their nutritive powers. It has been supposed that this modification would generally correspond to the developement of the nucleus, to the endospermous perisperm. 2nd, At other times, these cells acquire rather large dimensions, at the same time preserving a certain degree of softness, and the seed would then be called *fleshy* (*charnu*). In this case, within the cells oil is sometimes formed, as in the Castor-öil plant, and the seed is termed *oleaginous* (*oléagineux*) (*fig.* 454). 3rd, These cells may acquire along with great thickness a very great degree of hardness, almost that of horn, and the perisperm is *horny* (*corné*), as in the Date (*Datte*) (*fig.* 455), the Coffee (*Café*) and the IRIS. Then, generally, in a very thin section, we can easily perceive under the microscope these cells, the cavity of which is small, the wall very thick and formed of several layers fitting on one another, and frequently perforated with small canals of communication with one another.

The solution of iodine is very useful in determining the nature of the perisperm. It discovers the slightest traces of fecula by rendering it blue. Under the appearance of a small demi-solid mass which

453. A section of the perisperm of Maïze.—*c* Cells.—*f* Granules of fecula which they contain.

454. A section of the CROTON TIGLIUM.—*c* Cells.—*h* Drops of oil which they contain.

455. A section of the Date.

it renders yellow, we distinguish the azotised matter, the existence of which is so general in the seeds, inasmuch as it is necessary for the first developement of the tissues. These substances, which form almost the whole of the contents of horny perisperms, are not wanting in others, and, in the farinaceous are associated withth e fecula. It is the gluten in the Cereales.

It is clear that these characteristics must be looked for in the ripe seed. They are only established by degrees; and when the perisperm has begun to be organized in the fecundated ovule, the cellular tissue which composes it may present a few differences in its structure, but not yet in the consistence of its walls and of the substances formed within them.

§ 576. EMBRYO[r] (*Embryon*).—Whilst these different changes are going on in the envelopes of the seed, other changes are taking place in the embryo, its most essential part, to which all the others are necessarily subordinate. Let us now examine this developement of the embryo, by looking back to its first appearance at the time of fecundation, that which answers to the close connection established between the pollen-tube and the apex of the nucleus. At the corresponding point of the cavity of the latter, most commonly doubled by the embryonary sac, we observe a simple vesicle (*fig.* 456, 1, *v*), at first filled with a demi-fluid substance containing granules, in which we soon find a utricle formed, then several others (*fig.* 456, 2, *e*), all provided with a nucleus (§§ 21, 336), which is generally very apparent. They are commonly united end to end in a series, of which the whole of the upper portion forms the *suspensor* (*suspenseur*), the lower extremity of which forms the embryo, limited at first to a single utricle, soon composed of several united together into a small mass (*fig.* 456, 3, *e*). In this developement the mother-cell or embryonary vesicle has not been long in disappearing.

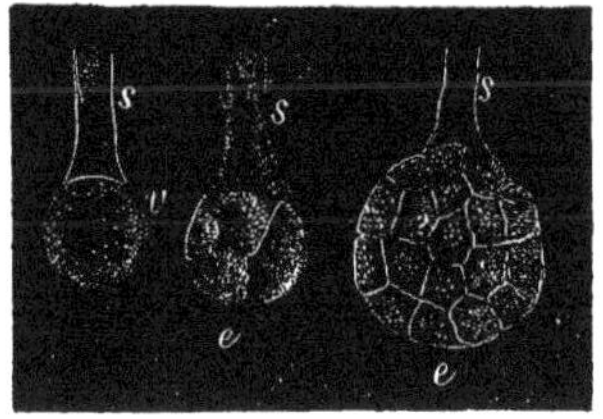

456

456. The developement of the embryo of the DRABA VERNA.—*s* Suspensor.—*v* Embryonary vesicle.—*e* Embryo.—1. The first stage, in which we only find the embryonary vesicle.—2. The second stage, in which several utricles are formed in the vesicle.—3. The third stage, when the embryo has become more manifest from the formation and conglomeration of a larger number of utricles.

r From ἔμβρυος, *the fœtus*.—TRANS.

The suspensor frequently does not exceed this degree of tenuity, at other times it is lengthened and strengthened by the addition of fresh cells; it almost always disappears, however, when the embryo, for sometime suspended by it from the apex of the sac, has acquired a certain size.

§ 577. We have already (§§ 27, 28, 29) explained the progressive changes, the constituent parts and the principal modifications of the embryo. We have seen that this small cellular mass, at first undivided, is afterwards separated so as to establish the distinction of several parts; we have found an axis and small lateral excrescences, sketches, as it were, of the first leaves; among these first leaves one or two, which we call cotyledons, present a peculiar form and structure, and, according to the unity or the plurality of the cotyledons, a fundamental difference is established between three great classes of vegetables. This difference increases in proportion to the developement of the plant. But, we have examined the embryo only as independent of the seed, and we have, besides, treated it in too general a manner to render it unnecessary to revert to it here more in detail.

The axis is formed the first, turning one of its extremities in the direction of the suspensor and the other towards the opposite side. Now, the former is always that, from which the root will afterwards spring, and assumes in the embryo the name of *radicle* (*radicule*); the second is that which will be lengthened into the stem, covered with leaves, and which, as a beginning, throws out the cotyledons. We distinguish, therefore, a radicular and a cotyledonous extremity. The radicular, being immediately continuous with the suspensor, is, consequently, turned towards the apex of the nucleus and the micropyle, which corresponds to it; the cotyledonous, directly opposite, is, therefore, turned towards the base of the nucleus, i. e., the chalaza: and these first relations will be almost always maintained, so that on the inspection of the seed, it is sufficient to be able to find the chalaza and the micropyle, to determine with great exactness the two corresponding extremities of the embryo still concealed under its envelopes.

Let us remark that this direction of the embryo or of the fresh plant is precisely inverse to that of the mother-plant, since we may consider the nucleus as forming the culminating point of the latter, and since the embryo is reversed with respect to the nucleus, turning upwards the point, which will at some future time be developed into a root, downwards the point, which will be developed into a stem.

This consideration establishes an essential distinction between it and the common buds (*bourgeons*) which we may compare to it, but which constantly continue the direction of the plant on which they are born.

§ 578. In the seed of a small number of vegetables, especially of several of the parasites, the embryo is limited to the axis, then undivided, as we may see, for instance, in the CUSCUTA or Dodder, (*fig.* 457); or if the cotyledons exist, it is in the rudimentary state, and they are often so small that we hardly recognise them, as in the PEKEA (*fig.* 458), and it sometimes even requires the aid of the microscope to see them, as in the Orchideæ. These cases are rare, and we commonly observe in the ripe embryo, besides the more or less voluminous cotyledons, the leaves which will follow, then amassed into one first extremely small bud, which we have called *gemmule*.

These different parts present rather marked differences according as the cotyledon is single or double. Let us examine them successively in both cases.

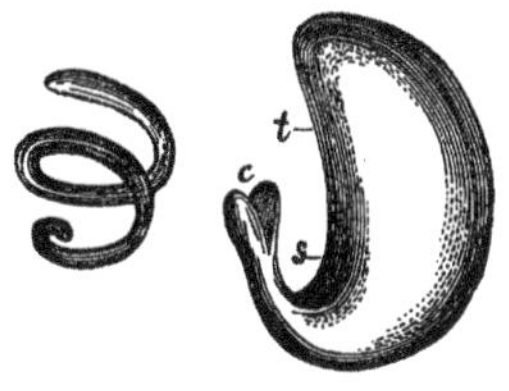

§ 579. MONOCOTYLEDONOUS EMBRYO (*Embryon Monocotylédoné*).—The most usual form of the Monocotyledonous embryo is that of a cylinder round at its two extremities or that of an ovoïd more or less lengthened (*fig.* 460). On the outside, it is difficult to distinguish the different parts; but, on cutting it vertically through the middle, we observe, at a variable height, a small nipple-like excrescence in a cavity immediately below the surface: this is the gemmule: the upper termination of the axis, to which the whole of the portion situated beneath belongs; a part, which is almost wholly composed of the small stem

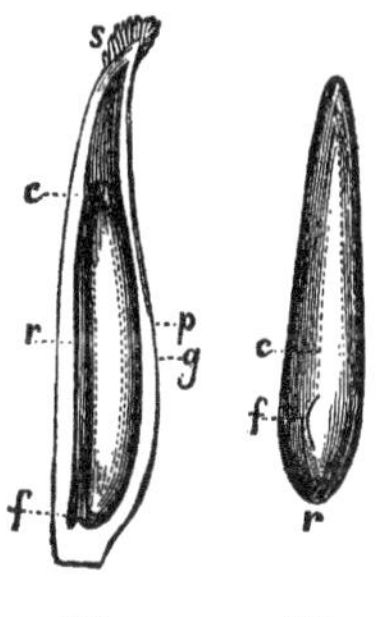

457. The embryo of the CUSCUTA.

458. The embryo of the PEKEA BUTYROSA.—*t* The large tigelle, forming almost the whole mass, bent back at its narrow extremity which lies close against the furrow *s*.—They have been separated in the figure to expose the two rudimentary cotyledons *c* which terminate it.

459. A vertical section of a carpel of the TRIGLOCHIN BARRETIERI.—*p* Pericarp surmounted by the sessile stigma *s*.—*g* Seed.—*f* Funiculus.—*r* Raphe.—*c* Chalaza.

460. The embryo.—*r* Radicle.—*f* Slit corresponding to the gemmule.—*c* Cotyledon.

or *Tigelle* of this shortened vegetable, but which is commonly called *Radicle* (*radicule*) (*fig.* 260, *r*), because it will be lengthened downwards into a root. The whole of the portion situated above the gemmule is the cotyledon (*fig.* 260, *c*). With strict attention, on examining with a lens of sufficient power the fresh or moistened embryo, it is possible to determine these different regions, even without dissection; for we may almost always discover a small slit (*fig.* 260, *f*) or external button-hole, which corresponds to the gemmule, most frequently also indicated by a slight projection on the surface of the embryo, and thence we know the limit between the radicular portion turned towards the micropyle and the cotyledonous portion turned towards the chalaza. To what do this cavity which contains the gemmule, this thin layer which covers it, correspond? We have stated that the cotyledon is nothing else than the first leaf of the little vegetable, and the gemmule the union of the leaves which will follow it. Now, if we compare a mass of common leaves extremely young with this, and take as a point of departure one of them which is already sufficiently developed for us to recognise a small sheath surmounted by a limb, we shall find the following ones enveloped by this sheath, beyond which they hardly project. These have absolutely the same relation to the first as the gemmule has to the cotyledon. The hollow part, which we observe at the base of the cotyledon, is, therefore, nothing else than its vaginal part; the slit, than the meeting of the edges of this part, close together or even covering one another. We are confirmed in this opinion by following the whole developement of the cotyledon, which at first appears under the form of a small nipple, is gradually lengthened, then acquires a vase-like form, at its base, whence there begins to project another nipple, the gemmule, at first free, then gradually covered by two small laminæ which grow from the two edges of the vase-like hollow. We find there the developement of a leaf (§ 147) the limb of which is the first to appear, then the vaginal part indicated at first by a simple swelling, and gradually forming a sheath for the other leaves situated nearer to the centre.

All this is very manifest in certain embryoes, as those of the Dioscorea and of other plants of the same family (*fig.* 461), the cotyledon of which (*c*) presents a thin dilated limb like that of a small true leaf, with a sheath (*g*) which surrounds the gemmule without entirely covering it. But, more generally, the form of the cotyledon is very far from that of the other common leaves, assuming the form of a cylinder, a cone or a club.

Sometimes the gemmule appears to be more or less free on the outside, whether the edges of the sheath do not join, or they are lengthened and expanded into a thin membrane. We can often recognise only a single leaf turned in the opposite direction to the cotyledon, so small are all the following ones; at other times we discover one or two more, rarely above this number, all of them successively decreasing.

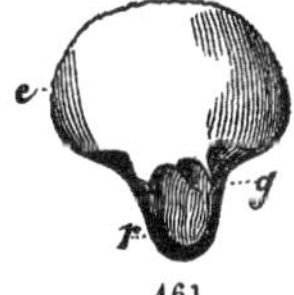

The radicle is, in some embryoes, as long and even longer than the cotyledon (*fig.* 76, *t*), and we then term them *macropodous* (*macropodes*) (from μακρός, *long;* πούς, ποδὸς, *foot*). Sometimes they are even extended laterally, so as to form a kind of excrescence. This may be extended so far as to constitute the greatest part of the embryonary mass. But more usually its radicle is, on the contrary, much shorter than the cotyledon; it is also, generally, thicker and of a more compact tissue. Frequently a small point appears at its extremity, where the suspensor ends and where at a later period we shall find the first root. But it is not the extremity itself which will be lengthened to form it, and we have already seen (§ 111) that it is a kind of internal swelling, which, piercing the external layer, will be thus developed.

§ 580. Dicotyledonous Embryo (*Embryon Dicotylédoné*).—The form of the Dicotyledonous embryoes is much too varied for it to be possible for us to express its shape in any general manner. Sometimes in the shape of a very long cylinder or ovoïd, they are similar to the Monocotyledonous; but they are always distinguished from them by the division of the cotyledonous extremity into two lobes; this division is more or less deep, according as the cotyledons are more or less developed with relation to the axis or tigelle which bears them. A very common form is that which we have already had occasion to point out and describe (*fig.* 77) in those of the Almond-tree, where two oval cotyledons (*cc*), pressed against one another, constitute the greater part of the embryo, whilst the axis is reduced to a much narrower and shorter body, so that it seems on the outside only like a small cone (*r*) projecting beneath the cotyledons; this lower portion below the cotyledons is the radicle, the extremity of which, as we have already said (§ 111), will be in a little time lengthened into a root. The other portion of the axis above their insertion,

461. The embryo of the Rajania cordata.—*r* Radicle.—*c* Cotyledon.—*g* Sheath, which conceals the gemmule.

the gemmule, sometimes hardly developed and concealed between them, is only seen after we have artificially separated them. It is frequently terminated itself by two small lobes, sometimes presents a much larger number of these lateral lòbes, the first sketches of the leaves, at other times appears still undivided.

§ 581. In a very great number of the embryoes, the cotyledons are equal. It is, nevertheless, rather frequent to observe between them a slight inequality, but too slight for us to take any notice of it. In some cases, however, it becomes very apparent and may even be so to such a point that one of the two cotyledons appears to be almost entirely wanting. We almost always find it, it is true, on looking carefully for it, but reduced to a simple rudiment, as in the TRAPA, a few HIRÆA (*fig.* 462), &c. &c.

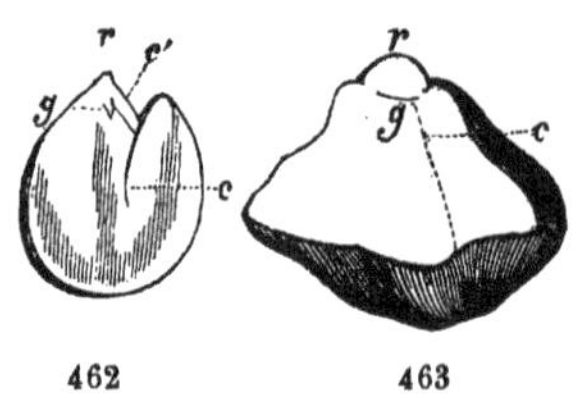

§ 582. At other times the appearance of unity in the number of the cotyledons is owing to another cause, viz. both are united more or less perfectly into a single body (*fig.* 463). But, then the gemmule (*g*) is not, as in true Monocotyledonous embryoes, situated near the surface and even in communication with the exterior by means of a small slit. It occupies a cavity situated quite in the inside and in the elongation of the axis. Besides, we may almost in every case recognise that the cotyledonous body is double from the traces, which the seam (*c*), as it were, leaves over the whole extent of the united faces; and if these traces are wanting, by studying the embryo in its younger state, before the cotyledons are thus joined and blended together, as in the TROPÆOLUM.

§ 583. But let us leave these unusual arrangements and let us take the most habitual, that in which the two cotyledons are equal and only contiguous. Sometimes they acquire great thickness, as in the Almond, the Haricot-bean, the Oak, &c., and we then say that they are fleshy; in these cases, the two faces in contact or the internal are generally plane; the free faces or external, more or less convex. Sometimes they are compressed into their laminæ, as in

462. The embryo of the HIRÆA SALZMANNIANA. A vertical section shewing the inequality of its two cotyledons, one of which *c* forms almost the whole of the embryonary mass.—*c* The small cotyledon.—*g* Gemmule.—*r* Radicle.

463. The embryo of the CARAPA GUIANENSIS. A vertical section shewing the union of the cotyledons, being only distinguishable from one another by a very fine line *c*.—*r* Radicle.—*g* Gemmule.

the Castor-oil plant, the Spindle-tree, &c., flattened on both of their faces, and we then call them foliaceous. The true nature of the cotyledons, the first leaves of the young plant, disguised in the first case, is more or less manifest in the second. Generally, their circumference is entire, even in those vegetables in which the subsequent leaves will be more or less profoundly cut; on the other hand, there are several in which they are already lobed, as in the Walnut, the Oak, the Lime-tree, (*fig.* 464). The foliaceous nature is again

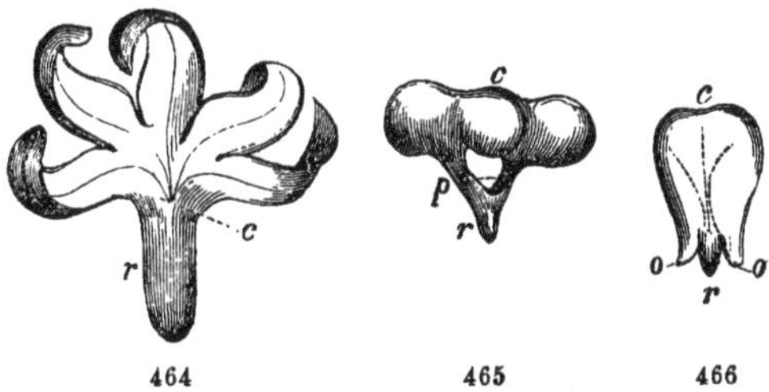

indicated by nerves already more or less evident, commonly very few on the fleshy cotyledons, and increasing in number in proportion as they decrease in thickness. We frequently observe at this early period stomata on the surface of the latter kind. Lastly, the cotyledons may be petiolate (*fig.* 465); that is, separated from the axis which bears them by a short contraction. More commonly they are sessile, formed only by an expansion or limb which is inserted immediately on the axis. It is not uncommon to see this scolloped base prolonged on both sides into a lobe below the point of insertion: if these two lobes are large and wide enough, the cotyledon is heart-shaped or cordi-form; if they are short and narrow like two small ears, the cotyledon is said to be ***bi-auriculate*** (*fig.* 466).

§ 584. We sometimes find more than two cotyledons. This may take place in certain seeds of plants, in which, nevertheless, the normal number is two; these are exceptional facts, analogous to those which exhibit leaves usually opposite to one another, in pairs in certain plants, becoming whorled in threes by some unwonted arrangement.

But there are other plants, in which the existence of more than

464. The embryo of the Lime-tree.—*r* Radicle.—*c* One of the cotyledons.

465. The embryo of the GERANIUM MOLLE.—*r* Radicle.—*c* Cotyledons which are attached by a foot or petiole *p*.

466. The embryo of the Elm.—*r* Radicle.—*c* Cotyledon.—*o o* Lobes, causing the cotyledon to become *bi-auriculate*.

two cotyledons in a whorl, is a constant and normal fact, as in several of the Coniferæ, principally in the Pines (*fig.* 467) and Abies, in several species of which we find the number of cotyledons raised as high as six, nine or even fifteen. In this case, their form is linear, as it will be afterwards in the leaves: and let us remark, that these leaves united into a fascicle on small, contracted, almost obliterated branches, will frequently present in their turn an analogous arrangement which we may study in Pines, the Larches, &c.

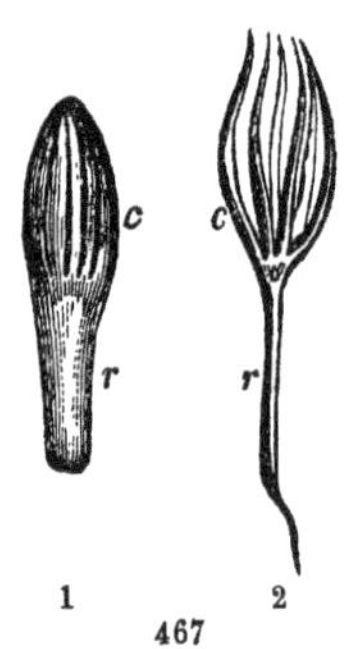

467

§ 585. This large number of cotyledons has caused some botanists to propose to substitute for the generic term of *Dicotyledonous*, that of *Polycotyledonous* vegetables. But the former answers for the majority, or rather for nearly the whole of these vegetables; it has been generally and for a long time received and ought, consequently, to be preserved. We shall only, therefore, be obliged to recall to our mind, that the essential difference of the embryoes in these two great classes of vegetables is, that these first leaves always grow alternately in some (monocotyledons), in others (dicotyledons) in whorls, either, most usually, in pairs, or, very rarely, in a larger number.

§ 586. We have said that the two cotyledons are most frequently found touching one another by their smooth plane faces; but they frequently also present other arrangements, analogous to those which we have pointed out in the leaves properly so called before their developement, when they are packed in the bud (*bourgeon*) in the state of vernation (§ 174). Thus, they may be bent into two halves, *reclinate* (*ré-*

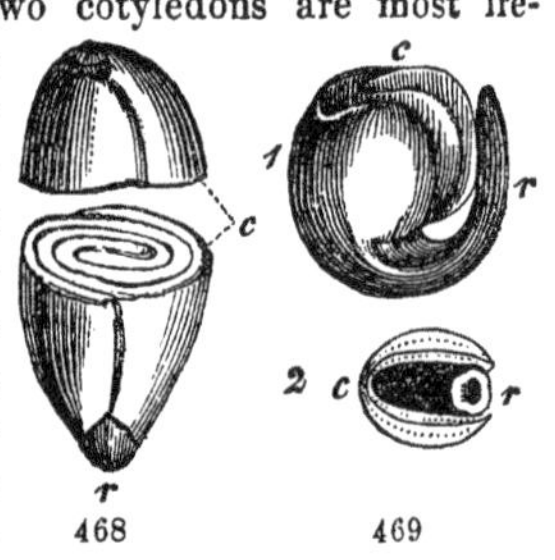

468 469

467. The embryo of the Pine.—1. In the seed.—2. When it has begun to germinate.—*r* Radicle.—*c* Cotyledons.

468. The embryo of the Pomegranate (*Grenadier*) (Punica granatum). It has been divided into two halves, and the upper raised, thereby shewing the way in which the cotyledons are rolled up *c*.—*r* Radicle.

469. The embryo of the Cabbage (*Chou*) (Brassica oleracea).—*r* Radicle.—*c* Cotyledons.—1. Undivided.—2. Horizontal section.

clinés) (*fig.* 164, 1) or *conduplicate* (*condupliqués*) (*figs.* 164, 2; 469), or *convolute* (*convolutés*) (*figs.* 164, 4; 468) or *circinnate* (*circinnés*) (*figs.* 164, 7; and 470). Most commonly the two cotyledons are thus bent and turned in the same direction, and parallelly, as if they only formed one single body; more rarely, it is in a contrary direction, as when they are *equitant* (*équitants*) (*fig.* 164, 9) or *demi-equitant* (*demi-équitants*) (*fig.* 164, 8). Sometimes they are also *plaited* or *rumpled* (*chiffonnés*). We can conceive that the foliaceous cotyledons may be folded and rolled in these different ways, sometimes very complicated, which cannot be defined by a single word and require a short, but very explicit description.

§ 587. After having examined the different positions, which the two cotyledons of the same embryo, may take one relatively to another, let us look at those which they may assume with regard to the other fundamental part of the embryo: the radicle. The latter very frequently follows the same direction as the cotyledons; the rectilinear direction if the embryo be straight, curvilinear if it be bent. This curve commonly assumes the shape of an arc of a circle, but sometimes becomes a true spire of several coils, arranged, either in one plane (*fig.* 470), or in several planes one above another (*fig.* 457). At other times, the direction

470

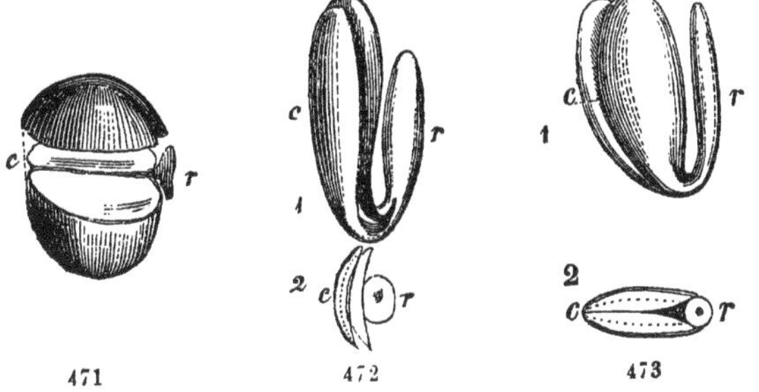

470. The embryo of the BUNIAS ORIENTALIS.

471. The embryo of the Pea, divided into two pieces; the upper one being raised shews how the fleshy cotyledons *c* are separated.

472—473. Embryoes of Cruciferous plants.—*r* Radicle.—*c* Cotyledons.

472. The embryo of the Woad (*Pastel*) (ISATIS TINCTORIA).—1. Undivided.—2. Horizontal section.

473. The embryo of the common Wall-flower (*Girofleé commune*) (CHEIRANTHUS CHEIRI).—1. Undivided.—2. Horizontal section.

of the radicle is not the same as that of the cotyledons, but forms an obtuse, acute or right angle with them; or even, bending itself completely, proceeds parallelly to the cotyledons, but in an inverse direction. The radicle thus bent may touch, either the face of

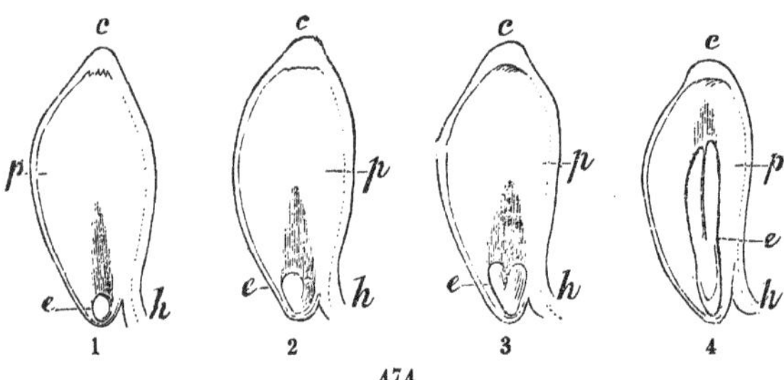

the cotyledons, or their edge. In the former case, they are said to be *incumbent* (*incombants*) (*fig.* 472); in the latter *accumbent* (*accombants*) (*figs.* 471, 473). These foldings of the radicles on the cotyledons may coincide with those of the cotyledons on themselves (*fig.* 469).

§ 588. Let us now study the different relations of the embryo to the different parts of the seed which encloses it, and at first to the perisperm, when this is developed.

We have seen that the embryo is in the beginning only a very small body suspended from the apex of the embryonary sac. We have seen that it gradually extends (*fig.* 474), and at last often fills the whole of it, absorbing all the juices which are accumulated there, and even a part of the envelopes which previously existed. Let us suppose all the intermediate degrees between this first and this last state of the embryo; let us suppose it arrested at each one of these degrees, and in each of these cases the place, which is not taken up by the embryo, occupied by the perisperm: we shall conceive all the possible relations of size between the two; relations, infinitely varied, of which nature presents us with all the

474. A seed of the Spindle-tree (*Fusain*) (Euonymus Europæus). We have given a vertical section of the seed at four different ages, shewing the relative developement of the embryo *e* and the perisperm *p*.—The arillus has been taken away.—*h* Hilum. —*c* Chalaza.—1. The first stage, when the embryo is of the form of a globule, as yet undivided, buried at the summit of the perisperm.—2. The second, when the cotyledons have begun to appear.—3. The third, when the embryo is longer, and its parts more distinct.—4. The fourth, when the embryo, as it lengthens, exceeds the half of the perisperm.

examples (*figs.* 475, 476, 477). Thus, the embryo may occupy

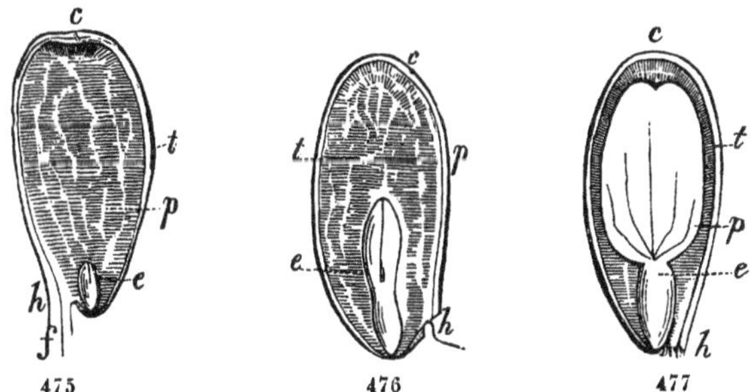

only a very small point at the summit of the perisperm, or be extended to its half, or more or less, or lastly equal it in length. It may be thinner or thicker, and this thickness will necessarily be in a direction inverse to that of the perisperm, the layer of which will be more and more attenuated in proportion as the embryo increases.

§ 589. The embryo may be directed in the same way as the seed, and is then said to be *axilary*. Then two cases will present themselves to our notice:—it either pushes the perisperm below it, with which it is in connection only by a part of its lower or cotyledonary extremity (*fig.* 478); or it is buried in the very substance of the perisperm, which then surrounds it on all sides, except quite at its radicular extremity (*fig.* 477). A junction rarely takes place between this extremity and the perisperm (in several Coniferæ, for instance), doubtless, by the means of the thickened suspensor.

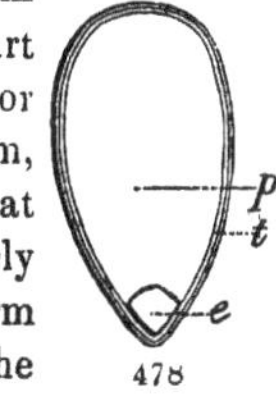

§ 590. At other times, the embryo does not follow the axis of the seed in its developement and is thrown on one side, generally on that which is opposite to the chalaza. In this case, even, it may be still completely enveloped by the perisperm, the layer of which is not so thick on one side as on the other. At other times, it is

475—477. Vertical sections of ripe seeds shewing the different relations in size of the embryo *e* with respect to the perisperm *p*.—*t* Tegument.—*f* Funiculus.—*h* Hilum.—*c* Chalaza.

475. The ripe seed of the HELLEBORUS NIGER.

476. ——————— DIPHYLLEIA PELTATA.

477. ——————— BERBERIS VULGARIS.

478. A vertical section of the seed of the CAREX DEPAUPERATA.—*t* Teguments.—*p* Perisperm.—*e* Embryo.

quite on the outside of the perisperm and is placed immediately under the teguments. This occurs especially in the curved seeds, resulting from campylotropous ovules; and then the chalaza occupying the concave part of the bend, the embryo, which is called *peripherical* (*périphérique*), follows the convex part, and appears to surround the perisperm instead of being surrounded (*fig.* 479): if the seed is not bent, if the embryo is small with regard to the perisperm, it is thrown on a point of its surface, as in the Gramineæ.

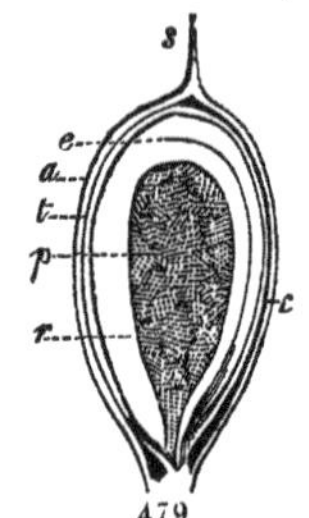

§ 591. Lastly, in a small number of cases, the developement of the different teguments may take place irregularly, so that the micropyle ceases to coincide with the apex of the nucleus, and, consequently, the axis of the seed (that is, the curved or straight line drawn from the micropyle to the chalaza) in reality no longer follows that of the embryonary sac. In this case the radicular end of the embryo, which is called *eccentrical* (*excentrique*), ends at a certain distance from the extremity of the seed. We see examples of it in the Primulaceæ, the Plantains, several of the Palms (*fig.* 480), &c.

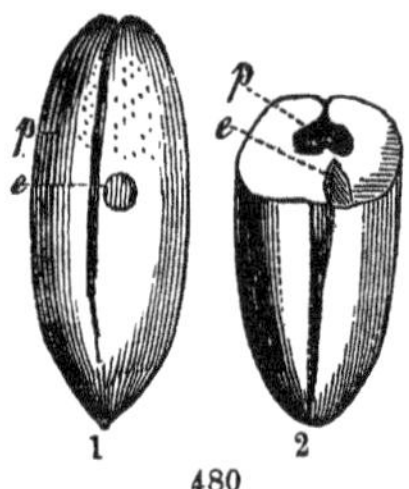

§ 592. We have just seen that the embryo, when it is accompanied with a perisperm, is most commonly surrounded by it; that at other times it is found on the outside, either on one of the extremities, or on the side. These *internal* and *external* positions induced Richard to call it EMBRYO INTRARIUS (*intraire*) in the first case, E. EXTRARIUS (*extraire*) in the second.

593. Let us examine, lastly, the relations of the embryo to the teguments of the seed, i. e., to the three principal points, the micropyle, the chalaza and the hilum. We know already that they are with very few exceptions very constant with regard to the first two, the cotyledonous extremity directed towards the chalaza, the radicular pointing to the micropyle. They only vary, then, with regard to the hilum. Now, the hilum coincides with the chalaza

479. A vertical section of a carpel of the MIRABILIS JALAPA, shewing the seed which it contains.—*a* Pericarp surmounted by the remains of the style *s*.—*t* The teguments of the seed.—*e* Embryo with its radicle *r* and its cotyledons *c*.—*p* Perisperm.

480. Stone of the Date.—*p* Perisperm.—*e* Embryo.—1. Undivided.—2. A horizontal section on a level with the embryo.

in the orthotropous ovules, whilst it is at the other extremity in the anatropous ovules. In the former case, the radicle takes a direction inverse to the hilum (RADICULA HILO CONTRARIA [*fig.* 481]); in the second, it points towards it (RADICULA HILUM SPECTANS

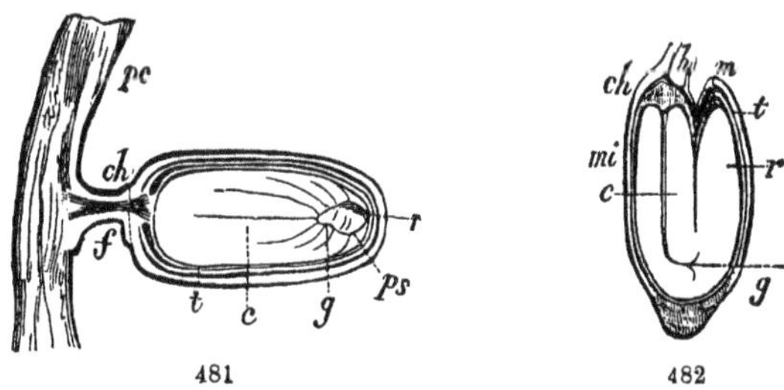

[*fig.* 483]). Richard terms the embryo in the former case *antitropous* (*antitrope*) (from ἀντί, *opposite*; τροπή, *the action of turning*), in the second *homotropous*, (*homotrope*) (from ὁμός, *same*). That which, when curved upon itself, brings the two ends together, he calls *amphitropous* (*amphitrope*) (from ἀμφί, *around*) (*fig.* 482); this, as we have seen, generally surrounds a part or the whole of the perisperm (*fig.* 479). It is clear, that the antitropous embryo will be formed in an orthotropous ovule; the homotropous embryo, in an anatropous ovule; the amphitropous embryo, in a campylotropous ovule[a].

§ 594. We have seen (§ 549) what are the different relations of the ovule to the ovary of the loculus which encloses it. They are apt to be modified by the changes which the ovule undergoes in its

481. A seed of the STERCULIA BATANGHAS. This is divided longitudinally along with the pericarp *p c*, to which it is attached.—*f* Funiculus.—*c h* Chalaza and hilum blended together.—*t* Teguments of the seed.—*p s* Perisperm, the top of which only can be seen.—*c* One of the cotyledons, the other has been removed to expose the gemmule *g*.—*r* Radicle.

482. The seed of the ERYSIMUM CHERANTOÏDES. It is cut lengthways.—*m* Micropyle.—*ch* Chalaza almost blended with the hilum *h*.—*t* Testa.—*mi* Internal membrane.—*r* Radicle.—*c* Cotyledons.—*g* Gemmule.

[a] The similarity of the sound of these words may, perhaps, cause some confusion before the ear be familiarized with them, so that it will be rather difficult to distinguish which belong to the ovule and which to the embryo. This confusion would be much increased by the adoption of two other terms proposed by Richard: that of *heterotropous* (*hétérotrope*) for the embryo, which does not follow the direction of the seed, and of *orthotropous* (*orthotrope*) for that, which is both homotropous and rectilinear: for the latter is the term of the anatropous ovule, and, consequently, when the ovule is not orthotropous, the embryo becomes so exactly. This induced us along with M. Brongniart to propose the epithets, straight, reflected, and curved, instead of orthotropous, anatropous, and campylotropous, introduced by Mirbel a long time after Richard had given these words the same ending.

developement; when it has arrived, however, at the state of a perfect seed, this, in its direction, cannot present other combinations than those which the ovules themselves present; it must be either erect (*fig.* 459), ascending, inverse or pendulous, either in the same direction as the funiculus, or in the opposite; it may be attached by its middle, and also be bent back upon itself. The figures 433 to 443, by which we have endeavoured to explain these different positions of the ovule, are applicable, therefore, to the ripe seed, as well as the words which designate them.

§ 595. But the identity of direction with relation to the loculus observed in two seeds belonging to different plants, does not imply the same identity in that of the embryoes. Thus, an erect ovule may be either straight or bent back, may turn its micrôpyle towards the top or towards the bottom of the loculus. The radicle, which nearly always corresponds to the micropyle, ought in the first case to be equally turned upwards; in the second, downwards. This is indicated by certain epithets applied to the radicle, which is termed *superior* (*supère*) when it is turned upwards (*fig.* 483, *er*); *inferior* (*infère*), when it is turned downwards (*fig.* 460); *ventral* or *centripetal* (*centripète*), when it is directed inwards: *dorsal* or *centrifugal* (*centrifuge*) (*fig.* 481, *r*), when it is turned outwards. It is clear, that, from this direction of the embryo combined with that of the seed, we may find the absolute direction of the ovule; in the same way as we may reciprocally foresee by this, what will be in a little time that of the embryo. An erect and straight (or orthotropous) ovule will announce early that the embryo will be antitropous, with the radicle superior; in the same way on meeting with the latter in the ripe seed, we thence conclude with certainty, what the direction of the ovule was. Since it is not always possible to observe these parts at every age, since the majority of plants that

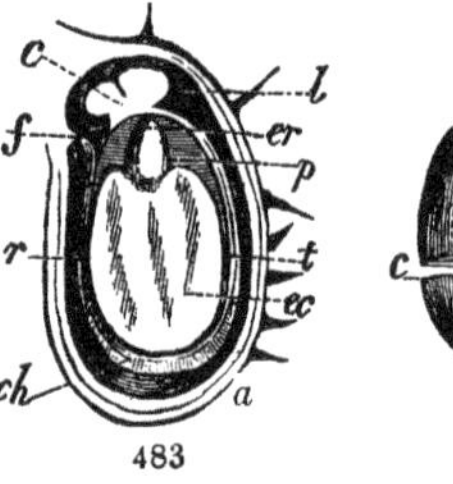

483. A vertical section of a carpel of the RICINUS COMMUNIS and the seed it encloses.—*a* Pericarp.—*l* Loculus.—*f* Funiculus.—*t* Teguments of the seed, the outer surmounted by the caruncle *c*, traversed by the small canal of the exostome, which no longer exactly corresponds to the endostome placed immediately above the radicle.—*r* Raphe.—*ch* Chalaza.—*p* Perisperm: the upper portion alone can be seen.—*e* Embryo with its radicle *e r* and its cotyledons *e c*.

484. The embryo divided into two parts by a horizontal section. The parts have been separated so as to shew the two cotyledons *c* applied to one another.—*r* Radicle.

we know, described by travellers and preserved in herbaria, are fixed a certain period of their developement and not in several successive stages, we can readily conceive the great importance of these characteristics, which may be substituted for one another, which allow us to divine from one isolated fact the conditions, that have preceded, or those that would have followed it.

§ 596. The micropyle is very visible in a certain number of seeds, as in those of the Iris, of the Windsor-bean, of the Haricot-bean, of the Pea and other Leguminosæ, in which it remains under the appearance of a small hole. But, in the majority of plants it disappears, and then the seed must be dissected to determine its situation and also the place where the point of the radicle is terminated.

As to the hilum and the chalaza they are generally more clearly marked on the ovule. The former is determined by the point at which the funiculus is fixed, or, when this attachment is broken and the seed is free, by the cicatrix which we find on the surface of the teguments. The latter is frequently recognised by its colour which is different from the rest of these teguments, paler, or, on the contrary and most generally, darker; at other times, of the same colour as they are, it is not so easily distinguishable (sometimes by the aid of dissection alone, which shews us in these teguments a thicker portion of a slightly different tissue, corresponding to the chalaza). Besides, it is always directed towards the cotyledonous extremity of the embryo. It also varies in its form, which is sometimes linear, sometimes and most frequently that of a circle, or, lastly, an intermediate shape between these two extremes. If the hilum is situated immediately on the outside of the chalaza (in the straight seeds or those with an antitropous embryo) these two points are confounded on the outside. If the hilum be far from the chalaza, the vascular fascicle, which, reaching the former along with the funiculus is terminated at the latter passing through the teguments, is marked under the latter as a line or a little band, generally a little buried in their substance, which we have called the *raphe* (*raphé*). This continuation of the funiculus may be considered as a part of itself, differing from the funiculus properly so called in adhering to the teguments instead of being free. When the ovule or the seed is turned in a

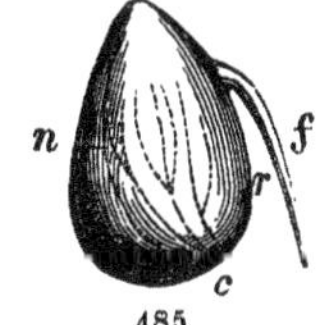

485. The seed of the Filbert (*Noisetier*).—*f* Funiculus.—*r* Raphe.—*c* Chalaza.—*n* Nerves, which are separated into rays expanding over the teguments of the seed.

direction inverse to that of the funiculus, when, for instance, we have an erect ovule on a pendulous funiculus (*fig.* 438), is the latter any thing else than a free raphe? Let us suppose it united to the seed, it will take one of its most common directions, it will become pendulous and curved. Nature shews us, in the different species of the genus ZYGOPHYLLUM, all the transitions from one state to the other. In the fruit of the GUAZAMA, when ripe, the raphe, very thick, is detached from the seed (*fig.* 487, *g*) which falls, and remaining under the shape of a small stiff thread attached to the placenta, assumes the very shape and appearance of a funiculus. It is evident, that the length of the raphe is always measured by the distance of the hilum from the chalaza; so that it is hardly developed in the majority of campylotropous seeds, whilst it acquires great length in the anatropous seeds. It corresponds nearly always to the ventral face of the seed, that which is turned towards the placenta. In a very small number of seeds, it follows the opposite or dorsal face (*fig.* 486).

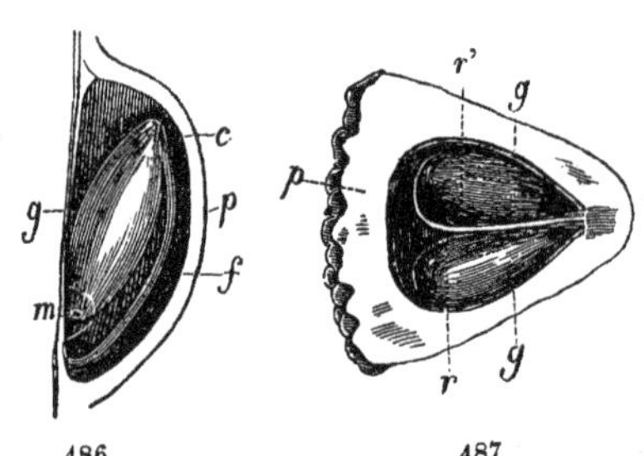

From the preceding facts, we find that from the external aspect of the seed, and from the determination of its different points or parts, the hilum, the chalaza, the micropyle, the raphe, we may discover the direction of the embryo which we do not see; but the reverse does not obtain, and if the embryo assists us to recognise these points on the teguments, it is not sufficient, since it has no necessary relations to the hilum, the position of which may vary.

§ 597. It now only remains for us to add a few facts to those which we have already stated (§ 570) with regard to the envelopes of the ripe seed, the number of which sometimes amount to three or four, like those of the ovule, but most commonly are reduced to two,

486. Horizontal section of a carpel of the FAGONIA CRETICA.—*p* Pericarp.—*f* Funiculus, which may be for the most part considered as a raphe detached from the seed.—*g* Seed.—*c* Chalaza.—*m* Micropyle.

487. A transverse section of one of the carpels of the fruit of the GUAZUMA ULMIFOLIA.—*p* Pericarp.—*gg* Seeds.—In one, the raphe *r* is still adherent to the seed: in the other, it is detached in the shape of a small hook *r'* projecting freely into the interior of the loculus.

an external or testa, and an internal membrane (ENDOPLEURA [*endoplèvre*] of De Candolle). The embryo, either deprived of the perisperm, or surrounded and accompanied by this posterior formation, becomes, with or without it, a body to which we give the name *nucleus* (*amande*); a body, externally covered by the internal membrane, which follows it throughout all its shapes. The testa also sometimes follows it, moulded on the nucleus and this intermediate membrane; this is what commonly occurs when the seed is straight or scarcely curved. But, if its curve is bent on itself, it is generally the internal membrane alone which is interposed in this fold, and the testa is only gradually buried in it. Sometimes even, instead of being regularly extended in a continuous manner over the internal face of the tissue, it forms wrinkles or numerous folds which are bent inwards, and thus divide more or less deeply the circumference of the cavity of the seed into a great number of compartments. The perisperm, which fills such a cavity, is, therefore, furrowed on its surface by wrinkles or grooves corresponding to all these folds; it is then said to be *ruminated* (*ruminé*) (RUMINATUM), as in the Anonaceæ, the Sago (*Sagou*), the Areca or Betel Nut (*Arec*), and several other Palms.

But, at other times, on the contrary, the testa may be formed on the outside of the elongations where the internal membrane does not follow it. These are small fleshy excrescences or caruncles, which most commonly circumscribe the micropyle (*fig.* 483, *c*); they are folds, membranes or wings, which (like those of the Samaræ) are sometimes extended on one or other extremity, sometimes grow from the circumference of the seed, either on one side only or on all sides, in number either one or more: the seed is then said to be *winged* (*ailée*) (ALATA).

The internal membrane most frequently merits its name on account of its thin and flexible tissue; sometimes, however, it is thickened, and even so much so as to form apparently a layer of the perisperm, to which its tissue (which is then fleshy) thus furnishes a transition more or less insensible. It is not always thus equally swollen; but it may be thickened only in some places, preserving in others its membranous texture. It is generally whitish or demi-transparent.

As to the testa, it may present the same appearance and the same colour; but is more commonly distinguishable by its deeper colour, as well as by its more compact tissue and greater thickness. Its consistence is sometimes soft and fleshy, sometimes coriaceous, frequently of a hardness which approaches more or less to that of the wood: then, if it is thin, it becomes fragile. But it frequently forms rather a thick layer adapted for the protection and preservation of the contained nucleus: it is then generally formed of small fibres transversely directed from the outside to the inside and pressed against one another, frequently arranged in two layers; the inner, formed of tapering cells of a fibrous texture; the outer, of a kind of epidermis the cells of which are broad, with larger cavities, sometimes secreting peculiar substances. The surface is smooth: or it is unequal, being covered with different projections, blunt or sharp, regular or irregular; or else, on the contrary, excavated into dots, small dimples, wrinkles, circles even, which form a kind of network. It is free from or covered with hairs of different kinds, analogous to those which we have seen on other parts. But there are some which present a peculiar and remarkable form; these are utricles more or less elongated, doubled in the inside with a spiral thread, as in the HYDROCHARIS, the COLLOMIA (*fig.* 488).

488

§ 598 We have said that it is in the external tegument of the ovule that the raphe proceeds. Now, since it is this tegument, which, either alone or blended with other more internal ones, forms the testa of the seed, it is also in the testa that we shall generally find the raphe, sometimes following a groove made on its superficies, more frequently a canal opened in its substance. The fascicle of the raphe is expanded towards the chalaza, as it bends back towards the inside, and, passing from the testa into the internal membrane, sometimes sends out branches, which rising again are dispersed and cover it (*fig.* 485, *n*). In the whole extent which corresponds to this expansion, the two envelopes are thickened and their tissue is modified in a remarkable manner.

§ 599. DIFFUSION OF THE SEED (*Dissémination*).—The ripening of the seed most generally coincides with that of the fruit. Then commences the diffusion of the seed, that is, the act by which the seeds, detached from the plant which has given them birth, are spread about at various distances from it to live for themselves, to grow into separate plants. The fruit is frequently detached with them by the disarticulation of its peduncle, and they then both fall

488. The utricles of the external layer of a seed of the COLLOMIA GRANDIFOLIA, very highly magnified and moistened with water while under observation.

to the ground, the one still containing the others. The funiculus is itself disarticulated at the point of the hilum, and the seed becomes free in the loculus. If the pericarp is dehiscent, the seed naturally falls out of it from the motions which may be frequently imparted to the dried fruit by the very pressure of the valves, which are contracted elastically as they separate; if it is indehiscent, the egress is slower through the pericarp, which, thenceforth deprived of life, is gradually decomposed and separated in shreds. Numerous causes favour the diffusion of seed; the weight, which has augmented in proportion as the strength of the adherence diminished; the motion imparted by the wind or the rain; the intervention of animals, which transport and sometimes even bury the seeds, either involuntarily and unwittingly, or voluntarily for the purpose of supporting themselves; and even when they have been nourished by the fruit, it frequently happens that the nucleus, protected by a thick ligneous testa, resists digestion and is restored intact to the earth with the excrements. Certain seeds assist the action of these external agents; those seeds, for instance, provided with tufts or wings, a kind of parachute, which sustain them in the air and allow the wind to carry them far away.

§ 600. Some seeds continue to grow without being detached from the plant. In some rather high trees, as in the Mangrove, the radicle piercing the testa and the pericarp, is sufficiently lengthened to reach the earth. In the plants which grow near the earth, the seed buries itself at the extremity of the branch, which it has not abandoned, and which frequently even assists it by bending at this time towards the soil, as in the Trifolium subterraneum.

§ 601. Several of the seeds escape from these influences, dry in the air, decay in the water, are eaten by animals; but there is always a certain number of them which from some cause or other, are preserved on the surface of the soil or are buried at a certain depth. Nature has secured the preservation of the species of vegetables by the number of the seeds which they bear, a number quite out of proportion to that of the individuals which live. We will mention the Poppy as an example, in which each fruit encloses such a multitude of seeds, that a very few years would suffice to cover the whole surface of the earth with Poppies, if all were developed during several successive generations.

§ 602. Germination.—A certain degree of heat and humidity is necessary for the subsequent life of the embryo, after the seed has fallen off with or without its pericarp. We have seen (§ 295), that it requires a certain proportion of oxygen, and, consequently, the free access of air: but in several seeds, when they are deprived of these conditions, life is suspended without being extinguished. We may thus preserve them for a great number of years by protecting them from the action of air and water: hence the custom of burying them to a great depth in cavities properly prepared. They are frequently found accidentally preserved. Freshly turned up earth, the edges of ditches of various depths in a soil which has long been uncultivated, are almost always covered with a fresh vegetation, sometimes different from that which we observed before, and it is by no means uncommon, for plants, long since extinct, but which indisputably once grew there, to reappear amongst them. Their appearance proves that their seeds, buried at this great depth, are preserved alive: withdrawn from the action of the air for so long a time, they begin to grow as soon as they have access to it.

§ 603. Let us suppose a seed to be placed in all the conditions favourable to its developement, and let us observe all the fresh changes which it undergoes. Sometimes they take place with inconceivable rapidity, sometimes very slowly: the Garden-cress germinates in a day or two, whilst there are seeds which require years. It is true, that these last are generally surrounded with teguments, which protect them from external agents, and which themselves resist their action; so that the germination, properly speaking only begins after a long interval has elapsed, since they were put in communication with these agents.

§ 604. We may distinguish two periods in the germination; the first, during which the embryo continues to grow within the seed which is now free; the second, during which, having come to light through the envelopes of this seed, but still attached to it there, it is developed outside of it. If we pursue a comparison which we have already mentioned (§ 573), that of the seed with the eggs of birds, we shall easily recognise that the first period corresponds to the changes, which take place in the inside of this egg during incubation, that is, during the process of sitting; that the second corresponds to the hatching.

§ 605. Let us examine at first what passes during the former period. Two cases may present themselves; the embryo either has or has not a perisperm.

If it has a perisperm, it is softened by the combined action of heat and humidity; its chemical nature changes at the expense of the elements furnished to it by the oxygen of the air and water (§ 295, 296). The embryo, in contact with it, absorbs these substances now become fit from their state of solution to penetrate and to nourish it by the modification which they have just undergone. Thus nourished, it increases in the same proportion as the perisperm decreases, and at last fills the whole of the inside of the seed, in which at first it only occupied a small space. The perisperm has now disappeared, and the embryo can extend itself only by breaking the teguments, which, softened, present a gradually decreasing resistance.

§ 606. If it has no perisperm, and the embryo already fills at the time of the diffusion the whole cavity of the seed, it is clear, that the time of germination will be considerably abridged, since its parts will have acquired a much greater developement than in the preceding case. It is the cotyledons, generally, which then form the greater part of the embryonary mass, and we ought to remark, that, in this case, their nature is analogous to that of the perisperm: it is a cellular mass, the cells of which are filled with fecula, as in the Haricot-Bean and Garden Pea, and frequently contain small drops of oil, as in the Nut, the Rape or Cole-seed (*Colza*). This mass performs with regard to the rest of the embryo the same functions as the perisperm, undergoes changes analogous to those, which we have just seen in it, and thus furnishes nourishment to the radicle and the gemmule, organs, in which is contained the whole force of the developement.

§ 607. Thus fortified, either at the expense of the perisperm or of its own cotyledons, the embryo, continuing to grow, presses its teguments, which are broken to afford it a passage. The radicle is almost always the first to appear on the outside (*fig.* 489, 1), as we might expect, since from the first its extremity was the nearest to the teguments, almost naked beneath them and corresponding to a natural solution of continuity,

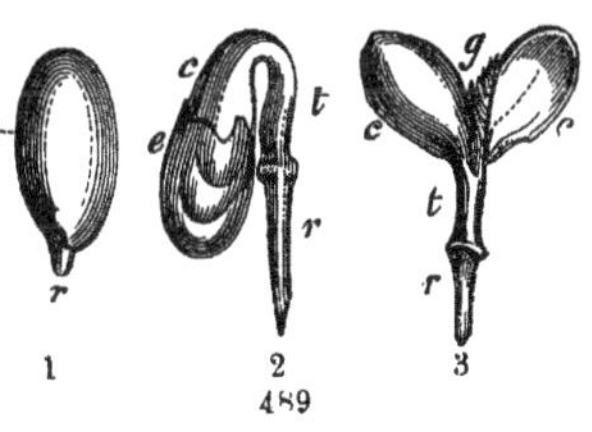

489. The germination of a Dicotyledonous seed, not perispermous, that of the Acacia Julibrissin.—*e* Envelope of the seed.—*r* Radicle of the embryo.—*t* Tigelle.—*c* Cotyledons.—*g* Gemmule.—1. The first stage, in which the radicle is seen outside the broken envelope.—2. The second stage, in which the parts developed, already very distinct from one another, are separated from the envelope, which still contains the top of the cotyledons.—3. The third stage, in which the embryo is wholly separated from the envelope, and in which the cotyledons, erect and separated, allow the gemmule to be perceived.

the micropyle. The radicle then forces its way to the outside. But, what we have termed radicle is almost entirely formed of the tigelle, at the summit of which is the gemmule, which, in its turn, is thus formed on the outside; its axis, up to this time contracted and almost wanting, is lengthened; its small lateral lobes, rudiments of the leaves, are developed, and all this system is vertically directed from bottom upwards, towards the sky. But in this germination the really radicular part, up to this time limited to the extremity alone of the radicle, has itself begun to be elongated (*fig.* 489, 2) and always in an inverse direction from top downwards, towards the centre of the earth. The cotyledon, single or double, is the last to be freed from the seed; sometimes, even, it is not disengaged and withers with it; sometimes it is liberated in its turn, and, becoming free, expands (*fig.* 489, 3) into a leaf at the point of the young stem, which separates the portion originally belonging to the radicle from that which belonged to the gemmule. Then all these parts begin to turn green under the influence of the air and light.

§ 608. Let us, nevertheless, remark that several embryoes have already appeared green within the seed, with a tinge sometimes pale or yellowish, but sometimes also very deep. We shall mention as examples, among the perispermous seeds, those of the Spindle-tree, of the Buckthorn, &c.; amongst the seeds without perisperm, those of the Pistachio, of the Sycamore, of the greater part of the Cruciferæ. But most frequently the embryo contained in the seed is whitish as well as the perisperm: the Miseltoe is the only plant we know in which the latter is green. The identity of colour between the embryo and the perisperm, confounding at the first glance these two bodies into a single mass, does not render their observation very easy. We may readily distinguish them by plunging the cut seed into boiling water, which, acting differently on the two tissues, clearly separates the bright white of the one from the more dusky white of the other.

§ 609. Let us add a few details on the differences which we have not yet described between the germination of the Monocotyledonous and that of the Dicotyledonous seeds.

The former are,.for most part, provided with a perisperm, most commonly very considerable, and in all these the cotyledon is not disengaged from the seed: only it sometimes forms on the outside an elongation of different degrees of length and thickness, by which it is attached to the axis, as in the Ephemereæ, the *Garlic*, the CANNA (*fig.* 490, 3): this elongation, which is produced by the act of germination, may be compared to the petiole, whilst the part (*c*) imprisoned in the inside is the limb of the cotyledon already quite formed. Sometimes it remains sessile on the axis, which is then a tangent to the seed. In every case the sheath, which surrounds the gemmule and which was indicated on the embryo by a small lateral slit (*fig.* 490, 2 *f*), has followed this gemmule on the

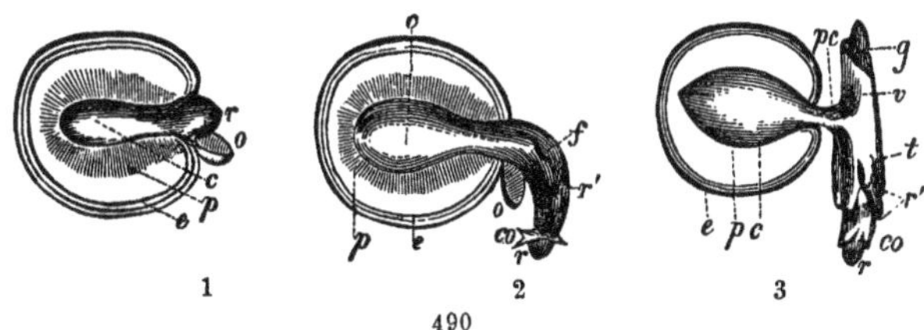

490

outside, and continues to follow it in its ascending direction, as it is lengthened along with it. Its slit is more and more pronounced and its two lips separate and allow the first leaves to pass (*fig.* 490, 3, *g*) through them, then the axis which bears them. The cotyledon shews us, therefore, in its evolution all the phases of the leaf; the limb is the first formed; then the sheath, then sometimes a petiole which separates one from the other. They only differ in the limb being stopped in its developement in the cotyledon, crowded by the body of the seed which continues to enclose it, from this very cause preserving a direction different from that of its sheath, which ascends and grows for some time.

In the small number of Monocotyledonous seeds which have no perisperm, as the Alismaceæ, Potameæ, &c., things do not take place quite in the same way. The cotyledon is generally freed from its teguments and raised vertically with the gemmule (*fig.* 79). We

490. The germination of a Monocotyledon, the CANNA INDICA. The seed has been cut so as to shew the perisperm diminishing in proportion to the increase of the embryo.—*e* Envelope of the seed.—*o* Its upper part, which is detached like an operculum to give egress to the radicle.—*p* Perisperm.—*c* Cotyledon.—*r* Radicle.—*r' r'* Secondary radicles.—*c o* Coleorhiza.—*f* Slit corresponding to the gemmule, afterwards opening to give egress to a long sheath *v*.—*p c* The contracted portion of the cotyledon (corresponding to the petiole), intermediate between its widened *c* (corresponding to the limb) and its vaginal part *v*.—*t* Tigelle.—*g* Gemmule.—1. The first stage, in which the radicle appears outside of the teguments.—2. The second stage, in which the slit *f* appears on the outside. The real radicle *r* has pierced the epidermis, with which it was surrounded, and which appears at its base in the form of a scolloped collar, or coleorhiza. We already see one of the secondary radicles, *r'* also surrounded by a coleorhiza.—3. The third stage, in which all these parts are more developed, and in which the gemmule *g* projects beyond the slit, the edges of which are prolonged into a sheath *v*.

have already treated (§ 111) of the peculiar mode of developement of endorhizal roots, and it is useless to revert to them here.

§ 610. As to the Dicotyledonous embryoes, sometimes their cotyledons also are confined in the seed, and then the egress of the gemmule presents some similarity to that of the Monocotyledonous ones; a resemblance, however, only apparent, since here the gemmule comes out of the interval of the cotyledons at their base and not out of the interior of a sheath. The two cotyledons are most generally separated from one another and the gemmule is freely lengthened in its direction, whilst the exorhizal radicle (§ 111) continues in its own.

§ 611. The cotyledons remain sometimes concealed under the earth, as in the ARACHIS, and are called *subterranean* (*hypogé*) (HYPOGÆUS) (from ὑπὸ, *under*; γῆ, *earth*). They are commonly raised above the earth to a certain height, depending upon the length of the tigelle; they are then called *epigeous* (*epigé*) (EPIGÆUS) (from ἐπί, *above*).

§ 612. The cotyledons continue, at the same time gradually exhausting themselves, to furnish the young plant with its nourishment which it now begins to draw directly from the earth. They wither and fall, the germination is finished, and the vegetable, thenceforth living on its own resources, begins that series of acts which we have tried to explain in as clear a manner as possible. We have thus traversed the entire circle of vegetation, and brought ourselves to the point whence we departed.

§ 613. SPORES OF ACOTYLEDONOUS VEGETABLES (*Spores de végétaux Acotylédonés*).—But we have treated of the ovary, which afterwards becomes the fruit, of the ovule, which afterwards becomes seed, of phanerogamous vegetables only, of those, in which there is a manifest fecundation by the action of a tube emanating from the pollen, or essential part of the stamen, to the nucleus, or essential part of the pistil, two kinds of organs completely distinct in their nature, with which we have become acquainted. Since the result of this fecundation is the production of an embryo furnished (if we lay aside the exceedingly rare exceptions) with one or more leaves of a particular structure, called cotyledons, we give the name of Cotyledonous to these vegetables indifferently, dividing them into two classes by the prefixes *Mono* and *Di*.

We know that there are others in which we do not find these two kinds of organs, the reciprocal action of which determines the formation of the reproductive bodies, which in these vegetables only appear to be a small homogeneous mass without distinction of parts and, consequently, without cotyledons. Hence the names of Cryptogamic and Acotyledonous vegetables which are given to them indifferently. We have already treated on the analogy of the stamens (§ 480), and we have seen that the bodies which we supposed to be such and termed ANTHERIDIA (*anthéridies*) differ completely and essentially from the true stamens, since we do not observe any pollen or even fovilla in them. Let us now enquire if we can find among cryptogamous plants a body analogous to the ovary or, at least, to the ovule.

§ 614. Several authors have fancied that they have recognised it. The Mosses and the HEPATICÆ being those vegetables in which the resemblance appears to be the least doubtful, we shall commence with them. In the HEPATICÆ, in the very substance of the expanded tissue which constitutes the plant, (RICCIA [*fig.* 493]), on its surface, or on other expansions distinct both in shape and situation (MARCHANTIA); in the Mosses, at the extremity of the branches or at the axils of the leaves, we observe small hollow bodies, the form of which (*fig.* 491) cannot be compared to any thing better than to that of a bottle. The walls of these bodies are formed by a layer of cellular tissue, and their cavity is filled by a mass of seeds, which we term *spores* (from σπορὰ, *seed*), each being developed into a small plant similar to that on which they grew. These spores are, therefore, analogous to the seed and the body enclosing them, which has been termed SPORANGIUM (*sporange*) (from σπορὰ, *seed*; ἀγγεῖον, *vase*), has been naturally compared to an ovary; the neck of the bottle, as it were, which surmounts this, to a style. This is very like a real style; for it withers in proportion as the spores approach maturity, and almost entirely disappears when this small fruit becomes completely ripe.

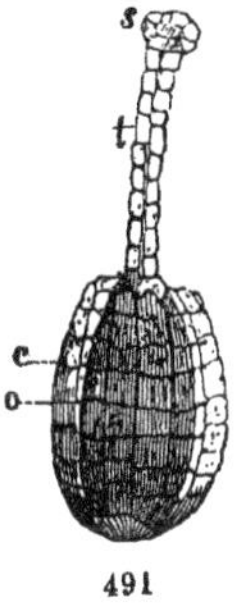

491

§ 615. But the comparison, when rigorously followed up, presents very important points of difference. The spores are free in the cavity which encloses them and are at no time continuous with its walls. In germination they are produced immediately by an elon-

491. Sporangium of the MARCHANTIA POLYMORPHA.—*o* The lower hollow swelling, which, containing the spores, has been compared to an ovary.—*t* The upper contraction in the shape of the neck of a bottle, which has been compared to the style. —*s* Terminal vase-like expansion, which has been compared to the stigma.—*c* Cellular tube, surrounding the sporangium like a calyx.

gation of themselves at a point of their circumference, and do not open to give passage to a fresh body formed in their interior. We may, therefore, compare them to naked embryoes, but by no means to seeds. Besides, in their structure we find nothing which reminds us of the complicated structure of the ovules, that combination of sacs fitting one to the other, in the innermost of which we at last find the embryo. They are simple utricles which, under a single or double membrane, enclose a liquid substance of an oleaginous consistence. If we examine the different changes which they undergo, before they reach this state, and if we follow all the phases of their formation, we see, that at first this so-called-ovary does not present any cavity in the inside, but a continuous cellular mass; that afterwards the cells, situated at the centre in the HEPATICÆ, round the centre in the Mosses, are much more largely developed than the external cells; that these cells thus developed are filled with a semi-fluid granular substance; that these granules at first scattered, then conglomerated, are at last separated into four small distinct masses (*fig.* 493); that it is, in short, each of these small masses which is organized into one of those seeds of which we have spoken, and that at the same time the cell, in which they are formed, is gradually absorbed, and thus disappears, as well as all similar cells; so that all the spores float freely in a common cavity the external cells of which form the wall.

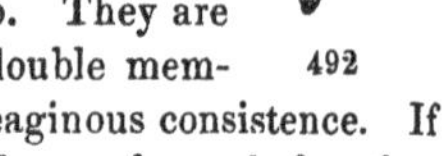

492

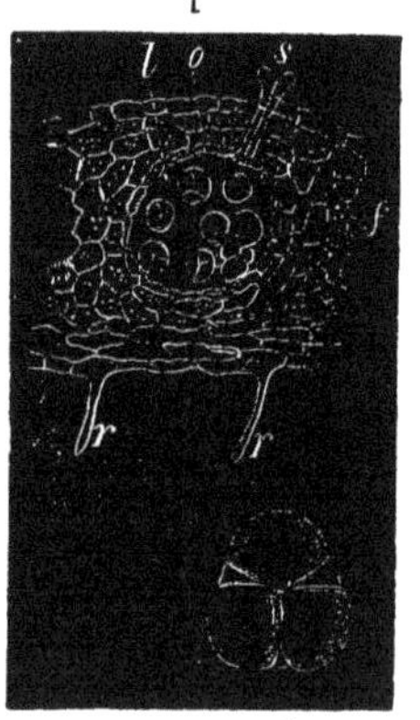

493

This cellular sac, enveloping a number of free utricles, does not, therefore, present to us the characteristics we have described in the ovary of the phanerogamous plants; no more than these utricles,

492. Spores of the MARCHANTIA POLYMORPHA. The germination of one is farther advanced than that of the other.

493. 1. The perpendicular section of a frond *f* of the RICCIA GLAUCA and of the sporangium *o* which is enclosed in its substance.—*s* Contraction or style, by means of which the sporangium communicates with the outside.—*l* Its cavity or loculus.—*s* Young spores, still united in fours in the mother-cells.—*t* Cells lengthened into the shape of roots.—2. One of these utricles more strongly magnified, with the four spores which it contains. We see three, which conceal the fourth behind them.

formed by fours in other mother-cells, present the characteristics of the ovules. But, we shall be struck by another analogy, which the whole of this formation of the spores and of the sporiferous sac presents to that of the pollen and of the anther (§ 463).

§ 616. In another family of Cryptogamic vegetables, that of Rhizocarpeæ, as in the *Pill-wort* and the MARSILEA, we find something resembling true fruits bearing bodies attached to their wall. But these bodies are themselves cellular sacs filled with spores, the formation of which presents in its successive phases all the changes which we have just described. In the Ferns, under the leaves; in the Lycopodiaceæ at their base; we find small sacs arranged in different ways in the former, solitary in the latter, but in both filled with free spores which are all formed in the same way. These sacs present still less analogy in their shape to ovaries, and are, besides, not furnished with the elongation, which in the Mosses and the HEPATICÆ we have compared to the style. In certain Ferns and LYCOPODIA, on the contrary, they bear a greater resemblance to the anther.

§ 617. As we descend to the vegetables which no longer present any distinction between the stem and the leaves, we see this apparatus still more simplified. We always find free spores in a cavity; but this cavity no longer appears to be any thing else than that of the mother-cell, which persists instead of disappearing by absorption, and the wall of which, then taking the name of THECA[1] (*thèque*) (*fig.* 494), forms that of the sporiferous sac. A demi-fluid granular mass fills it at first and then separates into a certain number of spores: only these are placed above one another instead of by the side of one another, and they are sometimes united end to end in twos (*fig.* 495), in fours, or in some higher number, being a multiple of two; so that each theca contains either several isolated spores, or several series of spores, which are themselves sometimes contained by another envelope or a common theca. These thecas are crowded in groups, either at the surface of the expansion, which forms the vegetable, or in its substance: we observe this in the Lichens and in some Fungi. But, in the latter we

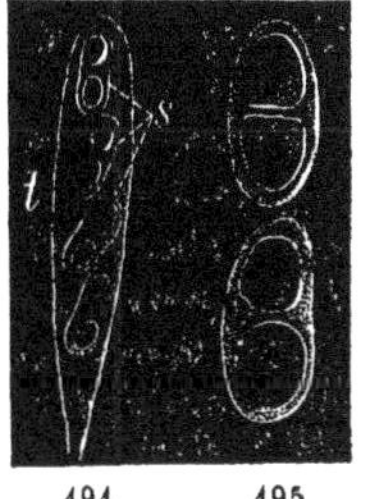

494 495

494. A theca of the SOLORINA SACCATA, enclosing eight spores united in couples.
495. Two of the preceding couples more strongly magnified.

[1] From θήκη, *a case to put anything in, a box, chest.*—TRANS.

find several others in which the spores become free in one or more internal cavities, by the absorption of the utricles in which they are at first formed, either frequently united in fours, or in a greater number, or, on the contrary, one only in each utricle.

§ 618. The spores appear in the same way either in fours or solitary in the ALGÆ, in which the mother-utricles, which continue to enclose them, are scattered in the substance of the tissue, or arranged in certain distinct or projecting places, either on the surface itself, or in the cavities which are formed in this surface. But, the more the structure of these vegetables is simplified, the less tendency have the sporiferous utricles to be distinguished from the others which form the rest of the tissue, so much so, in fact, that at last we find vegetables, each cell of which contains granules capable of reproducing them, the organs of reproduction being thus confounded with those of vegetation.

§ 619. A very remarkable phenomenon in the spores of these, the simplest, vegetables is the motion with which they are endowed at a certain period of their existence, that which immediately follows their egress from the mother-cell. These motions may be compared to those of the animals called Infusoria, and it has been recently discovered that they are carried on by means of similar organs, of vibratory ciliæ, i. e., of small threads growing from a part of the body and moving about in the water like fins. We have already represented two of these ciliæ at the extremity of the thread which forms the animalcule of the antheridium of the CHARA (*fig.* 353). M. Thuret, to whom we owe these observations, has discovered them also in the spores of certain fresh-water ALGÆ: two situated at one of the extremities in the spore of the CONFERVA (*Conferve*) (*fig.* 496); four in that of the CHÆTOPHORA (*fig.* 497); a complete circle in that of the PROLIFERÆ (*Prolifères*) (*fig.* 498); and, lastly, a great number scattered over the whole surface of that of the VAUCHERIA (*fig.* 499). This power of locomotion is transitory and only takes

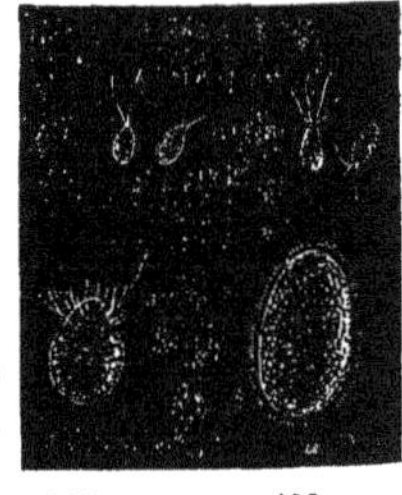

496—499. Spores of different fresh water ALGÆ.
496. Spore of a CONFERVA with two vibratory ciliæ.
497. ——— CHÆTOPHORA with four ciliæ.
498. ——— PROLIFERA with a circle of ciliæ.
499. ——— VAUCHERIA quite covered with ciliæ.

place (which is another equally curious fact) during the early part of the day. Then the motion is stopped, the spore passes from the animal into the vegetable life, and it is then that it may be said to begin to germinate.

§ 620. All the details, into which we have just entered, shew more and more how far the reproductive organs of cryptogamia differ from those of phanerogamia, and that we ought to recognise an ovary in the former only when we define as such the whole cavity enclosing bodies susceptible of being developed into a plant similar to that which gives them birth; a definition so general, that we should be compelled to rank under it a certain number of different parts without any true relations to one another.

§ 621. Theory of Schleiden.—The history of the developement of the spores which we have just given, whence we can trace a manifest analogy to the formation of the pollen in the anthers of phanerogamia, ought to precede the explanation of a theory which borrows from it a part of its proofs, and which has recently been expounded in Germany, proposed on one hand by M. Schleiden, on the other by M. Endlicher. It is most generally known by the name of the former, whose publications are anterior and much more explanatory, supported by numerous observations and figures. We shall, however, find the germ of the theory in the works of a French author: it was for a long time neglected and left in obscurity. Cl. J. Geoffroy, about the beginning of the eighteenth century, in a *Mémoire* on the structure and functions of the principal parts of the flower, in which we already find very just notions on the pollen and the seed, with the indication of the micropyle and of its destination, concluded from it that "the dusts of the flowers are the first germs of the plants, which, to be developed, require the juices which we meet with in the seeds, as animals require the egg before they are born." He shewed, taking his example from the Leguminosæ, that before the emission of the pollen we discover in the seeds "nothing else than its envelope or bark;" that after the emission we begin to perceive in the interior a small point or globule, which insensibly grows and consumes the liquid with which this interior is filled. He pointed out the micropyle as the opening through which would take place the introduction of this fresh body, the embryo, and shewed already the constant direction of the radicle towards this opening, through which it afterwards finds its egress during germination. He admitted that it is an entire grain of the pollen, which thus penetrates into the seed in order to be there developed into an embryo.

We have seen that this supposition is false, that the pollen-grain remains fixed to the stigma, and that it is the tube, formed by the elongation of one of its membranes, which is insinuated through the style to the nucleus of the ovule and conveys thither the fovilla or the contents of the substance contained in the pollen-grain into the interior of the seed.

§ 622. M. Schleiden follows it still further; he says that the extremity of the tube penetrates into the embryonary cavity by pushing before it the membrane which forms the apex of the nucleus, and that it is this part of the membrane thus bent back into a little pocket, which constitutes the embryonary vesicle; that the substance contained in the extremity of the tube forms the embryo; that the remaining portion, which has penetrated, forms the suspensor. According to this theory the ovule would furnish the embryo with the medium only, in which it is developed, and with the nourishment necessary for this developement, adapted to its nature, which it would not find elsewhere. M. Schleiden thus easily explains away the rather frequent existence of several embryoes in one single seed (as we observe, for instance, almost constantly in that of the Orange-tree); in this case, several tubes at once are, therefore, in contact with the same ovule. The cryptogamic plants would, consequently, differ from the phanerogamic in this particular, that their spores, true pollen-grains, are susceptible of arriving at their perfect state at the very place where they were formed, and it is not requisite for them to be modified by a preparatory sojourn in an ovule to acquire the germinative faculty.

The observation and verification of these facts is difficult; for they occur in the most minute parts, and the want of ordinary transparency around the micropyle and the apex of the nucleus only allows us to determine with a deal of trouble, whether the small body, which we see suspended from the summit of the embryonary cavity, really continues the tube, that runs into the opening of the teguments of the ovule. The author, however, mentions certain ovules, those of the Orchideæ, for instance, as exempt from this inconvenience and throwing great light on these facts.

§ 623. On the other hand, MM. Mirbel and Brongniart think they have several times determined the existence of the embryonary vesicle with the still rudimentary embryo, which it contains before the arrival of the pollen-tube. The extremity of the latter would not, then, furnish it, but would find it already prepared and would only serve to give it the vital excitement to cause its developement.

The objection would be still stronger, if we found seeds which were developed without the assistance of anthers. It has been more than once thought that some, in which this was the case, had been found, but almost always these facts resulted in recognising, by the side of the fecundated ovary, anthers, rudimentary it is true, even reduced to a few pollen-grains, the presence of which, however, was sufficient to bring back the phenomenon to its ordinary conditions. One genus of the Euphorbiaceæ, the CÆLEBOGYNE rather recently described, but for some years back cultivated in the conservatories of England, has several times fructified in them, and its seeds were evidently perfect, since we not only find there a well-formed embryo, but also on sowing the seed, this embryo has been developed into a similar plant. Now, the flowers are diæcious; we do not know and do not possess male plants, and the most minute researches prosecuted by our best observers, have not yet been able to determine the least trace of anthers or even of pollen. The embryo does not come from this pollen, which is entirely wanting: it must be perfectly formed in the ovule.

§ 624. However this may be, this fact and others of the same kind, which we could mention, up to this time cannot be explained, since the pollen, supposing that it does not directly furnish the embryo, is always at least indirectly necessary to call it to life in phanerogamic plants. This is proved by numerous experiments, some of which were known long ago; it was a well-known fact, that the Date-tree, to bear dates, required the plants of those trees, which bear ovaries only, to be placed close to those plants which bear stamens only, and they even knew how to supply this want of male plants by throwing on the female the dust of the male flowers gathered on distant trees and carried thither for that purpose. In all diæcious plants, if we take away the plants which bear the stamens, the ovaries of the others are not developed; in the monæcious, we obtain the same result by cutting off all the staminiferous flowers before they expand.

In nature, winds, insects and several other extraneous agents favour the transport of the pollen; but in our conservatories, which are protected from the wind and insects, this transport frequently does not take place spontaneously, then the gardener takes care to supply it by himself conveying the pollen to the stigma; and since this practice, several plants have fructified (especially the Orchideæ), which previously produced flowers in abundance, but never produced fruit.

§ 625. A no less decisive proof of the fecundation of the vegetable is the existence of hybrids or mules. We have remarked, indeed, that the pollen of a plant, in general, fecundates only the ovaries of all the plants of the same species; this faculty, however, is also extended to those of species very similar to one another. When two species, not quite identical, are thus fecundated one by the other, the seed, which results from this fecundation, produces a plant not exactly resembling either, but with some characteristics of both at the same time: this is what we call a *hybrid*. This mixture of characteristics, of which some belong to the plant which has furnished the stamen, others to the plant which has supplied the ovule, shews that there has been an action of both at the same time. This fact weakens the doctrines which have denied the fact of fecundation, by trying to explain the excitation and the developement of the embryo by theories which it would be too long to mention here. These theories, besides, are no longer required to explain the use and destination of these apparatus, so delicate and complicated, which we have endeavoured to describe, and the continuation of the functions, the accomplishment of which they protect and insure.

NECTARY.

§ 626. (*Nectaire*) (NECTARIUM).—We find in several flowers parts, which do not present the structure and the form of those, with the examination of which we have been occupied, the calycinal folioles, the petals, the stamens, the carpels: these are termed *accessory parts* (*parties accessoires*). They have already engaged our attention (§ 397), and we have stated that they were most commonly some of the essential parts disguised by degeneration and transformation, but, that under this disguise it was still possible to determine them from the position which they occupy in the flower and their relation of position to the neighbouring whorls: if they alternate with the parts of these whorls, they are transformed organs; if they are opposite to them, they are simple deduplications (§ 387). The stamens are especially subject to these transformations, and multiply the number of these accessory parts.

These are presented under very varied forms, under those of threads, of green or coloured laminæ, thick or membranous, of scales; so that they are frequently described under the different names which express their appearance. But, they very often take the shape of glands, and then they exercise their functions more or less manifestly, becoming the seat of a secretion, the honeyed product of which has received the name of NECTAR. Hence, that of *nectary* (*nectaire*) (NECTARIUM), which several authors give to these parts. And, since their analogy to the accessory parts of a different structure is incontestable, the example of Linnæus has been frequently followed, in extending this name to all these parts, even when they are not nectariferous organs.

But, on the other hand, these very secretions are frequently discovered on some part of the organs of the flower, in other respects perfectly formed, on true petals or true stamens; and Linnæus also called the depositories of these secretions, nectaries, so that by adopting his terminology we find ourselves led to apply the same denomination to parts which have no connection with one another: for instance, to some glandular portion of a petal, because it secretes; and to some filament or some scale, although it does not secrete.

§ 627. It seems, therefore, preferable to keep ourselves to the strict etymology and to reserve the name of nectaries for the parts of the flower, in which this formation of the nectar appears, whatever may be their position and their origin. It is in this more restrained sense that they are defined by the majority of authors, even by Linnæus himself, who has said: NECTARIUM, PARS MELLIFERA FLORI PROPRIA [u].

The formation of this sweet exudation is extremely frequent in flowers to which the bee comes to gather it for the sake of his honey. The afflux of sugar appears, indeed, to be necessary for the developement of the floral parts, and, if it be formed in several other parts of the vegetable, it seems to tend especially towards the flowers. Thus, we have recently remarked that the sap of Maize is loaded with a large proportion of sugar, but only before flowering; afterwards it has almost all passed into the flowers and disappeared from the rest of the plant.

§ 628. We know that the modified leaves, which form the different parts of the flower, present in their structure very well-defined differences from that of true leaves. These differences appear not only in their tissue, but in the surface itself from which they grow, and which forms the bottom of the flower or torus (§ 382);

[u] *Nectary, the honey-bearing part peculiar to the flower.*—TRANS.

a surface, which, instead of resembling bark, is frequently covered with a glandular layer especially in places ; now, we frequently find these increases of thickness at the base of the organs, and the latter becoming abortive, the thick part may still remain, and, in fact, be so much the more developed; hence, doubtless, the shape of glands to which the abortive parts are so frequently reduced. The glandular layer of the torus constitutes, doubtless, an apparatus adapted for modifying the juices which pass from the plant into the flower, and contributes to the formation of the nectar, generally so much the more abundant as the layer is itself more developed. This layer is increased not only by those projections, of which we have just spoken, but also by being extended over the surface of certain floral parts, which it doubles and covers.

§ 629. The glandular apparatus, moreover, are far from being totally dependent on the torus; we observe them at other parts of the flower more or less distant from its base; on the inner face of the perianth and calyx; on that of the petals, sometimes at their extremity, and frequently at that of the stamens, as in several Rutaceæ. We shall not add any more details on the varied forms of these nectaries, which resemble those of the glands we described in another part (§ 249). We shall content ourselves with mentioning as examples for study the pedicellate nectaries at the base of the stamens in the Laurel (*Laurier*) (*fig.* 316, *g g*) or those of the PARNASSIA (*fig.* 500, *n*), which seem to replace the anthers on filaments so regularly and so elegantly deduplicated; the projecting and sessile glands, whence grow the stamens of the Cruciferæ (*fig.* 325, *t*), or those, which appear around and beneath the pistil in the greater part of the Labiates: those, which crown the ovary in the Umbelliferæ; those, which form large cavities of a different colour, towards the base of

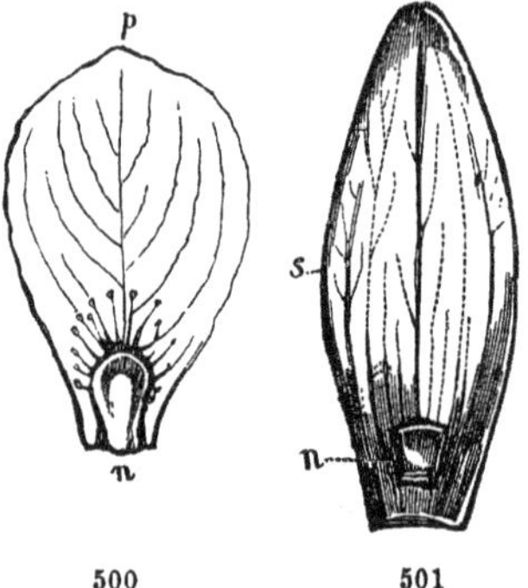

500. A nectary *n* of the PARNASSIA PALUSTRIS with the petal *p*: it is placed before the petal in the flower.

501. A division *s* of the perianth of the FRITILLARIA IMPERIALIS, with a nectary *n* at its base, under the form of a superficial cavity, differently coloured from the rest.

the inner face of the folioles of the perianth of the Crown Imperial (*fig.* 501), &c. &c.

We frequently find the nectary in the cavities of the appendiculate organs, especially in the spurs, and this cavity is, as it were, a reservoir, in which its produce is accumulated, as in the Melianthus, Tropæolum, Pelargonium.

It is by no means uncommon to meet a saccharine exudation without any appearance of a glandular surface; oozing out, for instance, from that of the petal, which apparently is not peculiarly modified. We may be convinced of this, by touching a great number at the time of flowering, the majority of which will shew, when touched, the presence of a colourless juice, which otherwise escapes the eye.

§ 630. Let us remark that this exudation follows the phases of the flowering, begins, grows, declines and ends with it; that it is very rare to see it precede the dehiscence of the anther and the expansion of the flower; that it is at its maximum during the emission of the pollen; that it ceases, when the stamen withers and the fruit sets. The nectaries frequently appear round the essential organs of reproduction (the stamens and the pistil), and there is hardly any doubt but that their result is connected with this function. Is it peculiarly with the functions of the stamen or with those of the pistil? It is most certainly not with one to the exclusion of the other, since in certain diclinous plants the male flowers present nectaries as well as the female. On the other hand, the action of the nectaries and that of the organs of the flower, if they have an evident influence over one another, do not appear, however, to be necessarily connected together. We may take away the petals, the stamens, the pistils; and the nectaries will continue to secrete, so long as we do not wound them; we may take away the nectaries or at least their produce, without hurting the fecundation or retarding the maturation of the fruit.

When we reflect in what proportion the nectar extravasates, flows out and is carried off by the insects during the flowering, and that then this oozing is stopped when the set fruit requires a large quantity of the juices, we have been tempted to consider the nectaries as much excretory organs as secretory, which excite the afflux of the juices by the waste they cause of the excess, which would be useless for the flower; and when the fruit in its developement requires a much larger proportion, these juices continue to arrive by the usual routes, and, there being no need of waste, are all employed in the process of ripening. We have seen, however, that the sugar, a large part at least, does not come quite formed to the fruit, but is there formed and mostly at a later period (§ 545).

§ 631. Whatever may be the function of the nectaries, they furnish very good characteristics in distinguishing plants, being very constant in number, shape, &c., in a given species. It is remarkable, that their developement at a particular part of the flower is frequently connected with its irregularity and seems to cause that side to be irregular in which the nectary is placed.

Some general phenomena of vegetation.

§ 632. After having considered in a general manner the life of the vegetable and the organs by means of which it is carried on, it remains for us to consider, how, by means of the modifications which these organs present, we may distinguish and classify the great number of vegetables that exist on the face of the earth. But, before beginning this long and important chapter, let us rapidly cast our eye over some points, that we have voluntarily omitted (§ 331) and have postponed for this place so as not to interrupt the preceding explanation.

Colouring principles of plants.

§ 633. The outward appearance of living vegetables is modified by various colours, among which green predominates: it is generally the colour of the young bark, of the leaves and of other organs which are most similar to them in their nature, such as the calyces, the carpels and the fruits still young. We have already had occasion to speak frequently of the principle to which this green colour is owing, the chlorophyll (§ 24), and when we were engaged with the respiration of the vegetables, we saw that it is most commonly formed by the combined action of the atmosphere and of light, whence result an accumulation of carbon and a loss of oxygen in the vegetable. But it is not impossible for the same effect to result from some other reason, when the parts protected from the action of the light become green, but they must be placed in some other medium than air. Thus, M. de Humboldt has determined that the plants carried into completely dark mines, into an atmosphere not fit for respiration, but strongly hydrogenized, not only preserve the green colour in their parts previously developed, but also shew it in the young shoots which are continually developed.

The change, therefore, owing to the absence of light is, without doubt, compensated by the fresh conditions from this change in the surrounding air, deficient in the oxygen, which the plant would have lost, and abounding in materials adapted for the formation of the chlorophyll. By analogous considerations, perhaps, we may explain the green colour, which we have pointed out in parts situated deeply in the interior of the plant, like the pith (§ 55), certain seeds and certain embryoes (§ 608), &c., &c. In every case, these are only young parts, enjoying a very active vital principle, which are thus coloured.

§ 634. But the leaves, as well as the organs which resemble them the most, are not constantly coloured green. There are some which habitually present other colours, either over the whole of their surface, or at several parts only (in which case they are said to be *variegated* [*panachées*] [VARIEGATA]), or over only one of their faces, commonly the under one. In the first case, the young bark commonly participates in this colouring; as in the red varieties of the Beech, the Beet-root, the Garden Orache. The AUCUBA presents us leaves spotted with yellow; the CALADIUM BICOLOR, spotted with red; certain species of Saxifrages, of CYCLAMEN, of TRADESCANTIA, of leaves, green above, red or brown beneath. It would be easy, but useless to multiply these examples.

§ 635. Several green leaves take at a certain period of their existence fresh tints and colours; that of a red more or less brilliant, sometimes heightened with brown; that of yellow more or less pale: and these changes are always reproduced in the same manner in the same species. We most generally see this in autumn, at that period of their life which immediately precedes their fall; the leaves then lose their green hues to assume others: those of the Poplar, of the Elm (*Orme*), of the Birch, for instance, to become yellow; those of the Sumac, to become a bright scarlet; those of the Cornel or Dogwood (*Cornouillier sanguin*), of the Guelder Rose (*Viorne*), a duller red; those of the Vine, spotted with yellow or purple, &c., &c.; we may just remark that in most instances, these new colours of the leaf correspond to that which the fruit of the same plant takes when ripe. Thus, the spots are yellow on the leaves of the Vine producing white grapes, purple, on those of the Vine producing black grapes.

§ 636. Yet these changes are by no means a necessary sign of approaching death. Several of our herbaceous plants, when they are developed late in the year, preserve their living leaves all the winter, and in a great number, we see at the latter end of autumn the tint of the leaves more or less modified, although they will fall the following spring. If we observe during winter those vegetables called evergreen, because they do not then lose their leaves, we shall see on those of several (the Rue, Firs, Houseleek, SEDUM, &c.), a tinge of a dirty yellow or more commonly slightly brown or red, entirely different from that which they had during the preceding fine season and which they will reassume in the following.

We find ourselves by this fact led to draw a natural conclusion: that the alteration of the colour is not owing to the changes caused in the tissues by old age, the precursor of death; but rather that it is allied to those which the season causes in their functions, which modifies their existence in the organs, where its life is not definitely suspended.

§ 637. Several plants, however, do not wait for the end of autumn to shew analogous changes, but begin to present them when their vegetation is in the strongest force, in circumstances which we consider the best adapted to render it active, as under the prolonged influence of a strong light. Let us compare those Gramineæ growing on walls or lawns exposed for a great part of the day to the direct action of the sun, with those which are only submitted to its action when less direct and constant (like those of our meadows, which pressed together mutually shelter one another:) we shall frequently find the former of a rusty red colour, whilst the latter are of a bright green. On high mountains this premature change in colour is frequently observed in a large number of vegetables; as may be seen by comparing plants gathered on the heights of the Alps with the same species taken from the valley. We must state that in these different circumstances, these plants, at the same time that they are for a longer period exposed to the light and the heat of the sun, are afterwards exposed to great coldness on account of the nocturnal radiation, and this cause may also have some influence on the phenomenon with which we have been occupied.

§ 638. Several plants present reddish or brownish tints in their leaves, which begin to expand out of the seed or buds, before they assume their green colour: a fresh proof that we ought not to assign these changes in colour to the modification caused by the decay of the vital functions.

§ 639. In those leaves nearest to the inflorescence the green colour is most frequently replaced by other more brilliant hues, generally passing gradually or rapidly to those of the flower. These leaves have been shortened and contracted into bracts or preserve their common form, as in the POINSETTIA. If they assume the hues of the corolla, the calyx equally participates in them.

§ 640. Green is the exception in the corolla, to which we must also add the coloured perianth of several Monocotyledonous plants (§ 418), whilst we there find the examples of other most varied and rich hues; so that it has been the particular study of the authors who have treated of the question which now occupies us. White, yellow, red, violet, blue are the colours, which we most frequently find in various degrees of intensity; or combined together in different ways, so as to furnish intermediate shades equally numerous. It is worthy of notice that yellow flowers may change into red or white, but never into blue; blue flowers, conversely, into red and white, but never into yellow; that in several very natural genera or even families, all the flowers affect the blue colour and its derivatives, or yellow and its derivates, but not both at once. Botanists have thence admitted two distinct series in the colours of flowers: the *Cyanic* (*cyanique*) (from κύανος, *blue*) and the *Xanthic* (*xanthique*) (from ξανθός, *yellow*). Green, composed of blue and yellow, is intermediate and, as it were, neuter between them, and both series ending with red seem to be blended at their two extremities. We have placed them here reduced to the smallest number of terms possible:

RED,—ORANGE,—YELLOW.—GREEN.—BLUE,—VIOLET,—RED.

White is, as we shall see, simply the absence or extreme dilution of these colouring principles; brown or black (always imperfect in the external parts of vegetables), merely the accumulation and concentration of the same principles.

§ 641. Now, instead of only studying the external phenomena, let us examine the interior to try and discover the seat of the colour, and how the combination is formed, from which results the sensation impressed upon our eye. The colouring matter is deposited in a liquid or semi-liquid state in the cells: we see it through their transparent walls. When we open these small reservoirs of the colour, we find its internal appearance much less varied than its external. It is always yellow, violet, red or green, either pure, or tending sometimes to yellow, sometimes to white. This substance is suspended in a colourless liquid, and according as it exists in a greater or less proportion, the colour is deeper or paler. When it is wanting, the colourless fluid remaining alone gives us the appearance of white; but it is very rarely perfectly pure, and there is almost always mingled with it some small quantity of one of the other colours: so that on pressing a white flower on perfectly white paper, it almost always leaves a slight yellowish or bluish stain.

Yellow or green substances are found in the more deeply situated cells; in the more superficial layers, even in those of the epidermis, we find the blue, red and violet substances. If yellow be beneath these layers of different tinges of red, the flower will be of an orange colour; if green, of a brown of various shades; if the green, blue or violet substance be accumulated in several layers of a compact cellular tissue, and be not much diluted, the colour then seems to approach black, but an attentive inspection always shews us one of these shades. The interposition of several rows of cells with a colourless fluid between the eye and the layers of the inner coloured cells, weakens the strength of the latter and seems to stretch over them a white layer; whence results a glaucous appearance, if these inner cells are of a green or blue colour.

§ 642. We generally observe the same colour only in the same cell. It is not, however, always and necessarily so: for instance, in those leaves the green colour of which is modified in winter, we sometimes find green granules of chlorophyll with a reddish juice in the same cells.

§ 643. We have already seen that the colour of the leaves may vary at different times of their existence. It is the same with the corollas: of a green colour, more or less pale, at their appearance in the flower-bud, they assume, as they expand, tints more and more brilliant, which generally attain their maximum at the time of the fecundation, then fade and become dull. But in some we observe changes of another nature: the formation of regular spots, which did not exist before their inflorescence, or a complete change in the general ground of the colour; the passage from one to the other is equally characteristic; from red to bright blue, for instance; as in several Boragineæ; from white to red, as in several Onagrariæ; from yellow to orange, to red, to violet, as in some species of LANTANA, in the CHEIRANTHUS SCOPARIUS; or, what is more rare and may appear singular, from yellow to the colour precisely opposite, to blue, as in the MYOSOTIS VERSICOLOR. What then takes place? Are fresh cells formed in which the fresh colour is produced, or else is the colouring substance modified in the already existing, enlarged cells? If this modification takes place, within what limits is it possible? Are there several kinds of colouring matter, or is there only a single one, susceptible of all the modifications which have hitherto been considered as different substances? Lastly, what are the causes, under the influence of which these modifications take place, if this hypothesis be true?

§ 644. Several authors are led to adopt it, and, founding it upon the fact, that the green colour is the most general in vegetables, that it is almost always that of the leaves, that the other organs are only modified leaves and almost always present at their first appearance this same green colour of a certain degree of intensity, they think that it is the green substance itself, which, as it is modified, determines the other colours found in several leaves at a certain period of their life, in all flowers during nearly the whole of their existence.

We know with what rapidity and facility vegetable colours are modified by contact with acids or alkalies: so much so, indeed, that the changes they undergo furnish the chemist with the most delicate tests for determining the acid or the alkaline nature of bodies. Red flowers change into blue, into green and even into yellow by the action of alkalies; by the action of the acids, they pass through these same colours in an inverse direction. It was, therefore, natural to explain by the influence of acids and of alkalis, which are introduced or formed in the interior of the plant in the action and process of life, the different modifications of colour, which the different parts assume externally, although originally green.

The question was afterwards carried farther, and it was pretty generally admitted that these changes were owing to the oxygen added or subtracted from the parts, which are coloured in such a manner or such another manner; and Schübler went as far as to give the name of oxygenized and of deoxygenized to the two series of colours, which De Candolle called xanthic and cyanic: the green principle, when it was combined with fresh quantities of oxygen, would pass to the yellow, then to the orange, then to the red; by losing, on the contrary, a part of the oxygen, which enters into its primitive composition, it would pass to blue, then to violet, then to red. By yielding oxygen, acids act; by attracting it alkalies act: thus, when the latter change the syrup of violet to green they would disoxygenize it by suroxygenizing themselves.

§ 645. This theory seems to be confirmed by an observation of M. Dutrochet, who, on putting the juices extracted from a leaf, green above and red beneath, in communication with the poles of a voltaïc pile, found the green substance accumulated at the negative pole, the red at the positive pole: the former, consequently, alkaline, the latter, acid. The author drew from this and other experiments made on the leaves or unicoloured petals conclusions somewhat different. He supposed all colouring matter to be composed of two principles, the one, electro-negative, the other electro-positive; every leaf or other foliaceous organ, a petal, for instance, to be analogous to a galvanic pile, having an upper disoxygenizing face, a lower oxygenizing one.

§ 646. The more complete and recent work of M. Marquart on the colours of flowers shews the weak points of Schübler's theory and substitutes a fresh one for it. It is also founded on the chlorophyll. He endeavoured to separate it by means of alcohol, which has the property of dissolving it, and to which the leaves, macerated for a few days, impart a greenish hue, at the same time losing their own colour. The residue, which we obtain by evaporation, is treated with sulphuric ether, and thus freed of a substance which is mixed with it and dissolved by the ether. M. Marquart considers the fresh residue, thus obtained, as the chlorophyll. Now this, treated with water changes into yellow, most probably on account of its combining chemically with this water; treated with concentrated sulphuric acid, it passes, on the contrary, to blue; and the author thinks that the acid produces this effect by absorbing the water, for which it has such a great affinity. Hence, he concludes that all these modifications of the chlorophyll ascribed to oxygen, are owing to the water; that one quantity more of water changes it into a yellow substance, which he terms Anthoxanthine (*anthoxanthine*), that one quantity less changes it into a blue substance, which he terms Anthocyane (*anthocyane*). The green matter is the neutral state between these two; it finds its proper elements in water and in carbonic acid, which combined to form it by losing the oxygen, furnished either by the one or by the other. We do not find in it any trace of azote or nitrogen.

The anthocyane is soluble in water or in spirits of wine diluted with water; this solution is turned red by acids, and reduced to its original green colour by alkalies. In its pure state it colours blue flowers; modified by the action of a weak acid, violet flowers; by a stronger acid, red flowers. The anthoxanthine colours yellow flowers; it is not much affected by acids or alkalies, and is in general only dissolved by concentrated alcohol or sulphuric ether; although in some few plants alcohol diluted with water, and even water alone will readily reduce it to solution. Concentrated sulphuric acid changes it into a fine indigo blue, then into a purple, doubtless absorbing its water; for in proportion as it attracts water from some other source, we find these colours grow weaker and disappear. These different properties lead us to consider the anthocyane to be one of those substances termed extractive: the anthoxanthine, to be a resinous body. But, we frequently find in the cells along with them other substances, which rather tend to disguise their nature; with the anthocyane, a white or slightly yellowish or greenish resin, which M. Marquart thinks is an intermediate state of the chlorophyll; with the anthoxanthine, a colourless liquid, probably the cellular juice. This liquid or this whitish resin is the only one in white flowers. We have already seen how other tints may result from the relative situation of the cells containing these differently coloured fundamental substances. It is easy to conceive, how the flower, at first green, passes to other colours, and how these may be modified in a certain order, since the action of life may add or subtract water combined with the colouring principles; the formation of acids acting with so much less intensity as they are weaker, in a transitory or constant manner, according as they are volatile or fixed; the formation of the alkalies, which act in an inverse direction, either by themselves without the intervention of any other agent, or by neutralizing the acids.

§ 647. But rather solid objections have been made to these theories, according to which the different colours are derived from a single substance modified in various ways. The chlorophyll is not found in the superficial layers, especially in those of the epidermis, and in these we mostly find the violet or red colouring principle; how then could it be formed at the expense of a substance which does not exist there? It is true, that in the cells most deeply situated and filled with chlorophyll (in those of the mesophyll) we see at a certain time red matter formed; that observation assisted by the microscope generally shews that it exists along with the green, that it does not replace it, but rather the juice, which was previously colourless; that it only sometimes disguises the chlorophyll by covering the green granules, which at other times remain perfectly distinct. The colourless juice of the cells by the prolonged action of a weak acid gradually becomes red, but without passing through blue, which would be the case if it arose from the anthocyane. The anthoxanthine and the chlorophyll assume in concentrated sulphuric acid a very deep blue, which M. Marquart attributes to the transition to the anthocyanic state. But, when the latter is submitted to the continual action of the acid in excess, it ought afterwards to become red, which, however, it does not. In what way then, do the leaves and flowers coloured yellow by the anthoxanthine pass immediately to red? M. Marquart himself admits, that it is by the superposition of more recently formed cells which are filled with anthocyane reddened by an acid; but we do not perceive the green and blue tints, which, according to this hypothesis, ought to have immediately preceded the red. We are led by these different considerations to doubt that the chlorophyll, the anthocyane, the red principle, can be different states of the same substance; and this doubt is confirmed by the weighty authority of M. Berzelius, who opposes this hypothesis, and admits red as a distinct substance, which he terms *erythrophyll* (*erythrophylle*) (from ἐρυθρὸς, *red* ; φύλλον, *leaf*). On the other hand, he has extracted a yellow principle from the leaves, which he terms *xanthophyll* (*xanthophylle*), and which seems to be different from the anthoxanthine: it is a fatty body of a peculiar nature scarcely soluble in alcohol. This substance, in summer joined to another rather blue than green, would give, when the other was gone, the yellow colour to the leaves we so frequently see them assume in autumn.

It is probable, therefore, that there are many different principles in the coloured parts of plants; and, perhaps, according to the distinct properties, which several of these principles have, at the same time only imparting to us the sensation of one and the same colour, we may suspect the existence of more than we have mentioned. Chemistry must decide these questions, and it is desirable for it to give to this study a certain foundation by determining the elementary composition of these colouring principles, beginning with the chlorophyll, as it did in the case of several other principles, extremely similar to one another, of which we have treated under the article of nutrition.

§ 648. We have seen brown shades arise from the superposition of cells filled with juices of a different colour (§ 641). Sometimes, however, the interior of the cells is filled with a colour really brown, which appears under the form of globules very like those of the chlorophyll. M. Berzelius does not think that there is any connection between these two substances, since the former is produced as a liquid, at first colourless, by the action of the oxygen.

§ 649. But this brown colour is frequently situated not in the inside, but in the very wall of the cells, which a colouring substance, varying according to the plant or its parts from yellow to the deepest brown, penetrates and impregnates in the same way as the lignine (§ 20). We sometimes also find globules of the same colour, and we may then have the same principle, partly free in the cavity of the cell, partly incorporated in the substance of its walls. This colour is the most rarely found in those cells, which are nearest to the superficies, as in certain species of JUNGERMANNIA or of AZOLLA; more commonly, it exists in cells deeply situated, those of parts which have not yet occupied our attention with regard to this subject; of the wood, for instance. These cells, which last much longer than those of the leaves or especially of the flowers, have walls thickened by the formation of several successive layers fitted to one another; and it is clear that these layers, in proportion as they are thus placed upon one another, will assume a colour becoming gradually deeper, even when each, considered separately, is of a very pale colour.

§ 650. The wood of Ferns and Palms commonly owes its colour to that, which the walls of the cells thus assume, and the same thing takes place in that of some Dicotyledons, as the Alder and the Mulberry; but the latter more frequently owe their hues to the intracellular matter, which is coloured in the inside either of the ligneous fibres, or of the utricles forming the medullary rays, as in the Ebony (*Ebénier*).

§ 651. We have seen (§ 67) the division of the wood into alburnum and heart-wood; we know that it is in the former, as the younger part, that liquids abound and vital phenomena are more actively at work, whilst they are moderated or even totally arrested in the heart-wood, the hard and dry part, old and, as it were, dead, being at the same time that part which is coloured. Here then the colour no longer appears to be united to life and comes almost exclusively under the province of chemistry, which, from the varied hues of the different kinds of wood, has doubtless to discover in its investigation here, as in flowers, several colouring principles, as well as numerous modifications of which they are susceptible. These changes take place slowly and protected from the air, from which the heart-wood is separated by the whole thickness of the alburnum and of the bark. The latter, which is directly exposed to the atmosphere, is placed in other conditions and undergoes more rapid and varied changes of colour. This action of the air is, moreover, perceived on all kinds of wood in contact with it and generally darkens them; Mahogany is a sufficient example of this.

§ 652. It frequently happens, that the combinations of the principles contained in the atmosphere and of those which colour the vegetable parts, take place more rapidly and, as it were, instantaneously. Thus, during life, the cells of the root of the Madder (*Garance*) are swelled with a yellow juice; when we pull them up, they assume on their surface in contact with the air that red colour which every one knows; if we wound or cut them, this fresh colour is produced on these successively fresh surfaces. They do not change their colour in pure oxygen, without there is a small quantity of water in a state of vapour in it. There are in our woods a great number of species of fungi of the genus BOLETUS, the flesh of which is perfectly white; if one of these be broken, the white soon passes to other colours: to a vinous tint in some, to a greenish blue in others, to the most intense indigo in the BOLETUS CYANESCENS. The flowers of several species of Orchideæ of a very pure white, such as those of the CALANTHE VERATRIFOLIA, of the BLETIA TANKERVILLÆ, &c., also change into a deep blue wherever they happen to be wounded or bruised, and, as they dry, they also assume this hue, which becomes so deep as to change into a black. Drying changes the whole surface of the plants of this family of Orchideæ into black: an analogous process is observable in several others, especially in the Rinanthaceæ, in which all, who are collecting a herbarium, must have observed this property, almost a characteristic of family. This modification is owing in some cases to the presence of the colouring principle of indigo, which, grey or whitish during life, on combining with oxygen takes the colour, which we are accustomed to see. Doubtless, the tannin, so frequent in certain parts of the vegetable, the gallic acid which is found in it, the salts of iron which are frequently recognised in rather notable proportions, and which, mingled with this acid, form ink, have something to do with other cases, in which we observe this change of colour after life. But the determination of these substances, the modifications and the fresh combinations of which they are susceptible, belong to another science.

HEAT LIBERATED BY VEGETABLES.

§ 653. Can vegetables, like animals, emit a heat from themselves, which, consequently, is higher than that of the atmosphere with which they are surrounded? Is this heat constant or intermittent, and under what circumstances does it acquire the greatest intensity? We can answer these questions to a certain extent by analogy, since the cause of heat in animals is known; for this effect arises from certain chemical combinations always accompanied by a disengagement of caloric, especially from that of carbon with oxygen; analogous combinations take place in the vegetable, but since these much weaker and slower actions are also performed in an intermittent manner, both under the influence of and protected from daylight, in different seasons and in different parts of the same plant: we thence conclude that vegetables will emit a degree of heat, produced by the action of life, but that it will not generally be very sensible; that it will increase only in certain phases of vegetation; that it will be very strong in certain parts and not traceable in others. This is indeed what we learn by experiment.

§ 654. The disengagement of heat is very strong in the flowers; it is the strongest in them at the moment of blossoming. In those of the family of the Aroïdeæ, it is powerful enough to be appreciated by a touch of the hand, and their arrangement is peculiarly adapted to observations of this kind. They are crowded on a long, thick axis, which is enveloped by a large spathe in the shape of a horn (*fig.* 185), which concentrates the heat produced by all the flowers at the same time and into which it is easy to introduce a thermometer. The pistils, each of which constitutes a female flower, are, moreover, separated from the stamens, each of which forms a male flower; we may thus, by alternatively cutting off both, determine in what proportion these two kinds of organs furnish the amount of heat.

When the spadix is expanded, it becomes for a few days the seat of a kind of fever, which establishes between it and the surrounding air, a difference of a greater or less number of degrees; this fever is intermittent and quotidian, for each day the heat begins to increase gradually: then, after having attained a maximum, to decrease until it returns to nearly its point of departure, a point which is always a little above that of the surrounding air. The maximum, small at the commencement of the inflorescence, becomes stronger for a few days, is then gradually weakened; and the change ceases at the end of a little time. It does not return each day precisely at the same hour, but is a little before or after that of the preceding day. In our common Lords and Ladies (Arum vulgare), the maximum of the heat thus developed, that is, above the heat of the atmosphere, is from 8° to 10°, it is still higher in the Arum Italicum and the Arum dracunculus, which belong to warmer countries; in the Colocasia odora, the Caladium pinnatifidum, which we cultivate in our conservatories: But the phenomenon seems to acquire great intensity in climates, in which these plants grow spontaneously, since in one of the observations made at Bourbon, five spadices of the ARUM CORDIFOLIUM, bound round the thermometer, made it mount to 25°; twelve made it rise to more than 30°.

§ 655. The spadix of the different Aroïdeæ, on which these observations have been made, is composed of an axis surrounded at the bottom by a certain number of pistils; a little higher up, by fertile anthers; above these, by anthers more or less completely barren; the whole enveloped by the spathe in the shape of a horn; the heat developed during the blossoming is the general result of the temperature to which these different parts are raised. But if we enquire what belongs to each of them in particular, we find that the heat is very unequally distributed; that the largest proportion generally arises from the fertile anthers; that the sterile ones sometimes appear to equal or even to surpass them, but remain much below them when their abortion and metamorphoses are complete; that the temperature of the pistils is always comparatively much weaker and that of the spathe still more so.

§ 656. During this blossoming, as during that of flowers generally, there is a certain quantity of oxygen taken up by them from the surrounding air; a certain quantity of carbonic acid gas exhaled, consequently, a proportional quantity of the carbon of the flower consumed. Now, M. de Saussure has determined, that this quantity is considerable in the Aroïdeæ during the disengagement of heat, much more than that which is consumed when the flower remains cold, and that it is very unequal in the different parts of the spadix, when isolated; that, for instance, in the ARUM MACULATUM and the DRACUNCULUS, whilst the anthers absorb more than one hundred and thirty times their volume of oxygen, the terminal mass, formed by the sterile anthers does not absorb more than thirty times its volume, the pistils only ten times, the horn-shaped spathe only from one-half to five. The heat developed by the different parts is, moreover, nearly in relation to these numbers. We hence necessarily conclude, that the heat is owing to this very active combination during certain phases of the blossoming.

§ 657. It was very naturally supposed that this phenomenon, so intense in the Aroïdeæ, ought to be found to a certain degree in other flowers, which, as we have just said, act in an analogous manner on the air, by combining a certain portion of the carbon they contain with its oxygen. Delicate observations have enabled us to recognise in certain flowers a decided developement of heat during the blossoming, but always very weak; it attained in some at the most one degree above the temperature of the atmosphere, less in a greater number of others, and was completely insensible in the majority. It is true, that the flower is frequently at this very moment, as we have seen, the seat of exhalation and, consequently, of evaporation, which, by the cold it produces, must tend to counteract some of the heat developed. By comparing the quantity of oxygen absorbed by a simple flower, that is, one with its stamens, and by the same flower doubled, that is, in which these stamens have been changed into petals, that consumed by the male flowers of a diclinous species, and by the female flowers of the same, we may be convinced that here, as in the Aroïdeæ, the anthers consume more than the petals and the pistils. It is, therefore, allowable to explain the same effect by the same cause. Let us remark, however, that we ought not to admit this consumption as the only reason of the feeble heat developed by these flowers, since the latter is not found in a constant ratio to the proportion of oxygen destroyed, and certain flowers consume more, although not manifesting any change of temperature, whilst others present a very appreciable one.

§ 658. We know that during germination the seed attracts the oxygen of the air and forms rather a large quantity of carbonic acid. We are led to conclude hence, that in this act of vegetable life, as in that of the blossoming, we ought to perceive a developement of heat. We may observe rather a considerable one, if we introduce a thermometer into the midst of a mass of germinating seeds; for instance, into the heaps of barley which are sprouting during the process of malting. Nevertheless, M. Dutrochet thinks that this heat is not a vital phenomenon, that it is owing to the same causes, which determine a large elevation of temperature in a mass of moist hay or of other vegetable matter heaped up in the same way, either living or dead, and which are only, according to him, chemical combinations between the organic vapours rising from these substances, condensed in the empty spaces which they leave between them, and undergoing a decomposition more or less speedy.

§ 659. The same author has directed his attention to determine the heat, which may be produced in the other parts of the vegetable, and in his researches has constructed for this purpose a thermo-electrical apparatus, that is, one measuring the heat by means of the developement of the electricity which accompanies it, and having this double advantage that the smallest quantities are indicated much more clearly than by the most delicate thermoscopes, and that every vegetable part may be easily examined, since it is only necessary to plunge into it two points forming part of this simple and very manageable apparatus.

We have already perceived a difference of temperature between vegetables and the surrounding air. These experiments were principally performed on the trunks of trees, into which it is easy by means of a gimlet, to insert a thermometer to the depth at which we wish to determine the temperature; which was sometimes a little higher than that of the air, sometimes, on the contrary, a little lower. But we may easily explain this difference without having recourse to a vital heat peculiar to the vegetable. The wood, a much worse conductor of heat than air, becoming, consequently, hot and cold much more slowly, will always have a tendency to maintain an even temperature with the air, and to be very rarely in equilibrium with it, since in our climate the heat does not remain constant very long, but changes perpetually according to the season, the day and the hour. The vegetable follows these variations, but only very slowly, and shews the previous, rather than the actual state of the atmosphere. Besides, in these experiments, that part into which the bulb of the thermometer was introduced is traversed by the ascending sap, which, protected from exposure, will have preserved the temperature of the soil at the depth at which it is absorbed by the extremities of the roots, cooler, when the air is warm, warmer, when it is cold. Lastly, if there be a developement of vital heat, it ought to be where these chemical combinations, of which we have spoken in another part of the work, are taking place actively under the influence of the vital principle, towards the periphery and not towards the centre, in the young parts and not in those which have already grown old.

These are the considerations which have guided M. Dutrochet; and the young stems which were not accessible to the thermometer, being so, at all events, to the points of his thermo-electrical apparatus, he has been able to determine that they are the seat of a production of caloric, very weak, it is true, since in the plants, which gave him his maximum, it attained at the most ·3° to ·4°, and in others it did not rise beyond a few hundredths. It varies, moreover, with the causes, on account of which the intensity of the vital phenomena is augmented or diminished; and we, therefore, see it so much the better as the young parts vegetate more vigorously, more marked during the day, in which these phenomena acquire the greatest activity; then growing gradually weaker it is sometimes even completely extinguished during night, although artificial darkness is a long time in effecting it. The leaves themselves, at least in certain vegetables in which the evaporation is almost inappreciable, and the fruits also, but in a much weaker degree, indicate an internal developement of heat.

Developement of Light.

§ 660. The daughter of Linnæus first observed this curious phenomenon of the light which escapes from the flowers of the Tropæolum. The observation was afterwards repeated by others and also on other flowers, more particularly on the flowers, whose colour is yellow or orange, with brilliant and golden tints, as the Sun-flower (*Soliel*), the Marygold (*Souci*), the Sannacetum patula and the erecta, &c., &c. This phenomenon is seen to the best advantage in those evenings which succeed a warm and stormy day; the flashes of this light are then produced with the greatest vivacity, but never when the atmosphere is humid. According to the account of a traveller, there is in Africa a species of Pandanus, in which the bursting of the spathe by the flowers which it envelopes is accompanied with noise and a flash of light. Botanists record phosphorescent fungi. In the Rhizomorpha, which have the appearance of blackish roots winding through dead wood found in cellars and other dark places, the light, frequently very intense, is produced at the extremities which are of a whitish colour and of a flaky texture, especially at those of the young and vigorous branches. M. Delile has observed that on the Agaric of the Olive; 1st, it is the lower surface of the cap, where the spores are accumulated, that frequently becomes luminous; 2nd, it is at the commencement and during the greatest activity of their growth; 3rd, that they are not so during the day, although placed in a perfectly dark place.

In all the preceding observations, the emission of light accompanies the exercise of the vital functions at the time when it is most active, in the parts which absorb oxygen and disengage carbonic acid gas. It is, therefore, somewhat similar to heat; and we should be tempted to infer thence, that these two phenomena may be referred to an analogous cause, a rather intense combustion.

This supposition is supported, besides, by experiments prosecuted on the RHIZOMORPHA, the light of which is extinguished in those gases, which animals cannot respire, and is revivified and increased in pure oxygen. It is, however, necessary to multiply and vary observations, before we can be authorized to affirm any thing on so delicate a subject.

§ 661. Several other facts are on record of phosphorescent lights emitted by vegetable substances, either in full or incipient decomposition, by fungi, by wood, generally when it has been exposed to moisture, after having been cut when the flow of the sap was in full force, &c. Here the phenomenon falls entirely under the science of Chemistry, since it is in being disorganized, in obeying chemical affinities, that these tissues assume these fresh properties, similar to those which we frequently observe in animal substances under the same circumstances. The light seems to be situated in a gelatinous substance, extended in a layer over the luminous surfaces, which friction spreads, in the same way as it does phosphorus.

PHENOMENA OF THE DIRECTION AND MOTIONS OF PLANTS.

§ 662. We have seen that the different parts of the plants are developed in the same direction[1]: the root towards the centre of the earth; the stem in an opposite direction towards the sky; the leaves towards the light. Their tendency to place themselves in connection with the medium adapted for the exercise of their functions will not alone explain the necessity of this constant direction, as a few very simple experiments clearly demonstrate. Cause a seed to germinate in an apparatus so constructed, that the humid and dark medium, earth or a sponge filled with water, be placed above it instead of being, as is usual, below: the radicle will not rise and bury itself there, but will descend and hang in the air; the tigelle will not seek the air and light which are beneath it,

[1] Some parasitical plants growing on trees, those of the family of the Loranthaceæ and, especially, the Miseltoe, which is its representative in our climate, form an exception to this rule. The seed of the Miseltoe is held to the branch by the glue or viscous fluid which surrounds it, germinates whilst thus attached and always throws out its radicle towards the centre of the branch, its gemmule in an inverse direction. On a globe, the position of which may be varied, the radicle is always directed in the same way towards the centre. On a glass window it has a continual tendency to send its roots into the interior of the apartment, whether it be placed on the inside or outside of the window. In a word, its gemmule always grows towards the light, its radicle towards the dark.

but will rise and force itself into the earth or sponge. Let us place at a little distance from the radicle, which is thus descending, a moistened sponge, parallel to it and vertical to the horizon, the radicle will not deviate, but will continue to proceed parallelly downwards through the air. This opposite tendency of the two portions of the vegetable axis is, therefore, a part of their nature. We know, moreover, that the primary axis alone is endowed with this property, which is weaker in the secondary, and which may even completely disappear, as is proved by the more and more oblique direction of these axes and even frequently by their horizontal progress; as, in the rhizomes.

§ 663. If we endeavour to change this natural direction of the parts, it is not long before they resume it of their own accord. A branch retained by force in a horizontal position is soon bent upwards at its extremity, as it grows in length, and recommences its ascent: under the same circumstances a root is bent downwards and recommences its descent. Knight proposed a very simple theory to explain this phenomenon; he supposes, that the juices by the effect of gravitation, are accumulated over the whole of the lower surface of the branch, which is placed horizontally; that the accumulation of the juices determines a more active developement in the whole of this half, the fibres of which are lengthened more during the same time than those of the upper half, and that, as a necessary result, it will form an arc with its convexity turned downwards and, consequently, directing its extremity upwards. As to the direction of the root, which is only lengthened at the end, we may also apply the theory of gravitation, which will cause the juices to accumulate at this lower end and make it descend. But this theory cannot be applied to the case in which the direction of the root is inverted, nor at that in which the secondary axes naturally follow a very oblique, horizontal or even descending direction, as, in certain weeping trees.

§ 664. M. Dutrochet supposes the reason of their assuming these directions lies in the very structure of the parts. If in a mass of cellular tissue, the cells gradually diminish in dimension in a certain direction, so that to one or more planes of large cells we find one or more planes of smaller cells applied, and if, by the effect of the endosmose, these cells happen to swell, the larger ones will fill quicker than the smaller ones, their plane will be more extensive than the plane of the latter, and since these planes are closely connected, they will necessarily bend, the larger occupying the convexity of the arc and the smaller the concavity. If, therefore, in an axis the cells gradually diminish from the outside inwards, there will be a tendency to incurvation inwards; if they gradually diminish from the inside outwards, there will be a tendency to incurvation outwards. These tendencies are not manifested so long as the stem is entire and equally strong throughout the whole of its substance, since the sides being equally distant from the centre and wishing to bend in the same way with regard to this centre neutralize each other. But let us cut the stem into two longitudinal parts, or let one of the sides be weaker than the opposite side, the equilibrium is broken and the incurvation takes place. The horizontal position tends to lessen the force of the endosmose in the side which is turned downwards, and, consequently, to determine the incurvation either to one side or to the other. Now, according to M. Dutrochet, in stems with a large pith the dimensions of the cells, considered as a whole, diminish from the centre to the circumference; in roots without pith, in which the cortical system is more largely developed, this takes place from the circumference to the centre. If we oppose their natural tendency, by placing them in a position more or less approaching the horizontal, the force of the endosmose will be weakened on that side which is turned downwards; it will cease, therefore, to neutralize the wish of the side which is turned upwards, and this will be bent in the direction, which the arrangement of its cellular planes enjoins, forming an arc, the concavity of which is turned upwards if it be a stem, downwards if it be a root, directing its unencumbered extremity upwards in the former case, downwards in the latter.

Whatever may be the value of this theory in elucidating this phenomenon of the direction of the axes of the vegetable, so difficult to explain, the principle, on which it is founded, may enable us to find out the reason of a large number of less complicated cases, where there is a movement in consequence of incurvation, by means of a power already known, the endosmose, which we have seen playing such an important part in the motion of the liquids, and which appears to perform equally important functions in that of the solids, on account of the unequal swelling of which the parts placed close together are susceptible. It was necessary, therefore, to enter into all the preceding details, although we have not hitherto touched upon motion properly so called, and the changes of direction, with which we have been occupied, are only the consequence of growth, by the addition of fresh parts to those already formed, and by no means necessitate a change of place in the latter.

§ 665. DIFFERENT KINDS OF MOTION.—The dehiscence of the anthers and that of the pericarps is accompanied by a change of form in these organs. It is frequently very slow, and only proceeds gradually, so as not to give to the observer the idea of motion; but not so at other times, when it takes place abruptly and almost suddenly. We have explained in a previous part (§§ 462, 472) the mechanism of this motion, which results from the structure of the parts: certain points or certain lines present less resistance than the rest of the walls; these are dilated or contracted, either by the action of vegetation itself, from which results the progressive extension of the tissues, the afflux of the liquid juices at one time and at another their diminution; or by the action of external physical causes, as variations in the temperature, in the hygrometric state of the atmosphere, &c. The abrupt dehiscence always pre-supposes a certain state of tension generally owing to an arrangement analogous to that which we have just described in the axes (§ 664), namely, to the unequal extensibility of planes of fibres or of cells placed close to one another, which are filled by the means of endosmose, or, on the contrary, are emptied by the gradual loss of the liquid parts. The valves up to maturity, have maintained their relations by their mutual antagonism, but, when once the equilibrium is broken, these valves now separated obey their own peculiar tendency, spring open and bend in different ways. This is what may be observed in the cocca of the Euphorbiaceæ, especially of the Sandbox-tree (*Sablier élastique*) (HURA CREPITANS), and in those of the Diosmeæ, in which, on account of these unequal tensions, even the different layers of the pericarp, the mesocarp and the endocarp, are separated from one another assuming different forms, the first remaining straight, the second being bent back forcibly.

Every one has seen the ripe capsule of the Balsam divided into five valves, each of which is then rolled spirally inwards. M. Dutrochet has shewn that this motion is increased, if we plunge the valve into pure water, that it decreases if we put it into syrup of sugar; that examined anatomically it presents cells decreasing from the exterior to the interior; that it is, therefore, a real incurvation by endosmose, these cells being occupied by a denser liquid than water, less dense than syrup of sugar.

The fruit of the MOMORDICA ELATERIUM is detached when ripe from its peduncle, causing at its base a hole through which a thic fluid contained in its loculus with the seeds, is thrown violently to the outside, and we may remark that then the pericarp is somewhat lengthened as it diminishes in diameter. Here, also, as in that of the Balsam, the cellular tissue of the pericarp decreases from the outside to the inside; the fluid contained in its interior, which is thickened in proportion as it approaches nearer maturity, acts in the manner of the syrup of sugar and tends to straighten the valves, and these, in this manner pressing on the liquid and repelling the summit of the peduncle which serves as a kind of cork, determine this singular kind of dehiscence.

These motions may be explained by physical and mechanical causes. Let us pass to others where the intervention of these causes is much less clear, and to explain which we must have recourse to the mysterious agent of life.

§ 666. We know that (§ 123) in a very large majority of plants the leaves present two faces: the upper turned towards the sun; the lower towards the earth. If this direction be inverted, the leaf has a tendency to resume its natural position by surmounting the obstacles, which we have interposed, and if it cannot attain its end, it pines away and at last dies. This return takes place at the petiole, or, when it is wanting, at the base of insertion; and we cannot attribute it to the elasticity of the fibres which have been twisted and tend to untwist, since, if the inversion of the leaves takes place naturally as in pendulous branches, the petiole twists itself to turn its upper face sunwards. This turning takes place as long as there is any life in the leaf, on branches detached from the plant, on leaves or even fragments of leaves, which we take care to suspend by a sufficiently movable support. We cannot explain the cause of this phenomenon by the action of the air and light on the leaf; since it also occurs in water and in darkness.

§ 667. Sleep of Plants.—But yet the position of the leaves is influenced by the light in an incontestable manner; this is proved by a mere glance and confirmed by the study of the functions. We do not allude here to the anxiety shewn by every part of the plant to get to the place that has the most light, and then developing there in the greatest abundance its branches and its leaves, &c., &c.: it is, indeed, a necessary consequence that it should be more developed on the side where it finds the most favourable conditions for the performance of the functions of these parts and, consequently, for their increase and their multiplication. We are now going to treat of the motions of the leaves, considered apart from the other parts of the plant, to put themselves in connection with the light, so that they frequently modify the other positions of the leaf, turning one face towards the sky and the other towards the earth, and in different ways, according to the state of the day being more or less clear, more or less advanced. This phenomenon cannot be studied better, than by prosecuting our examination in opposite conditions, when the light has full power and when it has none, during day and during night. Now, whoever, at night-fall, when it is just light enough to distinguish objects, or during the twilight, will fix his attention on a certain number of plants, he will be struck with the difference of appearance, presented by several of them, from that which is familiar to us, and will recognise that this change is owing to the fresh position that the leaves have taken.

This fresh state has been termed their sleep; but they are far from sleeping in the same way in different vegetables. They may, after having passed through a quadrant, direct their points downwards, as in the IMPATIENS NOLI-ME-TANGERE, upwards, as in the ATRIPLEX HORTENSIS, exposing in the former case their upper surface to the outside, their lower in the latter. But these are frequently only imperfectly raised or debased.

Compound leaves are especially subject to sleep and take it in the most varied positions, since in them not only the petiole can move on the branch which bears it, but the partial petioles on the common one, the folioles on the rachis. Hence, three kinds of motions may be combined together. In the compound leaves which present only one, that of the folioles, this last may during night be raised, as in the LOTUS, the Trefoil (*Tréfle*), &c., or be lowered, as in the OXALIS, the Milk-vetch (*Réglisse*), or laid flat on the rachis by directing its point either to front, as in the Sensitive plant and the most of the other MIMOSÆ and true ACACIAS, or backwards, as in the TEPHROSIA CARIBÆA. In these two last cases, the folioles are imbricated from back to front or from front backwards. When, moreover, the common petiole has its own motion, it is directed either upwards or downwards, as in the AMORPHA, thus making a greater or a less angle than during the day with the upper part of the axis which bears it. When, lastly, in a leaf several times compound, the petioles of different ranks are moved at the same time upon one another, as we may see in the Sensitive plant, the partial petioles are bent from the back to the front on the common one, and tend to become parallel to it, and the latter from top downwards on the branch. Intermediate directions may be observed between these, as that of the folioles being directed at the same time in front and upwards, of the petioles forming with the axis which supports them angles variable in the number of degrees. There are essential differences which serve as characteristics to distinguish the sleep of different plants from one another; and there are secondary ones which the same plant may present according to the greater or less intensity of its sleep. Thus, in the Sensitive plant, the common petioles do not always hang during night, and the partial ones are bent only after the shrinking of the folioles, so that we may distinguish by these three degrees, if this singular plant sleeps slightly or profoundly.

§ 668. The state of sleep presents remarkable connections with that which the young leaves, diversely folded, already presented in the prefoliation, and we may consider the first as being up to a certain point, the return of the second. Thus, to the modifications which we have already indicated, we may add sometimes one degree more, that of the bending of the limb of the leaves: we will cite, for instance, that of the foliole conduplicate on the outside in the OXALIS. The more the leaves are crowded at the time of the prefoliation, the more easily they return to the arrangement, which they presented at this period of their life, the more they are disposed to sleep, and they become less and less so as they grow old. The softness of the tissues, the general attribute of youth; the hardness which they unceasingly acquire with old age, are indications of the tendency they have to sleep. Thick, coriaceous or stiff leaves are not subject to it; those, which are subject to it, are always more or less soft and thin. Articulations are also very frequently remarked, and these are so well adapted to favour the play of the parts, that they are highly developed at the point of attachment of the leaves and of the folioles in the plants noted for this faculty, as in the MIMOSÆ and several other Leguminous plants.

§ 669. On a gloomy day, the leaves either do not waken or go to sleep much sooner, and the transition from light to darkness, when the sky is overcast at the approach of a storm, is perceived on these most sensitive of all plants. If we protect the plants from the influence of light by covering them or taking them into a perfectly dark room, they assume the position of sleep, some sooner, others later. If we throw sufficient light into the room, they begin to awaken and to arouse themselves by degrees. De Candolle, who prosecuted a series of ingenious experiments on this subject, succeeded in deceiving, as it were, the Sensitive plant and some other vegetables by causing them to sleep during the day in artificial darkness, and to waken at night by the light of lamps. There cannot, therefore, be any doubt as to the great influence that light exercises on this phenomenon.

§ 670. We find, however, some facts, which seem to prove that light is not the only agent in causing the sleep of plants. If we have plants which thus change their habits when we vary the circumstances around them, there are others, which, less complaisant, do not alter, which even in darkness continue to sleep during the night and to rouse themselves during the day. The Sensitive plant itself, deprived of all natural and artificial light presents alternations of the sleeping and waking states, though they necessarily become very irregular. The plants of the equinoctial regions preserve in our conservatories, in spite of the unequal distribution of day and night, the same habits of sleep which are natural to them in their native land, where the days and nights are equal. The hours of repose, moreover, vary in different vegetables. Their rest is not governed by the hours of the day, and some of them waken and go to rest for a greater or less time before the rising or the setting of the sun.

§ 671. Light acts also on flowers in an analogous manner, with some differences, however, owing to that of their structure and, consequently, of their functions. Certain flowers assume on their peduncles at different hours of the day various positions, so that they are always turned towards the sun; and hence they have received the name of *Heliotropes* (from *ἥλιος, sun; τροπή, turning round*). The same thing may be observed in the Sun-flower (*Soleil des jardins*) (HELIANTHUS ANNUUS). It may be easily observed on account of the size of the calathis which forms its compound flower: but, since we sometimes see on the same stem several flowers turned at the same time towards different points of the horizon, this property is at least problematical.

§ 672. A more constant fact is that of the expansion of certain flowers at certain hours, of their closing at certain others. We also say of them that their first state is that of wakening, their second of sleep; and, since these different flowers open and shut at different hours, their habits, when once known, allow us to determine the hour of the day by their transition from one of these states to the other. Linnæus called this *Flora's Clock*, and the tables, which he constructed to form it, have been since extended by other observations. But, in spite of the precision with which they have been formed, the clock is far from being a good one; it is very easy, in fact, to foresee that, in our climate especially, all days will not resemble one another, that the blossoming of certain plants is prolonged for a long time and is even renewed in different seasons, and that light so unequally distributed would frequently put the instrument out of order on account of its extreme delicacy.

§ 673. The petals, or the divisions which represent them, have with respect to one another a certain position which we have clearly seen, especially in the prefloration (§ 404). In order to expand, these parts are separated from one another by moving their extremity from the inside outwards; when they close, they converge afresh, passing through a contrary direction, and resume their former position more or less exactly; as we have seen the leaves during their sleep, grouped and covered in the same manner, as in the prefoliation.

§ 674. But in the life of the flowers, infinitely much more transitory than that of the leaves, the alternations of the waking and of the sleeping state are only shewn at most a very few times, most commonly only once. Those which open for a single day and are closed not to be reopened, have the term *ephemerous* (*éphémère*) applied to them: those, which are opened and closed for several days consecutively, are called *equinoxial* (*équinoxiales*). We also distinguish them into *diurnal* (*diurnes*) and *nocturnal* (*nocturnes*); if the greater part of plants expand during the day, there are others which, closed during the day, are expanded during night.

§ 675. This phenomenon of nocturnal expansion appears at first to be directly opposed to the habits of the leaves, the sleep of which is always occasioned by the absence of light; yet, if we remember that certain leaves rise up when they sleep, whilst others fall down, we shall recognise that we only find in flowers this same two-fold motion, and that the difference arises from the state of sleep not being defined strictly in the same manner in these two kinds of organs.

The variations in the degree of the light of day, moreover, influence flowers in the same way as leaves, and experiments, made with the aid of artificial darkness or light, have caused them to change their habits. We may, therefore, believe that phenomena so analogous may be referred to a common cause, and also that all these movements are performed by a similar mechanism. As they always result from incurvations, from flexions or from erections, M. Dutrochet naturally applies to them the theory, which we have previously explained (§ 664), that of the unequal extensibility of layers placed by the side of one another in the same tissue; we have found it to be caused by the afflux of liquids from endosmose. M. Dutrochet admits also the afflux of a gas, oxygen, to which another course would be open, that of the tracheæ and of the fibres, the action of which, opposed to that of the cells filled with liquids, going on during the night whilst the other takes place during the day, would thus cause the alternations of the sleeping and of the waking states.

§ 676. In studying the daily motions of the flowers, we must not ascribe an exclusive influence to light; heat, doubtless, exercises some little power; this is proved by the positions which they assume in very warm days. Does this act by its own force? Does it modify the hygrometric state of the atmosphere, which acts an important part in certain plants known for this reason by the name of *meteoric* (*météorique*). They indicate by the directions and the curves of their petals the dryness or the humidity of the air acting upon their tissues. The CALENDULA PLUVIALIS derives its name from closing its petals when the sky threatens rain; several Chicoraceæ are said in such a case never to open in the morning; and the SONCHUS SIBIRICUS heralds the approach of a wet day the evening before by not closing at night contrary to its custom. It has, therefore, been attempted to form a *Flora's hygrometer;* but this is to be depended on still less than her clock. We must be careful in coming to conclusions from the preceding observations; since there exist so many sources whence these causes derive their power of producing these effects; rain, for instance, may act in three ways at once: by obscuring the light, by making it cold and by saturating it with humidity.

§ 677. The motions observed in the stamens and the pistils ought to attract particular attention, as connected with the act of fecundation, which they favour and accomplish by bringing these parts together and scattering the pollen. This act takes place when the flower is fully expanded, and then these motions are very apparent in certain plants: they do not take place in the bud, they cease with the life of the flower, and cannot be excited either before or after the expansion. The stamens bring their anthers to the stigma by bending their filaments. This may be seen in several flowers, such as the Rue; and we shall remark that of the eight stamens situated in two rows, the outer ones, opposite to the petals, bend the first, the inner ones afterwards. This phenomenon is very manifest in the

PARNASSIA PALUSTRIS, the five stamens of which, bending inwards, are applied to the stigma one after the other as if they followed the order of their insertion; the ten stamens of the SAXIFRAGA TRIDACTYLITES move in the same way, but in pairs.

At other times it is the style which moves outwards towards the moveable stamens, as in the Passion flower, some Onagrarieæ and CACTI, the NIGELLA SATIVA, &c. The style of one flower may even deviate so much as to be put in connection with the stamens of another; as in a species of COLLINSONIA.

Lastly, these two motions may take place at the same time; the stigma and the anther may be carried towards one another by the inclination either combined or alternate of the filaments and of the styles, in the Mallows and other flowers of the family of the Malvaceæ, of the Onagrarieæ, &c.

Most commonly, this motion only takes place once during the life of the flower; it is very rarely renewed. Can they be compared to those of the leaves and of the corollas? Medicus, in the BOERHAAVIA DIANDRA, has seen them vary at the different periods of the day and modified during the night, so that, lying on the side of the flower during the morning, the pistil towards ten or eleven o'clock is gradually raised towards the centre until the stigma meets one of the anthers; that, on the contrary, in the evening, the stamens lying on their side rise in their turn to meet the pistil. There is, therefore, a certain connection between the progress of these organs and that of light, which exercises such a powerful influence on the greater part of vegetable phenomena.

§ 678. Nevertheless, other agents may put it in motion, and we now arrive at one of the most singular phenomena and up to this period the least easy of explanation; that of the motions more or less sudden excited by the touch of a foreign body. Several centuries ago, the motion of the filaments of the Pellitory (*Pariétaire*), when they were touched (on account of which the dehiscence of the anthers took place), was a well-known fact. If in a Barberry, we touch, even very slightly, the base of one of the filaments with a pin, it will bend over, so as to strike its anther against the style of the flower: and this experiment may be renewed a certain number of times. In the Cistineæ, in the SPARMANNIA, the irritation, acting in the same way on the base of the filaments, produces quite a contrary effect, since they jump from the inside to the outside thus moving from the pistil; but they afterwards return with greater force, like a spring bent in a contrary direction. In the LOBELIAS, the GRATIOLA, the Gentians, the stigmas move by the application of a foreign irritating body; the two laminæ of the stigma of several BIGNONIAS (*fig.* 396) separated at the moment of blossoming, approach and contract. That of the RUELLIA ANISOPHYLLA, bent in one direction, springs up straight when it is irritated, is then bent in an opposite direction and is placed in connection with the collector hairs of the corolla, which are quite laden with pollen. In the STYLIDIUM the filaments united with the style form a column usually outside the flower, but if it be touched ever so lightly, it starts up with a jerk and falls over on the other side. The florets of the Thistles and other plants of the same family, shew, when we happen to touch them, a kind of balancing owing to the contraction of their filaments inserted on the corolla, to which they also impart their motion.

This excitement, which we cause by artificial means, is frequently produced naturally by the force of the wind or of small bodies which it carries along with it, and especially by the agency of insects, which happen to alight on the flowers and agitate it in the inside by extracting the juices of the anthers and of the stigmas. Moreover, the motion sometimes takes place without any apparent cause of excitement, in the STYLIDIUM, for instance, about noon in very warm days, and it is remarkable that then it is much slower and more regular.

§ 679. We have chiefly studied the motions of the leaves produced by the excitement caused by foreign bodies; and every one knows the Sensitive plant, which we may truly call the type in this respect. Its bipennate leaves are composed of a slightly straightened general petiole, of four partial petioles, situated, two at the extremity of the preceding, and convergent, two a little lower, and growing almost at right angles; each of them bearing more than twenty pairs of small horizontal folioles. The petioles and the folioles are articulated at their base which is swollen into a small cellular mass, where the vascular fascicles are arranged in a circle near the circumference. Such is the position of the parts exposed to the light. If the plants happen to be shaken rather suddenly and with some force, the folioles suddenly rise up into an oblique position, so that those of the same pair touch one another at their upper face, and all those of the same series are imbricated from the bottom upwards; then the general petiole bends down and becomes pendulous; lastly, the four partial petioles are bent as they converge towards one another and thus become parallel to the general one, from the end of which they hang. This is exactly the position, which all these parts assume during their sleep. If instead of shaking the plant, we touch a single foliole, or its basilar swelling, it rises up, as if it were going to sleep; then the neighbouring ones imitate it, the nearest the first; if it is a foliole of the lower pairs, the motion proceeds from the bottom upwards; if it is one of the upper pairs, it proceeds from the top downwards. The sensibility is so much the more powerful as the plant is younger and more lively, as the weather is clearer, warmer and more humid. If these conditions be united, the slightest shock, the breath, the presence of a light insect suffice to excite these phenomena, the intensity and extent of which are always proportioned to those of the excitement: if it be rather powerful, the motion is communicated not only to the leaves close to that which has been touched, but from the partial petiole, which conveys it to the three others as well as to the common petiole, and frequently upwards even to the other leaves of the branch. The nature of the body causing the excitement seems to be indifferent. But it is not only mechanical excitement which determines these motions; a chemical one produces analogous effects; this may be done by placing a very small drop of a concentrated acid on a leaf, so softly as not to shake it, or by causing the rays of the sun concentrated by a lens to fall upon it. We may then very well follow the series of its changes, which take place very slowly, but more generally. The different parts do not transmit the excitement in an equal degree, which we hinted, when recommending the articulated swellings; even in these, an unequal distribution of sensibility is observable. Thus, on touching the upper part of the base of the petiole, we find no effect; on touching the lower part, it instantly bends.

§ 680. We have quoted the Sensitive plant as the most striking example of the curious faculty, which is now occupying our attention. Other plants of the same genus (MIMOSA), or of the same family (Leguminosæ), present it also in a remarkable degree although not so strongly, as well as others belonging to quite different families, as several Oxalideæ, and the singular plant, known under the name of Venus's fly-trap (*Attrape-mouche*) (DIONÆA MUSCIPULA) the leaf of which bending on its median nerve, when an insect touches it, thus makes it a captive till it has ceased struggling. We may, indeed, believe that this excitability is much more general than was at first supposed; and that it only escapes our observation in several other vegetables, even in those which we know most familiarly, because the excitement being much weaker and more slowly produced, is not so apparent. The leaves of the species of the OXALIS, so common in our fields (OXALIS CORNICU-

LATA, and in a much slighter degree in O. ACETOSELLA), if we strike them with quickly repeated gentle blows, assume in the course of a minute or two the position of sleep, bending their folioles on the median nerve and allowing their petiole to hang. The leaves of the common Acacia (ROBINIA PSEUDO-ACACIA), forcibly shaken, shew in the course of some little time the same arrangement as during night. Perhaps, the peculiar appearance, which several of our plants assume during high winds, is owing to analogous changes resulting from violent and reiterated shocks.

§ 681. All the motions which we have previously described were intermittent, only appearing at certain times either of the day or of the life of the vegetable, or excited by a known and foreign cause. In a very small number of plants, we may observe some which, from their spontaneity and continuity, deserve to be studied separately. We find these motions in some tropical species of the DESMODIUM (formerly confounded with HEDYSARUM or *Sainfoin*), and particularly in the D. GYRANS. Its leaves are composed of three folioles; the one terminal, large and only subject to the alternation of the waking and sleeping states; the two others lateral and very small. These, in warm weather, are in perpetual motion, alternately approaching and flying from the general petiole, rising and sinking by turns; and this motion is not interrupted during night. On examining them attentively, we find that these small folioles are fastened on swollen petioles: that these bend alternately in opposite directions, by slightly twisting themselves, then arching themselves inwards, then straightening themselves, then arching themselves outwards, a position in which they tend to stop from preference; that the limb has no motion of its own, but that having a tendency to fall from its own weight, it increases and renders more apparent that of the petiole, at the extremity of which it is balanced. We have, therefore, here again a result of the incurvation, and it is to be presumed that it is connected with the action of light, with the respiration and with the evaporation of the folioles. Indeed, it is undeniable that the motions of the latter are much lessened, and are even stopped in the dark; that they revive and are increased in light whether natural or artificial. We can suspend them by spreading over the surface of the limb a small pellicle of gum, which will put a stop to the performance of its functions, by rendering it impermeable; then we can frequently restore them by dissolving this coat, if it has not been suffered to remain too long. When we cut the limb, so as only to leave a small piece of the lower part, it continues to move rather a long time; but at last stops whilst the opposite intact foliole continues its gyration. When we cut it longitudinally into two halves, we see them move so long as they do not dry; and then they cease. The phenomenon is not, therefore, so different from that of daily motions, as would appear at first sight; and, perhaps, the smallness of the limbs, relatively to the petioles, favours its appearance here, whilst it cannot take place in other plants, where parts with an analogous structure would present different proportions.

§ 682. We have alluded to a continuous motion in some flowers also, those of certain Orchideæ, as the PTEROSTYLIS, the MEGACLINIUM FALCATUM. It is that one of the six divisions of the perianth, remarkable for a peculiar shape and known by the name of labellum, that moves in this manner. It is continued as it is disarticulated from the rest of the flower by a contraction in the shape of a thread, which is somewhat similar to the petiole of the foliole of the DESMODIUM GYRANS, and which in the same way determines the motion of the limb which it bears, causing it to rise and sink alternately at irregular intervals.

§ 683. We have explained the principal facts relative to the motions of plants. We have seen that the majority of their motions result from incurvation, the mechanism of which may be traced up to a certain point. But what cause sets this mechanism to work? We can see that the variations of the light, which determine corresponding variations in the manner in which the young parts supple and full of juices perform their functions, may modify, at the different states of the day, the proportion of the liquids in the cells; and if these happen to be in the turgescent state, this fact and that of their being of unequal dimensions, which compel the tissues to bend in one direction or another, shews good reasons for the phenomena of the sleeping and the waking states. We may also conceive again that powerful excitement prolonged for some time may cause the afflux of the fluids, the turgescence and the incurvation which follows it.

§ 684. But there remain many other facts which are inexplicable. Sensitive vegetables, placed in permanent darkness, ought to be kept in a certain state of equilibrium and of immobility, of sleep, or at least of half sleep, since their functions, if they be not completely interrupted, are then carried on in a continuous and uniform manner. The plants, however, although in fixed conditions which ought to cause an equally permanent state, are not long in resuming their habits, which only differ by not being so regular (§ 670). On the other hand, we mentioned those sudden motions caused by the action of a foreign stimulant, the effect of which is too instantaneous to be explained by an afflux of liquids, which would necessarily demand some time before it could produce these changes of position so marked, such as those which we observe in the Sensitive plant. These changes would, it is true, take place much more rapidly, when the gases, to which M. Dutrochet attributes a part of the theory of incurvation, are in equilibrium. But it remains to prove both the constant presence of these gases in the parts which are assigned to them, and how the external excitement provokes this sudden developement.

§ 685. Several naturalists, struck with the insufficiency of these mechanical or physical actions to explain the motions of the vegetables, have been led to admit a principle analogous to animal excitability. They found their theory on the rapidity with which the excitement may be carried from one point of the plant to another at a greater or less distance, as if by a kind of sympathy; on the excitability, much more lively in the young parts which are full of life, becoming weaker and disappearing in the old parts; on the fact, that when put actively in play and at several short intervals, it is weakened and ceases, to allow the plant to be refreshed and reanimated, as it were, after a sufficient period of repose; on this necessity for a sleeping alternating with a waking state and repairing its losses, much more necessary and greater in infancy, losing with age its duration and intensity, and being changed in old age into a kind of permanent demi-somnolescence; on the species of instinct, with which the parts of vegetables assume positions or directions favourable to the free exercise of their natural functions, and adapted to satisfy their wants, surmounting to attain them the obstacles, which may be placed in their way. These actions seem to them to be quite of the same order, as those which under similar circumstances are performed by the inferior animalculæ. A vital force which sets them in action is, consequently, recognised besides the mechanical and physical forces, which are only the means of execution. They adduce in support of their opinion, the action of the narcotics, which carried into the vegetable by absorption, are not long in lessening and suspending the motion, as in animals. But this last argument is overcome by other experiments which prove that other extracts, perfectly innocent, absorbed by the vegetable, suppress in the same way the excitability and that, consequently, they interrupt the phenomena of life not as poisons, but merely as foreign substances. Another objection, is that the sleep does not produce in vegetables, as in animals, a general state of relaxation; it is, on the contrary, a state of tension, opposite, it is true, to that of the waking state, but often equally characteristic, sometimes even more so, as in the leaves, which we have seen erect during night. The parts in this state resist everything, which tends to change their fresh position, and will be sometimes broken rather than assume any other.

§ 686. Moreover, how is the excitability transmitted? Some suppose it to be by the fibres; others, by the cellular tissue; others, by the parts contained in the cavities of the cells, fibres or vessels. But the experiments, made to demonstrate that is by such or such a way, contradict one another, and in the most sensitive vegetables, we find the same elements, arranged in the same manner as in those, which are by no means so; in their cavities we find the same substances; and if we admit under certain circumstances the fecula, the chlorophyll or other substances generally found in vegetables as conductors of the excitement, it would remain for us to enquire how these perfectly fresh properties have been communicated to them, properties which are usually wanting and would make of them a body of a different nature. We know what tissue in animals receives and transmits the excitement, what tissue is contracted as it receives it and thus determines the motion; we only know in vegetables the effect and some of the secondary causes.

§ 687. Concluding observations on the Vegetable and Animal Kingdoms.—Science is in doubt relative to some of the lowest organized beings, whether they are to be ranked among Animals or Vegetables. What argument then can be drawn from this mystery to enlighten that of the motion of vegetables? The very distinctions escape our notice. We have seen those bodies, which people the cells of the Antheridia (§ 480), the Sporangia of certain Algæ (§ 616), endowed with forms and motions analogous to those of the Infusorial Animalculæ. We have seen spores move by the aid of vibratory ciliæ, the ordinary organs of motion in these animalculæ. Even then, however, light seems to exercise its influence on their life, since it is only at certain hours of the day that we can observe this perfectly Animal locomotion, and then they pass into the immobility of the Vegetable, and are developed with all its characteristics. Let us again mention those Algæ, known by the name of OSCILLATORIÆ, because the filaments which compose them, formed of a series of cells attached end to end, short and swollen, are bent towards one side and towards the other like a finger or the extremity of a trumpet. One of their ends, free and frequently furnished at its point with a bunch of small irregular mucous threads, oscillates sometimes slowly, sometimes with sudden jerks; by the other end a great number of them are united together and form one common mass, from which the moveable extremities radiate.

§ 687 *b*. Having arrived at this limit, at which the two Kingdoms seem to be confounded, we shall now recognise the insufficiency of the definitions, which served as our point of departure in distinguishing Vegetables from Animals and were founded on the incapacity of feeling and of moving attributed to the former (§ 1). Can we then, with the more extended notions we have gained in the course of this work, lay down a more rigorous definition?

§ 688. This was once drawn from the chemical composition of the tissues, ternary in Vegetables, quaternary in Animals, in which azote is associated with oxygen, hydrogen and carbon. Yet we have seen (§ 305-311) that nitrogen also commonly exists in Vegetable substances. But, if we limit ourselves to the skeleton of the tissues the difference at first established holds good. The substance, which, under the form of utricles, of fibres or of vessels, essentially constitutes the Vegetable and serves as an envelope and, as it were, as a laboratory for all the other products, is always identical, always ternary: it is that which we have learnt to know already under the name of cellulose (§ 308.) On the contrary, Animal fibre reduced to its greatest state of purity, always contains a certain quantity of azote. Several characteristics, resulting for the most part from this different composition may enable us to distinguish a Vegetable from an Animal membrane. The former, where decomposed, furnishes acid products and residues; when burnt acetic acid and a carbonaceous residue, which is not altered in its shape; it is not coloured by the aqueous solution of iodine, is not very sensibly affected by diluted solutions of soda and potassa, by ammonia; not at all by the hydrochloric, acetic and tannic acids. The latter, when decomposed, gives both acid and ammoniacal products and residues; when burnt, carbonate of ammonia and a spongy carbonaceous residue; it is coloured yellow by iodine; is dissolved in soda, potassa and ammonia, in hydrochloric and acetic acids; is contracted by tannic acid at the same time combining intimately with it. Such are the distinctive characteristics clearly established by the numerous researches and works of M. Payen.

But, if instead of examining the Vegetable and Animal membranes, separated from the other matters, which fill their cavities and their interstices and most frequently penetrate and impregnate them, we consider them along with substances with which they form an organized body, especially when alive, the chemical composition and properties cease to furnish us with general characteristics of distinction. For we find in Vegetable cells on the one side quaternary products analogous, sometimes identical, to Animal substances; on the other, some purely mineral products (§ 323).

§ 689. We have pointed out another difference in the way in which the two Kingdoms are nourished. The Animal is nourished only by organic particles, the Vegetable only by inorganic particles.

But we must allow that these characteristics, subject, perhaps, to less exceptions than those which we employed at the beginning, become, however, very uncertain like them. We find this confusion when we wish to apply them to the beings situated on the limit of the two Kingdoms. M. Payen has determined the existence of a quaternary chemical compound, perfectly analogous to that of Animals, in the small moveable Vegetable bodies, which appear spontaneous, as the grains of the Fovilla, the corpuscules enclosed in the tubes of the CHARA. We may, from analogy, suppose it to exist in the spores of the ALGÆ ; and when they move, they form the whole Vegetable, which afterwards will merely be developed.

Are the means, by which these little beings are supported, really known? Do we know, whether these spores, which are just like the Infusorial Animalculæ, are of a class different from them, and whether the water, in which all these small bodies are developed, enters in a pure state into the one, laden with organic particles into the other.

§ 690. Let us in conclusion say, that, if we compare perfect Vegetables with perfect Animals, the differences taken together are great and furnish an exact definition founded on several characteristics at the same time; if we descend to the more imperfect or even to their parts, these very definitions, which we continue to apply, become incomplete, hypothetical or false; and our being unable to trace a clear line of demarcation between the two Kingdoms, to lay down a rule without exceptions, seems to prove the unity of the organic Kingdom, and confirms with regard to it at least this Linnæan axiom:

NATURA NON FACIT SALTUS.

THE PRINCIPLES OF CLASSIFICATION.

§ 691.—When we cast our eyes over the vegetables scattered around us, we see in each of them an *individual* (*individu*). This name itself indicates an undivided whole composed of parts continuously united to one another. Their appearance will frequently deceive us by representing above ground as distinct separate plants those, which belong to a common stem concealed under the earth. Thus the rhizomes of the CAREX ARENARIA overrun a very considerable extent in length, throwing up at certain distances stems, which rise above the ground and seem to form so many individuals, although, in reality, they are only so many parts of a single individual. It is clear, that all these will present a striking resemblance to one another; so that even considering them erroneously as so many different plants, we shall not hesitate to recognise them as the same plant and to apply the same name to them.

§ 692. SPECIES.—Now, this necessary resemblance of the different shoots of the same individual may be also found in several individuals really separate. A field of Rye or of Oats affords us thousands of examples, which we may easily separate, but, which we cannot distinguish from one another. In fields, in gardens, we recognise here and there plants which without the slightest hesitation we call by the same name. This collection of all the individuals thus resembling one another, has received in Natural History the name of SPECIES (*espèce*): their *common characteristics* (*caracterès*), those, the combination of which distinguish them from another set of individuals, are called *specific* (*specifiques*). We know, moreover, that by separating the shoots of an individual or by sowing its seeds, we shall obtain so many fresh individuals similar to the first. This fact completes the definition of the species; a collection of all those individuals, which resemble one another more than they resemble any others, and which by generation reproduce similar ones; so that we may from analogy suppose them all to have issued originally from the same individual.

§ 693. VARIETIES.—This fraternal resemblance, as it were, may be found, however, in different degrees. If two seeds taken from the same fruit are sown in different soils, in different climates, in different seasons, the two plants, developed in dissimilar conditions, will shew this inequality in the conditions of their nutrition by certain points of dissimilarity, so much the more strongly marked, as the causes have been more numerous and powerful. We cannot here take into consideration all the modifications of which a species is susceptible under the influence of these different conditions. The study of the different organs, of their structure, of their developement, of their means of nutrition, will up to a certain point enable us to presuppose these changes. Let us only remark, that they are so much the more frequent as they affect a less important organ and are less important in themselves. Thus, the changes of colour and, especially, of such a colour into such another, the developement or the absence of the hairs, the tissue being looser or more compact, are observed rather frequently in the same species and, what is more, in the same individual, when we diversify the circumstances in which the plants grow; these are then simple *variations* (*variations*). When the modification is more marked and permanent, it takes the name of *variety* (*variété*) (VARIETAS); then it appears with some degree of permanency in a certain number of individuals and will distinguish all the individuals presenting this change of form from the rest of the individuals of the same species, less clearly, however, than the latter are distinguished from those of a different species.

We have just seen that the accidental and individual variation may disappear with the cause which produced it, even in the very individual which was affected by it. At other times, the effect is permanent after the cause has vanished, and the individual preserves for its whole life its characteristics of variety. These may be still more permanent and be reproduced by extension in all the individuals obtained from the first by grafts, suckers or cuttings. But, if we sow the seeds, the new individuals, thus obtained, will no longer present these characteristics and will return to those of the primitive species.

Lastly, there are some in which the germs, contained in the seeds, preserve and transmit the characteristics of the variety on which they were formed. These hereditary varieties are frequently termed *races*.

A very powerful cause of variety is *hybridization*; that is, the fecundation of an individual of one species by that of a different one, when the pollen of the one is placed upon the stigma of the mother. This will not succeed when the plants are very different from one another; but it is undeniable that it takes place between plants of very nearly allied species, and that then the seeds, although having a general tendency to become abortive, are sometimes fertile. The plant, which springs thence, ought naturally to present intermediate characteristics between the two which have given it birth, and compared, either with one, or the other, differential characteristics which give it the appearance of a variety. But to which of the two shall we refer it? To that with which it has the most characteristics in common, if it presents very few with the other; if not, if they are nearly equal, it is simply called a *hybrid*[y] (*hybride*). But, after a few generations, the characteristics of one of the parents are more and more pronounced, especially, if a fresh cross takes place between the hybrid and one of the primitive species; it is then easily conceivable that we thus gain a very well defined variety. But hybrids are very rare in nature, in which the species most nearly allied by their characteristics are very rarely found to be neighbours in their situations. In our gardens, especially in botanical gardens, in which we endeavour to shew close to one another those species which most resemble one another, the crosses are much more frequent and multiplied.

Cultivation profits by all these properties to make vegetables vary, by changing the conditions of their nutrition, preserving and multiplying the produce which are the results, propagating them by seed, ameliorating them by fresh crosses. Hence, the prodigious number of the varieties of certain species of flowers and of fruits much esteemed by man. The species, thus acted upon during a long and continued series of generations, is represented by modifications, in which its primitive characteristics, changed in different degrees and shades, are recognised with difficulty, so much the greater as several are frequently borrowed from other species: a result very valuable for the cultivator, very puzzling for the botanist. But, if some domestic vegetables require such a complicated study, they are not numerous, and the majority of the species as they grow naturally, preserve their original characteristics intact and constant. The limits, in which they vary are exceedingly narrow, so that they are easily appreciated. We can then draw from each of them a description sufficiently general, to enable us to distinguish the species clearly. By this means we find the primitive

[y] From ὕβρις *an injury*.—TRANS.

types of some plants disguised and disfigured by the innumerable varieties of our gardens; of these we need only mention as examples, the DAHLIA, CALCEOLARIA, the GERANIUM, &c. Their study, a perfect chaos, presents little interest and attraction to the student of botany; it would possess much in the prosecution of physiology, if the cultivator could determine in what way chance led him to an end, with which he was previously unacquainted.

§ 694. GENERA (*Genres*).—If there were but a very small number of species our memory could without great difficulty retain the description of each and the particular name by which it would be designated. We find this among those people with whom the study of botany is limited to the distinguishing of the principal vegetables from one another, which grow around them in a very limited district; without being occupied with those which do not attract attention by their dimensions, their shapes, their beauty, their utility or some remarkable property, any more than with those which inhabit different countries. They would then be known in the same way as the other common words of the language, without any fixed order, as chance or want presented them; they would be defined by the real or imaginary characteristic, that recommended them to notice. Thus, in the most ancient works on Natural History we see a certain number of species described, the classification and description of which do not follow any fixed rule. The authors of these works thought more of pointing out the virtues and uses of the plants than of the characteristics, by the assistance of which they are recognised. These were, doubtless, considered to be superfluous, since the popular name, by which they were cited, was generally sufficient for their recognition.

At the revival of literature, the study of the Greek and Latin authors, in whose writings every thing was supposed capable of being found, absorbed for a long time the efforts and minds of the learned; and botany for a long time was limited to long and dry commentaries and disquisitions on Theophrastus, Pliny and Dioscorides. We can, however, understand, that for the clear comprehending of their works on Natural History, the study of the natural objects themselves would furnish a powerful ally: they were examined with respect to these works; they endeavoured to illustrate them, not only by writings, but also in a little time by figures. The obstinacy, with which they endeavoured to reconcile with these traditions of the fathers of science vegetables, observed in countries mostly different from that which had furnished them their materials, led without any doubt to several errors. But, nevertheless, this practice accustomed the annotators to know these vegetables as such, even though they frequently named them quite wrongly: they learned to distinguish many more than were mentioned in the writings of antiquity, and, this truth once known, they multiplied their researches and, consequently, the number of the known species of vegetables; so that a time at last came, when botanists felt themselves embarrassed in their researches by these fresh riches. The diversity of the things and of the words began to surpass the efforts of human memory.

§ 695. It was necessary, therefore, to assist the memory by establishing a certain order in this confused mass; and, in the same way as they at first naturally united into one species all the individuals similar to one another, they endeavoured to find all the species, with a certain resemblance to one another, which was wanting in other plants, and to unite them under the same name and under a common definition. Thus,from several of these units called species, units of a higher order were formed, to each of which was given the name GENUS (*genre*) (*γένος, race, stock, descent;* hence, *a people, nation*). This joining of several species into one group is a natural operation of the mind, although to a less degree than that of the individuals. Ancient authors furnish here and there examples of it, and the names which people, unacquainted with the science and even only half civilized, give to those vegetables for which their language has words, frequently prove, by the common termination of some of them, the idea of a connection between the objects which they served to distinguish. Such genera, doubtless, are frequently contrary to our actual rules which have been laid down, as well as those which resulted from the first attempts of botanists. But, it was still something to establish some rules, to recognise connections, and among the specific characteristics to raise several to a higher degree as common to a certain number of species, as *generic* (*génériques*).

SYSTEMS AND METHODS OF CLASSIFICATION.

§ 696.—It is evident that the genera would be multiplied along with the species, and their multiplication would render fresh divisions necessary, each of which would unite a limited number of these genera, which were similar to one another by some more general characteristics. This fresh operation greatly diminished the fatigue and difficulty of these researches by circumscribing them: let it be required either to find a genus already known, or to assign a place to a fresh genus, it would be no longer necessary to compare them with all known plants, for the greater part would be excluded from the comparison, as soon as the general characteristics were recognised, by which the plant under notice was attached to such a group or to such another; and the operation, thus divided, thus limited definitely to the study of a much smaller number of genera, becomes much more simple and at the same time much surer. The evident utility of these divisions augmented the number; the more general were divided in their turn; then these subdivided, and we thus obtain a continuation of subordinate groups of a higher rank than the genera and the species, the lowest rank of the classification.

This arrangement has been frequently compared to that of armies: a troop few in number may march without chief and without order; the want of these would soon be felt, if it became larger; they then would marshal the soldiers in squadrons, in companies, in batallions; large armies have their corps, their divisions, their regiments; the divisions are enlarged in the same proportion as they themselves increased, and in this manner enormous masses may be moved in perfect order, be managed with great facility, and the well defined place of the most insignificant soldier, permits us to find him, whilst it would be perfectly impossible to see him without this classification.

Thus have arisen the systems and methods of classification in Natural History. It is difficult to establish a clear distinction between the classification designated by these two terms. The former, it is true, is commonly defined as employing only the characteristics drawn from a single organ exclusively, the latter as employing several organs at once. But the examination of the greater part of the systems always shews them to be founded on the employment of several organs, as well as the methods; and, on the other hand, the latter generally make one more important than the rest. We shall, therefore, employ these two words almost indifferently in the following treatise.

§ 697. The most ancient authors of treatises on plants already divided into several categories the small number they mentioned, but merely taking account of their general aspect, and, above all, of their properties. In proportion as the number was extended, and botanists grew more skilled in the nature of the plants, they endeavoured to find in their characteristics the foundation of their divisions: and we ought to mention here Cæsalpinus, who, about the end of the sixteenth century, employed in his classification considerations drawn from the fruit and the seed. We shall not stop to consider the numerous attempts of early botanists at classification; because, each of them inventing his own method, only applied it to a very small number of plants. Others did not follow him; and, to compose a history of all these separate systems would be almost writing a review of all the works on botany published during a long series of years. Besides, those who wish to form an idea of them will find their contents given in an abridged form in more modern works, especially in the *Introduction des familles des plantes* of Adanson, and in the *Classes plantarum* of Linnæus: they will be easily understood, if the facts we have stated concerning the different organs and their principal modifications be kept in mind.

We must, however, mention at some length two, which have exercised more influence than the rest, as comprising all the plants known at the time of their publication, and as having been employed by others besides their authors, Ray and Tournefort.

§ 698. Ray's Method.—Ray, an English botanist, published his method towards the end of the seventeenth century, in which he examines and publishes more than eighteen thousand plants, an immense number for the time in which he lived, but which was much exaggerated on account of the quantity of varieties which he admitted. He commences by dividing plants into two classes, trees and herbs, and already knew how to distinguish Acotyledons (which he termed *imperfect*) from Cotyledons (which he termed *perfect*); then in the latter, he separates Monocotyledons from Dicotyledons, which he divides by considerations drawn from the compound or simple, apetalous or petalous flower, and from fruits, apocarpous (*naked seeds*) and syncarpous (*seeds enveloped with pulp*). He also knew how to distinguish in trees the adherent fruits (which he called *umbilicate*) from those which are not so. The genera are frequently grouped in a very correct manner, although at the same time we remark several false collocations; as much from the necessary imperfection of his method in itself, as from the incomplete or erroneous knowledge of the parts in several of these genera.

§ 699. Tournefort's Method.—Tournefort, a French botanist,

published about the same time his celebrated method; and, if he applied it to a much less number of plants, about ten thousand, at the same time including the varieties, he lessened the number by confining the species within more rigorous limits. Dividing vegetables like Ray into trees and herbs, he subdivided them according to the characteristics drawn from the envelopes of the flower, the absence or the presence of the corolla, with which he confounded the coloured perianth. He correctly places in the first rank as of most importance, the characteristic drawn from the distinction of the monopetalous and polypetalous corollas; in the second, that of their regularity or irregularity, then of their more peculiar forms, which we have considered in another part of the work (§ 437—440).

The composite flowers form a part of the petalous, but are clearly distinguished from the simple flowers. The Cryptogams form a division of the apetalous. This ingenious method served to the end of the eighteenth century as the foundation of botanical instruction in France. In this order were also planted the trees and plants in the *Jardin de Paris*, which, according to his fundamental division presented separately a garden of herbs and a garden of trees. Till within a few years, traces of this latter arrangement were still preserved; there were several aged trees scattered in a small wood on the very site of the galleries of botany and of mineralogy. There still remains a very small number, especially the first of the Acacias which was planted in our country (France).

But, on this subject, let us remark that if the distinction into trees and herbaceous plants presents some advantages in the planting of a garden, it is a radical fault in every method, which assumes it as a base, since the same genus (as the CORONILLA) presents ligneous species by the side of herbaceous ones; especially, since the same species (as the RICINUS) herbaceous in some countries, becomes ligneous in a more favourable locality.

LINNÆUS.

§ 700.—The system of Linnæus, published in 1734, caused botanists generally to abandon all those which had preceded it. It presented the grand attraction of novelty by being based on the organs of fecundation, which had up to that time been neglected, and the physiological functions of which, of a much higher value than those of the other parts of the flower, may be considered as a still recent discovery. Linnæus, moreover, connected this innovation with several others of great importance: he regulated the confusion that resulted from the multitude of the varieties, which, as well as the doubtful species, he reduced to those he could clearly describe; and thus, in spite of the addition of a large number of fresh plants unknown to his predecessors, he reduced the number of species to about seven thousand. He also diminished the number of genera, so well established by Tournefort, and completed their description by employing the stamens and certain parts of the pistil. But, especially, thanks to laws which are still and will probably remain in force, he introduced an admirable reformation in botanical language and nomenclature by rigorously defining each of the terms designed to express all the modifications of organs which he intended to employ as characteristics, and by reducing the name of the plant to two words; the former the *substantive*, which designates its genus; the second the *adjective*, which indicates its species. Before his time each genus bore a single name; but, for the species, this name must be followed by an entire sentence recapitulating all its distinctive signs; and the more species there are in a genus, the more signs were necessary to distinguish them; the sentences, thus growing longer by the very progress of botany, overloaded the memory beyond its strength and embarrassed conversation, in the course of which the mention of a plant would at each instant throw in a whole incidental sentence. This was a confusion similar to that which would be produced in society and language, if instead of distinguishing each one by a baptismal and a family name, the former were to be suppressed and the enumeration of several distinctive personal qualities substituted. The Linnæan nomenclature freed the memory for the profit of the other faculties, and simplified the language of botany. The works, in which the series of plants were explained according to his fresh system, presenting at the same time all these advantages as soon as they appeared, would, therefore, be almost universally in vogue. This is what really happened. Every point of the reformation was adopted on all sides: the system of Linnæus dethroned all the rest, and up to the end of the eighteenth century reigned almost without being challenged, except by a few minds behind or, on the contrary, more advanced than the generality. All the fresh plants were classified as they were discovered, and the *tableau* of the Vegetable Kingdom continued to increase, whilst its frames, as it were, were not changed. Since a considerable number of works have been arranged according to the Linnæan system, and that even during our own days; since on account of the binary nomenclature thenceforth adopted, they are frequently and easily consulted; since, on the contrary, the majority of the treatises of antiquity written in a language which is not the common one, are rarely so, and have only preserved for the most part an historical interest, it was better to omit or treat cursorily the other systems, with which the pupil is not obliged to be acquainted. But he ought to be familiar with that of Linnæus, and we must, consequently, explain it here more in detail.

§ 701. THE LINNÆAN SYSTEM.—It is generally the custom to define the system as founded on the number of the stamens, but, this is not correct; since Linnæus, although choosing his principal characteristics from the organs, first had regard to other considerations: that of their relations to the pistil, separated from the stamens in a different flower or both placed in the same flower; that of their relations to one another, whether of adherence by the filaments or by the anthers, or of size. Their absolute number was accounted as only the fifth or sixth importance. This may be clearly seen by a single glance at the accompanying table.

We have already had occasion under the head of the flower and of the stamens to explain all these names, which are again defined here by the table itself.

The twenty-four classes thus obtained are then each subdivided according to other characteristics derived either from the stamens or from the pistils. Thus, in the sixteenth, seventeenth, eighteenth, twentieth, twenty-first, twenty-second classes, we again find the absolute number of the stamens reappearing in order to furnish the secondary divisions; the MONADELPHIA DECANDRIA, for instance, will comprehend the plants which present ten stamens united to one another by their filaments; the GYNANDRIA HEXANDRIA, those which present six stamens consolidated with the pistil; the DIÆCIA PENTANDRIA, those the flowers of which with five stamens are deprived of pistils, which are only to be found in other non-stamniferous flowers placed on a different individual. The twenty-third class, from the distribution of the flowers of three sorts on the same individual or on two or three different ones, is itself divided into POLYGAMIA MONÆCIA, DIÆCIA, TRIÆCIA (*polygamie monœcie, diœcie, triœcie*). The nineteenth, the flowers of which united in the same capitulum, present five possible combinations of hermaphro-

THE LINNÆAN SYSTEM.

Stamens and pistils	visible	both in the same flower	not adherent to one another.	Stamens disunited	equal to one another	1 in each flower	1. MONANDRIA.
						2	2. DIANDRIA, (*fig.* 244).
						3	3. TRIANDRIA, (*fig.* 243).
						4	4. TETRANDRIA, (*fig.* 241).
						5	5. PENTANDRIA, (*fig.* 225).
						6	6. HEXANDRIA, (*fig.* 226).
						7	7. HEPTANDRIA.
						8	8. OCTANDRIA, (*fig.* 248).
						9	9. ENNEANDRIA, (*fig.* 240).
						10	10. DECANDRIA, (*fig.* 235).
						from 11 to 19	11. DODECANDRIA.
						20 or more inserted on the calyx	12. ICOSANDRIA, (*fig.* 369).
						20 or more inserted on the torus	13. POLYANDRIA.
					unequal	4 stamens, 2 long, 2 short	14. DIDYNAMIA, (*fig.* 326).
						6 ——, 4 —— 2 ——	15. TETRADYNAMIA (*fig.* 325).
				adherent to one another		their filaments being united into a single body	16. MONADELPHIA, (*fig.* 228).
						—— into two bodies	17. DIADELPHIA.
						—— into several	18. POLYADELPHIA, (*fig.* 322, 238).
						their anthers being united into a cylinder or tube	19. SYNGENESIA, (*fig.* 295).
			united to one another				20. GYNANDRIA, (*fig.* 551).
		not in the same flower.	Male and female flowers			on the same individual	21. MONŒCIA, (*fig.* 251).
						on two different individuals	22. DIŒCIA, (*fig.* 535, 536).
						and hermaphrodite, or on one or several individuals	23. POLYGAMIA.
	not visible						24. CRYPTOGAMIA.

dite, male, female, and neuter flowers, is divided into MONOGAMIA, POLYGAMIA, &c. As to the first fifteen classes, in which the absolute number of the free stamens has already been employed, Linnæus, to subdivide them has recourse to characteristics drawn from the fruit, short or long, (SILIQUA, SILICULOSA), in the fifteenth; *monospermous* (*monosperme*) (GYMNOSPERMIA [*gymnospermie*]) or *polyspermous* (*polysperme*) (ANGIOSPERMIA [*angiosperme*]) in the fourteenth; and in all the rest from the number of the styles, which, simple, double, triple, multiple, names the sections MONOGYNIA, DIGYNIA, TRIGYNIA. . . . POLYGYNIA (*monogynie*, *digynie*, *trigynie* . . . . *polygynie*) *:* the Chervil (*Cerfeuil*), for instance, which has five free stamens and two distinct styles, will be found in the PENTANDRIA DIGYNIA.

§ 702. It is evident that all these classes are far from having the same value, since some are founded on a characteristic which is only secondary in the rest: the absolute number of the stamens, for instance. This absolute number, moreover, is of much less importance than the number relatively to the other parts of the flower, from which results its general symmetry. The number of the styles is of less importance still, for it is only an apparent one; the real state of the case being frequently disguised either by the unions or by the deduplications; so that counting the styles does not give the number of the carpels, with which it would be much more important to become acquainted, and which would agree much more with the etymology of the name indicating the number of the female organs. Thus, PENTANDRIA MONOGYNIA (*pentandrie monogynie*) contains the Periwinkle (*Pervenche*), which has two carpels; the DIOSMA, which has three or five; and PENTAGYNIA (*pentagynie*), on the contrary, the STATICE, the ovary of which is unilocular.

It is true that these defects ought to be forgotten, if we are content to consider the system of Linnæus as a convenient and sure way of arriving at the determination of the plant. But we shall be convinced by trial, that it is far from being so much so as its exclusive partisans pretend; and although, as it left the hands of its author, it could be applied with success to the few genera, on which it was constructed, it no longer presents these advantages after having received the numerous additions of his successors. The variations in the number of the organs in the flowers of the same plant, those which result from their adherences to one another, from their abortions, throw doubt at each step over the place which it ought to occupy in the system. The exceptions are very numerous: the species of the most natural genera have been necessarily separated and arranged in different classes, and sometimes we should be obliged to do the same with the different flowers of the same species.

§ 703. Dichotomous Method.—Lamarck asserted that to solve the problem, which proposes to discover the name of an unknown plant, more convenient and surer methods than those of the Linnæan system might be found. He accepted the challenge, which was sent to him on this occasion and soon, as his answer, published the plan and the first sketch of a method, which is generally known by the name of the *Analytic* (*analytique*) or better still by that of the *Dichotomous* (*dichotomique*) *method*. In fact, it consists in putting to the student one first question, which divides vegetables into two classes, between which he must choose from a characteristic of the plant, which necessarily places it in one of the two to the exclusion of the other; then a second question, which divides this chosen class into two others, to one of which the plant will belong; then a third, a fourth, &c., &c.: so that at each question the circle is contracted, until the last one brings us by this series of successive exclusions to the one for which we were looking. This mode of procedure differs from that of other systems, because, much more clearly artificial, it employs almost indifferently all the characteristics without restraining itself to any order. As soon as there is an exception, doubtful, merely difficult to be perceived, it jumps to another, and even leads you to the same end in two different ways. Hence the dichotomous method can hardly be reduced to a table, like those which we have mentioned or explained before; for it borrows something from all at the same time, and its divisions, in which there is nothing fixed, vary according to the proposed end. All that we can do here then is to illustrate it by an example. Let us suppose that we have a Ranunculus to examine, being unacquainted with what genus we are engaged: the following is the series of questions which we shall have to answer, led by the answer of each to the following, which will be shewn by a number of reference. 1st, Has this plant flowers, or has it not? 2nd, Are the flowers under one common envelope, or are they separate? 3rd, Has the separate flower both pistils and stamens, or only one or the other of them? 4th, Furnished with both, has it also a calyx and a corolla; or else does it want one or the other only, or both?

5th, Is the corolla monopetalous or polypetalous? 6th, Has the polypetalous flower a free or adherent ovary? 7th, Is the free ovary simple or compound? 8th, If there be several ovaries, have the leaves stipules or not? 9th, In the latter case, is there or is there not a gland outside of each ovary? 10th, If there is none, is the fruit fleshy or not? 11th, If it is not so, are the leaves opposite or alternate? 12th, In this latter case, is the flower regular or irregular and spurred? 13th, Has the calyx of the regular flower three or five folioles? 14th, If there are five, is each petal deduplicated internally on the inside at its base by a small scale or not? If it is, the unknown plant belongs to the genus RANUNCULUS. We shall now be able to discover the species by a series of fresh questions.

We see what a variety of characteristics would be called into action; how we have taken, then abandoned one organ to pass to another, and sometimes to return again to the first.

This method is very convenient for beginners, whom it conducts, as it were, by the hand to the common end of their research, and they have need only of a very little knowledge of botany, but at the same time very clear and positive, since every error puts us on a wrong road, and thenceforth as we proceed we separate more and more from the end instead of approaching to it. It has the inconvenience of not resuming here and there, as is done by the classes and other successive divisions of the systems, the characteristics which it has employed; so that once arrived at our end, it is difficult to keep an account of all the intermediate points through which we have passed, and the memory hardly retains anything else than the name of the plant; which is of very little importance. Lastly, it has hitherto been applied only to the plants of some more or less limited countries, such as France, and, especially, the country round about Paris; it can, consequently, only be employed in these limits, and cannot be made subservient to the determination of any unknown plant.

It is true, that in a large work which is just now finished, M. Meisner, in explaining all the known phanerogamic genera, endeavours to facilitate the research by a series of distinctions, which enter into the spirit of the dichotomous method by presenting characteristics, if not always easy, at least distributed in such a manner as to present to the reader only two routes at each branch. But the use of this book presupposes a knowledge of botany already very advanced: that of the families, which serve him as his starting point, and concerning which it remains for us to treat.

NATURAL METHOD.

§ 704. We have seen all the vegetable individuals collected into species, or groups of all those which are similar to one another (§ 692), then the species arranged in genera, or groups of species similar to one another (§ 694). These latter groups are much more conventional than the former, since the species is furnished by nature, and in spite of doubts which may result from its variations, and according as we have a number of data of sufficient importance and have observed several generations under different conditions, we must either admit or reject its existence. Now, several of these units are joined in order to compose a higher order, or genus, of them. But our mind here furnishes the limits of the plants which it is to include. These limits consist of the sum of resemblances, greater or less, necessary to define it. The genus itself is no less natural, even when we change its limits. Let us suppose four species a, b, c, d, to be united together into one genus m, because they resemble one another more than every other plant; they have, nevertheless, peculiar differences sufficient to distinguish them: and let us suppose that these differences may be such that a resembles b more than c and d, that c resembles d more than a and b; we can from these facts make two genera: the one n including a and b; the other p comprising c and d; but the change thus introduced will not in any respect have altered the primitive relations of the species, if they were clearly determined at the beginning, and the four, having always some points of resemblance, wherever we place the boundary line between them, will always form a very natural whole. Let us now make $m = a + b + c + d \quad n + p = (a + b) + (c + d)$; for $n = a + b$ and $p = c + b$. It is evident, that we shall always have the same values. The multiplication of the genera, the necessary result of that of the species, which the discoveries of travellers are unceasingly increasing, does not, therefore, prove that they are not conformable to nature. It is however clear, that they are so much the more so as they are thus divided, since by this division they approach more and more to the species. It is sufficient to mention the name of certain well-known genera, to shew to what point they form natural associations. There is no need of botanical knowledge to arrange in one group the different species of Roses, of Willows, or of Trefoils, &c. The resemblance is frequently so great, that it is much more difficult to separate the species than to group them into a common genus: some time and study are requisite to distinguish such a Rose from such another; none, to pronounce that it is a Rose.

§ 705. Tournefort knew better than any of his predecessors how to define the genera, so that the majority only included species really similar, thus forming natural units. Linnæus reduced their number; but his reductions bearing on genera generally similar to one another, were only an inverse operation to that which we have supposed above, and, consequently, not very likely to change the natural relations or the genera, since by uniting *n* and *p* to form *m*, *a* and *b*, *c* and *d*, always preserve the same place relatively to one another. But, by pursuing the examination of the systems of these two great botanists, we shall find that, after having followed nature to the genera inclusively, they abandon her more or less completely in the continuation of their classification, when they wish to arrange the genera in a certain order. Let us take examples of the Linnæan system: a plant has six stamens, and one style; it will, therefore, be ranked in the class HEXANDRIA MONOGYNIA, which, will thus comprehend both the Reed (*Jonc*) and the Barberry (*Épine-Vinette*). Now there is not the slightest connection between these two plants; no more than between the Rice and the ATRAPHAXIS, which are brought together in the order DIGYNIA; between the Sorrel (*Oseille*) and the Saffron (*Colchique*), which are found together in TRIGYNIA; no more than between the Vine and the Periwinkle (*Pervenche*) in the PENTANDRIA MONOGYNIA; between the Carrot and the Gooseberry (*Groseillier*) in the order DIGYNIA, &c. What is the reason of this? Because Linnæus in uniting the *Grosulariæ* into one group, had regard to a series of characteristics drawn from all the parts of the plant; whilst in bringing them into the same class as the genus DAUCUS or Carrot, he only paid attention to the presence of the five stamens and of the two styles, relations, which are in no way connected with one another, and may be found in a multitude of plants essentially different from one another.

§ 706. FAMILIES.—An operation analogous to that which grouped the species was necessary for the classification of the species; their relations were to be studied, and those genera only to be brought together, which presented the greatest similarity to one another; from these units termed genera, by uniting those which resembled one another the most, whilst they did not resemble any others, fresh units of a higher order were composed. These are those natural collections of genera which are called *families* (*familles*), a happy term given to them by a French botanist Magnol; and the method, which would thus group the plants according to the relations derived from their structure, would be a *natural method* (*méthode naturelle*). But the discovery of these relations, which connect several genera into one common family, presented many difficulties. The resemblance of the individuals of the same species strikes us at the first glance; that of the species of one and the same genus, already much less evident and frequently more deceptive, requires a longer time and closer study, so much so in fact, that centuries elapsed before the science had reduced to order the chaos of species. That of the genera, being less apparent still, required that the observation should extend to other characteristics, that we should enter into a series of considerations different from those which had been deduced for the construction of species and of genera. Direct observation shews that the natural bodies, which most completely resemble one another externally, also generally resemble one another internally; and botanists were thus authorized in concluding that the relation of these external characteristics, from which the species and the genera had been established, would presuppose an analogous one in their internal characteristics. It was now necessary to study one comparatively with the other, to determine if there were any among both, which were in a constant state of dependence; and to work upon these as the bases of a classification, in which the genera should be brought together according to the presence of such a characteristic, indicating several others to be necessarily co-existent, and thus furnishing the warranty, as it were, of a real and close resemblance.

§ 707. The Families of Linnæus.—Linnæus was endowed with too sound judgment, with too fine a sense of appreciation, not to feel the defects of his own system; and he proved it by publishing under the title of *Fragments*, an essay on classification, in which the genera are distributed quite differently from his System. There are, nevertheless, entire series which are the same both in the system and in the fragments. In this case, the characteristic employed by the system to form the genera ought, therefore, to be connected with other important characteristics; it was not so in the contrary case. Thus, the greater part of the plants of the class Icosandria are divided into two natural neighbouring groups; all those of TETRADYNAMIA are united into one; the plants of PENTANDRIA, of HEXANDRIA, of POLYANDRIA, &c. &c., are scattered, on the contrary, into a number of different groups. The absolute number, therefore, of the stamens did not constitute in the eyes of Linnæus himself true relations so much as their position on the calyx or their relative proportion. But he did not publish the principles which guided him in the arrangement of his genera into natural orders; and, probably, he followed rather the inspirations of a fortunate genius and of consummate experience than a code of well digested laws, although in several of his works some few of these laws are found as axioms.

§ 708. THOSE OF BERNARD DE JUSSIEU.—A French botanist, whose name is connected with that of the natural method, Bernard de Jussieu, had received Linnæus into his hospitality, when he visited Paris before having published either his system or his fragments, but already learned, and well known by important works. Thence resulted an affectionate friendship and a correspondence, some passages of which prove that the attention of the two botanists was several times directed towards this great problem of the natural method, and that both were working towards the same point. It was not till twenty years afterwards, in 1759, that Bernard de Jussieu in a botanic garden established by Louis XV at the Trianon, attempted a natural arrangement of the genera, the fruit of his long studies and meditations. Like Linnæus he did not explain the principles which directed him; he did not even publish a list analogous to the *Fragments*. But the botanists of the time could go and study this learned enigma at the Trianon and, if they did study it, it was evident that they did not divine it.

§ 709. THOSE OF ADANSON.—We find a few years afterwards in 1763, the families of plants of Adanson published, who explained his principles as to their formation and laid down formulæ of their definitions like those of the genera. He found out that, in order to group the genera into families, attention ought to be paid to the whole of their characteristics and not to a single one; that an attempt at classification ought, consequently, to be preceded by a vast amount of labour, in which the organs of the vegetables, it was required to arrange, should be examined without neglecting any, the whole of their modifications should be determined in all the genera. Each point of their organization considered separately, would give us a separate system, which would present all of them in a certain order. If, in all these partial systems thus obtained the two same genera happen to be constantly brought into juxtaposition, it is evident, that they resemble one another in all the points of their organization, that they form part of the same natural group; if, on the contrary, they are constantly separated, they differ in all these points, and ought not, therefore, to form part of the same group. This principle cannot be denied, and the author thence deduced this rule: that we should be able to calculate by these means the different intervals, which in the general and natural order separate the different genera, intervals so much the less as they are brought oftener into juxtaposition in a greater number of particular systems. He constructed, in consequence, sixty-five systems in which he included all the considerations, by means of which he thought he could study and classify all plants: some general, such as the shape, the height, the breadth, the duration, the climate, &c., &c.; others drawn from the general organs, as the root, the branches, the leaves, the flowers, &c.; or partial, as the calyx, the corolla, the stamens, the pistil, the fruit, &c.; or of the component parts of these, as the anthers, the pollen, the seed, &c., as well as the modifications, which these parts may present in their number, their situation, &c. He then used these sixty-five combinations in the way we have stated above; bringing the genera together, or separating them, according as a larger number of his systems presented them close or far apart. A certain sum of resemblances existing between a certain number of genera constituted the characteristics of a family; a less sum indicated the greater or less interval which separated them from the rest. He obtained in this manner fifty-eight groups arranged in a certain order, which he called his natural order.

But, supposing his principles true, were they really applicable? His manner of proceeding was nothing else than an arithmetical calculation, in which every error in the figures would falsify the result, every fault in one of his practical systems would be multiplied in the general system. Now, the discovery of fresh plants would soon cause alterations in the numbers; the progress and discoveries of the science as to the organization of plants would modify most of these systems. This has, in fact, happened, for there are now described five times as many plants as were recorded then, and several facts, with which he was unacquainted, are now well known; the correctness of the principle would not, therefore, prevent the falsity of the deductions. But, it is not correct in principle. To make all the organs and the characteristics which we draw from them almost equal in importance, to make of them so many units of the same value entering into the calculation of the relations of the plants, is to give the same value to pieces of money of different metals and weights; it is, as it were, to issue a circulation of a purely fictitious value: it is, therefore, impossible not to perceive that the manner of proceeding followed by Adanson was quite artificial; and at the same time, whilst admiring his gigantic work as well as the variety and amount of knowledge which his attempts required, we must agree that his system would not conduct us to the desired end. Thus, his families, assemblages of genera most frequently without any real or close relation, have not been adopted by any of his successors and are themselves much less natural than those of his predecessors, Linnæus and, especially, Bernard de Jussieu.

The Method of A. L. de Jussieu.

§ 710. About the same time Antoine Laurent de Jussieu commenced the study of the science of botany under the tuition of his uncle Bernard, and there is no doubt but that the young man acquired in his close and intimate intercourse with the old man and in his learned lessons the germ, which he afterwards so well fecundated and developed. Ten years afterwards he explained to the *Académie des Sciences*, and applied to the planting of the botanical garden at Paris a fresh method, which, sixteen years afterwards (in 1789), matured by continual meditations and studies, received its definite form and expression by being extended over all the vegetables which were then known, in a fundamental work, the Genera plantarum. At the beginning he published the catalogue of the genera of the garden of the Trianon (which had never been edited) in the order in which Bernard had placed them; a precious monument for the history of the science, since it enables us to determine to what point the uncle had arrived, and from what point the nephew started. The latter having clearly laid down the principles which had guided him in the establishment of his classification, we may by comparing his work with those of his predecessors, discover which of his principles had been admitted or suspected by them, of which they had been ignorant. But we shall not endeavour here to determine the part of each in this grand work; this discussion, belonging wholly to the history and not to the elements of the science, does not enter into the plan of this work, which will only explain the laws without troubling itself with the legislators.

§ 711. A. L. de Jussieu admits, like Adanson, that the examination of all the parts of a plant is necessary for its classification; but, whilst he was pursuing this complete examination, he did not endeavour to deduce the order of the genera theoretically; in order to group them into families, he imitated the manner in which the genera themselves were formed. Botanists, struck with the complete and constant resemblance of certain individuals, had collected them into species; then, according to an equally constant resemblance, but much less complete, had collected the species into genera. The characteristics, which may vary in the same species, will depend on causes not innate in the plant, such as its height, the hardness or softness of its wood, certain modifications of shape and of colour, &c., which change with the soil, the climate and other purely accidental influences. The specific characteristics, on the contrary (those which ought to be presented by every individual, that is connected with others in forming a certain species, whatever may be the circumstances in which it is placed), will be inherent in the very nature of the plant. Amongst these characteristics there are some more important than others, less subject to vary in the different individuals; these, being always found in a certain number of species, impress upon them a resemblance sufficiently striking to allow us to constitute a genus. These will, therefore, have more value on account of their generality than the specific and the specific than the individual. But how can we appreciate these different values? Nature herself has indicated to the observer the species and several of the genera by the points of resemblance, with which she marks certain vegetables; beyond these genera this conducting thread was wanting, since all botanists, agreeing in almost every thing up to this point, differed after they reached it and followed each a separate route. There are, however, several large groups of vegetables connected with one another by characteristics of resemblance so evident, that they cannot escape the notice of the most casual observer, much less of a botanist. Besides these points of resemblance common to every species of one of these groups, there are some which are only common to a certain number among them; so that it may be subdivided into a large number of secondary groups. These had been recognised as genera by botanists.

There were, therefore, already a few collections of genera evidently more similar to one another than they were to those of any other group, or, in other terms, some families undeniably natural. Jussieu thought that this was the key of the natural method, since, by comparing the characteristics of one of these families with those of the genera which compose it, he would obtain the relation of one to the other; since, by comparing several of them with one another, he would see what characteristics, common to all the plants of the same family, varied in such a one and such another; since he would thus arrive at the value of each characteristic, and this value, once determined by means of these groups so clearly arranged by Nature herself, could in its turn be applied to the determination of those, on which she has not so clearly imprinted this family likeness, and which were the unknown quantities in the great problem. He chose, therefore, seven families universally admitted: those, which are known under the names of GRAMINEÆ (*Graminées*), LILIACEÆ (*Liliacées*), LABIATÆ (*Labiées*), COMPOSITÆ (*Composées*), UMBELLIFERÆ (*Umbellifères*), CRUCIFERÆ (*Crucifères*) and LEGUMINOSÆ (*Légumineuses*). He discovered that the structure of the embryo is identical in all the plants of one of these families; that it is Monocotyledonous in the Gramineæ and in the Liliaceæ, Dicotyledonous in the five others; that the structure of the seed is also identical; the Monocotyledonous embryo is placed in the axis of a fleshy perisperm in the Liliaceæ, on the side of a farinaceous perisperm in the Gramineæ; the Dicotyledonous embryo, at the summit of a hard and horny perisperm in the Umbelliferæ, without a perisperm in the three others; that the stamens, which may vary in their number in the same family, the Gramineæ, for instance, do not generally vary in the method of their insertion, hypogynous in the Gramineæ and in the Cruciferæ; on the corolla in the Labiatæ and the Compositæ; on an epigynous disk in the Umbelliferæ. He thus obtained the value of certain characteristics which would not vary in the same natural family. But, less in importance than these, there were others more variable, which he tried to appreciate in the same way, either by the study of other families formed by nature herself, or in those which he formed by applying these first rules and several others, also founded on his observations. We cannot here enter into the details of this long and arduous undertaking, from which resulted a hundred families containing all the plants known at that time.

§ 712. We see in all that precedes the employment of a

principle which had escaped the notice of Adanson: that of the *subordination of the characteristics*, which in Jussieu's system are, according to his own expression, weighed and not counted. They are considered as having unequal values: so that a characteristic of the first order is equivalent to several of the second, one of these to several of the third, and so on. This value is determined by observation and experience; and, in proportion as it descends in the scale, it becomes less and less constant. To employ the familiar comparison of coins of different metal to the different characteristics, which by their union would compose a certain sum of relations between the plants of the same family; the pieces of gold would have an invariable standard, higher than that of the silver; and those of copper would be worth a little more here, a little less there, destined in some kind to furnish the currency, in which the coin of a more precious metal forms the principal, and alone is rigorously controlled.

§ 713. The importance of the subordination of the characteristics results especially from a consideration, which we have not yet stated, but which necessarily arises from this combination of several of them in each family: a characteristic of a superior order necessarily carries along with it a certain number of a different order and excludes, on the contrary, a certain number of others; so that the simple enunciation of the former suffices to inform us of the co-existence or the absence of the latter, and one part of the organization of a plant is previously stated by a single fact which we have been able to determine. This wonderfully abridges and simplifies our researches as well as the language employed in recording them. Thus, we have seen in almost every chapter of this work, that the absence or the presence of the cotyledons, their unity or their plurality is manifested almost in every part of the plant, which presents important and striking differences, according as its germ has appeared to be differently constituted with regard to this relation. When we say that a plant is Monocotyledonous or Di-cotyledonous, we do not state this simple fact alone, but a large collection of facts; we have an idea of the general arrangement of the elementary organs in its tissues, of the manner in which it germinates and is ramified, of the structure and nervation of its leaves, of the symmetry of its flowers, &c., &c. From such secondary characteristics, we may in the same way deduce several others of a higher, equal or inferior order: to say, that the corolla is monopetalous, we mean that the plant is Dicotyledonous, that the stamens are inserted on the corolla in a definite number, equal to or less than that of its divisions. The knowledge of all these constant relations between the different parts, which allows us to draw conclusions both from a part to the whole and from the whole to a part, is the basis of the natural method; and, if this knowledge were perfect, we might say that the method is the science itself, since the place, which it would assign to each plant, would state its organization, and on its organization depends the exercise of the whole of its vital functions. Thus, we see that, in a family truly natural, great agreement in its economical or medical properties generally exists in the plants which compose it: this is by no means remarkable, since the similarity of organs will necessarily cause that of the products. This fact gives a great advantage to the Natural Method in its practical utility.

§ 714. Classes.—When the families are once constituted, it is only requisite to arrange them so as to bring together in their turn those, which resemble one another the most and to separate those which resemble one another the least. The manner of proceeding followed in the grouping of the genera naturally presents itself; the characteristics, common to several families at once, allowed several of them to be united in higher groups, and the established subordination of the characteristics indicated in what order they ought to be employed. That of the embryo evidently precedes all the rest and divides the Vegetable Kingdom into three great branches: the Acotyledonous, the Monocotyledonous, the Dicotyledonous. After this fundamental characteristic, but accounted of much less importance, A. L. de Jussieu placed the insertion of the stamens, hypogynous, perigynous or epigynous. But, in Dicotyledonous plants, these stamens are united by their filaments to the corolla, when it is monopetalous; so that in this case their insertion, instead of being immediately on the torus, on the calyx or on the ovary, takes place by the mediation of the corolla growing from one of these three points. The characteristic of the corolla, thus connected with that of the insertion, is accounted of equal value with it. The insertion is only the expression of the relative situation of the two orders of the organs of the flower, of the stamens with respect to the pistil, in the same envelope. But, if they are in separate flowers, this relation does not take place, and it is the fact of their separation which must be stated. Such are the principal considerations, from which the families were distributed into fifteen classes, which

KEY TO THE METHOD OF A. L. DE JUSSIEU.

Class	Division	Insertion	Anthers	Group
ACOTYLEDONS				1. ACOTYLEDONS, (*figs.* 502, 526).
MONOCOTYLEDONS	Stamens	hypogynous		2. MONOHYPOGYNOUS, (*fig.* 531).
		perigynous		3. MONOPERIGYNOUS, (*fig.* 546).
		epigynous		4. MONO-EPIGYNOUS (*figs.* 551, 555).
DICOTYLEDONS	apetalous	epigynous		5. EPISTAMINOUS, (*fig.* 599).
		perigynous		6. PERISTAMINOUS, (*figs.* 607, 612).
		hypogynous		7. HYPOSTAMINOUS, (*fig.* 616).
	monopetalous	hypogynous		8. HYPOCOROLLOUS, (*fig.* 678).
		perigynous		9. PERICOROLLOUS, (*fig.* 722).
		epigynous. Anthers	united to one another	10. EPICOROLLOUS SYNANTHEROUS, (*fig.* 730).
			distinct	11. EPICOROLLOUS CORISANTHEROUS, (*fig.* 707, B).
	polypetalous	epigynous		12. EPIPETALOUS (*fig.* 674).
		hypogynous		13. HYPOPETALOUS, (*figs.* 634, 642).
		perigynous		14. PERIPETALOUS, (*fig.* 663).
	Diclinous			15. DICLINOUS, (*figs.* 558, 560; 590, 592).

are shewn in the table, in which they will be clearly understood by a single glance of the eye. The terms employed in the first columns have been previously defined (§ 382, 390, 391); those of the last have been proposed at a more recent time for the sake of designating such class more conveniently.

§ 715. There are, therefore, two distinct parts to consider in the method of Jussieu: 1st, The grouping of the genera into families, 2nd, The arrangement of these families into classes and their series. It is generally this division into classes, such as is presented by the preceding table, that elementary books are content to present under the name of this method although it is only the least important part of this great work. The great step towards the establishment of the natural classification was that of the families which merited this name, and this is what A. L. de Jussieu has done. He seems himself to have pointed out this distinction of which we are speaking in the title of his work, which states the genera to be arranged into natural families according to a method employed in the *Jardin de Paris* (GENERA PLANTARUM SECUNDUM ORDINES NATURALES JUXTA METHODUM IN HORTO REGIO PARISIENSI EXARATAM.) He applied, therefore, the epithet to the families and not to the whole method. But in explaining the great principles, which ought to preside over the classification not only of plants but of all organized beings; in giving to the families, into which he distributed all vegetables, a solid basis, and at the same time forming a model for science, sufficient had been done to allow us to date from this moment the foundation of the natural method, which thenceforth was no longer to be discovered but to be perfected.

His families have all been preserved with the exception of the single changes which the progress of science necessarily causes, either by rendering us more conversant with plants, which were imperfectly known, or by discovering a great number of fresh ones, for which it is necessary either to make fresh limits or to enlarge the ancient ones. But, in these cases, if the conventional limits change, the actual relations are not at all changed; no more, for instance, than those of different points in an extent of country, in which a single province is divided into several departments.

As to the arrangement of the families, it has often been attacked and modified, not as to its fundamental division, which is universally admitted, but as to its secondary divisions, drawn from the insertion of the stamens. They are stated to admit several exceptions, to separate several natural connections and to join together some which are not so. These objections are frequently just; but, although half a century has passed away since this method was completed, and several attempts have been made to substitute a better one for it, we do not see that up to this time one has been found much better, nothing at least which justifies its adoption by the generality of botanists.

§ 716. De Candolle, who was the first to apply the natural method to the whole of the plants of a large country, France, and afterwards to every species of vegetable, has followed in his series of families an order, not essentially different from that of Jussieu. Indeed, separating Dicotyledonous plants into *thalamiflorous* (*thalamiflores*), which precisely answers to the hypopetalous; *calyciflorous* (*calyciflores*), which answers to the peripetalous; *corolliflorous* (*corolliflores*), which answers to the monopetalous; and *monochlamydous* (*monochlamydées*), which answers to the apetalous, he followed the rules drawn from the corolla and its insertions: his system only differing in each of the last two of his great classes comprehending several of those of Jussieu.

§ 717. An English botanist (whose name, if the limits of this book had not forbidden any historical digressions, would have appeared on several of the pages, since there are few important points of vegetable organization on which he has not thrown strong and fresh light), Mr. R. Brown, is also one of those, who have most contributed to the perfecting of the families; he has, moreover, indicated what remains to be done to arrive at the natural order. "A methodical and at the same time natural arrangement," said he, "is, in the actual extent of our knowledge, perhaps impracticable. "It is probable that the means of some day attaining it would be to "lay it for the present on one side, and to turn our whole attention "to the combination of those natural families, which are so suscep- "tible of being defined. The existence of several of these natural "classes is already recognised."

This plan has been followed by most botanists, who have been occupied with the solution of this important problem. The name of classes has been applied by them to much more limited groups than those to which A. L. de Jussieu gave this name, to several even of those of which he made single families, but which are singularly increased by modern discoveries. In short, the number of species known in his time, on the study of which he established his method, did not quite reach 20,000: we should not be far wrong in estimating those known at the present day at 100,000[1]. We may, therefore, assert without exaggeration that his families now represent collections of plants five times more extensive than at first, and they will have acquired much greater importance. Families, such as his Rosaceæ, Leguminosæ, Malvaceæ, Onagrarieæ, Euphorbiaceæ, Urticeæ, &c., &c., themselves alone evidently represent so many classes; whilst several others must be joined together to form groups of an equivalent value. Dr. Lindley has proposed the fresh name of *alliances* to designate these *classes*, reserving for the latter name the primitive signification, which we applied to the principal and less numerous divisions of the three great branches of the Vegetable Kingdom. He has distributed all the families into a very great quantity of these alliances, each of the latter comprises a small number of the former. In the most complete work which we now possess on the genera, M. Endlicher has also tried to unite the families into groups of a higher rank. When the Botanical Garden of Paris was recently planted, M. Adolphe Brongniart grouped 296 families into 68 classes, the characteristics of which he has traced. We hope that from these learned attempts and from the perfection which the study of organization will obtain, since it is now prosecuted farther than it was at the commencement of this century, a natural classification will be formed, as truly so at least, as we can possibly obtain from the multiplicity of the relations which connect vegetables with one another, and, especially, from the necessity of joining into one chain all the links which growing on every side, cannot be united in one place without being disjoined in another. But we must wait for this perfection and the sanction of time to fix as a definite thing this order so much desired: it is necessary that these groups, classes or alliances, as it may seem fit to the reader to term them, have been legitimised by general assent and their characteristics clearly defined in order that we may deduce a general system from their comparison.

§ 718. In spite of the multiplication of the families, however, their number is not so great as to prevent the memory from retaining their distinctive characteristics, especially when aided by the first grand division into three branches. The evident end of the method is to facilitate the acquirement of a complete knowledge of the different plants by substituting for these natural units, termed species

[1] Lindley in his *Vegetable Kingdom* (1846) p. 800, estimates the number of genera at 20,806 and of species at 82,606.—Trans.

or genera, which, on account of their number cannot all be present at once to the best endowed memory, other units of a higher order, sufficiently limited in number to enable the human mind to become acquainted with every one of them. This has been done by the establishment of families. When we know that a plant belongs to such or such a family, we have already some ideas of the principal points of its organization and of its relations to other plants. Whenever we want to extend our researches, thus confined to a narrower circle, they are more easily prosecuted, and the results become more tangible; hence, the incontestable progress which botany has made, since the system of families has replaced others, the secondary groups of which, uniting plants by a single point of their organization, only represent one single characteristic frequently very insignificant. Hence that statement uttered a few pages back, that the great step towards the discovery of the natural method has been the establishment of families worthy of this name, as well as of the principle of the subordination of characteristics.

In paying this tribute to the memory of the name I have the honour to bear, I am moved as much by sentiments of justice as by those of filial duty. It was, moreover, important to impress on the mind of the pupil this truth: that the work of Jussieu does not consist of the short table which we have presented to his notice as its summary; let it remain untouched even by reformation or let it be totally rejected, he must, at the same time knowing it perfectly, penetrate farther, if he wishes to have a clear idea of the natural method. The knowledge of the families is, doubtless, too vast, and demands too long a study for him to acquire it thoroughly; but it is well to study some of them with great care, to retain the characteristics on which they are founded. He will then be able from analogy to comprehend the description of others.

§ 719. The limits of this work do not allow us to explain them all even briefly. We shall, therefore, content ourselves with a series of tables shewing their principal characteristics. But from these families we shall choose a few to describe a little more in detail, although as succinctly as possible. These will be the most important or those, which, presenting some point of organization, which is not common or exceptional, will give us an opportunity of adding somewhat to the very general ideas to which we have tried to limit ourselves in the course of the preceding explanation. We shall also enter into some details of those, which present any peculiar properties in their produce, being either useful in arts and manufactures, proper for food or medicine, or, on the contrary, hurtful to man.

§ 720. After what we have just said, when we treat of the families of the Vegetable Kingdom, we shall still adopt the great divisions established by A. L. de Jussieu, in preference to those which have been more recently proposed, because the latter do not rest on fixed rules, and in spite of their merit, if we consider them one by one, they still want that systematic connection, by means of which the beginner may easily comprehend and fix them in his memory. We do not, however, believe that we ought strictly to follow the order, in which they were originally arranged. It remains for us to explain the fresh considerations that lead us to suppose this arrangement to be inverted in some of its parts.

Jussieu, when he laid down his families, wisely proceeded from the simple to the compound, commencing with the Acotyledons and finishing with the Dicotyledons. The truth of this progression has been generally admitted, not because of two cotyledons being more complex than one, and one than none; but, because, considered in all their parts, Acotyledons are evidently more simple than Cotyledonous plants, Dicotyledons than Monocotyledons: we find this from the examination of the organs, and there is no need of repeating the proofs here, which we have already explained under the article of each of those organs. This order cannot, therefore, be subject to any objection. Dicotyledons were divided into apetalous, monopetalous, polypetalous, and diclinous; for this series we think of substituting the following: 1st, diclinous, 2nd, apetalous, 3rd, polypetalous, 4th, monopetalous. We are going to examine in what the last appears to us to be of a higher degree of composition than the first, and why they, consequently, merit the fresh place that we assign to them.

§ 721. Every being is so much the more organized as its existence results from a greater number of functions, the performance of which requires a higher and more numerous class of organs. In general functions, some are of a higher order than others: these are not common to all, but become the peculiar attribute of a certain number of beings. Those, which are endowed with them, are indeed necessarily more elevated than any others; since, beside the same functions, they perform a certain number of different ones, for which the others were not adapted. It is, therefore, from the capacity to perform these acts, that we are able to determine the exact degree of the organization of a plant. This rule is included in that, which we previously drew from number only.

§ 722. It would be easy to prove from the same reasoning, that the same function may present different degrees of value in different beings, since it will not be performed in the same manner in all; but, in some, by certain parts; in others, by other parts added to the former. The organs which are the agents of it are multiplied and, therefore, perfected in the same proportion.

§ 723. The natural classification having for its purpose the representation of these different degrees of organization in their ascending progression, ought to determine in each being what are its most valuable characteristics, at first as functions, then as organs to perform them; and we shall call these organs the most important, not because they are the most indispensable to life, which is frequently preserved without them, but because they constitute the true nature of the being, that is, from the fact of its possessing them, it is that being and no other.

§ 724. Let us now apply these rules to vegetables. We have discovered two great functions in them: nutrition and reproduction. The latter is undeniably the most important, in the sense which we have attached to this word; since it necessarily supposes the former, since the plant is, during a part of its life, and may, during its whole life, be limited to the organs of vegetation, but it is only rendered complete by the rest being developed. By means of the gradual steps, therefore, towards the perfection of the latter, we must endeavour to establish the scale of rank in the Vegetable Kingdom: but to make the foundation stronger we must call to our assistance the facts arising from a comparative examination of the organs of vegetation, the gradations of which, as De Candolle has well explained, follow an almost parallel route, at least if they are considered in a very general manner.

§ 725. The plant is so much the more perfect as we see a much larger number of different organs concur in the reproduction. But which organ shall we place the first, as the most simple, as that which is to serve as a point of departure? We have examined the organs of the vegetable, some elementary, others compound. Among the former, the most simple is evidently a cell, since it is the first state of all the rest; and the most simple plant is that which is reduced to a cell or to a small number of cells exactly similar to one another: we observe this degree of simplicity in certain ALGÆ, which, consequently, will occupy the first place in a series proceeding from the simple to the compound. Each cell, when separated from the rest, is here perfectly adapted for propagating the plant: there is a complete confusion between the organs of vegetation and those of reproduction.

§ 726. We then find, in the same class, other vegetables, the tissue of which, although not presenting any distinction between the organs we have termed fundamental, is not, however, so homogeneous as the preceding. Some few of the cells are distinguished from the rest by a peculiar appearance, by peculiar products: so some are better adapted for reproducing, being separately developed into a plant similar to that of which they formed part. These portions of the tissue endowed with this peculiar property, but scattered and, as it were, lost in the midst of it, are more stationary in other vegetables, occupying a certain fixed place: the general form will be more regularly adhered to, and the individuality of the plant be better observed: for we can hardly recognise it in the lower degrees.

§ 727. In proportion as these parts, in which the faculty of reproduction is concentrated, are more distinguished and separated from the rest of the tissue, the latter assumes more fixed forms and begins to present some distinction of parts, the first sketches of the organs which we have termed fundamental: the one central, or the axis; the other lateral, or the leaves; this is what is observable, for instance, in the JUNGERMANNIA and the Mosses. Then the stems and the leaves are perfected. These leaves, with their shape more or less altered, are now called upon to bear on their surface (in the Ferns, for instance) the organs of propagation. But, in every case, these organs consist only of a portion of the cellular tissue, modified in a particular manner; so that in certain cells several others are formed which we have called spores. Sometimes, and always at the expense of the same tissue, there are developed others, again differing from the rest, the action of which, not yet well determined, according to several authors unites with the former in reproducing the plant.

§ 728. From this rapid examination of the Cryptogamia we may conclude, that the degree of confusion between the organs of vegetation and those of propagation depends on the degree of simplicity of the whole vegetable; that their distinction, more and more clear, generally demonstrates an organization more and more compound; this is proved by the perfection of the fundamental organs, which are complicated in the same ratio.

§ 729. Arrived at the Cotyledonous or Phanerogamic plants, we see the organs of reproduction assume a fresh and double form, that of the anther and of the ovule; and the reciprocal action of these two organs becomes necessary for the performance of the function. This necessity for their mutual action shews us a still more elevated degree of value in the function, which now assumes a fresh name: that of fecundation. It establishes a relation between the Vegetable and the Animal Kingdoms, the latter enjoying the higher organization. There can, therefore, be no doubt that Phanerogamic are more organized than Cryptogamic plants. It remains to discover how, in the former, we may establish this gradation, which we have tried to recognise in the latter.

§ 730. The organs of vegetation are, both in Phanerogamia and the higher Cryptogamia, axes and leaves; those of reproduction are comprehended under the general name of the flower, and we have seen that it is now generally agreed to consider the different parts of the flower as so many leaves more or less modified. The more complete the metamorphosis of one into the other becomes, the clearer and broader will be the distinction between the organs of vegetation and those of reproduction.

§ 731. The modification is always complete in the essential organs of fecundation, the anther and the ovule. The anther, (all the cells of which produce in their interior several others of them of a peculiar nature, immediate agents of the function [pollen-grains]), presents in this point of its organization an evident relation to sporiferous leaves of Cryptogamia; but the leaf in the latter is not completely changed, but still performs some of the functions of vegetation; in the anther, on account of its complete metamorphosis, it is confined to reproduction, and from this clear distinction of form and action it already bespeaks a higher organization. The ovule with its complicated structure appears to be not so much a single leaf as a small mass of leaves; but we can scarcely assign this origin to them from any facts, but mostly from reasoning and analogy. If they are really leaves, they can no longer be recognised and, moreover, perform entirely different functions. The ovules, besides, are generally concealed under an envelope formed by another leaf itself modified, although to a less degree (the carpel); so that we should be able to say that here the metamorphosis has been raised to its second power. There is nothing exactly parallel among Cryptogamia.

§ 732. But we have seen that very often other neighbouring leaves, laying aside their foliary appearances and functions, disguise themselves in order to form the envelopes of the flower: they separate the stamens and carpels farther from the vegetative organs and form along with them a more compound and distinct system. The accession of these fresh parts to the organs of reproduction appears, therefore, to be a fresh degree of organization.

§ 733. These different parts of the flower, however, frequently retain some vestiges of their foliaceous nature, without which we should never have been able to discover it; this is mostly when independent of one another. They then preserve on the shortened axis which bears them the relative positions which we find in the leaves. This characteristic of position, much more tenacious than that of the form, of the structure and, consequently, of the function, is the last which is effaced; when it, therefore, is effaced, the metamorphosis will have reached its highest state of developement. Now, this really does take place on account of the adherence between the different floral organs. It is clear that, it is much more difficult to recognise five leaves in a tube formed by the union of five anthers, in a quinquelocular ovary crowned by a simple style, in a calyx or a corolla with five teeth, than in so many sepals, petals, stamens, and carpels perfectly distinct: that in stamens regularly arranged in a spire on a flat or cylindrical torus (as in the Magnoliaceæ), we can admit modified leaves sooner than in these same stamens growing from the tube of the calyx or of the corolla, or from a disk covering the summit of an ovary blended with the calyx. Let us combine together into one flower these different degrees of adherence between its different parts, and we shall obtain a whole, in which any observer, if he be not previously instructed, will not suspect a succession of leaves, and in which the organs of reproduction will thus have become as distinct as they can be from those of vegetation by losing their last relations, those of position.

§ 734. It is now apparent why we have placed the monopetalous above the polypetalous, contrary to generally received custom. Besides, if according to another principle generally admitted, we estimate the value of the characteristics by their constancy, we shall see that of the monopetalous corolla, especially when connected with the insertion of the stamens, allows fewer exceptions than that of the polypetalous corolla. The majority of the polypetalous families contain a few apetalous genera, and several present an evident affinity to other families entirely without corolla. It is a fact so well known, that several authors propose to unite them into one large common class: M. Brongniart has done this by distributing Dicotyledons into two series; the one, the *gamopetalous* (§ 374), the other, the *dialypetalous* (from διαλύειν, *separate*), which comprehends plants either with separate petals or without any.

§ 735. Most authors would place at the head of the vegetable scale the thalamiflorous or hypogynous polypetalous plants, and amongst them the Ranunculaceæ, looking upon their flower as more perfect than any other, on account of the large number of essential organs (stamens and carpels) which it generally contains. The value here ascribed to the number does not seem to merit more consideration than it has obtained in the Linnæan system according to the judgment of all those who adopt the natural method. The very number of the floral parts frequently causes their relations of position on a continuous spire to re-appear, and thus gives them, although they belong to a single flower, the appearance of an inflorescence. It is thus that the carpels of certain Ranunculaceæ (as of the Adonis, of the Myosurus, &c.), represent a true spike and betray by this arrangement their foliary nature. The connection of the organs of the flower with leaves becomes also so manifest in some cases, that another of the Ranunculaceæ (the Hellebore) imbued the mind of Goëthe with his famous theory of metamorphosis. The inflorescence shews us the passage from the organs of vegetation to those of fecundation, and, mingling them together, belongs to both of these two great functions at the same time. The more insensible the transition is (and we have seen [§ 395] that it is quite so sometimes), the less clearly the system of the flower is distinguished, the more simple is its composition. It must have a lower place assigned to it according to the principles which we have laid down. The large number of parts of a single flower, which frequently renders it similar to an inflorescence, will, therefore, be far from indicating the highest degree of organization and would rather tempt us to look for it in an arrangement precisely inverse, that, in which an entire inflorescence resembles a single flower: as in the Compositæ. Besides, considering each flower of one of the Compositæ separately, in which the calyx blended with the ovary has assumed fresh forms, in which the monopetalous corolla is inserted in an epigynous disk and bears stamens united into a tube by their anthers, we should find nearly the maximum of adherences, and the organs of the flower resulting from leaves modified as completely as it would be possible for them.

§ 736. Monocotyledons may present different degrees in the composition of their flower, like Dicotyledons, and even reach, from the adherence between their parts, a state of complication almost as singular as that which has been just explained: the Orchideæ present an example of it. We do not, therefore, see any reason why in this respect they should be considered as inferior to them in organization; for if their envelopes are always limited to a simple perianth, it is also the case in several Dicotyledons, and even among the latter we find some of them which, reduced to a naked ovule, present a still greater degree of simplicity. These two great branches, considered with regard to the organs of fecundation, follow, therefore, two parallel lines rather than the same line, one behind the other. But when we compare the organs of vegetation, the parallelism disappears: Monocotyledons present a simpler structure, a more uniform tissue.

§ 737. We have prosecuted our researches into the principles to be observed in the establishment of the series of plants from the simplest to the most complex; but we have learnt from the differences of botanists, the difficulty of finding one, which will perfectly satisfy this condition and place all plants in their true relations with respect to one another. These affinities, indeed, are multiplied in Nature. Every species or every other collection of plants (genus, family, &c.) are related to several others at once by characteristics of an equal or almost equal value. Linnæus ingeniously compared the chart of the Vegetable Kingdom to a geographical map, in which each country touches at once several with which it is surrounded: if we were to draw a continuous line from one end to the other, it will only pass through a certain number of countries leaving a much larger number on the right and on the left. The series of families is this line, and we can only place them all in a line by transporting several out of their natural place. Mr. R. Brown explained this fact very happily, when he said that the connection between organized beings is a net and not a chain[a].

A third comparison, which we shall borrow from the Vegetable Kingdom itself, will aid us in understanding how this multiplicity of affinities does not exclude the idea of a general series, and how these lines, crossing one another in every direction, may be arranged

[a] JUSSÆANAM METHODUM SECUTUS SUM, CUJUS ORDINES PLERIQUE VERE NATURALES. . . . . NEC PRO ILLA ALIAM SUBSTITUERE TENTAVI, NEC DE ORDINUM SERIE ADMODUM SOLLICITUS FUI. IPSA NATURA ENIM CORPORA ORGANICA RETICULATIM POTIUS QUAM CATENATIM CONNECTENS, TALEM VIX AGNOVERIT.—FLOR. NOV. HOLL.—[*I have followed the method of Jussieu, whose orders are for the most part truly natural . . . . I neither have endeavoured to substitute any other system for it, nor was I very anxious about the series of the orders. For Nature herself, connecting organized beings in a net rather than in a chain, would hardly have recognised such a one*].—The quotation has been given at length to bring a superior authority to support what we have said (§ 626) and to justify the divisions we have adopted in the following part of the work (§ 710).

in a single continuous one. The families are like the branches of a large tree placed on a common trunk, each of which in its developement touches several others at the same time and may even cross them; some may grow beyond others which spring out of the trunk above them; but in spite of this divergence in one direction and this apparent confusion, they all converge towards the trunk and arise one after another on a single line extending from bottom to top. We may conceive without more details, in what way the metamorphosis may be continuous, and the ramification, differently modified, with its divisions of every rank and size may represent all those which we admit in the classification.

§ 738. The twigs, springing from the branches which represent the families, will give us an idea of the genera. Now, they may all grow successively one after the other on a simple branch, or very many together at the same height on a branch, which is itself ramified; thus forming in the first case a series, a group in the second. This double modification is also observed in the arrangement of the genera of the same family. There are *families arising from groups* (*familles par groupe*), the genera of which, very similar to one another and each touching several others at the same time, are united in some confusion. There are *families by connection* (*familles par enchaînement*), the genera of which, each being connected with that which follows and that which precedes it, form a real series, in which the last is bound to the first only by this series of intermediate links and may sometimes be rather dissimilar from it. The first are necessarily more natural than the second.

Before commencing the table and the explanation of the families, we must add a few more observations on their names and characteristics.

§ 739. 1st, ON THEIR NAMES.—Several of the most anciently and universally known families derive them from some one of their most apparent characteristics: as the UMBELLIFERÆ and CORYMBIFERÆ, from their method of inflorescence; the LEGUMINOSÆ and CONIFERÆ, from their fruit; the LABIATÆ and CRUCIFERÆ, from the form of the corolla; the PALMACEÆ and the GRAMINEÆ, from the appearance of the whole plant, &c., &c. But as to the rest, it is generally agreed to designate each family by the name of one of its principal genera, that, which we may consider as the type to which all the others approach, the termination of the Latin name of the genus being changed in another: ACEÆ (*acées*) (as in the RUBIACEÆ), INEÆ (*inées*) (as in the LAURINEÆ), IDEÆ (*idées*) (as in the CAPPARIDEÆ), ARIEÆ (*ariées*) (as in the ONAGRARIEÆ). The first of these terminations, that in ACEÆ, is most generally employed, and some authors, perhaps with reason, use it exclusively. It is generally agreed also to preserve the simple termination in EÆ (*eés*), which several names of families (JUNCEÆ) (*Joncées*) POLYGONEÆ) formerly assumed to designate divisions of an inferior order. Indeed, certain families are susceptible of being divided into several secondary groups, united by characteristics which are not yet regarded as sufficiently important to raise them to the dignity of a family: they are called *tribes* (TRIBUS). Thus, the MELIACEÆ (*Meliacées*) form a family, all the genera of which are united around the genus MELIA by certain common characteristics; but there are other characteristics, which are not common to all the genera of the family, and these present two combinations: the one set, which we find in the MELIA and some other genera; the other, which we observe in the rest and especially in the genus TRICHILIA. The MELIACEÆ may, therefore, be divided into two tribes, the MELIEÆ (*Méliées*) and the TRICHILIEÆ (*Trichiliées*). The tribes compose natural groups and are, consequently, like small families susceptible of being some day elevated to this rank, if it happens that by the discovery of rather a large number of fresh plants, the tribe, of which they form part, is able to assume sufficient size and importance to justify this dismemberment. Most of the tribes at first established by Jussieu in his families under the name of sections, have themselves since become families. It is, therefore, of little importance whether a group be a family or a tribe, provided that it is perfectly natural, so much the less, in fact, since all families are far from being of equal importance either on account of the number of the plants they contain, or on account of the value of the characteristics which distinguish them. In the subsequent enumeration of families, we have too little space to descend to the tribes, which we shall only indicate in some cases, when the characteristics, employed in our tables, separating two tribes of the same family from one another, will warrant our doing so. We shall also employ indifferently the several terminations we have just mentioned, preferring for each family that of the name under which it is best known. Let us advise those, who wish to study a few families from nature, always to choose well authenticated species of that genus, which gives it its name. They will be sure never to meet any of those exceptions which puzzle the student. Whatever possible changes take place in arrangement of the groups, it is very clear, for instance, that the Neem-tree (*Azedarach commun*) (Melia Azedarachta), the type of the genus Melia, will always be one of the Meliaceæ.

§ 740. 2nd, On their characteristics.—The characteristics of fructification (character fructificationis) are amongst the most important and are of essential service in defining the family. But we always unite with them those of vegetation, which as we have said, presenting some feature peculiar to each family, confirm those of reproduction and, in some cases, very much facilitate the enquiries. In this way, for instance, the simple opposite leaves with interpetiolar stipules aid us very much in recognising at a glance one of the Rubiaceæ. In the same way we employ in the description of the genera the characteristics of reproduction concurrently with those of vegetation. Linnæus only used the former reserving the latter for the distinction of the species.

Sometimes a family is described to its minutest details, so as not to omit any feature, this is its *natural character*. Sometimes we limit ourselves to the description of the characteristic features, those, the change of which distinguishes the family from every other: this is the *essential character*. We shall limit ourselves to the latter.

§ 741. But this characteristic results as we have just said, from the combination of several facts and not from a single isolated one. We should not, therefore, be contented with one of them, even if it be quite peculiar to the family: as, the tetradynamous stamens to the Cruciferæ. It would be trying to paint a portrait by representing a single feature of the countenance. We shall see from the tables, that it is necessary for understanding them, to retain in the mind, with the terms used to designate them, the organographic facts scattered throughout this book, especially those which we have related concerning the flower, concerning the symmetry of its parts and their insertions, concerning the situation of the seeds, and particularly concerning their structure, the different modifications of which furnish the most important characteristics.

§ 742. Let us, lastly, keep well in the mind the incompleteness of these tables, only intended to exhibit the prominent differences of organization in the families, but by no means to explain the whole of their organization. Constructed in the spirit of the analytical method (§ 703), they are necessarily more or less systematic, and do not always follow the natural order on account of the difficulty of making good and clear tables. Some families are, therefore, found a little out of the place that they ought to occupy. We have, nevertheless, endeavoured in this case to take them away from their proper place as little as possible, and to shew them at least in that group of the families, to which they have the most affinity; although even that has not always been allowable from the concessions which the establishment of certain large divisions necessitate: of that of the Diclinous plants, for instance. A few words, however, will be sufficient to point out these aberrations whenever they present themselves.

We shall not here repeat the characteristics which separate the three great branches or classes of the Vegetable Kingdom, since they have been explained in the different chapters of this book.

ACOTYLEDONOUS VEGETABLES.

§ 743.—We have already examined in a general way the organs of their vegetation (§§ 101, 109, 120, 152) and those of their reproduction (§§ 480, 613—620). It now remains for us to consider what divisions will be created by the different modifications of the organs in this great class. We have stated that some, very simple in their structure, only present cells, that others present fibro-vascular fascicles also; that some exhibit no distinction between the fundamental organs (stems and leaves); others have both stems and leaves. These preliminary ideas will assist us in understanding the following table; a few ulterior details will complete the explanation.

(Table I, page 550).

§ 744. Most of these groups are not so much families as classes, the numerous vegetables they contain being capable of being subdivided into secondary, and these into tertiary groups, which would correspond to so many families. We will not follow them as far as this degree of division, in as much as the simplicity of the organization would require a crowd of details, which this work would not contain, to explain the delicate characteristics, from which the distinction of these families would result. We shall content ourselves with some general observations on the most important of these classes and their principal divisions.

§ 745. ALGÆ[b] (*Algues*) (Algals).—The Algæ require an aquatic medium in which to grow: some few, it is true, are found on the surface of the earth, but only when it is extremely humid; almost all live immersed in water. We know under the general name of CONFERVA (*Conferve*) those, which inhabit fresh water; under that of FUCUS those, which inhabit salt water, and abound on the shore of the sea. But instead of this classification, which has obtained for a

[b] The English name of all these and subsequent divisions has been given on the authority of the *Vegetable Kingdom* by Dr. Lindley.—TRANS.

TABLE I.

FAMILIES.

ACOTYLEDONOUS VEGETABLES.

Structure	entirely cellular.	No axis,	no leaf, no foliaceous frond.	Aquatic plants		ALGÆ.
				Terrestrialplants.	No thallus	FUNGI.
					A thallus and thecæ	LICHENES.
			Leaves or foliaceous fronds			HEPATICÆ.
		An axis.	Leaves or foliaceous fronds,	capsule without operculum. Elaters. No Columella		
				capsule with operculum. A Columella. No Elaters		MUSCI.
			No leaves nor frond			CHARACEÆ.
	cellulo-vascular.	Reproductive organs	under scales forming terminal cones. No leaves. Sheaths around the stems			EQUISETACEÆ.
			solitary at the base of the leaves			LYCOPODIACEÆ.
			in groups on the surface of the leaves			FILICES.
			in receptacles under the form of fruits and situated near the origin of the roots			RHIZOCARPEÆ.

long time, we shall choose that, which is founded on the study of their structure and of their fructification; M. Decaisne has proposed such a one.

Some, as we have said, present the simplest organization that we can conceive, since they consist in a simple vesicle; in others, several vesicles are united end to end to form filaments, sometimes isolated, sometimes crowded, frequently with a certain regularity, so that they seem to radiate from a common centre. We have seen that these filaments are generally covered with a mucous covering, and this often forms an envelope common to the whole system of the filaments, so as to cause them to be united into a kind of individual. These cells isolated or united end to end, are filled with a green substance, each granule of which in the free vesicles may become a reproductive body. In certain cells of the most compound filaments, the green mass is separated at a certain time into several others (commonly four), and each of these small secondary masses represents a spore. The spores of these simple plants, escaped from the cell that has produced them, for some time enjoy movements analogous to those of animals (§ 619, [*fig.* 496—499]). We may, therefore, term these Algæ, ZOOSPOREÆ (*Zoosporées*) (from ζῶον, *a living being*).

In others much less numerous (which also consist of filaments formed of cells united end to end and filled with a green mass), one of the sides of these cells is lengthened at a certain time into a kind of pocket. The pockets belonging to two different filaments are united at their ends, then are pierced, so as to establish the communication of one cell with the other: the green mass of the one now passes into the other, is blended with that which it already contained, and thus forms the body which performs the functions of the spore. We have, therefore, a higher degree of complication since two distinct filaments are wanted for the formation of a spore, and we may designate those Algæ by the name of SYNSPOREÆ (*Synsporées*) (from σύν, which indicates *union*).

We then find a more intricate tissue: some, it is true, still consist of simple filaments; but in the others these cells and filaments are united to one another so as to form more complex bodies, which are lengthened into stems or are flattened into laminæ; these round or flat expansions, which we call the *frond* (*fronde*) (FRONS [*fig.* 502 *f*]), may be ramified a certain number of times, frequently by dichotomy. Some of their cells frequently project beyond the rest and are then seated on a kind of pedicel; and the substance con-

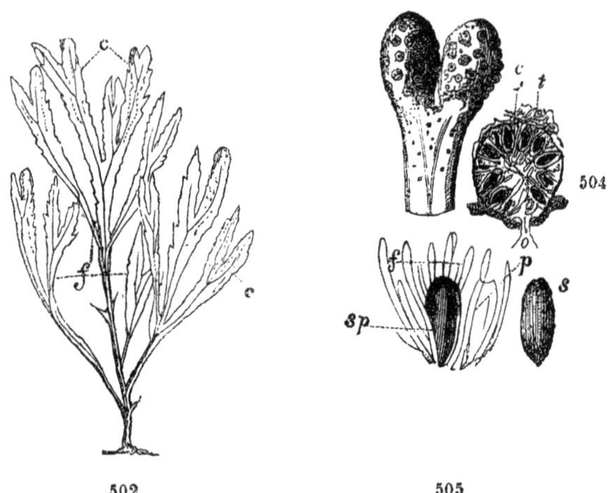

tained in these is organized into a spore to which the cellular membrane froms an envelope, *perispore* (*périspore*); the spore consists of a single nucleus covered with its own proper cellular membrane *epispore* (*épispore*) closely united to its substance, and continuing so after it has escaped from its mother cell. These Algæ may be called APLOSPOREÆ (*Aplosporeés*) (from ἁπλόος, *simple*). The spores are not always found at the surface of the frond, but they are frequently concealed in *conceptacles*, or cavities disseminated over this surface (*fig.* 503), which they continue by means of a small canal, called the *Orifice* or OSTIOLUM (*ostiole*) (*fig.* 504) by which they communicate with the outside.

This name is opposed to that of CHORISTOSPOREÆ (*choristosporées*) [from χωριστός, *separated*), applied to the following division which comprehends the most highly organized ALGÆ. In these, the

502. An Algal, the FUCUS SERRATUS. One of the APLOSPOREÆ. The whole plant is much smaller than nature.—*f* The Frond.—*c c* Conceptacles spread over the surface of its extremities.

503. The end of the frond, covered with conceptacles.

504. A vertical section of a conceptacle *c*, the inner surface of which is covered with spores.—*t* A part of the superficial tissue in which the conceptacle is buried.—*o* Orifice or ostiolum, by which it communicates with the exterior.

505. Spores, one *sp* still enveloped with its perispore; the other *s* has quitted the perispore *p*, which is represented on one side.—*f* Sterile filaments.

reproductive organs are of two kinds: some consist of a body projecting outwards, rather similar to the spore of the preceding, were it not for its forming a continuous mass and not being contained in a perispore, out of which it comes in order to germinate; others are formed in deeper cells at the expense of a mass at first simple, but afterwards divided into four spores. These commonly found in the CHORISTOSPOREÆ, really merit this name; the first although susceptible of germinating in the same way, are more analogous to bulbels. The whole plant presents the form of branches or of laminæ, and is always of a red colour, sometimes very brilliant; this colour changes into green, when the plant is exposed to the air. The APLOSPOREÆ are, on the contrary, green in the water, but lose their colour and become white, soon after they are taken out of it.

The simplest Algæ float freely in the water without being fixed to anything: the most highly organized may also exist under the same conditions, although more commonly they are fixed to the bottom and to the rocks by sucker-like expansions, resembling roots; but they are real grapples and not organs of absorption: for all these plants absorb through their surface the water that surrounds them and conveys their nourishment to them. They will, consequently, present in their composition the inorganic principles contained in the water. From this cause soda and iodine are found in great abundance in marine plants, which are collected and subjected to various processes in order to extract these substances from it. They secrete a mucilage, which in some species is sufficiently organic to be employed for the food of man. The nests of a kind of swallow, which are esteemed in China as very great delicacies, derive their properties from the FUCUS of which the bird constructs them.

§ 746. FUNGI (*Champignons*) (Fungals).—Whilst the Algæ live in water, the Fungi live in the earth or on its surface, very abundant in and upon animal and vegetable substances in decomposition. Although in some the organization is raised to a degree of complexity evidently superior to that of the Algæ, it descends in others as low as they do, to the very last degree of simplicity, as will be seen from the following classification, which is borrowed from Dr. Léveillé, whose researches have thrown so much light on the structure of these vegetables.

There are indeed some of them which consist of simple or

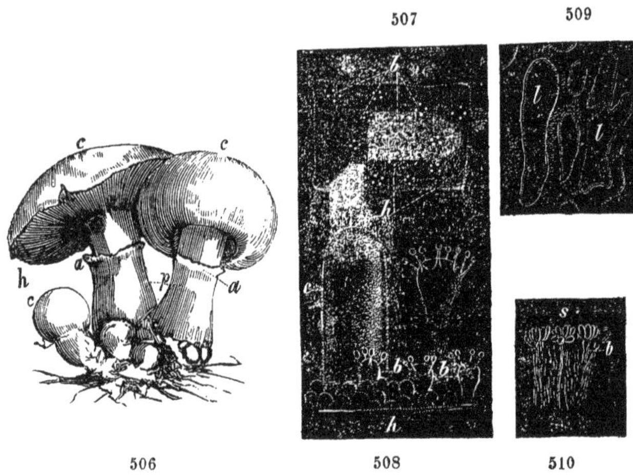

branched filaments, composed of articulations, which are last separated, sometimes throughout the whole length of the filament, sometimes only at its extremity. Each of these articulations is a spore, and, consequently, the vegetable appears to be composed only of organs of reproduction, which are confounded with those of vegetation. These Fungi may be termed ARTHROSPOREÆ (*Arthrosporés*) (from ἄρθρον, *a joint*).

Others, which we may call TRICHOSPORÆ (*Trichosporés*) (from θρίξ, τριχός, *hair*), present the same filamentose form, simple or ramified; but their spores, instead of composing the filament by being united end to end, are very distinct and are inserted either at its extremity, or lower down, sometimes singly, more frequently several together; arranged in a terminal fascicle or in whorls at

506. A group of Mushrooms (*Champignons de Couche*) (AGARICUS CAMPESTRIS), at various stages of developement.—*p* The stalk or STIPES.—*c* Cap or PILEUS.—*v* The Veil or VELUM, which at first unites the stalk to the cap, and afterwards, when the communication is broken, forms the annulus *a*.—*h* Lamellæ radiating under the inner face of the cap, covered by the HYMENIUM.

507. The upper part of the hymenium, on which we perceive the spores *s* in fours.

508. A small portion of the hymenium, very highly magnified and seen sideways. —*h* Its tissue.—*b* Basidia with their spores. One has been represented above, bearing a greater number.—*c* Cystidia.

509. A small portion of the cap of the CLATHRUS CANCELLATUS in the shape of trellis-work, with the hymenium which covers its lower face and is perceived on the open spaces *l* of the trellis.

510. The hymenium highly magnified, exhibiting the peculiar form of the basidia *b*.—*s* Spores.

equal distances, or lastly spread over the whole of the surface from the bottom to the top.

In others the spores are no longer placed on the outside, but are enclosed in membranous vesicles, which terminate simple or branched, continuous or articulated, capillary filaments. These vesicles, are, therefore, real sporangia (§ 614), which begin to indicate a much higher degree of organization; at a certain time, they open to allow the contained spores to escape. This may be easily observed in common Mould (*Moisissure*). We shall call these Fungi CYSTOSPOREÆ (*Cystosporés*) (from κύστις, *a bladder*).

We next find simple or branched filaments, each filament or each branch terminated by a single oval or round, simple or articulated, spore. But all these filaments are attached to a common body or receptacle, to which it has been agreed to give the name of STROMA (*στρῶμα, anything spread for resting upon*). Hence is derived that of STROMATOSPOREÆ (*Stromatosporés*) which is applied to these Fungi. The stroma, sometimes fleshy, is extended into a plane or concave surface, thus allowing the spores to project outwards; sometimes coriaceous or membranous, it is bent over them so as to enclose them in a cavity, which opens at the top by a pore. Sometimes the pores of several stromata grouped into a circle, end at the same circle, which thus seems to be an opening common to all. The stroma is sometimes raised on a footstalk narrower than it is, more commonly sessile[c].

Let us suppose, instead of the sporiferous filaments, a sac, either globular or lengthened into a club-shaped mass or into a cylinder and containing in its interior four or eight free spores, or, in a single word, what we termed (§ 618, *fig.* 494) the THECA, and these thecæ inserted on a common receptacle, which, as in the preceding case, either supports or completely envelopes them: we shall then have Fungi which are called THECASPOREÆ (*Thécasporés*). Here the receptacle, in general much more developed, no longer bears the name of stroma. In its relation to the thecæ, it presents the same series of modifications, which we described (§ 209) in the inflorescence of phanerogamia as taking place between the flowers and the axis which bears them. Thus, the receptacle of the THECASPOREÆ may be covered with thecæ over the whole of its external surface, as in the GEOGLOSSUM; or only at its swollen top; or over the upper face of this extremity hollowed out like a cup, as in the PEZIZA; or else this cup on the upper surface is closed over the

[c] The filaments attached to the stroma are called FLOCCI.—TRANS.

thecæ, which are then concealed in the cavity, till the spores escape from an opening in the top; or, as in the Truffle (*Truffe*), remain closed and allow the spores to escape only after decomposition. The thecæ are frequently mixed with very long empty cells or PARAPHYSES.

Lastly, we find the most perfect Fungi, and amongst them those, whose forms are the most familiar to us and are best known by the name of Mushrooms (*Champignons*). We still, however, observe something analogous to the preceding; the club-shaped, ovoïd, spherical or cup-shaped mass; the most common and remarkable (*fig.* 506) is that of a dome or cap (*c*), supported by a stand or stalk (*p*) more or less narrow, more or less lengthened. But that, which eminently distinguishes these Mushrooms, is the form of their reproductive organs. They are small, round bodies, terminated by two or more, frequently four points, each of which supports a spore at its extremity. These bodies have been termed BASIDIA (*basides* [*fig.* 508]), and the Fungi thus furnished, BASIDIOSPOREÆ (*Basidiosporés*). Rather frequently, but not constantly, along with these basidia are found a small number of other vesicular bodies commonly larger than they are, transparent, apparently filled by a liquid without granules or spores; they are then termed CYSTIDIA (*cystides*) (*fig.* 508 *c*). Some authors regard them as being designed for the fecundation of the spores, and performing with regard to them the functions of stamens; but, then we ought to find them in all the BASIDIOSPOREÆ, which is not the fact: they are probably analogous to the paraphyses. These basidia and cystidia like the thecæ in the preceding case, are situated either on the outside or in the inside. When situated in the inside, they are mingled (as in the SCLERODERMA) with the cells, to the walls of which they are glued, or cover the surface of the larger lacunæ (as in the LYCOPERDON); when situated on the outside, they are sometimes covered with a mucilaginous layer (as in the PHALLUS): but more frequently externally free, they are spread over the whole surface of the receptacle lengthened into a mass, and ramified like a tree (as in the CLAVARIA [*Clavaires*]), or else only over its lower face. The receptacle then generally assumes the form of an umbrella or cap, beneath which are situated laminæ radiating from the centre (as in the Agarics [*Agarics*]), or veins (as in the CANTHARELLUS), or tubes (as in the BOLETUS [*Bolets*]), or points (as in the HYDNUM), or lastly, a smooth surface or one bristled with short papillæ, (as in the THELEPHORA [*Téléphores*]). This surface or that of the points, of the veins, of the laminæ (otherwise called LAMELLÆ), in the interior of the tubes, is that which is covered with the basidia.

Various terms, in addition to those we have already mentioned, have been adopted to indicate all these different parts and thus abridge the descriptions of the Fungi. Thus, the layer formed by the reproductive bodies, the basidia or thecæ, is called the HYMENIUM. The simplest Fungi, such as we have described, are almost reduced to this hymenium and sometimes even to a fragment; in those which are more complicated, a layer of another tissue belonging to the system of vegetation is added and forms the receptacle; this receptacle now increases gradually and at last presents several parts to our notice. If it is perfectly closed, it is a PERIDIUM. But, even in the umbrella-shaped FUNGI, the *cap* (*chapeau*) (PILEUS), when quite young, forms a close cavity by a membrane called the *veil* (VELUM [*fig.* 507 *v*]), which connects its margin with the stalk or STIPES; when this is broken, it forms around the stalk a kind of little collar or an annular cicatrix only (*a*), ANNULUS (*anneau*); sometimes also, at its first appearance, a cellular sac, the *wrapper* or VOLVA envelopes the whole Fungus from its base, around which it is inserted; it is then irregularly torn and allows the Fungus to be developed.

What we have described does not constitute the whole of the Fungus; it is, as it were, only its inflorescence. Before this part was developed, we see filaments radiating from a centre (probably from the spore in germination) crossing one another in every direction; they at last agglomerate and are closed together at certain parts where the apparatus which we have described are formed. We term this filamentose net, of which rudimentary matter the Fungi is composed, MYCELIUM. It is generally concealed under the earth and escapes our notice on account of both its situation and its fragile texture. It is not rare to perceive it on the surface of humid and obscure places, on the floors and ceilings of our cellars, for instance. This mycelium is a kind of subterranean tree which exposes to day only its extremities loaded with the organs of reproduction, so that in general all the Fungi which grow in the neighbourhood of one another, really belong to a single individual: thence, the arrangement in a circle, which they frequently affect, the mycelium being regularly developed in a homogeneous medium and throwing out all its rays to the same distance.

The tissue of the FUNGI is a kind of felt composed of cells, some round, others lengthened and united end to end into tubes. The hymenium is often formed by the extremity of these tubes, some of which are terminated by thecæ, basidia or cystidia, so that these separate filaments really represent more simple Fungi, the CYSTOSPOREÆ, or TRICHOSPOREÆ.

The membrane of these cells is of the same nature as that of all other vegetable walls: it is cellulose. The tissue of the Fungi was formerly believed to be composed of quite a different substance, containing a large proportion of azote, which has been termed *fungine.* But this substance is foreign to the wall, and doubtless belongs to the substances which fill or penetrate it. They are found to be very superior to the Algæ in their secretions, among which we remark albumen, sugar, a fatty substance and several acids, without reckoning several peculiar to them, to which, doubtless, they owe their well known properties. It is a result of their composition that they should grow extremely fast, and, after a very transitory existence, should be decomposed, causing phenomena and producing substances very analogous to those we observe during the decay of animal matter.

They display very varied and sometimes very brilliant colours, but it is exceedingly rare to find green. They live and are coloured in darkness equally well as in light, and act on the atmosphere like all parts tinged with other hues than green. They vitiate the air very rapidly by absorbing its oxygen to form and exhale an equal quantity of carbonic acid gas. It is remarkable that in pure oxygen they absorb it, causing one part to combine with their carbon, and then restoring it under the form of carbonic acid gas; at the same time they preserve another part, which seems to replace in their tissue rather a large quantity of azote, which they then exhale. In an atmosphere of azote, they scarcely modify this gas. They, therefore, borrow this principle, so abundant among them, from the earth. This, however, is only what might be expected from seeing them grow on organic substances during the process of decomposition.

Every one knows that Mushrooms present both food esteemed for its delicacy and extremely dangerous poisons. There are not unhappily any characteristics by which we can distinguish the poisonous from the harmless; they ought to be used with more prudence, since the experience of others is not always decisive. It appears, indeed, that the mode of cooking several of them causes the effects they produce. We destroy the noxious qualities of certain species by cooking them in salt or infusing them in vinegar; this seems to prove that in a case of poisoning, it is necessary to refrain from salt or from vinegar, which, dissolving the venomous principle, would spread it with much more rapidity throughout the whole of the body.

§ 747. LICHENES, (*Lichens*) (Lichenals).—Lichens form those dry, foliaceous expansions, found on stones, the ground, the bark of trees, which they cover with varied tinges, which are peculiar to them. The expansion (which we term the THALLUS) of the Lichens has sometimes the consistence of fine dust and then it has no fixed form. At other times it forms a kind of crust of a more regular

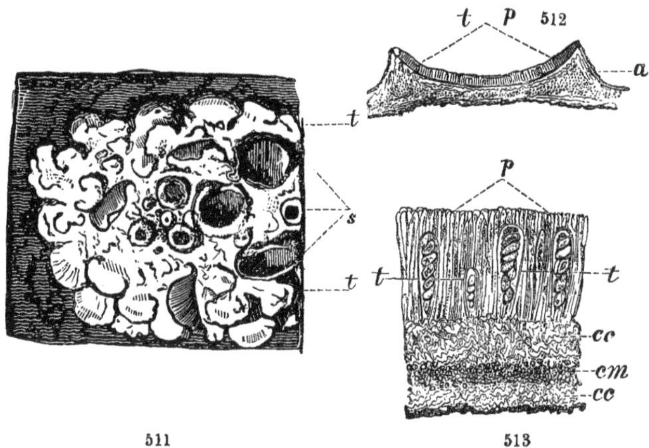

form, analogous in its consistence to the stroma of certain Fungi. Lastly, it may be extended into a lamina, the outline of which is clearly circumscribed, frequently divided into lobes, which are divided by a kind of dichotomy, or else are lengthened into simple or branched filaments. We recognise in the tissue two kinds of cells; the one, short, with thick walls, generally closely united to one another; the other lengthened into loosely felted filaments. The former alone can be observed in Lichens of a pulverulent or crustaceous consistence; in others, they form only the *central* or

511. A Lichen, one of the Hymenothalameæ, the PARMELIA ACETABULUM.—*t* Thallus.—*s* Apothecium in the shape of shields, at various stages of developement.

512. A vertical section of the Apothecium, magnified so as to exhibit the layer *t p* formed by the union of thecæ and paraphyses.

513. A small portion of the Apothecium, more highly magnified.—*c m* Medullary layer.—*c c* Cortical layer.—*t t* Thecæ at various stages of developement.—*p* Paraphyses.

medullary layer (*couche médullaire*) (*fig.* 513, *cm*) over both sides of which is extended a *cortical layer* (*couche corticale*) (*c c*) composed of filiform cells. Some descending from the lower surface like small filaments, serve to fix the lichen to the body on which it grows, as if they were roots. They, however, are like them in appearance only, since they perform none of their functions.

In the organs of reproduction, the Lichens have a close affinity to the THECASPOREÆ Fungals: for here they are also thecæ containing spores, in number either two or one of its multiples, most commonly four or eight, sometimes twelve or sixteen. They are congregated in groups sometimes immediately on the substance of the thallus, which thus in places forms the receptacle, sometimes on a peculiar intermediate substance (*fig.* 512, *a*). This receptacle is elevated around the groups into a projecting border, composed at the expense either of the thallus or of a peculiar substance, or of both at once. This sometimes forms a simple margin around it; sometimes, extending beyond the thecæ, it closes above them so as to retain them in a cavity, and then assumes the name of PERITHECIUM. It frequently envelopes them completely, but only when they are quite young, then bursts open and expands itself. Among the thecæ are mingled sterile filaments or paraphyses, which, longer and united at their top, connect all this system into one mass. This mass, with its receptacle, evidently represents that of the Fungals with their hymenium, and here assumes the name of APOTHECIUM. The rest of the thallus will, therefore, correspond to the mycelium, and thus establish an essential difference between these two classes.

We may divide the Lichens into several groups: those, in which the receptacle is formed by the thallus itself, CONIOTHALAMEÆ (*coniothalamés*)[d]; those, in which it is composed of a peculiar substance, IDIOTHALAMEÆ (*idiothalamés*)[e]; those, in which the perithecium is closed, GASTEROTHALAMEÆ (*gastérothalamés*)[f]; those, in which it is open, HYMENOTHALAMEÆ (*hyménothalamés*)[g]. The apothecium has frequently been designated by other names according to the different forms which it assumes: by those, for instance, of DISCUS, SCUTELLUM, TUBERCULUM, GLOBULUS, which are easily understood: or again by that of LIRELLA when, linear and flexible, it is opened by a longitudinal slit.

[d] κονία, *dust;* and θάλαμος, *an inner chamber, surrounded by other buildings.*—TRANS.

[e] From ἴδιος, *private, peculiar.*—TRANS.

[f] From γαστήρ, *the paunch.*—TRANS.

[g] From ὑμήν, ὑμένος, *a skin, membrane.*—TRANS.

The Lichens differ again from the Fungi in their being persistent for a very long time and, being attached to inorganic bodies, living or dead, but never in putrefaction, whilst they seem to require air and light. They rarely present however, the green colour, although all assume it without distinction when they are moistened or damp; and their tissue, properly dry, fragile or coriaceous, becomes moist, flexible, and easy to be torn.

The tissue of several Lichens is profitably employed as the food of man in certain cases and of animals in certain countries; the CENOMYCE RANGIFERINA is the principal food of the rein-deer during winter in Lapland. The CETRARIA ISLANDICA (*Iceland Moss*) ([*Lichen d'Islande*]), STICTA PULMONACEA and others furnish a wholesome and nutritious jelly, the use of which is advantageous in certain states of health. The cellulose, which forms the walls of the medullary layer, isomeric as we know (§ 299) with the fecula, approaches its properties as much as possible in these vegetables; it is even coloured blue by iodine in them. This, diluted into a jelly by a certain proportion of water, and mixed with a slightly bitter principle contained in the cells, furnishes in the Lichens a mild and slightly tonic aliment. Different species are remarkable for the abundance of colouring principles, which, however, require preparation to render them apparent. Naturally, in fact, their tissue is greyish; but after they have been acted upon by an alkali (potassa, or urine, so rich in ammonia) a red colour is obtained, then if we add potassa till it is in excess, blue. The Orchall (ROCELLA TINCTORIA) is very much employed in this branch of manufacture. Several other Lichens would furnish the same principle, but in a less proportion.

§ 748. HEPATICÆ (*Hépatiques*) (Liverworts).—The Hepaticæ form with the Mosses a natural class very distinct from all the preceding in the nature of the tissues, in which we see the chlorophyll appear in the interior of the cells. The surface is covered with openings or stomata adapted for putting them in communication with the atmosphere. By this entirely different structure the frond of the Hepaticæ is distinguished from that of the Lichens. Sometimes this frond bears the reproductive organs either buried in its substance near its superficies, as we have seen in the RICCIACEÆ (Crystalworts) (§ 615, *fig.* 493), or projecting beyond this superficies: at other times (in the MARCHANTIA) they are raised on a pedicel, which seems to be the first appearance of the axis, though not yet bearing leaves.

Lastly, in a large part of the JUNGERMANNIACEÆ (Scalemosses), we find an axis covered with small leaves of an entirely cellular texture, in the middle of each of which a series of other elongated cells begins to represent the median nerve.

The reproductive organs are frequently of two kinds: antheridia, which we have explained (§ 480), and sporangia, of which we have given some idea (§§ 614, 615, *figs.* 491, 493). A third kind may be added: these are small green, cellular bodies, attached by a contracted piece to the surface of the frond, which is raised around them something like an involucrum; they may be compared to bulbels.

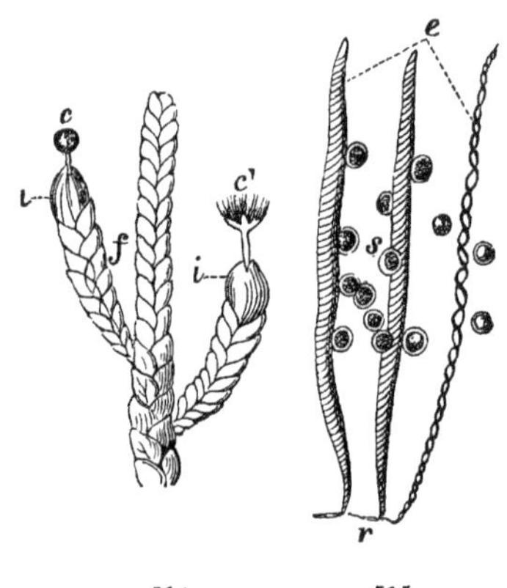

In the interior of the sacs or sporangia containing the spores we find utricles of two kinds: some, short, in which are formed the spores just like the pollen-grains (§ 615); these gradually disappear by absorption, leaving these spores (*fig.* 515, *s*) free in the interior of the cavity when they have arrived at maturity; the others (*e*), much longer and tapering like a spindle, in which only a few small green granules are found; their wall though at first continuous, is at last broken into a double spiral strap very similar to the thread of the tracheæ. It is in this state that they assume the name of *elaters* [h] (*élaterès*). By the motions which the atmospheric changes impart to these threads, so sensible of hygrometric variations, they assist in disseminating the spores arranged around them. Sometimes the sporangium is withered, sometimes by a real dehiscence it is separated into several valves (*fig.* 514, *c'*). This takes place in the JUNGERMANNIA, in which this sporangium,

514. A fragment of the JUNGERMANNIA TAMARISCI.—*f* Branches covered with imbricated, distichous leaves, the two lateral each bearing a capsule elevated on a thread-like support, surrounded at its base by an involucrum formed by the membranous envelope of the sporangium.—*c* Closed capsule.—*c'* A capsule opened by the dehiscence.

515. A point *r* of the receptacle bearing a few elaters *e*, one of which is separated into two spiral threads. Close to them we see free spores *s*.

h From ἐλατήρ, *one that drives away*, or *expels*.—TRANS.

at first developed in another sac, breaks it and is elevated on a pedicel more or less lengthened.

§ 749. Musci (*Mousses*) (Mosses).—Every one knows these small and elegant vegetables so abundant on the surface of the earth, of rocks, of the bark of trees, which they clothe with a green mantle; sometimes growing under water. On examining them closely, we find them to be formed of slender, simple or branched stems, covered with thin leaves, the texture of which is the same as that we have just described in the Hepaticæ. Their reproductive organs are also of

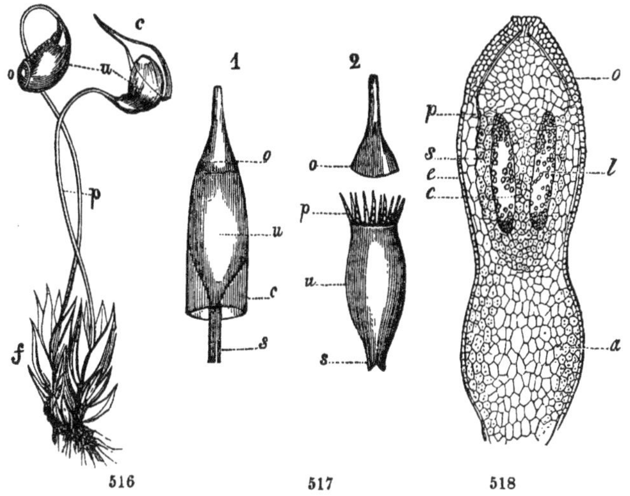

two kinds: 1st, antheridia (§ 480, *fig.* 352) grouped in the midst of terminal rosettes of leaves either situated on their axis, commonly

516. The Hygrometric Cord-Moss (FUNARIA HYGROMETRICA), slightly magnified. —*f* Leaves.—*u* An urn placed at the top of a long thread or pedicel *p* —*o* Operculum. —*c* Calyptra which is still in its place on one of the urns but has fallen from the other.

517. Urn of the ENCALYPTRA VULGARIS.—*u* Urn.—*o* Operculum.—*s* Top of the pedicel.—1. Before dehiscence, whilst still enveloped with the calyptra *c*.—2. After dehiscence, when the detaching of the operculum has liberated the peristome *p* terminated by 16 ciliæ or teeth.

518. The longitudinal section of a very young urn of the SPLACHNUM.—*a* Apophysis.—*c* Columella.—*s* Cavity or loculus around the columella filled with spores.—The tegument of the urn is formed from the outside to the inside by several different layers of cells: the first *c*, which forms the epidermis and is thickened at the summit to form the operculum *o*; two intermediate ones, whose top will afterwards be divided to form the peristome: an inner one *s*, which forms the wall of the loculus or sporiferous sac.

mixed with sterile filaments or paraphyses; 2nd, sporangia of a peculiar form. These, when quite young, single or several united together, sometimes separated from the antheridia on different plants or situated on some other part of the same plant, sometimes surrounded by these bodies, represent so many sessile sacs in the form of a bottle. Of several sporangia thus arranged, one is commonly developed whilst the others wither. This lengthens, and, as it does so, bursts open the external sac that envelopes it; this it carries along with it placed on its top like a night-cap, whence has been derived its name of CALYPTRA[i] (*coiffe* [*fig.* 516 *c*; 517 *c*]). We then distinguish two portions in the developed internal part: one lower and slender *pedicel*, sometimes termed the *fruit-stalk* or SETA (*soie* [*fig.* 516 *p*]); an upper swelling, globular or ovoïd, frequently in the shape of an urn, called the *capsule*, THECA (*thèque*) or *urn* (*fig.* 516 *u*). The interior of the capsule presents a cavity traversed at the centre by a kind of solid axis, the COLUMELLA (*columelle*) (*fig.* 518 *c*); the capsule is quite filled round this axis with a number of small spores become free by the absorption of their mother cells, the tissue of which at the beginning united the columella to the walls of the capsule. This, when ripe, is opened like a pyxidium by the separation of a conoid covering or OPERCULUM (*opercule*) (*o*), for a long time concealed under the calyptra; after the fall of the latter, however, this operculum is easily distinguished from the rest of the capsule by an annular furrow. When the dehiscence, as it were, takes place, it leaves this open at the top; this opening bears the name of PERISTOMIUM (*peristome*). The peristomium is surrounded by a border sometimes undivided or *naked* (*nu*), sometimes furnished with small teeth (4-64 in number) (*fig.* 517 *p*) frequently lengthened into straight or twisted hairs. These teeth are in one or two circles: whence the peristomium is said to be single or double; and these two circles terminate two cellular layers, which, under a thin epidermis, compose the wall of the capsule. It is very remarkable that these teeth are constant in their number in a given space, and always some multiple of 4, namely 4, 8, 16, 32, 64. From their texture and their hygrometric motions we may suppose they play the same part as the elaters in the Hepaticæ, which are totally wanting in Mosses. The peristomium is sometimes, though rarely, closed by a membrane stretched horizontally across it, the EPIPHRAGMA[j] (*épiphragme*). The sporiferous cavity does not

[i] From καλύπτρα, *a covering, especially a woman's veil.*—TRANS.

[j] Called also *Tympanum.*—TRANS.

occupy the whole of the capsule, the lower part of which, frequently solid, assumes the name of APOPHYSIS (*apophyse*).

The Mosses, like the Hepaticæ, do not secrete any remarkable product and are not employed for any purposes important enough for us to mention in this work.

§ 750. CHARACEÆ (*Characées*) (Charas).—We shall not stop here to treat of the Characeæ, the organs of vegetation of which we have already explained. They are remarkable for a degree of simplicity, which approaches closely to the Algæ, and for the decided rotatory motion of the juices contained in the cells (§ 282). The spore is composed of a mass of granules surrounded with several spiral tubes terminating at the top in five small teeth; the antheridium, situated below, is surrounded by fascicles of tubes (§ 480, *fig.* 353) united in the inside of a small globular box.

§ 751. EQUISETACEÆ (*Equisétacées*) (Horsetails).—These are readily distinguished from all other Acotyledons by the structure of their stem (§ 109), the arrangement of their branches outside a sheath embracing each articulation, around which they grow in whorls, and by that of their reproductive organs. The stem is terminated by a kind of cone formed by the union of a large number of scales in the shape of nails (*fig.* 520), perpendicular to the axis. Under the head of each of these nails there grows a circle of small sacs (*capsules* or THECÆ), each of which (*fig.* 521), split lengthways when ripe, allows a number of spores to escape. Each of these is in the form of a cellular mass, from the bottom of which spring four elastic threads (*fig.* 522), the motions of which assist the seed in its escape. At first the sac was filled by a continuous utricular tissue;

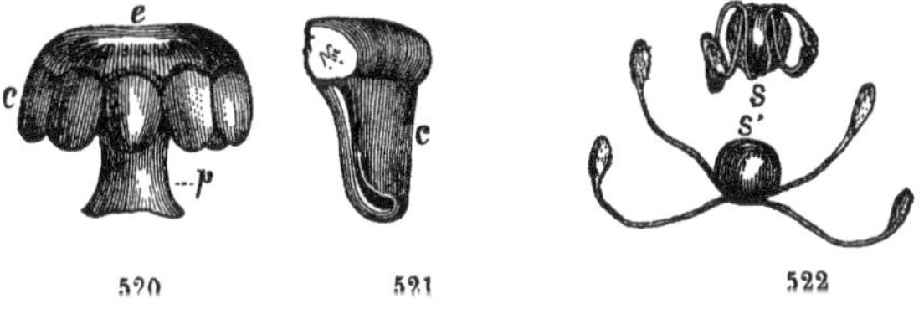

520. A scale *e* separated from the terminal cone of an EQUISETUM, with the whorl of capsules *c* which it bears and the contraction *p* by which it is attached to the common axis.

521. A capsule *c*. The inner side is next to us; it opens on this side by a slit.

522. A spore *s* with its four threads rolled in a spire around it.—*s'* The same with the threads unrolled.

then the utricles are separated and divided into spires, only adhering to the granular substance at one point of adherence so as to form these four threads. We have, therefore, a characteristic totally exceptional; the formation of a single spore in each mother cell, the wall of which, instead of being absorbed, remains to form the elater.

§ 752. Filices (*Fougères*) (Ferns).—We have already been occupied with the characteristics of vegetation of this large group

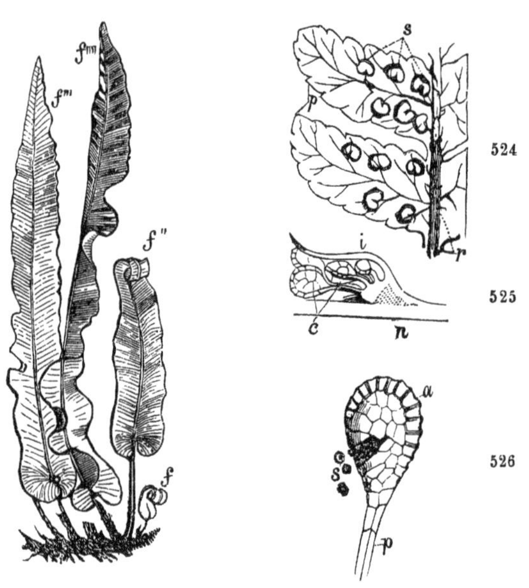

of Acotyledonous plants, with its stems (§§ 103—108), which in the species of our temperate climates grow under the earth, but which in several of those of tropical regions rise upwards into a

523. A plant of the Hart's-tongue (*Scolopendre*) (Scolopendrium officinale), with several fronds f f'' f''' f'''' at different stages of developement. On the lower surface of f'''' we find the sores in blackish transverse lines.

524—526. Fragments of the frond of another Fern (Nephrodium angulare). The under side.—*p* Two pinnules bearing sores *s*.—*r* Rachis which bears them.

525. A vertical section of a sore.—*n* The nerve which bears it.—*i* Indusium or fold of the epidermis which covers it.—*c* Capsules.

526. One of the capsules at the moment of dehiscence.—*s* Spores which are escaping from it.—*a* The cellular ring or annulus.

perpendicular trunk (*fig.* 117); with its adventitious roots (§ 120); with its leaves (§ 152), sometimes entire, but frequently very much divided. These leaves present the following constant characteristic: before they are developed, they are rolled up like a shepherd's crook, not only the general limb on the common petiole, but all the lobes (which are termed *pinnules*) on the partial petioles, so that, whilst the frond is quite young, the under surface is always concealed. We have already said a few words (§ 242) on the peculiar scarious hairs, that is, hairs expanded into scales or membranes, which are scattered in great abundance over the surface of the different parts: they also furnish useful characteristics for determining the genera and species. As to the reproductive organs, they are only small cellular sacs, or *capsules*, filled with spores and always situated on the lower surface of the leaves. The capsule-bearing leaves are sometimes of the same form as those which do not bear capsules; sometimes assume one a little different, in which the foliaceous parenchyma is much less developed and even almost completely disappears, leaving the nerves naked and covered with capsules.

These generally present in their cellular wall, one row of cells much larger and thicker than the rest, arranged end to end in a ring. The ring sometimes entirely surrounds the capsule, following a direction either vertical, as in the POLYPODIACEÆ (*fig.* 526), or horizontal or oblique, as in the HYMENOPHYLLEÆ. At other times surrounding it incompletely, it only forms a fragment of an oblique ring, as in the PARKERIACEÆ. Its physiological functions seem to be analogous to those of the elaters; that is, presenting more resistance than the rest of the walls, and contracting or expanding by the very process of growth or by its hygrometric changes, it determines the irregular rupture of these walls at some other part and by its motions forces the contained spores outwards. The dehiscence does not always take place in this manner; there is sometimes a regular split which opens the capsule either on one side, or quite round it, thus separating it into two valves. In this last case, we either observe an incomplete ring, as in the OSMUNDACEÆ, or none at all, as in the OPHIOGLOSSEÆ, in which these bivalved capsules are sometimes united in series by their sides. Lastly, they have a peculiar coriaceous consistence and are arranged in a circle, on the inner side of which they open, as in the MARATTIACEÆ.

The capsules are not scattered over the lower surface of the leaves singly, but in groups, which are called SORI (*sores*) (*fig.* 523 *f' s*).

These sori are found in different forms: sometimes they are circular, as in the POLYPODIUM (*Polypode*), sometimes more or less lengthened, as in the ASPLENIUM (*fig.* 523 *f' s*); sometimes separated from one another, sometimes arranged in a longitudinal series. Their position also varies with regard to the leaf under which they are scattered with more or less regularity either on the surface, or along the edge, as in the ADIANTUM; they may be arranged in continuous series on the margin, as in the PTERIS, or on the median nerve, as in the BLECHNUM.

They sometimes are found naked on the surface of the leaf, as in the POLYPODIUM: but more frequently a fine membrane, which seems to be a fold of the epidermis, is detached to cover them, and this is what is called the INDUSIUM. This sometimes forms a kind of collar (*collerette*) or cup (*cupule*) which surrounds the sori, as in the CYATHEA; but more frequently it covers them like the lid of a box fastened by hinges (*fig.* 525 *i*), and, in this case, continuous with the epidermis on one side, it presents on the other a free edge, which may be turned towards either the middle or the margin of the leaf (*fig.* 524). The indusium is attached by a single point, as in the NEPHRODIUM, or by a longer line, as in the ATHYRIUM. All these characteristics drawn from the shape of the sori, from that of the indusium, from the point of its attachment, from its shape and its direction, are employed in the distinction of the genera.

The capsules themselves considered separately, are either sessile or situated on a more or less elongated pedicel (*fig.* 526). The spores are formed in their interior in the same manner as those of the Cryptogamic plants we have already examined, that is, by fours in the mother cells, which at first form a continuous tissue, and afterwards being absorbed leave the spores free in the cavity of the capsule.

The spores by germination are lengthened into a filament composed of cells joined end to end, and this last, by the addition of cells formed on the side, is not long in being widened into a foliaceous expansion which may attain considerable dimensions. This expansion, close to the point where it begins, throws downwards radiculary fibres, upwards an axis with leaves. It has been compared to a Cotyledonous plant by several botanists, who, consequently, under the name of *Monocotyledonous Cryptogamia* separate the Ferns from the great branch of the Vegetable Kingdom, which is now occupying our attention. This mode of developement, however, will not sustain a rigorous comparison with the structure and germination of a true Monocotyledonous embryo; but is, on the contrary, perfectly analogous to that of the Acotyledonous, especially of the Hepaticæ, of which we have previously treated.

Antheridia are said to have been discovered in Ferns, but botanists do not agree on their nature and even on their existence. Some give this name to hairs scattered over different parts of the very young leaf; these hairs are swollen at their summit and filled with a granular substance; others, to small bodies which are sometimes found mingled with the capsules in the midst of the sori, and most frequently situated on the very pedicels of these capsules. They are lenticular and are filled with a substance, the granules of which, when thrown into water, exhibit active motions: but they have been discovered in a very few Ferns only, and, if they are organs really necessary for fecundation, we should of course find them in all.

In several Ferns of warm countries, the rhizomes contain a nutritious principle, which may be usefully employed as very wholesome food; but in those of our climate, the mucilage is mixed with another bitter principle, sometimes stimulant and even purgative, which renders them improper for food, but valuable for medicine. Certain species furnish good anthelmintics, that is, remedies for intestinal worms. This property is weakened or disappears in the leaves, in which an aromatic principle, being joined to the mucilage, communicates to it fresh properties.

§ 753. LYCOPODIACEÆ (*Lycopodiacées*) (Clubmosses). — These plants to a certain extent occupy a middle station between the Mosses, reminding us of the simple cellular texture of their leaves, and the Ferns, to the stems of which (§ 102) they bear great similarity. Their reproductive organs consist of small, yellowish, solitary sacs at the base of the leaves and are of two kinds: some filled with a number of small granules, which are arranged in fours in the mother cells forming at first a continuous tissue; the others OOPHORIDIA (*Oophoridies*), capsules enclosing four much larger bodies only. The former have been compared to antheridia, but their structure is in all respects similar to the sporangia of other Cryptogams; and, besides, they are found alone on a large number of LYCOPODIA (*Lycopodes*), which have no other means of propagation.

§ 754. RHIZOCARPEÆ (*Rhizocarpées*) (Pepperworts).—This family includes plants of very different appearances: as the Pill-wort (*Pilulaire*) (PILULARIA) with filiform leaves; the MARSILEA with long petioles terminated by four folioles; the SALVINIA with oval sessile leaves. These leaves are rolled up in the state of prefoliation into a shepherd's crook like those of the Ferns. The reproductive organs are small sacs, some enclosing very small granules, which have been considered to be antheridia: others, larger bodies, which have been considered as spores. These sacs are grouped with respect to one another in various ways in the same envelope or capsule (MARSILEACEÆ [*Marsiléacées*]), or separated in different capsules (SALVINIEÆ [*Salvinées*]); and these capsules, which remind us of small fruits, open by several valves and grow from or beneath the base of the leaves, in every case close to the commencement of the roots: whence comes the name of the family (from ῥίζα, *root*; καρπὸς, *fruit*). Very distinct from the Ferns, with which they were formerly confounded, they are, however, sufficiently allied to be regarded as forming part of the same class.

MONOCOTYLEDONOUS VEGETABLES.

§ 755. Their stems (§ 91—100), their roots (§ 119), their leaves (§ 150), the symmetry of their flower (§ 372), its envelope (§ 416, 417), their embryo (§ 579), and its method of germinating (§ 609), have been examined in a general manner, and in several other places the different points, which distinguish them from Acotyledons on one hand and Dicotyledons on the other, have been pointed out; we shall, therefore, refer the student to these passages to save us from repeating them. It now remains for us to explain those facts only which we learn from examining the different families in particular.

Jussieu divided them into hypogynous, perigynous and epigynous. We shall not follow this mode of division here, because the distinction between the first and the second of these ways of insertion of the stamens is not very clear in several of the families of Monocotyledons; in the Liliaceæ, for instance. The structure of the seed seems to us to furnish the first great distinction, being more constant and more important than any other. In the greater part this seed is provided with a perisperm, generally very thick, whilst in others it is entirely wanting; and these groups have very clear relations to one another. One of these relations is their living in water; and we may by that distinguish them from some other Monocotyledons without perisperm, although belonging to the former group: the Orchideæ, for instance. These are found in totally different habitats, living on the earth and on the trees. We have, therefore, the first division:—

SEED		
	without a perisperm. Aquatic vegetables	Table II.
	with a perisperm, except in some terrestrial vegetables	Table III.

Let us remark that these two groups do not follow one another in the natural series, but are rather parallel; in both, we gradually rise from the most simple flower, that is, one reduced to a stamen or a carpel, to the most compound, that is, to those which present all the whorls of the organs united together.

(**Table II, page 573**).

§ 756. We have elsewhere defined (§ 566, 580) the different epithets applied in this table to the embryo. This macropodous embryo, that is, one with a radicle very much developed with regard to the cotyledon, is, as we see, an almost general characteristic in the whole of this group of families with seeds without perisperm; for we also find it in the last three. The radicle (or rather the tigelle) thus elongated and swollen, commonly presents a tissue very rich in fecula and may thus, in supporting the young embryo, perform the physiological functions, with which either the cotyledons or the perisperm are commonly entrusted. We find the tigelle acquiring very remarkable dimensions in the Zoosteraceæ (*Zoostéracées*), forming most frequently a lateral excrescence, of which the greatest part of the substance of the embryo is composed. The same thing appears to take place in the Lemnaceæ (*Lemnacées*), commonly called Duckweeds (*Lentilles d'eau*), in which this mass surrounds on all sides the cotyledon concealed at the bottom of an internal canal, with which it is pierced at its centre.

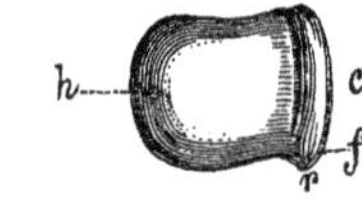

527

The envelopes of the flower are wanting in the greater part of these families; they begin to appear in the Juncagineæ (*Juncaginées*) (Arrowgrasses); we may here observe the passage from the inflorescence to the flower, as we explained (§ 385), when we examined two of their genera: the Lilæa and the Triglochin. In the latter, the parts of the embryo begin to shew their most general relations of size, the radicle being much shorter than the cotyledon (*fig.* 460). The tissue of these vegetables (like that of all aquatic plants in general) is very simple: the cellular tissue occupies a large extent, pierced with lacunæ filled with air or some other gas. These gases by diminishing the specific gravity of the plant, enable it to raise itself to the surface of the water. The vessels, on the contrary, are much less frequently found and are even completely wanting in some. From this arrangement will result very slight activity in the secretions and, consequently, the absence of peculiar properties as well as of any substances useful to man. Of all these plants, that

527. Embryo of the Ruppia maritima.—*c* Cotyledon.—*r* Radical.—*f* Slit corresponding to the gemmule.—*h* The lateral excrescence of the tigelle.

TABLE II.

FAMILIES.

MONOCOTYLEDONOUS VEGETABLES.

Aquatic. Seeds without perisperm.

Perianth	none, or scaly, or herbaceous.*			
	6 well-developed divisions, frequently (the 3 inner ones at least) petaloïd.**			
*Embryo	homotropous, macropodous. — . . . . Perianth none.	1 carpel. —	Fresh water plants.	NAIADEÆ.
	amphitropous, macropodous. — Perianth none or composed of 4 scales.	1 carpel or several distinct carpels. . .	Fresh water plants.	POTAMEÆ.
	homotropous, macropodous, and canaliculate in the interior.—Perianth none.	1 carpel. —	Fresh water plants.	LEMNACEÆ.
	antitropous, macropodous with a tigelle laterally developed. — Perianth none.	1 carpel or several distinct carpels. . .	Marine plants. .	ZOSTERACEÆ.
	homotropous, with a short radicle. — Perianth none, or herbaceous.	Carpels distinct or united into one single ovary.	Fresh water plants.	JUNCAGINEÆ.
**Ovules	A single one attached to the bottom of the loculus, campylotropous. — Ovaries free and distinct. .		Fresh water plants.	ALISMACEÆ.
	Several with parietal placentation — reflexed. — Ovaries free and distinct.		Fresh water plants.	BUTOMEÆ.
	Several with parietal placentation — straight. — Ovaries united into a single one adherent to the calyx.		Fresh water plants.	HYDROCHARIDEÆ.

most frequently mentioned is one of the HYDROCHARIDEÆ (*Hydrocharidées*), the VALISNERIA SPIRALIS, which covers certain branches of the Rhône and several canals and dykes of the south of France. It has often been narrated both in poetry and prose, how their male and female flowers, separated on different plants, approach at the time of blossoming; how the former are then detached by the rupture of their peduncle, float on the water, sustained by the small shell-like boat, which forms their convex perianth and approach the latter, fixed on their plant by a long thread, the spire of which is unrolled; how at last, after this approximation, the spire is collapsed and plunges under water, and how its seed ripens there.

§ 757. Among those Monocotyledons, which (with some exceptions) present seeds with a perisperm, the flowers of some, being less complex, have no real perianth; the envelope, which we do find, has not clearly the common characteristics as to the number and the structure of its parts, replaced by scales or by bracts; that of the others shews a real perianth with folioles whorled in threes. Thence we have a first division into APERIANTHEÆ (*Apérianthées*) and PERIANTHEÆ (*Périanthées*).

MONOCOTYLEDONOUS VEGETABLES.

Perisperimeæ. Aperiantheæ.

(Table III, page 597).

§ 758. These may be divided into SPADICEÆ (*Spadicées*) and GLUMACEÆ (*Glumacées*). The following are the reasons why these two divisions are so named: the former from its inflorescence being enveloped by a spathe (sometimes disguised by the small size and precocious fall of the general bract, which does not remain as a spathe); the second from the nature of the envelopes of the flower, which have received the name of *glumes* (GLUMÆ) (*glumes*) and represent small scaly bracts.

Among all the families enumerated in this table, we shall direct the notice of the student to the last two only, one of which especially, that of the Gramineæ, merits all the attention we can give it on account of its importance both with regard to its utility to man and its botanical relations.

§ 759. CYPERACEÆ (*Cyperacées*) (Sedges).—In common parlance under the name of herbs we confound all Monocotyledons, commonly green in all their parts even their flowers, with herbaceous stems, with entire leaves lengthened into narrow ribands which are traversed by longitudinal nerves; but these herbs really belong to several different families, more particularly to this and to the following.

The Cyperaceæ are easily distinguished from the Gramineæ by their solid stem, without any swelling at the origin of the leaves, frequently presenting the form of a triangular prism, a form, which is connected with the tristichous arrangement of the leaves. Their vaginal portion surrounds the stem without division to the very origin of the limb, or in other terms their sheath, is entire; the upper leaves have only the limb without sheath. The flowers are arranged in spikes towards the top of the plant. These sometimes, on account of their shortness, assume the name of spikelets (*épillets*): they are then arranged in different ways with respect to one another. These spikelets consist of a series of scaly bracts, at the axils of which are situated sometimes several stamens around a pistil, sometimes only stamens, or only pistils. These parts are pretty frequently wanting in the lower scales of the spikelet. These different combinations of hermaphrodite or of diclinous flowers and the various arrangements of the bracts of the axis which bears them are employed in distinguishing several tribes. Thus, distichous bracts along with hermaphrodite flowers are the characteristics of the Cypereæ (*Cyperées*); imbricate in every direction, of the Scirpeæ (*Scirpées*). When the stamens are separated from the pistils, the ovary may be concealed in a peculiar envelope or utricle, which is opened to give passage to the style, and by the two teeth of this opening shews that it is composed at the expense of two opposite bracts completely united, except at the top. We observe this in the Caricineæ (*Caricinées*), whilst in the Sclerieæ (*Sclériées*), also diclinous, the ovary is not closed. The number of the stamens extends from one to twelve; we most commonly find three, and their slender filaments bearing bilocular anthers are inserted below the ovary when they surround it. In this case, we sometimes find other sterile filaments having the appearance of hair or of scales. The ovary, surmounted with a style, bifid or trifid at its top, is hollowed into a single loculus containing an erect ovule. Afterwards its pericarp assumes a crustaceous or osseous consistence, as in the Sclerieæ. The seed (*fig.* 478) consists of a membranous sac, filled with a large farinaceous perisperm except at its lower end, under which is embedded a small embryo, turned, consequently, in the direction of the hilum. This embryo (*fig.* 528) has commonly the form of a spinning top and is slightly swollen (*c r*) on the side; this corresponds to the cotyledon and to the radicle, as is afterwards proved by the germination; and the remainder of the embryonary mass *a* is formed by the extremely dilated tigelle.

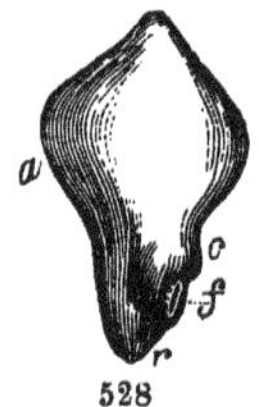

528

In speaking of the stem, we have considered that only, which appears above the soil, and which is frequently, in fact, only a branch springing from a horizontal rhizoma.

§ 760. GRAMINEÆ (*Graminées*) (Grasses).—This family is also called GRAMINALES. Their most general appearance is that known by the common name of Grass. We sometimes, however, find them of dimensions no longer agreeing with this name. The large Reed of the south of France, the ARUNDO DONAX [k], very greatly exceeds the height of a man, and under the tropics, Bamboos become large trees. Like the Cyperaceæ the Gramineæ have frequently a subterranean stem, whence spring those which are elevated above the soil. These are known by the name of *culm* or *straw* (*chaume*) (CULMUS) and are characterized by swellings at each knot, that is, at the origin of each leaf, as well as by their stem being hollow. The fibro-vascular fascicles, in fact, approach and press against one another towards the exterior, leaving the centre hollow except at the knots, where they are bent horizontally, cross one another, and by their lattice-work mixed with cellular tissue, form a kind of platform. The culm is, therefore, a hollow cylinder, the canal of which is interrupted by a series of partitions at the origin of the leaves. These surround the stem by a sheath, the insertion of which embraces the knot, and which is split for the greater part of its length on the opposite side, and is prolonged into a narrow limb or lamina above it. The separation of the limb from the sheath is most frequently marked by a small membranous prolongation, truncate, or sharp, or bifid, even scolloped and sometimes reduced to a tuft of hairs: this is the LIGULA (*ligule* [§ 150, *fig.* 151]). The leaves are commonly distichous, and from their axils frequently spring leaf-

528. The embryo of the CAREX DEPAUPERATA.—*r* Radicle.—*c* Cotyledon.—*f* Slit corresponding to the gemmule.—*a* The lateral swelling of the tigelle.

[k] So large does it grow, that in France and Italy it is cultivated for fishing-rods, fence-wood, vine-props and similar purposes.—TRANS.

buds (*bourgeons*), the developement of which determines the ramification of the plant.

This distichous arrangement is frequently found again in the bracts of the inflorescence, which consist of spikelets (*épillets*) (SPICULÆ), that is, of extremely short spikes, so short that for a long time they were each described as one single flower. Considered thus, these spikelets are grouped with regard to one another sometimes in panicles, as in the Oat (*Avoine*) (AVENA), sometimes in spikes, and in this latter case, it frequently happens that the axis, which bears them, is alternately hollow first on one side and then on the other, so as to receive their insertion. These spikes [as those of Wheat (*Froment*), of Rye (*Seigle*)], have become the most commonly quoted type of this method of inflorescence, although they are really compound, since each spikelet is a small group of flowers. Up to this point we find a great resemblance between the spikelet of the Gramineæ and that of the Cyperaceæ. The two lower bracts, in the same way bearing nothing at their axil, seem to form an envelope common to all the others, and take the name of *glumes* (*glumes*) (GLUMÆ [*figs.* 529, 530 *ge*, *gi*]). But each of the following ones present in their concavity not only the organs of reproduction, but also (and this distinguishes them from those of the Cyperaceæ) a second bract opposite to the first, a little more elevated and internal with respect to it. These bracts, which take the name of the *chaff of the receptacle* (*paillettes*) (PALEÆ [*fig.* 530 *pe*, *pi*]), opposite in pairs, thus form so many involucres between which are placed stamens and a pistil, and each of these small systems is a true flower. There may be found above the glumes one of these systems, or two, or three, or a much greater number, and according to these cases, the spikelet is said to be uniflorous, biflorous, triflorous, multiflorous. The stamens, amounting in number to six or more, sometimes reduced to three or even to one, but most commonly to three, are inserted beneath a central pistil (*figs.* 530, 531), which, in rarer cases, is wanting here and is only found in other separate flowers. Commonly we find also, on both sides and a little to the outside of the outermost stamen, two small membranous or scaly bodies which have been termed PALEOLÆ (*paléoles* [*figs.* 530, **B**, *l l*, 531 *p*]). Since the outer palea is marked with a median nerve, whilst the inner, on the contrary, has frequently none and is furnished with two lateral nerves one on each side, several authors consider this *parinervate palea* (*paillette parinervée*) to result from the union of two by their continuous margins; we should thus have three of them, before which would be placed the three stamens, and the paleolæ would form the intermediate whorl completed by a third which is abortive, but might have been observed in the very

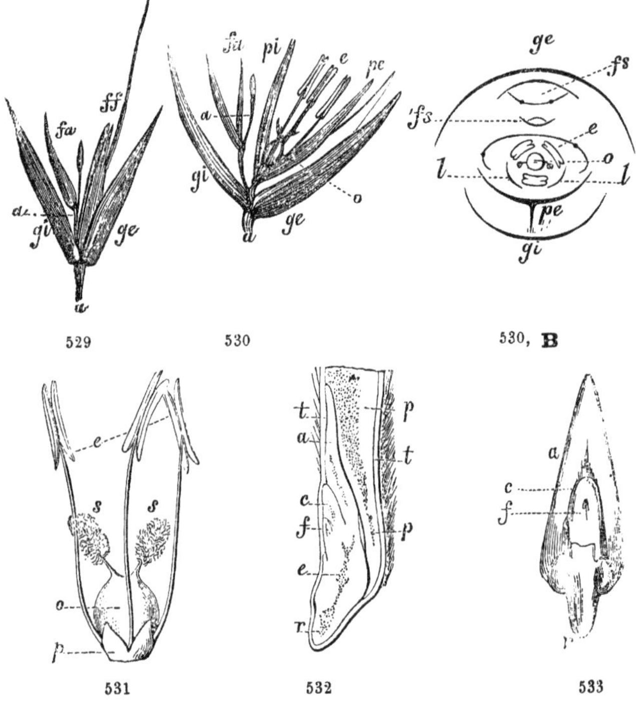

529. A spikelet of the Oat (*Avoine cultivée*) (AVENA SATIVA).—*a* Axis.—*ge* Outer glume.—*gi* Inner glume.—*ff* Lower fertile flower.—*fa* Two higher abortive ones.

530. The same, deprived of its envelopes so as to display the inner parts.—*pe* Outer palea of the fertile flower, terminated by a beard.—*pi* Inner palea.—*e* Stamens.—*o* Pistil.—The other letters have the same signification as in the preceding figure.

530, **B** Diagram of the spikelet.—The letters again refer to the same parts. —*ll* Paleolæ.

531. The fertile flower without its glume.—*e* Stamens.—*p* Paleolæ.—*o* Ovary.—*s* Stigmas.

532. The lower part of the vertical section of the cariopsis.—*t* Teguments of the seed and cariopsis blended together.—*p* Perisperm.—*e* Embryo.—The other letters refer to the same parts as in the following figure.

533. The embryo.—*r* Radicle.—*c* Cotyledon.—*f* Slit corresponding to the gemmule.—*a* Lateral swelling or hypoblast.

young bud. But if we admit this, the inner palea must have its origin on the same axis as the outer, and not on a secondary axis.

However this may be, the stamens consist of a thread-shaped filament and an anther with two loculi, which, joined at their middle (at the base of which the filament is attached), separated at their extremities, thus represent a kind of *x* (*fig.* 531 *e*). The pistil is an ovary surmounted by two styles (sometimes united into one), branching for a greater or less part of their length into flat shreds, which form the two hispid or pulmose stigmas *s*. It has only a single loculus filled with a single ovule longitudinally adnate to the inner wall. Afterwards the seed, as it ripens, is confounded by its tegument (*fig.* 532 *t*) with the pericarp, and thus forms a CARIOPSIS (*cariopse*) (§ 526). The greater part of its mass is composed of a farinaceous perisperm *p*; but on the outside and at the bottom, is perceived a small distinct body, buried in its surface, hardly projecting: this is the embryo (*figs.* 532 *e*, 533), which is supported on the perisperm by an enlarged part in the form of a shield *a*. At the bottom and on the outside of this we see a smaller body project which, continuous with the former at its middle, presents two extremities, the one upper and the other lower. It is between these two that we perceive the small gemmular slit *f*; the upper *c* is, therefore, the cotyledon, the lower *r* the radicle, and the shield (*hypoblaste* of Richard) is only a lateral excrescence of the tigelle analogous to what we have already seen in some Zosteraceæ. We have described the germination of the seed of one of the Gramineæ (§ 111, *fig.* 120).

All these parts, and especially those of the flower, have received from different authors a variety of different names, to state which would occupy more space than this book can afford. We shall content ourselves with adding that the name of glume, instead of being applied to each one of the lower sterile bracts of the spikelet is sometimes given to the whole, and then they are so many *valves of the glume*; that that of *bale* or GLUMELLA is given to the whole of the paleæ, which are then the *valves of the paleæ*. Let us add again, in order to understand generic characteristics and descriptions, that in the external bracts of the glume and of the bale the median nerve is frequently prolonged into a bristle of a greater or less length above the summit, or one at other times quitting the blade a little below the apex. This is called the *awn*, *beard* or ARISTA. The mode of inflorescence, the number of the flowers in each spikelet, their complete developement or the abortion of several (which constantly takes place in some of them), the union or the separation of the stamens and of the pistils in the same flower, the presence or the absence of the glumes, the consistence and the form of the paleæ, the united or distinct styles, the nature of the stigmas, the number of the stamens and that of the paleolæ,—these are the characteristics which vary in the family, and the combination of which is employed in order to distinguish the tribes and the genera.

This immense family, distributed over the whole globe, serves for purposes as varied as they are important. The abundance of fecula in its fruits causes man to cultivate a certain number of species, which assume the name of *cereal* (*céréales*); he prèfers of course those with the largest seed, Wheat (*Froment*) in temperate climates; along with it a little farther north, Barley (*Orge*), Rye (*Seigle*) and Oats (*Avoine*); farther south, Maïze (Maïs), Rice (*Riz*), and SORGHUM (*Doura*, *Sorgho*); some others under the tropics, as the POA ABYSSINICA, several species of PANICUM and of ELEUSINE. The farina or flour, which is extracted from the perisperm, is, when ground, a doubly nutritious aliment both from the fecula which it contains and from the gluten (§§ 23, 310) which is associated with it. The bran is the *débris* of the pericarp and owes its qualities to the amylaceous particles which are still attached to it. The sap of several Gramineæ contains sugar in solution: it is extracted in great quantities from the Sugar Cane (*Canne*) (SACCHARUM OFFICINALE). The presence of sugar determines fermentation, in consequence of which are produced different liquids of an alcoholic nature, so much sought after for the beverage and several other purposes of man. From this property rum is obtained by distillation from the juice of the Sugar Cane, arrach from rice, and beer from barley. The success in making the last, the process of which consists in submitting to fermentation in a large vat of water the barley to which the first impulse of germination has been given, depends on a certain part of the fecula of the young plant being converted into sugar during germination. These various nutritive principles being so abundant in the different parts of Gramineæ cause them to be employed as food for animals and as the staple of pastures and meadows. Lastly, we have seen (§§ 20, 325) that the Gramineæ have a peculiar affinity for silica, which penetrating with their sap and solidifying in the walls of their outer cells, frequently incrusts their epidermis and their knots: thence the rigidity and the incorruptibility of certain straws, properties which the ingenuity of man has turned to his advantage.

All the Gramineæ are not totally scentless; some, at the time of blossoming exhale a soft, but at the same time penetrating smell, with which all who are walking past cannot fail to be struck at this time, especially when the individuals are numerous. The Vernal Grass (*Flouve*) (ANTHOXANTHUM) is quoted as one of the most odoriferous of our indigenous species. Some of them are so to a much higher degree in warmer climates and an essential oil is thence extracted. The substance, called *Vetiver*[1] by the French, so much used as a perfume, is the root of one of the Gramineæ (ANDROPOGON MURICATUM).

§ 761. Jussieu formed his monohypogynous plants from the preceding families; from the following, his monoperigynous and his monoepigynous plants, between which it is not easy to draw a line of distinction.

We shall, therefore, divide them according to another characteristic, which is also generally allied to that of insertion: it is more useful, as being more easily determined; we refer to the adherence or non-adherence of the calyx to the ovary. The plants of these Monocotyledonous families, which it now remains for us to examine, present a perianth with folioles almost always arranged in ternary whorls, most commonly in two, which are either similar to one another, both presenting the appearance either of a calyx or of a corolla, or different from one another, the external being then like a calyx, the internal like petals.

MONOCOTYLEDONOUS VEGETABLES.

Seed with a Perisperm. Flower with a Perianth.

Table IV. p. 598.

§ 762. PALMACEÆ (*Palmacées*) (Palms).—We have already explained (§§ 93, 94) the structure of the stem of the Palms (*Palmiers*) (PALMÆ) and their general habits (*fig.* 114, 1). Although the stem is generally a tall, straight trunk without branches, there are some exceptions. Thus, it is divided at a certain height by regular dichotomy in the Doom Palm (*Doum*) (CRUCIFERA THEBAICA)[m]; in several others, it is reduced to a bulb or to a rhizoma. The trunk,

[1] This substance is known by the name of *Khus* in India. The natives employ its roots in the manufacture of covers for palanquins and other purposes, for which its fragrance renders it peculiarly desirable.—TRANS.

[m] A native of Upper Egypt. Paxton makes this a synonym of HYPOPHANE CORIACEA, whilst Lindley calls it an allied species.—[*Pax. Bot. Dict.*—*Lind. Vegt. King.*, p. 135].—TRANS.

when it is elongated, may be thick or slender; its internodes are sometimes very short, sometimes separated from one another by long intervals; its surface is sometimes smooth and even shining (as in the CALAMUS), frequently, on the contrary, very rough on

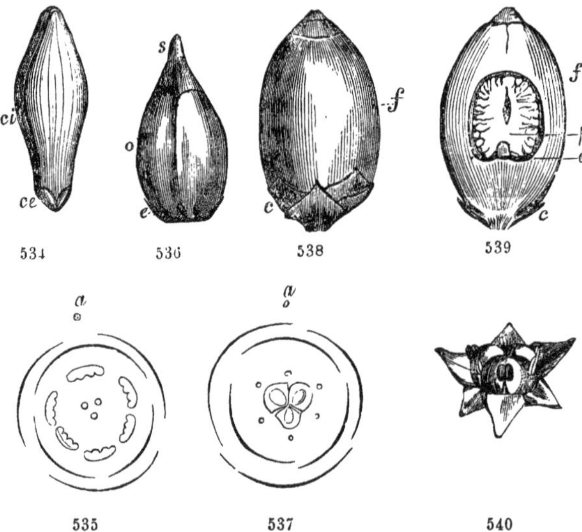

account of the remaining bases of the leaves, or even in the older parts where they are detached, unequal, knotty and hacked; it is not rare to find it armed with straight thorns. The adventitious roots, springing from above the soil and accumulated towards the base of the stem, frequently form around it a kind of net-work which thickens it into a kind of cone.

The leaves, which attain considerable dimensions, are supported by a long, strong, but very flexible petiole, to which their thick

534. The flower of the ARECA CATHECU, not yet expanded.—*c e* External perianth.—*c i* Internal perianth.

535. Diagram of this flower in which the stamens are developed and the ovaries abortive.—*a* Position of the axis of the inflorescence, with respect to the flower.

536. Another flower stripped of its perianth, in which the stamens *e* are partly abortive and the ovary *o* is developed —*s* Stigma.

537. Diagram of this flower with its perianth.

538. Fruit *f* of the same, surrounded at its base by the persistent perianth *c*.

539. Vertical section of this fruit.—*c* Perianth.—*f* Pericarp.—*p* Ruminate perisperm. —*e* Embryo.

540. Flower of the CHAMÆROPS HUMILIS.

limb is attached not in a straight, but a serpentine or zigzag line, so as to form a series of folds, which are exceedingly like those of a fan and are expanded exactly in the same way. These folds are, indeed, sometimes arranged like a true fan, being all inserted at the extremity of the enlarged petiole; sometimes they are arranged like the feathers on a quill, being inserted one above the other on both sides of the petiole, which then becomes the median nerve or rachis. The whole of this plaited limb is at first continuous, but it soon splits all along the folds, and is thus more or less deeply divided into a number of blades giving to the whole a palmatisect or pennisect appearance (*fig.* 114, 1). From the axils of these leaves, which, renewed by a terminal bud (*bourgeon*), form a kind of tuft at the top of the stem, the flowers grow in a spadix, either simple or more frequently branched; and the spathes, with which they are at first enveloped and which remain with them for a certain time after they have blossomed, have a thick, hard, sometimes ligneous tissue, almost in the shape of the keel of a ship. There may be one or more; they are complete or incomplete, and sometimes are even quite wanting. The flowers may be hermaphrodite, or polygamous, or monæcious, or diæcious (in the Date, for instance). Their perianth (*figs.* 538, *c*, 540) is composed of two whorls of coriaceous folioles, the three inner ones of which have not always the same form and the same length as the outer, and are sometimes united to one another. The stamens, most commonly six in number (*figs.* 535, 540), reduced rarely to three, are sometimes more numerous in the diclinous flowers; their filaments are free or monadelphous. The pistil is composed of three distinct (*fig.* 540) or united (*fig.* 536) ovaries (the styles also participate in this union), each enclosing one or two erect ovules; but frequently, especially in the case of the union of the ovaries, two loculi are sterile, and we then find only one. The fruit, which sometimes acquires enormous dimensions (in the Cocoa nut, for instance), under a thick, fleshy or fibrous envelope encloses a nucleus almost always reduced to a single loculus with a wall sometimes thin, but frequently acquiring the hardness of wood or even of stone. The seed, which fills it and is blended with it, is composed for the most part of a thick perisperm, generally very hard, horny or cartilaginous, frequently ruminate (*fig.* 539 *p*), at the bottom or on the sides of which we find a small superficial cavity, where a small embryo *e* is nestled: this embryo is, consequently, turned sometimes towards the hilum, sometimes quite in a different direction.

Several tribes of Palms have been distinguished according to the different modifications of their inflorescences and of the spathes which commonly accompany them, and according to those of the fruit, which varies in the consistence of the pericarp, composed of several distinct carpels or of only one, and in this case, containing several loculi and several seeds, or else only one of each. They may be subdivided according to the two distinct forms of the foliage: the divisions and the shape of the perianth; the number and appearance of the free or united stamens; the form of the anthers; the degree of coherence of the ovaries and of the styles and their abortions; the form, the size, the tissue of the fruit and of its parts, of the nucleus, of the perisperm; the position of the embryo; the distribution of the pistils and of stamens on the same or on different flowers, on the same or different plants. All these different modifications variously combined are employed in distinguishing rather a numerous assemblage of genera.

Several plants belonging to this beautiful family render the most varied and important services to the inhabitants of the countries to which they are indigenous. On the one hand, their wood is employed in the construction of huts, the roof of which is made of their large hard leaves without much labour; and their flexible and tenacious fibres, found in every part, are made into ropes, arms and different domestic utensils. On the other hand, different species furnish delicious food that hardly requires dressing. Every one knows that numerous tribes are fed almost exclusively on Dates, and that the Cocoa-nut contains an acidulated milky juice, a most grateful beverage in warm climates: this milk is nothing else than the perisperm still in a fluid state, which afterwards is gradually thickened and is at last solidified into a mass as hard as stone itself. The terminal bud of another valuable species, the ARECA OLERACEA, commonly known by the name of the West Indian Cabbage Palm is held in high estimation as delicate food. We find, moreover, in the products of Palms some of those substances, which we have seen to be so useful in the Gramineæ: an abundance of fecula in the inner cells of the trunk of a large number of species, especially of the SAGUS and PHŒNIX DACTYLIFERA; this is the substance known as *Sago* (*Sagou*); sugar mixed with the sap, from which are manufactured fermented liquors, as Palm wine, the most esteemed of which is extracted from the ELAIS GUINEENSIS, as Arrach, which is made from the juice of the ARECA CATHECU fermented with rice. A liquor is drunk in India under the name of *Toddy* which is obtained by the incision of the spathes of the Cocoa-nut tree (*Cocotier*) and others. The middle of the Cocoa-nut owes a part of its nutritious properties to the oily principle which is mixed with it, and it is remarkable that a similar principle is found in several other Palms of the same tribe: *Palm-oil* is extracted from the ELAIS which we have already named. We obtain also a kind of Wax (*Cire de Palme*) from this family; this wax exudes from the spaces between the insertion of the leaves of the CEROXYLON ANDICOLA (§ 327), and flows in great abundance down the trunk, covering it with a thick coating.

§ 763. JUNCACEÆ (*Joncacées*) (Rushes).—We do not refer to this family here merely as the family represented by the plants of our country, and generally confounded under the name of herb with the Gramineæ, in the same way as we frequently on the other hand confound several herbs growing in morasses under the name of Rushes. The structure of their flowers renders them easy to be distinguished both by the existence of a perianth with six parts and by that of an ovary with three loculi; but the scaly or herbaceous consistence of the former furnishes a kind of passage from the floral envelopes of the Glumaceæ to the coloured perianths of the following families.

§ 764. LILIACEÆ (*Liliacées*) (Lilyworts).—The perianth acquires its most brilliant colours in the Liliaceæ, so much so as to render them the object of our admiration when we meet them in the meadow or in the trim parterre of our gardens. We need only allude to the Tulip, the Hyacinth, the Lily, the Crown Imperial, the Asphodel, to give the student an idea of the beauty of this family. In our climate they are all herbaceous; their stems, often short and swollen into bulbs (the different modifications of which we have explained in another part of the work [§ 182]), are at other times very long, either creeping or erect, and sometimes even very much branched. But in warmer climates we find some that are truly arborescent, as the YUCCA, some ALOES, &c.; and among them we find examples of the largest Monocotyledonous trees, as the Dragon-tree (*Draconnier*), (§ 194). The leaves are long and generally very narrow, with parallel nerves. Their sheaths are largely developed around certain bulbs, which they thicken and partly form.

The flowers (*figs.* 226, 541) present the exact type of those of Monocotyledons, a perianth with six folioles in two concentric rows, similar to one another, sometimes distinct and sometimes united at the bottom into a tube, six stamens opposite to these folioles, also arranged in two whorls, inserted on their tube when they are united, if not quite at their base, low enough for them in some cases to be considered as hypogynous; three ovaries alternating with the three innermost stamens, united into a single one, as well as their styles and even sometimes their stigmas, which will be divided into three lobes; each loculus (*fig.* 543 *o*) encloses several ovules *g* in one or two longitudinal rows attached to the internal angle. The fruit is generally a loculicidal capsule. A certain number of genera with fleshy fruit now joined to them, were formerly separated under the name of ASPARAGINEÆ (*Asparaginées*). The seed (*fig.* 544) presents in a fleshy perisperm *p* an embryo *e* most frequently erect, sometimes bent, but directed in every case towards the point of attachment. The testa, which forms its tegument, is spongy in some (those of which the original family of the LILIACEÆ was composed), crustaceous and brittle in others, with which was formed that of the ASPHODELEÆ (*Asphodélés*), membranous in a certain number, the ALOINEÆ (*Aloïnées*).

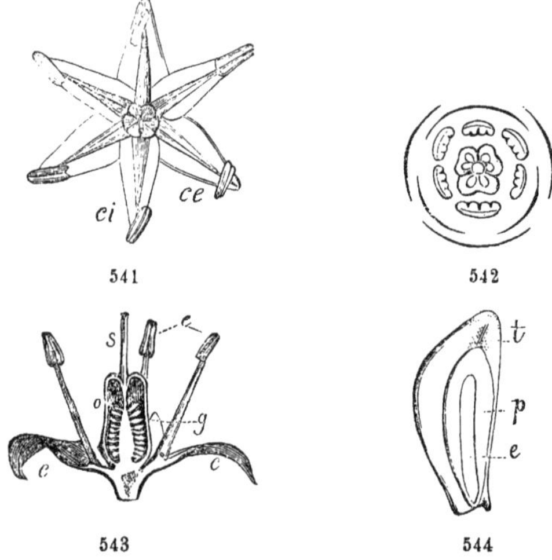

541. The upper side of the flower of the SCILLA AUTUMNALIS.—*ce* External Perianth.—*ci* Internal Perianth.

542. A diagram of the same.

543. A vertical section of the flower.—*c c* Perianth.—*e* Stamens.—*o* Ovary.—*s* Style and stigmas.—*g* Ovules.

544. A longitudinal section of the seed.—*t* Tegument.—*p* Perisperm.—*e* Embryo.

It is not only as ornamental plants that the Liliaceæ are cultivated. Several are esteemed for culinary purposes and belong in general to the genus Allium (*Ail*), (Onions, Leeks, Shallots, Garlic, &c.) They owe their properties to the juice, which has a very strong taste and is slightly stimulating in its effects; it abounds in every part and especially in their bulbiform stems. These properties may become very intense and the plants thus become useful in medicine on account of their bitter juices, as the Squill (*Scille*), Bitter Aloes (*Aloës*), and others which would take too much space to enumerate.

In the neighbouring family, that of the Melanthaceæ (*Melanthacées*) (Melanths), as the Colchicum (*Colchique*), Veratrum, we remark still much more energy, and we find strong poisons in the plants.

§ 765. Amaryllidaceæ (*Amaryllidácées*) (Amaryllids).—Let us suppose that in some of the Liliaceæ the divisions of the perianth are united at the bottom to one another and to the ovary, and we shall then have the definition of this fresh family, which presents the same characteristics of vegetation, the same symmetry in its flower. Its seeds present analogous modifications in the texture of their testa, especially if we unite with it the small neighbouring family of the Hypoxideæ (*Hypoxidées*) (Hypoxids), which differs from it in the same way as the tribe of the Asphodeleæ differs from the other Liliaceæ. In some genera, of which we form a distinct tribe under the name of Narcisseæ (*Narcissées*), the divisions of the perianth are deduplicated internally; they separate at the same time into a coloured, tongue-shaped fold, whence results a kind of entire or dentate general collar or coronet. We observe in some Amaryllidaceæ a singular modification of the seed, the tegument or the perisperm of which, losing its ordinary consistence, is swollen into a loose cellular, succulent and greenish tissue, and attains dimensions very much larger than those of their normal state. These seeds are termed *bulbiform* (*bulbiforme*), on account of their apparent resemblance to a large bulbel.

The Amaryllidaceæ are esteemed as ornamental plants as much as the Liliaceæ, and their juices have analogous and still more powerful properties.

§ 766. Iridaceæ (*Iridacées*) (Irids).—This family, though rather similar to the preceding, is easily distinguished from them by its having three stamens placed before the three external divisions of the perianth, the anther opening outwards (*fig.* 545). Their filaments are sometimes united into a tube. The three stigmas opposite to the anthers are expanded into as many petaloïd bodies (*fig.* 546 *s*); these are gathered in the Crocus (*Safran*) to make Saffron so well known by its taste and by the colouring principle which it contains in large quantities. The perisperm of the seeds (*fig.* 547) is sometimes thick and fleshy, sometimes horny. This consistence, something similar to that of Coffee, has suggested the idea of trying to replace it by the seed of a species of Iris (IRIS PSEUDO-ACORUS), and it is pretended that when roasted and prepared in the same way it can hardly be distinguished from it.

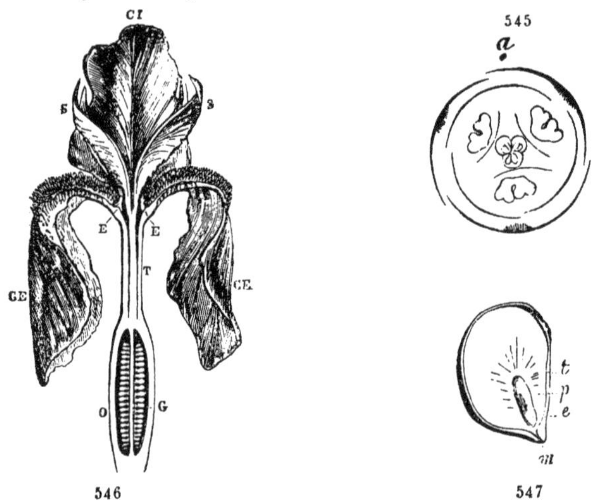

§ 767. BROMELIACEÆ (*Broméliacées*) (Bromelworts).—This family presents all the transitions from the free to the adherent ovary; for the TILLANDSIEÆ (*Tillandsiées*), which the sections established in our tables have obliged us to present at a certain distance from it, form really only one of its tribes. Amongst these we find parasitical plants living on other trees. The seeds are frequently remarkable for a raphe, which, almost as thick as the rest of the seed,

545. Diagram of the flower of the IRIS GERMANICA.—*a* The position of the axis in the inflorescence.

546. The vertical section of the flower.—C E External divisions of the perianth.—C I Internal divisions.—T Its tube, above the part adhering to the ovary.—O This ovary.—G Ovules.—E Stamens.—S Stigmas.

547. A longitudinal section of a seed.—*t* Teguments.—*p* Perisperm.—*e* Embryo.—*m* Micropyle.

is at last partly separated, and also for its embryo, which is situated at the top of the perisperm, one of its sides being prolonged into a point for some distance.

Several Bromeliaceæ are remarkable for the beauty of their flowers, but are much rarer than preceding families, because natives of tropical climates, they can only be cultivated in a conservatory. The best known is the Pine-apple (*Ananas*), so much esteemed for its fruit (§ 541, *fig.* 429).

§ 768. DIOSCOREACEÆ (*Dioscoréacées*) (Yams).—This small family merits some notice on account of several curious points in its organization. Along with the Aroideæ and the Smilacineæ, it presents an exception to the common nervation of the leaves in Monocotyledons, since the nerves of theirs by being ramified and anastomosed into a net, as well as the form of the limb, bear a strong resemblance to that of Dicotyledons. This singular difference is to be remarked in the embryo, in which the cotyledon is expanded into a real limb (§ 579, *fig.* 461), and in which the gemmule is almost uncovered. The stem is equally remarkable; for if it emits each year climbing branches, they do not appear, as in other vegetables, to grow from a series of successive knots, but rather from a subterraneous offshoot which is something like the first internode continuing to grow and acquire enormous dimensions. We have had occasion to mention the case of a species, the TAMNUS ELEPHANTIPES (§ 100), curious from the developement of a kind of suberous bark on its surface. In the DIOSCOREA ALATA, known as the Winged Yam (*Igname*), the rhizoma swelled into a fleshy, mucilaginous and rather saccharine tubercle, forms one of the commonest means of support of the inhabitants of the equatorial regions.

The fleshy or dry and winged fruit may serve to distinguish two tribes, TAMNEÆ (*Tamnées*) and the DIOSCOREÆ (*Dioscorées*).

§ 769. MUSACEÆ (*Musacées*) (Musads).—The typical genus of this family is the BANANA (*Bananier*) or MUSA, found almost over the whole of the torrid zone, in which its fruit, of which we have already spoken (§§ 512, 517), forms an important means of supporting life. We have also spoken of its large leaves (*fig.* 152), the long sheaths of which fitting to one another form the apparent stem, whilst the real one is concealed underneath the surface of the ground. We have described its tracheæ, remarkable for their size and for the number of the parallel spiral threads (§ 9, *fig.* 39).

These threads are said to be employed for purposes for which they are not adapted, as they break with the greatest facility. Those, which are obtained from plants of this family and especially from the MUSA TEXTILES to make cords and matting, are long straight fibres, analogous to those of the liber. In the flowers of the Musaceæ, some of which are extremely remarkable for their extraordinary shapes and brilliant colours (those of the STRELITZIA, for instance), one of the six stamens commonly proves abortive, but reappears in another genus, the RAVENALA. We find one or more seeds in each of the three loculi of the fleshy or dehiscent fruit; they commonly present a fringed arillus differently coloured from the rest of the fruit.

§ 770. CANNACEÆ [n] (*Cannacées*) (Marants).—This family borders on the preceding in the form of their distichous leaves with transverse nerves, situated on long sheaths, which, fitting one in the other, strengthen as well as lengthen the branches which grow from a subterranean stem. This becomes in many species a rich store of fecula of first-rate quality in some of them (especially in the MARANTA ARUNDINACEA) well known by the commercial name of *Arrow-root*. The flowers deserve all the attention of botanists, who for a long time were unable to understand them on account of the irregular transformation of the stamens into real petals, one only of which preserves a trace of its true nature by the presence of the loculus of an anther on one of its edges. The style and the stigma participate in this transformation, and the different ways, in which all these parts are united, complicated the difficulty, which is now clearly solved. We have explained the seed (§ 572) and its germination (*fig.* 490): sometimes instead of being straight, it is completely bent in two on itself.

§ 771. SCITAMINEÆ [o] (*Scitaminées*) (Gingerworts).—The appearance of the plants is the same in this family as in the preceding; but in the flower, the stamens, also assuming for the most part the form of other organs, have not quite the same appearances and the same relations. One of them retains its real nature and is terminated by a large bilocular anther; another, placed opposite to it, is changed into a large petal. The rest are found in various rudimentary states. Let us remark also that here the fertile stamen is placed before one of the internal divisions of the perianth and, con-

[n] Lindley calls this family MARANTACEÆ. *Veg. King.*, p. 168.—TRANS.
[o] ZINGIBERACEÆ. *Lind. Veg. King.*, p. 165.—TRANS.

sequently, belongs to the whorl of internal stamens; the contrary takes place in the Cannaceæ.

The seed presents a very rare and extraordinary fact, which, however, is easily explained by the actual theory of the ovule (§ 573): it is the existence of two concentric perisperms of different natures (*fig.* 548 *pe*, *pi*). The external is analogous to that of the Cannaceæ.

The rhizomes are very rich in fecula, but it is not generally employed on account of its being mixed with a very powerful and strongly smelling essential oil. On the other hand, this oil, found in every part, communicates to them an aroma, which gives them great value as spices. Almost all the Scitamineæ possess this property; we may mention as instances, the Grains of Paradise (*Amome*), the Zedoary (*Zédoaire*) and the Ginger (*Gingembre*). Turmeric, a very valuable colouring matter, is extracted from one of them, the CURCUMA LONGA.

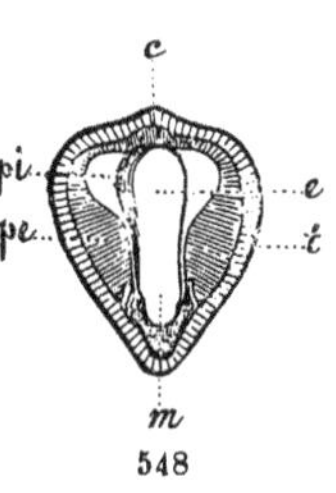

§ 772. ORCHIDEÆ (*Orchidées*) (Orchids).—The flowers of this family attract the attention of the superficial observer by their fantastic forms, of the botanists by their peculiar structure. Let us endeavour in order to understand it well, to bring it to the established type of Monocotyledons. The perianth adherent to the sessile ovary is divided above it into six pieces, three outer ones rather similar to one another, and three inner ones different from the former and from one another. In general, the former three and two of the latter are turned upwards towards the side of the axis of the inflorescence, the sixth falls down in a contrary direction, and in this way the perianth becomes, as it were, labiate, the upper lip being formed by the union of five divisions, the lower one by the sixth, which thence assumes the name of LABELLUM (*labelle*). In the very young flower, this labellum was situated by the side of the axis (*fig.* 550); but afterwards the ovary, turning on itself, inverts the position of the parts and carries them to that in which the expanded flower shews them to us (*fig.* 549). The labellum, by its form and colours, frequently totally different from the other parts, contributes the most to give such a singular aspect to the flower, and sometimes presents a rude resemblance to several other objects of nature,

548. A vertical section of the seed of the HEDYCHIUM CARNEUM.—*t* Tegument.—*m* Micropyle.—*c* Chalaza.—*pe* External perisperm.—*pi* Internal perisperm.—*e* Embryo.

especially to certain insects. We ought, consequently, to find three stamens opposite to the three external divisions; and we find them so in the Lady's Slipper (*Sabot de Venus*) (CYPRIPEDIUM); only one of the three (that which is situated towards the top of the flower) is sterile, and, instead of bearing an anther, is expanded into a kind of shield. In the greater part of the other Orchideæ, on the contrary, the third alone is antheriferous (*fig.* 550 *e*) and at the first glance we should suppose that the two others had disappeared; but a close examination will discover them under the form of two very small cellular lumps, which are termed STAMINODIA (*Staminodes*) (*fig.* 550 *s*), and in some monstrous flowers they have been found developed into real anthers. We are hindered from easily recognising this whorl of stamens, by their being supported on a body rising from the summit of the ovary at the centre of the flower, being a short obliquely truncated column, turning outwards a plane or slightly concave surface covered with a viscous covering, instead of being regularlyinserted between the internal divisions of the perianth. This surface is that of the stigma, and hence we see that the central body results from the union of this with the anthers; this union, in addition to the abortion of several parts, tends to mask their true nature. The name of *Column* (*Colonne*) or GYNOSTEMIUM (*Gynostéme*) has been given to this body composed of the stigma and of the anther which is inserted sometimes beneath it, remaining parallel to it (*fig.* 441 *a, s*), sometimes above it, projecting beyond it for the whole of its length (in which case it is said to be terminal), sometimes upright, sometimes bent back above the surface of the stigma. In this last case, the anther is at last frequently detached; in others, it remains in its place even after the emission of the pollen. The pollen presents an unusual structure; it has the appearance of several distinct masses having the consistence of wax, or of a larger number of smaller pieces in the shape of a wedge, united into one on an axis of a viscous substance (*fig.* 557); or at other times, the most common, of a mass of dust, with seeds frequently united in fours, most likely as they were so formed in their mother-cells. It has been discovered that in every case it is composed of grains analogous to those of pollen of any other plant, easily separated from one another, but glued together by some other substance. The anther (*fig.* 552) is divided into two loculi, which are opened on the side of the stigma, each of them being frequently subdivided by internal partitions into several locelli. Each loculus or locellus is filled with a *pollen-mass* (*masse pollinique*) (*fig.* 553, 557), which

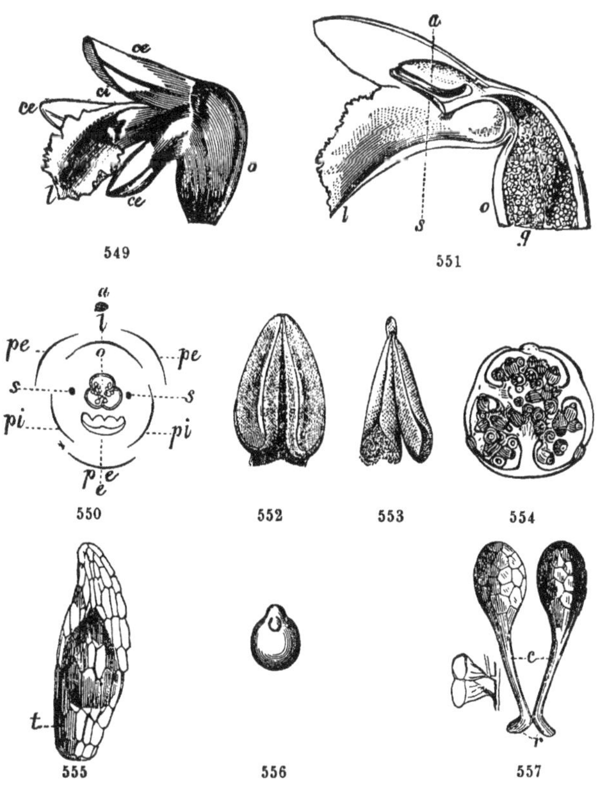

549. The flower of the SPIRANTHES AUTUMNALIS, after the turning.—*o* Ovary with the adherent perianth.—*ce* External divisions of the perianth.—*c i* Internal divisions, the lower one *l*, more highly developed, assumes the name of labellum.

550. Diagram of this flower before the turning.—*a* The axis of the spike.—*p e* External divisions of the perianth.—*p i* Internal divisions.—*l* Labellum.—*e* Fertile anther.—*s* Abortive anthers or Staminodia.—*o* Ovary.

551. The vertical section of the top of the flower.—*o* Adherent ovary with parietal ovules *q*.—*l* Labellum.—*s* Stigma.—*a* Anther.

552. The inner face of one of the anthers exhibiting its two loculi.

553. Granular pollen-masses taken from the anther.

554. Horizontal section of the ovary with its parietal placentas.

555. A seed with its external tegument *t*.

556. Embryo of the OPHRYS ANTHROPOPHORA stripped of its teguments.

557. Pollen-masses of the ORCHIS MACULATA, with grains united in small wedge-shaped masses, two of which have been represented at the side.—*c* Caudicle terminated at the bottom by the retinaculum *r*.

results from the grains of pollen being glued together, as we have just said. We always find, therefore, two of these masses or a larger number, some multiple of two. Each of them is sometimes contracted at the bottom into a kind of tail, *caudicle* (*caudicule* [*fig.* 557 *c*]), and this in some cases is terminated by a small glandular body, the RETINACULUM (*rétinacle*), which is placed in a kind of pocket, the BURSICULUS (*bursicule*), situated beneath the anther. It was necessary for us to detail these facts, because several tribes have been formed from the characteristics drawn from them. It is sufficient for the pupil, who does not wish to enter so fully into their examination, to remember that there is one anther with two loculi, each containing one or more masses of pollen.

As to the ovary, it is much more uniform throughout the whole of the family, twisted on itself, as we have said, and with only a single loculus which communicates with the middle of the surface of the stigma by a rather large canal. From this canal to the bottom there extend three longitudinal placentas opposite to the internal divisions of the perianth and quite covered with ovules in thousands (*fig.* 551 *g*). The ovary becomes a capsule, whose singular dehiscence we have described (§ 538, *fig.* 425), in which the three placentas remain attached at the bottom to the peduncles, at the top to the perianth, whilst three intermediate slips are detached and fall off. The VANILLA, with its indehiscent and pulpy fruit, forms an exception to this rule.

The seeds, innumerable and very small, are *scobiform*, that is, are very similar to fine sawdust. When we examine them more closely, we find that they are generally covered with an extremely loose, spindle-shaped, outer tegument (*fig.* 555 *t*), and another, much thicker, spheroïd or ovoïd, under which is a small cellular mass, apparently continuous; in this, however, we discover (*fig.* 556) with the aid of the microscope a small groove, the edge being raised a little on one side, from which during germination the axis of the plant will grow; we may, therefore, assume the raised edge of the groove to be the cotyledon, and its bottom to be the gemmule. We have here the tigelle enormously developed. This embryonary mass appears to be analogous to the tubercle found at the roots of several well-developed Orchideæ. The stem of the year rises from this tubercle: it then withers and another is formed on the side for the following year.

The true roots are fasciculated (*fig.* 125); the stems are simple or branched; the leaves are simple, entire, marked with longitudinal nerves, sometimes articulated at their base, and in several exotic species swollen beneath the articulation into a fleshy mass. Our Orchideæ grow in the earth; in tropical climates we find a large number of them growing on trees, *Epiphyte Orchids* (*Orchidées épiphytes*); they do not live on them as parasites, but establish their roots in the splits, the holes and the angles of the bark, extracting, doubtless, sufficient nourishment from the humus amassed in these places: their roots probably derive the greatest part of their food from the humidity of the air with which they are in contact; in fact, a humid atmosphere is indispensable to their growth. Hence the manner of cultivating them in baskets, surrounding their roots only with damp moss or broken potsherds, between which the air may circulate freely.

If we except the VANILLA, the slightly fleshy fruit of which contains such a delicious perfume and furnishes, consequently, so highly esteemed a condiment, we hardly find any other parts of the Orchids employed except the tubercles of certain species, from which a very useful nourishing diet is prepared, *salep*, a mixture of the fecula (found there in large quantities) and the teguments which enclose it, and another principle analogous to the gums, termed *bassorine* (*bassorine*), concentrated in small horny lumps scattered throughout the mass of these tubercles. In spite of such limited uses, the plants of this family are much esteemed on account of the beauty and the singularity of their flowers; their cultivation, which requires a stove, has become in some countries a perfect art. Linnæus was acquainted with merely a dozen exotic species, the catalogues of botanists describe more than fifteen hundred[p].

[p] Lindley doubtfully makes the whole number of species to be *three thousand*. From thirty to forty of these are natives of Britain.—TRANS.

DICOTYLEDONOUS VEGETABLES.

§ 773. Dicotyledons, which form the greater part of Phanerogamic plants, have necessarily occupied much of our attention and furnished us with the most of our examples. Their general characteristics and the principal points of their organization have, therefore, been previously explained at greater length, and several chapters have been more particularly dedicated to them. We have described their stems (§§ 50—90, 332—345), their roots (§ 118), their leaves (§§ 128—140, 151), the symmetry of their flower, the modifications of their embryo (§§ 29, 580—598), those of their seed and its germination (§§ 607, 610). The examination of their families will complete our knowledge of their characteristics by shewing us in what way they differ and combine. We shall be satisfied with giving the facts mentioned in the tables with respect to the greater part of these families; for room would be wanting on account of their number, and besides, the differences do not always turn on those points which should occupy our attention here.

At first we shall follow the grand division proposed by A. L. de Jussieu, but changing his order and successively examining the diclinous, the apetalous, the polypetalous and the monopetalous.

Dicotyledonous Vegetables.

Diclinous.

(Table V. page 599.)

§ 774. Among the families which belong to this division, there are two, the Coniferæ and the Cycadaceæ, which, by the peculiar characteristics of their general appearance and of some of their organs, have for a long time attracted the attention of botanists; their place, however, has been more clearly marked by recent obser-

TABLE III.

FAMILIES.

MONOCOTYLEDONOUS VEGETABLES.

Seed with perisperm. Flower without perianth.

Embryo	with a short radicle, not passing beyond the other parts. A largely developed bract at the base of a terminal spike	SPADICEÆ.
	with a macropodous radicle, laterally developed. Short, scaly bracts, corresponding with lateral spikelets . . .	GLUMACEÆ.

SPADICEÆ.

Flowers	naked, the males, consisting of one stamen each. Spathe . . .	enveloping, persistent. Embryo	apicilar, antitropous	PISTIACEÆ.
			axillary, antitropous. Leaves with branched nerves.	AROIDEÆ.
		shorter and caducous. Embryo	short, axillary. Several uni-ovulate loculi, or one with parietal placentation. Flowers diæcious	PANDANACEÆ.
			short, axillary. A single loculus with parietal placentation. Flowers monæcious, mixed on the same spike	CYCLANTHACEÆ.
			axillary, homotropous. A single loculus with one pendulous ovule. Flowers monæcious, separated on the spikes	TYPHINEÆ.
	surrounded with scales, similar to the perianth, but always alternate . .	Spathe short and caducous. Ovules half reflexed .	Embryo axillary. Ovary with several loculi.	ORONTIACEÆ.

GLUMACEÆ.

Seed	Erect. Embryo, extrarius, apicilar . . . .—1 scale for each flower.—Stems solid and angular.—Leaves tristichous	CYPERACEÆ.
	Adnate on the side. Embryo extrarius, lateral.—2 scales for each flower.—Stems, straw or culm . Leaves distichous	GRAMINEÆ.

vations and theories. We have described the ovules as being always enclosed in an ovary and have shewn that the *naked* seeds of ancient authors were not really so, and that in some cases they only appeared to be naked from the union of the teguments of the seed with those of the fruit. We have explained the structure of the ovules, consisting of a central body or nucleus in a single or double envelope, which adheres to it at one end and leaves a small opening at the other. Now, the bodies, which in the Coniferæ and the Cycadeæ, had been considered as ovaries with a style and stigma (according to some even with an adherent calyx), do not shew, when closely examined, this diversity of parts, but seem rather to present the simple structure of ovules, a nucleus in a double envelope opening at the top ; but the top, in this case, would have (*fig.* 562, *o*) a little longer point, something like a style, and the outline of the micropyle would be sometimes scolloped like a stigma. These are ovules, which the scales on which they are inserted, erect or pendulous, do not envelope as a pericarp. They are, therefore, naked ovules, and we may, therefore, call those vegetables *gymnospermous* (from γυμνὸς, *naked*, and σπέρμα, *seed*) in which they are found ; whilst all the rest with their closed ovaries are *angiospermous* (from ἀγγεῖον, *vase*): two words invented by Linnæus, but wrongly applied by him.

This characteristic of the organs of reproduction connected with others of vegetation, is doubtless sufficiently important to allow us to separate this small group of *gymnospermous Dicotyledons* from all the rest, which are *angiospermous*. It has not been done here, as it disturbs the established order less, and it would hardly have altered the place of our two families in the series, when we apply this mode of division to the diclinous alone.

§ 775. CYCADEÆ (*Cycadées*) (Cycads).—Let us first mention the great resemblance between this family and the Palmaceæ. This resemblance disappears, when we examine the interior, where we find several concentric layers of wood, formed, it is true, with great slowness, so that a single one may be the produce of several years and thus deceive observers. The pinnules of the leaves are also flat and not plaited, as in the Palms.

§ 776. CONIFERÆ (*Conifères*) (Conifers).—Those trees, so well known by the name of Evergreens, belong to this family, which does not comprehend any herbaceous plant. We have explained (§ 7,

TABLE IV.

FAMILIES.

MONOCOTYLEDONOUS VEGETABLES.

Seed with a Perisperm.—Flower with a Perianth.

Ovary
- free . . . . 1.
- adherent to the calyx . . . . 2.

1. Embryo
- excentric, or antitropous, or homotropous, intrarius. Flowers hermaphrodite, or diclinous, in a branched spadix. . . . Perianth consisting of six calyx-like, coriaceous divisions . . . Leaves, divided, palmate or pinnate . . . PALMACEÆ.
- antitropous, extrarius,
  - without a fold of the tegument,
    - 3 monospermous loculi. Axilary placentation . . Perianth consisting of two to six calyx-like divisions . . . Leaves grass-like . . . RESTIACEÆ.
    - 1 polyspermous loculus. Parietal placentation . . Perianth consisting of internal petal-like divisions . . . Leaves, sword-shaped, or grass-like . . . . XYRIDEÆ.
  - in a fold of the tegument . 3 dispermous loculi, with peltate seeds. Axilary placentation . . Perianth consisting of internal petal-like divisions . . . Leaves, wide, with longitudinal nerves . . . COMMELINACEÆ.
- homotropous
  - extrarius . . . . 3 polyspermous loculi. Axilary placentation . . Perianth consisting of internal petal-like divisions . . . Leaves, riband-, or sword-shaped . . . . . TILLANDSIEÆ.
  - intrarius. Perianth with six similar divisions,
    - all glumaceous, regular.—Style simple. . . . 3 stigmas.—Capsule . . . —Leaves, grass-like . . —JUNCACEÆ.
    - all calyx-like, or petal-like . .
      - irregular
        - quinque- or sex-partite.—Stamens on a tube.—Embryo, curved . . . —Leaves, grass-like . . —GILLIESIACEÆ.
        - tubular. . . . . —No stamen-bearing tube.—Embryo, straight . . . —Leaves, with a wide limb.—PONTEDERIACEÆ.
      - regular . . . Style
        - undivided. . —Stigmas united or distinct . . . Leaves, with parallel nervation . . . . . . LILIACEÆ.
        - divided. . . —Stamens
          - extrorse.—Capsule . . . Leaves, with parallel nervation . . . . . . MELANTHACEÆ.
          - introrse.—Berry . . . Leaves, with anastomosed nervation . . . . . SMILACINEÆ.

2. Flowers
- diclinous . . one or two whorls of fertile stamens.— . . . . Anthers introrse.—Climbing plants . . . Leaves, with branched nervation . . . . . DIOSCOREACEÆ.
- hermaphrodite
  - one or two whorls of fertile stamens. Embryo . . . .
    - intrarius.—Anthers
      - extrorse. Three; opposite to the external divisions, opening longitudinally . . . . . . Leaves, sword-shaped . IRIDACEÆ.
      - introrse.
        - Three; opposite to the internal divisions, opening transversely. Perianth coloured, with external winged divisions . . . BURMANNIACEÆ.
        - Six. Perianth
          - tubular, with flat divisions, the whole being petal-like. Seeds with a coriaceous testa . . . HÆMODORACEÆ.
          - regular, with equitant divisions, the whole being petal-like. Seeds, with a testa
            - crustaceous, shining . . . —HYPOXIDEÆ.
            - membranous or fleshy . . . —AMARYLLIDACEÆ.
        - Six—Five. Perianth irregular, the whole being petal-like . . . Leaves, with transverse nervation . . . . MUSACEÆ.
    - extrarius.—Stamens introrse . . . Six. Perianth with external, calyx-like, internal, petal-like divisions. Leaves, with longitudinal nervation . . . . BROMELIACEÆ.
  - all the stamens of one whorl and several others being abortive. Perisperm . . . . . .
    - single.—Axilary placentation. 3 Loculi.—Filaments, petaloid, one only bearing an anther with one loculus . . . Leaves, with transverse nervation . . . . CANNACEÆ.
    - double.—Placentation axilary. 3 Loculi. Filaments, petaloid or abortive, one only bearing an anther with two loculi . . . Leaves, with transverse nervation . . . . SCITAMINEÆ.
    - wanting. Cotyledon almost invisible.—Placentation
      - axilary. 3 loculi.—One to three epigynous anthers . . . . . . Leaves, with longitudinal nervation . . . . APOSTASIACEÆ.
      - parietal. 1 loculus.—One or two epigynous anthers . . . . . . Leaves, with longitudinal nervation . . . . ORCHIDEÆ.

[*To face page* 598.]

TABLE V.

FAMILIES.

DICOTYLEDONOUS VEGETABLES.

Diclinous.

I. GYMNOSPERMOUS, i. e., with naked ovules on scales.
- Embryo, with cotyledons
  - partly united together, with a free radicle. Leaves, pinnatisect.—Trunk, simple. —CYCADEÆ.
  - quite distinct, with a radicle adherent to the perisperm. Leaves, simple. Trunk, branched. —CONIFERÆ.

II. ANGIOSPERMOUS, i. e., with ovules enclosed in ovaries.
- Vegetables
  - supporting themselves, having a stem and leaves. Perianth
    - simple. 1.
    - double. 2.
  - parasitical, without green leaves and sometimes without stem. Perianth, simple. 3.

1. Ovaries
- a single one
  - monolocular
    - oligospermous. Ovules
      - erect. Seeds with perisperm
        - double and embryo antitropous,
          - several in each carpel. Several male flowers, monandrous, around several females —SAURUREÆ.
          - a single one in each carpel. Several male flowers, monandrous, around a single female —PIPERACEÆ.
        - Ovary simple or wanting
          - adherent. Embryo antitropous with lobed cotyledons. Perisperm wanting. Flowers in catkins —JUGLANDEÆ.
          - free. Embryo monospermous
            - antitropous, straight. Perisperm and calyx wanting. Flowers amentaceous —MYRICACEÆ.
            - homotropous, straight. Large ruminate perisperm. A calyx. Flowers separated. The males monadelphous. —MYRISTICEÆ.
            - antitropous
              - straight. Perisperm fleshy. A calyx. Sap aqueous —URTICACEÆ.
              - curved. Perisperm wanting. A calyx. Sap aqueous —CANNABINEÆ.
      - pendulous or campylotropous
        - solitary. Ovary
          - adherent. Embryo homotropous, straight, very small, in a large perisperm —GUNNERACEÆ.
          - free. Embryo
            - curved
              - antitropous. Perisperm wanting. Sap milky —ARTOCARPACEÆ.
              - amphitropous. Perisperm fleshy. Sap milky —MORACEÆ.
            - straight. Antitropous.
              - Perisperm wanting. Calyx multifid —CERATOPHYLLEÆ.
              - very small in a large perisperm. Calyx wanting. Leaves, opposite —CHLORANTHACEÆ.
              - Perisperm and calyx wanting. Leaves, alternate —PLATANACEÆ.
        - in pairs. Ovary
          - free. Embryo, straight
            - antitropous. Perisperm wanting —PLATANACEÆ.
            - homotropous, with a perisperm. Leaves, alternate. —STILAGINEÆ.
          - adherent. Embryo, straight, homotropous, much smaller than the perisperm. Leaves opposite —GARRYACEÆ.
    - polyspermous
      - adherent. Placentation parietal. Fruit opening at the top —DATISCEÆ.
      - free. Seeds
        - naked on the surface of a dissepiment or valves —PODOSTEMEÆ.
        - with tufts of hair, ascending from two parietal placentas, basilar. Flowers amentaceous —SALICINEÆ.
  - plurilocular
    - free. Ovules
      - 1 or 2, pendulous. Perisperm
        - wanting. 2 loculi. Flowers
          - amentaceous —BETULINEÆ.
          - not amentaceous, polygamous —ULMACEÆ.
        - fleshy. 3 loculi, rarely more or less. As many cocca —EUPHORBIACEÆ.
      - 6 to 8, peltate. Perisperm thin. 2 loculi. Male flowers, monandrous, in spikes —BALSAMIFLUÆ.
      - indefinite, attached to the dissepiments. Perisperm fleshy. 4 loculi. Male flowers, polyandrous, monadelphous —NEPENTHEÆ.
    - adherent. Seeds
      - 1 or 2 in each loculus. Perisperm wanting. Fruit indehiscent in an involucrum. Male flowers, amentaceous —CUPULIFERÆ.
      - indefinite. Perisperm thin. Fruit indehiscent without an involucrum, trilocular, three winged. Flowers in cymes. Calyx coloured. —BEGONIACEÆ.
- several in a common calyx. In the male flowers, many stamens on the calyx.—In the female*
  - *Ovules solitary
    - pendulous. Embryo, straight, homotropous in a fleshy perisperm. Anthers opening by a longitudinal slit —MONIMIEÆ.
    - erect. Embryo, straight, homotropous in a fleshy perisperm. Anthers opening by a valve —ATHEROSPERMEÆ.

2. Ovary
- free. Placentation
  - axilary. Several loculi; in each
    - 1 seed, ascending. Perisperm fleshy —EMPETRACEÆ.
    - 1 or 2 seeds, pendulous. Perisperm fleshy —EUPHORBIACEÆ.
  - parietal. Seeds indefinite in number, without perisperm. Anthers rectilinear —PAPAYACEÆ.
- adherent. Placentation, parietal. Seeds, indefinite, rarely definite, without perisperm. Anthers flexuous —CUCURBITACEÆ.

3. Ovary adherent. Loculi
- 1 or 2 with a pendulous ovule. Male flowers, mono- tri- tetr- androus. Perianth with as many divisions. 1 or 2 styles, free —BALANOPHOREÆ.
- 1, multiovulate, with parietal or free placentas. Perianth
  - with five divisions. Anthers, numerous, opening by a pore at the top. Styles united —RAFFLESIACEÆ.
  - with 3—6 divisions. Anthers, equal or double the number, opening by a slit. Styles united —CYTINACEÆ.

figs. 33, 34) the peculiar nature of their fibres marked with large pores regularly arranged. With the exception of some tracheæ scattered in the medullary sheath, these fibres alone constitute the wood. By this fact alone one of the Coniferæ may be easily distinguished almost without exception from one of any other family. The form of the leaves, reduced, as in the Firs, Larches, &c., to very narrow blades or even to a needle (*fig.* 133), is less characteristic; for we find them much wider in other genera (ARAUCARIA, CUNINGHAMIA), and even quite like an ordinary limb (DAMMARA, GINCKO). Let us remark that in several species the last branches are sufficiently shortened for these acicular leaves to be arranged in fascicles seeming to spring two or more from the same point, as in the Larch, the Pine, &c.

The flowers are monæcious or diæcious. The males consist of small catkins (*chatons*) (*fig.* 558) covered with anthers or more frequently with scales which bear one or more anthers (*fig.* 559). They are frequently grouped in a common inflorescence, a kind of close spike. Each anther or each stamen-bearing scale is considered as a flower. The females are those naked ovules of which we have spoken, and which in different shapes are placed on a scale, one, two or more in number (*figs.* 561, 562). These ovule-bearing scales are arranged on a common axis in the form of a more or less elongated cone (*fig.* 430), to which we sometimes give the name of GALBULUS (*galbule*), when it is very short and composed of a very small number of scales (*fig.* 431). At other times several scales are imbricated without bearing any ovule, but thus forming a kind of common involucrum around a single ovule or two at the most, which are then more or less completely enveloped by a cupula.

According to the various modifications of the fruit, we may divide this family into several tribes; they might be termed families, and in that case the Coniferæ will of course be considered as a class. In the ABIETINEÆ (*Abiétinées*), the membranous scales form a cone and each bears inverse ovules united at their base; in the CUPRESSINEÆ (*Cupressinées*), reduced to a small number, they form a galbulus, each scale of which bears free and erect ovules. One cupula surrounds or envelopes the ovule in the TAXINEÆ (*Taxinées*) (Taxads), the stamens of which are naked, and also in the GNETACEÆ (*Gnétacées*) (Joint Firs), the stamens of which have a small perianth at the base of each stamen and also articulated stems.

The seed of the Coniferæ (*fig.* 564) is very remarkable; first from the presence of several rudimentary embryoes in each, arranged in a whorl around one which is developed; this plurality of embryoes

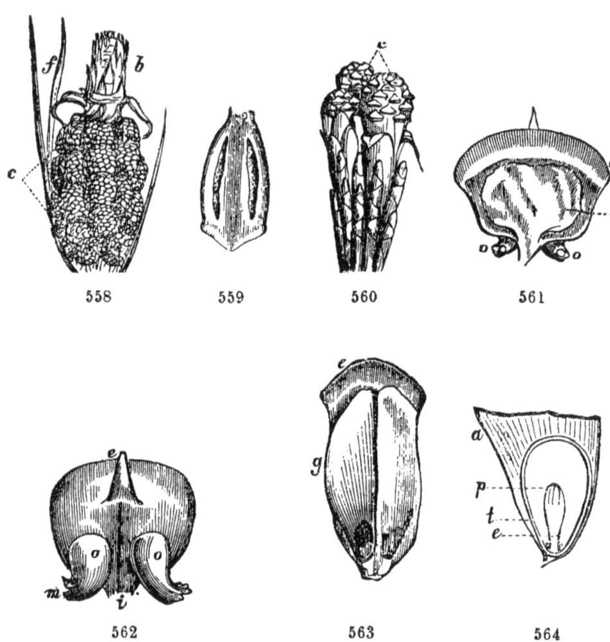

558—564. The organs of the fructification of the Common Pine (PINUS SYLVESTRIS).

558. The collection of male catkins *c.*—*f* Leaves.—*b* Terminal bud.

559. Male flower, or an anther-bearing scale.

560. Three collections of female flowers or young cones *c* at the extremity of a branch.

561. A scale detached from one of these cones. The outside is turned towards the observer.—*b* Bract.—*e* Scale.—*o o* The tops of the ovules.

562. The same, the inside being towards us.—*e* Scale.—*i* The part where it is inserted on the axis.—*o o* The naked, inverse ovules.—*m* Their upper opening or micropyle, which is described as a stigma by those who wish to call this organ an ovary instead of ovule.

563. The same taken from a ripe cone.—*e* and *i* have the same signification.—*g* One of the seeds with its wing. The other has been taken away.

564. A longitudinal section of the seed.—*a* Base of the wing.—*t* Tegument.—*p* Perisperm.—*e* Embryo. Close to the radicle we see two other bodies, which are abortive embryoes.

is still more clearly marked in the Cycadeæ. The developed embryo occupies the axis of a large fleshy perisperm; we have seen that it is frequently polycotyledonous (§ 584, *fig.* 467), and another characteristic still more exceptional is, that it is united to the surrounding perisperm by the extremity of its radicle: this does not take place in the Cycadeæ.

We see to what a degree of simplicity the organs of reproduction descend in this group, reduced to anthers and ovules, sometimes even to one of each. We do not find any plants more, or even as simple among Monocotyledons; this is another proof that these two great branches of Phanerogamic plants run parallelly rather than in a progressive series.

The wood of the Coniferæ is very useful for all kinds of carpenter's work. Its usefulness is owing to the large quantities of resin secreted in its tissues, which render it impermeable to water and capable of being preserved for a long time. This resin, liquid during life, becomes concrete after death by the evaporation of the essential oils which held it in solution. It is found in every part, but mostly in large lacunæ regularly distributed in the bark. It varies in different species and is mixed with different principles, and, according to these different states, assumes the name of pitch (*poix*), of balsams (*baumes*), of turpentine (*térébinthe*). From trees of this family we also obtain storax and sandarach (*sandaraque*). The resins produce a stimulating or even irritating effect on the animal economy; medicine takes advantage of this property and employs different parts and different products of several species belonging to the family which now occupies our attention. The galbuli of the Juniper (*Genièvre*) (which are improperly called berries on account of their fleshy scales being united and forming a body apparently simple), are employed in the manufacture of *gin*, which, doubtless, derives its flavour and some of its properties from them; other wild indigenous fruits, richer in sugared principles, also assist in imparting this flavour. The resinous principles do not exist in the kernel of the seed; soft and oily, it is eaten in some species in which it is large enough, as in the Stone pin.

The Cycadeæ are also filled with a juice throughout the tissue and accumulated in the lacunæ; but it is of a different nature, mucilaginous and nauseous, or else insipid.

§ 777. We shall notice some of the *angiospermous diclinous* families particularly.

AMENTACEÆ (*Amentacées*) (Amentals).—Under this name, several families were confounded. They were grouped together from the common characteristic of their male flowers being in a catkin, which connects them also with the Juglandeæ, which differ, however, in their leaves being compound and not simple as in the others. Along with the Coniferæ these families furnish almost all the great trees of our country and form the whole of our forests; the BETULINEÆ (*Bétulinées*) (Birchworts) furnish the Birch (*Bouleau*) and the Alder (*Aune*); the CUPULIFERÆ (*Cupulifères*) (Mast-

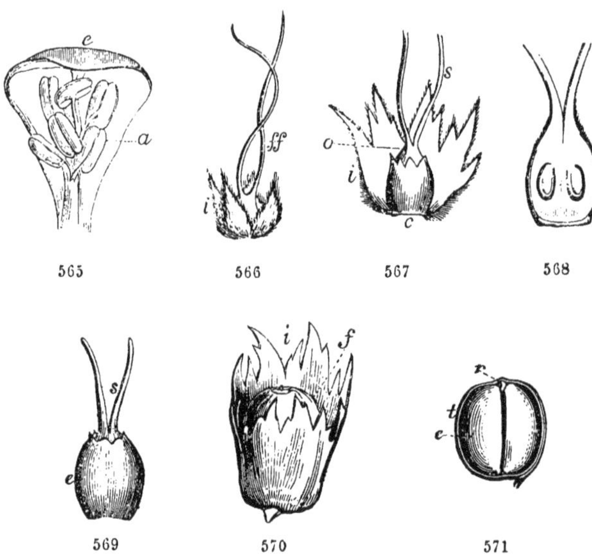

565—571. The organs of the fructification of one of the Cupuliferæ, the Hazel (*Coudrier*) (CORYLUS AVELLANA).

565. One stamen-bearing scale *e* or male flower taken from the catkin.—*a* Stamens.

566. Female flower *ff*, very young, with its involucrum *i*.

567. The same, more developed, the involucrum *i* allowing us to see the ovary *o* which is almost wholly covered by the calyx *c*.—*s* Styles.

568. A longitudinal section of the same shewing the two loculi, each containing a pendulous ovule.

569. The same, still more developed.

570. The ripe fruit *f* enveloped with the involucrum *i*.

571. The seed. Half of the teguments *t* have been removed shewing the embryo *e*. —*r* Radicle.

worts)[q], the Oak, the Chestnut, the Beech, the Hazel (*Coudrier*), the Hornbeam (*Charme*); the SALICINEÆ (*Salicinées*) (Willowworts), the Poplar and the Willow; the PLATANACEÆ (*Platanées*) (Planes), the Plane-tree; the ULMACEÆ (*Ulmacées*) (Elmworts), the Elm; the JUGLANDEÆ (*Juglandées*) (Juglands), the Walnut (*Noyer*). The MYRICACEÆ (*Myricées*) (Galeworts) are represented in our climate only by low shrubs, but in the Archipelagoes of Asia by large trees, the appearance of which is somewhat similar to that of certain Coniferæ, as the CASUARINA. This genus has become the type of a small distinct family in the opinion of some botanists. The great usefulness of these vegetables to man, of the wood of all of them, of the tanning property of the bark of several, of the seeds of some, is too well known for us to dwell a single instant on them. Let us remark in passing that the presence of fecula and of oil mixed in different proportions renders these seeds, as that of the Chestnut, of the Beech, of the Filbert, of the Walnut, capable of being employed for food and for extracting oil, or for both at the same time.

§ 778. URTICACEÆ (*Urticées*) (Nettleworts).—This old established family formerly contained several others which are now separated. 1st. That family which retains this name having for its type the genus URTICA (*Ortie*) (*figs.* 572, 577), so well known on account of the abominable stinging properties of its hairs, the structure of which we have explained (§ 247, *fig.* 213); this property is very intense in several tropical species, and causes violent inflammations, always very painful, sometimes mortal. 2nd. The CANNABINEÆ (*Cannabinées*) (Hempworts), to which among others belongs the Hop (*Houblon*) used in brewing beer, to which it imparts an agreeable bitter owing to the resinous principle contained in the small yellowish glands, covering the whole of the surface, especially of the calyx; this principle is the *Lupuline* (§ 246 *n*); also the Hemp (*Chanvre*), so useful from the tenacity of the fibres of its bark: this property is found in several other plants of this family and of the Nettleworts. The seed of the Hemp is the *Chènevis*. Its leaves contain a powerful narcotic principle; from those of the Indian Hemp the intoxicating condiment so much esteemed in Egypt and Arabia, the Hadschy (*hashish*) is prepared, of which so many marvellous accounts have been told. According to some authorities, the etymology of the word *assassin*

[q] CORYLACEÆ.—*Lind. Veg. Kingd.*, p. 290.—TRANS.

is referred to this plant, because the Old Man of the Mountain, that chief who so well knew how to find agents for all the murders which he wished to have committed, obtained the blind devotedness of his sectaries, by giving them with the drunkenness of the hadschy a foretaste of the celestial happiness, which he promised them in recompense for their perpetual obedience. 3rd. The ARTOCARPACEÆ

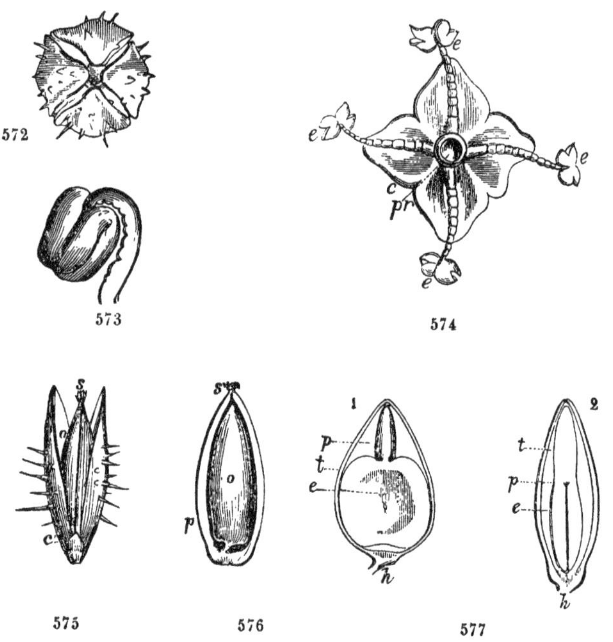

572—577. Organs of the fructification of the Nettle (*Ortie*) (URTICA URENS).

572. Bud of the male flower.

573. Stamen taken from the preceding, shewing the curve and structure of the moveable filament, and that of its anther before dehiscence.

574. Male flower expanded.—*c* Calyx.—*e e e e* Stamens lying back and open, hypogynous.—*p r* Rudiment of the central pistil.

575. Female flower.—*c* Calyx with unequal folioles, the two outer ones being the smaller.—*o* Ovary.—*s* Sessile stigma.

576. Vertical section of the pistil shewing the direction of the ovule *o*.—*p* Wall of the ovary.—*s* Stigma.

577. The seed, cut vertically, parallelly (1) and perpendicularly (2) to the cotyledons.—*t* Tegument.—*h* Hilum.—*p* Perisperm.—*e* Embryo.

(*Artocarpées*) (Artocarps), amongst which we find two plants well known as furnishing bread and milk already prepared by nature: the one, the ARTOCARPUS INCISA or Bread-fruit-tree; the other, the GALACTODENDRON or Cow-tree, a native of the Cordilleras of Venezuela; it furnishes by incision an enormous quantity of a white thick liquid, which has the taste and some of the qualities of real cow's milk. It contains more than one half of water, and, with a little sugar and albumen, a very large proportion of a fatty substance to which it seems to owe its principal properties. The presence of large quantities of milky juice is common to other plants of the same family; but, salutary or innocent in some, it becomes noxious, virulent and even poisonous in others. It is singular to find included in the same family two trees of such different properties, the Cow-tree and the ANTIARIS TOXICARIA of Java, which furnishes the *Upas*, a most deadly poison, which has been the subject of the most harrowing stories. Some of its properties, but not the fundamental, may be referred to the presence of *strychnine*, an alcaloïd now well studied and the subject of many experiments in chemistry and medicine. 4th. The MORACEÆ (*Morées*) (Morads), remarkable for some trees as the Mulberry, the Indian-rubber-tree, and the Fig-tree. The species of the last genus are extremely numerous and also contain, as well as most other plants of the same family, a milky juice commonly very acrid. Like that of the preceding, it merits attention on account of the presence of a peculiar principle employed in manufactures, Indian-rubber or *Caoutchouc*, which, however, frequently exists in juices of this nature, although extracted from vegetables belonging to very different families. 5th. The GUNNERACEÆ (*Gunnéracées*) and some other genera, which have become types of small families or placed in others.

§ 779. PIPERACEÆ (*Pipéracées*) (Pepperworts).—This family has been separated some little time from the Urticaceæ, to which it was at first united. Its type is the genus PIPER (*Poivre*) (Pepper), so well known from its being daily employed at our meals to promote digestion. Other genera have the same properties in their different parts, as is proved by the leaves of the Pepper betel (*Poivre betel*) (CHAVICA BETLE), which the inhabitants of certain parts of Asia delight in continually chewing on account of their intoxicating effects. But the Piperaceæ deserve more of our attention on account of several peculiar points of their organization, such as the presence of fibro-vascular fascicles in their pith, which imparts to their young stems the appearance of Monocotyledons, the existence of a double perisperm, the inner one reduced, as in the NYMPHÆA, to a small fleshy sac, to which the embryo remains attached by its suspensor, persistent and occupying the top of the seed (*fig.* 582 *pi*), the rest of which is filled by the external perisperm *pe*, remarkable for the abundance of its acrid and aromatic principles, and for being that part mostly employed. On the flower-bearing axis around each carpel, which we consider as a female flower, are inserted two or more stamens each of which may be considered as a male flower.

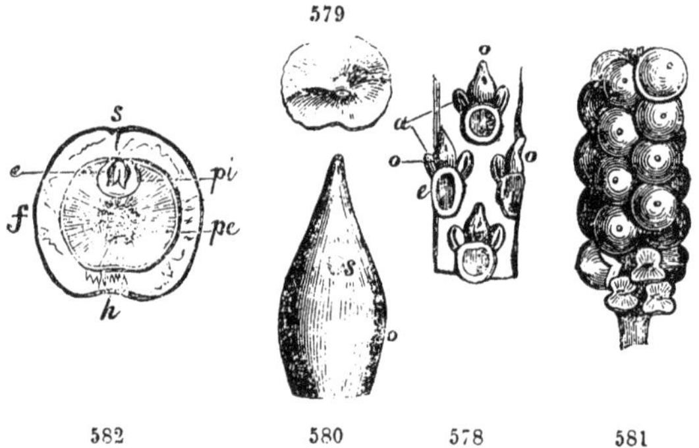

§ 780. MYRISTICEÆ (*Myristicées*) (Nutmegs).—This family also furnishes another very important spice, the Nutmeg (*Noix de muscade*), the perisperm of which contains great quantities of an aromatic oil in its tissue. The arillus of the seed is the spice we know as Mace. The bark and the pericarp of the Nutmeg-tree (*Muscadier*), however, abound in an acrid and viscous juice.

578—582. The organs of the fructification of the Black Pepper (PIPER NIGRUM).

578. Part of the flower-bearing spike.—*e* Scales, accompanied by two anthers or male flowers around a pistil *o* or female flower.

579. The inside of the scale.

580. The pistil.—*o* Ovary.—*s* Stigma.

581. The spike in fruit.

582. The vertical section of a separate fruit.—*h* The point of insertion of the fruit and the seed, corresponding, consequently, to the hilum.—*f* Pericarp.—*s* Stigma.—*pe* External perisperm.—*pi* Internal perisperm or fleshy sac which contains the embryo *e*.

§ 781. NEPENTHEÆ (*Népenthées*) (Nepenths).—The type of this family is the genus NEPENTHES, the median nerve of which is prolonged beyond the limb, bearing a fresh, foliaceous, pitcher-shaped expansion; to the top of this pitcher a kind of covering is fitted, attached as if by a hinge and susceptible of depression and elevation, so that the pitcher is sometimes found closed, sometimes open. We frequently find it full of an aqueous liquid which appears to be secreted in its interior.

§ 782. Before we finish, we will mention a few singular families, the plants of which, living as parasites on the roots of others, are hardly raised above the soil, and, sometimes without stem, are always without any leaves except a few scales. One of them, the CYTINUS [r], grows on the roots of the CISTUS in the South of Europe. But the most curious, without exception, is the RAFFLESIA [s], the flowers of which expand immediately on the surface of the earth. Those of the first species which was discovered, a true giant of vegetable life, puzzled the minds of botanists for some time. They could not define this extraordinary production of nature. The flower borne on a short subterranean stem which rises from that of a CISSUS, on which it is, as it were, grafted, was nearly a metre (French) or 39·333 inches (English) in diameter, and we may easily conceive that the very size of the parts, exaggerating the smallest details, would render them exceedingly difficult to be recognised. Long study of this and afterwards of other species of less unusual dimensions, has clearly demonstrated the nature of these plants. They are remarkable for other characteristics than that of their dimensions; as, for the dehiscence of their anthers, which open by a small pore at the summit, sometimes common to a number of locelli, which are formed in each anther; for their placentas covered with small seeds, which, applied to the walls, are detached to hang freely in the loculus or traverse it from the bottom to the top; for the undivided embryo, such as we observe rather frequently in plants parasitical on roots and deprived of leaves; this renders their being deprived of cotyledons merely a matter of course. These families form the passage to the Aristolochiaceæ; if we mention them, therefore, before some other diclinous plants with a double perianth, of which it now remains for us to speak, it is because these ought to be placed here rather on account of the systematical bond connecting them to the others in our tables than on account of their real affini-

[r] Belonging to the family CYTINACEÆ (Cistusrapes).—TRANS.
[s] Belonging to the family RAFFLESIACEÆ (Patma worts).—TRANS.

ties, which would doubtless assign their place to them in some other part.

§ 783. EUPHORBIACEÆ (*Euphorbiacées*) (Spurgeworts).—These are considered by several authors to belong to the polypetalous hypogynous plants, being situated not far from the Malvaceæ or the Rutaceæ ; this may be true, if we only examine those genera with flowers clearly provided with petals. But we see in Table V, that we have discovered in this family on the other hand the existence of flowers with a simple perianth or even none. This great family, in short, presents great variety in the composition of the flower, which, almost complete in certain genera (the JATROPHA [*fig.* 251], for instance), successively descends in others to the lowest degree (the EUPHORBIA [*figs.* 256, 583, 584, 585], for instance). We observe rather frequently some genera less complete than the others in the same family; they are members of these families by some essential characteristics ; but impoverished and degraded members, which represent it badly : and then, in general, it is to the most complete that

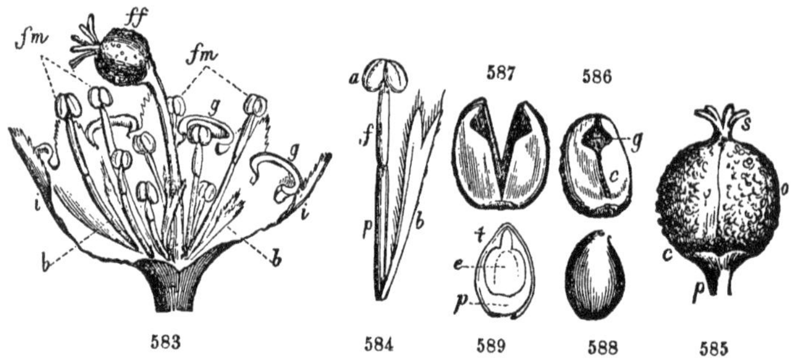

583—589. The organs of the fructification of the EUPHORBIA PALUSTRIS.

583. Inflorescence. The involucrum *i* has been opened and taken away so as to exhibit the situation of the flowers it contains.—*gg* Glandular lobes alternating with as many divisions.—*b* Membranous laminæ or bracts at the base of the flowers.—*fm*, *fm* Male flowers, each consisting of a stamen.—*ff* Central female flower.

584. A male flower.—*b* Bract.—*p* Pedicel.—*f* Articulated filament on the pedicel. —*a* Anther.

585. Female flower.—*p* Top of the pedicel which bears them.—*c* Calyx.—*o* Ovary. —*s* Stigma.

586. A coccum *c*. The inside. We perceive the seed *g* through the opening where the nourishing vessels enter.

587. A coccum after the dehiscence and the emission of the seed.

588. Seed.

589. The vertical section of the same.—*t* Teguments.—*p* Perisperm.—*e* Embryo.

we must apply in order to determine the true type of the family disguised by the reductions of the rest. But in the Euphorbiaceæ very few would present this more elevated type, whilst the greater part, particularly in the large genus EUPHORBIA whence the name is derived, present in their flower extreme simplicity, which, by assimilating their entire inflorescence to one flower (§ 385, *fig.* 583), brings them into close affinity with several Amentaceæ and Urticeæ. Thus, on the other hand, whatever may be the definitive place of the Euphorbiaceæ, marked lower down in the series by the structure of the greater number of its genera, higher by that of some, the flower of which is more perfectly composed, we may follow the insensible transition from one to the other; all, moreover, are united together by some common characteristic, as the constant separation of the stamens and of the pistils in different flowers, the hypogyny of the distinct or frequently united stamens, the free ovary with several loculi and one or at the most two ovules pendulous from the internal angle in each (these loculi, most frequently three in number, being separated at the maturity into so many cocca [*figs.* 586, 587]), the existence of a thick, fleshy, oleaginous perisperm, around an embryo with superior radicle, with large, flattened cotyledons (*fig.* 589). Their appearance is very varied from that of lofty trees to lowly herbs. The singular forms of some African species of EUPHORBIA remind us of those of the CACTUS.

Several plants of this family, and particularly those of its principal genus, have a peculiar milky and acrid juice. It is especially in this juice that the principle resides which gives to the Euphorbiaceæ uniform properties. These are of unequal strength in the different species, so that its action, reduced in some and causing slight irritation, produces in others violent inflammation; in some it becomes at last a violent poison. The different parts in which the tracheæ or proper vessels abound, the root, the leaves, the bark especially, will, therefore, act powerfully on the animal economy; the seeds also produce the same effects. It has been ingeniously discovered that the properties of their parts are not equal in strength, that those of the embryo, of the radicle particularly, are much more powerful than those of the perisperm. This unequal distribution of the most active principles in the different parts of the same plant accounts for the contradictory results to which experiments have often led observers, when they neglected to keep an account of which part they employed. Medicine has often taken advantage of these virtues of the Euphorbiaceæ in order to obtain emetics (as from the roots of the EUPHORBIA IPECACUANHA) or purges. But for these purgative purposes, the concentrated milky juice extracted from certain species of EUPHORBIA, especially of the succulent kinds, has for a long time been abandoned as too dangerous, and the oil extracted from the seeds has been employed in preference, from those of the Castor-oil plant (*Ricin*) or Palma-Christi, for instance, if we wish for a gentle action; from those of the CROTON TIGLIUM, if we wish for a very violent one. The JATROPHA are also much employed in medicine.

It is very remarkable that we find, along with the side of energic medicaments and even violent poisons, very nutritious food, as the *Cassava*, obtained from the JANIPHA, a genus which borders close on the EUPHORBIA; it furnishes the food of a large part of the population of South America. This is only an apparent contradiction; the thick and fleshy root, from which this farina is extracted, is very dangerous when raw, and the juice, with which it is then filled, causes terrible accidents and even speedy death. But cooking destroys the poisonous principle and, consequently, the root is only eaten after having been gratered and passed through a sieve, washed and subjected to the action of fire on an iron plate. Whilst it is being washed a very pure fecula is deposited which is then called *Tapioca*.

A tree of this family, the Manchineel (*Mancenillier*), has often been mentioned as presenting in the highest degree the toxical properties of the Euphorbiaceæ, since its shade alone may prove mortal to the imprudent wretch who reposes under it. The fact has never been well determined, and the experiment tried by courageous travellers has given no fatal result; which does not, like every other negative result, decide the question. The principle, which imparts these properties, being commonly volatile, as is apparent from several facts (among others its destruction by cooking in the CASSAVA), it is clear that the atmosphere around the Manchineel will, according to different meteoric circumstances, be charged with the evaporated parts in different degrees, if it is indeed ever so. This, however, is incontestable, that the milky juice of the same tree is very strongly impregnated with it.

Caoutchouc, which exists in the juice of the Figworts, is also found in that of the Euphorbiaceæ, particularly of the SIPHONIA ELASTICA, a tree of Guiana; this tree, in fact, is supposed to be the most fertile source of the supply. Others, in which the milky juice is wanting, have on the other hand a colouring principle, the Turnsole (*Tourne-*

sol), which we have already met with in another family, the Lichens. A small plant, common in the south of France, the Crozophora tinctoria, has been for a long time used for this purpose.

§ 784. Cucurbitaceæ (*Cucurbitacées*) (Cucurbits).—This family is further removed than the preceding from all those we have enumerated in this division, and ought rather to be placed among the perigynous polypetalous plants along with the Passion flowers and the Loasads, in spite of their diclinous flowers, although their internal perianth, when it exists, is not a real corolla and also is not divided into distinct petals. The Melon, the Water-Melon

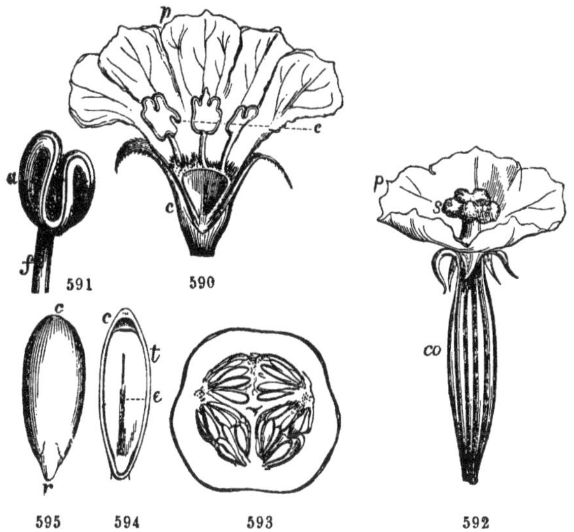

590—595. The organs of the fructification of the Cucumber (*Concombre*) (Cucumis sativus).

590. Male flower, the envelopes of which have been split longitudinally and are separated so as to shew the inside.—*c* Calyx.—*p* Inner coloured calyx or corolla. *e* Perigynous stamens.

591. A stamen.—*f* Filament.—*a* Anther.

592. Female flower.—*c o* Calyx united to the ovary.—*p* Corolla.—*s* Stigmas.

593. Horizontal section of the ovary, shewing its division into three loculi and the parietal insertion of its ovules.

594. Vertical section of the seed.—*t* Tegument swollen at the chalaza *c*.—*e* Embryo.

595. Embryo.—*r* Radicle.—*c* Cotyledons.

(*Pastèque*), the Gourd (*Citrouille*), the Cucumber (*Concombre*) are sufficient to shew the food it furnishes to man and the general aspect of the plants which compose it. They have herbaceous, creeping and climbing stems, covered with palmatinervate, lobed leaves as well as with tendrils, whose anomalous position on the side and not in the axil of the petiole deserves to be taken notice of. In the flowers, sometimes very large, the calyx, terminated by five teeth, is doubled internally by a second envelope which perhaps also belongs to it. It bears in the males five stamens with broad filaments crowned by a flexible anther (*fig.* 591), frequently grouped into threes (*fig.* 590). In the female, the ovary is completely united to it (*fig.* 592) and bears its ovules on the three fleshy parietal placentas, projecting into the loculus (*fig.* 593), so as almost to fill it entirely; it is terminated by a short style and a thick velvety stigma. We know from the examples we have mentioned the nature of the fruits, which, sometimes rather small, acquire at other times enormous dimensions and frequently assume extraordinary forms, as in the Calabashes (*Calebasses*). The numerous flat seeds contain under a coriaceous testa an embryo without perisperm, turning its radicle towards the point of attachment (*figs.* 594, 595).

§ 785. We also eat, but generally when cooked, the fleshy fruit of the Papaw (*Papayer*) (CARICA PAPAYER), the type of a small neighbouring family, indigenous to South America. It contains along with water and a little oil a large quantity of fibrine, to which it doubtless owes its nutritious properties.

TABLE VI.

FAMILIES.

DICOTYLEDONOUS VEGETABLES,

with hermaphrodite apetalous flowers.

Embryo
- straight, occupying the axis of the seed, in a fleshy perisperm, or without perisperm . . . 1.
- amphitropous, on the side of or quite round a farinaceous perisperm. Placentation central . . . 2.

1.—Ovary
- adherent
  - 3—6 loculi. Numerous ovules with axillary placentation. Embryo, very short, towards the end of a large fleshy perisperm. 6—12 epigynous stamens . . . ARISTOLOCHIEÆ.
  - 1 loculus. Ovules few in number
    - at the top of an erect central placenta. Seeds with a perisperm and short, flat cotyledons. Isostemony. Perigyny . . . SANTALACEÆ.
    - pendulous from the top of the loculus. Seeds without perisperm, with long, foliaceous, convolute cotyledons. Diplostemony. Perigyny . . . MYROBALANEÆ.
- free. Embryo
  - antitropous
    - Large perisperm. Several valves and as many placentas, opposite, parietal, polyspermous. Diplostemony. Monadelphia. Sometimes, appendices alternating with the stamens . . . SAMYDEÆ.
    - No perisperm. 2 valves and as many placentas, opposite, parietal, dispermous. Diplostemony or isostemony. Appendices alternating with the stamens . . . AQUILARINEÆ.
  - homotropous
    - 4 dispermous loculi. Seeds ascending. Calyx bellying out, 4- fid. As many alternate stamens . . .—PENÆACEÆ.
    - 1 loculus. Seeds
      - without perisperm
        - 1 or 2 upright. Radicle inferior. Calyx 4-partite. As many opposite stamens, inserted towards the top of these divisions . . . PROTEACEÆ.
        - 1 pendulous. Radicle superior. Calyx
          - 4-6- partite. Stamens double or triple the number, opening by valves . . . LAURINEÆ.
          - tubular. Stamens double, equal or half the number, opening by slits . . . THYMELACEÆ.
      - with a perisperm
        - 1 pendulous. Radicle superior . . .
        - 1 straight. Radicle inferior. Calyx enveloping the ovary. Isostemony. Stamens opening by slits . . . ELÆAGNEÆ.

2.—Loculi.
- Several, monospermous and as many distinct styles. Calyx, herbaceous or coloured, 4-5- partite. As many or more opposite stamens . .—PHYTOLACINEÆ.
- One. Embryo
  - lateral, hardly curved, antitropous with superior radicle. Calyx, herbaceous or coloured, 3-4-5-6- partite. Stamens equal or more in number. 2—4 styles . . . POLYGONEÆ.
  - annular or spiral.
    - No involucrum. Calyx
      - tubular, hardened, 4-5- dentate. Stamens, perigynous, equal, less or greater in number. 1 seed. 1—2 styles . . . SELERANTHEÆ.
      - 3-5- partite, herbaceous. Stamens, equal in number, opposite. 1 seed. 4-5 distinct stigmas . . . ATRIPLICEÆ.
      - 3-5- partite, scarious, with two bracteoles. Stamens equal in number, opposite; or double, the alternate sterile. One or more seeds. Style simple. Stigma simple or lobed . . . AMARANTACEÆ.
    - Involucrum, from 1- to multi-florate. Calyx, tubular, petaloid, its hardened base enveloping the fruit. Limb with 4—10 divisions. Stamens, hypogynous, equal, less or greater in number. 1 seed. Style and stigma simple . . . NYCTAGINEÆ.

Table VIII.

FAMILIES.

DICOTYLEDONOUS VEGETABLES.

Hypogynous, Polypetalous Plants, with parietal placentation.

Placentas | opposite to the valves . . . . . . 1.
| alternate . . . . 2.

1. Embryo | in the axis of a perisperm, to which it is almost equal. Stamens | definite in number | Style, bi-tri-fid. No stipules. Flowers, regular. Anthers, introrse . . . . } Frankeniaceæ.
| | | Style, simple. Stipules. Flowers, regular. Anthers, extrorse . . . . } Sauvagesieæ.
| | | Several styles. No stipules. Flowers, regular. Anthers, extrorse. . . . . } Droseræ.
| | | Style simple. Stipules. Flowers commonly regular. Anthers, extrorse . . . . . . } Violarieæ.
| | indefinite in number.—Embryo | antitropous, curved . . . . . . —Cistineæ.
| | | homotropous, straight . . . . . —Bixineæ.
| very small at the extremity of a large perisperm, straight. 5 petals and as many stamens . . . . —Pittosporeæ.
| without perisperm, straight, antitropous. 3 valves. Seeds with tufts of hair. Stamens, equal in number to the petals, or double of them.—Tamariscineæ.

2. Embryo | without perisperm, bent on itself. Flowers | irregular. Stamens definite or indefinite in number. Capsule gaping at the top . . . . —Resedaceæ.
| | regular. Sepals and petals 4. Stamens | indefinite in number. Capsule or berry . . . —Capparidaceæ.
| | | definite in number, didynamous. Siliqua . . —Cruciferæ.
| very small at the extremity of a large perisperm, straight. Flowers | irregular. Sepals, petals, stamens definite in number, placentaries, 2 and its multiples. Herbs with an aqueous juice . . . } Fumariaceæ.
| | regular. Sepals and petals, 2 and its multiples. Stamens indefinite in number. Herbs with a milky or coloured juice . } Papaveraceæ.

[*To face page* 612.]

TABLE X.

FAMILIES.

DICOTYLEDONOUS VEGETABLES.

Hypogynous, Polypetalous Plants, with axillary placentation.

Embryo
- very small, imbedded at the extremity of a large perisperm . . . . 1.
- surrounded with a perisperm, to which it is almost equal . . . . 2.
- without perisperm . . . . 3.

1. Carpels
   - distinct. Number of the parts of the flower
     - quinary. Perisperm
       - horny . . . Stamens indefinite in number. Seeds without arillus . . . . —RANUNCULACEÆ.
       - fleshy . . . Stamens indefinite in number. Seeds with an arillus . . . . —DILLENIACEÆ.
     - ternary. Perisperm, fleshy
       - ruminate. Stamens indefinite in number. No arillus . . . . —ANONACEÆ.
       - solid. . Stamens opening by
         - slits
           - indefinite in number. Seeds, attached to the internal angle, with an arillus.—MAGNOLIACEÆ.
           - definite in number. Seeds, scattered over the walls. No arillus . . —LARDIZABALACEÆ.
         - valves, definite in number. One carpel. . . . . . . No arillus . . —BERBERIDACEÆ.
   - united into a plurilocular ovary. Loculi
     - containing 1 or 2 erect seeds. . . . Stamens equal in number and opposite to the petals. Climbing shrubs . . . . —VINIFERÆ.
     - polyspermous . . . . Stamens indefinite . . . . . . Aquatic plants . . . . —SARRACENIEÆ.

2. Calyx with imbricate. prefloration Stamens
   - definite in number
     - opposite to the petals. Carpels, distinct, monospermous. Seeds, kidney-shaped. Flowers, diclinous by abortion . . . . —MENISPERMACEÆ.
     - alternating with the petals or double their number
       - free. Perisperm
         - fleshy. Flowers
           - diclinous by abortion . . . . —XANTHOXYLACEÆ.
           - hermaphrodite. Endocarp
             - separated from the mesocarp, bivalved .—DIOSMEÆ { of Europe. of Australasia.
             - united to the mesocarp . . . . . —RUTACEÆ.
         - horny. Styles united. . . . . Endocarp, united to the mesocarp . . . . . —ZYGOPHYLLACEÆ.
       - united
         - to one another. Flowers
           - regular. Perisperm
             - horny. Styles distinct / Inflorescence
               - terminal, definite. 3 to 5 loculi, 2-spermous. Petals simple . . . . LINACEÆ.
               - axillary. One loculus (by abortion), monospermous. Petals appendiculate. . . . . . ERYTHROXYLACEÆ.
             - fleshy. Styles
               - distinct. Inflorescence, terminal, definite . . . . . . OXALIDEÆ.
               - united. Seeds
                 - without wings .—MELIACEÆ (Melieæ).
                 - with wings . —CEDRELACEÆ.
           - irregular. Perisperm fleshy. Seeds, carunculate. Anthers, often monolocular . . . . . POLYGALACEÆ.
         - to the petals or free. Perisperm fleshy. Drupe. Placentation sometimes central. Anthers bi-locular.—OLACINEÆ.
   - indefinite in number
     - free, united to the base or polyadelphous. Calyx simple . . . . —TERNSTRŒMIACEÆ.
     - monadelphous. Calyx
       - with an involucre. Connective hardly developed in proportion to the loculi of the anther . . . . .—CHLENACEÆ.
       - simple. . . . Connective largely developed with very small loculi . . . . .—HUMIRIACEÆ.

valvate. Stamens { free or polyadelphous. Anthers, bi-locular opening by | pores at the top. Petals | entire with stamens opposite in pairs . . . —TREMANDREÆ.
scolloped. Stamens multiple . . . —ELÆOCARPEÆ.
longitudinal slits. Petals entire. . . . Stamens commonly indefinite in number . . . —TILIACEÆ.
monadelphous. Flowers | apetalous, diclinous by abortion . . . Pollen, smooth, globular.—STERCULIACEÆ.
petalous, hermaphrodite. Anthers | bi-locular . . . Pollen, smooth, globular.—BYTTNERIACEÆ.
uni-locular (with stamens indefinite in numb.) or bi-locular (with stamens def.) Pollen, smooth, trihedric. } BOMBACEÆ.

3. Calyx with prefloration | valvate. Pollen | trihedric, smooth . . . Anthers, sometimes bilocular, and then definite. Cotyledons, foliaceous, bent on the radicle } BOMBACEÆ.
globular, rough . . . Anthers, uni-locular, indefinite . . . Cotyledons, foliaceous, bent on the radicle . . . —MALVACEÆ.

imbricate. Stamens | indefinite in number. Styles | united or wanting. Leaves, simple | alternate. Cotyledons distinct { following the axis of the radicle. Calyx accrete, two lobes of which are very long . . . } DIPTEROCARPEÆ.
sometimes bent on the radicle, which is much shorter than they are. Calyx not accrete . . . —TERNSTRŒMIACEÆ.
straight, small, shorter than the radicle . . . —MARCGRAVIACEÆ.
opposite. . Cotyledons united to one another, straight . . . —GUTTIFERÆ.
separate. . . . Leaves, opposite | compound. Cotyledons, almost wanting, bent on the largely developed radicle . . . —RHIZOBOLEÆ.
simple. . Cotyledons, following the axis of the radicle, as large as it is . . . —HYPERICINEÆ.

definite in number, | united | by the anthers. Cotyledons straight and thick. Capsule with five monospermous loculi. Isostemony. Flowers, irregular . . . —BALSAMINEÆ.
by the filaments. Cotyledons | plaited lengthways and bent on the radicle. Five carpels, united to their styles, with an oblong receptacle. Flowers, regular or irregular . . . } GERANIACEÆ.
straight, following the axis of the radicle. | Diplostemony. | Hesperidium. Leaves, dotted . . . —AURANTIACEÆ.
Fleshy fruit or capsule. Leaves compound, not dotted.—MELIACEÆ (Trichilieæ).
Meiostemony (commonly triandria). Three samaras or trilocular berry. Leaves, simple, opposite . . . } HIPPOCRATEACEÆ.
curved, or bent on themselves. Generally diplostemony. Samaras, cocca or fleshy fruit, bi- or tri-locular. In each ovule one lycotropous ovule. Leaves, simple, opposite . . . } MALPIGHIACEÆ.

free. Embryo | curved. Ovules | 2 placed by the side of one another. Double samara. Leaves opposite . . . —ACERINEÆ.
1 erect . . . } Samaras, capsules or fleshy fruit. Leaves, commonly alternate . . . —SAPINDACEÆ.
placed above one another . | 2-3.
2, | 3 loculi. Capsule, coriaceous, uni-locular (by abortion). Leaves, opposite . —HIPPOCASTANEÆ.
5 loculi. Cocca with a bi-valved pericarp separated from the mesocarp. Leaves, alternate . . . } DIOSMEÆ (Cusparieæ).
straight. Carpels | united | into one capsule with several polyspermous loculi. Several styles . . . —ELATINEÆ.
until ripe, then separated, indehiscent, monospermous. One style . . . —TROPÆOLEÆ.
distinct. | Cocca with a bivalved endocarp, separated from the mesocarp. Leaves dotted . . . —DIOSMEÆ (African).
Drupes | several. Styles | distinct at the base, growing from the top of the ovaries. One pendulous ovule . . . } SIMARUBACEÆ.
united into one, inserted on a gynobasic ovule. One erect ovule . . . —OCHNACEÆ.
one. Stigma sessile. 2 pendulous ovules. Leaves, opposite, dotted . . . —AMYRIDEÆ.

[To face page 613.]

DICOTYLEDONOUS VEGETABLES,

with Hermaphrodite Apetalous flowers.

(Table VI. page 612.)

§ 786.—We know that Jussieu divided Apetalous plants into three classes; the Epistaminous, Peristaminous and Hypostaminous. Of the families enumerated in Table VI., the first alone composed the first division, the last two belonged to the third, all the rest to the second. We have not followed this method of dividing them in this table, because the perigynous insertion of the stamens, very evident, it is true, in the greater part of these families, becomes less so in the Polygonaceæ, especially in the Atripliceæ and Phytolacineæ, in which it sometimes passes to the hypogynous and merits this last name almost as much as in the two following, connected, besides, with them in a large very natural group characterized by the peculiar structure of the seed. Let us observe that in these apetalous families we very commonly observe in the parts of the flower, some other number than *five*, frequently *three*, which is more frequently found in Monocotyledonous vegetables.

§ 787. ARISTOLOCHIEÆ (*Aristolochiées*) (Birthworts).—These plants are remarkable for several characteristics, and especially for the really epigynous insertion of the stamens (which is rather a rare case) and the ternary number of the parts. The calyx adherent to the ovary (*fig.* 595) is prolonged above it into a swollen tube, which is terminated by three segments, sometimes equal, sometimes unequal, with valved prefloration. This calycinal limb frequently presents very lively colours and sometimes dimensions so large that the flower of a species is mentioned as being used in America for a child's cap. The stamens, from six to twelve in number, rarely indefinite, are generally reduced to anthers almost sessile, situated on an annular epigynous disk or united to the base of the style, with which they then seem to form one body (*fig.* 599). The short column-like style, which is crowned by a stigma divided into 6, 4 or 3 rays, terminates an ovary divided into as many loculi, each of which encloses a large number of ovules fastened in one or two

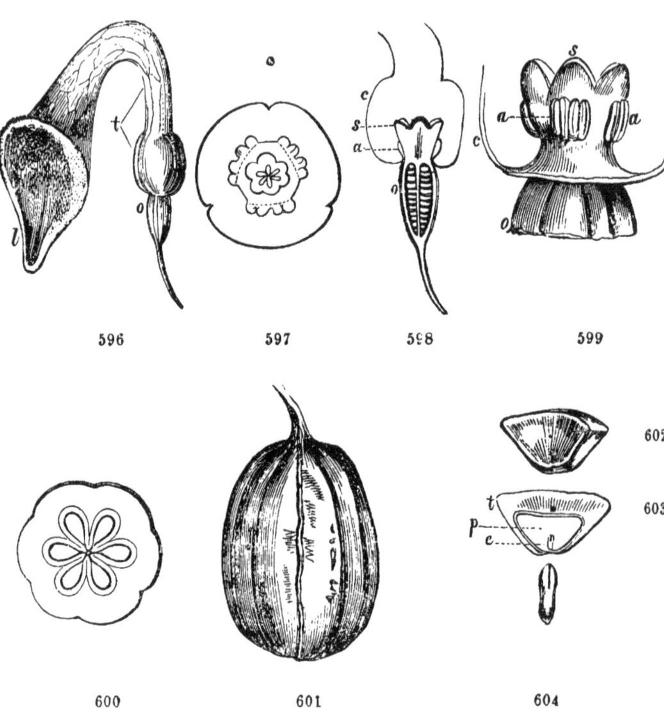

rows on the internal angle, ascending or horizontal. It becomes a fleshy or more commonly a capsular fruit (*fig.* 601) with loculicidal dehiscence, each loculus of which contains a large number

596—604. The organs of the fructification of the ARISTOLOCHIA CLEMATITIS.

596. The flower.—*o* Part of the calyx adherent to the ovary.—*t* Upper part of its tube swollen at the bottom.—*l* Its limb prolonged laterally into a tongue-shaped blade.

597. Diagram of this flower.

598. Vertical section of the lower portion of this flower.—*o* Ovary.—*s* Stigma.—*a* Anthers.—*c* Swelling of the calycinal tube.

599. Stigma *s* with the anthers *a a* united by pairs to the lobes.—*o* Top of the ovary.—*c* Swelling of the calycinal tube.

600. Horizontal section of the ovary.

601. Ripe fruit.

602. Seed.

603. Vertical section of the seed.—*t* Tegument thickened at the side of the chalaza.—*p* Perisperm.—*e* Embryo.

604. Embryo.

of seeds (*fig.* 602), flat or angular, presenting towards the top of a large fleshy or slightly horny perisperm a very small straight embryo, the radicle of which, longer than the cotyledons, is directed towards the point of attachment (*fig.* 603). The stems are herbaceous or frutescent. The stems, in the latter case frequently climbing, often present that anomalous structure which we have already pointed out (§ 84); the alternate or simple leaves are frequently furnished with two large stipules, which are united into a single one on the other side of the stem. The roots are quite bitter, possessing tonic and stimulating virtues, which causes several of them to be employed in medicine: the most important of these is the Virginian Snake-root (*Serpentaire*) (ARISTOLOCHIA SERPENTARIA).

§ 788.—We shall also mention several other families: SANTALACEÆ (*Santalacées*) (Sandalworts). The tree, so much esteemed as a perfume and known as *Sandal-wood* (*bois de Santal*), belongs to this family. The developement of their ovule deserves our attention as being quite exceptional. From the bottom of a single loculus a central column is raised, from the top of which are suspended several ovules consisting of as many naked nuclei. Only one of them is developed, and in this the nucleus is soon excavated into the embryonary sac, which is lengthened outwards and grows alone, thus forming the external tegument of the seed.

§ 789. PROTEACEÆ (*Protéacées*) (Proteads).—Here each of the four calycinal divisions, more or less deep, generally bears a stamen inserted at a greater or less height on its inner surface; a very rare arrangement for perigynous stamens, which commonly grow from the tube, i. e., below the place where the limb is divided.

§ 790. THYMELACEÆ (*Thymélœacées*) (Daphnads),—in which membranous appendices, frequently inserted at the top of the calycinal tube between its divisions, now begin to present some appearance of petals. Their bark is remarkable for two things: on one hand, for the extreme tenacity of the fibres of its liber, which renders it impossible to break the branches of several species; this property is employed to manufacture ropes; the fibres are detached in concentric layers, forming a thin and elegantly formed net-work in the LAGETTA, which for this reason has been termed the Lace-bark tree (*Bois dentelle*); on the other, for the extreme causticity of its juices, which act on the skin as local irritants, producing vesication. Mezereum bark is, consequently, employed under the name of *Garou* for this purpose.

§ 791. LAURINEÆ (*Laurinées*) (Laurels).—The anthers of this family present the singular mode of dehiscence by valves, which we have described in another part of the work (§ 452, *fig.* 316), and sometimes the existence of the four loculi superposed by twos (*fig.* 609): an extremely rare organization. The calyx has 4 or 6 divisions (*fig.* 605), alternating in two rows, and bears the stamens, which are opposite and double in number, consequently, in four rows. Those of the internal rows are frequently sterile: but, when fertile, they present this singular characteristic, that their anthers are turned and opened outwards, whilst those of the external rows

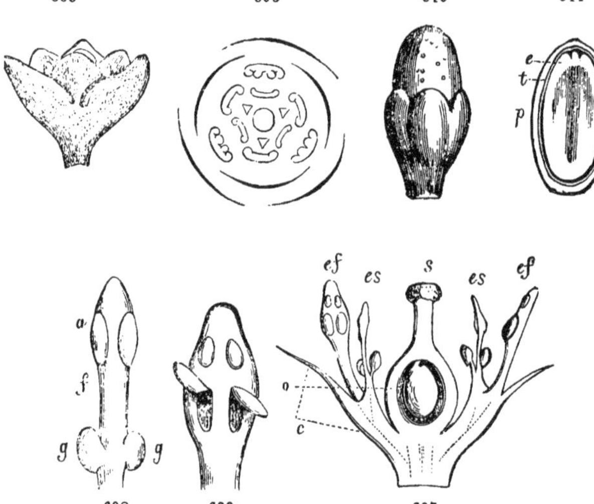

605—611. Organs of the fructification of the Cinnamon-tree (*Cannellier*) (LAURUS CINNAMOMUM).

605. The flower.

606. The diagram of this flower.

607. Vertical section of the flower.—*c* Calyx.—*ef* Fertile stamens.—*es* Barren stamens.—*o* Ovary with its single loculus and pendulous ovule.—*s* Style and stigma.

608. Stamen.—*f* Filament, with two glandular bodies *gg* at its base.—*a* Anther.

609. The side of one of the anthers just as it is opening.

610. Fruit accompanied by the persistant calyx.

611. A vertical section of the same without its calyx.—*p* Pericarp.—*t* Tegument of the seed.—*e* Embryo.

are turned and opened inwards. An ovary, terminated by a simple style and stigma, excavated into a single loculus in which hang one or two ovules (*fig.* 607 *o*); a fleshy fruit; an embryo without perisperm, the thick cotyledons of which conceal the very short and superior radicle (*fig.* 611): such are the other characteristics of this family, composed of trees frequently of a great height. Among them, the Laurel of the poets (Laurus nobilis) is, doubtless, the best known, both on account of its growing in our temperate climates and on account of its being the material from which the triumphal crowns of antiquity were made. But other species are of more positive utility, such as those furnishing the valuable spice, *Cinnamon* (*cannelle*). It is the bark of several species, principally of the Laurus cinnamomum: it owes its properties to a volatile oil found also, though less abundantly, in other parts, as well as in other vegetables of the same family. There is also another principle, *Camphor* (*camphre*); its presence in the plants, in which the volatile oil is abundant, is a fact confirmed by other families. In this it is principally produced by the Laurus Camphora or Camphor-tree (*Camphrier*). There exists along with it in the tissue of the Laurineæ another fixed oil, sometimes rather acrid, but sweet and very abundant in one of the most esteemed fruits of the tropics, that of the Avocado Pear (*Avocatier*) (Laurus persea).

§ 792. Polygoneæ (*Polygonées*) (Buckwheats).—These are for the most part herbaceous plants with alternate leaves, rolled outwards in the prefloration. We have already spoken of the singular stipules, which we have shewn united into a sheath (ochrea) surrounding the stem (§ 145, *fig.* 127). The number of the calycinal divisions is quinary (*fig.* 613), or ternary in two rows; the stamens, which they bear towards their base, are opposite and equal to them in number: in the latter case, they are placed in two rows, the inner one of which is incomplete, remarkable, moreover, as in the Laurineæ, for its extrorse anthers, whilst they are introrse in the outer row (*fig.* 613). The ovary, surmounted with 2, 3 or 4 free or united styles, sometimes extremely short, terminated by simple or plumose stigmas, has on the outside so many angles and in a single loculus contains a single erect ovule (*fig.* 612 *o*). It becomes a cariopsis or an achenium; and in its seed the straight or arched embryo, placed on the side of a farinaceous perisperm, turns its radicle upwards, that is, in a direction contrary to the point of attachment (*fig.* 614). The farina of this perisperm in the Buck-

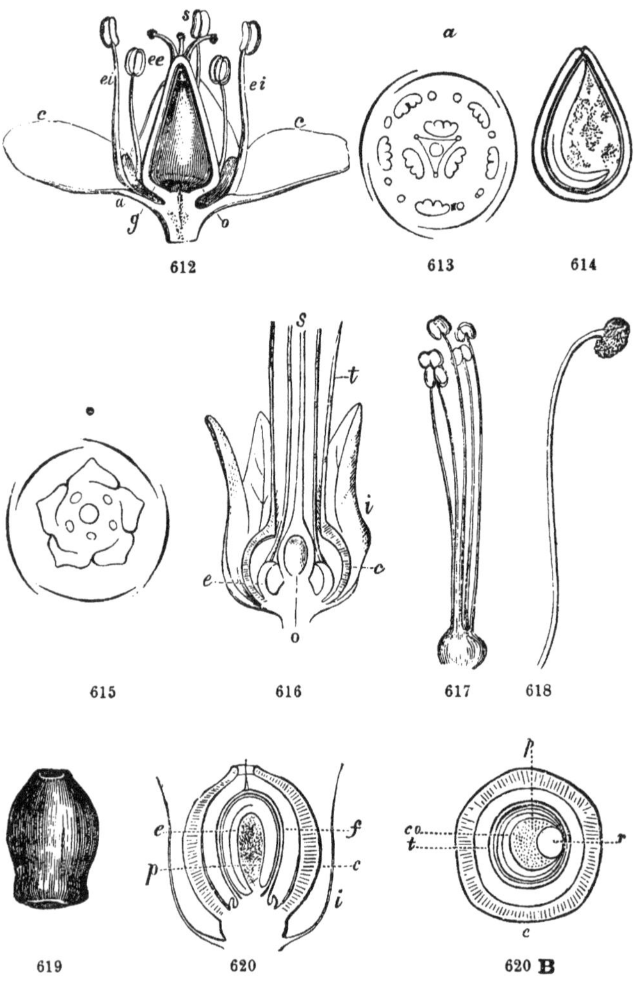

612. Flower of the Buckwheat (*Sarrasin*) (POLYGONUM FAGOPYRUM). A vertical section.—*c* Calyx.—*e i* Outer introrse stamens.—*ee* Inner extrorse stamens.—*a* Glandular appendices.—*o* Ovary with its erect ovule *g*.—*s* Styles and stigmas.

613. The diagram.—*a* Axis.

614. Vertical section of the seed.

wheat (*Sarrasin*) (POLYGONUM FAGOPYRUM) is employed as the food of man and animals. The leaves and the young shoots of several species of Sorrel (*Oseille*) (RUMEX) and of Rhubarb (*Rhubarbe*) (RHEUM) are eaten. The presence of oxalic acid in great abundance communicates to several of the species an agreeable acid taste. But other principles and, consequently, other properties are found in the roots, in which a resinous is associated with a gummy and an astringent substance. Hence, doubtless, their well-known virtues, at the same time both purgative and tonic, especially in the Rhubarb.

§ 793. NYCTAGINEÆ (*Nyctaginées*) (Nyctagos).—We have explained (§ 540, *fig.* 427), the fruit and the seed of the MIRABILIS JALAPA, the type of this family; we have seen that the base of the hardened calyx envelopes and seems to make part of it (*fig.* 620). Previously, from the upper contraction of this green base springs a coloured vase-shaped limb (*fig.* 616 *t*), which afterwards is divided and detached at this point. Around and beneath the ovary are inserted stamens definite in number, the free filaments of which traverse this upper defile (*fig.* 616), as it were, without adhering to it (although they appear to do so) and bear bilocular anthers. The ovule is single and erect (*fig.* 615 *o*) like the seed, the embryo of which, rolled around a farinaceous perisperm, turns its radicle downwards towards the point of attachment (*fig.* 620 *e*). We shall mention the purgative properties of the roots of this family only on account of the opinion formed on this circumstance, falsely attributing to the plant, which we have mentioned above, the origin of *jalap* and, consequently, the specific name of JALAPA.

615—620. Organs of the fructification of the MIRABILIS JALAPA.
615. The diagram of the flower.
616. The vertical section of the lower part of this flower.—*i* Involucrum.—*c* The base of the green calyx, swollen around the ovary.—*t* The coloured part of its tube.—*e* The lower part of its filaments.—*s* Part of the style.—*o* Ovary with its erect ovule.
617. Stamens with an arched swelling at the base of their filaments.
618. Style and stigma.
619. Fruit enveloped with the hardened, persistant base of the calyx.
620. Vertical section of the same.—*i* Involucrum.—*c* Calyx. *f* Pericarp.—*p* Perisperm.—*e* Embryo.
620 **B**. Horizontal section of the same.—*c* Calyx.—*t* Tegument of the seed with the pericarp.—*p* Perisperm.—*r* Radicle.—*c o* Cotyledons.

Polypetalous Dicotyledonous Vegetables.

§ 794.—Jussieu, applying to them his method of division derived from the three methods of insertion, distinguished them as Epipetalous (*Epipétalées*), Hypopetalous (*Hypopétalées*) and Peripetalous (*Péripétalées*). We shall adopt it with some slight modifications; we shall confound, in short, the epigynous with the perigynous, because in the very small number of families, of which the first of these classes was composed, the insertion of the stamens on the circumference of the disk (which covers, it is true, the top of the ovary, but which is attached by another part to the calyx), is really ambiguous; then we shall begin by separating, without regard to the insertion, a small group of families connected with the preceding by a very peculiar characteristic, the structure of the seeds with farinaceous perisperm surrounded by the embryo (*fig.* 625) and situated on a central placenta (*fig.* 624, 2). We ought, perhaps, also to neglect this last characteristic and join to this group, notwithstanding their parietal placentation, two other families, the one, the Ficoïdeæ, in which the arched embryo forms a half ring on the side of a farinaceous perisperm; the other, the Cactaceæ, which the former would draw along with it although deprived of perisperm, but indicating an analogous tendency by the general curve of its embryo.

(**Table VII. page 621.**)

The insertion and the presence of the petals appear to be of little importance in this group; for there are found in the first family some hypogynous plants, in the latter some perigynous genera, some apetalous ones in both, and as to the Paronychieæ (*Paronychiées*) (Knotworts) we may say that they are only the Scleranthеæ with the addition of a corolla. Sometimes in the same genus, oftener in the same species, the petals are present or wanting almost indifferently. They all, however, form together a group so incontestably natural, that all authors are disposed to admit it. We do not observe in it any remarkable property, any useful plant, if we except that of eating, when cooked, the fleshy leaves of some Portulaceæ (*Portulacées*) (Purslanes), especially of the Common Purslane (*Pourpier*), the type of the family.

TABLE VII.

FAMILIES.

POLYPETALOUS DICOTYLEDONOUS VEGETABLES,

with parietal placentation and with a farinaceous perisperm surrounded by the embryo.

Stamens	perigynous.	Sepals frequently reduced to two. No stipules. Plants commonly fleshy	PORTULACEÆ.
		Sepals, equal in number to the petals. Stipules, scarious. Plants commonly dry	PARONYCHIEÆ.
	hypogynous.	—4 or 5 sepals and as many petals. Plants commonly dry	—CARYOPHYLLACEÆ.

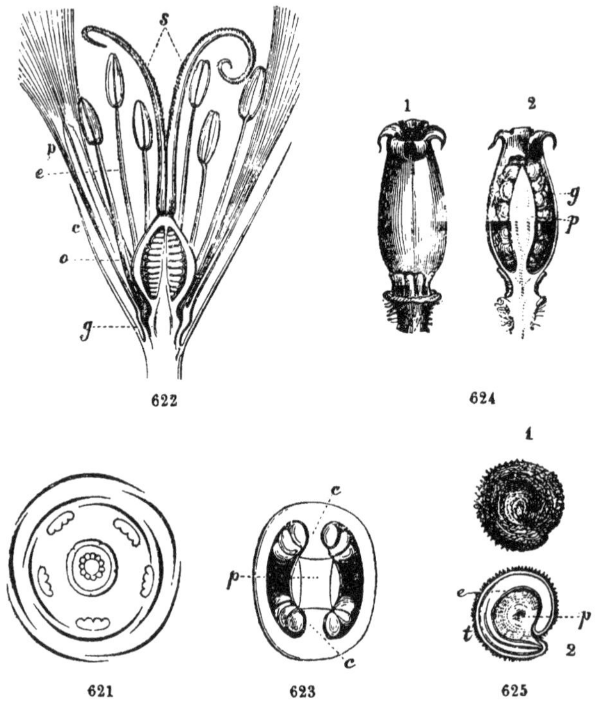

CARYOPHYLLACEÆ (*Caryophyllées*) (Cloveworts).—To the characteristics of the placentation (the nature of which we have previously explained [§ 505]) and of the seed, we shall add the following: petals furnished with claws; stamens double in number, the oppo-

621. Diagram of the flower of the ALSINE MEDIA.

622. Section of the flower of the Clove Pink (*Œillet à bouquets*) (DIANTHUS CARYOPHYLLUS).—*c* Calyx.—*p* Petals, united at the base to the opposite stamen.—*e* Stamens.—*g* Gynophore.—*o* Ovary.—*s* Styles covered with a papillose stigma along the whole of their inner surface.

623. Horizontal section of its very young ovary, whilst it is still separated into two loculi by the partitions *c*, which will afterwards be destroyed, leaving the central part or placenta *p*.

624. Capsule of the Corn Cockle (*Nielle*) (AGROSTEMMA GITHAGO), at the moment of the dehiscence, by which the pericarp is separated into several valves at the top only.—1. Entire.—2. Vertical section, shewing the seeds *g* grouped in a central mass on the placenta *p*.

625. Seed.—1. Entire.—2. Vertical section.—*t* Tegument.—*e* Embryo.—*p* Perisperm.

site ones sometimes united to them at the base (*fig.* 622); the ovary frequently raised on a column-shaped axis, which also supports the petals and the stamens (§ 386 *b*, *fig.* 233), surmounted with 2 to 5 stigmas elongated like styles, but covered with papillæ over the whole of their inner surface (*fig.* 622 *s*); the capsule (*fig.* 624) with as many valves, each of which is itself frequently split into two. All the species are herbaceous plants, very rarely assuming a ligneous consistence. At their swollen knots two simple and entire leaves are opposite to one another. Some botanists refer to the Paronychieæ the few genera in which they are accompanied with stipules.

Hypogynous Polypetalous Dicotyledonous Vegetables.

§ 795.—We shall subdivide these plants according to their parietal or axilary placentation: in the former category we shall place the fruits composed of carpels united either by their edges or reflexed sides into incomplete dissepiments; in the latter, the fruits, in which the reflexed sides of each carpel form a complete loculus, whether it remains separated from the rest as a distinct carpel or is laterally united to them in a plurilocular ovary. All the apocarpous fruits of the hypogynous plants will, therefore, be referred to the latter, even when the ovules, erect or pendulous from the bottom of the loculus or scattered over its walls, do not seem to be attached to the internal angle. Thus, our division may be stated as, 1st, a unilocular ovary with several placentas; 2nd, a plurilocular ovary or distinct carpels.

Hypogynous Polypetalous Dicotyledonous Vegetables,

with parietal placentation.

§ 796.—The placentas are sometimes on the edges of the valves of the fruit, and, consequently, alternate with them, sometimes occupy the middle of their length and are opposite to them. In some cases of indehiscent fruit, the other characteristics derived from the structure of the seed will allow us to supply tho absonoo of this one.

(Table VIII. page 612).

§ 797.—We shall mention among these families the Violarieæ (*Violariées*) (Violetworts), with flowers presenting sepals, petals and stamens, each five in number: the loculi of their anthers are situated on a large connective, which is prolonged into a kind of point above them (*fig.* 317) and is sometimes united to them, forming a kind of tube applied to the ovary. The style is simple, oblique, terminated by a thick stigma, turned inwards and pierced at its middle (*fig.* 181); the fruit is a capsule with three valves. This may be divided into two tribes, 1st, those having regular flowers (the ALSODINEÆ [*Alsodinées*]), 2nd, irregular flowers (the VIOLEÆ [*Violées*], which are by far the most numerous). We have given an example of the irregularity presented by two of the stamens (*fig.* 317). The roots in this family rather frequently possess emetic properties. Several species indigenous to South America are known and sold under the name of *Ipécacuanha* (*Ipécacuanhas*).

§ 798. CISTINEÆ (*Cistinées*) (Rock-Roses).—In this family the flowers are regular with the exception of the calyx, the two outer folioles of which are frequently shorter than the others; the stamens are indefinite; the placentas are 3 in number, sometimes 5 or even 10, sometimes project into the interior of the loculus, and the incomplete dissepiments, which they bound, may even be sufficiently advanced to meet towards the centre at a greater or less height and thus divide the cavity into as many demi-loculi. It cannot be well conceived, how, in the ovules, the micropyle, situated at the extremity opposite to the hilum, can be placed in relation (for the purpose of fecundation) with the placentas, to which they are connected only by a very long funiculus; if we examine the inside of the ovary at this period we see the pollen tubes, arrived along the conducting tissue at its surface, lengthen and hang in the empty space of the loculus. Several of the species of this family, herbaceous or frutescent, are covered with a resinous down, and this furnishes in the CISTUS CRETICUS and others the balsamic substance known as *Ladanum* (*Labdanum*).

§ 799.—We shall here allude to the BIXINEÆ (*Bixinées*) (Bixads) merely on account of the colouring matter, so well known as *Arnotto* (*Rocou*), supplied by the pulpy envelope of the seed of the BIXA ORELLANA, and which naturally red, becomes a golden yellow by the action of alkalies; to the RESEDACEÆ (*Résédacées*) (Weldworts), on account of that, which, under the name of *Weld* (*Gaude*) generally employed for dyeing yellow, is the produce of a plant common in our fields, the RESEDA LUTEOLA. Regretting that we cannot dwell on the interesting irregular flower of this last family, we shall call attention to the terminal separation of the walls of its ovary, which leaves its cavity exposed, as if the union of the carpellary leaves, commonly so complete, was here stopped half-way[t]. The flower of the CAPPARIDACEÆ (*Capparidées*) (Capparids), often so curious in its irregularity, would be the object of our examination if space would allow; *Capers* (*Câpres*) are flower-buds of the genus CAPPARIS, which is its type.

§ 800. CRUCIFERÆ (*Crucifères*) (Crucifers).—Four sepals in a cross, four petals alternating with them (*fig.* 284), six tetradynamous stamens on or within four glands, the whole of which form a hypogynous disk (*fig.* 627), an ovary with two parietal placentas, a siliqua (*figs.* 630, 631) for the fruit, and seeds without perisperm, are characteristics which easily and surely distinguish this very natural family so largely distributed throughout our country. We have defined the siliqua (§ 538, *fig.* 426) with its dissepiment so different from others, determining the co-existence of two characteristics commonly incompatible, the parietal placentation and the plurality of loculi (*figs.* 629, 632, 633); we have seen the different ways in which the radicle is bent on the cotyledons (*figs.* 472, 473, 469, 482). The species are almost without exception herbaceous; their leaves are alternate and without stipules, their flowers white or yellow, rarely reddish. They are remarkable for the presence in their tissues of a large proportion of azote and of a volatile oil. To the former, they owe not only their nutritious properties, of which the numerous varieties of Cabbage (*Chou*) will furnish the best examples, but also their readiness to putrify, and the infectious and animal odour, which they exhale as ammonia is formed. To the latter they owe their stimulating properties, so strong in Mustard (*Moutarde*): these, when weakened and tempered by a saccharine substance, impart valuable qualities to certain roots, as the Radish (*Radis*) and the Turnip (*Rave*). This weakening, which naturally results from their dwelling under the surface of the earth protected from the action of the light, is artificially produced in the external parts; by causing them to become abortive, as in the inflorescence of the Cauliflower (*Chou-Fleur*); by covering their young shoots, as in the Sea-Kale (CRAMBE MARITIMA), and only choosing the inner leaves of the buds, as in the Cabbage (*Chou pommé*). Medicine takes advantage of these exciting properties and employs them to restore the tone of the organs in certain debilitating

[t] To this family belongs the most fragrant ornament of our flower borders, the Mignionette (*Réséda*) (RESEDA ODORATA).—TRANS.

diseases, especially in the Scurvy (*scorbut*). The Cruciferæ, indeed, are eminently antiscorbutic ; so generally so, in fact, that in a cele-

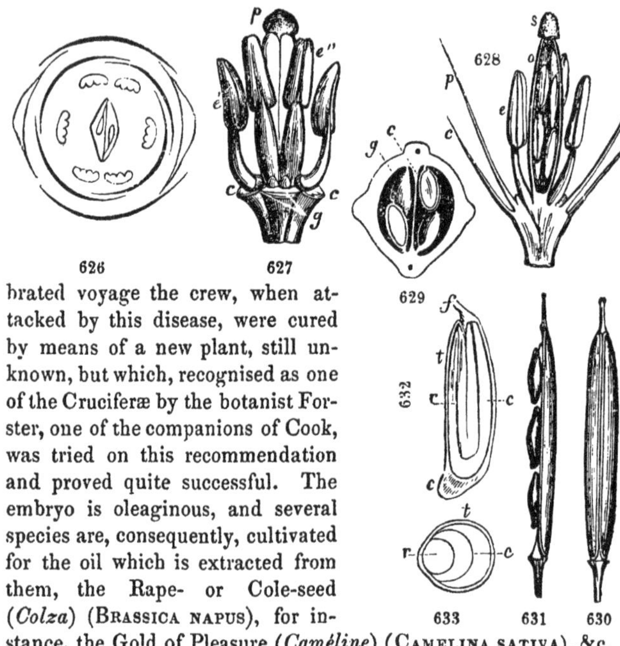

brated voyage the crew, when attacked by this disease, were cured by means of a new plant, still unknown, but which, recognised as one of the Cruciferæ by the botanist Forster, one of the companions of Cook, was tried on this recommendation and proved quite successful. The embryo is oleaginous, and several species are, consequently, cultivated for the oil which is extracted from them, the Rape- or Cole-seed (*Colza*) (BRASSICA NAPUS), for instance, the Gold of Pleasure (*Caméline*) (CAMELINA SATIVA), &c.

§ 801. PAPAVERACEÆ (*Papavéracées*) (Poppyworts).—We here again find in the flower the parts which grow alternately: the calyx

626—633. The organs of fructification of the ERYSIMUM MURALE.

626. The diagram of the flower.

627. The flower without its envelopes.—*c* Wounds resulting from the fall of the folioles of the calyx.—*g* Glands, which accompany the insertion of the stamens.—*e'* The two shorter stamens.—*e''* The pairs of longer stamens.—*p* Pistil.

628. Vertical section of the flower.—*c* Calyx.—*p* Petals.—*e* Stamens.—*o* Covered ovary.—*s* Stigma.

629. Horizontal section of the ovary.—*c* Dissepiment.—*g* Ovules.

630. Siliqua.

631. The same; one of the valves has been taken away shewing the seeds attached to the replum.

632. Vertical section of the seed.—*f* Funiculus.—*t* Tegument swollen at the chalaza *c*.—*r* Radicle.—*c* Cotyledons.

633. Horizontal section of the seed.—*t* Tegument.—*r* Radicle.—*c* Incumbent cotyledons.

of two caducous sepals (three is an exception); the petals four (or one of its multiples) in number; the stamens, double in number, or more commonly a higher multiple, and then opposite to the petals in fascicles. The style is short or wanting, the stigmas two or more in number: we have already explained the peltate shape and radiating arrangement which they affect (§ 513, *fig.* 397). The fruit presents in the inside as many projecting placentas under the form of incomplete dissepiments, and at maturity is split into as many valves either completely, or else only at the top, which, crowned by the stigma-bearing shield, then offers in its circumference a circle of openings, through which the seeds escape. These are extremely numerous with a very small embryo towards the extremity of a large, fleshy, oleaginous perisperm. The stems are commonly herbaceous, the leaves alternate, and all the parts distended with a peculiar juice, generally milky, rarely of any other colour. This juice has well-marked properties; some resulting from great acridity, as that of the Celandine (*Eclaire*); on this account the roots of several Papaveraceæ are employed as purgatives or emetics: others are narcotic, principally found in the Poppy (*Pavot*); they are owing to several alcaloids, which their sap conveys, *Meconine* (*Méconine*), *Codeine* (*Codéine*), *Narcotine* (*Narcotine*), and especially *Morphine* (*Morphine*). These substances along with several others compose *Opium*, which is only the concreted juice, after having been extracted from the capsules and their peduncles, where it is more abundant than elsewhere. These principles are not found in the seed, from which an oil is extracted; this oil was for a long time suspected of noxious qualities on account of its origin, but now recognised as innocent, is admitted in commerce and is extensively employed in adulterating olive oil.

§ 802. We shall here place a small intermediate group among polypetalous plants, between those which have the parietal and those which have axilary placentation; for it presents both at the same time, but is distinguished by the small fleshy sac, which envelopes its embryo. This sac originates in an inner perisperm, commonly accompanied with an external swelling into a farinaceous mass, more rarely isolated.

(Table IX. page 628).

Nymphæaceæ, (*Nymphéacées*) (Water-lilies).—We shall not linger

TABLE IX. **FAMILIES.**

Embryo in a peculiar sac.—Fruit	Monolocular, polyspermous.—Large farinaceous perisperm		—NYMPHÆACEÆ.
	Composed of several carpels.	in a fleshy disk. 1 or 2 pendulous ovules.—Perisperm wanting	—NELUMBONEÆ.
		on a receptacle scarcely enlarged. 2 or 3 ovules attached to the internal angle.—Perisperm fleshy	CABOMBACEÆ.

long on this family, the type of which, the White Water-lily, has already occupied us several times (§§ 366, 368, *figs.* 223, 452). The seeds, the structure of which is so remarkable from the existence of an internal perisperm forming a small sac around the embryo, may render some services by the mass of the external or farinaceous perisperm, to which recourse has been had in times of famine. In South America, they eat the seeds of another of the Nymphæaceæ, calling it Water Maize (*Maïs d'eau*): it is the handsomest of all these handsome flowers and has, consequently, been deemed worthy of being dedicated to the Queen of England under the name of VICTORIA REGIA [u]. The leaves and the flowers of these plants float on stagnant waters, under which are concealed their creeping stems, also rich in fecula. They may be employed for food, but only after prolonged washing, which deprives them of the bitter principles mixed with the fecula.

CABOMBACEÆ (*Cabombacées*) (Water-shields)—also live in water, as well as the NELUMBONEÆ (Water Beans). Their singular fruit, with its egg-shaped carpels, dispersed over and half buried in a large fleshy plate, as well as its flowers and leaves, may be seen in almost all Chinese paintings. Their rhizomes and fruits are also eaten: in their seeds, the farinaceous cotyledons supply the place of the absent perisperm.

HYPOGYNOUS POLYPETALOUS DICOTYLEDONOUS VEGETABLES,

with axilary Placentation.

§ 803.—As the families, which present this triple characteristic, are extremely numerous, we shall try to distribute them in their turn into several sections, and the structure of the seed will furnish us with the first division. The embryo is naked under the teguments; or rather

[u] To give the reader an idea of the magnitude of this splendid production of nature, the size of some of the parts is given, taken from the description of Sir W. J. Hooker in the Botanical Magazine. (It has also been published separately in a folio volume). The leaves are five to six feet across. The flower-stalk is an inch thick near the calyx, covered with elastic prickles about three quarters of an inch long. The four-leaved calyx is a foot in diameter, but is concealed by the hundred-leaved corolla, which diffuses an agreeable perfume.—TRANS.

it is surrounded by a perisperm which it almost equals in length: or else much shorter than it, it is buried at its extremity. But let us remark, that if this latter characteristic has a real value, the two others seem to have a much less one. The perisperm, when its mass is not proportionally much larger than the embryo, appears to lose a deal of its importance in the classification; it passes by insensible degradations from large to small, and even totally disappears in plants evidently nearly allied: also in our tables we shall be sometimes brought in two ways to the same family, generally, it is true, to different tribes.

(Table X. page 613).

§ 804. RANUNCULACEÆ (*Renonculacées*) (Crowfoots).—This family is an excellent subject for the study of those, who wish to understand clearly the meaning of a family, so much the more so, since it served in some way as a base for all the labour of A. L. de Jussieu, who by first examining this family perceived the early dawn of the real method of classifying plants according to their natural characteristics. A calyx composed of five sepals, five alternate petals, stamens, indefinite in number, free on a plane or projecting torus, to the bottom of which they are inserted (*fig.* 634 *e*); several independent carpels (*fig.* 624 *pi*), sometimes indehiscent and monospermous, sometimes dehiscent and polyspermous; seeds, in which the small embryo is buried in the side of the hilum towards the extremity of a large horny perisperm (*fig.* 639); such are its general characteristics, such is the type, the derivations from which may be followed in a certain number of genera: some, in which the quinary number of the parts gives way to the ternary, others, in which the petals change their shape, assuming the form of small laminæ or horns, or else disappearing. They are wanting, for instance, as in the CLEMATEÆ (*Clématidées*), and the calyx then assumes all the colours and the appearance of the corolla. Their prefloration is valvate with opposite leaves. In the ANEMONEÆ (*Anémonées*), the prefloration is imbricate with alternate leaves. The RANUNCULEÆ (*Ranunculées*) present the type described above, with achenia enclosing a single erect seed (*figs.* 638, 639), whilst it is pendulous in the two preceding tribes. The HELLEBOREÆ (*Helleborées*) have polyspermous follicles with rolled petals. In all these plants, the stamens are terminated by adnate extrorse anthers (*fig.* 634); but they become introrse in the PÆONIEÆ (*Pæoniées*), the fruit of which is composed

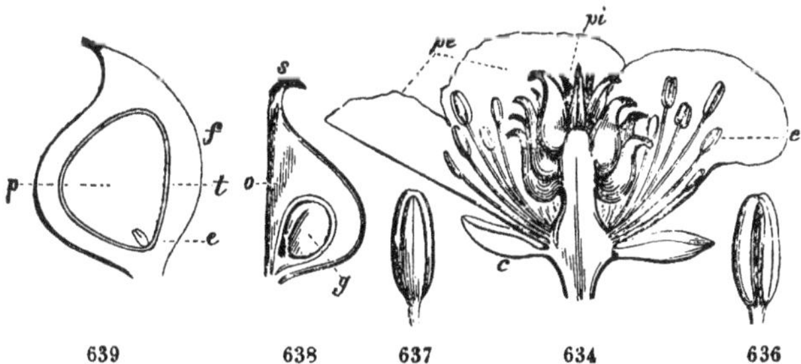

639 638 637 634 636

of several dehiscent or indehiscent carpels, enclosing several seeds. We have seen in an example taken from this family, an exception to the rule how free carpels may pass into a single and multilocular ovary. It shews us, moreover, from all that precedes, how certain characteristics may be modified in a very natural group, what those are, on the contrary, which are the most invariable; it teaches us their subordination, demonstrating the importance of the seed and assigning a more elevated position to the relations of situation or of adherence of the parts than to their number. The Ranunculaceæ are for the most part herbaceous plants; some few are shrubs most frequently climbing. The leaves without stipules are sometimes simple and even reduced to phyllods; but their limb is generally more or less deeply lobed. The juice, of an aqueous appearance, is extremely acrid and caustic; the principles to which it owes this

635

634—639. The organs of fructification of the RANUNCULUS ACRIS.

634. Vertical section of the flower.—*c* Calyx.—*p e* Petals.—*e* Stamens.—*p i* Pistil composed of several carpels on a lengthened axis.

635. The diagram of the flower.

636. The outer side of the anther, at which it opens.

637. The inner side of the same.

638. Vertical section of an ovary *o* shewing the ovule *g*.—*s* Stigma.

639. Vertical section of a ripe carpel.—*f* Pericarp.—*t* Tegument of the seed.—*p* Perisperm.—*e* Embryo.

property appear to be very volatile; they are also much more energetic in the roots than in the aerial parts, in which they are dissipated in the air or in the surrounding water, although certain parts manifest them in a very high degree: as in the Aconites (*Aconits*), such well-known poisons, from the flowers of which bees have been known to extract honey of a poisonous quality; as in different species of RANUNCULUS and of ANEMONE, the leaves of which have been employed at certain times and in certain countries as vesicatories on account of their action on the skin. Hence the name of Beggar's Weed (*Herbe aux gueux*) given to the CLEMATIS, because beggars rub themselves with it in order to produce on their bodies superficial and transitory ulcers. The Hellebore, so celebrated by the writers of antiquity, acts as a violent purgative. In the seeds the acridity still exists, but mixed with an aromatic principle; on account of this they have been employed by the poor as condiments instead of pepper, especially those of the NIGELLA SATIVA and of the DELPHINIUM STAPHYSAGRIA, in which, moreover, is found a peculiar alcaloid, the *Delphine*.

§ 805.—The quinary number is also observable in the flower of the DILLENIACEÆ (*Dilleniacées*) (Dilleniads), but its place is usurped by the ternary in the MAGNOLIACEÆ (*Magnoliacées*) (Magnoliads), of which we have already had occasion to speak (§§ 368, 396 [*fig.* 224]), and in which at the same time the whorls of the petals are multiplied; in the ANONACEÆ (*Anonacées*) (Anonads), in which we find them merely doubled. In all these families, we find the adnate stamens generally opening outwards or laterally, the small embryo at the extremity of a large perisperm, but a different consistence or shape in the last. The two last present analogous properties and an aromatic principle, which produces the Aniseed of Bordeaux from the fruits of the ILLICIUM ANISATUM. This, very frequently joined to a certain degree of bitterness, imparts to the barks of the DRYMIS and others tonic virtues, which have been compared to those of the quinquina or Peruvian bark. This aroma, dissolved in a sweet mucilage, causes the fruits of certain Anonaceæ to be esteemed as dessert fruit, such as the Custard Apple, &c.

§ 806.—We shall again find along with the apocarpous fruit the ternary number in the BERBERIDACEÆ (*Berberidées*) (Berberids), as well as in the LARDIZABALACEÆ (*Lardizabalées*) (Lardizabalads) and several of the MENISPERMACEÆ (*Ménispermacées*) (Menispermads); these two last families are closely connected with one another, both having flowers unisexual by abortion, enclosing in each of their carpels, the former a large number of ovules dispersed over the lateral walls, the latter, a single curved ovule attached to the side. But the latter family differs from all those, which we have previously mentioned, in the developement its embryo begins to assume with regard to the perisperm, to which it is almost equal. Their climbing stems are remarkable for the formation of their wood, the zones of which, separated by so many cellular zones, do not correspond to the succession of years; they are similar to that we have described and represented (*fig.* 110) with this difference, that the small fascicles of liber are only found in one single zone, the innermost. The roots of several species are bitter and tonic, especially that known as Kalumba- or Calumba-root (*Colombo*), and they are, consequently, employed in foreign countries as febrifuges. The fruits are frequently narcotic, and one of them, the Cocculus Indicus (*Coque du Levant*), is employed in poaching fish: the poachers have only to collect the fish, which, stupified by the poison, rise to the surface of the water into which the poison has been thrown. In this fruit has been discovered a peculiar alcaloid, the *picrotoxine*, in which this property probably resides.

§ 807.—The stamens are opposite to the petals in the Berberidaceæ, Lardizabalaceæ and several Menispermaceæ; but we have already shewn (§ 397) that it is a necessary result of the parts of the flower being arranged in ternary whorls and being doubled in each kind of organ. But it is not the same in the Ampelideæ (*Ampelidées*) or Viniferæ (*Vinifères*) (Vineworts), the whorls of which are ternary or quaternary, and in which, nevertheless, the four or five petals are opposite to the stamens (*fig.* 640). Here the opposite position is the result of the abortion of a whole row of stamens, which is proved by their rudimentary existence under the form of five lobes in the Leea. The ovary, seated in the middle of a large glandular disc (*fig.* 642), the circumference of which bears the stamens, is surmounted by a simple style and stigma, and excavated into from two to six loculi, at the bottom of which are one or more erect ovules (*fig.* 643). It becomes a berry (*baie*) and every one knows the seeds or small stones (*fig.* 645) in their interior; under their ligneous tegument we find a hard perisperm twice as long as the embryo, which occupies its axis turned towards the point of attachment (*fig.* 646). The Viniferæ are for the most part climbing shrubs, with swollen knots susceptible of being disarticulated, with alternate lobed leaves, which may be otherwise stated as composed of several pennate or palmate folioles. We have seen how the inflorescences opposite to these leaves may be changed into tendrils (§ 185, *fig.* 172). We have pointed out the size of the vessels which convey the sap and its force

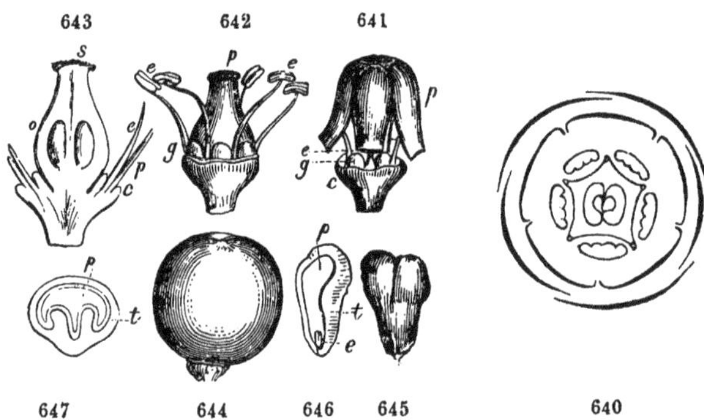

and abundance (§§ 257, 268) in the stems. Some stems, those of several species of Cissus, resemble in their zones, alternately ligneous and cellular, those of the Menispermaceæ; but we do not find any liber, which in the Vine (*Vigne*) is each year detached with the whole of the cortical layer. Is there any need of mentioning the great utility of the Grape to man? The abundance of sugar along with a vegetable acid (tartaric) in the pulp imparts to the fresh fruit its agreeable flavour, is concentrated in the dried fruit and, communicating to the juices the property of fermenting, enables us to make of them the most esteemed of all alcoholic beverages.

§ 808.—The Zygophyllaceæ (*Zygophyllées*) (Bean Capers), the Rutaceæ (*Rutacées*) (Rueworts), the Diosmeæ (*Diosmées*), the Xan-

640—647. The organs of fructification of the Vine (*Vigne*) (Vitis vinifera).

640. The diagram of the flower.

641. The flower at the time of blossoming, which detaches the petals *p* at the bottom, whilst they are united at the top.—*c* Calyx.—*g* Glands.—*e* Stamens, the filaments of which only are seen.

642. The flower after the fall of the petals.—*g* Glands. —*e* Stamens.—*p* Pistil.

643. The vertical section of the flower.—*c* Calyx.—*p* Petals.—*e* Filaments.—*o* Ovary with its two loculi and their erect ovules.—*s* Stigma.

644. Fruit (commonly called the *Grape*).

645. Seed (commonly called the *Grape-stone*).

646. Vertical section of the same.—*t* Tegument.—*p* Perisperm.—*e* Embryo.

647. Horizontal section of the same about the middle.—*t* Tegument.—*p* Perisperm.

THOXYLACEÆ (*Zanthoxylées*) (Xanthoxyls) and the SIMARUBACEÆ (*Simarubées*) (Quassiads) form together a very natural group, which is sometimes confounded in a single family under the common name of RUTACEÆ, of which these would be so many tribes. They are, however, separated in our tables, because the perisperm is wanting in some and is developed in others. We find this double structure in the same family, that of the Diosmeæ ; but it then characterizes sections so much the more distinct as the plants which compose them are indigenous to different parts of the globe: those of New Holland being furnished with a perisperm, those of Africa being without one, as well as a small portion of those of America (the CUSPARIEÆ [*Cuspariées*]), very remarkable also for their variously wrinkled cotyledons, folded on themselves and on the radicle which they envelope. The perisperm of the Zygophyllaceæ is of a horny consistence; they are herbs, or more frequently shrubs or trees with opposite leaves, one of which is frequently much less developed than the other, with an indefinite inflorescence. Their wood is sometimes extremely hard, as in the GUAIACUM (*Gaiac*), in which both it and the bark have exciting properties, owing, doubtless, to a peculiar principle, the *Guaiacine*, which has been described as a resin, although it differs in some points from such substances. In the three following families, which are generally represented only by ligneous plants, we almost always find an essential oil very abundantly diffused throughout every part. Its presence is indicated by transparent dots, which are perceived on the leaves when held between the eye and the light, these dots being the small reservoirs of the colourless volatile oil. Exciting properties, the general attribute of these oils, and a powerful aromatic odour, frequently perceived at a great distance, result from their presence in several of these plants, as in the Rue and the DICTAMNUS of our gardens, in the Diosmeæ or Bucku Plants of our green-houses. Sailors approaching the Cape of Good Hope, frequently perceive when far out at sea, breezes strongly impregnated with the odour arising from the numerous species of this family covering this region. In some Xanthoxylaceæ, the fruits participate in these qualities and are popularly called Peppers (*Poivriers*) where they are indigenous; but a bitter principle is also found along with this in others, which makes their bark good febrifuges, especially in the celebrated *Angostura bark* (*écorce d' Angusture*) (TICOREA FEBRIFUGA). In the Quassiads, the latter is the only principle, and their properties, consequently, are somewhat different, as is proved by the medical use of the bark of the SIMARUBA and of the QUASSIA AMARA. The presence of a resinous substance, the *quassine*, has been determined in them.

§ 809.—We have described the peculiar fruit of the GERANIACEÆ (*Géraniacées*) (Cranesbills) (§ 533, *fig.* 410). Those species, cultivated in our gardens under the name of GERANIUMS (the varieties of which have been multiplied to such an extent), all in reality belong to the genus PELARGONIUM, indigenous to South Africa and New Holland.

§ 810. We again find in the MALVACEÆ (*Malvacées*) (Mallow-worts)—another example of those large natural groups, which unite several families. That family, which at first bore this name, now comprehends the STERCULIACEÆ (*Sterculiacées*) (Sterculiads), BYTTNERIACEÆ) (*Byttnériacées*) (Byttneriads), BOMBACEÆ (*Bombacées*), and MALVACEÆ properly so called. The last, which are most familiar to us, of which the Mallow (*Mauve*) (MALVA) and the Marsh-Mallow (*Guimauve*) (ALTHÆA) may give us an idea, are characterized by their thick calyx with valvate prefloration (common, moreover, to all the group), rather frequently surrounded externally by an involucrum or calyculus (*fig.* 273); by their petals, in general large, oblique and obcordiform, twisted even after the expansion; by their stamens having their filaments united for a part of their length into a cylinder joined to the base of the petals, terminated at the top by an entire or quinque-lobed edge, divided externally into a number of filaments, each terminated by a kidney-shaped unilocular anther (*figs.* 311, 651), filled with large globular and bristled pollen-grains; by their carpels, verticillate around a large central axis in the shape of a column (*figs.* 408, 652), from the top of which grow the styles united together except at their extremity (*figs.* 389, 650 *s*); each contains one or more seeds, in which the embryo without perisperm bends back its radicle between its plaited cotyledons (*figs.* 655, 656). The alternate leaves with stipules are more or less deeply lobed and then are generally very subject to vary. The different parts are commonly quite impregnated with a mucilaginous substance imparting to them the emollient properties for which they are renowned. The GOSSYPIUM belongs to this family; the hairy covering of its seeds is the *Cotton* (*Coton*) of commerce.

The BOMBACEÆ, some species of which also have their seeds

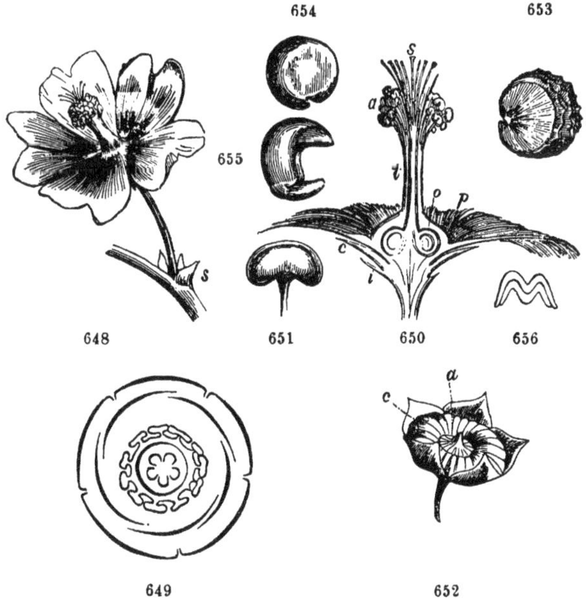

covered with a kind of hair, form a family very nearly related to the Malyaceæ, distinguishable by means of their general habit and appearance rather than by any clearly defined characteristics of fructification, among which the different form of the pollen is, perhaps, the most constant. We mention them here only as including the largest trees that grow on the face of the earth. The best known

648—656. The organs of the fructification of the Mallow (*Mauve*) (MALVA SYLVESTRIS).

648. The flower, with its peduncle accompanied with two stipules *s*.

649. Diagram.

650. Vertical section of the flower.—*i* Calyculus or involucrum.—*c* Calyx.—*p* Petals.—*t* Tube of monadelphous stamens, widened above the ovary *o* and united at its base to the petals, divided at its top into a large number of filaments bearing as many anthers *a*.—*s* Styles separate at the top, united below into one.

651. An anther with the top of the filament.

652. Fruit surrounded by the persistant calyx.—*c* A whorl of cocca, united by the axis *a*.

653. A coccum; the side is turned towards us.

654. The seed.

655. The embryo.

656. A section of the embryo near its middle, shewing the arrangement of its cotyledons.

is the Baobab Tree (ADANSONIA), which inhabits the western coasts of Tropical Africa. But other Bombaceæ may almost enter the lists as rivals of this giant among vegetables. Their branches, which extend to a great distance and touch the earth at their extremities bending under their own weight, cover spaces of ground so large that a single tree seen at a little distance appears to be a large thicket. Their developement is less remarkable for its height than for its width; and their stem instead of taking the form of a straight column is more commonly swollen into the shape of an immense barrel. When we examine the illustrations of botanical works, in which a representation of this tropical vegetable has been given, we shall soon remark several trees of this family remarkable for their dimensions or even for their extraordinary shapes. As an example we shall mention the CHORISIA VENTRICOSA, the trunk of which, widened about the middle of its height, contracts insensibly towards its base and its top and thus presents the appearance of a gigantic spindle.

The tree, which furnishes the Cocoa or Cacao, the THEOBROMA, belongs to the BYTTNERIACEÆ (*Byttneriacées*). Its fleshy, oleaginous, brownish embryo of the consistence of wax, after having been roasted is employed in the fabrication of chocolate, the rather intense bitter of the substance being tempered by the addition of sugar. The oily pulp, which fills the loculus and surrounds the seeds, has a little of their taste and yields an ardent spirit.

§ 811. The TERNSTRÖMIACEÆ (*Ternstrœmiacées*) (Theads)—are divided into several tribes, one of which, the CAMELLIEÆ (*Camelliées*), is of sufficient importance to arrest our attention on account of two shrubs which it contains; the one, the CAMELLIA, the beauty of whose flowers (sometimes known by the name of the Rose of the World [*Rose du Japon*]) has brought it in vogue; its cultivation has, consequently, produced an immense number of rich varieties; the other, the flowers of which few persons know, although they also are very elegant, but the leaves of which have become one of the most important objects of commerce, the Tea-plant (*Thé*) (THEA). We know that it comes originally from China, which grows it for the consumption of the world, although its cultivation has been attempted in other countries, especially in Brazil, and more recently in Assam. The leaf, gathered when young, is slightly roasted and pressed in order to dispel a rather plentiful and slightly corrosive juice; it is then rolled and dried more or less rapidly, according as green or black tea is wished for; for the latter, leaves rather older and, consequently, of a more ligneous consistence are employed. Thus prepared they contain, in addition to several substances common to all leaves three others which give them peculiar properties: 1st, an essential oil, which communicates the aroma to the tea; 2nd, *theïne* (*théine*), a quaternary substance, rich in azote, since it is composed of 8 atoms of carbon, 10 of hydrogen, 2 of azote, 2 of oxygen; 3rd, *caseine*, another substance, the component parts of which have been already explained (§ 310). The last is insoluble in warm water which dissolves the other two, the only ones, consequently, which are found in the infusion of the tea as we drink it. It is not an exciting beverage only, but it is also nourishing, since it may contain six per cent of the theine, generally a little less, according to its quality and the facility of being dissolved. This property of the tea, which was formerly far from being suspected, accounts for its general use in all countries except France and the strongly concentrated state in which it is drunk. But the Chinese and other Asiatic nations do not confine themselves to the infusion: they eat the leaves when boiled. Now, since, after having been deprived of the soluble principles, the caseine still remains and in such a proportion that this residue may contain 28 per cent of it, the leaves furnish food very much richer in azote than the infusion.

§ 812. The GUTTIFERÆ [x] (*Guttifères*) (Guttifers)—derive their name from the presence of a gummy, resinous juice, commonly yellow and bitter, known in several of the species of this family under the name of *Gamboge Gum Gutta* (*Gomme-Gutte*), very much used in painting. Its energy and activity, which in over-doses make it a strong poison from the inflammation which it provokes, has caused its internal use to be proscribed, although certain very active purgative remedies are said to borrow a part of their power from it. This juice is not found in the pulp of the fruits; for there are several of them which are eaten, and even one of them, the Mangosteen (*Mangostan*), passes for the most delicious production of the tropics.

§ 813. We shall mention the ERYTHROXYLACEÆ (*Erythroxylées*) (Erythroxyls)—on account of the ERYTHROXYLON COCA, one of its species, the leaves of which are very much used in Peru; they are eaten with a little powdered chalk. It is used continually by the miners in Peru. It is reported, that they can by its aid pass a long

[x] CLUSIACEÆ. *Lind. Veg. King.*, p. 400.—TRANS.

time without any kind of nourishment even when working very hard; in this case it would seem that this family, like that of Tea, contains a very nutritious principle. But other travellers attribute its effect to a totally different cause; this principle, according to them, is a powerful narcotic, the power of which seems to surpass that of opium itself. It would be very interesting if Chemistry would enlighten us on the component parts of this plant.

§ 814. All the MALPIGHIACEÆ (*Malpighiacées*) (Malpighiads)—present a modification of the ovule, which M. Grisebach has called *lycotropous* (*lycotrope*) (from λύκος, *a horse's bit*). It presents, indeed, the form of a bit, or rather of a fish-hook, the pendulous funiculus of which would form the half of one of its sides, the ovule curved in an opposite direction being the curve of the hook and the other the ascending half. Also the embryo, modelling itself on this curve of the ovule in which it is formed, has its cotyledons almost always bent on themselves. Let us remark that in this, as well as in the following and several other polypetalous families, some of the genera have fleshy fruits, others dry fruits, the greatest number samaras: the Malpighiaceæ, in fact, will furnish us with the greatest number of examples of the most varied modifications of this kind of fruit. We have represented one of them (*fig.* 404). They are all trees or shrubs, and several are climbers remarkable for an abnormal structure (§ 86, *fig.* 106, 107). We have also explained (§ 251, *fig.* 218) the structure of the glands which commonly exist on their leaves or on their petioles, the presence of which is so characteristic on their sepals, at the back of which they are united in pairs, as well as the hairs peculiar to the genus MALPIGHIA (§ 248, *fig.* 214), whence they borrow their name.

§ 815.—The SAPINDACEÆ (*Sapindacées*) (Soapworts) are also remarkable for the structure of their climbers (§ 88, *fig.* 109). They are remarkable for the frequent want of symmetry between their stamens, often reduced to 8, and their envelopes, which belong to the quinary series; for the petals, each being frequently deduplicated as if by a second internal petal. The fleshy fruits of several species are eaten; and those of the EUPHORIA, known under the names of *Litchi* and of *Longan* (*Longan*), are much esteemed in China. Those of the SAPINDUS SAPONARIA and some others are remarkable for melting slowly in water like soap, whitening it and forming suds, and, therefore, being proper for washing. A white substance is extracted from them by alcohol. This is very soluble in water, in which it produces the effect we have just spoken of, being neuter, not volatile, composed only of carbon, hydrogen and oxygen: it is the *saponine*, the same as that obtained from the root of the SAPONARIA and other Caryophyllaceæ.

§ 816. In the MELIACEÆ (*Méliacées*) (Meliads)—we observe a remarkable example of those deduplications of the stamens of which we have spoken (§ 445). The filament is wholly or partly united to and placed on the outside of rather a wide plate, commonly bifid at the summit, and these plates are joined to one another at their edges, so as to form a tube bearing the stamens. Another shorter tube frequently surrounds the ovary. The seed either has (in the MELIEÆ [*Méliées*]) or has not a perisperm (in the TRICHILIEÆ [*Trichiliées*]). The leaves most frequently are one or more times compound. The properties of these bitter, astringent and tonic plants may acquire such an intensity as to cause vomiting, purging and even poisoning. This bitterness is found in the fleshy seeds of some species, especially of the CARAPA, with which the natives of Guiana cover their naked bodies in order to protect themselves from the attacks of insects.

§ 817. In the CEDRELACEÆ (*Cedrélacées*) (Cedrelads),—a nearly allied family formerly confounded with the preceding, we find the same principles, but otherwise combined, so that the exciting ones almost disappear, whilst the bitter ones, predominating, impart febrifugal properties to several species: their bark has, therefore, been employed with great success in the countries where they are indigenous. They also contribute, doubtless, to preserve the wood from the ravages of insects, assist it to last a long time and make the seed fine-grained and dense. The trees of this family, indeed, supply us with the most esteemed woods for cabinet work; it will be sufficient for us to mention among them the Mahogany (*Acajou*) (SWIETENIA MAHOGONI), although several others, less known in commerce, are as much and more esteemed in their own country.

§ 818. The AURANTIACEÆ (*Aurantiacées*) (Citronworts)—have for their type the Orange-tree, the fruit of which has received from several authors the name of HESPERIDIUM (*hespéridie*) and has already several times occupied our attention (§§ 515, 522, 531). The fruit is almost the same in the greater part of the other genera, with the exception of the modifications of form, of size, of colour and of flavour; the difference of these is immense and is a striking example of the influence of cultivation on domestic fruits. Every part is perforated with small vesicular glands or cavities filled with a volatile oil, the nature of which varies according to the organs and which are spread over the leaves as dots, transparent when held up to the light. These leaves are simple or frequently compound, and we shall allude in this respect to those of the Orange, which seem to be simple, but the presence of two foliaceous edges on the petiole and the articulation of the leaf beneath them clearly indicates a trifoliate leaf. Their wood is hard and compact, and is, therefore, employed in cabinet work, as that of the Citron (*Citronnier*).

PERIGYNOUS POLYPETALOUS DICOTYLEDONOUS VEGETABLES.

§ 819.—We shall divide the perigynous, like the hypogynous, according to the axillary or parietal placentation. The seed has a perisperm in a certain number of families and is without one in others; this enables us to establish two sections in the perigynous plants with axillary placentation, amongst which we shall place those with parietal placentation, in order to obtain a series which is better connected with the previous and subsequent parts of the general series.

(See Table XI. page 642).

§ 820.—Several families the SPONDIACEÆ (*Spondiacées*), the BURSERACEÆ (*Burséracées*) (Amyrids), the CONNARACEÆ (*Connaracées*) (Connarads), the TEREBINTACEÆ (*Térébinthacées*) (Terebinths or Anacards) were at first confounded together in a single one under the last name. They present, indeed, some common characteristics, but have some others very distinct, especially in the fruit, which is composed of separate carpels with a homotropous embryo in the Terebintaceæ, antitropous in the Connaraceæ, united into a drupe with several kernels in the Burseraceæ, into a single plurilocular one in the Spondiaceæ. The abortions are rather frequent in the flowers of several of these families, so that some few of their plants seem (as exceptions) to enter into the number of the diclinous or the apetalous. But they are necessarily placed here on account of more numerous and complete plants, the type of which they pre-

TABLE XI.

FAMILIES.

PERIGYNOUS POLYPETALOUS DICOTYLEDONOUS VEGETABLES.

1. Placentation axillary. Seed without perisperm.

Calyx free. Carpels united. Seeds definite in number. Flowers regular. Drupe bi-tri-locular. 1 style. Cotyledons flat. Isostemony —CHAILLETIACEÆ.
with a quinquelocular kernel. 5 styles. Cotyledons flat. Diplostemony —SPONDIACEÆ.
with 2—5 kernels. Style simple or wanting. Cotyledons scolloped. Diplostemony —BURSERACEÆ.
free. Embryo antitropous. 5—1 follicles. One erect seed in each. Diplostemony. Monadelphia —CONNARACEÆ.
amphitropous. Carpel, indehiscent, commonly fleshy. One seed on a funiculus erect from the bottom. Flowers regular. No stipules —TEREBINTACEÆ.
Legume. Flowers, papilionaceous. Stamens diadelphous. Stipules. PAPILIONACEÆ.
Flowers, irregular, frequently 3-1- petalous. Stamens free. Stipules. SWARTZIEÆ.
homotropous, straight. Legume.—Flowers irregular with imbricate prefloration. Stipules. CÆSALPINIEÆ.
regular with valvate prefloration. Stipules. MIMOSEÆ. } LEGUMINOSÆ.
Carpels, one or more, mono- or di-spermous. Cotyledons flat. Flowers rosaceous. Anthers introrse. Stipules —ROSACEÆ.
Several achenia. Cotyledons contorted. Petals, indefinite in number, similar to the calyx. Anthers extrorse. No stipules —CALYCANTHACEÆ.
Several follicles, whorled, polyspermous, each accompanied by a scale. Embryo cylindrical. Cotyledons flat —CRASSULACEÆ.
united. Flowers irregular. Petals and stamens, very frequently reduced to one. Calyx spurred —VOCHYSIACEÆ.
Petals equal in number to the divisions of the calyx. Stamens, in number equal to, double or triple of the petals. Calyx tubular }
regular. Seeds indefinite in number. Anthers opening by splits } LYTHRARIEÆ.
pores } MELASTOMACEÆ.
adherent to the 1- pluri- locular ovary. Anthers opening by terminal pores. Stamens definite in number. One style }
splits. Stamens indefinite in number. Several styles. Fruit fleshy. Cotyledons straight —POMACEÆ.
One style. A double whorl of loculi placed above one another. Cotyledons contorted }
Leaves not dotted } GRANATEÆ.
A regular whorl of loculi*
*Stamens monadelphous with a hood-shaped tube. Fruit, ligneous. Leaves not dotted —LECYTHIDEÆ.
with a straight tube. Berry. Leaves not dotted —BARRINGTONIEÆ.
free. Fruit fleshy. Leaves dotted —MYRTACEÆ.
free or polyadelphous. Fruit dry. Leaves dotted —LEPTOSPERMEÆ.
definite in number. Ovules one or more, erect. 1 loculus. Leaves dotted —CHAMÆLAUCIEÆ.
pendulous **

** Anthers curved, opening near the base. Several loculi. Fruit fleshy. Cotyledons contorted. No stipules —MEMECYLEÆ.
straight, opening lengthways. Embryo with a radicle very long. Cotyledons flat. One or more loculi. Stipules —RHIZOPHOREÆ.
very short. Cotyledons contorted or wrinkled. One loculus with several ovules pendulous from the top. Pollen egg-shaped. No stipules } COMBRETACEÆ.
flat. Several loculi. Pollen three-horned. No stipules —ONAGRARIEÆ.

[To face page 642.]

TABLE XI. Continuation.

FAMILIES.

2. Placentation parietal.

Embryo in the axis of a fleshy perisperm . . . . Stamens opposite to the petals in fascicles, alternating with multifid scales. One style. . . . Leaves commonly opposite . . . Herbs. . . . . —LOASACEÆ.
. . . . or alone, alternating with glands. Several styles. Leaves alternate . . . . Trees. . . . . —HOMALINEÆ.
alternate with the petals, equal to them in number . . . . united in a central column. Ovary free. Styles . . . terminal. Arillus . . . Embryo with foliaceous cotyledons. Plants climbing. Stipules. —PASSIFLORACEÆ.
lateral. Arillus wanting. Embryo cylindrical. . . . Plants not climbing. No stipules.—MALESHERBIACEÆ.
free. Ovary free Capsule, tri-valvate . . . . . Embryo with foliaceous cotyledons . . . . —TURNERACEÆ.
Capsule bi- or tri-valvate . . . Embryo cylindrical. Frequently diplostemony . . . . . . —SAXIFRAGACEÆ.
adherent. Berry. 2 placentaria. . . Embryo very small at the end of a large perisperm . . . . . —GROSULARIACEÆ.
without perisperm. Stamens definite, double the number of the petals, monadelphous. Anthers unilocular. Ovary free . . . Style and stigma single. . . Capsule tri-valvate . . —MORINGACEÆ.
indefinite in number, as well as the petals . . . . . . Anthers bi-locular. Ovary adherent. A long style. Several stigmas. Berry. Succulent plants —CACTACEÆ.
surrounding a farinaceous perisperm. Stamens and petals indefinite . . . . . . Anthers bi-locular. Ovary semi-adherent. Several sessile stigmas. . . Capsule with loculicidal dehiscence. Succulent plants . . . . FICOIDEÆ.

3. Placentation axillary. Seed with a perisperm.

Seeds indefinite in number. Ovary plurilocular free. . . . . Stamens, double the number of the petals, alternating with sterile appendices. Capsule 4-locular. . . . . —FRANCOACEÆ.
free or adherent. Stamens without alternate appendices equal to or double the petals. Carpels separating at the top. As many distinct styles. —SAXIFRAGACEÆ.
at the base. Style simple . . . —ESCALLONIACEÆ.
or multiple. Capsule with apicilar dehiscence, or breaking on the sides. Styles distinct or united . . . . PHILADELPHACEÆ.
indefinite. Anthers opening at the top. Capsule bi-locular. Styles distinct . . . . . . —BAUERACEÆ.
definite in number. Ovary adherent. Ovules pendulous. Embryo in the axis of a perisperm, to which it is almost equal, *
* with foliaceous cotyledons. One or more ovules. Carpels separating at the top. 2 styles. Stamens double the number of the petals . . . . HAMAMELIDEÆ.
One ovule. Drupe. Style, one. Stamens equal or multiple . . . . . . —ALANGIEÆ.
cylindrical . . One ovule. Three or four indehiscent carpels. Styles distinct. Stamens equal or double . —HALORAGEÆ.
very small, near the end of a large perisperm.**
** horny. One ovule. Achenium, double. 2 styles. . . . . . Stamens equal. Prefloration imbricate.—UMBELLIFERÆ.
fleshy. One ovule. Berry, 2-pluri-locular As many styles . . . . Stamens equal. Prefloration valvate —ARALIACEÆ.
Style single . . . . —HEDERACEÆ.
Drupe, with tri-locular kernel. Style, one . Stamens equal. Prefloration valvate . —CORNACEÆ.
. . . . 1-2 ovules. Fruit, dry, 3-4-locular. . . Styles, free or united. Stamens equal. Prefloration imbricate.—BRUNIACEÆ.
ascending . . . .
free. Embryo equal to the perisperm, having wide, foliaceous cotyledons, with a very short radicle. ***
*** Ovules 1—2 ascending . . . Fruit fleshy or capsule with septicidal dehiscence. Stamens equal and opposite to the petals. . . Prefloration of the calyx valvate . . . . RHAMNACEÆ.
One or more ascending. Fruit fleshy or capsule with loculicidal dehiscence. Stamens equal to and alternate with the petals. Prefloration of the calyx imbricate . . . . CELASTRINEÆ.
cylindrical, with very short cotyledons. One erect ovule. 3—5 indehiscent carpels. . . . . . Stamens equal and alternate. Claws of the petals united . . . . STACKHOUSIACEÆ.

[To face page 643.]

sent with some of those reductions of which we have spoken in another part of the work (§ 388).

The BURSERACEÆ (*Burséracées*) are trees or shrubs full of resinous juices, several of which are known in commerce under the names of balsams, gums and incenses. We shall mention only the best known, as Beshan or Balm of Mecca, (*baume de La Mecque*), furnished by the BALSAMODENDRON OPOBALSAMUM; that of Gilead by the B. GILEADENSE; Myrrh (*Myrrhe*) by the B. MYRRHA; Gum Elemi (*Gomme Elemi*) by the ICICA HEPTAPHYLLA. In India the BOSWELLIA SERRATA produces the true incense, under the name of which several other resinous substances are sold, some of which belong, and others do not belong to this family. In tropical countries where these different trees are indigenous, it is commonly the branches themselves, thoroughly impregnated with their juices, that are burnt in the temples. It is clear that these substances have in different degrees the stimulating properties, which generally belong to the resins, and it is on this ground that they are employed in medicine.

We shall find them again in the TEREBINTACEÆ; but the volatile oil, which holds their resin in solution, is frequently of a very acrid nature, and their juice applied to the skin, and especially when taken internally, causes serious accidents, as that of several *Sumacs*: they are said to occur even from the emanations alone of some trees of this family. But these juices render great services to the arts by supplying some of those beautiful varnishes known by the name of lacquers; these are at first white, so long as the innumerable particles of its organic substance, as yet separated, disperse light in every direction; afterwards, when these particles, decomposed by contact with the air, are connected into a homogeneous mass, they pass into a fine red or black colour. The Japan Lacquer (*laque du Japon*) is produced by the STAGMARIA VERNICIFLUA; RHUS VERNIX yields the black varnish. Two species of the Pistachio nut (*Pistachier*) (PISTACIA LENTISCUS, P. ATLANTICA) furnish the substance we know as Gum Mastich, and another (P. TEREBINTHUS) the Scio turpentine (*Térébinthe de Chio*); hence the origin of the name given to the whole family, although the source of the greater part of turpentines is different, as we have seen (§ 776). In certain fruits, the pulp of the sarcocarp is sufficiently developed to allow the oil to aromatise it, and they become not only innocent but agreeable: those of the Mango tree (*Manguier*), for instance. Let us remark that in one of them, the ANACARDIUM OCCIDENTALE (commonly known as the Cashew-nut, [*Noix d' acajou*]), the peduncle is swollen into a much larger mass than the fruit itself. The seed is fleshy and commonly oily without any mixture of other exciting principles, as in the Pistachio nut (*Pistachier*) (P. VERA). The leaves of a Sumac (RHUS CORIARIA), rich in tannin, are employed by the tanners.

§ 821. LEGUMINOSÆ (*Légumineuses*) (Leguminous plants).—The pod or legumen (§ 530, *figs.* 406, 407) characterizes all these plants, to which this name has, consequently, been applied, and the group may be considered, less as a single family than as a collection of several. The most numerous and the best known to us, as being the only one represented in our country, is that of the PAPILIONACEÆ (*Papilionacées*) thus termed from its butterfly-shaped flower, which we have explained (§ 440, *figs.* 283, 658) ; the flower is characterized, besides, by ten stamens sometimes free, more commonly monadelphous or diadelphous, whether they are united by fives or the tenth is detached from a tube formed by the nine others (*figs.* 687, 658) ; lastly, by a curved embryo with a radicle bent on the accumbent cotyledons (*figs.* 471, 661). The flowers, still irregular in the CÆSALPINEÆ (*Cæsalpinées*), preserve the papilionaceous, though with a tendency to the rosaceous form: the ten stamens are most frequently free: the embryo straight. The petals are reduced in number or are even quite wanting in another group (the SWARTZEÆ [*Swartziées*]), in which the number of the stamens sometimes exceeds ten, and in which the embryo is curved. The group of the MIMOSEÆ (*Mimosées*) is very considerable; in it the corolla as well as the calyx becomes regular, the prefloration valvate, whilst it is imbricate in all the rest: the stamens are equal in number to the petals, or more frequently multiple, so much so as to become indefinite; the embryo is straight. Let us remark that in these two last families the insertion of the stamens, clearly perigynous, has a tendency in the others to approach the bottom of the calyx and to pass to hypogynous. Let us also remark, that sometimes the internal membrane of the seed is very thick and almost resembles a perisperm. We have several times had occasion to speak of the leaves, which in a large number of Leguminosæ are one or more times compound and frequently articulated, always with stipules at the origin of their petiole.

When we reflect on the large number of species contained in this group, which comprehends plants of all dimensions and of many

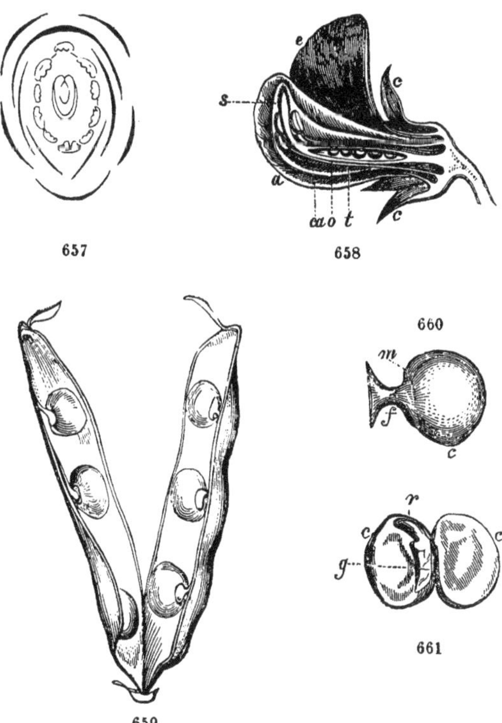

varied habits, from the highest trees to the most humble herbs, we must expect to find as great a difference in the products and properties. To recount them all would be too long a task, so that we must be satisfied with mentioning the most remarkable.

Several trees of this family are employed for carpenter's work in

657—661. The organs of the fructification of the Sweet Pea (*Pois de senteur*) (LATHYRUS ODORATUS).

657. Diagram of the flower.

658. Longitudinal section.—*c* Calyx.—*e* Standard.—*a* One of its wings.—*c a* Half of the keel.—*t* Tube of the stamens.—*o* Ovary with its ovules.—*s* Stigma.

659. The pod (*gousse*) opening in two valves, shewing the way in which the seeds are inserted.

660. A seed.—*f* Funiculus.—*c* Chalaza.—*m* Micropyle.

661. Embryo. The cotyledons *c c* have been separated, so as to exhibit the gemmule *g*, which was concealed between them.—*r* Radicle.

the countries of which they are natives, and we may refer to the Locust tree (*Faux-Acacia*) (ROBINIA PSEUDO-ACACIA) for its hardness, durability and resistance to moisture. The compact grain, the deep colours of the heart-wood of a large number cause them to be much esteemed for cabinet-work. Amongst these are the Sissoo-wood (*Palissandre*), whose origin was for a long time unknown, but is now reported to be from a species of DALBERGIA, the Pernambuco-wood (*bois de Fernambouc*) (CÆSALPINIA ECHINATA), Brazil-wood (*de Brésil*) (C. BRASILIENSIS), the Bukkum or Sappan-wood (*de Sappan*) (C. SAPPAN), the wood of the SWARTZIA TOMENTOSA, Camwood (BAPHIA NITIDA) and a great many others, among which an indigenous tree, the Laburnum (*Faux Ebénier*) (CYTISUS LABURNUM, may be mentioned. Several Leguminosæ are climbers and present an abnormal structure, an example of which we have already explained (§ 85, *fig.* 105). We remark sometimes the mixture of the cortical with the ligneous layers, so that we frequently have alternate zones of both; at other times, a kind of network formed by the cortical substance in the midst of the wood, in which it then forms elegant veins. This is very beautifully marked in the stem of a plant now very commonly cultivated on account of its handsome flowers, the GLYCINE SINENSIS.

Several herbaceous species of Papilionaceæ are rich in nutritious principles and are, therefore, cultivated as food for cattle. They form, in fact, the richest part of artificial meadows: as the Clover, the Lucerne, the Saintfoin, &c., &c. They are full of azotised substances, a portion of which, we know (§ 288), they can imbibe directly from the atmosphere.

This property is frequently found in the foliaceous pericarp of the fruits, the young pods of which are thereby rendered fit for food.

As to the seeds, they are of several kinds: some with thin and foliaceous cotyledons, not fit for food; others with thick cotyledons, which are frequently so. These, as they ripen, are filled with great quantities of fecula, such as the Kidney-bean, Windsor-bean, Peas, Vetches, &c., and several others less common or exotic, the names of which would not recall to our readers such familiar objects. This fecula is mixed with a large quantity of azotised matter, which renders them still more substantial food; it is gradually formed and accumulated in the seed, which, when young and limited for the most part to its teguments, would present cells filled with these principles and with a saccharine mucilage, and, consequently, would afford at this time a kind of nutriment differing from that it will afterwards afford. Peas, for instance, small and young, or old and large, are as different in their nutritive properties as in their taste. In others the cotyledons are fleshy and oily, as in the ARACHIS HYPOGÆA (the Underground Kidney-bean [*Pistache de terre*]); they contain a very large quantity of oil and have now become an important object of speculation. At other times an essential oil aromatises the seed, as that of the COUMAROUNA ODORATA (the Tonquin-bean [*Fève de Tonka*]), used for perfuming tobacco. The seeds with foliaceous cotyledons have frequently contrary properties and become purgative ; as those of the Bladder-Senna (*Baguenaudier*), of several species of Broom (*Genêts*) and Laburnum (CYTISUS), &c., &c. It is necessary, therefore, to use great precaution in the trials which we might be tempted to make, from the external resemblance of the fruits to the most familiar of our culinary leguminous vegetables.

But these purgative properties are also found in other parts : in the leaves, in the pericarps, especially in those which are foliaceous. The medicine best known in this respect is the Senna (*Séné*) (the leaves and fruit of the CASSIA SENNA and C. ACUTIFOLIA, which come to us from the East) : a peculiar substance is extracted from it, the *Cathartine*, which appears to be its active principle ; but, doubtless, another kind is found in the pulp which fills the cavity of the fruit in the CATHARTOCARPUS FISTULA, in the Tamarind (*Tamarin*), in the Carob-bean (*Caroubier*), the action of which is much milder. The preceding properties are especially observed in the Cæsalpinieæ. In the Mimoseæ, they are tonic and astringent, of which we shall only mention one example, the Catechu (*Cachou*), the juice of an Acacia (A. CATECHU), which is obtained by boiling the brown heart-wood and then allowing the solution to evaporate, thicken and dry. The large quantities of tannin found in it account for these properties and render the bark of several species of great value to tanners in the preparation of leather.

Among other products of certain Leguminosæ, we find some resins, such as Dragon's blood (*Sang-dragon*), extracted from the PTEROCARPUS DRACO ; some of them are liquid, because they retain a portion of the volatile oil which holds them in solution in the vegetable, as the Copaiva balsam (*Baume de Copahu*), furnished by several species of COPAIFERA, especially the C. OFFICINALIS ; some are joined with benzoïc acid and, consequently, form real balsams, as the balsam of Peru (*baume de Pérou*) (MYROSPERMUM PERUIFERUM), of Tolu (M. TOLUIFERUM).

This family also produces the most esteemed *gums* (*gommes*): Gum Arabic (*G. Arabique*), obtained from several species of ACACIA, especially the A. NILOTICA, Gum Senegal (*Sénégal*) from the ACACIA VEREK and ACACIA ADANSONII, Gum Tragacanth (*Adragante*) (falsely attributed to a small shrub of the south of Europe, the ASTRAGALUS TRAGACANTHA, but really obtained from oriental species of the same genus; the ASTRAGALUS GUMMIFER, VERUS, CRETICUS).

Lastly, valuable colours for painting and dying are obtained from this family, as Log-wood (*bois de Campêche*) (HÆMATOXYLON CAMPECHIANUM), of a reddish brown, water or alcohol easily dissolving its colouring matter which is owing to a principle called *hematine;* and especially Indigo, the colouring matter of which (*indigotine*) we have already pointed out in very dissimilar families, but which is mostly extracted from several species of the genus INDIGOFERA. These biennial plants are gathered the first year and plunged into water, in which they are allowed to ferment: when fermented, they are taken out and agitated in contact with the air until the water becomes blue by the combination of the indigotine with its oxygen; then the precipitation of the colouring principle is aided by a mixture of lime-water, and the precipitate is dried by evaporation.

§ 822. ROSACEÆ (*Rosacées*) (Roseworts).—This is also a family that may be considered as an association of several others, which it is impossible to separate from one another. It is a very instructive family to study, shewing us how certain characteristics may vary in the same natural group; how, by following these variations from one extreme to the other through a continuation of intermediate ones, we can retain no doubts as to the tie which unites them; how, lastly, seeing one characteristic unchangeable by the side of one which is continually changing, we learn to attribute to it more relative value. The pistil of an Apple-tree (*Pommier*) is composed of an ovary adherent to the calyx throughout the whole of its extent, enclosing five small loculi in the midst of a thick mass of flesh; that of a Strawberry (*Fraisier*), of a crowd of small separate carpels on the surface of a thickened axis, projecting above the free calyx: we have

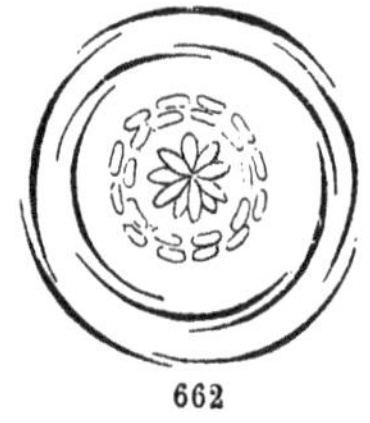

662

662—669. Th organs of the fructification of a Bramble (*Ronce*) (RUBUS STRIGOSUS).

662. Diagram of the flower.

an example of a syncarpous fruit in the former, of an apocarpous in the latter. But, if we examine a SPIRÆA, in which five distinct

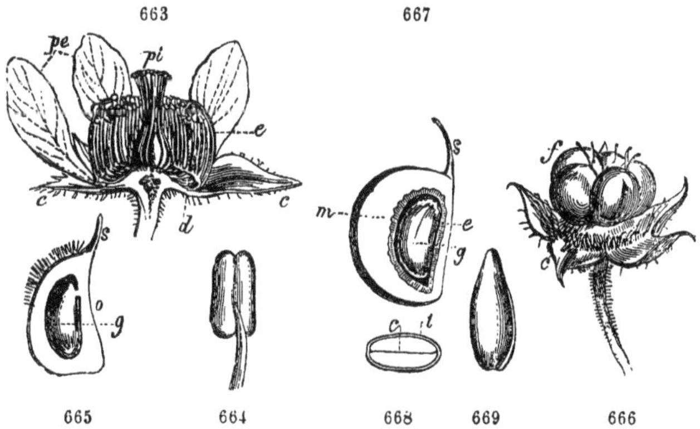

carpels are fixed on a plain torus at the bottom of the free calyx; then a Cherry-tree (*Cerisier*), in which there is only a single carpel, the cup-shaped calyx being raised around it; then an ALCHEMILLA, in which the calyx always free, contracts its tube above the carpels, in number from one to four; then a Rose (*fig.* 369), in which the carpels, more numerous and divided, seem to grow from the inner surface of the tube, which, swollen at their middle, closes above them, merely leaving sufficient room for the passage of the styles; if we go a step further and suppose that all these distinct parts are placed close together and united into one body, we shall have returned to the pistil of the Apple-tree. Yet the way, in which the stamens are inserted, has not been altered; it was constant in a circle towards the summit of the calycinal tube. The perigyny, therefore, of the stamens has more stability and importance than

663. Vertical section of the same.—*c* Calyx.—*p e* Petals.—*e* Stamens.—*d* Disk covering the bottom of the calyx, on which the stamens are inserted.—*p i* Pistil composed of several carpels.

664. An anther with the top of its filament. We see that part turned outwards in the flower.

665. Vertical section of the ovary *o* shewing the position of the ovule *g* —*s* Style.

666. Fruit.—*f* Fleshy carpels with the persistent calyx *c*, on which the withered filaments remain.

667. Vertical section of a carpel.—*s* Style.—*m* Fleshy mesocarp or sarcocarp.—*e* Endocarp.—*g* Seed.

668. Horizontal section of the seed.—*t* Tegument.—*c* Cotyledons of the embryo.

669. Embryo.

the relation of the calyx and ovary, free or adherent, to one another.

Let us add to what we have said, that the petals are inserted below the stamens on the calyx, alternating with the lobes of the calyx equal to them in number, most frequently five and expanded like a Rose; that the embryo has no perisperm, is straight and has fleshy cotyledons and a short radicle turned towards the point of the attachment of the seed; that the leaves are simple or compound but always have stipules: and we shall have the general characteristics of the Rosaceæ. The adherent ovary, with two ovules, rarely more or less, ascending in each loculus, changing into a fleshy fruit, will clearly distinguish the POMACEÆ (*Pomacées*) (Appleworts). Several distinct achenia, enveloped by and inserted on the bottom of a fleshy calyx, each enclosing a single pendulous seed, will characterize the ROSACEÆ (*Rosées*) (Roseworts) properly so called. The DRYADEÆ (*Dryadées*) have several achenia on a receptacle projecting into the centre of the flower, each with a pendulous or erect seed; the SANGUISORBACEÆ (*Sanguisorbées*) (Sanguisorbs) have achenia almost always reduced to one or two, covered with the contracted tube of the hardened calyx, which frequently does not bear petals; the SPIRÆACEÆ (*Spiræacées*), have five whorled carpels at the bottom of a calyx with a short tube, each enclosing two ovules at the most, pendulous or ascending, opening along an internal suture; the AMYGDALEÆ (*Amygdalées*) (Almondworts), a single free ovary with collateral pendulous ovules and afterwards a drupe; the CHRYSOBALANEÆ (*Chrysobalanées*) (Chrysobalans), the same with the exception of the two ovules being erect. Among the trees of our temperate climates, if almost all our forest trees belong to the Amentaceæ, the Rosaceæ supply us with almost all our orchard plants and furnish us with the greater part of the fruits we eat. Apples (*Pommes*), Pears (*Poires*), Quinces (*Coings*), Medlars (*Nèfles*), Service-apples (*Cormes*) are produced by the Pomaceæ; Cherries (*Cèrises*), Plums (*Prunes*), Apricots (*Abricots*), Peaches (*Pêches*), Almonds (*Amandes*), by the Amygdaleæ; Raspberries (*Framboises*) and Strawberries (*Fraises*), by the Dryadeæ. But we must remark that in these fruits, although all belong to the same family, we do not in every case eat the same part of the fruit; since in the Pomaceæ we use the thickened calyx, blended with the pericarp; in the Amygdaleæ, the sarcocarp only, excepting the Almond, the pericarp of which we reject and eat its embryo; in the Strawberry, the fleshy receptacle which bears the carpels, and in the Raspberry, the carpels without the receptacle.

Another fact worthy of attention is the presence of hydrocyanic acid, found in the leaves and kernels of the Amygdaleæ. It enters, therefore, but in an extremely small proportion, into the liquors which are made from the fruit of certain kinds of Cherry; as the Kirschenwasser from the Wild Cherry (CERASUS AVIUM).

§ 823. MELASTOMACEÆ (*Mélastomacées*) (Melastomads).—This family is composed almost exclusively of ligneous plants. Their opposite leaves are remarkable for the arrangement of the nerves, of which the lateral, one, two, three or four in number on each side, projecting almost as much as the median, are directed like it from the base to the summit of the leaf, preserving throughout the whole of their passage a uniform thickness, being united to one another by thinner transversal ones, so as to resemble the arrangement which we have pointed out in the leaves of several Monocotyledons. The ovary, entirely free in a small number, is more commonly adherent to the calyx, but in a peculiar and incomplete manner; for it adheres only by the longitudinal nerves on the surface of the ovary, and their intervals leave between this surface and that of the calyx so many lacunæ, in which the young anthers are imbedded; they are afterwards disengaged, straightening themselves on the filament. These anthers are very remarkable for their long and arched form: their two loculi are opened at the top (frequently prolonged into a kind of beak) by one common pore or by two distinct pores, and are united by a connective, which is frequently prolonged lower down, is articulated with the summit of the filament, and may present at the point of the articulation appendices of various forms.

§ 824. MYRTACEÆ (*Myrtacées*) (Myrtleblooms).—This family is subdivided into several secondary ones: 1st, the CHAMÆLAUCIEÆ (*Chamælauciées*) (Fringe Myrtles), the unilocular ovary of which encloses one or more erect ovules and becomes a dry, monospermous fruit, sometimes opening into two valves; the stamens are definite in number, being double or quadruple of that of the petals: several of them, commonly sterile, have free filaments, more rarely grouped in threes. 2nd, the LEPTOSPERMEÆ (*Leptospermées*), with a bi- or multilocular ovary becoming a dry fruit, most frequently capsular with numerous seeds, very rarely solitary; the stamens are indefinite in number and in groups opposite to the petals. 3rd, the MYRTEÆ (*Myrtées*) properly so called, which differ from the preceding by their fleshy fruit and their always free stamens. In all the preceding plants, the leaves are covered with transparent dots indicating so many reservoirs of an essential oil, which are not found in the following, namely: 4th, the BARRINGTONIEÆ (*Barringtoniées*) (Barringtoniads), the fruit of which is a berry or several oligospermous loculi, the stamens being numerous and most commonly monadelphous: 5th, the LECYTHIDEÆ (*Lécythidées*) (Lecyths), the indefinite filaments of which are united into a tube bent like a hood; the ovary with several multiovulate loculi becomes a fruit of rather considerable size, with a woody pericarp, indehiscent, or opening with a lid. It frequently happens in these last two families and in some of the genera belonging to the others that the different parts of the embryo are united into a homogeneous mass. At other times the cotyledons are plaited and foliaceous; they are most commonly flat in the first mentioned families.

The volatile oil, which exists in such abundance in these, imparts to them their tonic and stimulating properties and aromatic odour. Every one knows the Clove (*clou de gérofle*), which is the dried flower-bud of one of them (CARYOPHYLLUS AROMATICUS). It communicates an agreeable perfume to several of their fruits, which are, therefore, much esteemed as dessert fruit. The best known is the Guava (*goyaves*) (PSIDIUM). An astringent principle is also found in the bark of the root and of the fruits before they are ripe.

The Pomegranate (*Grenadier*) (PUNICA GRANATUM) is united to the Myrtaceæ by several authors, separated by others, who make it the type and hitherto the only species of a family, the GRANATEÆ (*Granatées*). It is especially distinguished by the peculiar arrangement of its loculi, which, instead of being arranged in a single whorl as in the greater part of the multilocular ovaries, form two of them: the one at the bottom, the other at the top and thrown to the outside (§ 506). Their unequal developement causes the interior of the ripe fruit to be divided into several irregular compartments by obliquely transversal dissepiments. The external envelope of the seeds is enlarged into a succulent pulp, that part which we eat in the Pomegranate, the whole of the pericarp or bark being rejected.

§ 825. ONAGRARIEÆ (*Onagrariées*) (Onagrads).—This family, such as it was primitively instituted, comprehended, besides that which we have preserved under this name, that of the COMBRETACEÆ (*Combretacées*) to which the MYROBALANEÆ (*Myrobalanées*) (Myrobalans) mentioned among apetalous plants and the HALORAGEÆ, (*Haloragées*)

(Hippurids) are closely allied. Some of the same genera, still left in the three polypetalous families, are destitute of a corolla, and this exception, which is also observed in very many others, proves the necessity of bringing in the series the polypetalous close to the apetalous, if we do not quite confound them together. Our tables shew the characteristics drawn principally from the structure of the seed, from which this group has been divided into several families. The Onagrarieæ properly so called include some plants, remarkable not only for their properties but also for the elegance of their flowers, causing them to be so much esteemed in our gardens. We must mention the extreme rarity of the quinary number in the parts of these flowers. The quarternary is nearly always found and the whorls are even reduced to two parts in the ŒNOTHERA (*fig.* 244): this is found in the greater part of the Halorageæ. Another point, which deserves our attention, is the constant and well characterized form of the pollen grains (§ 477, *figs.* 350, 351). An abnormal genus of this family is the TRAPA, an aquatic plant, the fruit of which is commonly known under the name of Water Chestnut (*Châtaigne d'eau*) on account of the thorny excrescences of its pericarp. The farinaceous embryo of several species is eaten both in Europe and Asia. One only of the two cotyledons forms nearly the whole mass of this embryo, the other being only rudimentary and even smaller than the radicle itself and but little developed (§ 581).

§ 826. PASSIFLORACEÆ (*Passiflorées*) (Passionworts).--The beauty of the flowers, well known by the name of *Passion-flowers* (*Fleurs de la Passion*), along with their strange shapes resulting from the presence and arrangement of a large number of appendices, attracted attention a long time since. We shall only stop here to explain why they are classed among perigynous plants, although apparently hypogynous. The stamens and the pistils are generally situated on the extremity of a column, rising from the very centre of the flower and apparently continuing the axis. But, on the other hand, the petals, the place of the insertion of which is commonly the same as that of the stamens, grow from the edge of the calycinal tube; and from its surface, a little lower, spring one or two circles of coloured filaments which can hardly be considered as any thing else than abortive stamens. The glandular layer of the torus, which, covering the whole of the surface of the tube, serves as a base to these different parts, is continuous with that of the stamen-bearing column and thus evidently connects the insertion of the anther-bearing stamens with those of the petals and of the coloured filaments. In the edible fruits of some Passionworts the fleshy and thick arillus enveloping the seeds is the part eaten. Its pulp on account of the abundance and limpidity of its juices, their acid and agreeable flavour, is much esteemed as dessert food, especially in the warm climates where the plants are indigenous. Rather a large number of them are cultivated in our greenhouses. They are for the most part herbaceous and climbing, with tendrils growing from the axils of their leaves and, consequently, representing transformed branches.

§ 827. GROSSULARIEÆ (*Grossulariées*) (Currantworts).—The external tegument of the seed, developed into a pulp and filled with large quantities of delicious juice, imparts to the Currantworts their qualities. It is easy, when the berry is cautiously opened by gently separating the seeds from one another, to perceive this arrangement. But, at the first sight the fleshy teguments seem to form a continuous pulpy mass between these seeds pressed against one another; this false notion has suggested the habitual expression of *seeds imbedded in the pulp* (*graines nichés dans la pulpe*) SEMINA NIDULANTIA.

§ 828. CACTACEÆ (*Cactées*) (Indian Figs).—In the fruits of these plants, on the contrary, it is the thickened pericarp which forms the pulp. Some of these plants are eaten in southern countries, especially that of the OPUNTIA, best known as the Indian-fig (*Figue d' Inde*). The Indian Figs, so remarkable for their singular shapes, which, as well as the beauty of the flowers of several of them, causes them to be much esteemed, are very succulent. The cellular tissue, extraordinarily developed, swells their stems and leaves and thus imparts to them dimensions and figures more or less different from those to which we are accustomed in the greater part ofother vegetables. This developement is generally connected with the scarcity of the stomata, from which result the defect of evaporation, the accumulation of the juices retained in the interior of the plant and the capability of living in extremely dry climates, in which plants otherwise constituted could not exist. The ligneous fascicles are in much less proportion in this tissue: regular zones of wood are only observed in a few species. The fibres or vessels which form them present a very remarkable structure; instead of a thread in a perfect spire, or divided into rings, we find a lamina of some breadth and thickness (*figs.* 670, 671). In this family there are plants with cylindrical stems with leaves or without leaves; but the majority depart far from these forms to assume those of variously fluted columns, to become flat like leaves, or, on the contrary, to be enlarged like spheroïd or ovoïd fruits, sometimes of enormous dimensions, frequently covered with regular nerves more or less projecting and sharp. In this last case, there is frequently a suppression of leaf buds (*bourgeons*) and, consequently of ramification. In the most of them they are replaced by small bundles of thorns arranged in regular natural series, frequently spiral, close to and above which grow the flowers. In these, there is almost an insensible transition from the calycinal folioles already coloured to the petals; and above the ovary to which they adhere, they are sometimes separated directly, sometimes they form a tube by their union prolonged for a greater distance.

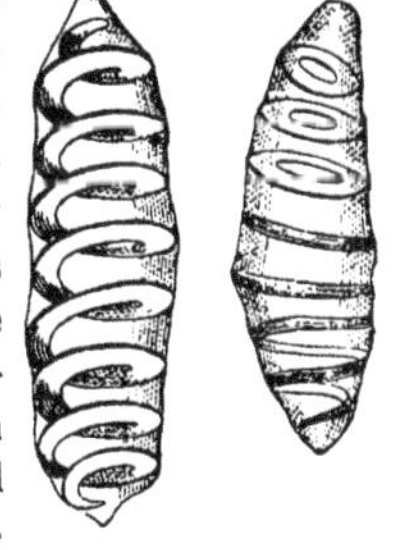

§ 829. Crassulaceæ (*Crassulacées*) (House-Leeks.)—These are also succulent herbs or shrubs; but the ordinary forms are thickened and not altered. The affinities of this family with that which follows give it a claim for its place here, although in our tables it is stationed at some distance from it, on account of the perisperm being entirely wanting in its seeds or only forming an extremely thin lamina. We have frequently had occasion to mention its flowers, as offering, perhaps, the most perfect type of a Dicotyledon (§§ 371, 386, *figs.* 225, 234, 235), and we have shewn the existence of the small fold or peculiar body, which is constantly found on the outside of each of the carpels. Several genera, by the union of the edges of their petals into a tube of greater or less length, to which even the bases of the filaments adhere, seem to be quite monopetalous. But all their characteristics are too clearly similar to those of the family for them to be separated. The juices of the Crassulaceæ are generally acrid and corrosive; for this reason that of several of them, of the House-Leek (*Joubarbe*), for instance, is popularly employed instead of caustic.

670—671. Two lengthened utricles taken from the Echinocactus coptonogonus, with a lamina completely spiral in one, partly spiral and partly divided into rings in the other.

§ 830. SAXIFRAGACEÆ (*Saxifragées*) (Saxifrages).—This considerable group comprehends several secondary ones besides the BAUERACEÆ (*Baueracées*) and the ESCALLONIACEÆ (*Escalloniacées*) (Escalloniads), which we have mentioned in the table; the CUNONIACEÆ *Cunoniacées*) (Cunoniads), trees or shrubs with opposite leaves, most frequently compound, accompanied with large interpetiolar stipules; the HYDRANGEACEÆ (*Hydrangeacées*) (Hydrangeads), trees or shrubs with simple opposite leaves, without stipules, among which we may mention the HORTENSIA of our gardens. As to the SAXIFRAGACEÆ properly so called, they are herbs with alternate leaves without stipules. The HAMAMELIDACEÆ (*Hamamelidées*) (Witch-Hazels) and the PHILADELPHACEÆ (*Philadelphacées*) (Syringas) seem to be nearly allied to them and, perhaps, also several other families now rather distant, but characterized in the same way by carpels most frequently two in number, polyspermous and distinct at the summit.

§ 831. UMBELLIFERÆ (*Umbellifères*) (Umbellifers).—This group, so natural and so easily recognised by several prominent features, has been long and generally known. Its name, which it at first received, has been preserved; it was derived from its manner of inflorescence which we have explained (§§ 208, 230, *fig.* 187). It is composed of plants for the most part herbaceous, annual or perennial, the aerial stem of which, being thus developed in the course of a year, during which it frequently acquires rather large dimensions, (a developement in which the pith cannot for a long time take part), becomes fistulous and presents a tympanum at each of its knots like that of the Gramineæ. The alternate leaves with a limb almost always deeply divided embrace its knots by a long and wide sheath, which is prolonged to a greater or less height and forms nearly the whole leaf in the upper ones. The flowers (*fig.* 674) are composed of an adherent calyx terminated by five small teeth, sometimes hardly visible, with which alternate as many petals inserted on the circumference of a large glandular disk covering the whole of the top of the ovary, and bearing the five alternate stamens, with filaments frequently curved inwards, always so in the flower-buds. From the centre of the disk grow two short styles, each terminated by a simple stigma and turned, one towards the centre of the umbel, the other towards its circumference; this arrangement corresponds to that of the two loculi, each enclosing a pendulous ovule and forming two achenia, which are at last separated, and are only kept together by the axis or fascicle of the nourishing vessels, which is divided into two filaments, each bearing the corresponding achenium suspended (§ 533, *fig.* 413). The seed, the teguments of which are nearly always blended with the pericarp, is almost entirely formed of a perisperm, generally horny, towards the upper extremity of which is imbedded a small cylindrical embryo (*fig.* 676).

We must consider some of these parts more in detail, if we wish

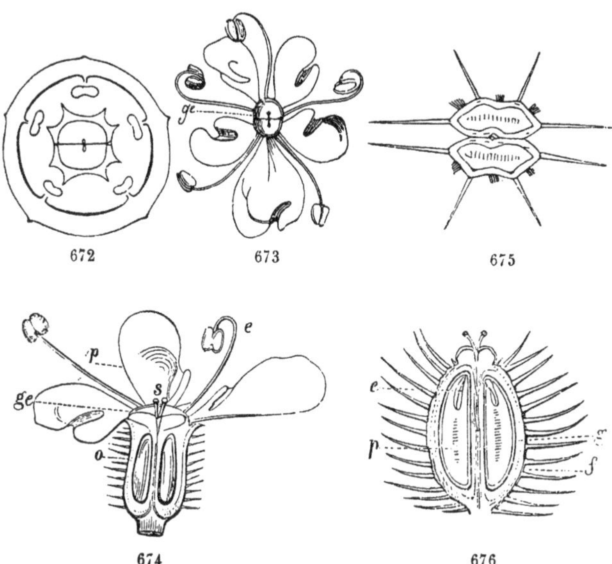

to comprehend the characteristics now employed in the distinction and arrangement of the genera of the Umbelliferæ. The petals are entire, crenate or bilobed, plane at the summit or prolonged into

672—676. The organs of the fructification of the Carrot (*Carotte*) (DAUCUS CAROTA).

672. The diagram of the flower.

673. The flower.—*g e* Epigynous disk. The observer is supposed to be looking at its top.

674. Vertical section of the flower.—*p* Petals.—*e* Stamens.—*o* Ovary blended with the adherent calyx.—*s* Styles and Stigmas.—*g e* Glandular epigynous disk.

675. Horizontal section of the fruit.

676. Vertical section.—*f* Pericarp.—*g* Seed.—*p* Perisperm.—*e* Embryo.

an inflexed point (*figs.* 282, 673): this corolla is frequently not quite regular, but the petals situated near the outside of the umbel are much more developed than the rest. The characteristics drawn from the fruit are very important. They are as follows; there are projecting nerves on its surface (*fig.* 672), along which they describe *ribs* or *ridges* (*côtes*) (JUGA) more or less developed, sometimes under the form of superficial lines, sometimes under that of a cock's comb. Now the adherent calyx is composed of five folioles, as the five free teeth at the summit proves; each of them presents a median nerve, and their edges united in pairs determine as many angles alternating with the former, so that the whole of the fruit presents ten of them, alternately corresponding to the median nerves (JUGA CARINALIA) and to the united edges (JUGA SUTURALIA); so that each of the two carpels presents five, one median, two intermediate and two lateral, which are joined to the similar ones of the opposite carpel. Between the five ridges, thus formed on the surface of a carpel, may be found four retreating angles or *vallecules* (VALLECULÆ). Sometimes a secondary nerve, double in each foliole, divides each vallecule lengthways and thus doubles the number. In the substance of the pericarp and along each vallecule are frequently situated one or more lacunæ filled with a peculiar resinous juice. They are gradually enlarged from top to bottom, where they are terminated in a CŒCUM and describe as many coloured linear receptacles, VITTÆ. The form and the number of the ribs, of the valleculæ, and the arrangement of the vittæ furnish the principal characteristics now employed, and we must, therefore, learn how to determine them. The internal faces, by which the two carpels are at first united and which are at last separated, are sometimes plane (UMBELLIFERÆ ORTHOSPERMEÆ [*Ombellifères orthospermées*] [*figs.* 675, 676]), at other times concave, either by the inflexion of their edges or lateral ribs (U. CAMPYLOSPERMEÆ [*O. Campylospermées*]), or more rarely because they are bent at their two extremities (U. CŒLOSPERMEÆ [*O. cœlospermées*]). The perisperm, forming the greatest part of the mass of each carpel, closely adherent to its teguments, presents the same modifications of shape.

The juice accumulated in the vittæ is an aromatic oil which communicates its properties and perfume to the seeds of several species, such as Aniseed (*Anis*), CORIANDER (*Coriandre*), FENNEL (*Fenouil*), Cummin (*Cumin*) &c., &c. This oil is frequently joined with a narcotic principle in the other parts of the vegetable, especially in the bark and leaves, where the peculiar juices are found. These present different qualities according to the proportion of the predominating principle. They sometimes form stimulating or antispasmodic gum-resins usefully employed in medicine, as *Assafœtida*, *Opopanax*, *Sagapenum*, *Galbanum*, *Gum ammoniac ;* they sometimes become more or less violent poisons in the CONIUM MACULATUM, the CICUTA VIROSA, the ÆTHUSA CYNAPIUM, the PHELLANDRIUM AQUATICUM, &c., plants which are vulgarly called Hemlock, Water Hemlock, Cowbane, Fool's Parsley, &c.; there is such confusion in the names of antiquity, that we cannot determine with precision that plant from which the famous Athenian death-drink was prepared; they are sometimes diluted so as merely to impart an agreeable flavour to the parts in the midst of which they are found, and which are then used for food, as the leaves of the Parsley (*Persil*), of the Chervil (*Cerfeuil*), the stems of the Angelica (*Angelique*). But these properties are especially diluted in the parts protected from the action of the light, particularly in the roots, as those of the Carrot (*Carotte*), of the Parsnip (*Panais*), &c., &c., in daily use: gardeners also produce this effect artificially by earthing up certain portions destined for food, as the stem of the Celery (*Céléri*). It has also been observed that the energy of the properties is augmented or diminished according to the heat of the climate: thus, the CONIUM MACULATUM, a dangerous poison in the South of Europe, may be eaten without hurt in Russia. The fleshy roots we have mentioned contain also rather a large proportion of saccharine matter.

§ 832. RHAMNACEÆ (*Rhamnées*) (Rhamnads).—Under this name were formerly confounded plants with stamens opposite to the petals (to which it is now exclusively applied) and the CELASTRINEÆ (*Célastrinées*) (Spindle-trees), in which the stamens alternate according to the more general law. These two families, nevertheless, have some properties in common: the green and yellow colouring principles of several of their species ; the acrid, purgative principle in several of their fruits, especially those of the Buckthorn (*Nerprun*); the astringent and stimulating principle that the herbaceous parts sometimes contain; hence the infusion of the leaves of some of them is used as tea. The Arabs chew the fresh leaves of the Khât (*Kat*) (CELASTRUS EDULIS) and thus procure an excitement analogous to that which results from the use of narcotics. It is remarkable that the Rhamnaceæ, along with these genera producing poisonous fruit, present others whose pericarp is swollen with a saccharine mucilage, which, imparting to it contrary properties, causes it to be esteemed as food. Every one knows that of the Jujube-tree (*Jujubier*) (ZIZYPHUS JUJUBA). Another of the same genus, the ZIZYPHUS LOTUS, gave its name to the ancient Lotophagi and is still collected by the inhabitants of the same country for food. In the HOVENIA DULCIS, the edible part is not the fruit, but its extremely thick and succulent peduncle, a modification which we have already found in a few other plants (§ 820). The Celastrineæ are separated into two tribes, which might even be raised to the rank of families: the STAPHYLEACEÆ (*Staphylées*) (Bladder-Nuts) with compound leaves, with seeds without an arillus, in which the perisperm forms a very thin lamina; the EVONYMEÆ (*Euonymées*), with simple leaves, having seeds with a fleshy arillus (*fig.* 451) and a thick perisperm: the type of these is the Spindle-tree (*Fusain*) (EUONYMUS), the light and porous charcoal of which is used for crayons, which are also made from several other kinds of wood possessed of the same qualities.

MONOPETALOUS DICOTYLEDONOUS VEGETABLES.

§ 833. These were divided by Jussieu (p. 533) into HYPOCOROLLEÆ (*Hypocorollées*), PERICOLLEÆ (*Péricollées*) and EPICOROLLEÆ (*Epicorollées*); the last are divided into two classes according as their anthers are distinct or united to one another. Whilst we follow this mode of classification, we shall confound the monopetalous plants having perigynous with those having epigynous insertions, on account of the practical difficulty we frequently experience in distinguishing one from the other.

HYPOGYNOUS MONOPETALOUS DICOTYLEDONOUS VEGETABLES.

§ 834. We shall commence with a certain number of families, which may be considered as establishing the passage from polypetalous to monopetalous plants. Several, indeed, present this double characteristic in their genera, connected as to the rest to one another by evident affinities; such are the Styracineæ, the Ebenaceæ, the Ilicineæ. Although in the rest the petals are united to one another to a certain height, it is sometimes for a very little distance; several characteristics, moreover, peculiar to plants essentially monopetalous are here wanting. In these the stamens are situated on the corolla; their number does not reach or is at most equal to that of its divisions, and lastly this number is not found in the carpels, which in most cases are reduced to three or more commonly to two. The subsequent families present, on the contrary, carpels frequently equal in number to the petals, stamens frequently double or some other multiple of their number, and very frequently also quite independent of the corolla. Several, it is true, following the general law, have their stamens inserted on the tube of this corolla, equal in number to its lobes: but they are then most commonly opposite to them, and other bodies, even sterile filaments, alternating with them occupy their normal place; this is a sufficient proof of the existence of a second whorl of stamens dissimilar up to a certain point on account of a more or less complete abortion. These different considerations have led us to present these families in a table by themselves; and if in some of their genera, even in a small number of whole families which we have thought it right to comprehend in it, we do not find these exceptional characteristics, this will be found to be their proper place from all their other characteristics to which we ought to have regard. The very mode of insertion itself seems to lose some of its importance in the group thus formed, which includes some few cases, very rare it is true, of perigyny: a fresh connection with the polypetalous families with which we have finished.

(See Table XII. page 666).

§ 835. ERICINEÆ (*Éricinées*) (Heathworts).—The first five families of this table have been all confounded under the name of ERICÆ or Heaths (*Bruyères*), and present, indeed, the most intimate relations in spite of the differences, which have caused them to be separated into several groups. The fruits, when they are fleshy like those of the Strawberry-tree (*Arbousier*), of the VACCINIUM, are eaten raw or cooked in the countries where they grow; the most used in Europe is that of the Bilberry (*Airelle*) (VACCINIUM MYRTILLUS), that of which Virgil has said,

"VACCINIA NIGRA LEGUNTUR."

That of the Common Strawberry-tree (ARBUTUS UNEDO) contains a very small portion of a narcotic principle which renders eating it in excess dangerous. This last principle is found, also, in several RHODORACEÆ (*Rhodoracées*), some species of RHODODENDRON, KALMIA, AZALEA, &c. It seems to be demonstrated that the honey which enervated a large number of the soldiers in the retreat of the ten thousand Greeks across Asia Minor under Xenophon had been collected by bees from the nectaries of the flowers of the RHODODENDRON PONTICA. The leaves of the LEDUM PALUSTRE, employed in the preparation of beer, render it extremely heady. Very astringent properties have also been found in several of the Ericineæ; the presence of tannin and of gallic acid has been clearly determined in the ARBUTUS UVA-URSI, the leaves of which are consequently employed in the preparation of leather.

§ 836. The STYRACINEÆ (*Styracinées*) (Storaxworts),—which it would be more convenient to place among the perigynous polypetalous plants, deserve to be mentioned as producing Gum benzoin or benjamin (*benjoin*), obtained from the STYRAX BENZOIN. Another drug well known in commerce, Storax, was supposed for a long time to come from the STYRAX OFFICINALE; but its origin is still doubtful.

§ 837. The EBENACEÆ (*Ebénacées*) (Ebenads)—are remarkable for the hardness of their wood, which has given to several the name of Iron-wood (*Bois de fer*). The best known is the Ebony (*Ebène*) the wood of the DIOSPYROS EBENUS, several species of which genus have the same qualities. The fruits of some of them are eaten when half decayed; the extreme acerbity of their flesh before it arrives at that state is a sufficiently clear proof of the existence of astringent principles.

§ 838. ILICINEÆ (*Ilicinées*) (Hollyworts).—(Called also AQUIFOLIACEÆ.)—This acerbity is again found in the fruits and the bark of several of these plants. We may remark in passing that this property is generally met with in the majority of the vegetables employed as substitutes for tea. We cannot pass over in silence the ILEX PARAGUENSIS, the infusion of the leaves of which is so much used in South America under the name of *Maté* or Paraguay Tea (*Thé du Paraguay*); it would be interesting to compare its chemical composition with that of real Tea[y]. The bark of the Common Holly (*Houx commun*) (ILEX AQUIFOLIUM) contains a peculiar substance called Birdlime.

§ 839. The JASMINACEÆ (*Jasminées*) (Jasminworts)—are well known from the abundance of the volatile oil which imparts such a delicious perfume to several of their flowers. Those of the OLEACEÆ (*Oléinées*) (Oliveworts), amongst which it is sufficient to cite the Lilac (*Lilas*) and the OLEA FRAGRANS, which are employed in China for flavouring tea, have generally a pleasant odour. But the latter family is especially useful from the presence of the oil in the pericarp of the Common Olive (*Olivier*), so generally cultivated all round the Mediterranean; this oil is also found in a less degree in the pericarp of some other drupaceous genera in which it is neglected. To this family is also assigned the Ash (*Frêne*), from several species of which, on their bark being wounded, exudes Manna, a saccharine and slightly purgative substance, the properties of which appear to be owing to a principle distinct from sugar, *mannite*, which is also found in several very different vegetables, even, it is said, in some Fungals.

[y] Lindley says that Mr. Stenhouse has discovered Theine in it.—*Lind. Veg. Kingd.*, p. 598.—TRANS.

§ 840. Sapotaceæ (*Sapotacées*) (Sapotads).—The pulpy fruits of rather a large number of these plants are esteemed in the tropical countries where they are indigenous, especially that of the Sapodilla Plum (*Sapotilier*), which gives its name to the family. Their seed, particularly that of the Bassia butyracea is rich in a thick fatty oil concreting and assuming the consistence of butter. We know another species of this genus by the name of the Butter-tree (*Arbre au beurre*) in Africa, and its produce by that of Galam butter (*Beurre de Galam*). Considered in its botanical relations, this family has real interest for us, because it shews the passage from those, in which the stamens are some multiple of the divisions of the corolla, to those, in which they are equal and opposite to them; we always find fertile stamens in this last position and sterile filaments in the interval between them.

§ 841. The Primulaceæ (*Primulacées*) (Primworts),—by their stamens opposite to the lobes of the corolla (*figs.* 677, 678), by the central placentation of their seeds (*figs.* 678—680), and by the situation of the embryo, which turns its side instead of its extremity towards the point of attachment (*fig.* 682), are easily distinguished from every other monopetalous family, if not from the myrsineæ (*Myrsinées*) (Ardisiads). But these last are, as it were, the Primulaceæ of tropical regions, where they grow exclusively and are only represented by trees or shrubs; whilst the Primulaceæ properly so called, indigenous to cold or temperate climates, are always herbaceous. They are only esteemed for the elegance of their flowers, which in several species appear at a time of the year when our fields and gardens are still destitute of the gifts of Flora. This precocity has given the name to the principal genus, the Primrose (*Primvère*) (Primula). The properties of this family are not very marked, but have a certain degree of energy, especially in the Pimpernel (*Mouron*) (Anagallis). The extract of the Anagallis arvensis is an acrid poison.

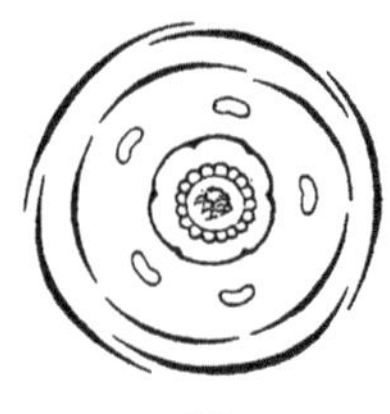

677

§ 842.—The families included in the following tables, which form the majority of monopetalous plants, constantly present those cha-

677—683. The organs of the fructification of the Oxlip (Primula elatior).
677. The diagram of the flower.

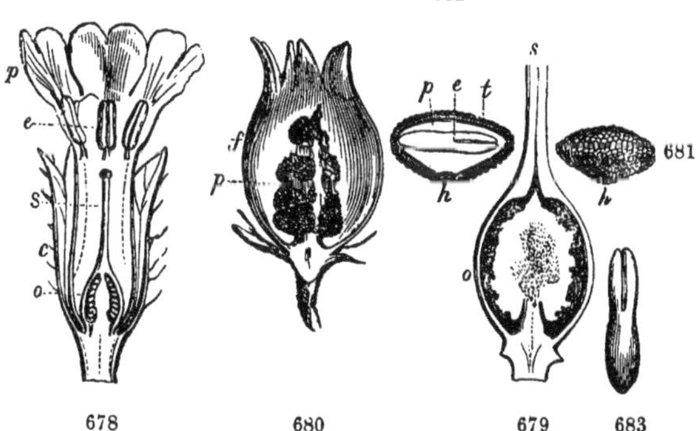

racteristics, which we have several times pointed out as connected with this modification of the corolla, viz., those of the number, the position and the insertion of the stamens, as well as of the number of the carpels, generally not so many as those of the petals, although in a few cases they are, on the contrary, more. Several plants of the families enumerated in the preceding table, in which we meet with these characteristics, ought to be arranged in one of the subsequent ones, if we had regard only to their systematic place; but we have preferred to leave them in that which their natural relations assign to them. We shall not be able in any respect to rank them with those of Table XIII, in which the corolla is irregular, and as to those of Table XIV, the examination of the other characteristics will easily decide the question in the small number of cases in which it would be doubtful.

(See Tables XIII. & XIV. pp. 666, 667.)

§ 843.—Before examining a few of the families mentioned in the two following tables, it is necessary to look at several general points of their organization. Those, whose unequal petals form

678. The vertical section of the flower.—*c* Calyx.—*p* Corolla.—*e* Stamens.—*o* Ovary.—*s* Style and stigma.

679. A vertical section of the ovary *o*, shewing the central placenta covered with ovules.—*s* Base of the style.

680. Vertical section of the fruit.—*f* Pericarp.—*p* Central placenta covered with seeds, some of which have been detached.

681. Seed.—*h* Hilum.

682. A vertical section of the same.—*t* Teguments.—*h* Hilum.—*p* Perisperm.—*e* Embryo.

683. The embryo.

an irregular corolla by their union, will occupy us first. One of these petals is generally opposite to the bract, *i. e.*, is turned outwards and united to the two adjoining ones, whilst the two others hang down from the opposite or inner side, so that the limb is divided into two parts or lips, the upper bi-lobed, the lower tri-lobed; and on cutting the corolla parallelly to the axis, we obtain two unequal halves of different forms, each of which constitutes one of these lips; cutting it perpendicularly to the axis, we have two symmetrical halves. The calyx may itself be regular or participate in this irregularity; in this last case, it will itself be bi-labiate. Of the five stamens which alternate with the five petals that, which is inserted in the interval of the two lobes of the upper lip, is rarely developed; it is most frequently abortive, either incompletely, then shewing itself by a rudimentary filament (as in several Scrophulariaceæ and Bignoniaceæ), or wholly so. In the latter case, of the four other stamens the two lower ones, those which alternate with the lobes of the lower lip, are very largely developed; the lateral ones, those which alternate with the two lips, are also developed but not so largely (in which case we have tetradynamous stamens), or are only incompletely developed and appear in the rudimentary state (in which case the flower is diandrous).

§ 844.—We have remarked in the families with both regular and irregular flowers, how frequent is the binary number of the carpels, and an attentive examination leads one to think that in reality it is still more so than our tables indicate. We know that in certain families the number of the loculi is frequently reduced from four to two, but that then that of the seeds is double in each; that in others the number (4) of the carpels is constant (as in the Labiatæ and the Boraginaceæ); but then even the single style is bifid or terminated by two stigmas, and each of the stigmas is opposite to a couple of carpels. Besides, the insertions of the four ovules do not commonly grow regularly, but are approximated in pairs opposite to the two stigmas. Certain monstrosities have the carpels disjoined, but in pairs, each of which bears a style with its stigma; and a genus of Dichondreæ presents two distinct styles, each serving for a pair of carpels. It would be allowable, perhaps, to conclude thence that each of these couples represents a single bi-lobed or bi-ovulate carpel. This is confirmed by the frequent existence of two ovules in each of the loculi of the clearly bilocular ovaries, and by the tendency which these loculi have to divide into two compartments on account of the recurrence of a median dissepiment. This in some

TABLE XII.

FAMILIES.

MONOPETALOUS DICOTYLEDONOUS VEGETABLES,

with a regular corolla;

with their stamens commonly hypogynous, frequently independent of the corolla, multiple, double or opposite, very rarely equal and alternate, or less; with carpels often equal in number to the lobes of the corolla.

Anthers
- uni-locular, without appendices. Stamens free or placed on the corolla, most frequently equal in number to its divisions. Ovary free. Fruit, fleshy or capsular . . . . } EPACRIDEÆ.
- bi-locular, opening
  - at the top, frequently lengthened into horns. Stamens most frequently independent of the corolla and double the number of its divisions*.
    - * Ovary
      - free. Seeds
        - winged. Capsule. Anthers, pointless, two-celled, with terminal pores.—PYROLACEÆ.
        - not winged.
          - Capsule with septicidal dehiscence. Anthers, pointless.—RHODORACEÆ.
          - Fruit fleshy, commonly capsular with loculicidal dehiscence. Anthers generally bearded . . . . . } ERICINEÆ.
      - adherent. Berry. Anthers bearded . . . .—VACCINIEÆ.
  - along the whole of their length by a slit, without prolongations or appendices**

** Stamens
- double or multiple. Ovary
  - adherent, with 3—5 multiovulate loculi. . . . . Fruit fleshy. Perisperm fleshy. Plants ligneous—STYRACINEÆ.
  - free, with 3 or more loculi. 1 or 2 pendulous ovules. Fruit fleshy. Perisperm horny. Plants ligneous—EBENACEÆ.
- less in number (2) than the divisions of the corolla. Ovary, free, bi-locular. 1—2 ovules
  - pendulous. Prefloration valvate. Fruit fleshy. Perisperm thick. Plants ligneous } JASMINACEÆ.
  - erect. Prefloration imbricate. Fruit fleshy, a capsule or samara. Perisperm thin. Plants ligneous . . . . } OLEACEÆ.
- equal in number,
  - alternate. Ovary free, 2—3— pluri-locular. 1 pendulous ovule. Placentation axillary. Drupe. Embryo, very small, towards the extremity of a large perisperm. Plants ligneous . . . . . . } ILICINEÆ.
  - opposite.
    - Placentation axillary. Ovary, free, pluri-locular. 1 ascending ovule. Fruit fleshy. Perisperm wanting or fleshy. Plants ligneous . . . . } SAPOTACEÆ.
    - Placentation central. Ovary, free, uni-locular . . . .
      - Several ovules.
        - No perisperm. Folliculus. Plants ligneous . . . } ÆGICERACEÆ.
        - Embryo, excentric in a fleshy or horny perisperm.
          - Drupe. Plants ligneous . . . . . } MYRSINEÆ.
          - Capsule or pyxidium. Plants herbaceous . . . . } PRIMULACEÆ.
      - One ovule. Embryo in the axis of a fleshy perisperm. Capsule. Plants herbaceous . . . . } PLUMBAGINEÆ.
  - alternate. Ovary, free, 1 or 2 loculi. Placentation axillary. Seeds with an excentric embryo in a fleshy perisperm. Fruit, dry, or a pyxidium . . . . } PLANTAGINEÆ.

TABLE XIII.

FAMILIES.

HYPOGYNOUS MONOPETALOUS DICOTYLEDONOUS VEGETABLES,

with an irregular corolla, bearing the alternate stamens, reduced to 4 didynamous or to 2 by the complete or partial abortion of the rest.

Ovaries | one, with a terminal style | uni-locular. Placentation | lateral. Seed, one, pendulous. Perisperm thick. Stamens, 4 . . . . . .—GLOBULARINEÆ.
central. Seeds, numerous. Perisperm wanting. Stamens, 2 . . . . . .—UTRICULARINEÆ.
parietal in several lines. Seeds numerous. Perisperm . . } wanting. . . Stamens, didynamous or 2—CYRTANDRACEÆ.
} thick. | Plants leafy. Ovary sometimes semi-adherent, with 2 placent. Stamens didynamous or 2 . . }—GESSNERIACEÆ.
| Leaves scaly. Frequently 4 placent. Stamens didynamous } OROBANCHEÆ.

bi-locular. Placentation axillary. Seeds | indefinite in number. | not winged. Perisperm, thick, fleshy. Stamens didynamous or 2 . . . . } SCROPHULARIACEÆ.
winged. . . Perisperm wanting. . Stamens didynamous . . . . } BIGNONIACEÆ.
definite . . | one or more, retinaculate. Perisperm wanting. Stamens, didynamous or 2, fertile . . . . } ACANTHACEÆ.
One, pendulous. Perisperm fleshy. Stamens didynamous. Anthers | bi-locular . . . . } MYOPORACEÆ.
uni-locular . . .—SELAGINEÆ.
One erect. Perisperm fleshy. 4 or 5 stamens .—STILBINEÆ.

2—4—locular. Seeds, definite. Fruit spiny. Perisperm wanting. Stamens didynamous . . . . .—PEDALIACEÆ.
2—4—8—locular. Seed, one, erect. Fruit unarmed. Perisperm wanting. Stamens didynamous . . . . .—VERBENACEÆ.

4 distinct, with a gynobasic style. . . . . Seed, one, erect. . . . . . . Perisperm wanting. Stamens didynamous, very rarely 2.—LABIATÆ.

[To face page 666.]

Table XIV.

FAMILIES.

HYPOGYNOUS MONOPETALOUS DICOTYLEDONOUS VEGETABLES,

with a regular corolla, bearing alternate stamens, equal in number to that of the lobes.

Ovaries

- several, distinct,
  - with one gynobasic style.
    - 4 achænia. . . . . Seeds pendulous. . . Perisperm wanting . . . . —Boraginaceæ.
    - Drupes, 1—6—locular. Seeds, solitary, erect. Embryo, amphitropous around a fleshy perisperm . —Nolanaceæ.
  - with two basilar styles. 2—4 achænia. . . . . Seeds erect. . . . Perisperm wanting. Cotyledons wrinkled . . . . . —Dichondreæ.
- one, with one or more terminal styles. Perisperm . . .
  - wanting. Seeds
    - 1—2 erect in each loculus. Radicle inferior. Cotyledons
      - wrinkled. Fruit, fleshy or capsular, 2—3—4—locular . . . —Convolvulaceæ.
      - wanting. Pyxidium, bi-locular . . . . —Cuscuteæ.
    - one, pendulous in each loculus. Radicle superior. Cotyledons bent. Drupe, having a kernel with 4—8—loculi . . . . Cordiaceæ.
  - thin . . . . Radicle superior. Cotyledons flat. . Drupe with four kernels . . . . —Ehretiaceæ.
  - thick. Loculi
    - 3. Placentation axillary. Ovules definite or indefinite in number. Capsule with loculicidal dehiscence. Seeds
      - winged . . . . —Cobæaceæ.
      - not winged . . . . —Polemoniaceæ.
    - 1. Placentation parietal. Ovules definite or indefinite in number. Capsule with loculicidal dehiscence. Inflorescence . . . . .
      - scorpioid . . . . —Hydrophylleæ.
      - straight . . . . —Gentianaceæ.
    - 2. Placentation axillary. Ovules indefinite in number. Leaves .
      - alternate. Styles
        - two, distinct. Capsule with loculicidal dehiscence. Embryo straight . . . . Hydroleaceæ.
        - one. Berry or capsule with septicidal dehiscence. Embryo curved . . . . Solanaceæ.
      - opposite
        - without stipules. Prefloration of the corolla contorted.—Gentianaceæ.
        - with stipules. Prefloration
          - valvate. Capsule with two cocca . . . . . . Spigeliaceæ.
          - imbricate
            - Capsule . . —Loganiaceæ.
            - Berry . . . —Potaliaceæ.
          - contorted. Berry or capsule . . . . Apocynaceæ.
- two, distinct, with terminal styles united by the stigma. Pollen
  - pulverulent. Perisperm fleshy or horny . . . . Apocynaceæ.
  - in solid masses in each loculus of the anther, attached to the stigma by a caudiculus. Perisperm thin . . . . Asclepiadaceæ.

Table XV. **FAMILIES.**

PERIGYNOUS MONOPETALOUS DICOTYLEDONOUS VEGETABLES,

with adherent ovary; with a regular or irregular corolla, bearing commonly the alternate stamens equal in number to the lobes, rarely less.

- Anthers
  - distinct. Leaves
    - opposite
      - with interpetiolar stipules. Two or more loculi, mono- or poly-spermous. Fruit fleshy or capsular. Perisperm fleshy or horny. Prefloration of the corolla, valvate or contorted . . . RUBIACEÆ.
      - without stipules.
        - Two or more loculi, mono- or poly-spermous. Berry. Perisperm fleshy or horny. Prefloration imbricate—CAPRIFOLIACEÆ.
        - One monospermous loculus. Stamens . .
          - equal in number and opposite. Prefloration valvate. Plants ligneous, parasitical. Berry. Perisperm fleshy . . . LORANTHACEÆ.
          - less in number and alternate. Prefloration imbricate. Plants herbaceous. Fruit indehiscent. Perisperm
            - wanting. Flowers in a cyme . . . VALERIANACEÆ.
            - fleshy. Flowers in contracted cymes—DIPSACEÆ.
    - alternate without stipules.
      - Pyxidium, bi-locular. Seeds indefinite in number. . . . Perisperm wanting—SPHENOCLEACEÆ.
      - Capsule, 2-8-locular. Seeds indefinite in number. Stamens independent of the corolla. Perisperm thick —CAMPANULACEÆ.
  - united
    - to the style. Flowers separate. . . . Seeds indefinite. Capsule bi-locular. . . . Perisperm fleshy .—STYLIDIEÆ.
    - to one another, forming a tube. Flowers
      - separate. Perisperm
        - indusiate. Prefloration induplicate. Seeds
          - definite in number. Fruit, a drupe, or dry and indehiscent, 1—4— locular . . . SCÆVOLACEÆ.
          - indefinite in number. Capsule 2—4—locular .—GOODENIACEÆ.
        - naked. Prefloration valvate. Seeds indefinite in number. Flowers . . .
          - irregular. Fruit indehiscent or a capsule with two or three loculi . . . LOBELIACEÆ.
          - regular. A capsule with two, three, five or ten loculi . . . CAMPANULACEÆ.
      - united in a common involucrum into a calathis or composite flower. Achænium monospermous. Seed
        - erect. . . Perisperm wanting . . . COMPOSITÆ.
        - pendulous. Perisperm fleshy . . . —CALYCERACEÆ.

[*To face page* 667.]

cases raises the apparent number of the loculi to 8; there are really four, but each cut into two by a dissepiment. In this case (in certain Verbenaceæ) instead of 8 unilocular we find 4 bilocular nuclei.

The position of the two loculi relatively to the axis of the flower is, on the contrary, constant and important. In the Scrophulariaceæ, Solanaceæ, Acanthaceæ, &c., one of the loculi is the upper of the two, *i. e.*, turned towards the side of the axis; the other, the lower, *i. e.*, turned towards the side of the bracts. In the Gentianaceæ, Apocynaceæ, Asclepiadaceæ, &c., they are both lateral ones, situated, with respect to the axis, the one to the right and the other to the left.

§ 845. Bignoniaceæ (*Bignoniacées*) (Bignoniads).—The plants of this family are shrubs or trees remarkable for the beauty of their flowers, some of which are frequently cultivated in parks and gardens: as the Catalpa, which has been known for a long time and, as it were, naturalized. Several of these shrubs are creepers, and the greater part present in their ligneous system the singular arrangement which we have already pointed out (§ 87, *fig.* 108). The wood forms a column with four very deep flutes, so that a horizontal section represents a kind of Maltese cross. The interval of the four ligneous lobes is entirely filled by the cortical substance, which regularly covers the whole of the circumference, itself preserving almost a cylindrical form, the modification of the interior not being at all manifested on the outside and becoming apparent only by dissection. More rarely the number of these ligneous lobes is double, and each of them is itself afterwards bilobed, thus raising as high as sixteen the number of the lobes of the ligneous body alternately deeper or shallower.

§ 846. Acanthaceæ (*Acanthacées*) (Acanthads).—We have mentioned retinaculate seeds as a distinctive characteristic of this family. The word, retinaculum, is here applied to a prolongation of the placenta which is advanced under and subtends each seed; it is a small groove terminated in a point frequently bent into a hook. After the fall of the seed, we perceive these retinacula persistent and projecting on the inner edge of the dissepiments, which are separated from one another and frequently also from the valves to which they are opposite. In a small number of genera they are wanting, then replaced by a small horny cupula (Thunbergieæ [*Thunbergiées*]) which encircles the hilum, or by a papillus (Nelsonieæ [*Nelsoniées*]).

§ 847. LABIATÆ (*Labiées*) (Labiates).—The labiate corolla (*figs.* 296, 685), the didynamous stamens rarely reduced to two (as in the Sages [*Sauges*]) by the almost complete abortion of the intermediate two, and the four ovaries with a single gynobasic style (*fig.* 372) bifid at its top (*fig.* 686 *s*), readily distinguish this family from every other. Add to these their quadrangular stem and their opposite leaves; also the existence of a large number of small reservoirs

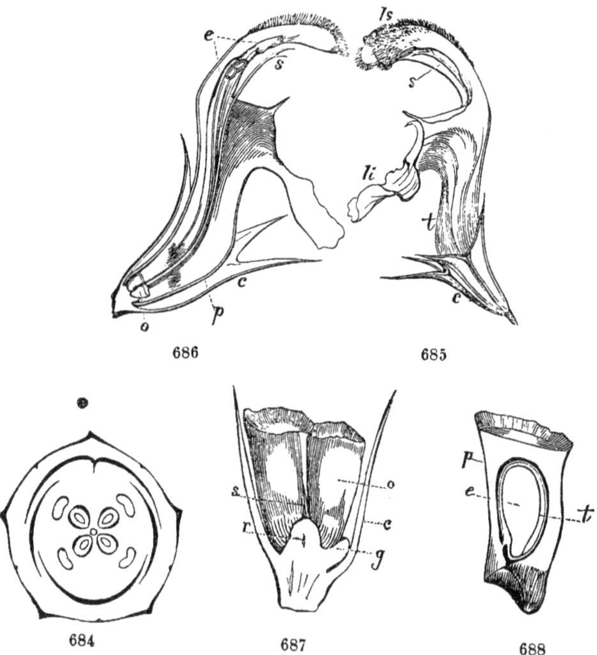

684—688. The organs of the fructification of the White Dead Nettle (LAMIUM ALBUM).—*c* Calyx.—*p* Corolla.—*t* Tube.—*ls* The upper lip.—*li* The lower lip.—*e* Stamens.—*s* Styles and stigmas.

684. The diagram of the flower.

685. The flower. One of the sides is turned towards the observer.

686. Vertical section of the same.

687. Vertical section of the fruit. Two of the carpels have been taken away.—*c* Persistent calyx.—*g* Gland.—*r* Gynobasic receptacle, *i.e.*, the part bearing the style *s*. —*o* Two carpels.

688. A vertical section of a carpel.—*p* Pericarp.—*t* Tegument of the seed.—*e* Embryo.

of an essential oil, with which the leaves are covered. The Labiatæ owe their aromatic odour to these oils, as varied as the species, and so agreeable in some, as in the Sage (*Sauge*), the Thyme (*Thym*), the Balm (*Mélisse*,) the Lavender (*Lavande*), the Mint (*Menthe*), the Rosemary (*Romarin*), the Patchouli (*Patchouly* [a species of COLEUS]), &c., &c. Sometimes the perfume is used in the form of the extracted oil; mostly in the form of distilled waters. Certain leaves, those of the Savory (*Sarriette*), of the Marjoram (*Marjolaine*), of the Basil-Thyme (*Basilic*) &c., are introduced into our food as condiments. The infusion of several already named, as the Sage, Balm, &c., and some others, as the Ground Ivy (GLECHOMA), slightly tonic, is sometimes taken as tea. Along with these essential oils, the exciting principles of which we know, we frequently find a resinous gummy principle, slightly bitter, from which will result these tonic properties. Several of these infusions are recommended on account of this as being stomachic: and even if the last principle abounds, they may become febrifuges (as the Germander [*Germandrée*], CHAMÆPITYS, SCORDIUM). A substance resembling Camphor (which we have already pointed out as found in a very different family, that of the Laurineæ) is associated with the volatile oil of the Labiates, in greater abundance in some, as the Sage, the Lavender, and especially the Rosemary. The roots of some species present tuberculous swellings, the fecula of which would furnish nutritious food; and among them we may mention one of our own country, the STACHYS PALUSTRIS.

§ 848. The BORAGINACEÆ (*Borraginées*) (Borageworts),—by their four distinct ovaries with a single gynobasic style approach the Labiates; but their alternate leaves on a round stem, and their corolla, almost constantly regular, and, even when it is not so (in the ECHIUM), bearing five antheriferous stamens, distinguish them at the first glance; and this distinction would be easy even if we had only one leaf, for it is readily known by its soft consistence, its surface covered with asperities which result from the swollen and hardened bases of simple hairs, its tissue totally without oil; and in spite of the resemblance of the pistil or of the fruit, they would still be sufficient without any other characteristic, on account of the ovules being pendulous, instead of being erect, and from the direction of the radicle which is its necessary consequence, and which, inferior in the Labiates (*fig.* 688), is superior in the Borageworts (*fig.* 691). The properties of the latter, abundant in mucilage, which merely imparts to their infusion emollient properties, are equally distinct. The root of several species, especially the Alkanet (*Orcanette*) (ANCHUSA TINCTORIA), for which we may substitute those of the ONOSMA ECHIOÏDES, and of the LITHOSPERMUM TINCTORIUM, is

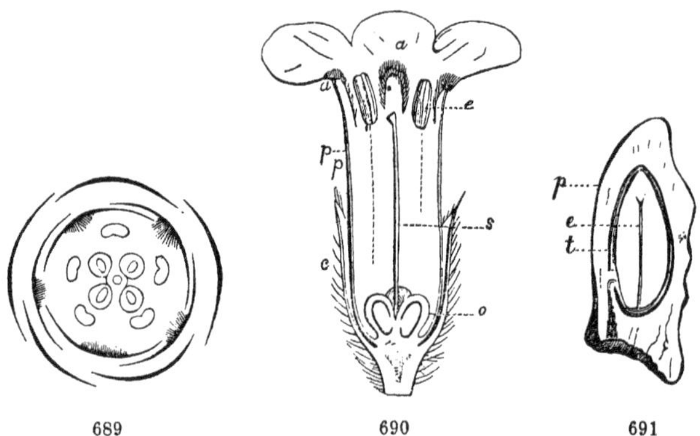

employed in dyeing. It is red on the outside in contact with the air, white within, and contains a substance insoluble in water, but soluble in alcohol, ether, oils and fatty bodies in general; it forms a blue colour by the addition of alkalies. Its alcoholic solution throws down variously coloured precipitates when metallic solutions are added to it.

The EHRETIACEÆ (*Ehrétiacées*) (Ehretiads) and CORDIACEÆ (*Cordiacées*) (Sebestens) were originally confounded with the Boraginaceæ, and are still associated with them by several authors as simple tribes. We have seen in the table that they differ from this family by the insertion of the style, to which is added the union of the carpels into a single ovary which frequently becomes a fleshy fruit. Those of the Sebestens (*Sébesteniers*) (CORDIA SEBESTEN and C. MYXA) have a mucilaginous pulp employed in medicine.

§ 849. SOLANACEÆ (*Solanacées*) (Nightshades).—The plants of this family deserve to be mentioned for the energy and at the same

689—691. The organs of the fructification of a Bugloss (*Buglose*) (ANCHUSA ITALICA).
689. The diagram of the flower.
690. The vertical section.—*c* Calyx.—*p* Corolla.— *a* Its appendices.—*e* Stamens. —*o* Ovaries, two of which are divided.—*s* Style.
691. A vertical section of one of the carpels.—*p* Pericarp.—*t* Teguments of the seed.—*e* Embryo.

time for the diversity of their properties. The commonest is the narcotic principle which resides in the juices of the roots, leaves and fruits of certain well-known species: of the Belladonna (*Belladone*) (ATROPA BELLADONA), the Mandrake (*Mandragore* (A. MANDRAGORA), formerly so much renowned, the Henbane (*Jusquiame*) (HYOSCIAMUS NIGER, and other species of the same genus), the Stramony or Thorn Apple (*Pomme-épineuse* or *Stramoine*) (DATURA STRAMONIUM),

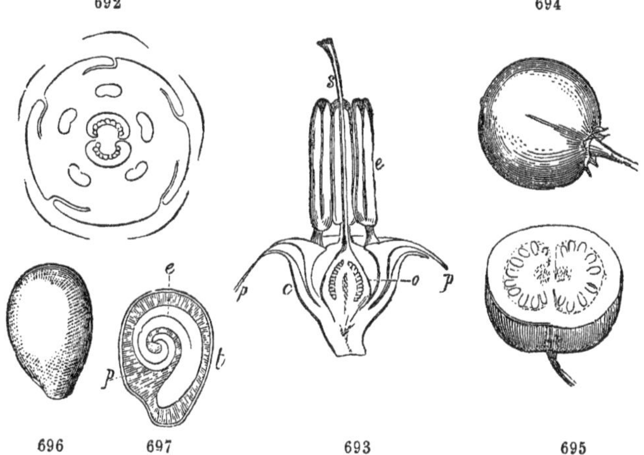

different species of the genus SOLANUM (as the Nightshade [S. NIGRUM] so common in our climate). Chemistry has discovered peculiar and at the same time analogous principles in these different plants. The names of the plants have been given to them, as the *atropine*, *hyoscianine*, *daturine*, *solanine*, and to these their qualities seem to be owing. The PHYSALIS SOMNIFERA and the NICANDRA PHYSALODES produce similar although less intense effects. Those of the leaves of the Tobacco (*Tabac*) are very violent, when they are taken internally; but it is administered in this way only as a medicament;

692—697. The organs of the fructification of the Potato (*Pomme de terre*) (SOLANUM TUBEROSUM).

692. The diagram of the flower.

693. The vertical section of the flower.—*c* Calyx.—*p* The lower part of the corolla.—*e* Stamens.—*o* Ovary.—*s* Style and Stigma.

694. The fruit.

695. Horizontal section of the same.

696. Seed.

697. The vertical section of the same.—*t* Tegument.—*p* Perisperm.—*e* Embryo.

in common use, it is put in connexion with the outermost parts only of the membrane which covers the intestinal tube; with the nostrils in powder; with the mouth chewed or smoked, a form in which its effect will be much weakened, and yet is very powerful to any one not accustomed to it. It is a native of America: the inhabitants of Haïti call it *Yati:* the name of *Tobacco*, applied to the pipe, has been given by the Europeans to the plant. Sir Walter Raleigh introduced it into England in 1586; but it was cultivated as early as 1560 in Portugal, whence it was carried into France by the ambassador Nicot, whose name has been given to the genus (NICOTIANA). Its use was at first strictly forbidden by several sovereigns, but was established in spite of their menaces and punishments, and has at last become an important source of public revenue. Spread over the surface of the whole earth, its culture has also been generalized, and it is surprising to find a plant originally a native of tropical countries successfully cultivated in Scotland and in Sweden; but it is easy to explain this, when we reflect that it is an annual herb, which requires only a few hot months to come to perfection, and in its native country grows on mountains and, consequently, in a more temperate climate. Several species are cultivated; most generally the NICOTIANA TABACUM with rose-coloured flowers; the N. RUSTICA with yellow flowers, cultivated by preference in western Africa and Egypt, as well as in the south of Europe, and from it is prepared the Salonica Tobacco, and probably also the Latakia. That of Shiraz is the N. PERSICA, perhaps, a native of that country; this, however, is far from being certain. We find in this family, by the side of these poisonous products, others of a perfectly different nature. The fruits of the Cayenne Pepper Tree (*Piment*) (CAPSICUM) are extremely *piquant* to the taste and even acrid, but are eaten with impunity; and those of the Love Apple or Tomato (*Tomate*) (LYCOPERSICUM ESCULENTUM), of the Egg Apple (*Aubergine*) (SOLANUM MELONGENA) and of some others are common articles of cookery. But the use of the Potato (*Pomme de terre*) (SOLANUM TUBEROSUM) is the greatest contrast to that of the narcotics we at first mentioned. It is true that this food so universally employed is furnished by another part of the plant very much modified, by the subterranean stems (§ 190, *fig.* 178), which, as they swell, form rich deposits of fecula. This very useful vegetable also comes from America; but from what part of it? It has been found growing wild on the mountains of Chili, towards the 33° of south latitude; on those of Peru, where, perhaps, it had been introduced by the Incas; recently on the peaks of Mexico, where, however, it was not known in the time of Montezuma, and Raleigh introduced it into England from Virginia. But it is very difficult to determine, whether a plant so easily propagated has always grown spontaneously in a certain place or has been left by man at some previous time. However this may be, the Potato was established in Europe with more difficulty than Tobacco, and it may be considered to have come into general use only during the last century. It was cultivated sooner in the South of France; but its adoption in the North was brought about by the most persevering efforts of an enlightened philanthropist, Parmentier.

§ 850. SCROPHULARIACEÆ (*Scrofularinées*) (Figworts).—These are very closely related to the Solanaceæ, from which they differ only in the irregularity of their corollas and that of their stamens, reduced to four didynamous ones by the abortion of the fifth, or to two by the abortion of the three others. Several genera have, consequently, been attributed to both families in turn: as the Mullein (*Bouillon blanc*) (VERBASCUM) at first included in the *Solanaceæ* on account of its five stamens; now in the *Scrophulariaceæ* because these same stamens five in number, it is true, but unequal and different from one another, as well as the lobes of the corolla, thus manifest their tendency to sterility. Jussieu distinguished the SCROPHULARIACEÆ with septicidal dehiscence from the PEDICULARINEÆ (*Pedicularinées*) or RHINANTHACEÆ (*Rhinanthacées*) (Rhinanths) with loculicidal dehiscence; they are now united, because the former method of dehiscence has been remarked in plants which cannot be separated from the former family, and they constitute together one vast group subdivided into several tribes. The septicidal dehiscence is sometimes also found. They are generally, as well as the Solanaceæ, acrid and bitter, and we sometimes discover in them narcotic properties, especially in the Foxglove (*Digitale*) (DIGITALIS), a strong poison when it is administered in a large dose. Its action is very singular, being chiefly felt in the circulation, which it lowers to a remarkable degree, after having momentarily accelerated it; whence its employment in those diseases in which it is important to moderate the speed of the blood, in palpitations of the heart and aneurisms.

§ 851. CONVOLVULACEÆ (*Convolvulacées*) (Bindweeds). The CUSCUTEÆ (*Cuscutées*) (Dodders) and DICHONDREÆ (*Dichondrées*) are generally united to them as simple tribes. Several of the distinctive characteristics of the true Convolvulaceæ have been pointed out in Table XIV: we will add to them the decidedly quincuncial

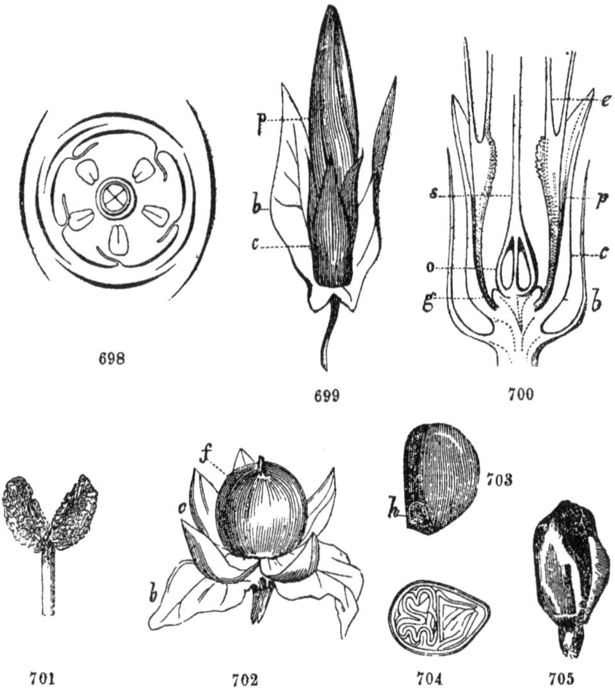

arrangement of the five sepals of the calyx inserted at unequal heights, the corolla with five folds which are twisted in the flower-bud (*figs.* 259, 698, 699), the loculicidal dehiscence of the capsule. The greater part of the species of this family are climbing plants: when

698—705. The organs of the fructification of the CONVOLVULUS SEPIUM.
698. The diagram of the flower.
699. Flower-bud.—*b* Bracts.—*c* Calyx.—*p* Corolla.
700. Vertical section of the lower part of the flower.—*b* Bracts.—*c* Calyx.—*p* Tube of the corolla bearing the filaments of the stamens *e*—*o* Ovary.—*s* Style.
701. Top of the style and stigma.
702. Fruit *f* surrounded by the calyx *c* and of the bracts *b* which remain.
703. Seed.—*h* Hilum.
704. Its section, shewing the plaited cotyledons.
705. Embryo.

they are ligneous as well as climbing, we find the covering of the cortical by the ligneous system. This forms a central column around which are arranged other ligneous fascicles in concentric rings more or less numerous according to the age of the plant. The cortical substance is interposed between these rings and, in each ring, between the different fascicles which compose it. It is clearly marked in a fresh section, so much the more manifestly, as in the plants of this family it is generally traversed by lactiferous vessels full of a milky juice, and this juice, flowing over the section, renders the cortical tissue surrounding these ligneous fascicles more distinct. This juice is generally very purgative, a quality depending on its resinous nature. It has been found in a large number of species of the genus Convolvulus or Bindweed (*Liseron*), some of which have been specially employed in medicine, as the Jalap (C. jalapa), the Scammony (*Scammonée*) (C. scammonia), the Turbith (*Turbith végétal*) (C. turpethum) and others. This principle abounds principally in the roots from which they are extracted. It is remarkable that we find in the same genus other plants, which, by the purgative principle being almost completely suppressed and fecula being largely developed, become good and wholesome food. Amongst these the best known is the sweet Potato (*Patate*) (C. batatas). The C. dissectus contains a large proportion of hydrocyanic acid ; this liquor, is, consequently. used in the preparation of Noyau.

§ 852. Gentianaceæ (*Gentianées*) (Gentianworts).—They have been mentioned (Table XIV) among the families with parietal and also axillary placentation. This is on account of the edges of their two carpels covered with seeds, being sometimes almost directly united, sometimes joined and bent towards the interior of the loculus, so as to form there two complete or incomplete dissepiments. We have pointed out among their characteristics the opposition of the leaves; this occurs in all, except in two genera abundant in our waters, the Villarsia and the Menyanthes ; the leaves of the latter are not only alternate but compound, whilst they are simple in all the rest ; so that the small tribe of the Menyantheæ (*Menyanthées*) has been made of them. The different parts of all the Gentianaceæ are extremely bitter, which renders them tonic, stomachic and febrifugal.

§ 853. Apocynaceæ (*Apocynées*) (Dogbanes) and Asclepiadaceæ

(*Asclépiadées*) (Asclepiads).—These two families, at first united into one, of which they were simply considered as two tribes, presenting indeed the most intimate connections and only different in the arrangement of their stamens, separate in the former, in which the pollen presents its usual pulverulent structure, whilst in the latter it is united into masses either granular or more frequently of a consistence similar to that of wax, generally ten in number, *i. e.*, one in each of the loculi of the five bilocular, extrorse anthers which are fastened to the circumference of a large pentagonal stigma. At a less advanced period of the developement of the flower, in five furrows of this stigma which alternate with the anthers, two small granduliform bodies are organized; afterwards confounded, each is prolonged into a kind of gelatinous tail. This tail, at the moment of dehiscence, is united to the extremity of the corresponding pollen mass and attracts it out of the loculus, so that examined at this time this mass, the gland borne on the stigma and its prolongation, seem to make only one body. This pollen body (*fig.* 709) is formed by a cellular tissue with closely united cells each enclosing a grain with a simple membrane, the surrounding cellular wall of which may, perhaps, be considered as the external membrane. However this may be, a longitudinal slit is at last established on one of the sides of the mass, and from the cells thus opened escape the pollen grains *p*, which are only applied to the lower part of the large stigma (*fig.* 708 *p p*) near the insertion of the style, into which the pollen tubes also penetrate. This organization of the pollen can hardly be compared to that we have previously explained in a certain number of Orchideæ, and it is singular enough to justify the distinction of the Asclepiadeæ. There is another characteristic of several of their genera: this is the existence of as many appendices of various shapes opposite to each of the stamens, which form within the corolla a whorl˙ also as much developed as it is and described under the name of CORONA (*Couronne*) (*figs.* 707, 708 *a*). We have explained the arrangement of the two distinct ovaries as well as the styles which terminate them (*fig.* 708 *o*, *s*) united only by means of the large stigmatical body we have just described. They are afterwards changed into two polyspermous follicules (*fig.* 711) in all the Asclepiadaceæ and in all the true Apocynaceæ; but in one tribe of the latter (OPHIOXYLEÆ [*Ophioxylées*]) they become two drupes, and in another (CARISSEÆ [*Cariséæs*]) are united at first into one, which, most commonly, becomes a berry. It is in these last two

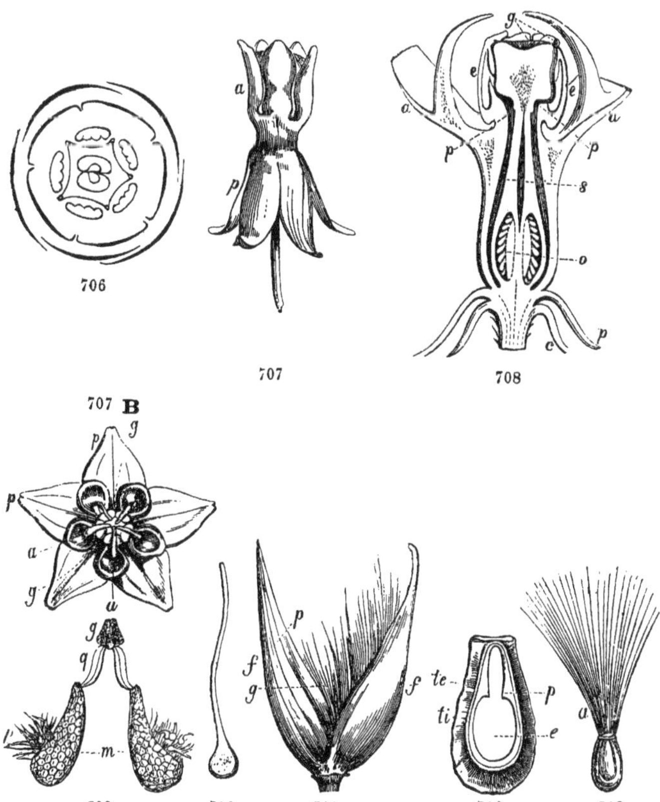

706—713. The organs of the fructification of the ASCLEPIAS NIVEA.—*c* Calyx. —*p* Corolla.—*a* Appendices forming the corona.—*g* Glandular bodies situated on the stigmas and bearing the pollen masses.

706. The diagram of the flower

707. The whole flower.

707 **B**. The same. The observer is supposed to be looking at the top.

708. Vertical section of the same.—*e* Stamens.—*o* Ovaries.—*s* Styles united at the top by the large stigmatical body, to the base of which the pollen tubes penetrate towards the points *p p*.

709. Two pollen masses *m* attached by two prolongations in the shape of a tail *q* to another body *g*, formed by the union of two glands.—*p* Pollen grains beginning to escape from the masses.

710. One of them highly magnified.

cases that we sometimes find the ovules definite in number or even solitary.

The plants of the two families are frequently climbing and frutescent, the ligneous body is more or less deeply separated into several lobes by so many solutions of continuity, which are filled with the cortical substance. The juice, generally milky, is acrid and bitter, and from the excitement it produces different effects arise, according to the part of the body on which it acts; vomiting or purging, the abundant secretion of perspiration and of urine. Thus, the leaves of the CYNANCHUM ARGEL act like those of Senna, which are frequently adulterated with them, but they are much more dangerous; the juice of C. MONSPELIACUM is also known under the name of Montpellier Scammony, (*Scammonée de Montpellier*) and purges violently, whilst the root of the C. IPECACUANHA, one of those which are confounded in commerce under the latter name, causes vomiting. The C. VINCETOXICUM owes this name to the evacuations which it provokes and which are so useful in cases of poisoning. But we may say as a general thing that these dangerous principles are less marked in the Asclepiadaceæ than in the other family, and there are some, the milk of which would be innocent and might be employed as food. It is also rich in Caoutchouc which is extracted from some species for commerce.

We have just said that the juice of the Apocynaceæ presents acrid properties in a more intense degree. We shall mention, among a number of examples, only the NERIUM OLEANDER (*Laurier-rose*), the extract of which is a very violent, acrid, narcotic poison, and the emanations alone of which may, especially in the southern countries where it grows spontaneously, cause most serious accidents. We shall linger a little longer on the seeds with a large horny perisperm of the genus STRYCHNOS which contains one of the most active poisons that we know, the alcaloïd, which we know by the name of *Strychnine*. It causes (doubtless, by acting on the spinal marrow) contractions in the muscles, which after a few convulsions are succeeded by stiffness and immobility, then by asphyxia or the suppression of the respiratory action. This is what may have sometimes been observed with regard to the wandering dogs poisoned by balls of meat thrown for that purpose in our public

711. Fruit at the time of the dehiscence.—*ff* Follicles.—*p* Detached placenta.—*g* Seeds with tufts of hair.

712. A seed.—*a* The tuft of hair.

713. Vertical section of a seed stripped of its tuft.—*t e* External tegument.—*t i* Internal tegument.—*p* Perisperm.—*e* Embryo.

promenades and prepared with the Nux vomica (*Noix vomique.*) It is from this plant (STRYCHNOS NUX-VOMICA) and from the Bean of St. Ignatius (*Fève de Saint Ignace*) (S. IGNATIANA) that Strychnine is extracted, which also imparts its properties to the bark of the False Angostura (*écorce de Fausse Angusture*), which also appears to come from a STRYCHNOS, perhaps, from the Nux vomica itself. There is also another celebrated poison in which the Javanese dip their arrows, called by them the Upas tieuté (*Tieuté*), another species of the same genus (S. TIEUTÉ). But medicine has skilfully applied these formidable properties to salutary purposes and has employed strychnine in those cases where the paralysed muscular action has to be roused by a very energetic agent: but it administers a very weak dose, that of a small fraction of a grain. The fleshy pericarp of several species of Carisseæ does not share in these dangerous qualities and is eaten in the countries, in which they grow: such are the fruits of the CARISSA EDULIS and C. CARANDAS, of the MELODINUS MONOGYNUS, of the WILLUGHBEIA EDULIS, &c., &c.

PERIGYNOUS MONOPETALOUS DICOTYLEDONOUS VEGETABLES.

(See Table XV. page 667.)

§ 854. RUBIACEÆ (*Rubiacées*) (Stellates).—This group, one of the largest and most natural of the Vegetable Kingdom, may be subdivided into several secondary ones according to different considerations. At first into two great sections: that of the COFFEACEÆ (*Cofféacées*,) with one or more rarely two ovules; that of the CINCHONACEÆ (*Cinchonacées*) (Cinchonads) with multiovulate loculi. They are then divided into tribes according to the nature of their fruit, which is either fleshy, (a berry or a drupe with several kernels); or dry, indehiscent or dehiscent; its carpels then remain united or are separated when ripe (*fig.* 718); the loculi, most commonly reduced to two, are at other times more numerous; according to the fleshy or horny consistence of the perisperm: according to the inflorescence, the flowers of which are frequently compressed into a spike and are sometimes even blended together and united to one another by their ovaries: according to the developement of the bracts, sometimes united to one another in the interval of the two petioles and thus sometimes forming a kind of sheath of different forms. In the *Rubiaceæ of Europe* these stipules are developed into leaves similar to the real ones, and augment their number

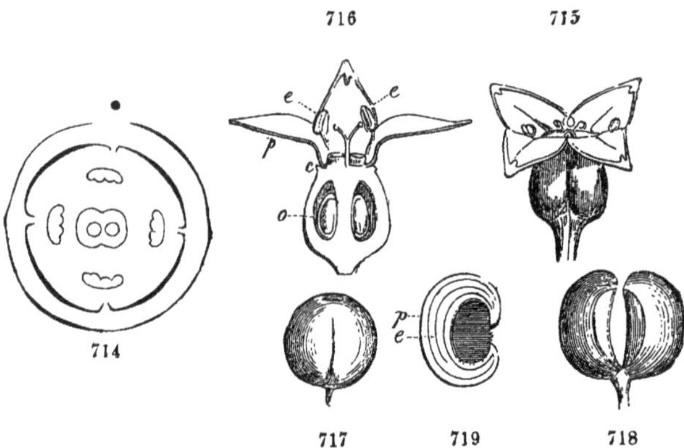

to a greater or less degree, according to the different kinds of union or of the deduplication of the accessory parts. There then results a whorl of these commonly straight leaves arranged like the rays of a star, whence these plants have obtained the name of Stellates (*Etoilées*) (STELLATÆ): but we only find two opposite buds at each knot. The adherent ovary is frequently crowned with a fleshy disk (*fig.* 716), which is pierced by the simple style, but frequently divided to a great depth into as many branches as there are loculi.

This family presents in rather a large number of its species remarkable properties which it now remains for us to examine. The bark of several is astringent and bitter to a high degree, and has, therefore, febrifugal virtues, especially that of the CINCHONA, more commonly known by the name of Peruvian bark (*Quinquina*). These are owing to alcaloïds which we have already had occasion to mention (§ 314), cinchonine and especially quinine. There are species, the bark of which contains both at the same time, others which contain only one: their medical action also is not quite the same. Formerly they were administered, either naturally, or in an extract after having dissolved their active prin-

714. The diagram of the flower of the GALLIUM MOLLUGO.
715. The flower.
716. The vertical section of this flower.—*c* Calyx blended with the ovary *o*.—*p* Corolla.—*e* Stamens.
717. The fruit of the Madder (*Garance*) (RUBIA TINCTORUM).
718. The same after two carpels have been taken away.
719. The vertical section of the seed.—*p* Perisperm.—*e* Embryo.

ciples in water, or rather in alcohol, which is better adapted for making the solution. Now that we know how to extract the active principles, we employ these directly and, consequently, with a much greater degree of certainty as to the effect, which will be produced, and the dose which ought to be administered. We may conceive, therefore, how the compound medicament, which was obtained from the bark, differs from the simple medicine, which is furnished by the alcaloïd now employed: there are other Rubiaceæ, the PORTLANDIA HEXANDRIA, for instance, in the bark of which both quinine and cinchonine have also been found. But there are some, which, although employed as febrifuges, contain neither of them, the EXOSTEMA, for instance. This property resides, therefore, in bitter principles which may vary; it is not an attribute peculiar to quinine, which possesses it only in a more energetic degree, better known and, consequently, worthy of more confidence. The name of Peruvian bark, vulgarly applied to the bark of several plants both of this family and of others totally different, by no means implies, therefore, the existence of quinine or of cinchonine, but only that of some bitter, tonic and astringent principle, the efficacy of which in curing fevers has been proved.

The roots of other Rubiaceæ are celebrated as emetics, and among them especially the CEPHÆLIS IPECACUANHA: this name has been also given to others, either of the same family (PSYCHOTRIA EMETICA, several species of RICHARDSONIA and of SPERMACOCE), or of entirely different families, as we explained when treating of them. Chemistry has extracted the active principle of the CEPHÆLIS, the *emetine*, sixteen per cent of which enters into the composition of its roots; it is now administered by itself in doses of from four to six grains. Will it also be discovered in all the other emetic roots thus known by the name of Ipecacuanha?

Other roots are esteemed for their colouring principles and are employed in dyeing, especially that of the Madder (*Garance*) (RUBIA TINCTORUM), of which we have previously spoken (§ 652). Several species of the same genus (R. CORDIFOLIA and R. ANGUSTIFOLIA), natives of another country, have the same properties: these appear to be common to others of our own country belonging to the same tribe, as the ASPERULA TINCTORIA, &c., or to other tribes, as several MORINDÆ, the HYDROPHYLAX MARITIMA and the OLDENLANDIA UMBELLATA, the root of which is commonly known under the name of *Chaya-vair*. But, not so rich in colouring matter as the Madder, they are neglected or else much less generally employed.

Coffee (*Café*) is the seed of a plant of this family, the COFFEA ARABICA, and nearly the whole of its mass is formed by the horny perisperm to which it owes its properties; these are developed by roasting the seed; a concrete oil is thereby volatized and we gain the much esteemed aroma. We also find in it an oil fusible at 25°, a bitter and an azotised principle termed *Caffeine*, but which, singular enough, appears to be identical with theine (§ 811); it is, therefore, nourishing to a certain point for those, who, not content with the infusion, do not separate the grounds. This plant, the cultivation of which is now extended almost everywhere under the tropics, comes from Ethiopia, whence it was, towards the end of the fifteenth century, conveyed to Mocha, where it has been so well acclimatised that it was for a long time believed to be a native and is still considered to be the best in quality. Coffee brought by the Venetians, was known in France and England in the middle of the seventeenth century; the plants were afterwards brought into Europe by the Dutch who had cultivated it at Batavia and the Mauritius. They were first introduced into the Jardin de Paris in 1713; and thence, four years afterwards, into the French colonies of the Antilles. It has frequently been a subject for the historian to recount how all the plantations, now so extensive at Martinico, Cayenne, Bourbon, Berbice, came from a single plant, saved during the passage by the care of Captain Declieux, who even went so far as to divide his ration of water with it. It is probable that the seeds of other Rubiaceæ with a horny perisperm would present some analogy; and some trials made on those of the GALIUM, at the time when the continental system prevented the arrival of colonial Coffee into France, authorize this supposition. They have not, however, been pursued, for it is the same with the substitutes for coffee as with all substitutes in general, the worse has been thrown aside when the better could be easily procured.

§ 855. CAPRIFOLIACEÆ (*Caprifoliacées*) (Caprifoils).—This family is divided into two tribes: the one, whose type is the Elder (*Sureau*) (SAMBUCUS), that of the SAMBUCINEÆ (*Sambucinées*), the corolla of which is regular, the ovary surmounted by three sessile stigmas, the seed traversed by a raphe, which, according to custom, follows its inner side; the other, whose type is the Honey-suckle (*Chèvrefeuille*), that of the LONICEREÆ (*Lonicérées*), with a corolla sometimes irregular, with a filiform style, the raphe traversing the outer side of the seed. To this singular arrangement common to all the genera, we in some may add another which is deserving of mention; of the different loculi of the same ovary, one or two (as in the SYMPHORICARPUS) contain a single seed which comes to maturity, whilst the others contain several of them and become abortive with them. The fruits of two flowers close to one another are frequently afterwards united so as to appear to be only one, as in the CHAMÆCERASUS (*Chamérisiers*).

§ 856. LORANTHACEÆ (*Loranthacées*) (Loranths).—This family has been doubtfully placed here; for, if its petals are frequently united so as to form a staminiferous tube, they are often also completely independent; they are even wanting in several genera and these last are generally diclinous. According to those genera, which have been most carefully examined, they should be ranged in the polypetalous, near the ARALIACEÆ (*Araliacées*) and CORNACEÆ; in the apetalous, near the SANTALACEÆ or, perhaps, the PROTEACEÆ. The vegetation of these plants, some idea of which may be gained from our Mistletoe (*Gui*) (VISCUM ALBUM), is singular; for, as they germinate, they are implanted on the bark of other trees and thence extract their nutriment, sometimes by roots which are insinuated between the bark and the wood, more frequently by a wide base, buried in the wood of the tree in proportion as it grows by the addition of fresh annual layers, a direct union being thus established between it and that of the parasite. This is generally composed of a series of articulations, at each of which it ramifies by dichotomy, and preserves its colour, most commonly green, the whole year. The anthers of the Mistletoe are very singularly formed: each is fastened to the surface of a division of the calyx, perforated like a sponge with a great number of pores, which are the openings of so many pollen-bearing lacunæ. Only one of its ovules is developed, or rather two or three are united together, so that the seed then presents so many embryos divergent at their radicles, convergent and even blended at their cotyledons. The perisperm which envelopes them is green and the pericarp is swollen by a viscous substance of the nature of glue, from which Birdlime is made.

§ 857. VALERIANACEÆ (*Valérianées*) (Valerianworts.)—We have placed them among those families whose fruit presents a single loculus; but we find in the ovary three, two of which are always abortive and traces only of which are to be found when the fruit is ripe. The stamens are rarely equal in number to the divisions of the corolla, more generally reduced to three or even to one (CENTRANTHUS). The roots of some Valerians (*Valérianes*) (VALERIANA OFFICINALIS, V. PHU, V. CELTICA) are bitter and aromatic with a strong odour, which is considered to be disagreeable amongst us, whilst other analogous ones are much esteemed as perfumes in the East.

§ 858. DIPSACEÆ (*Dipsacées*) (Teazelworts).—The flowers are remarkable for the presence around each of an involucrum or caliculus, which causes the calyx to appear double (*figs.* 276, 277). The true or inner calyx is frequently only incompletely adherent to the ovary and that only in the upper part of its contracted tube. This characteristic and that of the four stamens sometimes didynamous connect this family with other monopetalous hypogynous ones, especially the GLOBULARINEÆ (*Globularinées*) (Selagids). On the other hand, the arrangement of the flowers compressed into a capitulum, which is enclosed by an elongated involucrum (*figs.* 188, 200), gives them a very great resemblance to the calathis of the Compositæ, and the frequent passage from the divisions of the calycinal limb to the state of tufts of hair which we also observe in the Valerians, confirms this affinity, which has, perhaps, been somewhat exaggerated.

§ 859. CAMPANULACEÆ (*Campanulacées*) (Bellworts.)—This family is one of the remarkable exceptions to the general rule of monopetalous plants, which we have found only in some of those enumerated in Table XII: the stamens are not inserted on the corolla, but directly on the calyx (*fig.* 722). It is true that this corolla is of a peculiar tissue, dry and membranous (like that of several Ericineæ), and, instead of falling in one piece like the greater part of the staminiferous corollas, it remains attached to its place where it is persistent and withered above the fruit (*fig.* 416 *e*, 724). The fruit is opened at the top, either by several valves, which remain coherent for the rest of their extent, or by lateral openings (*fig.* 416, *t*) corresponding to so many loculi, the number of which is sometimes equal to that of the other parts of the flower, is sometimes reduced to three or to two. The Campanulaceæ, by their appearance, by their prefloration, by their styles covered with collector hairs, by their anthers sometimes united into a tube (in the JASIONE), are very nearly related to several of the Compositæ, and particularly to the Cichoraceæ, by their milky juice. This is slightly acrid, but not sufficiently so to

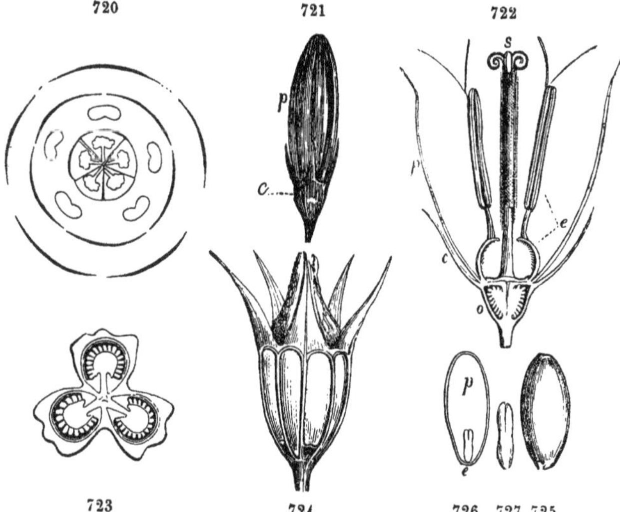

prevent the young roots of several species, as that of the Rampion (*Raiponce*) (Campanula Rapunculus) from being eaten.

§ 860. Lobeliaceæ (*Lobéliacées*) (Lobeliads).—Closely allied to the preceding, they serve to connect them more closely to the Compositæ by the constant union of their anthers, their corolla, frequently split laterally with its five divisions thrown on one side, reminding us of the semi-flosculous flowers. They have also collector hairs arranged beneath the stigma in a circle, which may, doubtless, be considered as analogous to the Indusium of the neighbouring families. Their juice is also milky, but extremely acrid, which imparts energetic and even poisonous properties to several species and renders their use suspicious.

720—727. The organs of the fructification of several species of Campanula.—*c* Calyx.—*p* Corolla.

720. The diagram of the flower of the Campanula medium.

721. The flower-bud of the Campanula rotundifolia.

722. Vertical section of the flower.—*s* Stigmas.—*o* Ovary with the tube of the adherent calyx.—*e* Stamens.

723. Horizontal section of the ovary.

724. Fruit crowned with the limb of the calyx.

725. Seed.

726. Vertical section of the same.—*p* Perisperm.—*e* Embryo.

727. The Embryo.

§ 861. COMPOSITÆ (*Composées*) (Composites).—This group of plants, in which there are 9,000 known species, must be considered less as a family than as a class. We have seen that it forms in fact the tenth (the EPICOROLLEÆ SYNANTHERÆ) (*Epicorollées synanthères*) of the method of Jussieu, and almost every author has agreed to admit it as a class in their different classifications under one name or another, Linnæus (§ 701) under that of SYNGENESIA. As to the subdivisions which they then established, in order to understand them perfectly we must first examine the structure and the arrangement of the flowers. They are crowded at the extremity of a peduncle more or less widened on a capitulum or calathis (§ 209), surrounded by an involucrum of one or more rows of folioles (§ 230). They present by this arrangement the appearance of a single flower, the involucrum of which would be the calyx, and thence the name of *Common Calyx* (*Calice Commun*) which was formerly given to it. The small flowers or florets may be of two kinds: some regular, the limb of which is divided into five equal teeth or lobes (*fig.* 731); the others irregular, the limb of which, split for a great length, is thrown outwards as a tongue-shaped body composed of five united parts and, consequently, terminated by five small teeth (*figs.* 295, 729): the former are termed FLOSCULI (*fleurons*), the latter, SEMI-FLOSCULI or LIGULÆ (*demi-fleurons* or *ligules*). These florets are sometimes hermaphrodite, sometimes only male or female, sometimes neuter. The divisions of the group have been founded on the various combinations which may thus be presented in the same capitulum. Linnæus has distinguished them according to the distribution of the sexes in the florets of the same capitulum, which may be all hermaphrodite, POLYGAMIA ÆQUALIS (*Polygamie égale*), hermaphrodite mixed with females, P. SUPERFLUA (*P. superflue*), or with neuters, P. FRUSTANEA (*P. frustanée*), some males and others females, P. NECESSARIA (*P. nécessaire*), or according to that of the involucres, several being close together on one capitulum, P. SEPARATA (*P. séparée*). Tournefort, who has been much more generally followed, separated them into SEMI-FLOSCULEÆ ([*Semi-flosculeuses*] those in which the capitulum is composed exclusively of SEMI-FLOSCULI) and RADIATÆ ([*radiées*] those in which it is composed of both): the latter name was derived from the fact of the semi-flosculi occupying the circumference of the calathis, arranged in a circle (RADIUS) whence the ligulæ radiated outwards; the flosculeæ, the centre, where all of them formed a *Disk* (*Disque*) (DISCUS). Afterwards, Vaillant, and after him Jussieu, modified this arrangement a little, preserving the

Semi-flosculeæ under the name of CICHORACEÆ (*Chicoracées*), uniting under that of CORYMBIFERÆ (*Corymbifères*), the whole of the Radiatæ with some Flosculeæ, the rest of which form the CYNAROCEPHALÆ (*Cynarocéphales*), distinguishable by their habits and appearance and by their style swollen beneath the stigmas.

This last classification has been respected to a certain extent, although modern botanists have greatly multiplied the divisions and the subdivisions of the Compositæ, which are now divided into three great series: 1st, The LIGULIFLORÆ (*Liguliflores*) (*fig.* 729), which answer to the Chicoraceæ or Semi-flosculeæ. 2nd, The LABIATIFLORÆ (*Labiatiflores*) (*fig.* 730), the corollas of which present an irregularity different from that of the preceding, being divided into two lips; the one turned inwards, formed with one or two divisions; the other turned outwards, formed of four or three others. These plants were formerly scarcely known, and this is why we find them omitted in the ancient classifications. 3rd, The TUBULIFLORÆ (*Tubuliflores*) (*fig.* 731), the florets of which, either all of them, or those of the disks only, are tubular and regular. These, consequently, comprehend the Radiatæ and the Flosculeæ, but among them there is one tribe (that of the CYNAREÆ [*Cinarées*]) which answers to the Cynarocephalæ. Besides this, four others have been admitted and have been founded principally on the differences in the structure of the style and stigmas; a characteristic, very important in this group, because its presence is always an evidence of that of several others. Now this style, simple in the male florets, is always divided in the female and the hermaphrodite into two terminal branches, partly covered with collector hairs, and traversed on the edge of their internal face by two small glandular bands which are considered as the real stigmas, although this name is frequently applied to the entire branches. We have already seen that in the CYNAREÆ (*fig.* 734) there is beneath these branches a swelling or knot frequently covered with hairs; the stigmatical bands traverse the whole length of the branch and are joined in one at its top. In the SENECIONIDEÆ (*Sénéçionidées*) (*fig.* 735), the style is perfectly cylindrical, the branches are truncate at their top, which is frequently crowned with a pencil of hairs, beyond which they are elongated at other times into a cone or some other appendix; but it is always at this point that the stigmatical bands are stopped without running into one. In the ASTEROIDEÆ (*Astéroïdées*) (*fig.* 736), the linear branches are continued without change to their top, except their being flattened and covered with very fine hairs on the outside, the

738

729

730

739

728

731

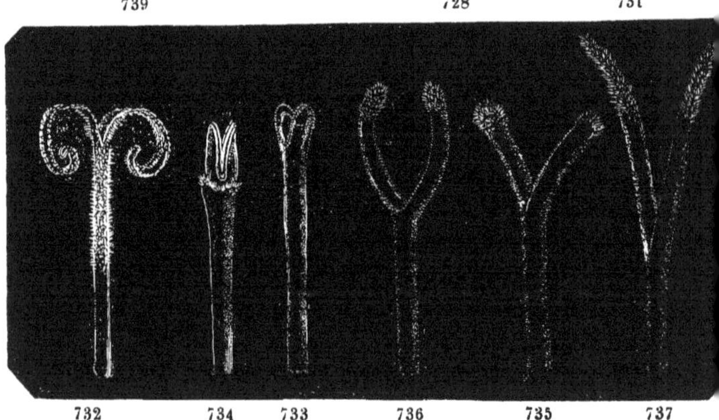

732 734 733 736 735 737

bands ceasing at the same height. The branches are long, widened into club-shaped masses, covered with papillæ on the outside, in the EUPATORIACEÆ (*Eupatoriacées*) (*fig.* 737); they are either lengthened and subulate, or short and obtuse, bristled with long, equal hairs in the VERNONIACEÆ (*Vernoniées*) (*fig.* 738); in both, the marginal bands stop before they attain to the middle of the branch. These seven tribes, the Ligulifloræ, the Labiatifloræ and the five we have just mentioned in the Tubulifloræ, have been subdivided into a large number of sections, which we have not room to explain here. It is necessary, nevertheless, to add a few more details on the principal points of the structure of the organs of this very important class, and to explain the peculiar terms by which we designate their different modifications; terms destined to abridge descriptions, which could not be understood, if we did not know the value of the words employed for this purpose.

The top of the peduncle, widened into a kind of plate, which, bearing the florets of the capitulum, is called the *Receptacle* (PHORANTHIUM or CLINANTHIUM of some authors [§ 209]), is flat or concave, or, on the contrary, convex or even conical. The florets may spring immediately from its smooth surface, or else their insertion is more or less buried in it thus causing it to be filled with shallow (RECEPTACULUM AREOLATUM) or even deep holes (R. ALVEOLATUM), the edges of which are raised around the base of each ovary or

728—739.—The organs of the fructification of several Compositæ.

728. The diagram of a floret of the Groundsel (*Séneçon*) (SENECIO).—The outermost dotted circle shews the tuft of hair or the limb of the calyx.

729. A Semi-flosculus of the Cichory (*Chicorée*) (CICHORIUM INTYBUS).—*o* Ovary adherent to the calyx.—*e* Tube formed by the stamens and traversed by the bifid style *s*.

730. The floret of one of the Labiatifloræ (CHÆTANTHERA LINEARIS).—*o* Adherent calyx and ovary.—*t* Tube of the corolla.—*ls* Its upper lip.—*li* Its under lip.—*e* Tube of the anthers.—*s* Top of the style.

731. A Flosculus of one of the Flosculeæ (ASTER RUBRICAULIS). This is a vertical section shewing the erect ovule *o* in the ovary blended with the calyx, and the tube *e* of the anthers, situated on the corolla *p* and traversed by the style *s*.—*a* Tuft of hairs.

732—738. The tops of the styles of Compositæ belonging to different tribes.—The two bands of the stigmas may be seen on the inner surface of the two branches which terminate each of these styles.—Several are furnished with collector hairs, on the outside, inside or at the top.

732. The top of the style of one of the Cichoraceæ, (the Cichory).
733. ——————— Labiatifloræ (CHÆTANTHERA LINEARIS).
734. ——————— Cynareæ (THEVENOTIA).
735. ——————— Senecionideæ (SENECIO DORIA).
736. ——————— Asteroideæ (ASTER ADULTERINUS).
737. ——————— Eupatoriaceæ (STEVIA PURPUREA).
738. ——————— Vernoniaceæ (VERNONIA ANGUSTIFOLIA).
739. Vertical section of the ripe fruit of a Groundsel (*Séneçon*) (SENECIO).

achenium in laminæ, sometimes continuous, sometimes divided into irregular, membranous, tongue-shaped bodies, or frequently into fringe or hairs (R. FIMBRILLIFERUM). This assemblage of florets is surrounded by an involucrum of folioles or bracts of various shapes, frequently reduced to that of scales, and then bearing this name, sometimes terminated in spines (as in the Thistles [*Chardons*]), arranged in a circle in one or in two concentric rows (§ 230), or most frequently in an imbricate spire. They are united together at the bottom in some cases, but frequently remain distinct. These bracts for several rows do not bear any florets at their axil excepting frequently those of the inner; but each floret may be accompanied by a bract of its own, which grows with it from the receptacle and, concealed among the florets out of the reach of the light, assumes the consistence and the appearance of a whitish scale or of a membrane, *Bracteoles* (*bractéoles*). When these bracteoles cover the receptacle, it is said to be *paleaceous* or *surrounded by chaffy scales* (*paléacé*) (RECEPTACULUM PALEACEUM); when they are entirely wanting, it is said to be *naked* (*nu*) (R. NUDUM VEL EPALEACEUM). The involucrum contains nothing but hermaphrodite florets (CAPITULA HOMOGAMA) (from *ὁμὸς, same; γάμη, a marriage*), or else florets of two kinds (C. HETEROGAMA) (from *ἕτερος, other*). In the latter case, the neuters or the females occupy the circumference, the hermaphrodites or the males occupy the centre. When it unites the male and female florets, it is *monœcious* (*monoïque*) (C. MONOÏCA). When the capitula are composed some of male florets only, others of female, but situated on the same plant, they are *heterocephalous*[z] (*hétérocephale*) (C. HETEROCEPHALA); if the male florets are situated on other plants than the females, they are *diœcious* (*dioïque*) (C. DIOÏCA). We have already seen that the capitula may contain only flosculous florets (C. FLOSCULOSA SEU DISCOIDEÆ) or only semi-flosculous (C. SEMI-FLOSCULOSA SEU LIGULATA), or both at once (C. RADIATA). They may all be bilabiate also (C. FALSO-DISCOIDEA), or those of the circumference ligulate and those of the centre labiate, (C. FALSO-RADIATA SEU RADIATIFORMIA). Sometimes, in flosculous or false discoidous capitula, as in those of the Corn Blue-bottle (*Bleuet*) and of several other CENTAUREÆ (*Centaurées*), the external florets, whilst they preserve the form of the internal, may assume a much greater developement (C. CORONATA).

[z] *ἕτερος, other; κεφαλή, head.*—TRANS.

The calyx is adherent to the ovary, which it completely covers, and is sometimes terminated with it so as to present no trace of limb; at other times it is prolonged a little higher into a kind of small crown, more frequently into several divisions which rarely present the form of folioles, more frequently that of scales, most generally that of a tuft of hair beginning at the very summit of the ovary (Sessile pappus [*aigrette sessile*] [PAPPUS SESSILIS]), or elevated on a prolongation of the calycinal tube in the form of a filament (Stipitate pappus [*aigrette stipitée*] [P. STIPITATUS]). We have examined in another part (§ 427) the nature and the different modifications of the hairs of the pappus.

To the principal forms of the corolla we have recorded, we will add a few more remarkable characteristics, such as that of its nervation. We know that in general in the flowers of other plants the median nerve is the most largely developed; therefore, in the tube of a monopetalous corolla, the five principal nerves are opposite to the five lobes, in the centre of which they terminate. It is not thus in the florets of the Compositæ; the five nerves alternate with the lobes; when they have reached them they are parted into two, which are prolonged on their corresponding edges, so that each division of the limb is bounded by two projecting nerves confluent at its top (*figs.* 729—731). The lateral nerves are, therefore, most developed here; united in pairs in the tube, separated in the limb. As to the median nerves, they are sometimes also found, but are more frequently wanting. It has been proposed to name the Compositæ from this very remarkable characteristic, NERVAMPHIPETALEÆ (*Nervamphipétalées*). The valvate prefloration is connected with this arrangement. We find corollas of all colours; sometimes all the florets of the same capitulum are of the same (CAPITULA HOMOCHROMA[a]), sometimes of different colours (C. HETROCHROMA[b]), those of the disk being then always yellow, those of the radius white or some shade of the cyanic series.

The stamens, situated on the tube of the corolla, have their filaments free or united, but their anthers are always united by their edges, and thus form a tube of themselves (*figs.* 729—731, *e*,) being separated from one another only at their tops, commonly lengthened into an appendix of a greater or less length, and frequently also at their bases elongated into a tail (ANTHERÆ CAUDATÆ), which at other times is wanting (ANTHERÆ ECAUDATÆ). The style we have already described; it traverses the tube formed by the anthers and,

[a] ὁμὸς, *same*; χρῶμα, *colour*.—TRANS. [b] ἕτερος, *other*.—TRANS.

as it lengthens, sweeps off the pollen, which is there retained, by means of its collector hairs buried in the slits of their loculi.

The ovary has a single erect ovule in a single loculus (*fig.* 731, *o*). Yet, from the double number of the stigmas and the existence of two little cords, which sometimes traverse (beginning from the origin of the style) the inner wall of the loculus from top to bottom as far as the insertion of the ovule, would it not be allowable to suppose that it really was formed by the union of two carpels? It becomes an achenium, which, by the absence or the presence and the nature of the tuft of hairs furnishes useful characteristics. The pericarp is at last sometimes blended with the teguments (composed of a double membrane) of the seed as it increases. The embryo turns its short radicle (*fig.* 739) downwards towards the point of attachment.

The CICHORACEÆ or TUBULIFLORÆ have a milky juice analogous to that of the Campanulaceæ. It is bitter, slightly astringent and even narcotic. These properties are found in almost all the wild species to a more or less intense degree; they are remarked especially in the species of the Wild Lettuces, the LACTUCA SYLVESTRIS, and L. VIROSA, the extract of which is employed instead of opium, but without causing the same accidents. But these properties are weaker in other species, especially in those which we cultivate, and of which we eat either the roots, as those of the Salsify (*Salsifis*) and of the Scorsonera (*Scorsonère*), or the young shoots or the leaves, as those of the Lettuce (*Laitue*), of the Cichory (*Chicorée*), of the Dandelion (*Pissenlit*), of the Goats' Beard (*Barbe-de-Bouc*), &c., &c. Let us remark that we employ in this manner the parts which are quite young or are etiolated either naturally, like those which grow under ground, or artificially, so that the peculiar juices have not yet been completely elaborated, and possess only that weak degree of astringency or of bitterness which pleases the palate.

We may say as much of the different CYNAROCEPHALEÆ or CYNAREÆ, which are fit for food; as the leaves of the Cardoon (*Cardon*) (CYNARA CARDUNCULUS), that have been caused to blanch or etiolate; the receptacles of the Artichoke (*Artichaut*) and others, which are gathered before the expansion of the flower and are eaten even raw, whilst they are still extremely young. Every one knows the extreme bitterness of the other parts of the Artichoke, and it is a characteristic common to all the other plants of this tribe. On this account several of them have been employed as stomachics.

It is again found in those which were confounded under the name of CORYMBIFERÆ, but is modified in them by the coexistence of a resinous principle which commonly increases its properties. If this resin, instead of being concentrated and solidified, remains in the state of a volatile oil, the plant will be at once tonic, aromatic, and antispasmodic, as the Chamomile (*Camomille*), the Wormwood (*Armoise*), the Tansy (*Tanaisie*), the Yarrow (ACHILLEA), &c. It has even been proposed to use the infusion of several of them in the shape of tea. The predominance of the bitter principle imparts to them febrifugal virtues, as to the different species confounded under the name of Chamomile, &c. That of the resin will augment the stimulating properties and will provoke perspiration, salivation and an abundant secretion of urine; it is, doubtless, to these effects that several exotic species owe their renown as antidotes against the bites of serpents; such are a species of EUPATORIUM, the E. AYAPANA, and the Guaco (MIKANIA GUACO).—We find in a small number of Corymbiferæ deposits of fecula which are turned to the use of man and animals; the Jerusalem Artichoke (*Topinambour*) (HELIANTHUS TUBEROSUS) will quite bear comparison in this respect with the Potato. We eat the lower subterranean branches changed into tubercles, covered with eyes and filled with fecula. In another, the Elecampane (*Aunée*) (INULA HELENIUM), there has been discovered a bitter principle, which has been termed *inuline*, analogous to fecula, of which it has almost the composition (43·72 of carbon, 6·20 of hydrogen, 50·8 of oxygen by weight) and all its properties, with the exception of not being very soluble in warm water, which dissolves $\frac{1}{4}$, or cold water, which dissolves $\frac{1}{50}$; and being coloured yellow, not blue, by iodine. It is found also in a large number of other vegetables in which it takes the place of the fecula.

The seeds of most Compositæ are oleaginous, of which we may be easily convinced by examining those of the Sunflower (*Soleil*) (HELIANTHUS ANNUUS). Several species are even cultivated for the purpose of extracting oil, as the MADIA SATIVA, GUIZOTIA OLEIFERA, &c.

BOTANICAL GEOGRAPHY.

---

§ 862. We know that every plant is not found uniformly over the whole surface of the globe, but appears only on this or that part of its surface. These limits, assigned to each of them, depend on several causes. The organization, variously modified in different vegetables, imposes on them several different conditions of existence, and they can only live and multiply where they find the conditions peculiar to each of them. Observation also proves that every plant has not started from a common centre, whence they would then be dispersed over the earth, but that there existed a number of original centres of vegetation, each species having its own, although, on the other hand, many seem to have been common to several centres at the same time. If the conditions are different in two places, the vegetation will be so; but the similarity of the one does not necessarily cause that of the other, especially at great distances, since the plants have generally not been able to pass from one of these points to the other, where they would have prospered very well. Thus, the distribution of vegetables on the earth is ruled by complicated causes, some physical, depending on their nature and on the agents which surround them, some hidden from our researches in the mysterious origin of created beings.

§ 863. Botanical Geography is that part of the science which treats of this distribution of vegetables. The fact of their existence in such or such a medium presenting a certain number of physical conditions constitutes their *station;* the fact of their existence in such or such a country constitutes their *habitat, habitation* or *locality.* When we say that a plant grows in marshes, on the sand of the sea shore, on the rocks of mountains, on the edges of the glaciers, we thereby indicate its *station.* When we say that it grows in Europe, in France, in Auvergne, around Paris, we indicate its *habitation* in terms gradually more and more precise. These ideas may be applied to units of a higher order than species; we may enquire into the distribution of whole genera, or even of tribes or families. These large associations of species, in the organization of which we may presuppose great uniformity, frequently present very remarkable similarity in their stations, or their habitations or in both at the same time.

§ 864. But the cause ought to occupy our attention before the effect; and before entering into more details and illustrating what precedes by examples, it will be better to state some general considerations on the manner in which we find on the surface of the earth these external agents playing such an important part in vegetation, such as heat, light, air, water, &c., which in each place are combined in certain proportions to form the climate.

Heat gradually decreases from the equator towards the poles, and that rather regularly, if we consider the same meridian separately. But if we notice this decrease on several at the same time, we are struck with the differences which it presents in this respect. Each place, in the course of a year, receives a certain quantity of heat; and if we compare these quantities for a long series of years, we thence deduce the mean temperature of the place. The line, which would pass through a series of places having the same mean temperature, is called *isothermal* (*isotherme*) (*ἴσος, equal; θερμὸς, heat*). We should be led to think at the first glance that these isothermal lines are only the expression of the greater or less distance from the great source of heat, the sun; that each of them, consequently, cuts the meridians at equal distances from the equator and corresponds to a certain degree of latitude. But, on instituting a comparison between those isothermal lines we have been able to determine by direct observation, we perceive that they form on the globe, instead of regular circumferences parallel to the equator, curves unequally distant from it at the different points. The line of the maximum temperature does not coincide exactly with the equator, but departs a little from it, here southwards, there northwards. The points of the maximum cold do not appear to coincide with the poles, but in our hemisphere to stop at 12° or 15° on this side, concentrated at the north of the two great continents so as to form, as it were, two poles of cold. The isothermal lines present in their inflexions around these poles a certain resemblance to one another, although they are very far from being exactly parallel. In the Northern Hemisphere (the only one, in which an opportunity for making these observations has been afforded, and in which they have been repeated in a very great number of points so as to trace these lines in a less incomplete manner) following the isothermal lines from west to east, we see them decline southwards into the interior of the two great continents and especially of America; rise northwards in the large seas, which are interposed between them, especially in the Atlantic Ocean. The temperature of the ancient continent is, therefore, generally much higher than that of the new one; that of the continents lower in the interior than near the sea, and much higher on the western than on the eastern side. These differences at the same latitudes may be very considerable, and so much the more as we depart from the equator, so that on approaching the north they at last attain as much as 20°. Thus, the northern part of the United States towards 44° north latitude and Drontheim on the western side of Norway towards 63° are included in the same isothermal line (that in which the mean temperature is 5° centigrade.)

§ 865. It does not follow from several places being situated on the same isothermal line, from their having received in the course of the whole year the same amount of heat, that their climate should be identical. Indeed, this sum may be divided in different ways between the different months and, consequently, between the seasons ; or with a certain degree of equality, so that both winter and summer may be very temperate; or, on the contrary, very unequal, so that the summer may be very warm and the winter very cold. These differences of extreme temperature have much more influence on vegetation than the mean temperature. The line, which would pass through all the places in which the temperature of the winter (in an average year) descends to the same point, is called *isochimenal* (*isochimène*) (*χειμὼν, winter*) ; and that, which would pass through the places in which that of the summer is raised to the same degree of heat, *isotheral* (*isothère*) (*θέρος, summer.*) These fresh lines, not coinciding in their turn with the isothermal, do not comprehend the same series of places.

§ 866. The mass of waters is much more constant in its temperature than the earth, so that at a given moment the difference between two points of different latitude is less, and in a given place the difference between winter and summer is also less. The adjacent land participates in this uniformity; and thence the distinction of climates into *marine* and *continental;* the former, that of the shores and islands, more temperate, and so much the more so in proportion to the smallness of the island, to its distance from the mainland; the latter, in which the difference between the summer heat and the winter cold is so much the more marked as it is situated nearer the middle of the continent. Thus, in the Feroë Isles, about 62° north latitude, the heat does not attain 12° in summer, but hardly descends below 4° in winter, thus giving between these two seasons a difference of 7° ; about the same latitude, on the contrary, at Yakouzk in Siberia, the thermometer descends in winter to more than 37° below zero and rises in summer higher than 17° above, thus passing over an interval of 46°.

§ 867. We have not yet taken into consideration another cause, which exercises a powerful influence on the unequal distribution of heat over the surface of the earth, of which we have spoken as if it presented everywhere the same level, that of the sea. But every one knows that it is otherwise, and that this surface is far from being equal over any part of its extent, but that it is raised into table land of various heights and is studded with mountains forming chains of a greater or less length, crowned here and there by still more lofty peaks. Now, in proportion to the elevation, we find that the temperature is lowered, and in such proportion that an ascent of a few hours is sufficient to afford us an opportunity of passing through all the degrees of decreasing temperature. A very high mountain, situated under the line and covered at its summit with eternal snows, as the Chimborazo in the great Cordilleras of the Andes, represents, therefore, in a very limited space all the changes we should more gradually experience, if we travelled from the equator towards the pole. Some authors have, consequently, compared the two hemispheres of our globe to two enormous mountains joined together at their base: an ingenious comparison, but yet not exact in several respects. For the distribution of water, which covers such a large extent on the two hemispheres, and which we have seen to be so powerful in modifying the climates; that of the air, the density of which does not decrease from the equator towards the pole, as it does from the bottom towards the top of the mountain; that of light, so dissimilar at the poles and at the top of an equatorial mountain, establish so many clearly defined differences.

If the law, according to which the heat decreases from the equator to the pole, varies according to the different meridians, that law, according to which the height decreases in proportion as we rise in height, appears for its part to vary according to different circumstances, as the season, the hour of the day, the inclination and the exposed situation of the slope. The decrease is more gradual in winter, during night, on a very gentle slope or on table lands. A difference of 655 feet English (200 *mètres F.*) more or less, gives according to these circumstances a mean of one degree of difference in the temperature, almost as much as two degrees of latitude would give. At a certain height, the cold is such that the heat of summer-days cannot dissolve the ice and snow formed during the rest of the year; and there begins the limit of eternal snows; a limit necessarily so much the less elevated as the climate is less warm at the base of the mountain, or, in other terms, as it approaches nearer the poles: this at a certain distance from them, about 75°, is found, after being gradually lowered, to descend to the level of the sea. Thus, this limit is found at almost 16,400 feet in the Cordilleras between the tropics, at about 8,850 feet in our Alps, below 3,300 feet in Iceland. The glaciers are prolongations, which descend lower than the limit from the accidents of the ground and mark the natural way assigned for the escape of the water which arises from the melting of the snow.

§ 868. The humidity of the atmosphere exercises an important influence on vegetation, because the water, either volatised to the state of a light vapour, frequently even invisible, or to that of a fog of a greater or less density, touches the aerial parts of plants; or condensed, it falls again in the form of rain and so, after having bathed these same parts, penetrates into the soil. The atmosphere is naturally so much the drier as the surface on which it reposes, contains less water, as it is at a greater distance from every reservoir which supplies this defect, and also as it is more heated, so as to rarify quickly all the vapour that arises. A temperature low enough to diminish evaporation and to condense the vapour into a fog or into rain, not enough so to render it solid, is the most favourable to moisture, which will, consequently, be maintained more habitually at certain latitudes and heights. But an elevated temperature also favours it to a remarkable degree, when, on the one hand, it acts on a sufficient quantity of water (a part of which it converts into vapour), and, on the other, these vapours, once formed, meet a cause which tends to maintain them at this degree of density or to bring them back to a much greater. Hence the great rains, which, in certain seasons, fall regularly each day in countries situated between the tropics. Thence, the constant and tepid moisture of their large forests, by the shade of which it is preserved and renewed. The influence of trees crowded together in great numbers on the state of the atmosphere, in which they prevent dryness by opposing evaporation, may be easily verified on a small scale in our own climates: and whole countries are completely changed on account of great fellings and clearings of wood. The neighbourhood of the sea, combined with the prevailing direction of the winds, determining that of the vapours formed on its surface, is a more or less abundant source of humidity, which is, consequently, more constant on islands than on continents. Humidity is, therefore, a condition, which very frequently accompanies those causing uniformity of temperature. The presence of smaller reservoirs, lakes, marshes, great and little watercourses acts in an analogous manner but in proportional limits. The nature and the height of the mountains also contribute much to modify the hygrometric state of the atmosphere. If their summits are high enough, their slopes sufficiently gentle to be the seat of eternal snows and of glaciers, they are so many vast reservoirs destined to feed numberless streamlets of water, which, after having furrowed the sides in all directions, are united lower down, to form larger brooks and become the origin of great rivers flowing at their feet through the valleys and plains. But from the top of peaks, either too low or too steep to retain the snow, transitory torrents only flow. The dryness, which reigns there, is frequently extended to some distance around them, and so much the more as they are less clothed with wood. The chains of mountains exercise their influence again by the lowering of the temperature of the soil on account of their elevation, and tend to condense the vapours which are blown about in large masses by certain winds, and which, stopped by this barrier, fall down in a liquid state, so that such a side of a mountain may generally be very moist, whilst the opposite one remains dry.

§ 869. We have previously seen that Light plays an important part in the majority of the chemical phenomena from which results the composition of the vegetable tissues, and that their ripening, their colouring, their motions, take place for the most part under its influence, combined with that of heat. We may conceive without there being any need of here entering into long explanations, how light is unequally and variously distributed at different parts of the globe: that it is a necessary consequence of their various positions with respect to the sun. Situated near the equator, they undergo the alternate action of nights equal to the days, during which time the rays of the sun are almost perpendicular to the surface of the earth. In proportion as we leave it, the alternation of the seasons is felt, causing the inequality of the day and night; this submits them to a longer privation of light during one part of the year, to its prolonged presence during another part; at the same time the rays become more and more oblique and, consequently, weaker and weaker as far as the polar regions: here this obliquity and this inequality acquire their maximum, so that total darkness obtains during one period of the year, and during the other continual, but very feeble light. The analogy, which we have observed between the latitudes, in proportion as we go from the equator, and the heights, as we rise above the level of the sea, totally disappears, therefore, in the distribution of light; since on the mountains the highest parts are the longest illuminated and enjoy a greater length of day, whilst their bulk, intercepting the rays of the sun, retards day and advances night in the lower parts. The plants, however, of the polar regions and those of high mountains are to a certain point in the same conditions with regard to light, if, concealed under the snow for the greater part of the year, they see the day during a few weeks of summer only.

Let us add again that the proximity of large sheets of water by the production of vapours, which are interposed between the earth and the sun, proportionally diminish the intensity of light. This cause, contributing so effectually to equalize the temperature and generally to raise the mean temperature, has, therefore, an inverse influence on light, which it tends to weaken.

§ 870. All the preceding facts belong to Meteorology. This science investigates the causes, which by the combination of various conditions form the different climates. It teaches us how they emanate from one first source, the solar action, which, on account of the regular motion of our planet, the varied form of the land and its relations with the water as well as the inequalities of its height, influences every part directly with a certain force, and indirectly, by causing the various currents of the atmosphere and of the sea, the one regular, the other variable on account of perturbations resulting from secondary, but analogous causes; how this solar action is, consequently, unequally distributed over the surface of the globe. All these considerations are foreign to the subject which now occupies our attention: the general results alone ought to be explained here; they could not be omitted, so closely is Botanical Geography connected in this respect with Meteorology, so powerful an influence does the climate exercise on vegetation.

§ 871. Let us now examine the general modifications the latter presents in relation to the climates we have just mentioned.

However short a time we may have looked for plants, we shall soon be struck with the inequality with which their different species are distributed. Some are met with in a very limited space, others, on the contrary, are dispersed over a large number of places at the same time. This difference, which our botanical excursions shew us on a small scale, is also perceived, when we compare the results of those which have made us acquainted with the vegetation of many vast countries: certain plants are peculiar to certain countries, others common to several. These limits, in which the habit of each species is contracted or extended, constitute what is termed its AREA (*aire*). Those, the area of which is very much circumscribed, may, therefore, be considered as characteristic of the vegetation of this space, which they do not outstep; but we shall not treat of them in this place, where we intend to enter into the most general points only. Those, the area of which is very large either in latitude or in height, cannot by this very fact of their distribution, be employed to characterize a particular region; we ought, therefore, to reject these also and to choose others, which are found abundantly in several distant parts of the globe, but not out of a certain zone of a greater or less width, of which they thus form one of the distinctive features. The more the list of these characteristic vegetables can be increased, the more exact will be the description of the area. But this multiplicity of detail can be found only in a complete treatise; in an abridged description, we must limit ourselves to a small number of vegetables chosen from those, which by their height, their remarkable appearance or their utility are most likely to arrest our attention, and for that reason have not escaped that of travellers, even of those who are totally ignorant of botany. Trees are in general very useful in this respect, so much the more so as they may be considered as being closely connected with the climate, to the vicissitudes of which they are more exposed during the changes of the whole year than herbaceous plants, which may be partly protected from their action during a portion of the year, and especially than annual plants, which only live one season. Certain regions are also characterized by the presence of groups of a more elevated order, genera, families or their tribes, whenever their area is thus circumscribed; we can then understand that the description of the region gains importance from its including a larger number of characteristics. Besides, it is not necessary for the whole of the species of the group in question to be exclusively found in the region we wish to delineate; it is sufficient that the greater number of them be found there. Without the natural method, Botanical Geography would necessarily be lost in endless details, and we may say that it has been founded by the establishment of families, and that it will be perfected as they are improved.

§ 872. Let us now cast a glance at the principal regions thus characterized, whether by the existence of certain peculiar and remarkable vegetables, or by the exclusive presence or the great abundance of those of certain families. We will examine them, proceeding from the equator towards the poles, and we shall compare to each of the successive zones belonging to a latitude more and more elevated those places under lower latitudes which correspond to it, as they are situated at a much greater height and, consequently, submitted to a similar temperature.

§ 873. The zone, which is limited on the two hemispheres by the tropics and from time immemorial has been designated by the name of *Torrid,* presents a vegetation very distinct from that, in the midst of which we live, in the vigour, in the variety, in the peculiar forms and characteristics of a large number of the plants which compose it. The proportion of the ligneous vegetables growing there appears to be large; and if the moistness and richness of the soil are added to the high temperature, we shall find that these are large trees growing in vast forests of an aspect quite different from ours; for, instead of the uniform repetition of a very limited number of species, they present an infinite diversity, whether we examine their *tout ensemble* or only a small part. These species, moreover, for the most part belong to other genera, to other families than those of the temperate zones. In vast countries only thinly inhabited, where the trees have not yet been destroyed to supply the wants of man, and their existence has no other limits than those which nature assigns to them, these *Virgin Forests* (*Forêts Vierges*) (SYLVÆ PRIMÆVÆ) have acquired the most magnificent developement. The strength of the vegetation is manifested not only by their stems of so very remarkable a thickness and height, but also by the production of other more humble plants, some ligneous, others herbaceous, which, under the shelter of their loftier brethren, flourish luxuriantly in this warm and humid atmosphere; by that of the parasitical plants, which cover and partly conceal these trunks; especially by that of the climbers which run from one to the other, mounting to their very tops only to fall and remount again, binding them round and round and clasping the giants of the forest with their net-work of branches. One of the distinctive features of this tropical vegetation depends on its being exposed to influences scarcely varying during the whole course of the year. In more temperate climates, in which the seasons are decidedly different from one another, one brings with it the blossoming, the other the ripening; so that we see the majority of trees after a repose, during which they have remained more or less naked, become clothed with leaves, with flowers at the same time and afterwards with fruits. Under the equator all these phases are confounded; and since this extreme activity causes a very great production of the leaves, which do not fall annually, we are struck by the much less proportion of flowers and, consequently, of fruits in a given time, although we find them always on the tree.

§ 874. But if the soil, although rich enough for the developement of the arborescent species, is not, from its nature and from the inadequate distribution of water on its surface and in its substance, constantly humid, if the moisture be renewed only at intervals by means of rains, themselves dependant on a certain regular alternation in the state of the atmosphere, we observe changes more analogous to those of our seasons, only they are inverted; dryness puts a stop to vegetation, and despoils the trees of their leaves, which afterwards flourish again as soon as the great periodical rains begin to water them. This may be observed, for instance, by comparing the Virgin Forests with those woods, consisting of less lofty trees with an intermittent vegetation, which bear at Brazil the name of *Catingas.*

§ 875. Lastly, sandy and irregularly irrigated soil can only produce frutescent and herbaceous plants; its vegetation, suspended during the dry season, revivifies during the rainy and covers with a rich but transitory carpet of verdure and of flowers the earth, which appeared naked and sterile during the rest of the year. We see this in the vast plains of the tropical regions, plane or undulated and deprived of the natural and continuous irrigation, which results from the neighbourhood of great mountains. These plains, some covered with many species, others, on the contrary, with a uniform vegetation, bear according to these variations different names in different countries. They form the *Campos* of Brazil, the *Pampas* of Paraguay, the *Llanos* of the Orinooka. The alternation of repose and of activity causes an effect analogous to that of our seasons, the complete absence of flowers during one time, but during another their multiplicity and diversity.

§ 876. The *Palms* and other *Arborescent Monocotyledonous Plants* (*Pan-*

danaceæ, Dragon-trees, &c.), as well as the *Tree Ferns*, contribute especially to impress a peculiar appearance on tropical vegetation. Another equally characteristic form, is that of the *Scitamineæ*, always remembering that it has been agreed to comprehend under this name not only the plants of this family, but also those of the *Musaceæ* and of the *Cannaceæ*. The BANANA (which acquires it's full developement in the stove houses of Europe) will give us some idea of them. We will mention those families which may be termed tropical, either because they do not appear beyond the tropics, or because the maximum of their species are found there. Such are the *Bromeliaceæ, Aroideæ, Dioscoreaceæ, Piperaceæ, Laurineæ, Myristiceæ, Anonaceæ, Bombaceæ, Sterculiaceæ, Byttneriaceæ, Ternströmiaceæ, Guttiferæ, Marcgraviaceæ, Meliaceæ, Ochnaceæ, Connaraceæ, Anacardiaceæ, Chailletiaceæ, Vochysiaceæ, Melastomaceæ, Myrtaceæ, Turneraceæ, Cactaceæ, Myrsineæ, Sapotaceæ, Ebenaceæ, Jasminaceæ, Verbenaceæ, Cyrtandraceæ, Acanthaceæ, Gesneriaceæ.* Several large families, which, in our climate have a greater or a less number of species, are found to be represented between the tropics by others still more numerous, (as the *Euphorbiaceæ, Convolvulaceæ*, &c.); but some are of different forms, as the Bamboos (*Bambous*), or other *Arborescent Gramineæ*, the *Epiphyte Orchids;* and others are distinguished by peculiar features sufficient to form entire tribes, as the *Mimoseæ* and the *Cæsalpineæ* in the LEGUMINOSÆ, the *Cordiaceæ* in the BORAGINACEÆ, the *Rubiaceæ properly so called.* Let us lastly mention several characteristic families, because among their species we find those parasites of a very curious vegetation, (the *Loranthaceæ, Rafflesiaceæ, Balanophoreæ*); and especially several of those climbing plants, which we have more than once mentioned, (the *Malpighiaceæ, Sapindaceæ, Menispermaceæ, Bignoniaceæ, Apocynaceæ, Asclepiadaceæ.*)

§ 877. Up to this time we have spoken of the intertropical zone as enjoying the same climate throughout the whole of its extent. But it can by no means be quite so. The progress of the earth round the sun, which for us brings about the extremes of winter and summer, brings back on the contrary, for the regions situated immediately under the equator, exactly similar conditions, and every difference has a tendency to be effaced more and more in the passage of the sun from one tropic to another. There does not exist, therefore, any distinction in the seasons; the mean temperature is found to be the same as that of the whole year; it is also the temperature of the soil at a certain depth, where the phenomena of life are taking place in the subterranean parts of vegetables. The length of the days and nights, being always equal, tends to complete this constant uniformity in the conditions to which they are subjected. A few degrees of latitude hardly alters these conditions; but in proportion as we leave the equator, the distinction of the seasons will become more and more easily perceived. This difference, it is true (if we content ourselves with the general appearance and do not take account of certain parts in which local influences cause rather remarkable variations), is nearly always imperceptible, and the isothermal lines, whilst they are a few degrees of heat lower, are not far from the isochimenal and the isotheral lines, all preserving a certain degree of parallelism with the equator, and the interior of the soil maintaining at a certain depth a constant temperature, which is no other than the mean. However this may be, appreciable differences in vegetation result from it; and we may under this head subdivide this large zone into the *Equatorial* (*Equatoriale*), comprising almost 15° on both sides of the equator, and the *Tropical* (*Tropicale*), extending from 15° to 24°. To be content with a few principal features chosen from those we have mentioned above, the first is characterized by the more exclusive presence of the *Palms* and of the *Scitamineæ;* the latter by that of the *Tree Ferns*, of the *Melastomaceæ* and the *Piperaceæ.* The former extends from the level of the sea up to a height of about 2,000 feet; if we ascend on these mountains as high as 4,000 feet, we shall find a zone corresponding to the latter. It is clear, that there can be no defined separation between the two, either of temperatures or of natural productions, and that the differences are never distinctly seen except we are placed at points sufficiently distant in latitude or height.

§ 878. The large zones, which are commonly termed temperate and extend from the tropics to the polar circles, necessarily present from one of these limits to the other differences in climate and in vegetation, more clearly defined than those we have hitherto mentioned. We shall, therefore, divide them into several regions, the boundaries of which are determined less by latitudes than by isothermal lines, the latter, as we have shewn, becoming gradually independent of the former.

§ 879. Our first zone, extending from the tropics to 34° or 36°(which would be better defined as being traversed near its middle by the isothermal line of 20° and might be termed the *juxta-tropical* (*juxta-tropicale*) *region*), shews us the transition from the Tropical Flora to that of climates essentially temperate. We still observe several plants and forms which we have previously enumerated, but much more rarely and joined to a large number of those of our country. The *Palms*, the great *Monocotyledons* and the *Tree Ferns* are still found there; the *Melastomaceæ* are also numerous; the *Myrtaceæ*, *Laurineæ*, *Diosmeæ*, *Proteaceæ*, *Magnoliaceæ* now acquire their largest numerical developement. Along with them we find some representatives of the families of the following zone, naturally increasing in number in proportion as we approach it; we find there European genera and even a certain number of identical species. This zone is placed in peculiarly favourable conditions for the habitation of man by this mixture of very different productions and the possibility of borrowing from very different climates the greater part of those plants which might be agreeable or useful to him. It comprehends the countries which were first inhabited by the human race, and those islands called the Fortunate Islands by the ancients.

§ 880. The portion of the temperate zone situated on the outside of the preceding, may itself, in a general manner, be divided in each hemisphere into three secondary zones; the first or *warm temperate* (*temperée chaude*), traversed by the isothermal lines of 15° to 10°; an intermediate one, or *cold temperate* (*temperée froide*), by those of 10° to 5°; the last one by that of 5° to 0°. This does not merit the name of *temperate* and may assume that of *sub-arctic* (*sous-arctique*) on account of the proximity of the polar circle, to which it reaches and beyond which it even advances at a small number of places, those, which correspond to the Western shores of Europe and of America, whilst in the other continents it remains more or less distant from it. Paris, where the mean temperature is 10° 8′; London, where it is 10° 4′; Vienna, where it is 10° 1′; are situated nearly on the common boundary of the first two.

§ 881. The examination of these three secondary zones, and even of those which follow them, no longer presents the same difficulties as that of the preceding, to explain which we were obliged to mention vegetables, whose name brings to our mind nothing but rather vague ideas, since we generally know them only as dwarfs in our greenhouses, reduced to fragments in our herbaria, and we can most frequently obtain only a very slight idea of their appearance from descriptions or paintings. Once arrived at climates really temperate we find ourselves in well known places, and we can pursue our studies on Nature herself, which is much more valuable than reading all the books ever published. For we have no need of travelling to the poles and of leaving France, since the South belongs to the warm zone and our mountains will furnish us with all those, which grow on the line of eternal snows where all vegetation ceases. He, who can climb the Pyrenees starting from the plains of the Roussillon, or ascend from Provence to the summit of the Alps, which extend close to the sea-shore, will see in this short excursion under his eyes all the changes he would observe, if he were to traverse Europe from the South to the North, even to the very confines of Lapland. We shall, therefore, prefer to follow this route. We shall still point out on our way the families which furnish the principal and characteristic features of the vegetation; but we shall also take advantage of a few remarkable vegetables, familiar to the greater part of our readers, to serve us as land marks; then we shall cast a glance over the other parts of the globe comprised in the same zone, when the modifications of their vegetation will be more easily understood, as we shall only have to compare them with those we already know.

§ 882. We have mentioned Provence and Roussillon. All the countries, bathed by the waters of the Mediterranean, present the most striking relations to them in their vegetation to a certain distance from the shore and form together an almost uniform botanical region. Some few of the tropical families advance as far as this, but are represented only by a small number of species; as the *Palms* by the Date-tree (*Dattier*) and the CHAMÆROPS; the *Terebintaceæ* by the Mastic-tree (*Lentisque*) and the Pistachio (*Pistachier*); the *Myrtaceæ* by the Myrtle and the Pomegranate; the *Laurineæ* by the Laurel; the *Arborescent Apocynaceæ* by the Oleander (*Laurier-rose*). On the other hand, other families, before this not numerous, multiply their representatives, as the *Caryophyllaceæ*, the *Cistineæ*, the *Labiatæ*, which, covering all the dry uncultivated ground, fill the air with their aromatic exhalations. The *Cruciferæ* also begin to appear. Among the *Coniferæ* we find the Cypress and the Stone Pines; among the *Amentaceæ*, the Holm Oak, the Cork-tree, the Planes, &c. A cultivated tree, the *Olive*, is a very good one to choose as characteristic of this region, since we find it almost everywhere in this and scarcely at all in any other zone.

§ 883. The vegetation of the environs of Paris will give us a general idea of a large part of the cold temperate zone. The families we have just named are also found in a great proportion, but fewer of the *Labiatæ* and *Caryophyllaceæ*, more, on the contrary, of the *Umbelliferæ* and the *Cruciferæ*. The families of the trees are still the same, but represented by other species; the *Coniferæ* by the common Pine, the Larch, &c.; the *Amentaceæ* by the Oak, the Hazel, the Beech, the Birch, the Alder, the Willow, all losing their leaves in winter; and thence there arises great difference in the appearance of the landscape according to the season. These different vegetables themselves vary, either in their proportionate number or in their species, according to the part of the zone under consideration.

§ 884. Let us suppose the spectator to be placed at the foot of the Alps, opposite to one of those high peaks crowned with eternal snows. On looking at the mountain he will easily see that the vegetation, which immediately surrounds it and imparts its features to the Centre and the North of France, disappears at a certain height to give place to another, which itself undergoes successive changes as he ascends the hill. Now, since at a certain distance his eye will be able to appreciate only the masses formed by the large vegetables, in the midst of which are concealed others more humble, he will be inclined to compare them to a series of ribands placed above one another: the first is that of the trees with caducous leaves distinguished by their light-coloured verdure; then that of the *Coniferæ* with their deep-coloured and almost black foliage; then, lastly, a band, the more undecided green of which is interrupted here and there by masses of another colour; the trees gradually lowering in height, till they reach the sinous line where the snow commences. This change in colour is owing to the trees, whose tops were close together and uniformly coloured, being replaced by shrubs or herbs.

If, from the point where the objects are thus gathered into masses, he advances towards the mountain and ascends it, he will at first be able to gather the plants of our fields, then, on the first slopes he will find others more or less different, which are designated by the name of *Alpine* (*Alpestre*), as the *Aconite*, the ASTRANTIA, certain species of ARTEMISIA, of *Groundsel* or SENECIO, of PRENANTHES, of ACHILLÆA, of SAXIFRAGA, of POTENTILLA. After having kept close to the border of the Walnuts, traversed woods of Chesnut-trees (*Châtaigniers*), he will find the latter cease and give way to woods composed of Oaks, of Beeches and of Birches. But the Oaks will cease the first (about 2,800 feet), the Beeches (*Hêtres*) a little later (about 3,300 feet). Then the woods will be formed almost exclusively of evergreen trees (the Larch, the Common Pine), which themselves disappear at successive stages (about 5,900 feet). The Birch mounts a little higher still (about 6,560 feet). One of the *Coniferæ*, the PINUS CEMBRA, is still sometimes observed for about 300 feet higher. Beyond this limit, the trees are not so high, but form humble shrubs, as a species of Alder, the ALNUS VIRIDIS. About this place, we find that shrub so characteristic of a region of the Alps, of which it is termed the Rose, the RHODODENDRON, which ceases in its turn higher up to give place to still more humble plants, very little elevated above the surface of the ground, called *Alpine* (*Alpine*): they are species of some of those families, which the observer found at his starting point: *Cruciferæ*, *Caryophyllaceæ*, *Ranunculaceæ*, *Rosaceæ*, *Leguminosæ*, *Compositæ*, *Cyperaceæ*, *Gramineæ*, but of different species; there are also numerous and fresh representatives of other families, which appear very rarely on the plains: *Saxifragaceæ*, *Gentianaceæ*, &c. &c. Annual plants are almost entirely wanting: this may easily be accounted for, since a single unfavourable year is sufficient to destroy the whole race by preventing the complete ripening of their seeds; this will be frequently the case in such a rigorous climate. Perennial or ligneous plants, on the contrary, are preserved under the soil maintained at a much higher temperature, are thus shielded from the destructive influence of the atmosphere, and are developed whenever it becomes milder and warmer: but it is only for a very short time and in certain places only once in several years. Hence the stems are hardly raised above the ground, hence those, which are commonly frutescent, touch the soil, sometimes creeping, sometimes short, stiff and entangled, forming nere and there thick and compact masses, like a shrub much cropped every year. The appearance peculiar to each family is in some degree destroyed and is replaced by the general appearance of the Alpine plant: this is found even in genera of several species commonly arborescent, as the Willow, which here creeps close to the soil. On the edge of pieces of water, where the brows of mountains form gentle slopes, or are flattened into steps on which a layer of humus may be collected, the vegetation forms extended masses; but most frequently this mass is broken by the accidents of the ground, and the verdure appears only in shreds in the intervals, the splits or the clefts of the rocks. The higher we mount the more it is impoverished, until at last these rocks are clothed with no other vegetation than that of the *Lichens*, the scurf of which somewhat varies the monotonous tint of the rocky surface. We have now arrived at the line of eternal snows, where organized beings can no longer accomplish the functions and the purposes of their life.

§ 885. Let us now compare what we shall observe in our progress from the Centre of France towards the pole with what we have observed in the ascent of the Alps. We see the absolute number of the species and the relative number of those of certain families gradually diminishing in the same way: the *Labiatæ*, the *Umbelliferæ*, the *Rubiaceæ*, &c.: those of several others completely disappearing: *Malvaceæ, Cistineæ, Euphorbiaceæ.* By assuming characteristic vegetables as points of comparison, such as those trees we have followed on the slopes of the Alps, we shall find their distribution almost analogous, if we consider it in a general manner, a little different, however, if we subject it to a more detailed and rigorous examination. Thus, in the West of Norway and Sweden the Beech stops at 60°, a little sooner than the Oak, which is found at 61°: it is the northern limit of the cold temperate zone. We shall now enter into the sub-arctic zone, in the midst of forests of evergreen trees, of Firs, which cease about 68°, of Pines, which cease about 70°, but the Larch is entirely wanting. The Common Birch is found a little farther still. They are, therefore, the same vegetables, the whole of which characterize the different zones determined by the different heights of the mountains; but they here cease in a different and sometimes inverse order. We do not afterwards meet with any thing else but low shrubs, and near the extremity of Lapland we enter into the polar circle. But this may itself be divided into two: the one, the *arctic* (*arctique*), analogous to that of the Alps, which is without trees but still clothed with humble shrubs. Here the Dwarf Birch, up to 71°, replaces the Evergreen Alder of the mountains, and the Rhododendron is represented by a peculiar species, the R. Laponicum. At Spitzbergen, lastly, we are in the region of Alpine Plants, in the other zone which we may properly term *polar* (*polaire*), where the vegetation, roused for a few weeks only, sleeps buried under the snow for the rest of the year, and produces nothing but a few perennial, sub-frutescent, puny vegetables; the same, for the most part, which we pointed out near the limits of eternal snow. In the preceding parallel of the different zones of vegetation, formed according to the altitude and according to the latitude, we have for the latter chosen the most favoured portion of the earth, that in which the isothermal lines rise a little towards the pole, viz. the West of Europe. By following other meridians, we should have found the successive zones stopping at much less elevated latitudes, so much the less as we should have been more closely approaching those, which traverse the centre of the great continents or approach their eastern shores.

§ 886. Let us also recall what we have stated (§ 865); that the mean temperature exercises less influence on vegetation than the extreme temperature of winter, and especially than that of summer as well as the duration of both. Because several vegetables, escaping under the earth or the snow which covers them and protects them from the action of the atmosphere, may thus brave the effects of the most rigorous winters and reappear during the summer, even passing through all the phases of blossoming and of fructification, if it be warm and long enough. These same conditions also allow the preservation of a certain number of annual species. We may, therefore, find remarkable differences between the vegetation of two points on the same isothermal line; between that point, where there is little difference between the summer and winter temperature, and that, where they differ much, as in the West and in the Interior of the continents, each of them excluding a certain number of plants which the other admits. The isothermal lines, consequently, cannot (any more than those of latitude or those of altitude) rigorously define a vegetable zone: the isochimenal and the isotheral lines would not be more than enough. The vegetation of a country more or less limited, is a result of these and several other influences combined. It is much more complex than the climate, to which it is subordinate only in a general manner. We cannot, therefore, pretend to circumscribe its numerous variations in certain continuous lines, or to bring them into the formulæ of a small number of laws. We may thereby conceive how incomplete and imperfect is the sketch we have traced, being obliged to confine ourselves to a few pages and to shun a multitude of details, here, however, so necessary: in this explanation, also, we have had recourse less to theories than to examples. We have naturally taken ours from Europe and especially from France, so that the reader can get some little advantage from his familiarity with the examples. Let us, however, state a few more important facts.

§ 887. In this comparison we shall follow an inverse process; we shall descend from the top of the mountains to their base, from the pole towards the equator.

If we consider the highest zone of vegetation on the peaks situated in different latitudes and on very different parts of the globe, that zone, which borders on the limit of eternal snow and has been termed polar, we shall find that it every where presents the same appearance, that of which we have tried to give some idea, very incomplete, it is true, in the Alpine plants (§ 884). On the heights of the Caucasus, of the Himalaya, of the Mexican Andes and of the Peruvian or Chilian Andes, botanical travellers have described to us the same aspect of vegetation arrested at a little distance from the soil, formed by the herbaceous shoots of perennial plants developed during a short summer, by the stiff branches of the ligneous species (the direction of which has a tendency towards the horizontal instead of the vertical) entangled in compact masses, which sometimes can be penetrated only by the aid of the hatchet. The species, which we mentioned as existing on the principal range of Europe, the Alps, are found for the most part on its other mountains, those of Norway and Sweden, of Spain, of Turkey, the Apennines, the Carpathians, the Pyrenees. They are mixed, doubtless, in each of these countries with some that are peculiar to the place, but the general features of vegetation are the same. In Asia, the Caucasus and the Himalaya present the greatest analogy: we generally find the same families, the same genera, but represented by different species and so much the more so as we go farther from the parts which we have chosen as examples. In America, these plants, which by extension are also called *Alpine*, but which ought properly to be termed *Andine*, again belong to the same families, some to the same genera, but the greater number to fresh genera, especially to several of the Compositæ and Umbelliferæ. Others at this height represent other families, as the Oxalis, the Calandrinia (*Portulaceæ*), and there are a few *Malvaceæ* which approach this limit.

§ 888. The vegetation of the polar artic circles is not very different in the old and the new worlds. In this respect we will compare two well-known countries: Lapland, botanised and described by M. Vahlenberg; and Melville Island by Mr. R. Brown. The latter place presents great interest, inasmuch as it borders on one of the poles of cold (§ 864) and may, therefore, be considered as the extreme limit of vegetation at the level of the sea, with a mean temperature of 18° below zero, with winters, in which the thermometer descends below 33°, with summers, in which it does not rise above 3°. We find altogether 116 plants, 19 Cryptogams and 67 Phanerogams. We have given a list of the families and the number of the species of each: *Fungi* (2 species), *Lichenes* (15), *Hepaticæ* (2), *Musci* (30), *Cyperaceæ* (4), *Gramineæ* (14), *Juncaceæ* (2), *Amentaceæ* (1), *Polygoneæ* (2), *Caryophyllaceæ* (5), *Cruciferæ* (9), *Papaveraceæ* (1), *Ranunculaceæ* (5), *Rosaceæ* (4), *Leguminosæ* (2), *Saxifragaceæ* (10),

Ericineæ (1), *Scrophulariaceæ* (1), *Campanulaceæ* (1), *Cichoraceæ* (1), *Corymbiferæ* (4). Now, 70 of these species (26 Dicotyledons, 8 Monocotyledons, 36 Acotyledons) are common to this place and to the North of Europe, 45 (20 Dicotyledons, 12 Monocotyledons, 13 Acotyledons) are peculiar to the North of America. Ramond, on the other hand, has described among 133 plants from the top of one of the Pyrenees 35 species (15 Cryptogams, 20 Phanerogams) identical with those of Melville Island. As to the recently discovered polar antarctic countries, they are as regards the botanist as if they did not exist. Navigators have not been able to see the soil under the thick layer of ice and snow that covers them and almost always bars approach to them.

§ 889. In this Hemisphere, the zone, which we have termed the arctic, covered by the ocean, is interesting to the botanist only from its species of FUCUS. As to the Northern Hemisphere, in which the sea occupies only a very small part, we may be content with the glance cast over Lapland, so nearly does the vegetation of the arctic zone resemble that of the polar. It mostly presents the same parts as it does, to which are added several others of form and habits already superior, although not yet elevated to the dignity of trees. But we find more clearly marked difference if we compare these two zones on the Alps and on the Andes. On the Chimborazo, for instance, between 9,840 and 14,000 feet, along with those humble species which characterize the upper region, we find shrubs of a higher rank multiplying, and even a few trees towards the bottom. Certain *Compositæ* assume this form. Two species of this family, ESPELETIA and CHUQUIRAGA, may from their abundance over the whole zone be employed to characterize it; some of the plants belong to the tribe of the *Labiatifloræ*. Other families (*Escalloniaceæ*, *Araliaceæ*, *Ebenaceæ*) have their representatives there, and that of the *Ericineæ* has also several genera and tribes. One of them, the BEFARIA, seems to replace the RHODODENDRON of the Alps.

§ 890. The temperate zone, which we have only considered as it relates to Europe, must now be followed into other parts of the globe, first into the Northern, then into the Southern Hemisphere. It comprehends in Asia a vast extent of country bounded on the north by a part of Siberia, by the northern declivity of the Altai, containing in the south those countries known by the common name of the Levant, and ending at the southern slopes of the Himalaya. The greater part of this extent is enclosed between the two great chains of mountains we have just mentioned; it has been explored by botanists very sparingly; we cannot, therefore, lay claim to sufficient knowledge to describe the general aspect. The botany of its boundaries only is well known; the vegetation of the northern part of the Levant is very similar to corresponding latitudes in Europe, and runs into that of the tropical regions towards the south; in a long band of Siberia (where a considerable diminution of temperature brings us back in several parts to the sub-arctic region, although the latitude is not so high) we find, on the contrary, several new species of European families, developed, doubtless, under the influence of comparatively very warm summers. The vegetation of the tropics is lost on the slopes of the Himalaya, and that of different temperate climates is established according to the height at which the observer is placed. Lastly, this Asiatic zone is terminated on the east by the North of China and Japan, where the physiognomy of European vegetation is not yet effaced; this is clear from several plants belonging to the same families and the same genera; but it is modified by the admixture of other families (*Magnoliaceæ*, *Menispermaceæ*, *Byttneriacaceæ*, *Ternströmiaceæ*, *Hippocastaneæ*, *Sapindaceæ*, *Xanthoxylaceæ*, *Calycanthaceæ*, *Bignoniaceæ*, *Commelinaceæ*, *Dioscoreaceæ*) strangers to Europe though also found in America. Two very remarkable trees, the Tea-tree in China and the CAMELLIA in Japan, will characterize the warm zone.

§ 891. In North America, the immense territories of the United States form almost the whole of the temperate zone. The warm zone comprised almost between 30° and 36° is characterized by the developement of trees belonging to some of the families we have just mentioned and especially to the *Magnoliaceæ*. The cold, compared with the corresponding European zone, is distinguished by the scarcity of the *Cruciferæ*, *Umbelliferæ*, *Cichoraceæ* and *Cynareæ*. Other *Compositæ* (as the ASTER and SOLIDAGO), on the contrary, are very abundant as well as the trees of the *Coniferæ* and *Amentaceæ*. These species belong to the same genera as those of Europe, but are different and much more numerous, as PINUS, ABIES, LARIX, THUJA, JUNIPERUS, TAXUS, CARPINUS, BETULUS, ALNUS, JUGLANDA, FRAXINUS, SALIX, ACER and especially QUERCUS.

§ 892. Passing now to the other Hemisphere, we see the comparatively small extent occupied by the land in the temperate zone. A single glance at a map will clearly shew us this; the different continents are the widest under the tropics, then grow narrow at first gradually, but with some little rapidity as they approach the antarctic pole, at a great distance from which they end. Thus, the greater part of South America, of Africa and nearly the half of New Holland, belong to the tropical region. Africa, ending at 35° south latitude, New Holland at about 42°, do not pass beyond the warm temperate zone, to which the southern extremity alone of the former belongs. America only, extending to 55°, enters into the cold temperate zone.

Its extreme limit, at the land around Magellan's Straits, presents a remarkable analogy to the other hemisphere, also characterized by the presence of certain trees (SALIX and FAGUS) attaining rather large dimensions. But the American characteristic is shewn by the presence of a DRYMIS, an evergreen tree belonging to the *Magnoliaceæ*, of an ESCALLONIA, of a FUCHSIA, &c. Re-ascending on the one hand to the mouth of the Rio de la Plata, on the other, towards the northern frontiers of Chili which border on the juxta-tropical regions, we gradually pass through all the modifications of the temperate zone. The plants of Chili in nearly 100 families contain about fifteen not indigenous to Europe, some even which seem peculiar to this region, as the tribe of the *Labiatifloræ* among the *Compositæ*, the *Loaseæ*, *Gilliesaceæ*, *Francoaceæ*, *Malesherbaceæ*, *Nolanaceæ*, &c. Among trees the ACACIA CAVEN, a tropical form, along with the CACTUS PERUVIANUS, abounds in the North; about the Middle, singular *Rhamnaceæ* with thorny branches (COLLETIA), one of the *Homalineæ* (ARISTOTELIA NAQUI), peculiar genera of *Rosaceæ* (QUILLAIA and KAGENECKIA), a Laurel, some species of the ESCALLONIA, which grow as far down as the edge of the sea: in the South, along with the Beech (FAGUS) and the DRYMIS, various kinds of *Myrtaceæ*, two genera of *Monimiaceæ*, some *Cunoniaceæ*, *Bixineæ* (AZARA) and a few *Proteaceæ*, not very numerous, it is true, in genera (LOMATIA, EMBOTHRIUM, QUADRARIA) and species, but a very large number of their individuals form whole forests. Between these trees climb some species of CISSUS and LARDIZABALA, representatives of the Climbers.

§ 893. If under the very equator we consider the zone of the Andes, which corresponds by its height to this temperate region, we shall find from 3,000 to 9,000 feet a DRYMIS and an ESCALLONIA (those genera we have found in the land about Magellan's Straits), as well as trees of peculiar interest: the QUINQUINA, the different species of which are met at various heights and even descend so low as the limit of the Tree Ferns. But tropical plants encroach more on the limits of the temperate zone along the slopes of the mountains than on those of the latitude, and Palms, Epiphyte Orchids, Sensitive Plants, Melastomads, &c., are found in great abundance even in the midst of the region of the QUINQUINAS.

§ 894. Southern countries, of which New Holland forms the larger part, are very singular in their vegetation. More than $\frac{8}{10}$ of their species are peculiar to them; several form quite distinct families; others, the greater part, families scarcely represented at all in the other parts of the globe. Even those, which belong to extensively spread and well-known families are so disguised that they were not recognised when they were first discovered. A witty botanist said, whilst looking over a herbarium full of them, 'Here we are at a *bal masqué*.' The disguises are now well known. But they have been mostly collected and studied between 32° and the southern extremity; this, therefore, belongs to the temperate zone; and it is this part of the vegetation, moreover, which bears a peculiar stamp, for towards the equator we find the characteristics approach more nearly to the general vegetation of the tropics and especially to that of the East Indies. The species of two genera, the one a Myrtle-bloom, the other a Leguminous plant, the EUCALYPTUS and the ACACIAS with leaves reduced to phyllods, are most generally found, and by their number and their dimensions form, perhaps, the half of the vegetation which covers these countries. We have already described the singular position of their leaves (page 117, note o), which causes such extraordinary effects of light and shade in the woods. The *Leguminosæ*, *Euphorbiaceæ*, *Compositæ*, *Orchideæ*, *Cyperaceæ* and *Filices* are found in great numbers here, but not more so than in any other part of the earth; whilst four others, the *Myrtaceæ*, *Proteaceæ*, *Restiaceæ* and *Epacrideæ* have more representatives here than in any other country. Most of the species of the *Goodeniaceæ*, *Stylidieæ*, *Myoporaceæ*, *Pittosporeæ*, *Dilleniaceæ* and *Halorageæ*, a certain tribe of the *Diosmeæ* and the two small families, *Tremandreæ* and *Stackhousieæ*, are found only in this country.

§ 895. The Islands of New Holland are nearly in the same latitude as this zone which we have just examined and are the nearest land to it. They are of more interest, since a little to the south of them we find the antipodes of Paris, so that they would seem to represent on the other side of the globe a part of our Mediterranean or Olive-tree region. Yet their vegetation has very different characteristics, some being common to that of New Holland, a greater number to that of Polynesia and, consequently, to the tropics. We find in it Palms (CORYPHA AUSTRALIS), Tree Ferns and Tree Dracænas, forests of *Coniferæ* with wide leaves (DAMMARA) and habits very different to ours, along with *Myrtaceæ* (METROSIDEROS). Let us remark, however, that these forests are fast disappearing before the axe, and the culinary vegetables of Europe are speedily taking their place, so that the aspect of vast tracts of country will soon be totally changed.

§ 896. The Cape of Good Hope, lastly, presents a very distinct physiognomy, analogous in some points to that of the Austral countries on account of the presence of the *Proteaceæ*, *Diosmeæ*, *Restiaceæ*, as well as of the Heaths (*Bruyères*), which here seem to replace the absent *Epacrideæ*. But, on the other hand, the *Dilleniaceæ*, the ACACIAS with phyllods and the EUCALYPTUS are wanting, whilst other plants, which are rare or not found in New Holland, have become abundant here and characterize the country, as the *Iridaceæ*, the *Ficoideæ*, the PELARGONIUM, the ALOE, the STAPELIA (a genus of the *Asclepiadaceæ*), the *Brunaceæ*, the *Selagineæ*, &c., &c. Certain *Compositæ*, especially those known as Everlasting Flowers (*Immortelles* [GNAPHALIUM, ELICHRYSUM]), are very numerous.

At the Cape as well as in New Holland there are no very lofty mountains in which we can follow the degradation of the vegetation peculiar to these two points of the globe. New Zealand has mountains high enough to retain the snow always on their tops; but botanists have not yet explored them.

§ 897. Having arrived at this point, we find ourselves brought back to the juxta- and inter-tropical zones, which served us as the point of departure. We have generally referred merely to the large continents and have mentioned only a very small number of islands. We must now say a few words concerning the differences the islands may present when compared with the continents. Those which are of large size may themselves be considered as small continents, but they always, however, present from the great extent of their shore a larger proportion of land submitted to the action of a moister and more temperate climate, which may be termed *Marine.* This difference has necessarily some influence on their vegetation, on which it imprints some peculiar characteristics, mingled with those which it presents in common with parts of the neighbouring continents situated in the same latitude. One of these characteristics is the relative abundance of the Cellular Acotyledons, especially of the Ferns, to which this climate appears to be peculiarly favourable, being both moist and warm. They are found, then, in a greater proportion with respect to the rest, as the island is less in size and, consequently, more completely placed in these conditions of temperature. Thus, in the great island of Jamaica, the number of Ferns, compared with phanerogamic species is as 1 : 10. The proportion is as 1 : 8 in the Isle of France and the Mauritius, as 1 : 6 in New Zealand, as 1 : 4 in Otaheite, as 1 : 3 in Norfolk Island, as 1 : 2 in Tristan d'Acunha. Another characteristic of the island vegetation with respect to that of the continents is, that the whole number of species is less in an equal extent, and so much the less as the island is smaller and farther from the main-land: an almost necessary result of the obstacles interposed by the sea to the transition of a species at first foreign to the soil, which, on the contrary, on an equal extent of continent may arrive and at last be established as it advances nearer and nearer to the neighbouring zones. A marine climate in several places, especially as it leaves the tropics, appears to be hurtful to arborescent vegetation. Its ill effects are most probably increased by the action of frequent and violent winds. This may be seen on our coasts. Iceland, the Shetland and Feroe Isles, have no trees or only some stunted shrubs in a few sheltered spots, whilst these trees in as high and even higher latitude acquire great vigour and form extensive forests on the coast of Norway. We have also seen in the Southern Hemisphere large trees as far as Terra del Fuego, but in the Malonines, although nearer the equator by several degrees, at the most only a few humble shrubs with a flora in other respects almost identical.

§ 898. The fact we have stated at the beginning of this part of the book clearly arises from the details into which we have just entered; a great number of places on the earth present differences in their vegetation independent of the various conditions in which they are placed, as if each of them at first had been the object of a separate creation. Two distant places with an analogous and even identical climate and all the other circumstances, which ought to cause similarity between the natural productions, may nevertheless produce different plants. Each of them at first received its own and not the other plants, though both could have lived there very well. So true is this, that several species carried from one to the other live and flourish there in as great perfection as in their native country. We have mentioned New Zealand as an example (§ 895); we have several in our own country, as the Canada Flea-bane (ERIGERON CANADENSE), which, formerly introduced into Europe as an ornamental plant, has now become the commonest and most troublesome weed of our gardens; there are also several annual plants, which, by the fortuitous sowing of their seed in corn brought from other countries, are so well naturalized in ours that it is now difficult to distinguish them from our indigenous plants. Let us mention two more instances, the AGAVE (commonly known by the name of Aloe) and the Prickly Pear (*Raquette*) (CACTUS OPUNTIA), which cover Algeria, Sicily, a part of the shore of Spain, of Italy, of Greece, so that travellers, struck by the peculiar aspect of these countries, regard them as types of African vegetation; neither, however, were found in Europe or Africa before the discovery of America, both being indigenous to that continent. Our Milk Thistle (*Chardon-Marie*) and Cardoon (*Cardon*) have invaded the plains of the Rio de la Plata; the Chickweed (*Mouron des oiseaux*), the Herb-Robert (*Herbe-à-Robert*), the Cow-bane or Water Hemlock (*Ciguë*), the Common Nettle (*Ortie dioïque*), the Viper's Bugloss (*Vipérine commune*), the White Horehound (*Marrube commun*), are now flourishing in great abundance in the environs of the towns of the Brazils and grow luxuriantly even in their very streets. Almost every country could furnish similar examples of the emigration of certain plants following the footsteps of man. If they did not grow there before, it was not because conditions adapted to their existence were wanting, but because the Almighty, Who planted the earth, placed these vegetables in one place and not in another.

We may conceive that a species, growing in this way from any centre, is propagated as it radiates around that centre so long as it finds conditions suitable for its growth. Different latitudes, chains of mountains, deserts, and especially the sea, are so many natural barriers opposed to its indefinite extension, and enclose it in bounds narrower than are assigned to it by the conditions adapted to its peculiar organization. According to the differences in the constitution, as it were, which allow some and prevent others from living in various habitats, some are scattered over a vast extent, others are concentrated in more or less contracted limits; but we meet with some in very distant parts, separated by natural obstacles of which we have just pointed out some, which they were not able to pass alone. They might have been, as in the cases we have just mentioned, transported by man or by some one of those agents which favour the dispersion of the seed (§ 599). There are some cases, in which this agency cannot be supposed to act; we are from this led to admit that several must have belonged to different centres of vegetation at the same time, and that each of these centres is composed of vegetables, a large portion of which are peculiar to itself, a smaller portion common to several others at the same time. The name of *sporadic* (*sporadiques*) (σποραδικός; *scattered, wandering*) has been given to those vegetables spread over large spaces and in several different countries, of *endemic* (*endémiques*) (ἔνδημος; *dwelling at home*) to those which we find in a single country. Among the former, some exist at very different parts of the same zone, but without passing beyond it (as the SAUVAGESIA ERECTA, which grows in the Antilles, in Guiana, in Brazil, in Madagascar, in Java;) others on several zones at once (as the SCIRPUS MARITIMUS, which grows in Europe, in North America, in Western India, in Senegal, at the Cape, in New Holland; the SAMOLUS VALERANDI, almost as widely distributed). The same epithets may be applied to the genera and families as well as to the species, necessarily in more extended limits. The *Cactaceæ*, confined to Intertropical America (beyond the north of which they scarcely pass), the QUINQUINAS, to a certain zone of the Andes, are examples of an endemic family and genus.

§ 899. If two parts rather far from one another, but placed in analogous conditions, do not present the same vegetation, there exist, nevertheless, between the two kinds of vegetation relations which cannot be misunderstood. The plants, on the one hand, differ inasmuch as they belong to two different centres; on the other hand, are assimilated inasmuch as they are destined to live in similar conditions: thus, the same genera may be represented by different species, the same families by different genera or neighbouring families. Examples may be quoted by thousands; it will be sufficient to mention some, nearly all of which have been already mentioned, as the *Amentaceæ* and the *Coniferæ* of temperate Europe represented by other species of the same genera in the same zone of North America; those of the *Coniferæ* by other genera (ARAUCARIA, PODOCARPUS) in that of South America; the Common Beech, placed near the northern limit of the temperate zone in our Hemisphere; the Antarctic Beech (*Hêtre antarctique*), placed near the south boundary of the Southern Hemisphere: two species of CHAMÆROPS marking the northern limit of the Palms, the C. HUMILIS in Europe, the C. PALMETTO in America; the RHODODENDRON of the Alps, replaced in Lapland by another species, on the Andes by another genus, the BEFARIA; the presence of the *Diosmeæ* in the Austral countries, at the Cape of Good Hope, in the South of Europe, but at each of these points presenting genera different enough to form as many distinct tribes; the *Ericineæ* of the Cape, replaced in Australia by the neighbouring family of the *Epacrideæ*; the *Selagineæ* by the *Myoporaceæ*, &c. &c. We may, therefore, borrowing a word from chemistry, say that in these combinations of families, of genera, of species forming the vegetation of a country, there exist equivalents, there take place substitutions in order to cause that of another country to be analogous though different.

§ 900. In order to prosecute the comparative study of all the divisions of the vegetation comprised in the science of Botanical Geography, it is necessary to determine and describe all the plants of every country. The books written for this purpose have received, since the time of Linnæus, the name of *Floras*, a term also employed in the acceptation in which we have used that of vegetation. The FLORA GALLICA of De Candolle is the work written by that author on the plants of France; the French Flora is a general term for the whole of these plants. Unhappily botanists are generally obliged to confine themselves to the political and not to the natural divisions of the countries which they describe. To arrive, therefore, at more general results we are obliged to connect the Floras of several authors, generally composed with different intentions and on different plans, not containing facts of the same value, and causing doubt as to the identity of the species on account of the difference of nomenclature. That unity is wanting, which we should obtain if each Flora comprised a natural division of the globe.

§ 901. But how to determine these botanical regions clearly? There are some which Nature herself has separated by surrounding them with impassable barriers, Islands, for instance, surrounded by the ocean, as Saint Helena, the Sandwich Islands, Madagascar, &c. &c. The difficulty is the division of the Continents from the Archipelagoes and Islands which are not far from them. There are, doubtless, certain parts surrounded by boundaries, which arrest vegetation as it radiates from its centre, such as seas, deserts, high chains of mountains. But it is rare to find them thus completely imprisoned, for some break, some means of communication is sure to exist, by which plants pass from one place to another and extend into neighbouring countries, thus tending to assimilate their Floras. De

Candolle has proposed a certain number of these botanical regions, and they were correctly founded on the state of science at that time, before so many facts were known as at the present day. Travellers had generally herborized only round certain parts distant enough from one another for each to present a peculiar appearance in its Flora. The botanist, who collected plants around Rio Janeiro, then around Buenos Ayres, then in the country around Magellan's Straits, would find three very distinct centres. But if he pursued his herborizations by land and searched all the parts intermediate between Rio, on one hand, and proceeded northwards to the sea of the Antilles, on the other, southwards to Cape Horn, he would have seen the Flora of Patagonia insensibly blend with that of the Argentine Republic, the latter with that of the central provinces of Brazil, this with that of the northern provinces and of Guiana, so that it becomes impossible to assign any fixed limits to each of these regions. The same thing would take place as he advanced from the East to the West from any point of the shore of the Atlantic to the Great Cordilleras. The southern end of Africa, that well characterized region, so long as we do not go far from the Cape of Good Hope, has been the less clearly defined as discoveries have increased, as we ascend from the colony towards the equator. We thus find that these regions are not so decidedly marked as was supposed. This is so true that in 1820 twenty regions only were formed, whilst fifteen years afterwards the son of De Candolle, adopting those by his father, was obliged to form forty-five.

M. Schouw, one of the authors who has written the most largely on the Geography of plants and has contributed the most to its advancement, has tried to form better founded rules to determine regions, which according to him ought not to be elevated to this dignity, except one half of the whole number of the species found in each of them, as well as one fourth of the genera and some families, be peculiar to it. If we again find in any other part a few species of several of these characteristic genera and families, they are only few and far between, whilst they are the most numerous in that region, to which they give their name. According to this principle he at first established eighteen regions and afterwards twenty-five, which he named some like De Candolle from the geographical situation, the greater part from the vegetation, which form their distinctive feature from their greater numerical proportion or their remarkable appearance. Some may be subdivided into provinces, which themselves must be distinguished from one another by a fourth of the species and a few genera which are peculiar to each. Thus, the *region of the Labiatæ and of the Caryophyllaceæ* (corresponding to that we have called the *region of the Olive-trees*) is divided into several provinces, that *of the Cistus* (*Ciste*), (the Spanish

Peninsula), that of the *Scabious and of the Sages* (the South of France, Italy, and Sicily), that of the *frutescent Labiatæ* (the Levant), &c. &c.

§ 902. We have reviewed the different countries of the earth in a superficial manner. We may in the study of Botanical Geography instead of this order follow one somewhat inverse, when Botany in its turn guides Geography; in this we take the families one by one and examine how each is dispersed over the globe. By this general comparison we verify some of those truths we have already mentioned as to the concentration or the dispersion of certain species, genera and families, and determine their relative proportion either over the whole earth, or over great divisions or parts, or in each of its well-known countries. The term, *Botanical Arithmetic*, has been applied to the determination of these proportions by M. de Humboldt, who, laying aside a few earlier attempts, deserves to be called the founder of the science of Botanical Geography, upon which he has cast so much light by his labours and discoveries in Meteorology, by the rich and learned results of his protracted travels and by his example leading so many minds to prosecute the enquiries commenced by him. We may compare in any Flora, supposed to be perfect, the number of the species of one family in particular either with that of another, or with the total number obtained from the whole of the families. When this calculation has been made at a certain number of places conveniently chosen, we recognise a certain constancy in the relations between Floras placed on the same isothermal line, so that the knowledge of the number of the plants of a family at any place would give us to a certain extent some idea of the rest of the vegetation, if the isothermal line is known; and reciprocally, of the isothermal line, if the whole number of the plants is known. We are, doubtless, very far from possessing that degree of knowledge which would enable us to draw up tables, explaining the Botany and Meteorology of the different parts of the globe by means of one another. The facts of both sciences require verifying with great precision, although the results already obtained may throw some light on the questions they do not decide. We will only mention a few general connexions of the numbers in this distribution of vegetables on the surface of the earth.

§ 903. It is an established truth that the absolute number of the species augments progressively from the poles to the equator, where we find the greatest number. We must not, however, suppose that this greater proportion necessarily results from the fact of their being under a lower latitude. The scanty vegetation of large tracts of country situated between the tropics compared with the rich Flora of temperate countries (that of Arabia compared with that of France or of the Cape of Good Hope, that of the North with that of the South of New Holland) would disprove this assertion. But it is evident, that if a tropical country is intersected by valleys and mountains, it will correspond to a larger number of zones beginning from the foot of these mountains, and that the diversity of the plants will be increased in proportion to that of the conditions in which they are found. The Flora of the East Indies has been remarkably increased by exploring not only the Ghaut and the Nilgherry but also the sides of the Himalaya mountains. Intertropical America has been called the promised land of botanists, on account of the wonderful and almost inexhaustible variety of the plants, which are produced by the various accidental conformations of the country. Whilst the great chains of Asia, running from east to west, will for the most part be in nearly the same latitude, the Cordilleras of America, running from north to south, not only present in the same latitude the succession of every vegetable zone, but also at each point a very different latitude and, consequently, fresh details in the vegetation. The secondary chains, which are detached from them, others which cross them in every direction, the numerous water-courses which pour down them, the extensive valleys traversed by the largest rivers in the world, are so many powerful causes of fertility and variety; and we must not be astonished at Mexico, Columbia and especially Brazil containing in an equal extent of land more species totally different from one another than the greater part of other places on the earth.

§ 904. These more numerous species, dispersed over the land between the tropics, necessarily correspond to a greater number of families and of genera, which diminish gradually as we approach the poles. But, as each genus is then represented by a less number of species, the number of genera becomes larger in proportion to that of the species. Thus, the French Flora now contains more than 7,000 species forming more than 1,100 genera; that of Sweden rather more than 2,300 species forming 566 genera; that of Lapland rather less than 1,100 species forming 297 genera; so that the mean number of the species in each genus in France is about 6·36363, in Sweden 4·063, in Lapland 3·703.

§ 905. The absolute number of ligneous species and their proportion to the herbaceous species is also augmented as we approach the equator. The number of annual or biennial species increases, therefore, in a contrary direction, but not so far as the poles. Temperate climates seem to be best adapted for their delicate nature. They are found there in the greatest abundance and then again begin to decrease towards the poles. We have seen that they disappear in the coldest zones, whether of latitude or of height, where the greater part of the plants are perennial or subfrutescent.

§ 906. A corollary of the preceding proposition is that the height of plants also increases from the poles towards the equator. But this rule seems to be inverted in the Algals, which, though very small in the tropical, acquire enormous dimensions in the polar seas. At Cape Horn a species has been found that attains a length of nearly 330 feet[a].

§ 907. Let us now find the relative proportions of the species belonging to the three great branches of the Vegetable Kingdom under different latitudes. If we examine the numbers given by the Floras, we shall be led to admit the following law, that the number of Cryptogams or Acotyledons relatively to that of Phanerogams or Cotyledons increases as we leave the equator. According to the tables given by M. de Humboldt for the middle parts of the three great terrestrial zones, the Cryptogamic species would be equal in number to the Phanerogamic in the glacial zone (from 67° to 70°), one half as many less in the temperate zone (from 45° to 52°), to nearly eight times less in the equatorial zone (from 6° to 10°), the proportion for the plains being as 1 : 15 and for the mountains as 1 : 5. The latter numbers confirm the former. But we must state that in the Floras the number of Cryptogams is far from being so precisely fixed as that of the Phanerogams; that of the former increasing much in researches that add little to that of the latter (as in the Flora of Paris); that the different countries of Europe have been better explored by resident botanists than foreign countries could by travellers, whose notice would not be greatly attracted by such obscure plants as the greater part of Acotyledons; that Acotyledons near the poles attract more attention from botanists than Cotyledons, because the latter are quickly described, and botanists sooner turn their attention to the former: that the proportions will necessarily be somewhat modified by these partial investigations, which, prosecuted with the same zeal and industry in the tropics, would doubtless furnish us with results somewhat different from those we now have, either over the whole earth or in each zone, principally in the warmest. Moreover, all that we have just said is particularly applicable to the Cellular Acotyledons. We shall see that the distribution of the Vascular ones follow other better known and surer laws.

§ 908. On comparing the relative proportion of the two great branches of Cotyledonous Vegetables with one another, we find that the relative proportion of Monocotyledons increases as we leave the equator. As far

[a] This plant is the MACROCYSTIS PYRIFERA. Captain Cook mentions specimens 120 feet long. But they are very frequently found 300 feet in length, and in the Flora Antarctica some are mentioned, which were estimated to be upwards of 700 feet in length; they were observed in a strait between two of the Crozet Islands. It forms natural barriers, so strong that a boat can hardly be forced through them. It is also useful in making natural breakwaters, as the water inside these barriers is perfectly smooth. The stem is round, slimy and smooth, and seldom has a diameter of so much as an inch.—TRANS.

as 10°, it was relatively to the whole of the Phanerogams nearly as 1 : 6 in the New World, as 1 : 5 in the Old. Increasing progressively it becomes as 1 : 4 towards the middle of the temperate zone, as 1 : 3 near its

Groups or Families.	Proportion to the whole of the Phanerogams. Equatorial Zone, lat. 0°—10°.	Temperate Zone, lat. 45°—52°.	Glacial Zone, lat. 67°—70°.	
Juncaceæ	As 1 : 400	A 1 : 90	As 1 : 25	The proportion increases from the equator towards the pole.
Cyperaceæ	Old World 1 : 22 New World 1 : 50	1 : 20	1 : 9	
Gramineæ	1 : 14	1 : 12	1 : 10	
Amentaceæ	1 : 800	Europe 1 : 45 America 1 : 25	1 : 20	
Ericineæ	1 : 130	Europe 1 : 100 America 1 : 36	1 : 25	
Euphorbiaceæ	1 : 32	1 . 80	1 : 500	The proportion increases from the pole towards the equator.
Rubiaceæ	Old World 1 : 14 New World 1 : 25	1 : 60	1 : 80	
Leguminosæ	1 : 10	1 : 18	1 : 35	
Malvaceæ	1 : 35	1 : 200	——	
Cruciferæ	1 : 800	Europe 1 : 18 America 1 : 60	1 : 24	The proportion diminishes from the temperate zone towards the pole and the equator.
Umbelliferæ	1 : 500	1 : 40	1 : 80	
Labiatæ	1 : 40	Europe 1 : 25 America 1 : 40	1 : 70	
Compositæ	Old World 1 : 18 New World 1 : 18	1 : 8 1 : 6	1 : 13	
Ferns	Flat Countries 1 : 20 Mountainous Countries 1 : 3 to 1 : 8	1 : 70	1 : 25	

boundaries. But it rather decreases in the glacial zones, as in Greenland. It is clear that the proportion of Dicotyledons is inverse and may be expressed by the reciprocals of the preceding fractions. The increase of certain families and the diminution of certain others bring about these results; this will be understood from the opposite table borrowed from M de Humboldt. It indicates with respect to the whole of the Phanerogams (taking the average of the three great zones) the proportions of some of the commonest families, the most important from the number of their species, the contingent of which, varying according to the zone, will of course cause the greatest difference in the relations.

§ 909. These plants, belonging to various families whose species vary according to the country, impart a peculiar appearance on their native land by their different combinations. But this depends also on another cause to which we have not yet alluded;—this is the number of the individuals of a species living on a certain extent of ground. In every country, if any one considers with attention the surrounding vegetation and, not contented with a glance cast over the whole, endeavours to analyze its details, he will immediately find that among the plants composing it, some exist in great abundance, that such a species covers a large space, with its crowded individuals, whilst those of such another are only few and far between. Hence the number of certain species at one part or the increase of one at the expense of every other is the cause of the variety or monotony imparted to the mind through the eye. The plants which live close together in this manner are called *Gregarious Plants* (*Plantes Sociales*), a term borrowed from Zoology; if we find here and there a specimen separated at a great distance from any other, it is a very rare exception. Their presence always indicates the same nature in the soil which they cover; the line at which they stop, a change in the soil. This is easily verified on the banks of certain streams. Along canals, in which the level is nearly constant, the banks at different heights are in different conditions of moisture and frequently also of soil: so we see certain species of *Juncaceæ*, of *Cyperaceæ*, of *Gramineæ*, placed above one another in regular, narrow, parallel bands, each composed of the same species. This regular position is observed on a large scale along the magnificent streams of Equatorial America, where the navigator for several whole days has the monotonous aspect of continuous lines of great trees, each species of which invariably occupies a seat of a different height. Certain species of JUNCUS, of CAREX cover whole marshes; and on the edge of our stagnant waters crowd the ARUNDO PHRAGMITES, the SCIRPUS LACUSTRIS, forming a kind of girdle beyond which the depth becomes too deep on the one side, too shallow on the other, for them to prosper. Our readers are familiar with the Furze

(*Ajonc*) (ULEX EUROPÆUS,) which covers the commons, and the Heath which, giving its name to the hills of the North, clothes the soil with its fragrant, reddish-brown carpet. This mass of vegetation, formed by a single species, indicates great facility of reproduction and tenacity of life in the plant, as well as great sterility in the soil. If any other plants grow there, they are soon smothered by the gregarious plant, whose domain they have usurped, or else are very rarely met with. We have mentioned some of the commonest species of our own country, but every other possesses some plants which invade certain districts to the exclusion of others; these districts are known by different names according to the plant and the country. Several are sometimes found together, and there are many, though forming the staple of the vegetation, which allow a large number of other species to be nourished by the same soil.

§ 910. We are now naturally led to examine another influential agent in the distribution of plants, THE SOIL; we have hitherto neglected it, since we have considered the great regions of the globe as a whole, and since the variations resulting from the differences of the ground are much more local and numerous in each of these regions, frequently in very limited spaces. Under the general name of *soil*, we must understand *every medium in which a plant can grow*, so that we shall sometimes find water designated as such.

§ 911. Let us begin with the sea, in which we have seen growing a part of the Algæ (§ 745), those which we commonly know as FUCUS, and which, clinging to the bottom or to the rocks, but not rooted to them, absorb their nourishment from the salt water which surrounds them. Some even float freely, not being attached to anything. Among Phanerogams, the *Zosteraceæ* alone (TABLE II) are marine plants.

§ 912. Among fresh water plants, we find another part of the *Algæ* (§ 745), some floating freely, the greater part rooted in the bottom, the *Characeæ*, *Rhizocarpareæ*, some *Mosses* and *Hepaticæ;* among Phanerogams, almost all the species of Monocotyledonous plants having a seed without a perisperm and a perianth, either altogether wanting or herbaceous (TABLE II); others having a seed with a perisperm, as the *Pistiaceæ* and certain *Typhineæ;* among Dicotyledons, the *Ceratophylleæ*, *Nymphæaceæ*, *Nelumboneæ*, *Cabombaceæ*, the greater part of the *Halorageæ*, *Utricularineæ*, &c., &c.

§ 913. The greater part of these plants elevate their flower-bearing tops above the surface of the water and thus form an almost insensible transition to the plants of the marshes or banks, which have their lower part only under water, their inflorescences and frequently a part of their leaves above: as the *Juncaceæ*, *Alismaceæ*, *Butomaceæ*. The *Gramineæ* and *Cyperaceæ* furnish numerous examples. Let us also mention the

Orontiaceæ, *Pontederiaceæ*, some *Lycopodiaceæ*, the *Iridaceæ*, *Orchideæ*, *Polygoneæ*, *Caryophyllaceæ*, *Cruciferæ*, *Ranunculaceæ*, *Lythrarieæ*, *Rosaceæ*, *Onagrarieæ*, *Umbelliferæ*, *Plantagineæ*, *Scrophulariaceæ*, *Labiatæ*, and *Compositæ*. Some prefer stagnant water, some extensive sheets, as lakes or large ponds, others crowded together inhabit shallow pools and ditches; some delight in running streams; others, in the cold water supplied by the eternal snows and glaciers, as the pretty species of *Saxifragaceæ* and other Alpine Plants, that inhabit the banks of brooks in those high regions.

Salt water, destructive to most plants, is, on the contrary, necessary for the existence of several, which flourish on the sea shore, some of which even throw out their roots to be bathed by the tide, as the AVICENNIA and the Mangrove-tree (*Manglier* [*Rhizaphoreæ*, TABLE XI]); these trees are exceedingly gregarious and are common to the shores of all the tropical seas, to which they impart a singular appearance from their strong roots arising above the water and forming, as it were, so many arched buttresses, from the centre of which rises the stem.

Peat bogs are certain marshes of a peculiar nature covered with gregarious plants, the roots of which are closely matted together and at last form a kind of spongy moveable sod, the principal part of which is mostly composed of the species of a genus of Moss[b], the SPHAGNUM, in which flourish certain plants (DROSERA, OXYCOCCUS, some species of SALIX, &c.; and a few Ferns, as the OSMUNDA REGALIS). The vegetation of each year raises the surface, and the sods of previous years are more and more buried, at the same time ceasing to live. The remains of these plants, protected from the action of the air, are not decomposed, but at last constitute with the mud, which connects them during life, rather a compact mass; this is used for fuel under the name of *Peat* or *Turf*.

Certain plants are found both in places covered with water and on dry ground. Several marsh plants have this faculty of accommodating themselves to their position and are called *amphibious*. Some, designated by the peculiar epithet of *inundated*, grow on land alternately covered with water and laid dry. The leaves of amphibious plants are subject to vary in form, according as they grow in water or air: the Water Crowfoot (RANUNCULUS AQUATILIS) is a good example for exemplifying this fact.

§ 914. We have already spoken (§ § 324, 325) of the influence which the nature of the soil exercises on vegetation; but we have only considered it with regard to its being the food of plants, and we have now to see how it can affect the distribution of their species, genera and families. Soils,

[b] We may mention as a singular fact on the authority of Darwin, that neither SPHAGNUM nor any other species of *Moss* was found to compose any portion of the peat in South America.—TRANS.

different in their chemical composition, present some slight differences in their spontaneous productions. Thus, calcareous, siliceous or argilaceous soils have doubtless a few plants which are peculiar to each of them, but not numerous or constant enough to form a clear distinction between their Floras. It is otherwise with salt soils: they are covered with certain species, and several of them assume characteristic shapes in their short, thick foliage, as the SALSOLA, SALICORNIA. Other *Atripliceæ*, some *Cruciferæ* (CRAMBE and CAKILE), some *Primulaceæ* (SAMOLUS and GLAUX), some species of STATICE, are also abundant on the sea shore, and we have remarked that the same or analogous species (§ 324) are found inland on saline soil.

But generally the composition of the soil modifies its physical properties, rendering it more friable or more compact, more or less permeable to air and water, better adapted for retaining or allowing the latter to escape; so that the same soil may be favourable or adverse to the growth of the same plant in two opposite climates, and reciprocally the same plant will require soil of a different nature in these two climates. Thus, Kirwan has shewn that in a dry climate Wheat prefers aluminous soil because it is more affected by the moisture of the atmosphere; in a wet climate, siliceous soil, because it is less so.

§ 915. We may say almost as much concerning the relations between the geological formation of the soil and its vegetation. Since the soil is prepared and elaborated to a little depth in the superficial layers of the earth, Geology, teaching us the origin as well as the nature of this layer and of that on which it reposes, gives us facts very valuable in the majority of cases; but it neither can nor will enter into purely local details, which are frequently changed by physical accidents. Thus, on geological maps we paint with the same colour several of the plains around Paris, over which a layer of Millstone grit (*Meulière*) extends. Yet, if we compare the vegetation of Montmorency covered with cornfields, with that of Sannois covered with a short, sterile sod, or with that of Meudon covered with hard wood, principally with Spanish Chestnut-trees, amongst which flourish the AIRA FLUXUOSA, the MELAMPYRUM SYLVATICUM, the PTERIS AQUILINA, we shall be struck with the great difference between the plants of these places; this difference results sometimes from the Millstone grit being accompanied with clay, sometimes from its very thin layer reposing on sand, and sometimes from its being frequently uncovered. There can be no doubt, however, but that the excellent geological maps, such as those possessed by several countries of Europe and especially by France, will prove of great use in herborization and assist us in determining at some future time relations which are now but vaguely understood.

§ 916. The quantity of water contained by the soil plays the most important part in vegetation; if there be none, we shall find no plants. Thus, the interior of Africa is occupied by large deserts without vegetation at any season: for water-courses are wanting and under this latitude the vapour of the atmosphere, suddenly rarified when placed in contact with the burning sand, is not condensed into rain. But at a few places where some water moistens the soil, it is covered with vegetables and forms an *Oasis*, a kind of island in the midst of a sea of sand. In climates at a greater distance from the equator or tempered by the neighbourhood of large ranges of mountains, rain may form and furnish the wide plains with water, which are not otherwise irrigated. The country, therefore, after having presented the appearance of a desert during the dry season, is suddenly covered with a rapidly developed vegetation, generally composed of gregarious herbaceous plants.

We have mentioned (§ 875) the Pampas and Llanos of the centre of South America. The Savannahs or Prairies of North America, the Steppes of Siberia and Tartary may be compared to them, excepting the differences arising from their being situated in the temperate zone, which subjects them to the changes of our seasons, and from their vegetation originating from such distant centres. Among these deserts of the centre of Asia, there are vast expanses of country impregnated with salt, and these produce peculiar vegetables analogous to those of the sea shore: the sea at some anterior epoch has, doubtless, covered them. The Lands in France and the Heaths in England represent, happily on a less scale, these dry, sterile places. On certain low coasts, the wind which is continually blowing over the sea, pushes landwards the sand, which collects in little heaps in parallel chains: these gradually advance and gain every year on the vegetable soil, which is buried: thus are formed the DUNES (Appendix B). But their sterility is not irremediable, thanks to the moisture of the soil kept up by the sea wind. Trees, such as species of the Pine, may prosper there and render a double service by preventing the further encroachment of sand and rendering the land useful. To prevent this encroachment (as in Holland) Creeping Gramineæ are employed, those being preferred which increase very fast, as the ARUNDO ARENARIA. When the land is once fixed, several cultivated plants will be produced.

§ 917. We know that the mineral elements of the soil, with the water which penetrates it, are associated with the *débris* of organized beings to form the real vegetable soil, that soil, the richness of which exercises the greatest influence on vegetation. The presence of vegetables at a given place is a guarantee (and so much the more so as they leave a quantity of *débris* themselves) of the succession and multiplication of other individuals, which will be again favoured by the presence of animals enticed thither for food and shelter. But, before forming this layer of humus, it was necessary for some few plants to be established and to deposit the first manure; they thus prepare the soil for others, which in their turn enrich the first deposit, augmented by the successive generations of the same or different plants, the variety of which increases in the same proportion. At whatever point this progression may be arrested, the quality of the original soil influences the admission of the first colonies of plants and, consequently, the general nature of the vegetation in the end.

§ 918. The nature of the soil determines a large number of the stations of plants. They inhabit the sea, its shore impregnated with saline deposits, or land distant from the sea but still impregnated with the same deposits: fresh water, stagnant in large or small quantities, flowing in brooks or rivers; their shores, marshes, morasses, peat-bogs; rocks; sandy places, the chemical composition of which may vary, but which is mostly siliceous: places sterile on some other account, because the soil is hardened, on the contrary, by the heat into a mass too compact for the roots of other plants to pierce: soil, in which clay, chalk, gypsum or some other element predominates, formed in the place either by inundations, by disintegration, by volcanic or some other action. At other times the indication of the station is borrowed from the association of the plant with others combined together in a certain manner. We thus distinguish those growing in forests, in meadows, in hedge-rows, in cultivated land (PLANTÆ ARVENSES), &c. We here find the influence man exercises over the distribution of vegetation, since he has artificially determined these last combinations. But he also distributes them involuntarily. Certain wild plants, certain noxious weeds, which he would rather extirpate than propagate, accompany him everywhere and multiply around his dwelling, such as the Nettle, several species of CHENOPODIUM and RUMEX, the Mallows, the Chickweed, &c. Their presence in the midst of a desert country, at the top of high mountains, indicates that he has once passed, and that some shepherd has, perchance, raised his lonely hut there. There are plants, which crown the tops of walls; others (like the Pellitory [PARIETARIA]) establish their roots in the fissures and slight projections; some run at their feet, seizing upon and covering the rubbish (PLANTÆ RUDERALES).

§ 919. Civilized man, for whom the spontaneous productions of the country are not sufficient, and who delights in collecting around him vegetables and animals useful to him or pleasing to his sight, and in destroying such as are noxious or displeasing to him, necessarily modifies to a certain extent the distribution of these beings and the appearance of Nature. We see her thus altered mostly in Europe, where a place must be either totally sterile or quite inaccessible to be wholly abandoned to itself. Forests in a state of nature have a great tendency to run over the soil, as may be seen in the South of Chili, where thickets, once established on the edge or in the middle of the prairies, advance every year, contract the space occupied by the Gramineæ and at last completely replace them. The contrary takes place in cultivated countries. Forests, which at first cover the greatest extent, disappear gradually before the footsteps of man; and those which are preserved, subjected to his rule, have neither the same appearance nor the same influence on the surrounding vegetation. The conditions of the climate have thus been changed; those of the soil are unceasingly altered by culture, which, moreover, regulates what species shall cover it. Several of those forming the indigenous Flora are destroyed, at least in places; others, on the contrary, are introduced; these are generally Cereal and other agricultural vegetables brought from distant countries. But whatever may be these modifications, they are not so great as to deprive nature of the power of exercising her rights; she is the directrix of man: the abundant growth of indigenous, the flourishing of cultivated plants are a double indication, by which she may be known. The latter even furnish excellent materials for the study of Botanical Geography: when we use them, however, we ought to remember that human industry can push the cultivation of plants beyond their natural limits; but these limits thus extended preserve their relations with regard to the different species. We must also remember that the absence of any plant does not always suppose the impossibility of its being cultivated there, but that other places are more advantageous. A plant is always cultivated with most success in its native country; it is probable that it was cultivated there the first. The analogous climates are the next in rank, and in proportion as we leave this zone, it becomes more and more difficult to rear it. Having regard to these facts Agriculture and Botanical Geography will throw mutual light on one another: the latter will borrow from the former well-defined points of reference. If we see certain indigenous vegetables accompanying such or such a kind of cultivated plants, when we find them in some other part we might thence conclude that this same kind of cultivated plant would also succeed there.

§ 920. In the short space now remaining for the rapid examination of the rules of the distribution of cultivated vegetables, we shall confine ourselves to a small number, to those used for the food of man and, consequently, the most spread over the globe. We shall borrow most of the following details from the excellent work of M. Schouw.

The cultivation of *Cereals* (§ 700) is extended northwards in Norway and Sweden as far as 70°, close to the limit where trees cease. It is the only part at which it passes the polar circle; it stops on this side of it at every other part of the globe, about 60° in the West of Siberia, about

55° towards the East; near the western side, it does not reach Kamtschatka, i. e. as far as 51°. In America, it may reach 57° on the western coast, as we find in Russian America; but on the eastern it will not rise higher than 50° or at the most 52°. The line, which circumscribes it on the north in both continents, follows the same inflexions as the isothermal lines.

Barley ripens at this limit, to which *Oats* also approach; but the harvesting of the latter is less sure and only succeeds once in several years. Their seeds form the food of man in the North of Scotland, Norway, Sweden and Siberia.

Farther south, *Rye* is cultivated; it is also found as far north as Oats in Norway and Sweden. It is the most esteemed in this part of the cold temperate zone, which is formed by the South of Sweden and Norway, Denmark, almost all the land bordering on the rivers flowing into the Baltic, the North of Germany and a part of Siberia. We now begin to find *Wheat*, and then Oats are cultivated only for the food of horses, Barley, for the brewing of beer and ale.

Then begins a large zone in which *Wheat* is grown almost to the exclusion of Rye. This zone comprehends the South of Scotland, England, the Centre of France, a part of Germany, Hungary, the Crimæa and the Caucasus, and those parts of Central Asia under cultivation. Since the Vine flourishes in a part of this zone, wine replaces beer and, consequently, Barley is not so much grown.

Wheat extends farther south; but then we find along with it *Rice* and *Maize:* these are found in Spain and Portugal, a part of the South of France (especially that part which borders on the Mediterranean), Italy, Greece, Asia Minor and Syria, Persia, the North of India, Arabia, Egypt, Nubia, Barbary and the Canary Isles. In the last mentioned countries Maize and Rice are most generally cultivated towards the South, in some along with the *Sorghum* and the Poa Abyssinica. *Rye*, in this double zone of corn, is confined to rather lofty situations on the mountains; *Oats*, also; but they soon give way to *Barley*, on account of the preference given to it as food for horses and mules. At the eastern extremity of the ancient continent, in China and Japan, on account of some reason which appears inherent in the habits of the country, our seeds are abandoned for the almost exclusive culture of Rice. It also predominates in the southern provinces of the United States; but that of Maize is more general in the rest of this part of America than in our continent.

In the Torrid Zone, *Maize* in America and *Rice* in Asia are the principal grains: this distribution is doubtless connected with the original centres of these two Gramineæ. Both of them are also cultivated in Africa.

In the Southern Hemisphere, the temperate regions of which would doubtless admit the greater part of these plants, they will be scarcer on account of the less civilized state and the thinness of the population, and depend partly on the seed brought by the colonists. In the South of Brazil, in Buenos Ayres, Chili, the Cape of Good Hope, New South Wales and New Holland, *Wheat* predominates, and *Barley* and *Rye* are found farther south and in Van Dieman's Island.

Let us examine the distribution of Cereal plants over the different zones in height on mountains; we shall find it analogous to that over the different zones of latitude. Let us take an example presenting both phases at once, the Andes under Equatorial America. *Maize* predominates to the height of 3,000 or 6,000 feet, but it grows about 1,000 feet higher. Between 6,000 and 9,000 feet, the Cereals of Europe are most generally found: *Rye* and *Barley* towards the top, *Wheat* lower down.

It is clear that we must give most importance to the extreme limit of height or latitude. The other limit proves nothing, except that the cultivation of one kind of grain is abandoned as soon as we find conditions adapted for the production of a superior one. Nevertheless, from some experiments of MM. Edwards and Collin, it would seem that, besides the limit assigned to the different species by the minimum of heat necessary for their fructification, there exists an inverse one assigned by the maximum of heat, which, when passed, hinders their developement. According to these authors, this limit is a mean temperature of 18° for certain species, a little more or even 22° for certain others: and the observations of the heights under the tropics, at which this culture ceases, would verify this conclusion. Would not the few exceptions, that occur in climates with a temperature superior to this maximum, depend on the Cereals having been cultivated during a season, whose mean temperature has been lower then the usual one? However this may be, on examining the northern limits of the different Cereals and following them over the whole series of the places where they have been established, we shall see that we may say in a general manner that they are parallel to one another and almost follow the inflexions of the isotheral lines, *i. e.* of lines traced through the places, where the mean temperature of the summer is the same. The ripening of all these annual plants is regulated both by the length and heat of the summer.

§ 921. The *Potatoe* (§ 849) at a recent period has spread into every cultivated country and has been added to the farinaceous aliments derived from the seed of the Cereals, so much so as in some countries to replace them. It may be cultivated as far north as the Cereals, and even farther, if the early varieties, capable of being brought to maturity by a short summer, are planted. It is, therefore, cultivated in Iceland and on high mountains in Europe, where the Cereals would not come to perfection. In hot countries, on the contrary, the *Potatoe* easily degenerates and is not cultivated except where the height is sufficient to produce a suitable climate. According to M. Humboldt it is cultivated on the Equatorial Andes at a height of from 9,800 to 13,110 feet.

§ 922. In Upper Peru, the *Quinoa* was commonly cultivated before the arrival of the Europeans, and even now, though more rarely, for the sake of its farinaceous seeds. It is a species of the genus CHENOPODIUM, which belongs to the family of the Atripliceæ.

§ 923. The seeds of several species of the genus POLYGONUM, the type of the neighbouring family of the Polygoneæ (§ 792), are used as food by the inhabitants of the mountains and high table lands of the North of Asia, of which these plants are natives. Their seeds are also farinaceous. One of them, the Buck Wheat (*Sarrazin*) (POLYGONUM FAGOPYRUM), is widely distributed through the North of Europe, particularly in Brittany, where it forms the principal food of the peasants.

§ 924. The inhabitants of some mountainous districts, of the Apennines in Italy, of Cevennes and Limousin in France, are supported for a part of the year by Chestnuts. The *Chestnut-tree* (*Châtaignier*) (§ 777) grows spontaneously in all the mountainous countries of the South of Europe, in Asia Minor and the Caucasus, and is cultivated very far beyond its natural limits. But, it requires for the maturation of its fruit a certain degree of height for rather a protracted time. Beyond London and Belgium, about 51° north latitude, the fruit does not ripen, and the tree is cultivated only for the sake of its wood or as an ornament. Since it must as a tree undergo the whole force of the winter, its boundary in the north is most probably marked by an isochimenal line. But it is also impatient of heat: even in Italy it only grows on the slopes of the mountains and is never found on the Atlas.

§ 925. Between the Tropics, in all those countries which are situated very little above the level of the sea, other vegetable products form the food of man, because the quantity of the nutritious principle furnished by them is generally much larger in a given space, thus indulging the love of ease and the aversion to toil of the inhabitants of these scorching regions. We have mentioned: 1st, the BANANA (§ 769), which is cultivated for its fruits as far as Syria; it can hardly ripen its fruits on the Andes at a height of 6,500 feet, when the mean temperature falls to 18°—20°: 2nd, the *Date* (§ 762), a Palm of North Africa, where certain nations feed on its fruit, which cannot ripen beyond a line drawn from Spain to Syria, about 29° or 30° north latitude, although the tree may vegetate a few degrees farther north: 3rd, the *Cocoa-nut-tree* (§ 762) a native of South Asia, now spread like the Banana over the whole intertropical zone, but flourishing only on the sea shore, far from which it will not grow. It requires a mean temperature of more than 22°, and is, therefore, arrested nearly at the commencement of the Cereals; it furnishes several countries, such as Hindoostan and Ceylon, with important articles of food and commerce: 4th, the *Bread-fruit-tree* (§ 778) is the food of the greater part of the inhabitants of the South Sea Islands, to which it is indigenous; it has been transported to the Antilles, to Brazil, to Guiana and to the Mauritius, but as it is exceedingly tender and easily affected by cold it cannot pass beyond the 22° or 23° of latitude.

§ 926. We have mentioned a few plants cultivated for their farinaceous roots: the Winged Yam (*Igname*) (§ 768), a native of the Indian Archipelago, whose culture hardly extends beyond ten degrees on each side of the equator in the Old World: the *Batata* (§ 951), originally from India, succeeds in our temperate climates, although not cultivated in large quantities beyond the warm zone, *i. e.* beyond 41° or 42°; the *Cassava* (§ 783), extending from Brazil to the western shores of Africa, is cultivated in America as far as thirty degrees on each side of the equator and cannot exist on mountains at a greater height than 3,300 feet.

§ 927. We have seen, whilst studying the families, to what an extent fermented and alcoholic liquors are consumed by man, who procures them from some vegetable or other in nearly every country of the world. We shall here treat only of the most important, the *Vine* (§ 807), with respect to the limits of its cultivation in large quantities for the making of wine. This limit appears to have extended farther northwards than at present, since wine was made in Brittany and Normandy, where they no longer make it. This cessation seems to have arisen less from the deterioration of the climate, as some pretend, than from civilization substituting better vintages by improving the means of conveyance. The line, where the cultivation of the Vine ends at present, begins on the western coast of France, towards Nantes (47° 20′); thence it rises as far as Paris (49°), a little higher still in Champagne and on the Moselle and Rhine, as far as 51°; then, after several undulations passes into Silesia about the same degree; then descends southwards into Hungary at 48° or 49°, where it is sustained at the same latitude as far as the Crimea and the North of the Caspian, where it disappears. The southern limit of the Vine is at the Canaries, about 27° 48′, then it follows the shore of Barbary, is interrupted and reappears in a very small part of Egypt and then is very abundant in Persia at 29° and even at 27°. It does not ripen in Japan, nor is it cultivated in China, where it would doubtless come to maturity, were not that vast empire devoted to the consumption of tea.

In the Southern Hemisphere and in America it has been cultivated with success; but only at a few separated places according to the whims and customs of the colonists, and not on a scale large enough for its actual limits to be fixed by nature. In North America, where the first navigators found several indigenous species of the Vine, the northern limit of its cultivation does not rise beyond 37° on the banks of the Ohio, 38° in New California; its southern limit, 26° at New Biscay, 32° at New Mexico. In the Southern Hemisphere, where at no part does it reach 40°, we observe it at Chili and at Buenos Ayres; about 34° in New Holland and at the Cape of Good Hope, so celebrated for its wine.

On the mountains of Europe, it ascends to 990 feet at the most in Hungary; in the North of Switzerland to 1,700 feet; does not pass beyond 2,100 feet on the southern declivities of the Alps, and may approach 3,100 on the Southern Apennines and in Sicily, though it does not rise higher than 2,600 in Teneriffe.

From all these facts we may conclude that the Vine requires a temperate climate, and depends less on the mean temperature of the year than on that of the summer, which must have a certain strength that the plant may ripen its fruits, and a certain length that it may receive rather a high temperature in the autumn, at which time this ripening is finished. Is there no place in the tropics where it can meet with the favourable conditions? Modern observations seem to decide the question in the affirmative, since, beside several places already mentioned (as one of the Cape Verd Islands, St. Thomas, near the coast of Guinea and Abyssinia), wine, which is much praised by travellers, is made on the western coast of South America about 18°, 14° and even 6°. We might suppose that the height would compensate for the latitude; but this could not be true everywhere, especially when the Vine is planted near the sea shore. The chief thing seems to be, that the climate be very dry, for moisture totally precludes its cultivation.

It is cultivated in various ways. Sometimes, the plants are left to themselves, sometimes they are made to climb up poles or very low trellis work; or up trees, either very low and pruned closely as in the North of Italy, or tall and natural as in the kingdom of Naples, where the Vine ascends lofty poplars and runs in festoons from one to another at various heights. These latter methods have the double advantage of multiplying the surfaces of the leaves and of slowly ripening the grapes, sheltered by the foliage from the great heat that would otherwise act too quickly or too unequally upon them. Nevertheless, there and even farther south, as in Sicily, we find them trained up poles; whilst they are caused to climb up trees in Dauphiny. It is true that the quality of the juice does not thereby gain much; at least we find that in our own latitudes, when the Vines are thus left to themselves and allowed to enlace the trees, the grape is rarely ripened. It seems also capable of growing on all kinds of soil, but of acquiring all those qualities, which form the basis of good wine grapes, on those only that are dry and strong. Moreover, we know that neighbouring vineyards, placed in identical circumstances with regard to climate and soil, furnish wines of totally different qualities; but this may arise from the various ways in manufacturing and adulterating it: it is, therefore, difficult to determine how far nature has any influence on it. Acids generally predominate in those grapes which approach the Northern Limit; Saccharine principles and, consequently, alcohol in those of the Southern.

To be perfectly satisfied with the history of this geographical distribution, we must pay attention to the differences of species and of varieties, which flourish and predominate in each different latitude; but the determination of the species of the Vine has become one of the most complicated questions of Agricultural Botany, so greatly have they been multiplied and hybridised.

§ 928. The limits of this work, already too much extended, will not allow us to notice the distribution of several other cultivated plants, used for food or manufactures, and we shall be obliged to request our reader to turn for a short account of them to the description of the family of each: —such are the *Olive* (§ 839), the *Sugar Cane* (§ 760), the *Coffee-tree* (§ 854), the *Cacao* (§ 810), the *Tea-tree* (§ 811), the *Indigo* (§ 821), and several other plants from which we obtain thread, ropes, cloths and dyes.

We shall now, in finishing, content ourselves with calling the attention of the reader to the intimate relation of the different branches of the science to one another, and of theories to facts. The classification has light thrown on it by the study of the organization, which in its turn is explained by the classification; it puts in order the chaos of the innumerable species of vegetables, enables us to determine which are peculiar to each part of the globe, concludes from the natural association of plants, from which results the Flora of each country, which man may try to introduce, and thus becomes one of the most useful handmaids of Agriculture.

FOSSIL PLANTS.

§ 929.—We have endeavoured to give the reader a general notion of the distribution of vegetables on the surface of the globe, as far as our present knowledge of facts will allow us. But it may be asked, 'Has this distribution been the same at all times?' This has to be determined: and we can only refer to the study of the nature of FOSSILS as the means of withdrawing the veil from this mysterious subject. Fossils are remains, animal or vegetable, which have been buried in the different layers of the globe as they were formed in their turn on its surface. Now it is evident, that the results we shall hence obtain will be far from being as exact and general as those derived from the study of our contemporary vegetation; because, on the one hand, several plants must have existed of which we cannot find any trace; and on the other, the excavations, which expose these fossil remains, have only been made at a few places, almost exclusively in Europe, and that only with a view to commercial pursuits and not with that care and precaution requisite for discovering and preserving intact these *débris* of the organic world. Besides, we only find them in the state of fragments, from which it is excessively difficult to determine the species, genus or even the family of the vegetable. The characteristics of flowers and fruits, from which existing plants are classified, are almost constantly wanting in Fossils. On account of this the characteristics of vegetation are obliged to be studied much more closely and attentively. Nevertheless, modern researches, and principally those of M. Adolphe Brongniart, have surmounted several of these difficulties. There are now a large number of Fossils accurately described, and with such precision that they may be classified in families, genera and even species. We cannot here give a complete catalogue of them, limited as it is, and we must be contented with general statements, similar to those given on Botanical Arithmetic (§ 907). They belong more properly to the science of Geology [c].

§ 930. We first begin to find the traces of this lost vegetation in the TRANSITION PERIOD, but as yet they are very rare; they then increase in the CARBONIFEROUS SYSTEM, in the upper layers of which they acquire their maximum (Appendix C). During this long period vegetation seems to have undergone very important changes as to the species, at the same time preserving the same essential characteristics as a whole. These characteristics are the numerical predominance and the large developement of Vascular Cryptogams and an exceedingly small number of FUCUS, which are referred to the most distant period: afterwards amongst the Phanerogams, a few Monocotyledons and no Dicotyledons except those belonging to the Gymnospermous Plants (§ 774), *i. e.* to the *Cycadeæ* and the *Coniferæ*, or at least to families which would be analogous to them. The former contain the species of SIGILLARIA (Appendix C, *fig.* 745) and of STIGMARIA (Appendix C, *fig.* 746), the latter, those of the WALCHIA (Appendix C, *fig.* 747), which are somewhat related to our

[c] De Jussieu refers his readers to a treatise on Geology, mentioning the places referred to. The Translator has given a short description of the different formations in the Appendix at the end of the work.—TRANS.

ARAUCARIA. Several of the Fossils comprehended under the name of CALAMITES seem also to belong to this class of vegetables, whilst others belong to the *Equisetaceæ*, some of which were large trees instead of being as at present weak and humble stems. We may say as much of the *Lycopodiaceæ;* we have discovered whole trunks of the genus LEPIDODENDRON (Appendix C, *fig.* 744), which are sometimes sixty or seventy feet long. But the Ferns are the most abundant in this Ancient Flora, of which they form nearly the half; and it is remarkable that several of them were large trees, although flourishing mostly in the temperate zone, far from the climates where the Arborescent Ferns now grow. Besides, all the species are analogous to those now living under the tropics and not to those in our zone; we may thence conclude that it enjoyed a higher temperature. This large proportion of the Vascular Cryptogams with respect to the rest of the plants would seem to indicate a climate, not only warmer but also moister and more uniform. These forests would, therefore, occupy land intersected by arms of the sea and hardly raised above its level, rather than large continents left dry (§ 877). We cannot doubt that the Coal was formed by the accumulated masses of all these vegetables, altered and modified in the same way as the peat-bogs of our time might be, if they were covered by large deposits of mineral substances, were compressed by their weight and exposed at the same time to a more elevated temperature: it thus seems probable that its manner of formation had some analogy to that of our peat bogs (§ 913).

§ 931. This luxuriant, though uniform vegetation disappears in the strata which cover the Carboniferous System; and in the formations of the SECONDARY PERIOD which follow, the number of vegetable fossils that are found is much less. This is explained, doubtless, by the fact of the greater part of the strata of this formation being deposited in the sea, so that the terrestrial vegetables, which they may contain, must have been transported from great distances into this enormous mass, and have been thus preserved, whenever they were not completely destroyed. Yet some places, where these vegetable remains have accumulated, present a comparatively rich Flora. Thus, after the PERMIAN or MAGNESIAN LIMESTONE SYSTEM, where there have been found a few marine plants only, the UPPER RED SANDSTONE or TRIASSIC SYSTEM is remarkable for the presence of terrestrial plants of the same families which we have mentioned; the VARIEGATED SANDSTONE FORMATION for almost equal proportions of Vascular Cryptogams and of Phanerogams, amongst which we must mention several species of a genus of the *Coniferæ*, the VOLTZIA (Appendix D, *fig.* 748); the *Cycadeæ* are wanting here, whilst they reappear in the MUSCHELKALK and abound in the VARIEGATED MARLS where they form the half of the Flora; this large proportion is very remarkable for a family, the known number of whose living species does not exceed thirty. This proportion hardly decreases in the richer Flora of the OOLITIC SYSTEM or JURA LIMESTONE GROUP, (Appendix E), where the *Cycadeæ* are again associated with the *Coniferæ*, where the Ferns form about one half, and where a gigantic EQUISETUM makes its appearance (Appendix E, *fig.* 749). The Secondary Period is terminated by the CRETACEOUS SYSTEM (Appendix F), in the first deposits of which we find several species of *Cycadeæ*, (MANTELLIA [Appendix F, *fig.* 750]), *Coniferæ*, *Equisetaceæ*, Ferns, but very few marine plants. We see, therefore, in the whole period which follows the Carboniferous System and precedes the Tertiary Period, that the few existing monuments of the terrestrial vegetation continue to shew us the predominance of the Vascular Cryptogams and of the Gymnospermous Dicotyledons (the latter, especially the *Cycadeæ*, always in an increasing ratio), the absence of every other Dicotyledonous Phanerogam and rather an insignificant number of Monocotyledons.

§ 932. The general character of the vegetation completely changes in the TERTIARY PERIOD, during which were deposited the strata, that now form the soil of the principal capitals of Europe, London, Paris and Vienna. Henceforth, the external conditions seem to tend towards the equilibrium in which we now see them: for the relations of the large classes of vegetables to one another gradually approach those we have already mentioned as those of the actual present state. Thus, in the whole of the known plants of that period, the Gymnospermous Dicotyledons do not comprise more than $\frac{1}{10}$, whilst the rest of the Dicotyledons, which have not hitherto appeared, comprise more than $\frac{7}{10}$ and the Monocotyledons $\frac{1}{6}$. The Vascular Cryptogams are less than $\frac{1}{20}$. The soil of Europe was then covered as at present with Pines, Firs, THUJAS, Birches, Beeches, Elms, Poplars, Walnuts, Sycamores and other nearly identical trees, which are now growing in our climates. The climate must have been that of the temperate zone with rather an elevated temperature. This is proved by the presence of some Palms in the North of France, very different from those which now exist on the shores of the Mediterranean, as well as several other plants that now are actually confined to warmer regions. Another point worthy of our attention is the fact of these fossil species seemingly presenting greater analogy to the trees of Northern America than to those of Europe (Appendix G).

§ 933. This glance cast over the phases of vegetation revealed to us by the fossils, shews us a curious fact: that the progression from the simple to the compound, which the natural classification has tried to establish in the series of the Acotyledons to the Cotyledons, of the Gymnospermous to the Angiospermous Dicotyledons, is realized in a general manner in their successive appearance on the surface of the globe[d].

[a] Some doubt may be cast on this assertion of De Jussieu by a recent experiment conducted by Dr. Lindley. One hundred and seventy plants were thrown into a vessel containing fresh water. Amongst these were species of all the families of which the Coal Measures seem to consist, and some others, which might be supposed to have coexisted with them. In the course of two years, one hundred and twenty one species had disappeared, being entirely decomposed. Of the fifty-six which still remained, the most perfect were species of the *Coniferæ*, Ferns, Palms, *Lycopodiaceæ* and the like, to which the plants of the Coal Measures are most analogous. We must at the same time, if we admit the similarity of Coals to Peat, remember the properties of the Peat of our day, which enable Vegetables and even *Animals* to resist the effects of decomposition.—TRANS.

APPENDIX.

(APPENDIX A, page 228).

§ 934.—A very singular fact connected with the flow of the sap is mentioned by Darwin. The italics are the Translator's.

"In a few places there were Palms, and I was surprised to see one at " an elevation of at least 4,500 feet. These Palms are, for their family, " ugly trees. Their stem is very large and of a curious form, being " thicker in the middle than at the base or top. They are excessively " numerous in some parts of Chile and valuable on account of a sort " of treacle made from the sap. On one estate near Petorca they tried " to count them, but failed, after having numbered several hundred " thousand. Every year in the early spring, in August, very many are " cut down, and when the trunk is lying on the ground, the crown of " leaves is lopped off. The sap then immediately begins to flow from " the upper end and continues so doing for some months: it is, how-" ever, necessary that a thin slice should be shaved off from that end every " morning, so as to expose a fresh surface. A good tree will give ninety " gallons, and all this must have been contained in the vessels of the " apparently dry trunk. It is said that the sap flows much more " quickly on those days when the sun is powerful; *and likewise, that it is* " *absolutely necessary to take care, in cutting down the tree, that it should* " *fall with its head upwards on the side of the hill; for if it falls down the* " *slope, scarcely any sap will flow; although in that case one would have* " *thought that the action would have been aided, instead of checked, by* " *the force of gravity.* The sap is concentrated by boiling, and is then " called treacle, which it very much resembles in taste."—*Darwin's Naturalist's Voyage*, p. 256.

(APPENDIX B, page 729).

Dunes.

§ 935.—The following figures (740, 741) will illustrate the way in which the Dunes are formed by the constant encroachment of sand on cultivated soil. Whole countries are thus being covered with drift sand, such as Egypt, the North Coast of Cornwall, &c.

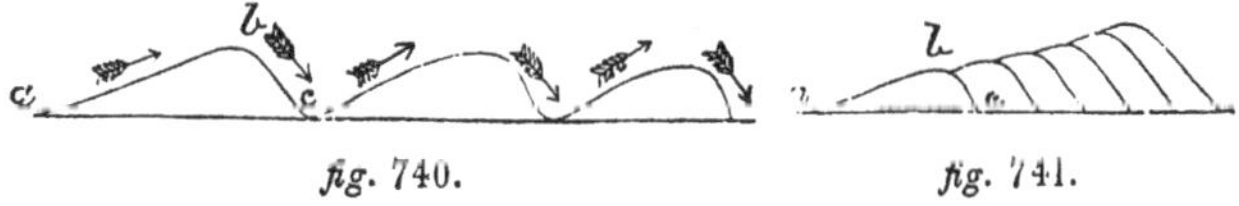

fig. 740. *fig.* 741.

These figures are supposed to represent hillocks of sand placed perpendicularly to the direction of the prevailing winds. Now the wind pushes the sand from the bottom *a* to the top *b*; it then falls in the direction *b c*, and thus forms a talus always steeper than the one before it. A hillock *a b c*, taken by itself (*fig.* 741) will increase continually at the back, if fresh sand be still supplied to it in the front, or will be removed, provided the original sand be kept constantly in motion. Again, if the wind act on all the hillocks at once, the whole mass is advanced farther inland, whilst, if fresh sand be supplied in front, new hillocks will be formed.

(APPENDIX C, page 738).

SECONDARY PERIOD.

CARBONIFEROUS SYSTEM.

§ 936.—In its complete state this System consists of the *Lower Carboniferous shales, Carboniferous limestone, Millstone grit, Coal measures, and Upper Coal grits*, the limestones alternating with the sandstones and shales.

The *Carboniferous limestone* is also called *Mountain* or *Metalliferous limestone.* It receives the last name from the great quantity of its mineral riches in Derbyshire, &c. It is largely developed in England, Belgium and the North of France.

The *Grit* or *Sandstone* is found of all degrees of fineness and coarseness. In it are found quartz pebbles of all sizes under an egg, though pieces even larger than this have been seen. Mica and feldspar also form an ingredient of these strata. The *Coal measures* exist in this part of the formation, and are frequently alternate with layers of sandstone, shale, &c.

The vegetable remains of this formation. Besides the Coal itself, formed by the accumulation of decomposed vegetables, whose *débris* may be recognised by the microscope, the Carboniferous System presents a large number of plants that have preserved their organic characteristics: stems and trunks of trees are found in the Sandstones; leaves of various kinds have left their impressions on the Shales which accompany the Coal. These remains are related to the Ferns, to the *Equisetaceæ*, to the *Lycopodiaceæ*, to the *Coniferæ* and to various families now totally unknown, which are very similar to the *Cycadeæ*.

The impressions of Ferns are exceedingly numerous; amongst them we find the PECOPTERIS (*fig.* 742), the SPHENOPTERIS, NEUROPTERIS, &c. The folioles of the PECOPTERIS are hardly detached from the peduncle and are sometimes united into a single leaf very deeply lobed, having a principal nerve, to which the secondary ones are perpendicular. There are also found several plants of doubtful families, as the SPHENOPHYLLUM, ANNULARIA, ASTEROPHYLLITES, PHYLLOTHECA, &c.

As was stated in the text (page 738), several fossils have been confounded under the general term of CALAMITES (*fig.* 743). It is the opinion of many very learned Botanists that the Calamites belong to a family now totally extinct. These remains are stems, originally cylindrical but now crushed and flattened, fluted lengthways, presenting here and there articulations, whence grow the branches. Though they are termed CALAMITES, they have no analogy to the CALAMUS of the present day, one of the Palm family. They are generally found converted into solidified clay, or into carbonate of iron, rarely into a siliceous substance. The external vegetable tissue, which has left its impression on the mineral mass, is frequently discovered in the form of Coal.

The genus LEPIDODENDRON (*fig.* 744) of the family of the *Lycopodiaceæ* has stems characterized by the rhomboidal projections arranged in spirals, which clearly shew the cicatrices of the leaves near the top.

The SIGILLARIA appears to be nearly allied to the Cycadeæ. These plants are generally stems, which seem to have been flattened by the pressure of the earth (*fig.* 745), and are marked by longitudinal flutes, but not articulated like the CALAMITES. They are covered with cicatrices arranged in longitudinal series, and not in spires as in the LEPIDODENDRON. The STIGMARIA (*fig.* 746) in the opinion of M. Ad. Brongniart can be nothing

742—750. Various species of fossil plants.
742. One of the Ferns, the PECOPTERIS AQUILINA.
743. The CALAMITES CANNÆFORMIS.
744. One of the Lycopodiaceæ, the LEPIDODENDRON ELEGANS.
745. A species of a family nearly allied to the Cycadeæ, the SIGILLARIA PACHYDERMA.
746. The STIGMARIA FICOIDES. These fossils in the opinion of M. Brongniart are nothing but the roots of other plants.

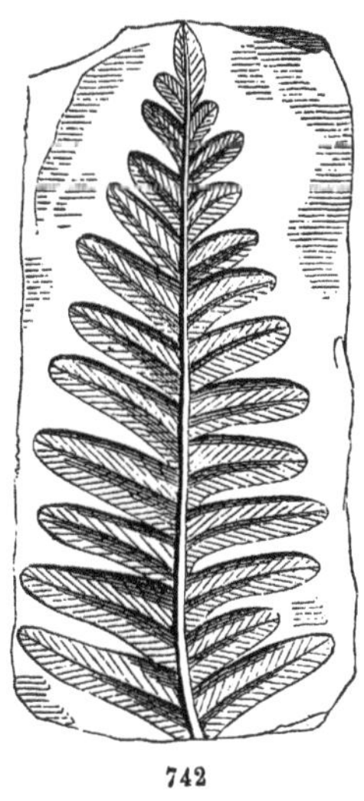

742

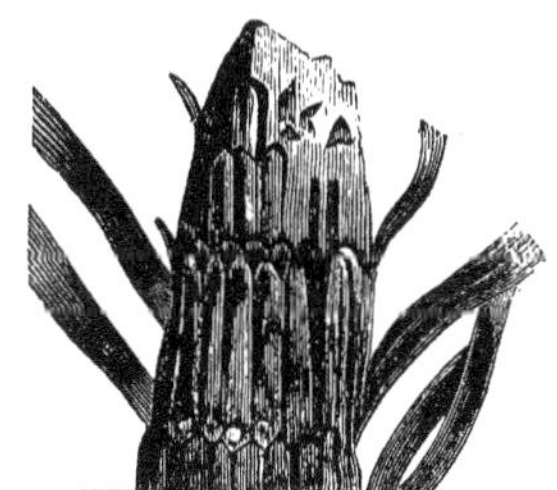

743

744

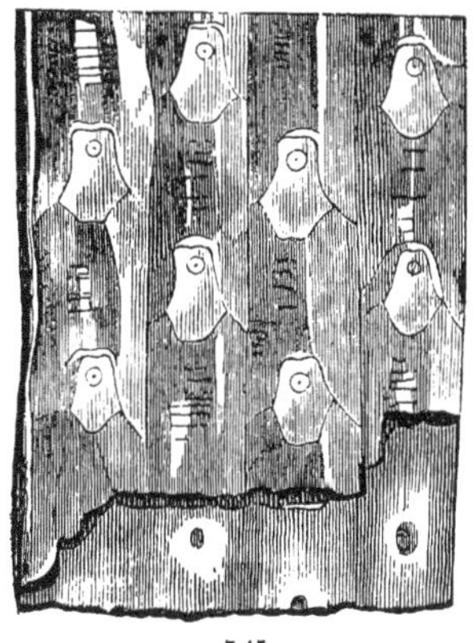

745

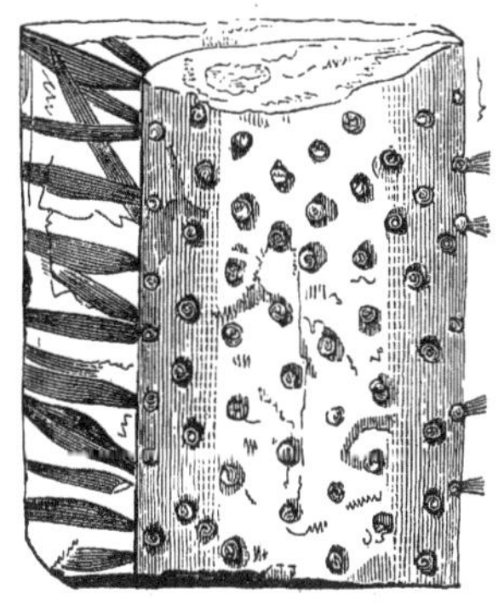

746

but the roots of these plants; they were traversed by a ligneous axis which was surrounded by soft parts.

The upper layers contain large quantities of the *Coniferæ*. Their remains seem to approach our genus ARAUCARIA. They have sessile leaves arranged in spires. M. Adolphe Brongniart seems inclined to refer them all to the genus WALCHIA (*fig.* 747).

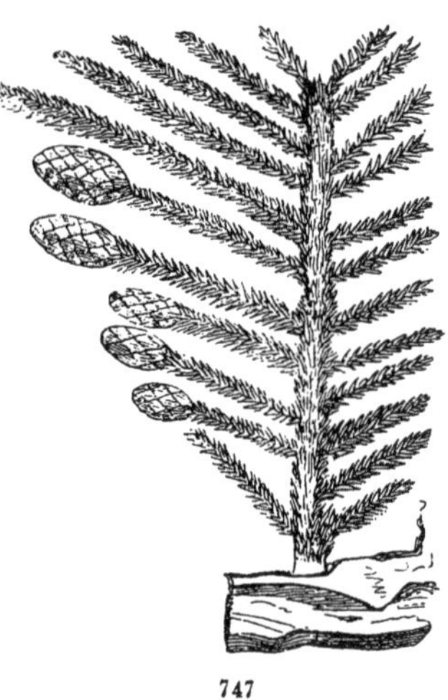

747

Before we quit the Carboniferous System, we will say a few words on the general features of its vegetation. As we said before, the Vascular Cryptogams greatly exceed in number the Phanerogams, so much so, indeed, that out of 260 species 220 are Cryptogams. The Equisetaceæ, now small herbaceous plants, were then trees more than ten feet high and five or six inches in diameter, and the Arborescent Lycopodiaceæ attained a height of sixty or seventy feet. The earth was then covered with a rich vegetation. All the species and some of the families are at the present day totally unknown.

(APPENDIX D, page 739).

THE PERMIAN OR MAGNESIAN LIMESTONE, AND THE UPPER NEW RED SANDSTONE OR TRIASSIC SYSTEMS.

§ 937.—These two systems may be termed the SALIFEROUS FORMATION, because salt very commonly lies in beds or nodules in this series of rocks. It is composed of the *Lower new red sandstone*, *Muschelkalk*, *Magnesian limestone*, *Variegated sandstone*, and *Variegated marls*.

747. One of the Coniferæ, the WALCHIA HYPNOIDES.

The first strata of this formation not containing many fossil plants we shall pass on to the *Variegated sandstone* (*Grès bigarré*) and to the *Variegated marls* (*Marnes irisées* or *Keuper* of Germany). In the former the Vascular Cryptogams and Phanerogams are nearly equal. We find in it several species of a genus of the *Coniferæ*, the VOLTZIA (*fig.* 748). The *Cycadeæ* or analogous families begin to appear again in the

748

Muschelkalk (*Calcaire Conchylien*), and in the *Variegated marls.* The former containing species of the genus MANTELLIA and the latter of the genera PTEROPHYLLUM and NILSONIA, &c.

The *Lower new red sandstone* consists of conglomerates, sands and marls: the *Magnesian limestone*, of marl slate, shelly limestone and yellow magnesian limestone; it is a marine formation: the *Variegated sandstone*, of red, white, blue and green argillaceous sandstone containing siliceous matter, and very frequently gypsum and rock-salt: the *Muschelkalk*, of grey, greenish-grey limestone, with siliceous nodules; it contains large quantities of the remains of shells and other zoological fossils: the *Variegated marls*, of mottled marls of a red, blue, blackish or greenish grey colour, slaty clays and fine grained sandstone.

748. One of the Coniferæ, the VOLTZIA HETEROPHYLLA.

(APPENDIX E, page 740).

JURA LIMESTONE GROUP.

§ 938.—This group comprehends the *Portland beds*, *Kimmeridge clay*, *Coral rag*, *Oxford clay*, *Cornbrash*, *Forest marble* and the *Great Oolite*. It derives its name from its constituting the main body of the Jura chain of mountains.

The *Portland beds* consist of shelly limestone and beds of chert; the *Kimmeridge clay* of slaty clay of a bluish or greyish yellow colour, containing gypsum; the *Coral rag* of limestones full of corals, of yellow-sands, and freestones full of shells. The *Oxford clay* is a marine formation consisting of a stiff blue clay mixed with calcareous matter, iron pyrites, gypsum, bituminous shale, &c. The *Cornbrash* is formed by grey or bluish rubbly limestone separated by layers of clay; it caps the escarpment of the Lower Oolites: its name is derived from the land being so well adapted for the growth of corn. The *Forest marble* is a carbonate of lime, sometimes crystalline, sometimes marly. The *Great Oolite* is composed of limestone crowded with remains of land animals, birds, amphibia, plants, shells of extreme beauty and in a most excellent state of preservation.

The *Coniferæ*, which had nearly disappeared, now become rather more numerous under peculiar genera, TAXITES, BRACHYPHYLLUM, &c. We have also *Cycadeæ*, Ferns of various species (every one different from those we meet with in the older series) and lastly a real EQUISETUM (*fig.* 749).

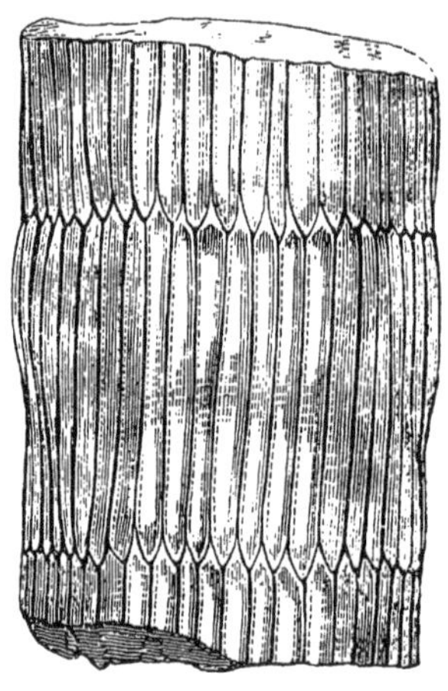
749

These fossil *Cycadeæ* appear to be very closely allied to the existing species. This is proved by a microscopical examination of the fossils. They resemble one another in the internal and external structure of the trunk, and in the method of increase of the plants by leaf-buds or bulbels.

749. One of the Equisetaceæ, the EQUISETUM COLUMNARE.

(APPENDIX F, page 740).

The Cretaceous System.

§ 939.—This is the last formation of this period. It consists of the *Maestricht beds*, *Chalk*, *Green-sand* and *Gault*. The *Chalk* consists of two strata, both of marine origin. The upper may be described as *Chalk with flints*, from its containing numerous layers of flints at intervals of from four to six feet. It is white, soft, and exceedingly full of Zoophytic remains. Pyrites are also found in the upper Chalk. It becomes Oolitic in some parts of the South of France. It covers the greater part of Our Island, of Ireland and a large part of Europe, being found in France, Sweden, and even Russia. The lower layer is also of marine formation. It may be described as the *Chalk without flints*. This, however, must not be taken absolutely, as flints are found in it, though not in such abundance as in the upper layer.

The *Upper Green sand* is immediately subjacent to the Chalk. It consists of grey, irony sands full of green particles of silicate of iron, and of layers of calcareous limestone. It is split with numerous fissures thus allowing the water which collects in the Chalk to pass downwards to the *Gault*. It is hardened into stone at some places, as at Godstone, where it is called 'Godstone Firestone.'

The *Gault* is a stiff blue clay, the lower part filled with iron pyrites and the upper part containing green particles of silicate of iron.

We now come to the *Lower Green sand*, which is composed of yellow ferruginous sand. In it are also found layers of flat bedded grey sandstone, full of fossils, also a good deal of ochre. It also contains much calcareous matter and the green silicate of iron.

The vegetable remains of this group are by no means extensive. They consist mostly of *Algæ* and other marine plants. We sometimes find the wood of Dicotyledons. These stems, however, appear to have been nothing but drift-wood, for they are covered with the scratches of rocks and the perforations of boring worms. It is also a singular fact that none of the plants of the epoch of the Cretaceous System appear to have been in existence either before or since. We shall conclude our notice of this system by mentioning the MANTELLIA NIDIFORMIS (*fig.* 750), of which we have already spoken. It belongs to the family of the *Cycadeæ*.

750. One of the Cycadeæ, the MANTELLIA NIDIFORMIS.

(APPENDIX G, page 740).

THE TERTIARY PERIOD.

§ 940.—We shall not dwell long upon this period, as the limits of our book, long since exceeded, will not allow us to do so. It may be divided into four Systems, the *Newer Pliocene*, the *Older Pliocene*, the *Miocene*, the *Eocene*.

The *Newer Pliocene* constitutes the Sicilian deposits, and contains limestone, sands, clays, conglomerates, marls, lignites, gypsum, and *marine*, *land* and *fresh-water* fossils. These three kinds of fossils will separate this and all other members of the Tertiary period into two subdivisions, the *Marine* and the *Freshwater*.

The *Older Pliocene* contains Sub-apennine marl, English crag, and some other deposits that occurred in the Newer Pliocene, along with *marine*, *land* and *fresh-water* fossils.

The *Miocene* contains nearly the same deposits as the preceding though they vary in their mineral composition. It also contains the three kinds of fossils. The Vienna basin is of the Miocene formation.

The *Eocene* is found in the London and Paris basins, and consists of London clay, calcareous sand, conglomerates, &c. The three classes of fossils are again found in this system.

The vegetable fossils of this period are exceedingly abundant. It is very singular that no fossils of the Tertiary period are found in any of the earlier formations. As we have already stated (§ 932) the plants bear a great resemblance to those now in existence. In fact we find such genera as JUGLANS, ULMUS, POPULUS, ACER and COCOS. A singular group of fossil fruits is imbedded in the London clay and in the Isle of Sheppey. The fragments of wood, which are taken from these deposits, are not unfrequently silicified and opalised.

FINIS.

OXFORD: PRINTED BY I. SHRIMPTON.

For EU product safety concerns, contact us at Calle de José Abascal, 56–1°,
28003 Madrid, Spain or eugpsr@cambridge.org.

www.ingramcontent.com/pod-product-compliance
Ingram Content Group UK Ltd.
Pitfield, Milton Keynes, MK11 3LW, UK
UKHW042207080726
473066UK00007B/283